Lippincott's Illustrated Reviews: Biochemistry

Lippincott's Illustrated Reviews: Biochemistry

Pamela C. Champe, PhD
Department of Biochemistry
University of Medicine and Dentistry of New Jersey–
Robert Wood Johnson Medical School
Piscataway, New Jersey

Richard A. Harvey, PhD
Department of Biochemistry
University of Medicine and Dentistry of New Jersey–
Robert Wood Johnson Medical School
Piscataway, New Jersey

Denise R. Ferrier, PhD
Department of Biochemistry
Drexel University College of Medicine
Philadelphia, Pennsylvania

INDIAN EDITION Distributed in India by
J.P. BROTHERS MEDICAL
PUBLISHERS (P) LTD.

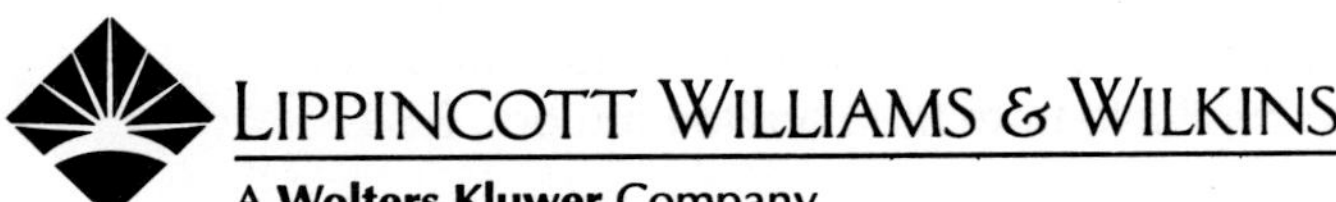

LIPPINCOTT WILLIAMS & WILKINS
A **Wolters Kluwer** Company
Philadelphia • Baltimore • New York • London
Buenos Aires • Hong Kong • Sydney • Tokyo

Acquisitions Editor: Neil Marquardt
Managing Editor: Sonya Seigafuse
Marketing Manager: Scott Lavine
Production Editor: Jennifer D. Glazer

351 West Camden Street
Baltimore, MD 21201

530 Walnut Street
Philadelphia, PA 19106

Reprinted 2005
Printed in India by Gopsons Papers Ltd. Noida

Distributed in India by J.P. Brothers Medical Publisher (P) Ltd.
This Book is not Permitted to be sold outside in India.

First Edition, 1987
Second Edition, 1994

Library of Congress information is available (0-7817-9983-X)

The publishers have made every effort to trace the copyright holders for borrowed material. If they have inadvertently overlooked any, they will be pleased to make the necessary arrangements at the first opportunity.

To purchase additional copies of this book, call our customer service department at **(800) 638-3030** or fax orders to **(301) 824-7390.** International customers should call **(301) 714-2324.**

Visit Lippincott Williams & Wilkins on the Internet: http://www.LWW.com. Lippincott Williams & Wilkins customer service representatives are available from 8:30 am to 6:00 pm, EST.

04 05 06 07 08
1 2 3 4 5 6 7 8 9 10

Contributing Authors:

Cal McLaughlin, PhD
Department of Biological Chemistry
University of California, Irvine
Irvine, California

Vernon E. Reichenbecher, PhD
Department of Biochemistry and Molecular Biology
Marshall University School of Medicine
Huntington, West Virginia

Computer Graphics:

Michael Cooper
Cooper Graphics
www.cooper247.com

This book is dedicated to
Marilyn Schorin,
whose generously shared insights
and unwavering support guided
unfocused words into coherent
ideas.

Acknowledgments

We are grateful to the many friends and colleagues who generously contributed their time and effort to help us make this book as accurate and as useful as possible. The support of our other colleagues at the University of Medicine and Dentistry of New Jersey–Robert Wood Johnson Medical School is highly valued. We (RAH and PCC) owe special thanks to our Chairman, Dr. Masayori Inouye, who has encouraged us over the years in this and other teaching projects. We are particularly indebted to Dr. Mary Mycek of the University of Medicine and Dentistry of New Jersey–New Jersey Medical School, who participated actively in this project. We also appreciate the many helpful comments of Dr. William Zehring and Dr. Jeff Mann.

Without talented artists, an Illustrated Review would be impossible, and we have been particularly fortunate in working with Michael Cooper throughout this project. His artistic sense and computer graphics expertise have greatly added to our ability to bring biochemistry "stories" alive for our readers.

The editors and production staff of Lippincott Williams & Wilkins were a constant source of encouragement and discipline. We particularly want to acknowledge the tremendously helpful, supportive, and creative contributions of our editor, Neil Marquardt, whose imagination and positive attitude helped us bring this complex project to completion. Final editing and assembly of the book has been greatly enhanced through the efforts of Jennifer Glazer.

Contents

UNIT I: Protein Structure and Function

Amino Acids

1

I. OVERVIEW

Proteins are the most abundant and functionally diverse molecules in living systems. Virtually every life process depends on this class of molecules. For example, enzymes and polypeptide hormones direct and regulate metabolism in the body, whereas contractile proteins in muscle permit movement. In bone, the protein collagen forms a framework for the deposition of calcium phosphate crystals, acting like the steel cables in reinforced concrete. In the bloodstream, proteins, such as hemoglobin and plasma albumin, shuttle molecules essential to life, whereas immunoglobulins fight infectious bacteria and viruses. In short, proteins display an incredible diversity of functions, yet all share the common structural feature of being linear polymers of amino acids. This chapter describes the properties of amino acids; Chapter 2 explores how these simple building blocks are joined to form proteins that have unique three-dimensional structures, making them capable of performing specific biologic functions.

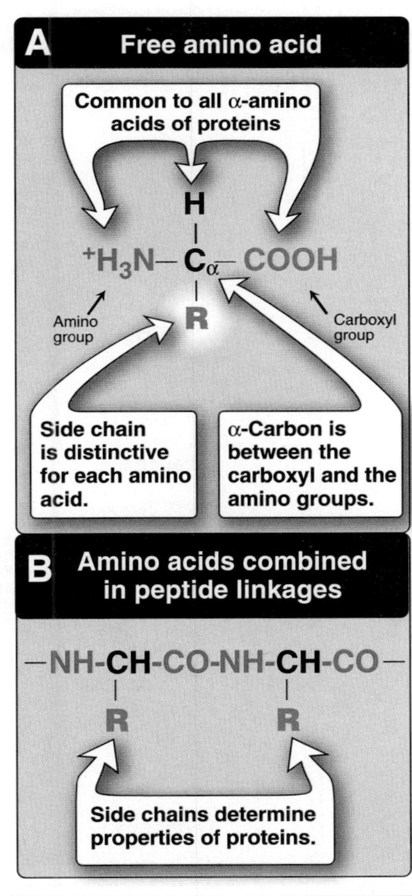

Figure 1.1
Structural features of amino acids (shown in their fully protonated form).

II. STRUCTURE OF THE AMINO ACIDS

Although more than 300 different amino acids have been described in nature, only twenty are commonly found as constituents of mammalian proteins. [Note: These are the only amino acids that are coded for by DNA, the genetic material in the cell (see p. 393).] Each amino acid (except for proline, which is described on p. 4) has a **carboxyl group**, an **amino group**, and a distinctive side chain ("**R-group**") bonded to the α-carbon atom (Figure 1.1A). At physiologic pH (approximately pH = 7.4), the carboxyl group is dissociated, forming the negatively charged carboxylate ion ($-COO^-$), and the amino group is protonated ($-NH_3^+$). In proteins, almost all of these carboxyl and amino groups are combined in peptide linkage and, in general, are not available for chemical reaction except for hydrogen bond formation (Figure 1.1B). Thus, it is the nature of the side chains that ultimately dictates the role an amino acid plays in a protein. It is, therefore, useful to classify the amino acids according to the properties of their side chains—that is, whether they are nonpolar (that is, have an even distribution of electrons) or polar (that is, have an uneven distribution of electrons, such as acids and bases; Figures 1.2 and 1.3).

A. Amino acids with nonpolar side chains

Each of these amino acids has a nonpolar side chain that does not bind or give off protons or participate in hydrogen or ionic bonds (see Figure 1.2). The side chains of these amino acids can be thought of as "oily" or lipid-like, a property that promotes **hydrophobic interactions** (see Figure 2.9, p. 18).

1. **Location of nonpolar amino acids in proteins:** In proteins found in aqueous solutions, the side chains of the nonpolar amino acids tend to cluster together in the interior of the protein (Figure 1.4). This phenomenon is the result of the hydrophobicity of the nonpolar

NONPOLAR SIDE CHAINS

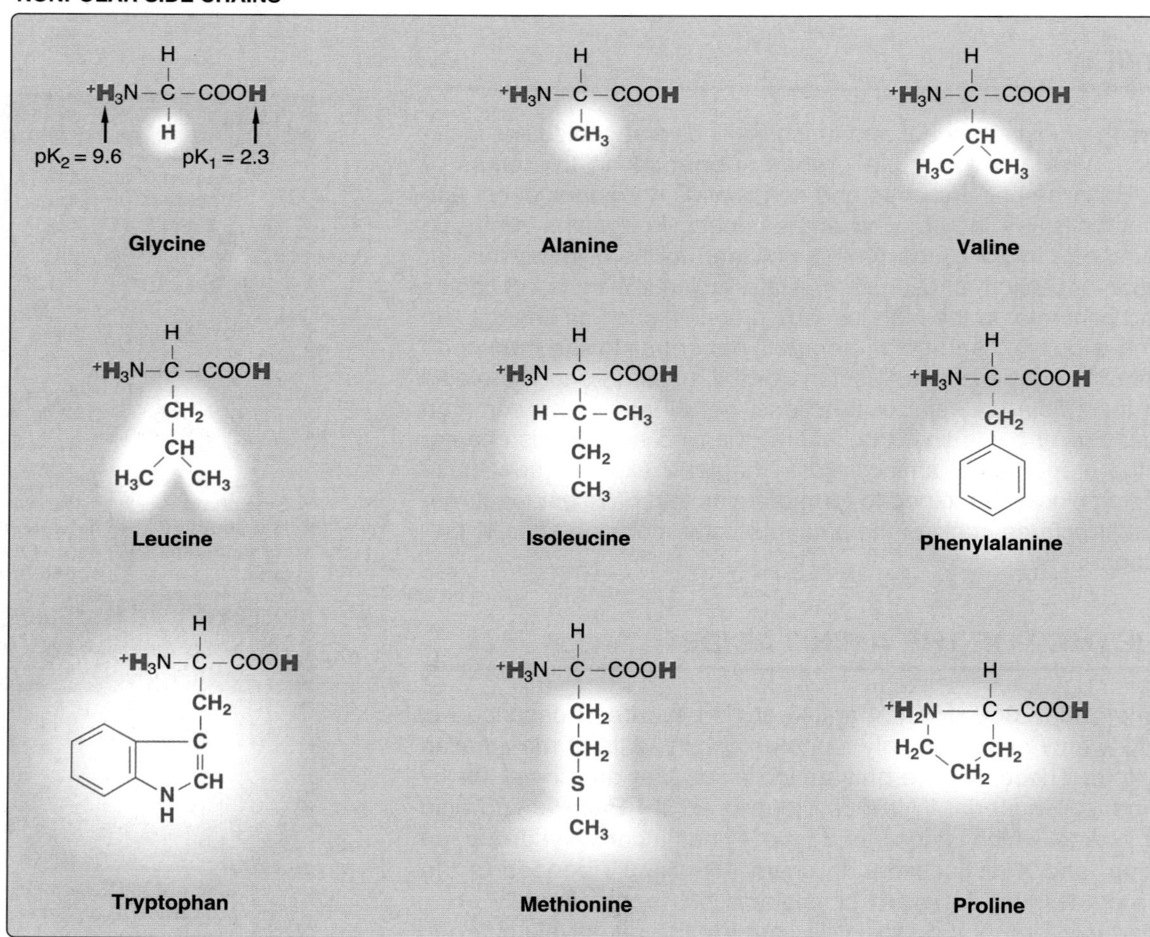

Figure 1.2
Classification of the twenty amino acids found in proteins, according to the charge and polarity of their side chains, is shown here and continues in Figure 1.3. Each amino acid is shown in its fully protonated form, with dissociable hydrogen ions represented in red print. The pK values for the α-carboxyl and α-amino groups of the nonpolar amino acids are similar to those shown for glycine. (Continued on Figure 1.3.)

UNCHARGED POLAR SIDE CHAINS

Serine

Threonine

Tyrosine: $pK_2 = 9.1$, $pK_1 = 2.2$, $pK_3 = 10.1$

Asparagine

Cysteine: $pK_1 = 1.7$, $pK_3 = 10.8$, $pK_2 = 8.3$

Glutamine

ACIDIC SIDE CHAINS

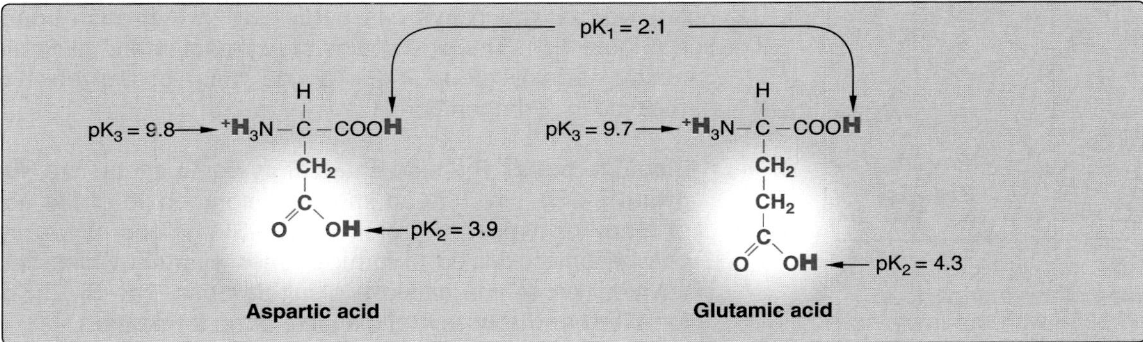

BASIC SIDE CHAINS

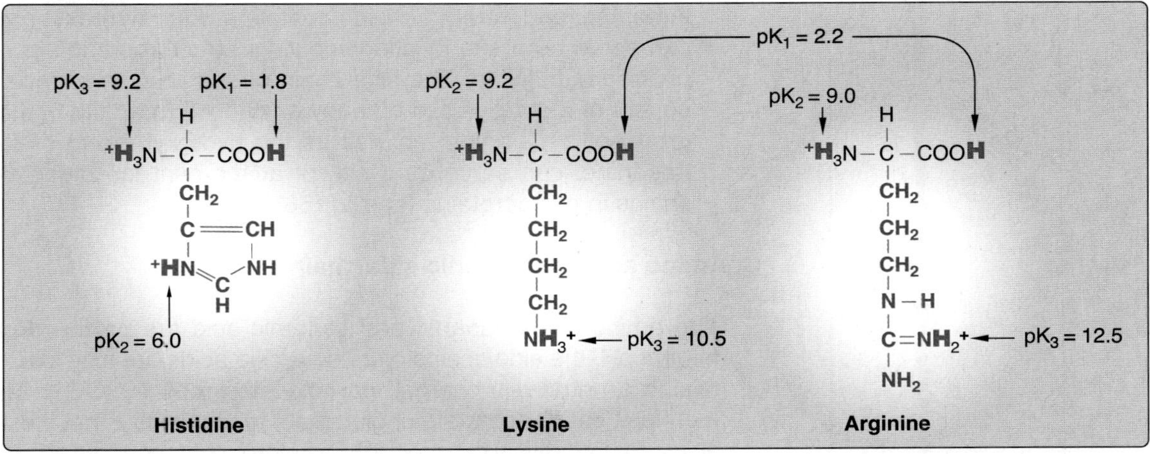

Figure 1.3
Classification of the twenty amino acids found in proteins, according to the charge and polarity of their side chains (continued from Figure 1.2).

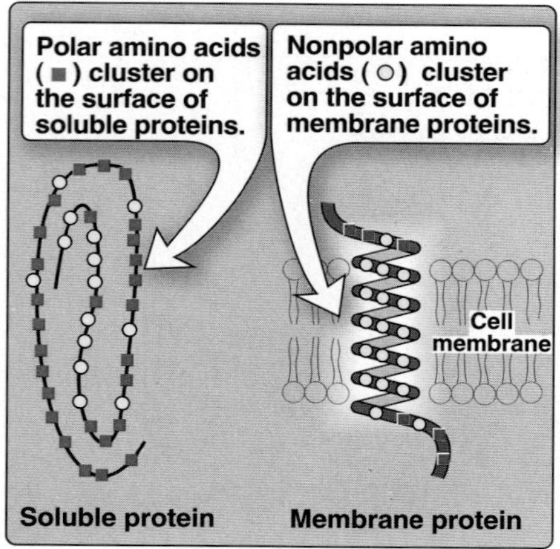

Figure 1.4
Location of nonpolar amino acids in soluble and membrane proteins.

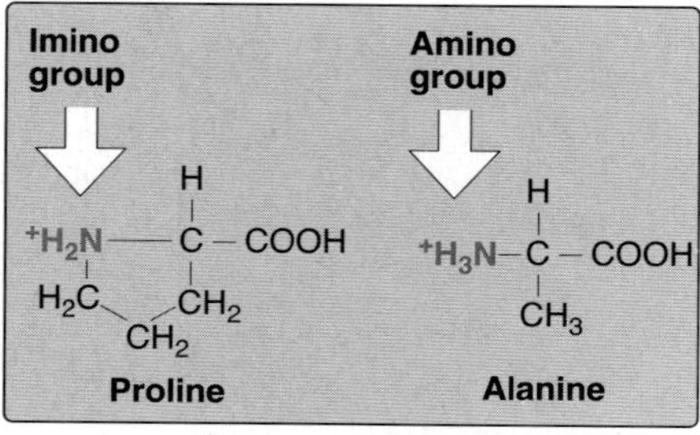

Figure 1.5
Comparison of the imino group found in proline with the α-amino group found in other amino acids, such as alanine.

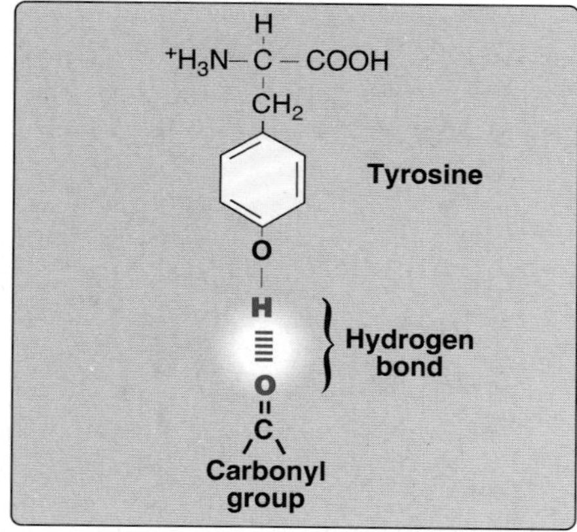

Figure 1.6
Hydrogen bond between the phenolic hydroxyl group of tyrosine and another molecule containing a carbonyl group.

R-groups, which act much like droplets of oil that coalesce in an aqueous environment. The nonpolar R-groups thus fill up the interior of the folded protein and help give it its three-dimensional shape. [Note: In proteins that are located in a hydrophobic environment, such as a membrane, the nonpolar R-groups are found on the outside surface of the protein, interacting with the lipid environment (see Figure 1.4).] The importance of these hydrophobic interactions in stabilizing protein structure is discussed on p. 19.

2. **Proline:** The side chain of proline and its α-amino group form a ring structure, and thus proline differs from other amino acids in that it contains an **imino group**, rather than an amino group (Figure 1.5). The unique geometry of proline contributes to the formation of the fibrous structure of collagen (see p. 45), and often interrupts the α-helices found in globular proteins (see p. 26).

B. Amino acids with uncharged polar side chains

These amino acids have zero net charge at neutral pH, although the side chains of cysteine and tyrosine can lose a proton at an alkaline pH (see Figure 1.3). Serine, threonine, and tyrosine each contain a polar hydroxyl group that can participate in **hydrogen bond** formation (Figure 1.6). The side chains of asparagine and glutamine each contain a carbonyl group and an amide group, both of which can also participate in hydrogen bonds.

1. **Disulfide bond:** The side chain of **cysteine** contains a **sulfhydryl group** (–SH), which is an important component of the active site of many enzymes. In proteins, the –SH groups of two cysteines can become oxidized to form a dimer, **cystine**, which contains a covalent cross-link called a disulfide bond (–S–S–). (See p. 19 for a further discussion of disulfide bond formation.)

2. **Side chains as sites of attachment for other compounds:** Serine, threonine, and, rarely, tyrosine contain a **polar hydroxyl group** that can serve as a site of attachment for structures such as a phosphate group. [Note: The side chain of serine is an important component of the active site of many enzymes.] In addition, the **amide group** of asparagine, as well as the hydroxyl group of serine or threonine, can serve as a site of attachment for oligosaccharide chains in glycoproteins (see p. 156).

C. Amino acids with acidic side chains

The amino acids aspartic and glutamic acid are **proton donors**. At neutral pH, the side chains of these amino acids are fully ionized, containing a negatively charged **carboxylate group** ($–COO^-$). They are, therefore, called aspartate or glutamate to emphasize that these amino acids are negatively charged at physiologic pH (see Figure 1.3).

D. Amino acids with basic side chains

The side chains of the basic amino acids **accept protons** (see Figure 1.3). At physiologic pH the side chains of lysine and arginine are fully ionized and positively charged. In contrast, histidine is weakly basic, and the free amino acid is largely uncharged at physiologic pH. However, when histidine is incorporated into a protein, its side chain can be either positively charged or neutral, depending on the ionic environment provided by the polypeptide chains of the protein. [Note: This is an important property of histidine that contributes to the role it plays in the functioning of proteins such as hemoglobin (see p. 26).]

E. Abbreviations and symbols for the commonly occurring amino acids

Each amino acid name has an associated three-letter abbreviation and a one-letter symbol (Figure 1.7). The one-letter codes are determined by the following rules:

1. **Unique first letter:** If only one amino acid begins with a particular letter, then that letter is used as its symbol. For example, I = isoleucine.

2. **Most commonly occurring amino acids have priority:** If more than one amino acid begins with a particular letter, the most common of these amino acids receives this letter as its symbol. For example, glycine is more common than glutamate, so G = glycine.

3. **Similar sounding names:** Some one-letter symbols sound like the amino acid they represent. For example, F = phenylalanine, or W = tryptophan ("twyptophan" as Elmer Fudd would say).

4. **Letter close to initial letter:** For the remaining amino acids, a one-letter symbol is assigned that is as close in the alphabet as possible to the initial of the amino acid. Further, B is assigned to Asx, signifying either aspartic acid or asparagine, Z is assigned to Glx, signifying either glutamic acid or glutamine, and X is assigned to an unidentified amino acid.

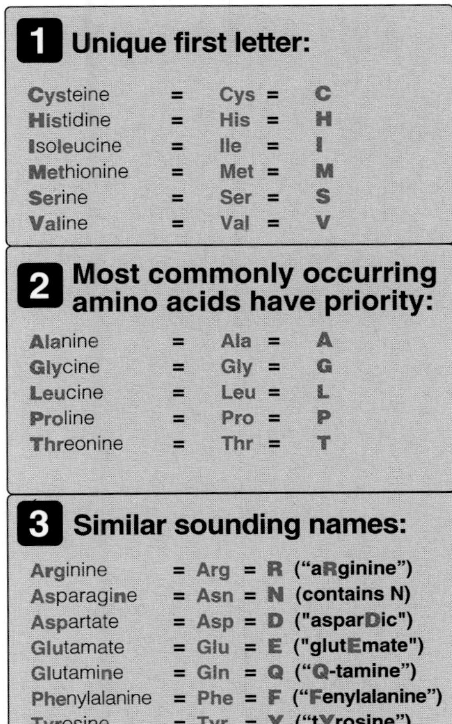

Figure 1.7
Abbreviations and symbols for the commonly occurring amino acids.

F. Optical properties of amino acids

The α-carbon of each amino acid is attached to four different chemical groups and is, therefore, a **chiral** or **optically active** carbon atom. Glycine is the exception because its α-carbon has two hydrogen substituents and, therefore, is optically inactive. [Note: Amino acids that have an asymmetric center at the α-carbon can exist in two forms, designated D and L, that are mirror images of each other (Figure 1.8). The two forms in each pair are termed **stereoisomers**, **optical isomers**, or **enantiomers**.] All amino acids found in proteins are of the L-configuration. However, **D-amino acids** are found in some antibiotics and in bacterial cell walls. (See p. 250 for a discussion of D-amino acid metabolism.)

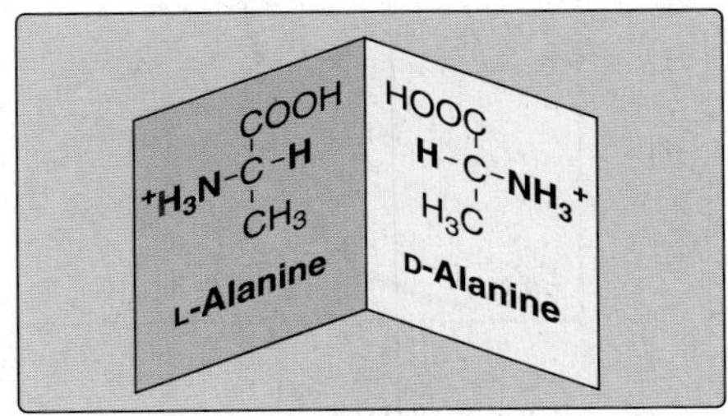

Figure 1.8
D and L forms of alanine are mirror images.

III. ACIDIC AND BASIC PROPERTIES OF AMINO ACIDS

Amino acids in aqueous solution contain weakly acidic α-carboxyl groups and weakly basic α-amino groups. In addition, each of the acidic and basic amino acids contains an ionizable group in its side chain. Thus, both free amino acids and some amino acids combined in peptide linkages can act as **buffers**. The quantitative relationship between the concentration of a weak acid (HA) and its conjugate base (A^-) is described by the **Henderson-Hasselbalch equation**.

A. Derivation of the equation

Consider the release of a proton by a weak acid represented by HA:

$$\underset{\text{weak acid}}{HA} \rightleftarrows \underset{\text{proton}}{H^+} + \underset{\text{salt form or conjugate base}}{A^-}$$

The "salt" or conjugate base, A^-, is the ionized form of a weak acid. By definition, the dissociation constant of the acid, K_a, is

$$K_a = \frac{[H^+][A^-]}{[HA]}$$

[Note: The larger the K_a, the stronger the acid, because most of the HA has been converted into H^+ and A^-. Conversely, the smaller the K_a, the less acid has dissociated and, therefore, the weaker the acid.] By solving for the $[H^+]$ in the above equation, taking the logarithm of both sides of the equation, multiplying both sides of the equation by –1, and substituting $pH = -\log [H^+]$ and $pK_a = -\log K_a$, we obtain the Henderson-Hasselbalch equation:

$$pH = pK_a + \log \frac{[A^-]}{[HA]}$$

B. Buffers

A buffer is a solution that resists change in pH following the addition of an acid or base. A buffer can be created by mixing a weak acid (HA) with its conjugate base (A^-). If an acid such as HCl is added to such a solution, A^- can neutralize it, in the process being converted to HA. If a base is added, HA can neutralize it, in the process being converted to A^-. Maximum buffering capacity occurs at a pH equal to the pK_a, but a conjugate acid/base pair can still serve as an effective buffer when the pH of a solution is within approximately ± 1 pH unit of the pK_a. [Note: If the amounts of HA and A^- are equal, the pH is equal to the pK_a.] As shown in Figure 1.9, a solution containing acetic acid ($HA = CH_3\text{–}COOH$) and acetate ($A^- = CH_3\text{–}COO^-$) with a pK_a of 4.8 resists a change in pH from pH 3.8 to 5.8, with maximum buffering at pH = 4.8. [Note: At pH values less than the pK_a, the protonated acid form ($CH_3\text{–}COOH$) is the predominant species. At pH values greater than the pK_a, the deprotonated base form ($CH_3\text{–}COO^-$) is the predominant species in solution.]

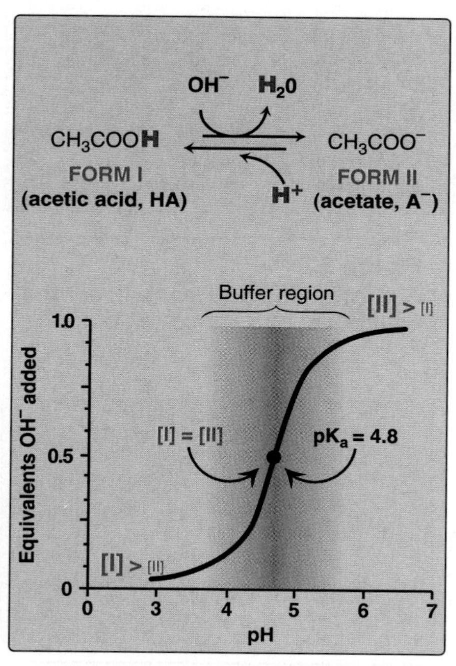

Figure 1.9
Titration curve of acetic acid.

C. Titration of an amino acid

1. **Dissociation of the carboxyl group:** The titration curve of an amino acid can be analyzed in the same way as described for acetic acid. Consider alanine, for example, which contains both an α-carboxyl and an α-amino group. At a low (acidic) pH, both of these groups

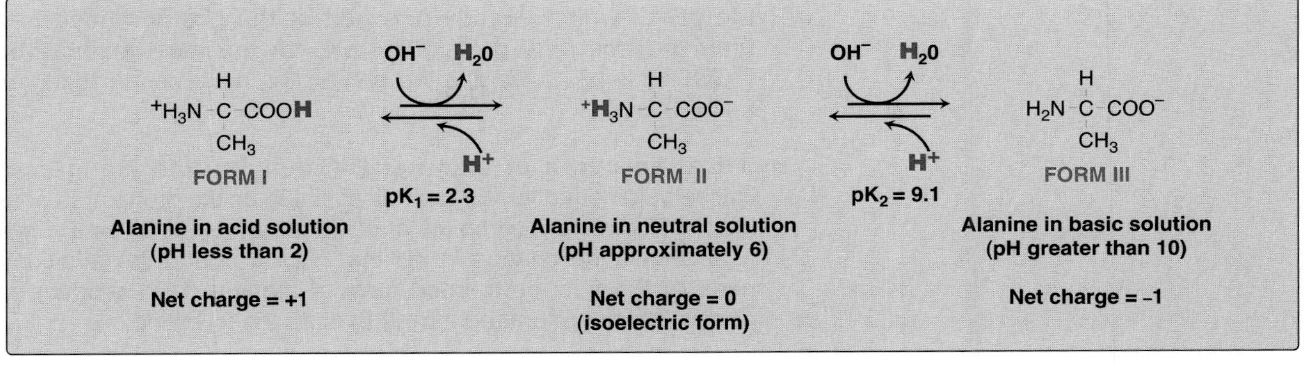

Figure 1.10
Ionic forms of alanine in acidic, neutral, and basic solutions.

are protonated (shown in Figure 1.10). As the pH of the solution is raised, the –COOH group of form I can dissociate by donating a proton to the medium. The release of a proton results in the formation of the carboxylate group, $-COO^-$. This structure is shown as form II, which is the **dipolar form** of the molecule (see Figure 1.10). [Note: This form, also called a **zwitterion**, is the **isoelectric form** of alanine—that is, it has an overall charge of zero.]

2. **Application of the Henderson-Hasselbalch equation:** The dissociation constant of the carboxyl group of an amino acid is called K_1, rather than K_a, because the molecule contains a second titratable group. The Henderson-Hasselbalch equation can be used to analyze the dissociation of the carboxyl group of alanine in the same way as described for acetic acid.

$$K_1 = \frac{[H^+][II]}{[I]}$$

where I is the fully protonated form of alanine, and II is the isoelectric form of alanine (see Figure 1.10). This equation can be rearranged and converted to its logarithmic form to yield:

$$pH = pK_1 + \log \frac{[II]}{[I]}$$

3. **Dissociation of the amino group:** The second titratable group of alanine is the amino ($-NH_3^+$) group shown in Figure 1.10. This is a much weaker acid than the –COOH group and, therefore, has a much smaller dissociation constant, K_2. [Note: Its pK_a is therefore larger.] Release of a proton from the protonated amino group of form II results in the fully deprotonated form of alanine, form III (see Figure 1.10).

4. **pKs of alanine:** The sequential dissociation of protons from the carboxyl and amino groups of alanine is summarized in Figure

1.10. Each titratable group has a pK_a that is numerically equal to the pH at which exactly one half of the protons have been removed from that group. The pK_a for the most acidic group (–COOH) is pK_1, whereas the pK_a for the next most acidic group ($–NH_3^+$) is pK_2.

5. **Titration curve of alanine:** By applying the Henderson-Hasselbalch equation to each dissociable acidic group, it is possible to calculate the complete titration curve of a weak acid. Figure 1.11 shows the change in pH that occurs during the addition of base to the fully protonated form of alanine (I) to produce the completely deprotonated form (III). Note the following:

   a. **Buffer pairs:** The –COOH/–COO⁻ pair can serve as a buffer in the pH region around pK_1, and the $–NH_3^+/–NH_2$ pair can buffer in the region around pK_2.

   b. **When pH = pK:** When the pH is equal to pK_1 (2.3), equal amounts of forms I and II of alanine exist in solution. When the pH is equal to pK_2 (9.1), equal amounts of forms II and III are present in solution.

   c. **Isoelectric point:** At neutral pH, alanine exists predominantly as the dipolar form II in which the amino and carboxyl groups are ionized, but the net charge is zero. The isoelectric point (pI) is the pH at which an amino acid is electrically neutral—that is, in which the sum of the positive charges equals the sum of the negative charges. [Note: For an amino acid, such as alanine, that has only two dissociable hydrogens (one from the α-carboxyl and one from the α-amino group), the pI is the average of pK_1 and pK_2 (pI = [2.3 + 9.1]/2 = 5.7, see Figure 1.10). The pI is thus midway between pK_1 (2.3) and pK_2 (9.1). It corresponds to the pH at which structure II (with a net charge of zero) predominates, and at which there are also equal amounts of form I (net charge of +1) and III (net charge of –1).]

6. **Net charge of amino acids at neutral pH:** At physiologic pH, all amino acids have a negatively charged group ($–COO^-$) and a positively charged group ($–NH_3^+$), both attached to the α-carbon. [Note: Glutamate, aspartate, histidine, arginine, and lysine have additional potentially charged groups in their side chains.] Substances, such as amino acids, that can act either as an acid or a base are defined as **amphoteric**, and are referred to as **ampholytes (amphoteric electrolytes)**.

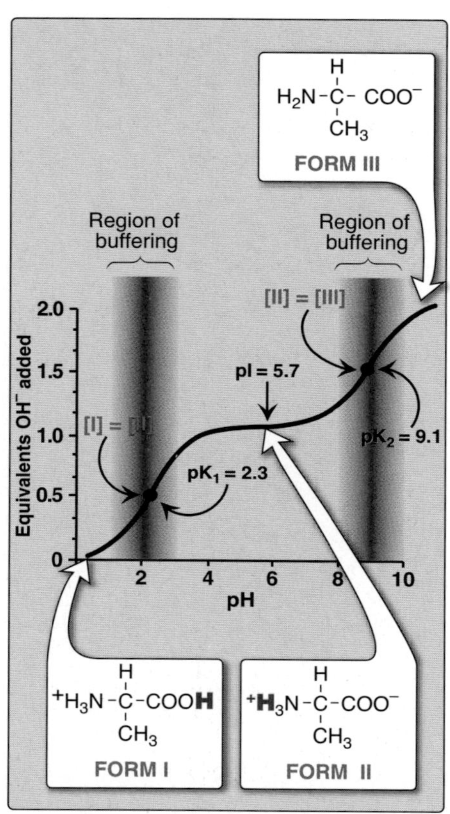

Figure 1.11
The titration curve of alanine.

D. Other applications of the Henderson-Hasselbalch equation

The Henderson-Hasselbalch equation can be used to calculate how the pH of a physiologic solution responds to changes in the concentration of weak acid and/or its corresponding "salt" form. For example, in the **bicarbonate buffer system**, the Henderson-Hasselbalch equation predicts how shifts in [HCO_3^-] and pCO_2 influence pH (Figure 1.12A). The equation is also useful for calculating the abundance of ionic forms of acidic and basic drugs. For example, most drugs are either weak acids or weak bases (Figure 1.12B). Acidic drugs (HA) release a proton (H^+), causing a charged anion (A^-) to form.

$$HA \rightleftarrows H^+ + A^-$$

Weak bases (BH^+) can also release a H^+. However, the protonated form of basic drugs is usually charged, and the loss of a proton produces the uncharged base (B).

$$BH^+ \rightleftarrows B + H^+$$

A drug passes through membranes more readily if it is uncharged. Thus, for a weak acid, the uncharged HA can permeate through membranes and A^- cannot. For a weak base, such as morphine, the uncharged form, B, penetrates through the cell membrane and BH^+ does not. Therefore, the effective concentration of the permeable form of each drug at its absorption site is determined by the relative concentrations of the charged and uncharged forms. The ratio between the two forms is, in turn, determined by the pH at the site of absorption, and by the strength of the weak acid or base, which is represented by the pK_a of the ionizable group. The Henderson-Hasselbalch equation is useful in determining how much drug is found on either side of a membrane that separates two compartments that differ in pH, for example, the stomach (pH 1.0–1.5) and blood plasma (pH 7.4).

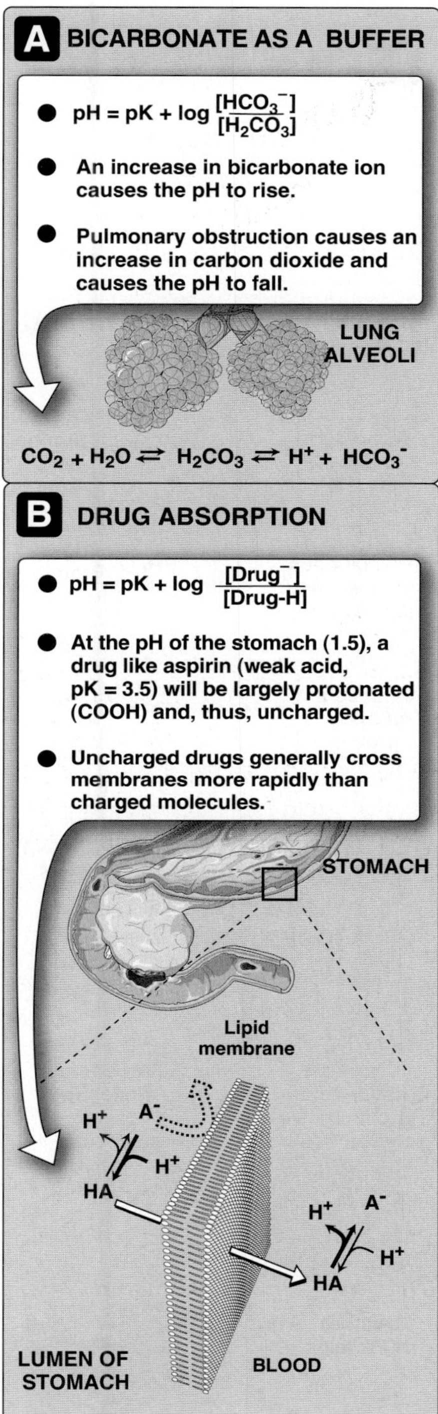

Figure 1.12
The Henderson-Hasselbalch equation is used to predict: A, changes in pH as the concentrations of HCO_3^- or CO_2 are altered; or B, the ionic forms of drugs.

IV. CONCEPT MAPS

Students sometimes view biochemistry as a blur of facts or equations to be memorized, rather than a body of concepts to be understood. Details provided to enrich understanding of these concepts inadvertently turn into distractions. What seems to be missing is a road map—a guide that provides the student with an intuitive understanding of how various topics fit together to make sense. The authors have, therefore, created a series of **biochemical concept maps** to graphically illustrate relationships between ideas presented in a chapter, and to show how the information can be grouped or organized. A concept map is, thus, a tool for visualizing the connections between concepts. Material is represented in a hierarchical fashion, with the most inclusive, most general concepts at the top of the map, and the more specific, less general concepts arranged beneath.

A. How is a concept map constructed?

1. **Concept boxes and links:** Educators define concepts as "perceived regularities in events or objects." In our biochemical maps, concepts include abstractions (for example, free energy), processes (for example, oxidative phosphorylation), and compounds (for example, glucose 6-phosphate). These broadly defined concepts are prioritized with the central idea positioned at the top of the page. The concepts that follow from this central idea are then drawn in boxes (Figure 1.13A). The size of the box and type indicate the relative importance of each idea. Lines are drawn between concept boxes to show which are related. The label on

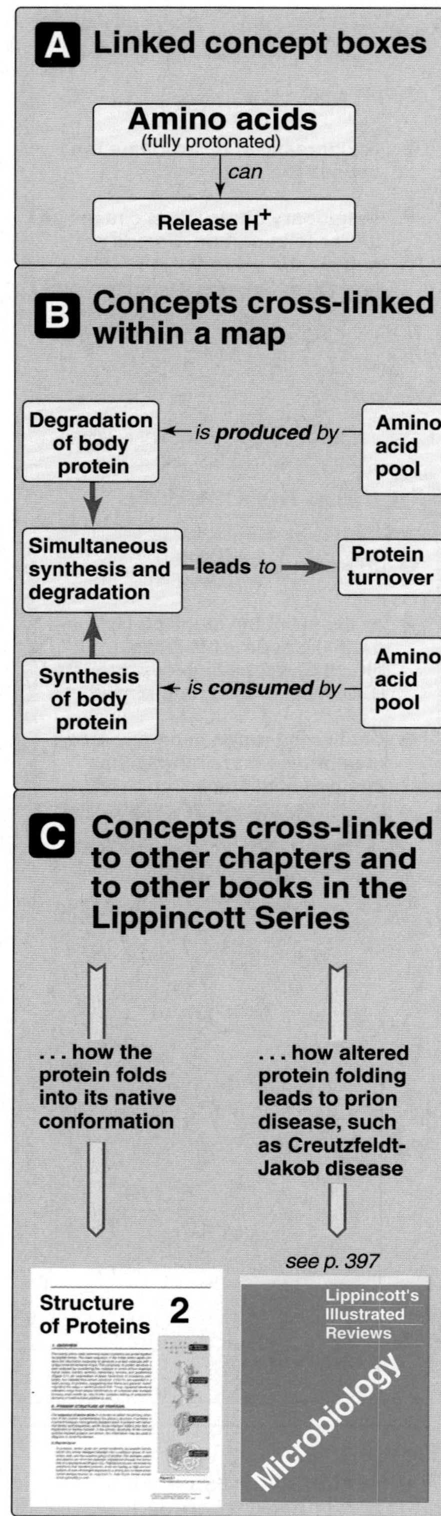

Figure 1.13
Symbols used in concept maps.

the line defines the relationship between two concepts, so that it reads as a valid statement, that is, the connection creates meaning. The lines with arrowheads indicate which direction the connection should be read.

2 **Cross-links:** Unlike linear flow charts or outlines, concept maps may contain cross-links that allow the reader to visualize complex relationships between ideas represented in different parts of the map (Figure 1.13B), or between the map and other chapters in this book, or companion books in the series (Figure 1.13C). Cross-links can thus identify concepts that are central to more than one discipline, empowering students to be effective in clinical situations, and on the United States Medical Licensure Examination (USMLE) or other examinations, that bridge disciplinary boundaries. Students learn to visually perceive non-linear relationships between facts, in contrast to cross referencing within linear text.

B. Concept maps and meaningful learning

"Meaningful learning" refers to a process in which students link new information to relevant concepts that they already possess. To learn meaningfully, individuals must consciously choose to relate new information to knowledge that they already know, rather than simply memorizing isolated facts or concept definitions. Rote is undesirable because such learning is easily forgotten, and is not readily applied in new problem-solving situations. Thus, the concept maps prepared by the authors should not be memorized. This would merely promote rote learning and defeat the purpose of the maps. Rather, the concept maps ideally function as templates or guides for organizing information, so the student can readily find the best ways to integrate new information into knowledge they already possess.

V. CHAPTER SUMMARY

Each amino acid has an **α-carboxyl group** and an **α-amino group** (except for proline, which has an **imino group**). At physiologic pH, the α-carboxyl group is dissociated, forming the negatively charged carboxylate ion ($-COO^-$), and the α-amino group is protonated ($-NH_3^+$). Each amino acid also contains one of twenty distinctive **side chains** attached to the α-carbon atom. The chemical nature of this side chain determines the function of an amino acid in a protein, and provides the basis for classification of the amino acids as **nonpolar**, **uncharged polar**, **acidic**, or **basic**. All free amino acids, plus charged amino acids in peptide chains, can serve as **buffers**. The quantitative relationship between the concentration of a weak acid (HA) and its conjugate base (A^-) is described by the **Henderson-Hasselbalch equation**. Buffering occurs within ±1 pH unit of the pK_a, and is maximal when pH = pK_a, at which $[A^-] = [HA]$. The α-carbon of each amino acid (except glycine) is attached to four different chemical groups and is, therefore, a **chiral** or **optically active** carbon atom. Only the L-form of amino acids is found in proteins synthesized by the human body.

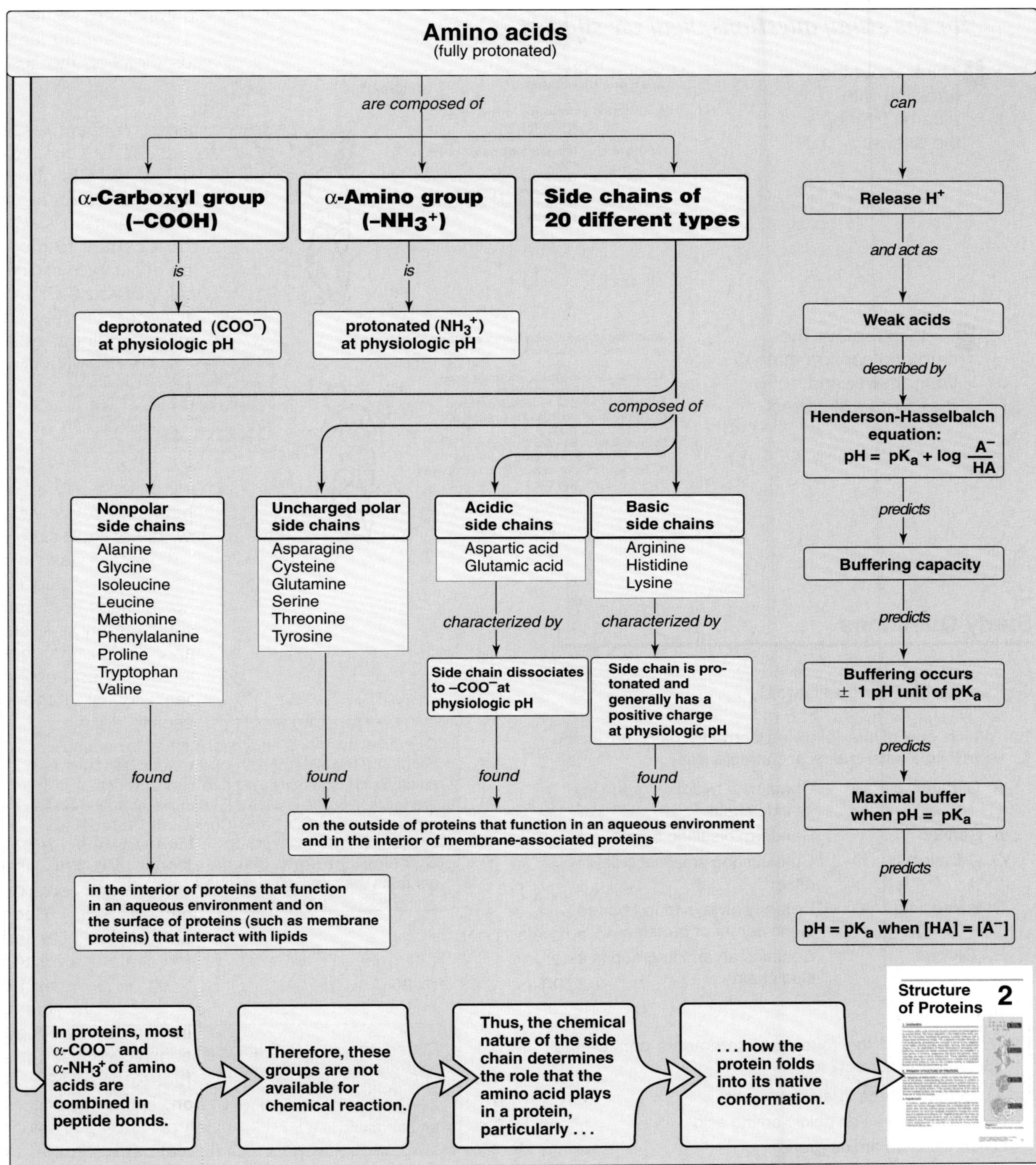

Figure 1.14
Key concept map for amino acids.

For the study questions, may we suggest...

1 Think about the question with a card covering the answer . . .

1.1 Which one of the following correctly pairs an amino acid with a valid chemical characteristic?

A. Glutamine: Contains a hydroxyl group in its side chain
B. Serine: Can form disulfide bonds
C. Cysteine: Contains the smallest side chain
D. Isoleucine: Is nearly always found buried in the center of proteins
E. Glycine: Contains an amide group in its side chain

2. . .then remove the card and confirm that your answer and reasoning are correct.

1.1 Which one of the following correctly pairs an amino acid with a valid chemical characteristic?

A. Glutamine: Contains a hydroxyl group in its side chain
B. Serine: Can form disulfide bonds
C. Cysteine: Contains the smallest side chain
D. Isoleucine: Is nearly always found buried in the center of proteins
E. Glycine: Contains an amide group in its side chain

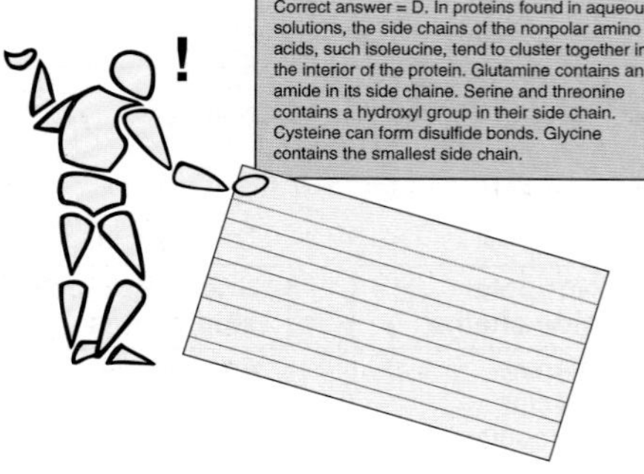

Study Questions

Choose the ONE correct answer

1.1 Which one of the following correctly pairs an amino acid with a valid chemical characteristic?

A. Glutamine: Contains a hydroxyl group in its side chain
B. Serine: Can form disulfide bonds
C. Cysteine: Contains the smallest side chain
D. Isoleucine: Is nearly always found buried in the center of proteins
E. Glycine: Contains an amide group in its side chain

Correct answer = D. In proteins found in aqueous solutions, the side chains of the nonpolar amino acids, such isoleucine, tend to cluster together in the interior of the protein. Glutamine contains an amide in its side chain. Serine and threonine contain a hydroxyl group in their side chain. Cysteine can form disulfide bonds. Glycine contains the smallest side chain.

1.2 Which one of the following statements concerning glutamine is correct?

A. Contains three titratable groups
B. Is classified as an acidic amino acid
C. Contains an amide group
D. Has E as its one-letter symbol
E. Migrates to the cathode (negative electrode) during electrophoresis at pH 7.0

Correct answer = C. Glutamine contains two titratable groups, α-carboxyl and α-amino. Glutamine is a polar, neutral amino acid that shows little electrophoretic migration at pH 7.0. The symbol for glutamine is "Q."

Structure of Proteins

2

I. OVERVIEW

The twenty amino acids commonly found in proteins are joined together by peptide bonds. The linear sequence of the linked amino acids contains the information necessary to generate a protein molecule with a unique three-dimensional shape. The complexity of protein structure is best analyzed by considering the molecule in terms of four organizational levels, namely, primary, secondary, tertiary, and quaternary (Figure 2.1). An examination of these hierarchies of increasing complexity has revealed that certain structural elements are repeated in a wide variety of proteins, suggesting that there are general "rules" regarding the ways in which proteins fold. These repeated structural elements range from simple combinations of α-helices and β-sheets forming small motifs (p. 18) to the complex folding of polypeptide domains of multifunctional proteins (p. 18).

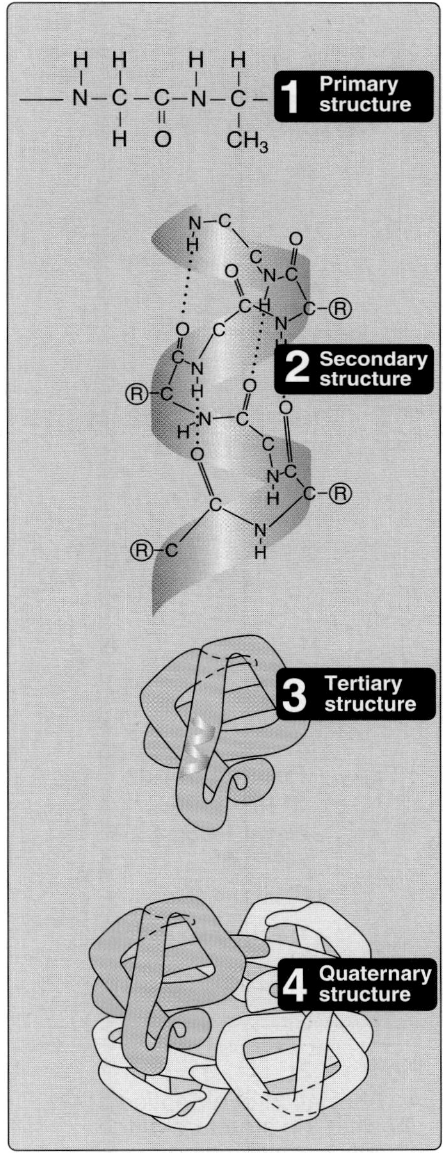

Figure 2.1
Four hierarchies of protein structure.

II. PRIMARY STRUCTURE OF PROTEINS

The **sequence of amino acids** in a protein is called the primary structure of the protein. Understanding the primary structure of proteins is important because many genetic diseases result in proteins with abnormal amino acid sequences, which cause improper folding and loss or impairment of normal function. If the primary structures of the normal and the mutated proteins are known, this information may be used to diagnose or study the disease.

A. Peptide bond

In proteins, amino acids are joined covalently by peptide bonds, which are amide linkages between the α-carboxyl group of one amino acid, and the α-amino group of another. For example, valine and alanine can form the dipeptide valylalanine through the formation of a peptide bond (Figure 2.2). Peptide bonds are not broken by conditions that denature proteins, such as heating or high concentrations of urea. Prolonged exposure to a strong acid or base at elevated temperatures is required to hydrolyze these bonds nonenzymically.

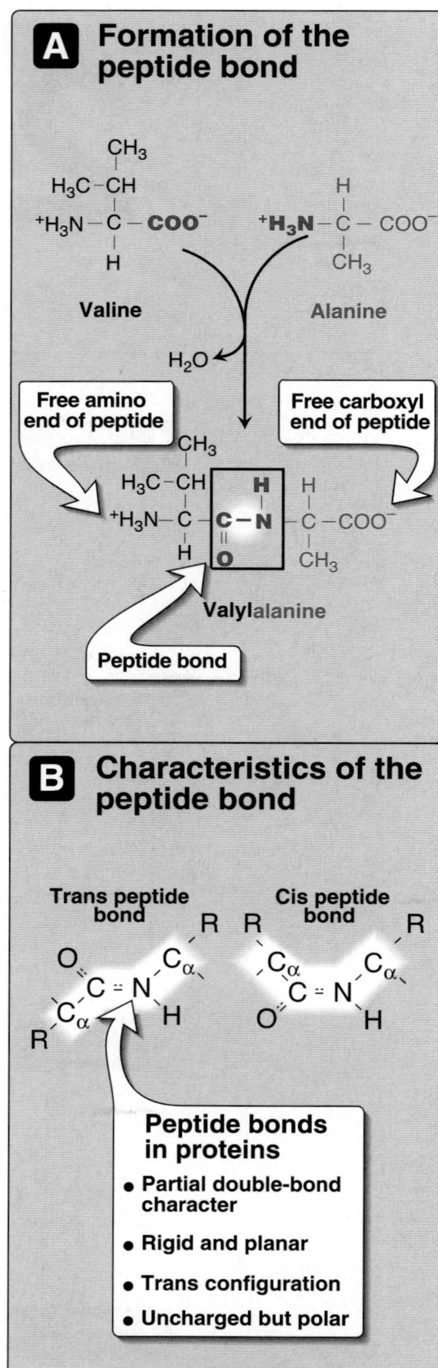

Figure 2.2
A. Formation of a peptide bond, showing the structure of the dipeptide valylalanine.
B. Characteristics of the peptide bond.

1. **Naming the peptide:** By convention, the free amino end of the peptide chain (N-terminal) is written to the left and the free carboxyl end (C-terminal) to the right. Therefore, all amino sequences are read from the N- to the C-terminal end of the peptide. For example, in Figure 2.2A, the order of the amino acids is "valine, alanine" not "alanine, valine." Linkage of many amino acids through peptide bonds results in an unbranched chain called a **polypeptide**. Each component amino acid in a polypeptide is called a "**residue**" or "**moiety**." When a polypeptide is named, all amino acid residues have their suffixes (–ine, –an, –ic, or –ate) changed to –yl, with the exception of the C-terminal amino acid. For example, a tripeptide composed of an N-terminal valine, a glycine, and a C-terminal leucine is called valylglycylleucine.

2. **Characteristics of the peptide bond:** The peptide bond has a **partial double-bond character**, that is, it is shorter than a single bond, and is **rigid** and **planar** (Figure 2.2B). This prevents free rotation around the bond between the carbonyl carbon and the nitrogen of the peptide bond. However, the bonds between the α-carbons and the α-amino or α-carboxyl groups can be freely rotated (although they are limited by the size and character of the R-groups). This allows the polypeptide chain to assume a variety of possible configurations. The peptide bond is generally a **trans bond** (instead of cis, see Figure 2.2B), in large part because of steric interference of the R-groups when in the cis position.

3. **Polarity of the peptide bond:** Like all amide linkages, the –C=O and –NH groups of the peptide bond are uncharged, and neither accept nor release protons over the pH range of 2 to 12. Thus, the charged groups present in polypeptides consist solely of the N-terminal α-amino group, the C-terminal α-carboxyl group, and any ionized groups present in the side chains of the constituent amino acids. [Note: The – C=O and –NH groups of the peptide bond are polar, and are involved in hydrogen bonds, for example, in α-helices and β-sheet structures, described on pp. 16–17.]

B. Determination of the amino acid composition of a polypeptide

The first step in determining the primary structure of a polypeptide is to identify and quantitate its constituent amino acids. A purified sample of the polypeptide to be analyzed is first hydrolyzed by strong acid at 110°C for 24 hours. This treatment cleaves the peptide bonds, and releases the individual amino acids, which can be separated by **cation-exchange chromatography**. In this technique, a mixture of amino acids is applied to a column that contains a resin to which a negatively charged group is tightly attached. [Note: If the attached group is positively charged, the column becomes an **anion-exchange column**.] The amino acids bind to the column with different affinities, depending on their charges, hydrophobicity, and other characteristics. Each amino acid is sequentially released from the chromatography column by eluting with solutions of increasing ionic strength and pH (Figure 2.3). The separated amino acids contained in the eluate from the column are quantitated by heating them with **ninhydrin**—a reagent that forms a purple compound with most amino acids, ammonia, and amines. The amount of each amino acid is determined spectrophotometrically by measuring the amount of light absorbed by the ninhydrin derivative. The analysis described above is performed using an **amino acid analyzer**—an automated machine whose components are depicted in Figure 2.3.

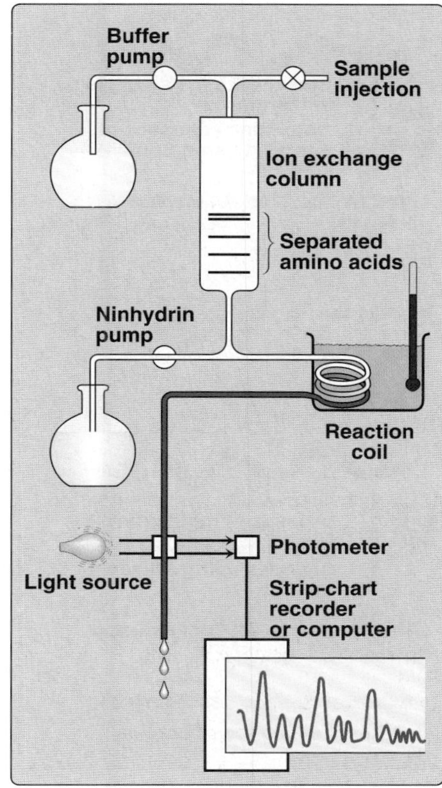

Figure 2.3
Determination of the amino acid composition of a polypeptide using an amino acid analyzer.

C. Sequencing of the peptide from its N-terminal end

Sequencing is a stepwise process of identifying the specific amino acids at each position in the peptide chain, beginning at the N-terminal end. **Phenylisothiocyanate**, known as **Edman's reagent**, is used to label the amino-terminal residue under mildly alkaline conditions (Figure 2.4). The resulting phenylthiohydantoin (PTH) derivative introduces an instability in the N-terminal peptide bond that can be selectively hydrolyzed without cleaving the other peptide bonds. The identity of the amino acid derivative can then be determined. Edman's reagent can be applied repeatedly to the shortened peptide obtained in each previous cycle. This process has been automated and, currently, the repetition of the method can be employed by a machine ("sequenator") to determine the sequence of more than 100 amino acid residues, starting at the amino terminal end of a polypeptide.

D. Cleavage of the polypeptide into smaller fragments

Many polypeptides have a primary structure composed of more than 100 amino acids. Such molecules cannot be sequenced directly from end to end by a sequenator. However, these large molecules can be cleaved at specific sites, and the resulting fragments sequenced. By using more than one cleaving agent (enzymes and/or chemicals) on separate samples of the purified polypeptide, overlapping fragments can be generated that permit the proper ordering of the sequenced fragments, thus providing a complete amino acid sequence of the large polypeptide (Figure 2.5).

E. Determination of a protein's primary structure by DNA sequencing

The sequence of nucleotides in a coding region of the DNA specifies the amino acid sequence of a polypeptide. Therefore, if the nucleotide sequence can be determined, it is possible, from knowledge of the genetic code (see p. 429), to translate the sequence of nucleotides into the corresponding amino acid sequence of that

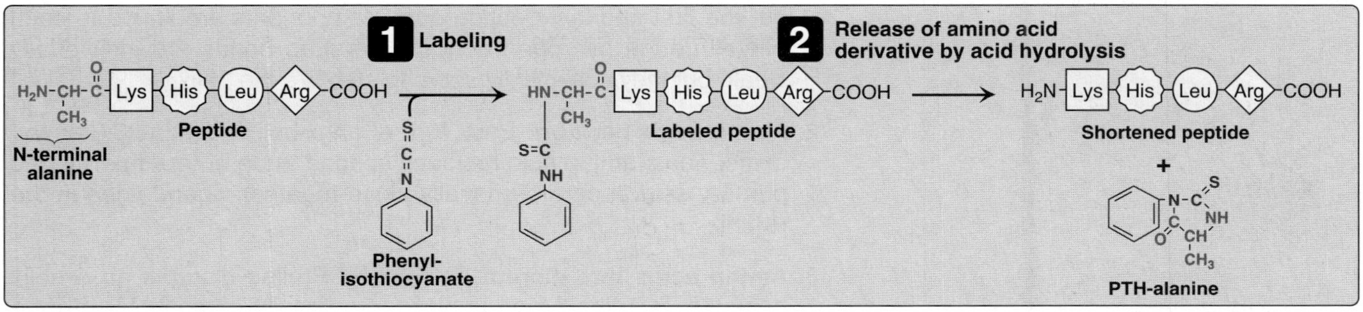

Figure 2.4
Determination of the amino-terminal residue of a polypeptide by Edman degradation.

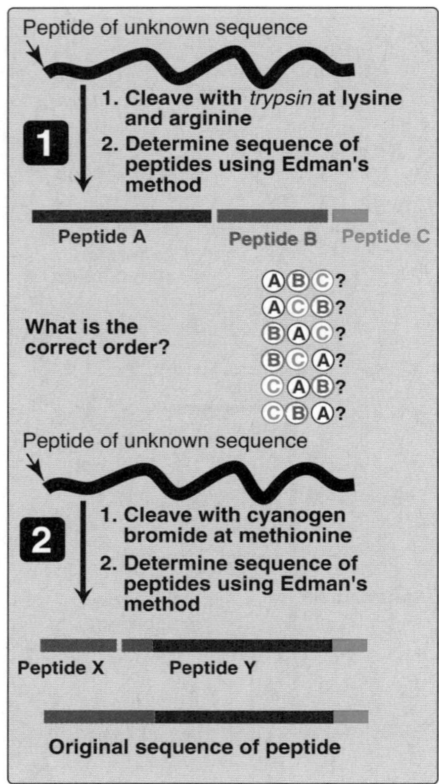

Figure 2.5
Overlapping of peptides produced by the action of *trypsin* and cyanogen bromide.

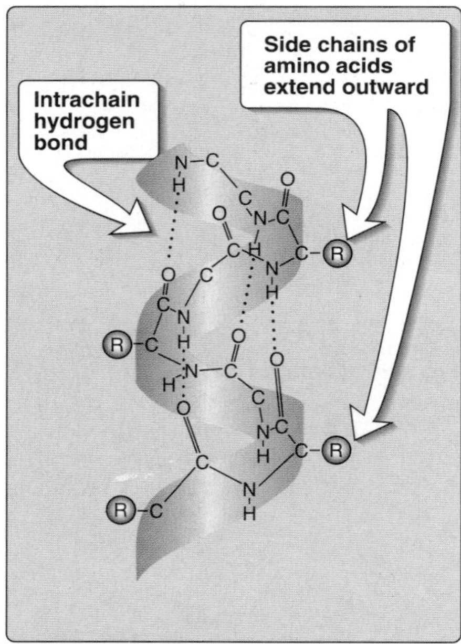

Figure 2.6
α-Helix showing peptide backbone.

polypeptide. This process, although routinely used to obtain the amino acid sequences of proteins, has the limitations of not being able to predict the positions of disulfide bonds in the folded chain, and not identifying any amino acids that are modified after their incorporation into the polypeptide (post-translational modification, see p. 440). Therefore, direct protein sequencing is an extremely important tool for determining the true character of the primary sequence of many polypeptides.

III. SECONDARY STRUCTURE OF PROTEINS

The polypeptide backbone does not assume a random three-dimensional structure, but instead generally forms regular arrangements of amino acids that are located near to each other in the linear sequence. These arrangements are termed the **secondary structure** of the polypeptide. The α-helix, β-sheet, and β-bend are examples of secondary structures frequently encountered in proteins. [Note: The collagen helix, another example of secondary structure, is discussed on p. 43.]

A. α-Helix

There are several different polypeptide helices found in nature, but the α-helix is the most common. It is a spiral structure, consisting of a tightly packed, coiled polypeptide backbone core, with the side chains of the component amino acids extending outward from the central axis to avoid interfering sterically with each other (Figure 2.6). A very diverse group of proteins contains α-helices. For example, the keratins are a family of closely related, fibrous proteins whose structure is nearly entirely α-helical. They are a major component of tissues such as hair and skin, and their rigidity is determined by the number of disulfide bonds between the constituent polypeptide chains. In contrast to keratin, myoglobin, whose structure is approximately eighty percent α-helical, is a globular, flexible molecule (see p. 26).

1. **Hydrogen bonds:** An α-helix is stabilized by extensive hydrogen bonding between the peptide-bond carbonyl oxygens and amide hydrogens that are part of the polypeptide backbone (see Figure 2.6). The hydrogen bonds extend up the spiral from the carbonyl oxygen of one peptide bond to the –NH– group of a peptide linkage four residues ahead in the polypeptide. This ensures that all but the first and last peptide bond components are linked to each other through hydrogen bonds. Hydrogen bonds are individually weak, but they collectively serve to stabilize the helix.

2. **Amino acids per turn:** Each turn of an α-helix contains 3.6 amino acids. Thus, amino acid residues spaced three or four apart in the primary sequence are spatially close together when folded in the α-helix.

3. **Amino acids that disrupt an α-helix:** Proline disrupts an α-helix because its imino group is not geometrically compatible with the right-handed spiral of the α-helix. Instead, it inserts a kink in the chain, which interferes with the smooth, helical structure. Large

numbers of charged amino acids (for example, glutamate, aspartate, histidine, lysine, or arginine) also disrupt the helix by forming ionic bonds, or by electrostatically repelling each other. Finally, amino acids with bulky side chains, such as tryptophan, or amino acids, such as valine or isoleucine, that branch at the β-carbon (the first carbon in the R-group, next to the α-carbon) can interfere with formation of the α-helix if they are present in large numbers.

B. β-Sheet

The β-sheet is another form of secondary structure in which all of the peptide bond components are involved in hydrogen bonding (Figure 2.7A). The surfaces of β-sheets appear "pleated," and these structures are, therefore, often called "**β-pleated sheets**." When illustrations are made of protein structure, β-strands are often visualized as broad arrows (Figure 2.7B).

1. **Comparison of a β-sheet and an α-helix:** Unlike the α-helix, β-sheets are composed of two or more peptide chains (β-strands), or segments of polypeptide chains, which are almost fully extended. Note also that in β-sheets the hydrogen bonds are perpendicular to the polypeptide backbone (see Figure 2.7A).

2. **Parallel and antiparallel sheets:** A β-sheet can be formed from two or more separate polypeptide chains or segments of polypeptide chains that are arranged either antiparallel to each other (with the N-terminal and C-terminal ends of the β-strands alternating as shown in Figure 2.7B), or parallel (with all the N-termini of the β-strands together as shown in Figure 2.7C). When the hydrogen bonds are formed between the polypeptide backbones of separate polypeptide chains, they are termed **interchain bonds**. A β-sheet can also be formed by a single polypeptide chain folding back on itself (see Figure 2.7C). In this case, the hydrogen bonds are **intrachain bonds**. In globular proteins, β-sheets always have a right-handed curl, or twist, when viewed along the polypeptide backbone. [Note: Twisted β-sheets often form the core of globular proteins.]

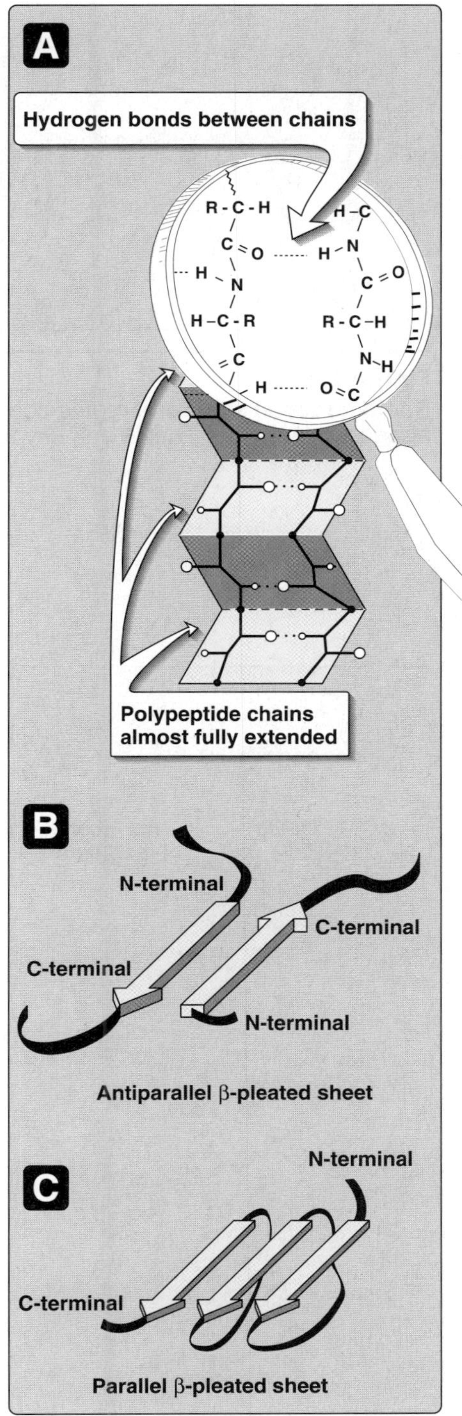

Figure 2.7
A. Structure of a β-sheet. B. An antiparallel β-sheet with the β-strands represented as broad arrows. C. A parallel β-sheet formed from a single polypeptide chain folding back on itself.

C. β-Bends (reverse turns)

β-Bends reverse the direction of a polypeptide chain, helping it form a compact, globular shape. They are usually found on the surface of protein molecules, and often include charged residues. [Note: β-Bends were given this name because they often connect successive strands of antiparallel β-sheets.] β-Bends are generally composed of four amino acids, one of which may be proline—the imino acid that causes a "kink" in the polypeptide chain. Glycine, the amino acid with the smallest R-group, is also frequently found in β-bends. β-Bends are stabilized by the formation of hydrogen and ionic bonds.

D. Nonrepetitive secondary structure

Approximately one half of an average globular protein is organized into repetitive structures, such as the α-helix and/or β-sheet. The remainder of the polypeptide chain is described as having a loop or coil conformation. These nonrepetitive secondary structures are not

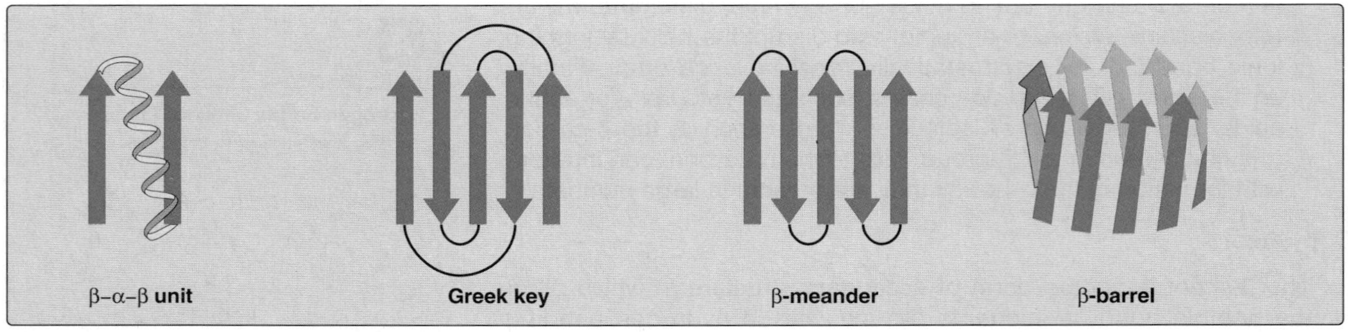

Figure 2.8
Some common structural motifs combining α-helices and β-sheets. The names describe their schematic appearance. [Note: The Greek key takes its name from a design often found in classical Greek pottery.]

"random," but rather simply have a less regular structure than those described above. [Note: The term "random coil" refers to the disordered structure obtained when proteins are denatured (see p. 21).]

E. Supersecondary structures (motifs)

Globular proteins are constructed by combining secondary structural elements (α-helices, β-sheets, nonrepetitive sequences). These form primarily the core region—that is, the interior of the molecule. They are connected by loop regions (for example, β-bends) at the surface of the protein. Supersecondary structures are usually produced by packing side chains from adjacent secondary structural elements close to each other. Thus, for example, α-helices and β-sheets that are adjacent in the amino acid sequence are also usually (but not always) adjacent in the final, folded protein. Some of the more common motifs are illustrated in Figure 2.8.

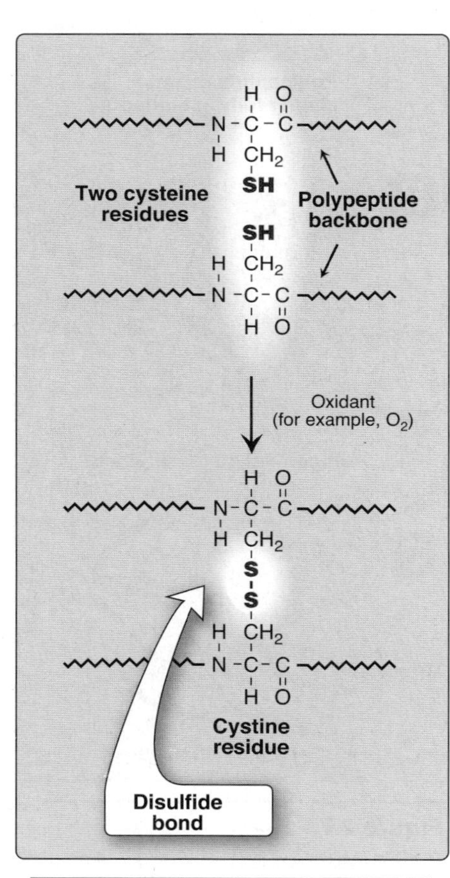

Figure 2.9
Formation of a disulfide bond by the oxidation of two cysteine residues, producing one cystine residue.

IV. TERTIARY STRUCTURE OF GLOBULAR PROTEINS

The primary structure of a polypeptide chain determines its **tertiary structure**. [Note: "Tertiary" refers both to the folding of domains (the basic units of structure and function, see discussion below), and the final arrangement of domains in the polypeptide.] The structure of globular proteins in **aqueous solution** is compact, with a high-density (close packing) of the atoms in the core of the molecule. **Hydrophobic side chains** are buried in the **interior**, whereas **hydrophilic groups** are generally found on the **surface** of the molecule. All hydrophilic groups (including components of the peptide bond) located in the interior of the polypeptide are involved in hydrogen bonds or electrostatic interactions. [Note: The α-helix and β-sheet structures provide maximal hydrogen bonding for peptide bond components within the interior of polypeptides. This eliminates the possibility that water molecules may become bound to these hydrophilic groups and, thus, disrupt the integrity of the protein.]

A. Domains

Domains are the fundamental functional and three-dimensional structural units of a polypeptide. Polypeptide chains that are greater than 200 amino acids in length generally consist of two or more domains. The core of a domain is built from combinations of **super-secondary structural elements (motifs)**. Folding of the peptide chain within a domain usually occurs independently of folding in other domains. Therefore, each domain has the characteristics of a small, compact globular protein that is structurally independent of the other domains in the polypeptide chain.

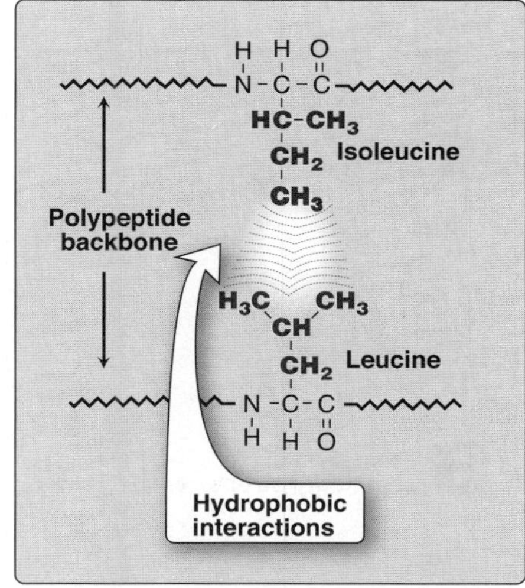

Figure 2.10
Hydrophobic interactions between amino acids with nonpolar side chains.

B. Interactions stabilizing tertiary structure

The unique three-dimensional structure of each polypeptide is determined by its amino acid sequence. Interactions between the amino acid side chains guide the folding of the polypeptide to form a compact structure. Four types of interactions cooperate in stabilizing the tertiary structures of globular proteins.

1. **Disulfide bonds:** A disulfide bond is a covalent linkage formed from the sulfhydryl group (–SH) of each of **two cysteine residues**, to produce a **cystine** residue (Figure 2.9). The two cysteines may be separated from each other by many amino acids in the primary sequence of a polypeptide, or may even be located on two different polypeptide chains; the folding of the polypeptide chain(s) brings the cysteine residues into proximity, and permits covalent bonding of their side chains. A disulfide bond contributes to the stability of the three-dimensional shape of the protein molecule. For example, many disulfide bonds are found in proteins such as immunoglobulins that are secreted by cells. [Note: These strong, covalent bonds help stabilize the structure of proteins, and prevent them from becoming denatured in the extracellular environment.]

2. **Hydrophobic interactions:** Amino acids with nonpolar side chains tend to be located in the interior of the polypeptide molecule, where they associate with other hydrophobic amino acids (Figure 2.10). In contrast, amino acids with polar or charged side chains tend to be located on the surface of the molecule in contact with the polar solvent. [Note: Proteins located in nonpolar (lipid) environments, such as a membrane, exhibit the reverse arrangement—that is, hydrophilic amino acid side chains are located in the interior of the polypeptide, whereas hydrophobic amino acids are located on the surface of the molecule in contact with the nonpolar environment (see Figure 1.4, p. 4).] In each case, the segregation of R-groups occurs that is energetically most favorable.

3. **Hydrogen bonds:** Amino acid side chains containing oxygen– or nitrogen–bound hydrogen, such as in the alcohol groups of serine and threonine, can form hydrogen bonds with electron-rich atoms, such as the oxygen of a carboxyl group or carbonyl group of a peptide bond (Figure 2.11; see also Figure 1.6, p. 4). Formation of hydrogen bonds between polar groups on the surface of proteins and the aqueous solvent enhances the solubility of the protein.

4. **Ionic interactions:** Negatively charged groups, such as the carboxyl group ($-COO^-$) in the side chain of aspartate or glutamate, can interact with positively charged groups, such as the amino group ($-NH_3^+$) in the side chain of lysine (see Figure 2.11).

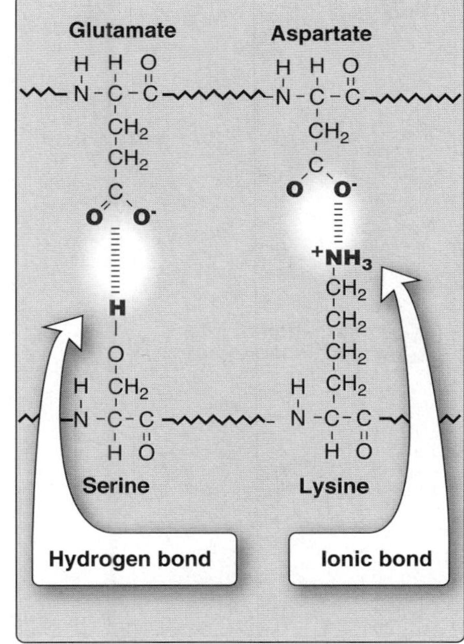

Figure 2.11
Interactions of side chains of amino acids through hydrogen bonds and ionic bonds.

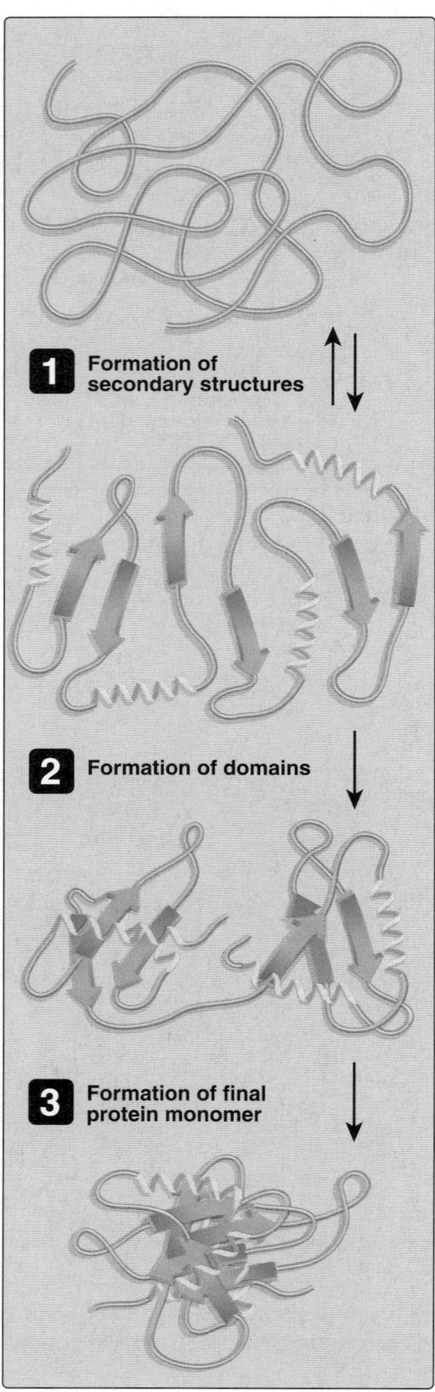

Figure 2.12
Steps in protein folding.

C. Protein folding

Interactions between the side chains of amino acids determine how a long polypeptide chain folds into the intricate three-dimensional shape of the functional protein. Protein folding, which occurs within the cell in seconds to minutes, employs a shortcut through the maze of all folding possibilities. As a peptide folds, its amino acid side chains are attracted and repulsed according to their chemical properties. For example, positively and negatively charged side chains attract each other. Conversely, similarly charged side chains repel each other. In addition, interactions involving hydrogen bonds, hydrophobic interactions, and disulfide bonds all seek to exert an influence on the folding process. This process of trial and error tests many, but not all, possible configurations, seeking a compromise in which attractions outweigh repulsions. This results in a correctly folded protein with a low energy state (Figure 2.12).]

D. Role of chaperones in protein folding

It is generally accepted that the information needed for correct protein folding is contained in the primary structure of the polypeptide. Given that premise, it is difficult to explain why most proteins when denatured (see below) do not resume their native conformations under favorable environmental conditions. One answer to this problem is that a protein begins to fold in stages during its synthesis, rather than waiting for synthesis of the entire chain to be totally completed. This limits competing folding configurations made available by longer stretches of nascent peptide. In addition, a specialized group of proteins, named "**chaperones**," are required for the proper folding of many species of proteins. The chaperones—also known as "**heat shock**" **proteins**—interact with the polypeptide at various stages during the folding process. Some chaperones are important in keeping the protein unfolded until its synthesis is finished, or act as catalysts by increasing the rates of the final stages in the folding process. Others protect proteins as they fold so that their vulnerable, exposed regions do not become tangled in unproductive encounters.

V. QUATERNARY STRUCTURE OF PROTEINS

Many proteins consist of a single polypeptide chain, and are defined as **monomeric proteins**. However, others may consist of two or more polypeptide chains that may be structurally identical or totally unrelated. The arrangement of these polypeptide subunits is called the quaternary structure of the protein. [Note: If there are two subunits, the protein is called "**dimeric**", if three subunits "**trimeric**", and, if several subunits, "**multimeric**."] Subunits are held together by noncovalent interactions (for example, hydrogen bonds, ionic bonds, and hydrophobic interactions). Subunits may either function independently of each other, or may work cooperatively, as in hemoglobin, in which the binding of oxygen to one subunit of the tetramer increases the affinity of the other subunits for oxygen (see p. 29).

VI. DENATURATION OF PROTEINS

Protein denaturation results in the unfolding and disorganization of the protein's secondary and tertiary structures, which are not accompanied by hydrolysis of peptide bonds. Denaturing agents include heat, organic solvents, mechanical mixing, strong acids or bases, detergents, and ions of heavy metals such as lead and mercury. Denaturation may, under ideal conditions, be reversible, in which case the protein refolds into its original native structure when the denaturing agent is removed. However, most proteins, once denatured, remain permanently disordered. Denatured proteins are often insoluble and, therefore, precipitate from solution.

VII. PROTEIN MISFOLDING

Protein folding is a complex, trial and error process that can sometimes result in improperly folded molecules. These misfolded proteins are usually tagged and degraded within the cell (see p. 441). However, this quality control system is not perfect, and intracellular or extracellular aggregates of misfolded proteins can accumulate, particularly as individuals age. Deposits of these misfolded proteins are associated with a number of diseases including amyloidoses.

A. Amyloidoses

Misfolding of proteins may occur spontaneously, or be caused by a mutation in a particular gene, which then produces an altered protein. In addition some apparently normal proteins can, after abnormal proteolytic cleavage, take on a unique conformational state that leads to the formation of long, fibrillar protein assemblies consisting of β-pleated sheets. Accumulation of these spontaneously aggregating proteins, called **amyloids**, has been implicated in many degenerating diseases—particularly in the neurodegenerative disorder, **Alzheimer disease**. The dominant component of the **amyloid plaque** that accumulates in Alzheimer disease is **Aβ**, a peptide of 40 to 43 amino acid residues. X-ray crystallography and infrared spectroscopy demonstrate a characteristic β-pleated sheet conformation in nonbranching fibrils. This peptide, when aggregated in a β-pleated sheet configuration, is neurotoxic, and is the central pathogenic event leading to the cognitive impairment characteristic of the disease. The Aβ amyloid that is deposited in the brain in Alzheimer disease is derived by proteolytic cleavages from the larger **amyloid precursor protein**—a single transmembrane protein expressed on the cell surface in the brain and other tissues (Figure 2.13). The Aβ peptides aggregate, generating the amyloid that is found in the brain parenchyma and around blood vessels. Most cases of Alzheimer disease are not genetically based, although at least five to ten percent of cases are familial. A second biologic factor involved in the development of Alzheimer disease is the accumulation of neurofibillary tangles in the brain. A key component of these tangled fibers is an abnormal form of the **tau protein**, which in its healthy version helps in the assembly of the microtubular structure. The defective tau, however, appears to block the actions of its normal counterpart.

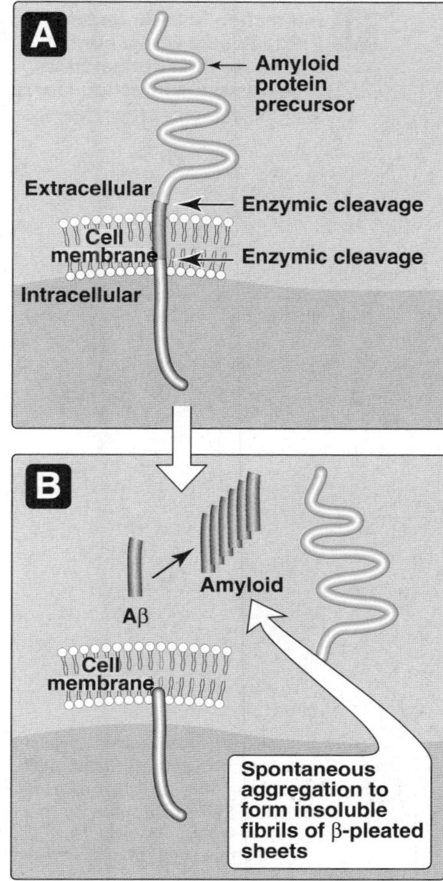

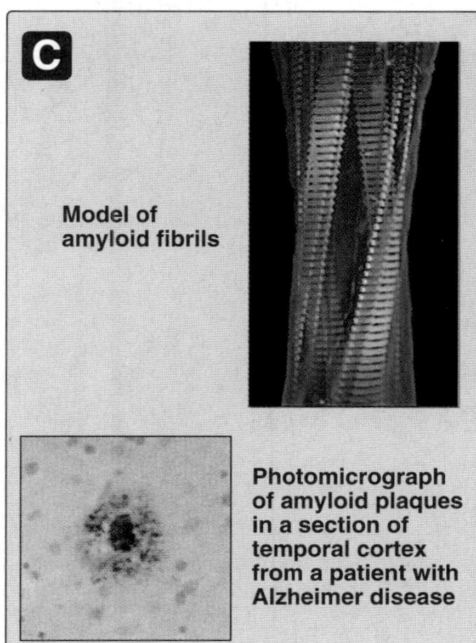

Figure 2.13
Formation of amyloid plaques found in Alzheimer disease.

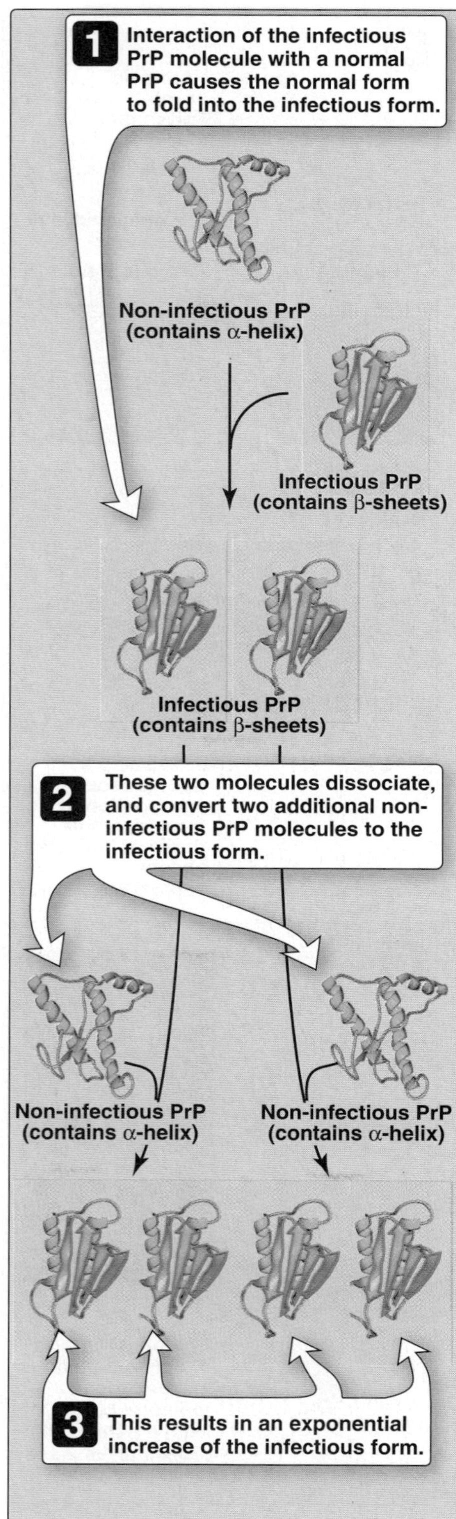

Figure 2.14
One proposed mechanism for multiplication of infectious prion agents.

B. Prion disease

The **prion protein** (**PrP**), has been strongly implicated as the causative agent of **transmissible spongiform encephalopathies (TSEs)**, including **Creutzfeldt-Jakob disease** in humans, **scrapie** in sheep, and **bovine spongiform encephalopathy** in cattle (popularly called "**mad cow disease**").[1] After an extensive series of purification procedures, scientists were astonished to find that the infectivity of the agent causing scrapie in sheep was associated with a single protein species that was not associated with detectable nucleic acid. This infectious protein is designated the prion protein. It is highly resistant to proteolytic degradation, and, when infectious, tends to form insoluble aggregates of fibrils, similar to the amyloid found in some other diseases of the brain. A noninfectious form of PrP, having the same amino acid and gene sequences as the infectious agent, is present in normal mammalian brains on the surface of neurons and glial cells. Thus, PrP is a host protein. No primary structure differences or alternate posttranslational modifications have been found between the normal and the infectious forms of the protein. The key to becoming infectious apparently lies in changes in the three-dimensional conformation of PrP. It has been observed that a number of α-helices present in noninfectious PrP are replaced by β-sheets in the infectious form (Figure 2.14). It is presumably this conformational difference that confers relative resistance to proteolytic degradation of infectious prions, and permits them to be distinguished from the normal PrP in infected tissue. The infective agent is thus an altered version of a normal protein, which acts as a "template" for converting the normal protein to the pathogenic conformation. The TSEs are invariably fatal, and no treatment is currently available that can alter this outcome.

VIII. CHAPTER SUMMARY

Central to understanding protein structure is the concept of the **native conformation** (Figure 2.15), which is the functional, fully-folded protein structure (for example, an active enzyme or structural protein). The unique three-dimensional structure of the native conformation is determined by its **primary structure**, that is, its amino acid sequence. Interactions between the amino acid side chains guide the folding of the polypeptide chain to form **secondary**, **tertiary**, and (sometimes) **quaternary structures**, which cooperate in stabilizing the native conformation of the protein. In addition, a specialized group of proteins named "**chaperones**" is required for the proper folding of many species of proteins. **Protein denaturation** results in the unfolding and disorganization of the protein's structure, which are not accompanied by hydrolysis of peptide bonds. Denaturation may be reversible or, more commonly, irreversible. Disease can occur when an apparently normal protein assumes a conformation that is cytotoxic, as in the case of **Alzheimer disease** and the **transmissible spongiform encephalopathies (TSEs)**, including **Creutzfeldt-Jakob disease**. In Alzheimer's disease, normal proteins, after abnormal chemical processing, take on a unique conformational state that leads to the formation of neurotoxic amyloid protein assemblies consisting of β-pleated sheets. In TSEs, the infective agent is an altered version of a normal prion protein that acts as a "template" for converting normal protein to the pathogenic conformation.

INFO LINK [1]See p. 397 in ***Lippincott's Illustrated Reviews: Microbiology*** for a more detailed discussion of prions.

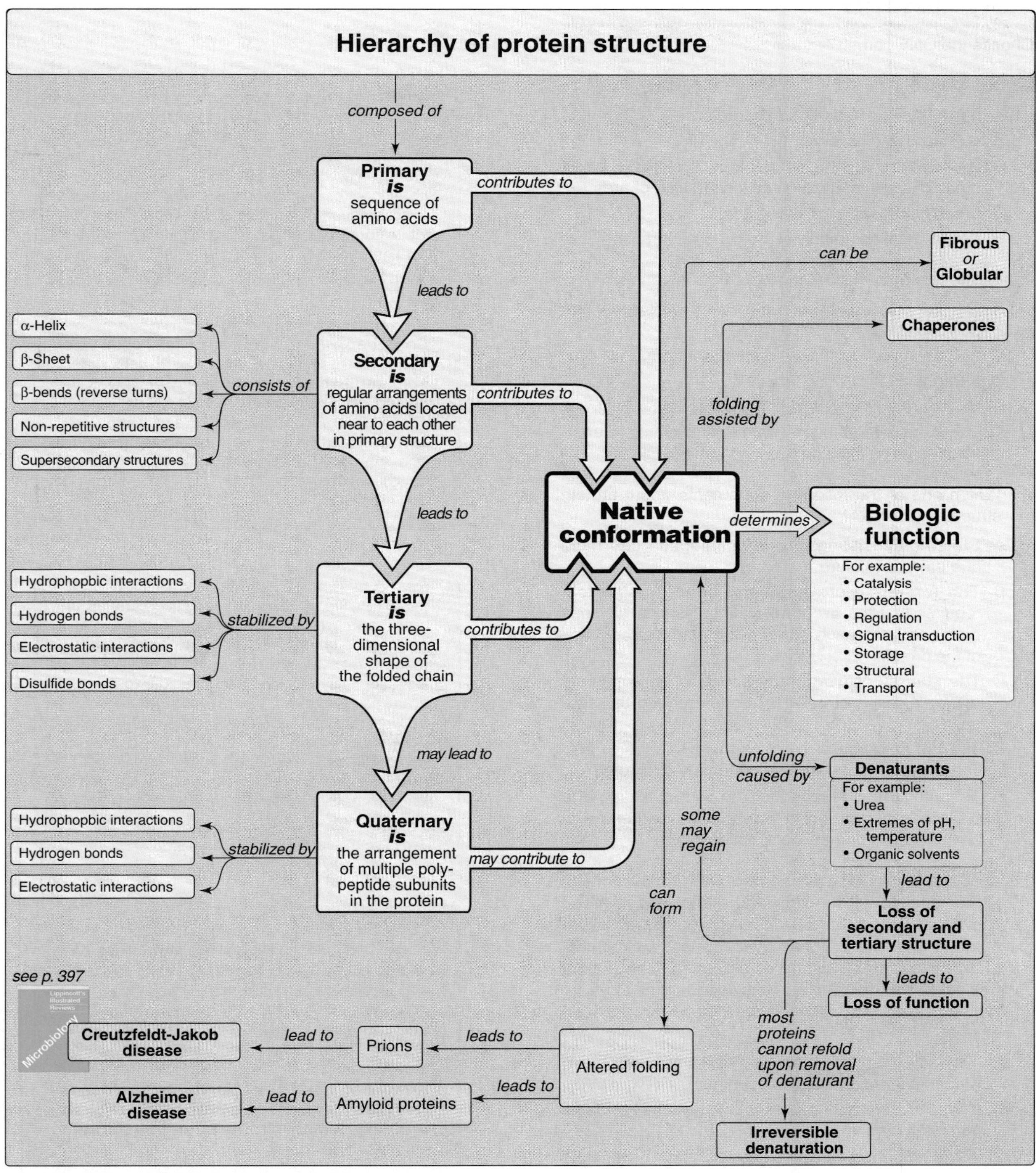

Figure 2.15
Key concept map for protein structure.

Study Questions

Choose the ONE correct answer

2.1 A peptide bond:

A. has a partial double-bond character.
B. is ionized at physiologic pH.
C. is cleaved by agents that denature proteins, such as organic solvents and high concentrations of urea.
D. is stable to heating in strong acids.
E. occurs most commonly in the cis configuration.

Correct answer = A. The peptide bond has a partial double-bond character. Unlike its components—the α-amino and α-carboxyl groups—the components of the peptide bond do not accept or give off protons. The peptide bond is not cleaved by organic solvents or urea, but is labile to strong acids. It is usually in the trans configuration.

2.2 Which one of the following statements is correct?

A. The α-helix can be composed of more than one polypeptide chain.
B. β-Sheets exist only in the antiparallel form.
C. β-Bends often contain proline.
D. Motifs are a type of secondary structure.
E. The α-helix is stabilized primarily by ionic interactions between the side chains of amino acids.

Correct answer = C. β-Bends often contain proline, which provides a kink. The α-helix differs from the β-sheet in that it always involves the coiling of a single polypeptide chain. The β-sheet occurs in both parallel and antiparallel forms. Motifs are elements of tertiary structure. The α-helix is stabilized primarily by hydrogen bonds between the –C=O and –NH– groups of peptide bonds.

2.3 Which one of the following statements about protein structure is correct?

A. Proteins consisting of one polypeptide can have quaternary structure.
B. The formation of a disulfide bond in a protein requires that the two participating cysteine residues be adjacent to each other in the primary sequence of the protein.
C. The stability of quaternary structure in proteins is mainly a result of covalent bonds among the subunits.
D. The denaturation of proteins always leads to irreversible loss of secondary and tertiary structure.
E. The information required for the correct folding of a protein is contained in the specific sequence of amino acids along the polypeptide chain.

Correct answer = E. The correct folding of a protein is guided by specific interactions among the side chains of the amino acid residues of a polypeptide chain. The two cysteine residues that react to form the disulfide bond may be a great distance apart in the primary structure (or on separate polypeptides), but are brought into close proximity by the three-dimensional folding of the polypeptide chain. Denaturation may either be reversible or irreversible. Quaternary structure requires more than one polypeptide chain. These chains associate through noncovalent interactions.

2.4 An 80-year-old man presented with impairment of higher intellectual function and alterations in mood and behavior. His family reported progressive disorientation and memory loss over the last six months. There is no family history of dementia. The patient was tentatively diagnosed with Alzeheimer disease. Which one of the following best describes the disease?

A. It is associated with β-amyloid—an abnormal protein with an altered amino acid sequence.
B. It results from accumulation of denatured proteins that have random conformations.
C. It is associated with the accumulation of amyloid precursor protein.
D. It is associated with the deposition of neurotoxic amyloid peptide aggregates.
E. It is an environmentally produced disease uninfluenced by the genetics of the individual.

Correct answer = D. Alzheimer disease is associated with long, fibrillar protein assemblies consisting of β-pleated sheets found in the brain and elsewhere. The disease is asssociated with abnormal processing of a normal protein. The accumulated altered protein occurs in a β-pleated sheet configuration that is neurotoxic. The Aβ amyloid that is deposited in the brain in Alzheimer disease is derived by proteolytic cleavages from the larger amyloid precursor protein—a single transmembrane protein expressed on the cell surface in the brain and other tissues. Most cases of Alzheimer disease are sporadic, although at least five to ten percent of cases are familial.

Globular Proteins

3

I. OVERVIEW

The previous chapter described the types of secondary and tertiary structures that are the bricks-and-mortar of protein architecture. By arranging these fundamental structural elements in different combinations, widely diverse proteins can be constructed that are capable of various specialized functions. This chapter examines the relationship between structure and function for the clinically important globular hemeproteins. Fibrous structural proteins are discussed in Chapter 4.

II. GLOBULAR HEMEPROTEINS

Hemeproteins are a group of specialized proteins that contain **heme** as a tightly bound **prosthetic group**. (See p. 54 for a discussion of prosthetic groups.) The role of the heme group is dictated by the environment created by the three-dimensional structure of the protein. For example, the heme group of a **cytochrome** functions as an electron carrier that is alternately oxidized and reduced (see p. 75). In contrast, the heme group of the enzyme **catalase** is part of the active site of the enzyme that catalyzes the breakdown of hydrogen peroxide (see p. 146). In **hemoglobin** and **myoglobin**, the two most abundant hemeproteins in humans, the heme group serves to reversibly bind oxygen.

A. Structure of heme

Heme is a complex of **protoporphyrin IX** and **ferrous iron (Fe^{2+})** (Figure 3.1). The iron is held in the center of the heme molecule by bonds to the four nitrogens of the porphyrin ring. The heme Fe^{2+} can form two additional bonds, one on each side of the planar porphyrin ring. For example, in myoglobin and hemoglobin, one of these positions is coordinated to the side chain of a histidine residue of the globin molecule, whereas the other position is available to bind oxygen (Figure 3.2). (See p. 276 for a discussion of the synthesis and degradation of heme.)

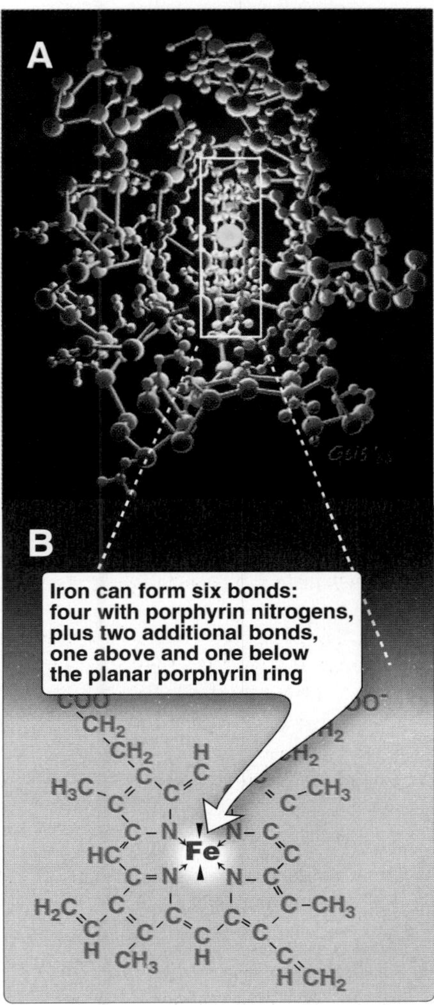

Figure 3.1
A. Hemeprotein (cytochrome c).
B. Structure of heme.

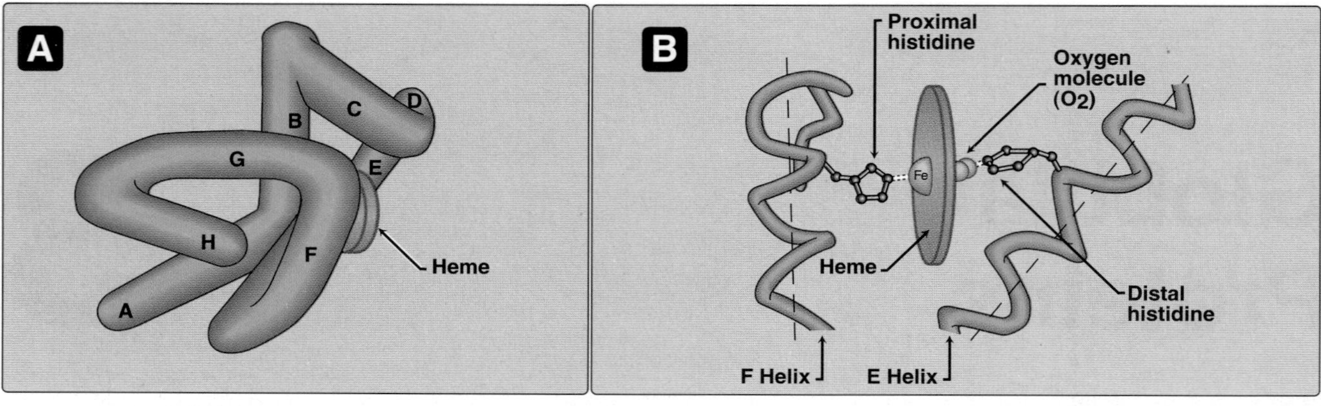

Figure 3.2
A. Model of myoglobin showing helices A to H. B. Schematic diagram of the oxygen-binding site of myoglobin.

B. Structure and function of myoglobin

Myoglobin, a hemeprotein present in **heart** and **skeletal muscle**, functions both as a reservoir for oxygen, and as an oxygen carrier that increases the rate of transport of oxygen within the muscle cell. Myoglobin consists of a single polypeptide chain that is structurally similar to the individual subunit polypeptide chains of the hemoglobin molecule. This homology makes myoglobin a useful model for interpreting some of the more complex properties of hemoglobin.

1. **α-Helical content:** Myoglobin is a compact molecule, with approximately eighty percent of its polypeptide chain folded into eight stretches of α-helix. These α-helical regions, labeled A to H in Figure 3.2A, are terminated either by the presence of proline, whose five-membered ring cannot be accommodated in an α-helix (see p. 16), or by β-bends and loops stabilized by hydrogen bonds and ionic bonds (see p. 17).

2. **Location of polar and nonpolar amino acid residues:** The interior of the myoglobin molecule is composed almost entirely of nonpolar amino acids. They are packed closely together, forming a structure stabilized by hydrophobic interactions between these clustered residues (see p. 19). In contrast, charged amino acids are located almost exclusively on the surface of the molecule, where they can form hydrogen bonds, with each other and with water.

3. **Binding of the heme group:** The heme group of myoglobin sits in a crevice in the molecule, which is lined with **nonpolar amino acids**. Notable exceptions are two histidine residues (Figure 3.2B). One, the **proximal histidine**, binds directly to the iron of heme. The second, or **distal histidine**, does not directly interact with the heme group, but helps stabilize the binding of oxygen to the ferrous iron. The protein, or globin, portion of myoglobin thus creates a special microenvironment for the heme that permits the reversible binding of one oxygen molecule (oxygenation). The simultaneous loss of electrons by the ferrous iron (oxidation) occurs only rarely.

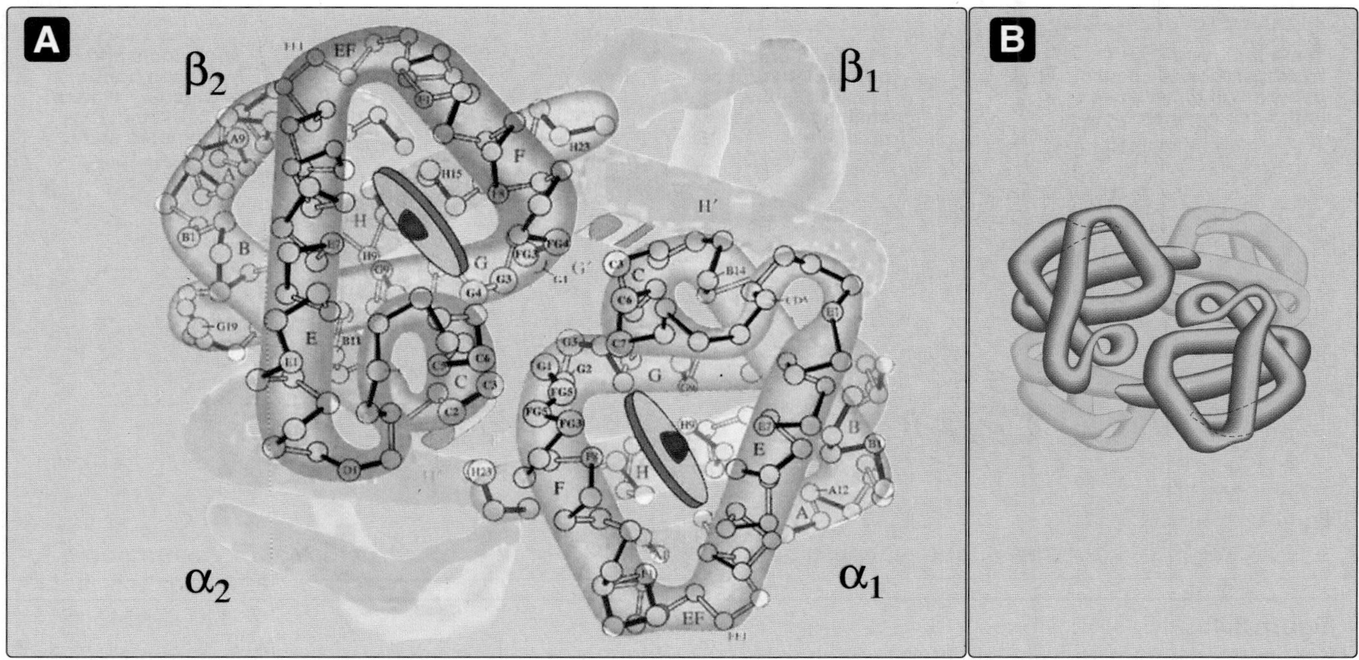

Figure 3.3
A. Structure of hemoglobin showing the polypeptide backbone. B. Simplified drawing showing the helices.

C. Structure and function of hemoglobin

Hemoglobin is found exclusively in **red blood cells**, where its main function is to transport oxygen from the lungs to the capillaries of the tissues. **Hemoglobin A**, the major hemoglobin in adults, is composed of four polypeptide chains—two alpha (α) chains and two beta (β) chains—held together by noncovalent interactions (Figure 3.3). Each subunit has stretches of α-helical structure, and a heme-binding pocket similar to that described for myoglobin. However, the tetrameric hemoglobin molecule is structurally and functionally more complex than myoglobin. For example, hemoglobin can transport CO_2 from the tissues to the lungs, and carry four molecules of O_2 from the lungs to the cells of the body. Further, the oxygen-binding properties of hemoglobin are regulated by interaction with **allosteric effectors** (see p. 62).

1. **Quaternary structure of hemoglobin:** The hemoglobin tetramer can be envisioned as being composed of two identical dimers, $(\alpha\beta)_1$ and $(\alpha\beta)_2$, in which the numbers refer to dimers one and two. The two polypeptide chains within each dimer are held tightly together, primarily by hydrophobic interactions (Figure 3.4). [Note: In this instance, hydrophobic amino acid residues are localized not only in the interior of the molecule, but also in a region on the surface of each subunit. Interchain hydrophobic interactions form strong associations between α-subunits and β-subunits in the dimers.] Ionic and hydrogen bonds also occur between the members of the dimer. In contrast, the two dimers are able to move with respect to each other, being held together primarily by polar bonds. The

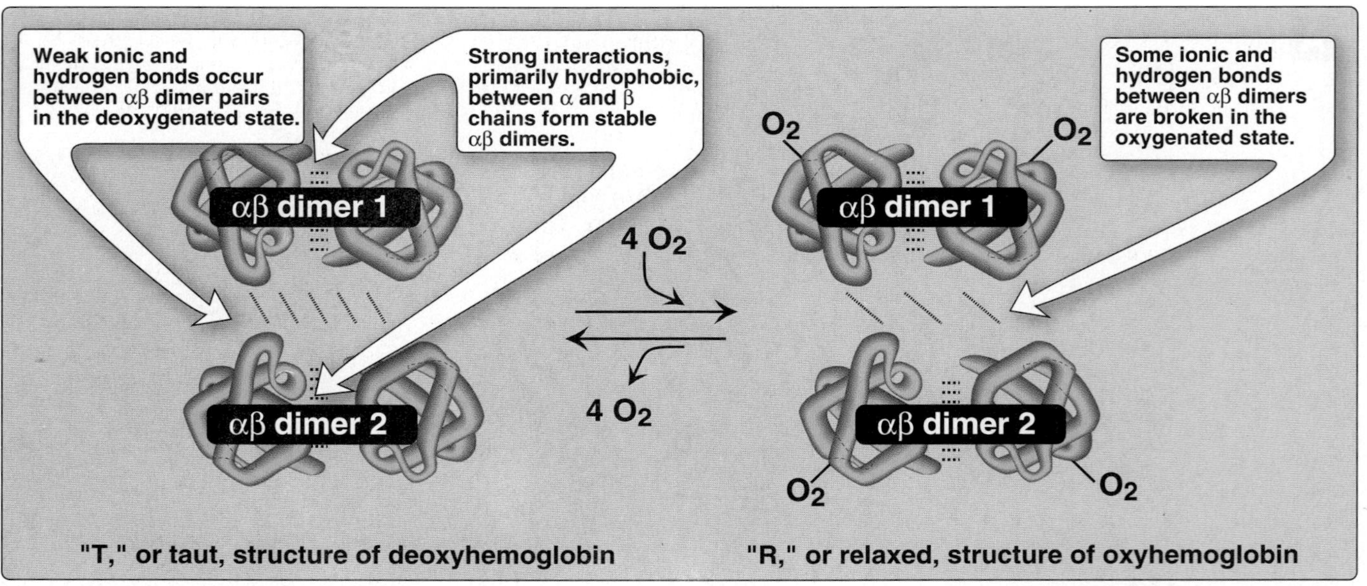

Figure 3.4
Schematic diagram showing structural changes resulting from oxygenation and deoxygenation of hemoglobin.

weaker interactions between these mobile dimers result in the two dimers occupying different relative positions in deoxyhemoglobin as compared with oxyhemoglobin (see Figure 3.4).

a. **T form:** The deoxy form of hemoglobin is called the "T," or **taut** (**tense**) form. In the T form, the two αβ dimers interact through a network of ionic bonds and hydrogen bonds that constrain the movement of the polypeptide chains. The T form is the **low oxygen-affinity form** of hemoglobin.

b. **R form:** The binding of oxygen to hemoglobin causes the rupture of some of the ionic bonds and hydrogen bonds between the αβ dimers. This leads to a structure called the "R," or **relaxed** form, in which the polypeptide chains have more freedom of movement (see Figure 3.4). The R form is the **high oxygen-affinity form** of hemoglobin.

D. Binding of oxygen to myoglobin and hemoglobin

Myoglobin can bind only one molecule of oxygen (O_2), because it contains only one heme group. In contrast, hemoglobin can bind four oxygen molecules—one at each of its four heme groups. The degree of **saturation** (**Y**) of these oxygen-binding sites on all myoglobin or hemoglobin molecules can vary between zero (all sites are empty) and 100 percent (all sites are full, Figure 3.5).

1. **Oxygen dissociation curve:** A plot of Y measured at different partial pressures of oxygen (pO_2) is called the oxygen dissociation curve. The curves for myoglobin and hemoglobin show important differences (see Figure 3.5). This graph illustrates that myoglobin has a higher oxygen affinity than does hemoglobin. The partial

pressure of oxygen needed to achieve half-saturation of the binding sites (P_{50}) is approximately 1 mm Hg for myoglobin and 26 mm Hg for hemoglobin. [Note: The higher the oxygen affinity (that is, the more tightly oxygen binds), the lower the P_{50}.]

a. **Myoglobin:** The oxygen dissociation curve for myoglobin has a **hyperbolic** shape (see Figure 3.5). This reflects the fact that myoglobin reversibly binds a single molecule of oxygen. Thus, oxygenated (MbO_2) and deoxygenated (Mb) myoglobin exist in a simple equilibrium:

$$Mb + O_2 \rightleftarrows MbO_2$$

The equilibrium is shifted to the right or to the left as oxygen is added to or removed from the system. [Note: Myoglobin is designed to bind oxygen released by hemoglobin at the low pO_2 found in muscle. Myoglobin, in turn, releases oxygen within the muscle cell in response to oxygen demand.]

b. **Hemoglobin:** The oxygen dissociation curve for hemoglobin is **sigmoidal** in shape (see Figure 3.5), indicating that the subunits cooperate in binding oxygen. **Cooperative binding** of oxygen by the four subunits of hemoglobin means that the binding of an oxygen molecule at one heme group increases the oxygen affinity of the remaining heme groups in the same hemoglobin molecule (Figure 3.6). This effect is referred to as **heme-heme interaction** (see below). Although it is more difficult for the first oxygen molecule to bind to hemoglobin, the subsequent binding of oxygen occurs with high affinity, as shown by the steep upward curve in the region near 20 to 30 mm Hg (see Figure 3.5).

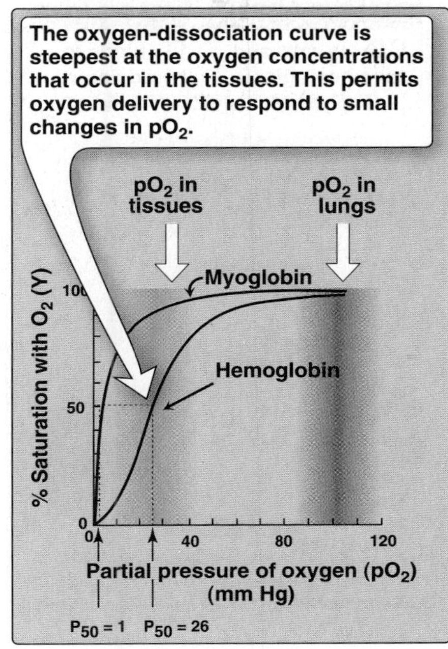

Figure 3.5
Oxygen dissociation curves for myoglobin and hemoglobin.

E. Allosteric effects

The ability of hemoglobin to reversibly bind oxygen is affected by the **pO_2** (through heme-heme interactions as described above), the **pH** of the environment, the **pCO_2**, and the availability of **2,3-bisphosphoglycerate**. These are collectively called allosteric ("other site") effectors, because their interaction at one site on the hemoglobin molecule affects the binding of oxygen to heme groups at other locations on the molecule. [Note: The binding of oxygen to myoglobin is not influenced by the allosteric effectors of hemoglobin.]

1. **Heme-heme interactions:** The sigmoidal oxygen-binding curve reflects specific structural changes that are initiated at one heme group and transmitted to other heme groups in the hemoglobin tetramer. The net effect is that the affinity of hemoglobin for the last oxygen bound is approximately 300 times greater than its affinity for the first oxygen bound.

a. **Loading and unloading oxygen:** The cooperative binding of oxygen allows hemoglobin to deliver more oxygen to the tissues in response to relatively small changes in the partial pressure of oxygen. This can be seen in Figure 3.5, which indicates the partial pressure of oxygen (pO2) in the alveoli of the

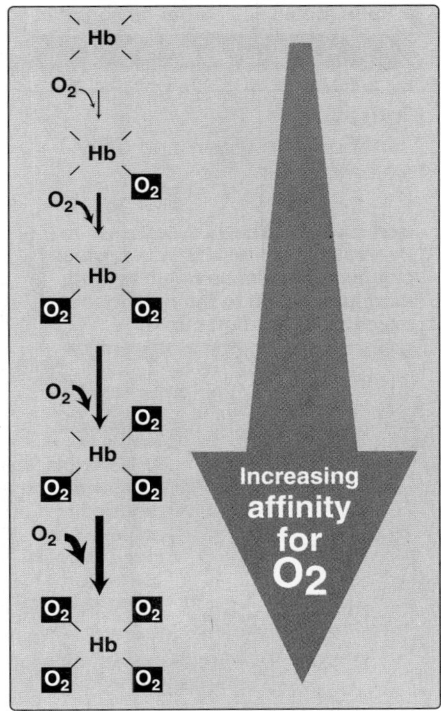

Figure 3.6
Hemoglobin binds oxygen with increasing affinity.

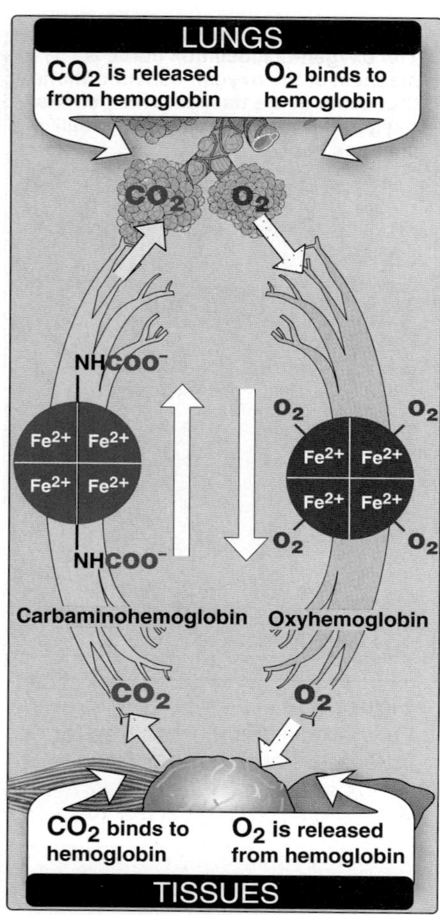

Figure 3.7
Transport of oxygen and CO_2 by hemoglobin.

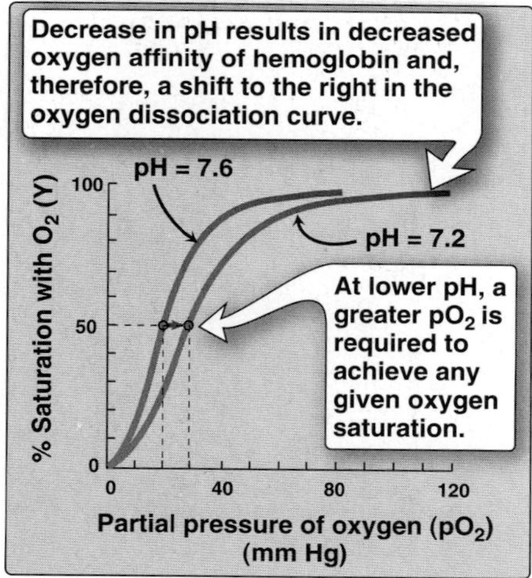

Figure 3.8
Effect of pH on the oxygen affinity of hemoglobin.

lung and the capillaries of the tissues. For example, in the lung, the concentration of oxygen is high and hemoglobin becomes virtually saturated (or "loaded") with oxygen. In contrast, in the peripheral tissues, oxyhemoglobin releases (or "unloads") much of its oxygen for use in the oxidative metabolism of the tissues (Figure 3.7).

b. **Significance of the sigmoidal O_2-dissociation curve:** The steep slope of the oxygen-dissociation curve over the range of oxygen concentrations that occur between the lungs and the tissues permits hemoglobin to carry and deliver oxygen efficiently from sites of high to sites of low pO_2. A molecule with a hyperbolic oxygen-dissociation curve, such as myoglobin, could not achieve the same degree of oxygen release within this range of partial pressures of oxygen. Instead, it would have maximum affinity for oxygen throughout this oxygen pressure range and, therefore, would deliver no oxygen to the tissues.

2. **Bohr effect:** The release of oxygen from hemoglobin is enhanced when the pH is lowered or when the hemoglobin is in the presence of an increased partial pressure of CO_2. Both result in a decreased oxygen affinity of hemoglobin and, therefore, a shift to the right in the oxygen dissociation curve (Figure 3.8). This change in oxygen binding is called the Bohr effect. Conversely, raising the pH or lowering the concentration of CO_2 results in a greater affinity for oxygen, and a shift to the left in the oxygen dissociation curve.

   a. **Source of the protons that lower the pH:** The concentration of both CO_2 and H^+ in the capillaries of metabolically active tissues is higher than that observed in alveolar capillaries of the lungs, where CO_2 is released into the expired air. [Note: Organic acids, such as lactic acid, are produced during anaerobic metabolism in rapidly contracting muscle (see p. 101).] In the tissues, CO_2 is converted by *carbonic anhydrase* to carbonic acid:

$$CO_2 + H_2O \rightleftarrows H_2CO_3$$

   which spontaneously loses a proton, becoming bicarbonate (the major blood buffer):

$$H_2CO_3 \rightleftarrows HCO_3^- + H^+$$

   The proton produced by this pair of reactions contributes to the lowering of pH. This differential pH gradient (lungs having a higher pH, tissues a lower pH) favors the unloading of oxygen in the peripheral tissues, and the loading of oxygen in the lung. Thus, the oxygen affinity of the hemoglobin molecule responds to small shifts in pH between the lungs and oxygen-consuming tissues, making hemoglobin a more efficient transporter of oxygen.

b. Mechanism of the Bohr effect: The Bohr effect reflects the fact that the deoxy form of hemoglobin has a greater affinity for protons than does oxyhemoglobin. This effect is caused by ionizable groups, such as the N-terminal α-amino groups, and specific histidine side chains that have higher pK_as in deoxyhemoglobin than in oxyhemoglobin. Therefore, an increase in the concentration of protons (resulting in a decrease in pH) causes these groups to become protonated (charged) and able to form ionic bonds (also called salt bridges). These bonds preferentially stabilize the deoxy form of hemoglobin, producing a decrease in oxygen affinity.

The Bohr effect can be represented schematically as:

$$\underset{\text{oxyhemoglobin}}{HbO_2 + H^+} \rightleftarrows \underset{\text{deoxyhemoglobin}}{HbH + O_2}$$

where an increase in protons (or a lower pO_2) shifts the equilibrium to the right (favoring deoxyhemoglobin), whereas an increase in pO_2 (or a decrease in protons) shifts the equilibrium to the left.

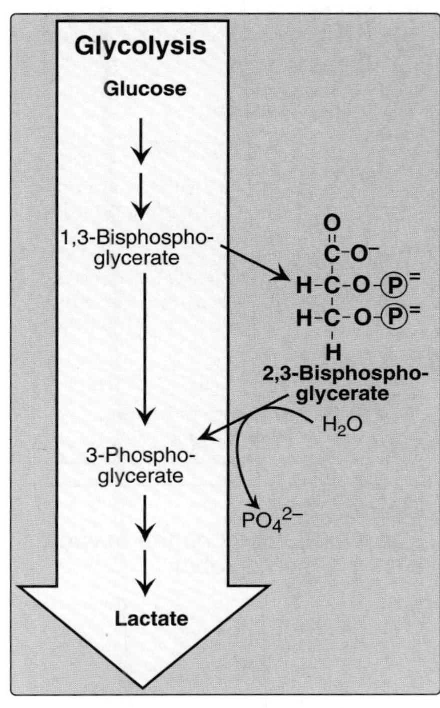

Figure 3.9
Synthesis of 2,3-bisphosphoglycerate. [Note: Ⓟ is a phosphoryl group.]

3. Effect of 2,3-bisphosphoglycerate on oxygen affinity: 2,3-Bisphosphoglycerate (2,3-BPG) is an important regulator of the binding of oxygen to hemoglobin. It is the most abundant organic phosphate in the red blood cell, where its concentration is approximately that of hemoglobin. 2,3-BPG is synthesized from an intermediate of the glycolytic pathway (Figure 3.9; see p. 99 for a discussion of 2,3-BPG synthesis in glycolysis).

a. Binding of 2,3-BPG to deoxyhemoglobin: 2,3-BPG decreases the oxygen affinity of hemoglobin by binding to deoxyhemoglobin but not to oxyhemoglobin. This preferential binding stabilizes the taut conformation of deoxyhemoglobin. The effect of binding 2,3-BPG can be represented schematically as:

$$\underset{\text{oxyhemoglobin}}{HbO_2 + \text{2,3-BPG}} \rightleftarrows \underset{\text{deoxyhemoglobin}}{\text{Hb–2,3-BPG} + O_2}$$

b. Binding site of 2,3-BPG: One molecule of 2,3-BPG binds to a pocket, formed by the two β-globin chains, in the center of the deoxyhemoglobin tetramer (Figure 3.10). This pocket contains several positively charged amino acids that form ionic bonds with the negatively charged phosphate groups of 2,3-BPG. [Note: A mutation of one of these residues can result in hemoglobin variants with abnormally high oxygen affinity.] 2,3-BPG is expelled on oxygenation of the hemoglobin.

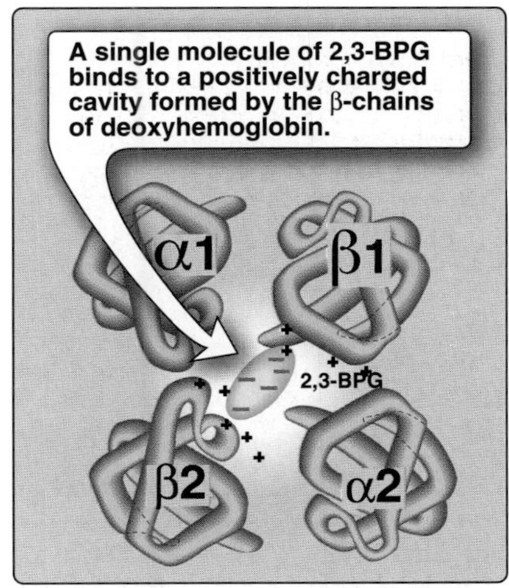

Figure 3.10
Binding of 2,3-BPG by deoxyhemoglobin.

c. Shift of the oxygen-dissociation curve: Hemoglobin from which 2,3-BPG has been removed has a high affinity for oxygen. However, as seen in the red blood cell, the presence of 2,3-BPG significantly reduces the affinity of hemoglobin for oxygen, shifting the oxygen-dissociation curve to the right (Figure 3.11). This reduced affinity enables hemoglobin to release oxygen efficiently at the partial pressures found in the tissues.

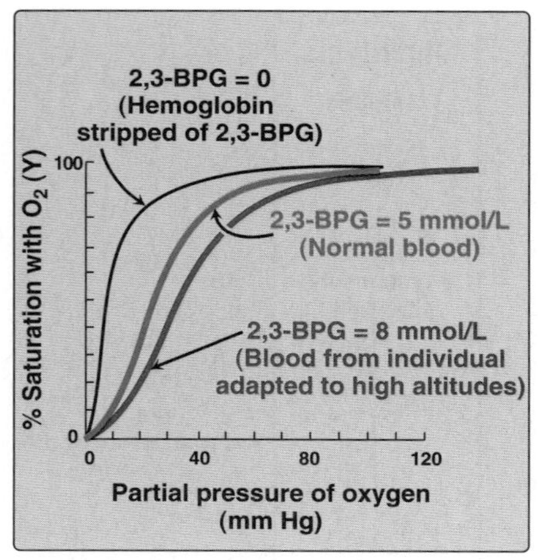

Figure 3.11
Effect of 2,3-BPG on the oxygen affinity of hemoglobin.

d. Response of 2,3-BPG levels to chronic hypoxia or anemia: The concentration of 2,3-BPG in the red blood cell increases in response to chronic hypoxia, such as that observed in obstructive pulmonary emphysema, or at high altitudes, where circulating hemoglobin may have difficulty receiving sufficient oxygen. Intracellular levels of 2,3-BPG are also elevated in chronic anemia, in which fewer than normal red blood cells are available to supply the body's oxygen needs. Elevated 2,3-BPG levels lower the oxygen affinity of hemoglobin, permitting greater unloading of oxygen in the capillaries of the tissues (see Figure 3.11).

e. Role of 2,3-BPG in transfused blood: 2,3-BPG is essential for the normal oxygen transport function of hemoglobin. For example, storing blood in acid-citrate-dextrose, a formerly widely used medium, leads to a decrease of 2,3-BPG in the red cells. Such blood displays an abnormally high oxygen affinity, and fails to unload its bound oxygen properly in the tissues. Hemoglobin deficient in 2,3-BPG thus acts as an oxygen "trap" rather than as an oxygen transport system. Transfused red blood cells are able to restore their depleted supplies of 2,3-BPG in 24 to 48 hours. However, severely ill patients may be seriously compromised if transfused with large quantities of such 2,3-BPG—"stripped" blood. The decrease in 2,3-BPG can be prevented by adding substrates such as inosine (hypoxanthine-ribose, see Figure 22.7, p. 292) to the storage medium. Inosine, an uncharged molecule, can enter the red blood cell, where its ribose moiety is released, phosphorylated, and enters the hexose monophosphate pathway (see p. 145), eventually being converted to 2,3-BPG.

4. **Binding of CO_2:** Most of the carbon dioxide produced in metabolism is hydrated and transported as bicarbonate ion (see p. 9). However, some CO_2 is carried as **carbamate** bound to the uncharged α-amino groups of hemoglobin (**carbamino-hemoglobin**; see Figure 3.7), which can be represented schematically as follows:

$$\text{Hb–NH}_2 + \text{CO}_2 \rightleftarrows \text{Hb–NH–COO}^- + \text{H}^+$$

The binding of CO_2 stabilizes the T (taut) or deoxy form of hemoglobin, resulting in a decrease in its affinity for oxygen (see p. 28). In the lungs, CO_2 disassociates from the hemoglobin, and is released in the breath.

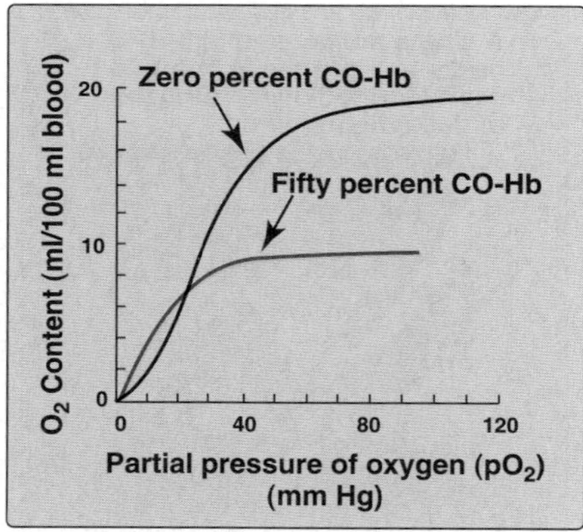

Figure 3.12
Effect of carbon monoxide on the oxygen affinity of hemoglobin. CO-Hb = carbon monoxy-hemoglobin.

5. **Binding of CO:** Carbon monoxide (CO) binds tightly (but reversibly) to the hemoglobin iron, forming **carbon monoxyhemoglobin (HbCO)**. When carbon monoxide binds to one or more of the four heme sites, hemoglobin shifts to the relaxed conformation, causing the remaining heme sites to bind oxygen with high affinity. This shifts the oxygen saturation curve to the left, and changes the normal sigmoidal shape toward a hyperbola. As a result, the affected hemoglobin is unable to release oxygen to the tissues (Figure 3.12). [Note: The affinity of hemoglobin for CO is 220 times greater than for oxygen. Consequently, even minute

concentrations of carbon monoxide in the environment can produce toxic concentrations of carbon monoxyhemoglobin in the blood. Carbon monoxide toxicity appears to result from a combination of tissue hypoxia and direct carbon monoxide-mediated damage at the cellular level.] Carbon monoxide poisoning is treated with 100 percent oxygen therapy, which facilitates the dissociation of CO from the hemoglobin.

F. Minor hemoglobins

It is important to remember that human hemoglobin A (**HbA**) is just one member of a functionally and structurally related family of proteins, the hemoglobins (Figure 3.13). Each of these oxygen-carrying proteins is a tetramer, composed of two α-globin-like polypeptides and two β-globin-like polypeptides. Certain hemoglobins, such as **HbF**, are normally synthesized only during fetal development, whereas others, such as **HbA$_2$**, are synthesized in the adult, although at low levels compared with HbA. HbA can also become modified by the covalent addition of a hexose. For example, addition of glucose forms the glucosylated hemoglobin derivative, **HbA$_{1c}$**.

Form	Chain composition	Fraction of total hemoglobin
HbA	$\alpha_2\beta_2$	90%
HbF	$\alpha_2\gamma_2$	<2%
HbA$_2$	$\alpha_2\delta_2$	2–5%
HbA$_{1c}$	$\alpha_2\beta_2$-glucose	3–9%

Figure 3.13
Normal adult human hemoglobins. [Note: The α-chains in these hemoglobins are identical.]

1. **Fetal hemoglobin (HbF):** HbF is a tetramer consisting of two α chains identical to those found in HbA, plus two gamma (γ) chains ($\alpha_2\gamma_2$, see Figure 3.13). The γ chains are members of the β-globin gene family (see p. 35).

   a. **HbF synthesis during development:** In the first few weeks after conception, **embryonic hemoglobin** (**Hb Gower 1**), composed of two zeta chains and two epsilon chains ($\zeta_2\varepsilon_2$), is synthesized by the embryonic yolk sac. Within a few weeks, the fetal liver begins to synthesize HbF in the developing bone marrow. HbF is the major hemoglobin found in the **fetus** and **newborn**, accounting for about sixty percent of the total hemoglobin in the erythrocytes during the last months of fetal life (Figure 3.14). HbA synthesis starts in the bone marrow at about the eighth month of pregnancy and gradually replaces HbF. (Figure 3.14 shows the relative production of each type of hemoglobin chain during fetal and postnatal life.)

   b. **Binding of 2,3-BPG to HbF:** Under physiologic conditions, HbF has a higher affinity for oxygen than does HbA, as a result of HbF's binding only weakly to 2,3-BPG. [Note: The γ-globin chains of HbF lack some of the positively charged amino acids that are responsible for binding 2,3-BPG in the β-globin chains.] Because 2,3-BPG serves to reduce the affinity of hemoglobin for oxygen, the weaker interaction between 2,3-BPG and HbF results in a higher oxygen affinity for HbF relative to HbA. In contrast, if both HbA and HbF are stripped of their 2,3-BPG, they then have a similar affinity for oxygen. The higher oxygen affinity of HbF facilitates the transfer of oxygen from the maternal circulation across the placenta to the red blood cells of the fetus.

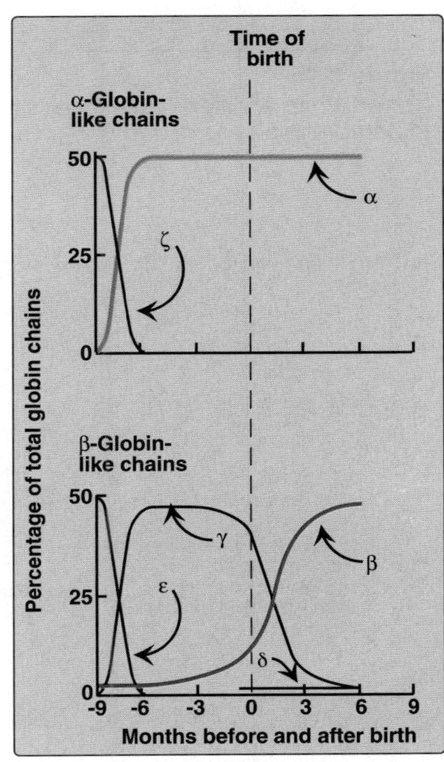

Figure 3.14
Developmental changes in hemoglobin.

2. **Hemoglobin A$_2$ (HbA$_2$):** HbA$_2$ is a minor component of normal adult hemoglobin, first appearing about twelve weeks after birth and, ultimately, constituting about two percent of the total hemo-

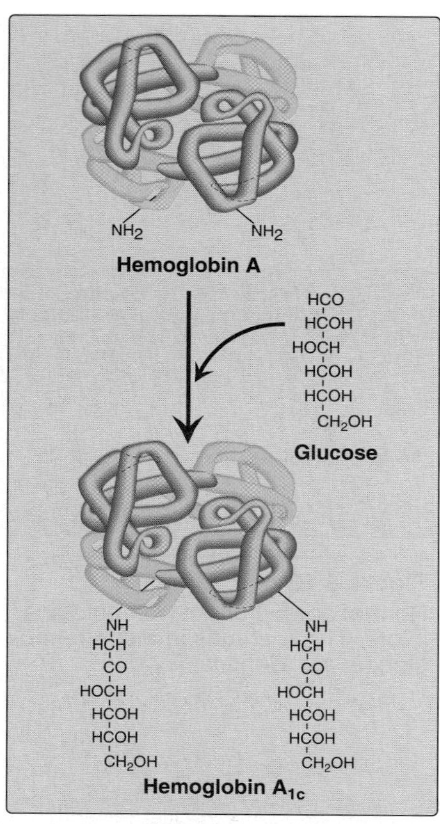

Figure 3.15
Nonenzymic addition of glucose to hemoglobin.

globin. It is composed of two α-globin chains and two delta- (δ-) globin chains (**$\alpha_2\delta_2$**, see Figure 3.13).

3. **Hemoglobin A_{1c}:** Under physiologic conditions, HbA is slowly and nonenzymically glycosylated, the extent of glycosylation being dependent on the plasma concentration of a particular hexose. The most abundant form of glycosylated hemoglobin is HbA_{1c}. It has glucose residues attached predominantly to the NH_2 groups of the N-terminal valines of the β-globin chains (Figure 3.15). Increased amounts of HbA_{1c} are found in red blood cells of patients with **diabetes mellitus**, because their HbA has contact with higher glucose concentrations during the 120 day lifetime of these cells. (See p. 339 for a discussion of the use of this phenomenon in assessing average blood glucose levels in persons with diabetes.)

III. ORGANIZATION OF THE GLOBIN GENES

To understand diseases resulting from genetic alterations in the structure or synthesis of hemoglobins, it is necessary to grasp how the hemoglobin genes, which direct the synthesis of the different globin chains, are structurally organized into gene families and also how they are expressed.

A. α-Gene family

The genes coding for the α-globin-like and β-globin-like subunits of the hemoglobin chains occur in two separate gene clusters (or families) located on two different chromosomes (Figure 3.16). The α-gene cluster on **chromosome 16** contains two genes for the **α-globin chains**. It also contains the **zeta (ζ) gene** that is expressed early in development as a component of embryonic hemoglobin, and a number of globin-like genes that are not expressed (that is, their genetic information is not used to produce globin chains). These are called **pseudogenes**.

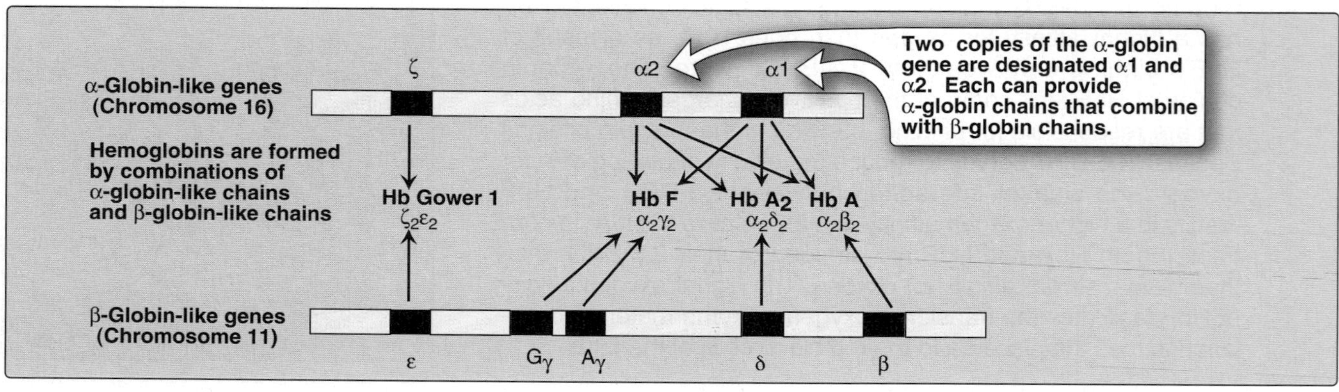

Figure 3.16
Organization of the globin gene families.

B. β-Gene family

A single gene for the **β-globin chain** is located on chromosome 11 (see Figure 3.16). There are an additional four β-globin-like genes: the **epsilon (ε) gene** (which, like the ζ gene, is expressed early in embryonic development), **two γ genes** (G_γ and A_γ that are expressed in fetal hemoglobin, HbF), and the **δ gene** that codes for the globin chain found in the minor adult hemoglobin HbA_2.

C. Steps in globin chain synthesis

Expression of a globin gene begins in the nucleus of red cell precursors, where the DNA sequence encoding the gene is transcribed. The RNA produced by transcription is actually a precursor of the messenger RNA (mRNA) that is used as a template for the synthesis of a globin chain. Before it can serve this function, two noncoding stretches of RNA (introns) must be removed from the mRNA precursor sequence, and the remaining three fragments (exons) reattached in a linear manner. The resulting mature mRNA enters the cytosol, where its genetic information is translated, producing a globin chain. (A summary of this process is shown in Figure 3.17. A more detailed description of protein synthesis is presented in Chapter 31, p. 429.)

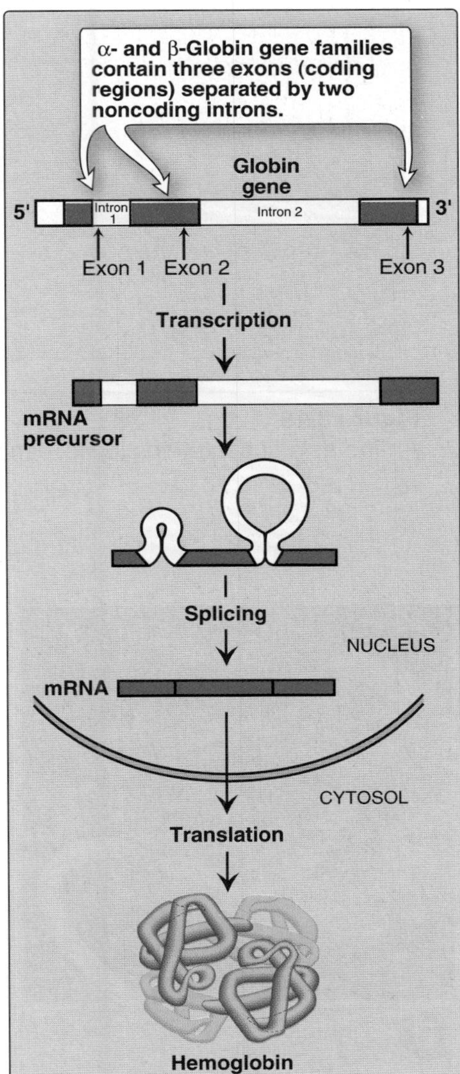

Figure 3.17
Synthesis of globin chains.

IV. HEMOGLOBINOPATHIES

Hemoglobinopathies have traditionally been defined as a family of disorders caused by production of a structurally abnormal hemoglobin molecule, synthesis of insufficient quantities of normal hemoglobin, or, rarely, both. **Sickle-cell anemia (HbS)**, **hemoglobin C disease (HbC)**, and the **thalassemia** syndromes are representative hemoglobinopathies that can have severe clinical consequences. The first two conditions result from production of hemoglobin with an altered amino acid sequence, whereas the thalassemias are caused by decreased production of normal hemoglobin.

A. Sickle cell disease (hemoglobin S disease)

Sickle cell disease (also called **sickle cell anemia**) is a genetic disorder of the blood caused by a single nucleotide alteration (a **point mutation**) in the **β-globin gene**. It is the most common inherited blood disorder in the United States, affecting 80,000 Americans. It occurs primarily in the African-American population, affecting one of 500 newborn African-American infants in the United States (Figure 3.18). Sickle cell disease is a homozygous, recessive disorder. It occurs in individuals who have inherited two mutant genes (one from each parent) that code for synthesis of the β-chains of the globin molecules. [Note: The mutant β-globin chain is designated β^S, and the resulting hemoglobin, $\alpha_2\beta^S_2$, is referred to as **HbS**.] An infant does not begin showing symptoms of the disease until sufficient HbF has been replaced by HbS so that sickling can occur (see below). Sickle cell disease is characterized by lifelong episodes of pain (**"crises"**), **chronic hemolytic anemia**, and increased susceptibility to infections, usually beginning in early childhood. Other symp-

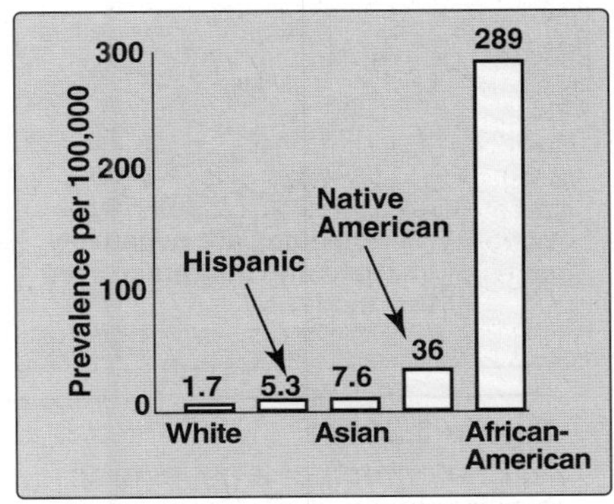

Figure 3.18
Sickle cell disease in the United States.

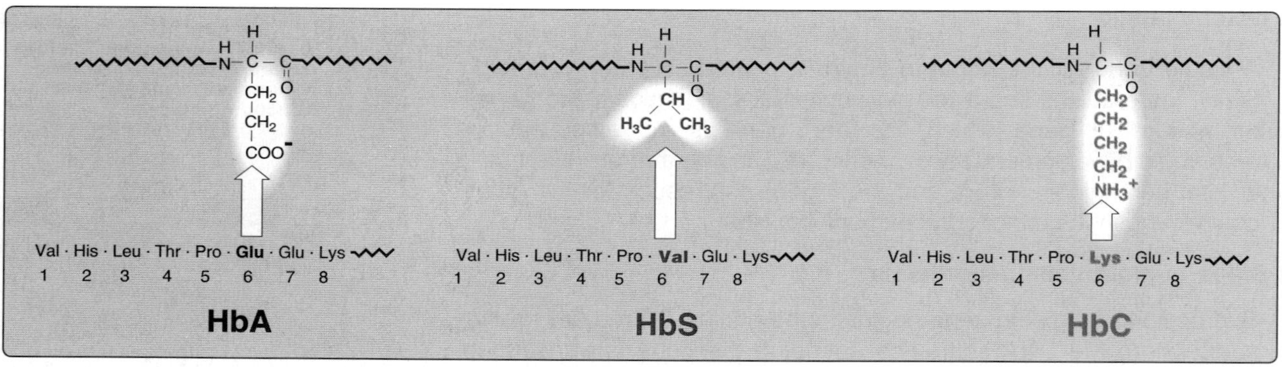

Figure 3.19
Amino acid substitutions in HbS and HbC.

toms include acute chest syndrome, stroke, and splenic and renal dysfunction. The lifetime of an erythrocyte homozygous for HbS is approximately twenty days, compared with 120 days for normal red blood cells. Heterozygotes, representing one of ten African-Americans, have one normal and one sickle cell gene. The blood cells of such heterozygotes contain both HbS and HbA. These individuals have **sickle cell trait**. They usually do not show clinical symptoms and can have a normal life span.

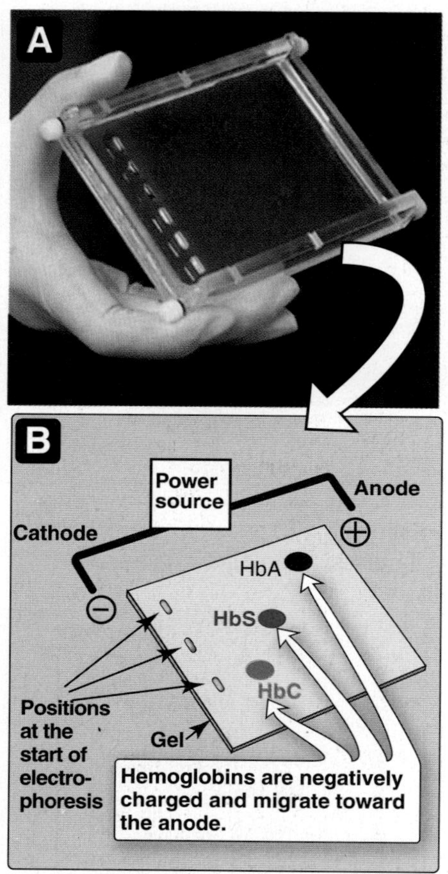

Figure 3.20
A. Photograph of a gel prior to electrophoresis. B. Diagram of hemoglobins A, S, and C after electrophoresis.

1. **Amino acid substitution in HbS β chains:** A molecule of HbS contains two normal α-globin chains and two mutant β-globin chains (β^S), in which glutamate at position six has been replaced with valine (Figure 3.19). Therefore, during electrophoresis at alkaline pH, HbS migrates more slowly toward the anode (positive electrode) than does HbA (Figure 3.20). This altered mobility of HbS is a result of the absence of the negatively charged glutamate residues in the two β-chains, thus rendering HbS less negative than HbA. [Note: Electrophoresis of hemoglobin obtained from lysed red blood cells is routinely used in the diagnosis of sickle cell trait and sickle cell disease.]

2. **Sickling causes tissue anoxia:** The substitution of the nonpolar valine for a charged glutamate residue forms a protrusion on the β-globin that fits into a complementary site on the α-chain of another hemoglobin molecule in the cell (Figure 3.21). At low oxygen tension, HbS polymerizes inside the red blood cells, first forming a gel, then subsequently assembling into a network of fibrous polymers that stiffen and distort the cell, producing rigid, misshapen erythrocytes. Such sickled cells frequently block the flow of blood in the narrow capillaries. This interruption in the supply of oxygen leads to localized anoxia (oxygen deprivation) in the tissue, causing pain and eventually death (**infarction**) of cells in the vicinity of the blockage.

3. **Variables that increase sickling:** The extent of sickling and, therefore, the severity of disease is enhanced by any variable that increases the proportion of HbS in the deoxy state (that is, reduces the affinity of HbS for oxygen). These variables include decreased oxygen ten-

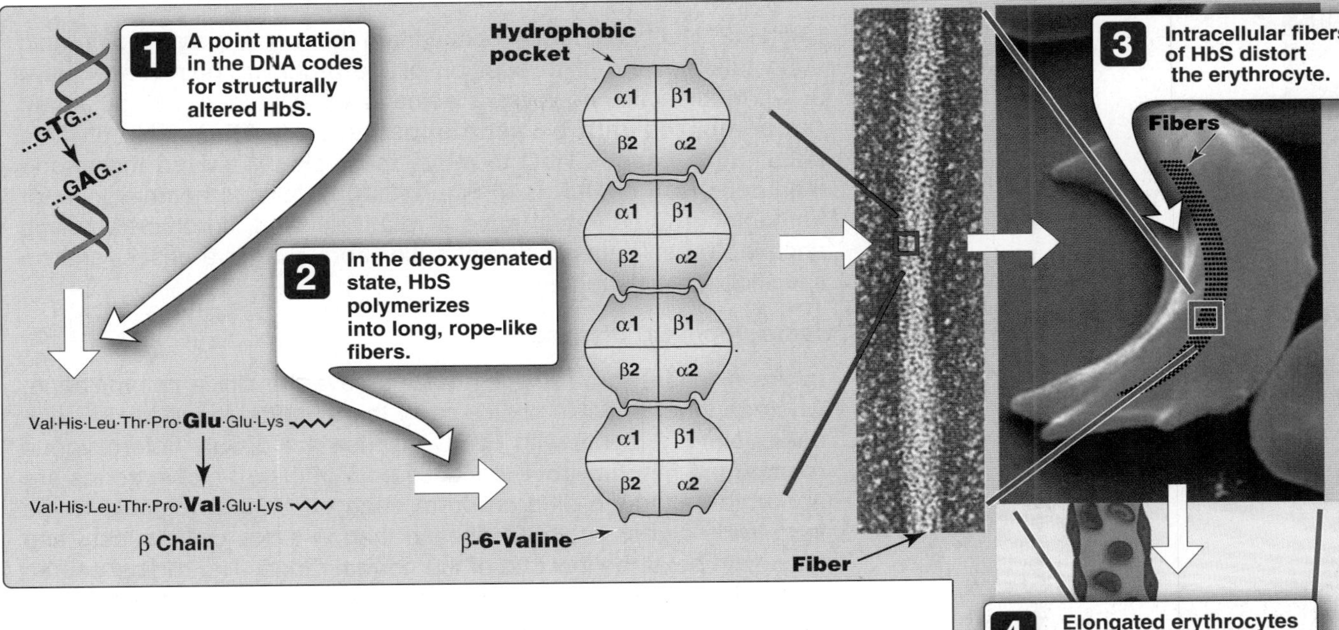

Figure 3.21
Molecular and cellular events leading to sickle cell crisis.

sion as a resu;t of high altitudes or flying in a nonpressurized plane, increased pCO_2, decreased pH, and an increased concentration of 2,3-BPG in erythrocytes.

4. **Treatment:** Therapy involves adequate **hydration**, **analgesics**, aggressive **antibiotic therapy** if infection is present, and **transfusions** in patients at high risk for fatal vasocculsions. Intermittent transfusions with packed red cells reduce the risk of stroke, but the benefits must be weighed against the complications of transfusion, which include iron overload (**hemosiderosis**), blood-borne infections, and immunologic complications. **Hydroxyurea**, an antitumor drug, decreases the frequency of painful crises and reduces mortality. The mechanism of action of the drug is not fully understood, but may include a modest increase in fetal hemoglobin, which decreases sickling.

5. **Possible selective advantage of the heterozygous state:** The high frequency of the HbS gene among black Africans, despite its damaging effects in the homozygous state, suggests that a selective advantage exists for heterozygous individuals. For example, heterozygotes for the sickle cell gene are less susceptible to malaria, caused by the parasite Plasmodium falciparum. This organism spends an obligatory part of its life cycle in the red blood cell. Because these cells in individuals heterozygous for HbS, like those in homozygotes, have a shorter life span than normal, the parasite cannot complete the intracellular stage of its development. This fact may provide a selective advantage to heterozygotes living in regions where malaria is a major cause of death. Figure 3.22 illustrates that in Africa, the geographic distribution of sickle cell disease is similar to that of malaria.

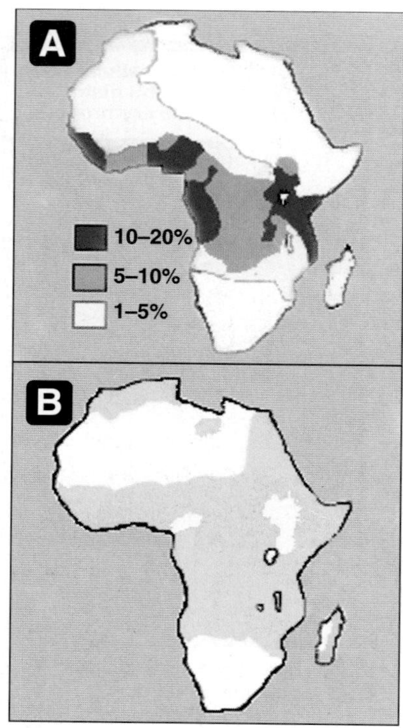

Figure 3.22
A. Distribution of sickle cell in Africa expressed as percentage of the population with disease.
B. Distribution of malaria in Africa.

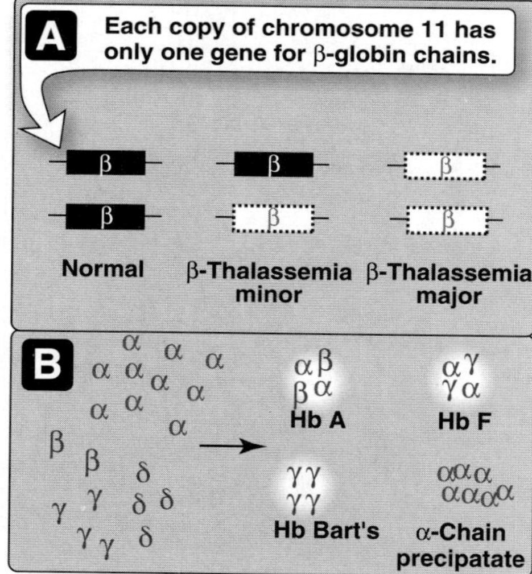

Figure 3.23
A. β-Globin gene deletions in the β-thalassemias. B. Hemoglobin tetramers formed in β-thalassemias.

B. Hemoglobin C disease

Like HbS, HbC is a hemoglobin variant that has a single amino acid substitution in the sixth position of the β-globin chain (see Figure 3.19). In this case, however, a lysine is substituted for the glutamate (as compared with a valine substitution in HbS). [Note: This substitution causes HbC to move more slowly toward the anode than does HbA or HbS (see Figure 3.20).] Patients homozygous for hemoglobin C generally have a relatively mild, chronic hemolytic anemia. These patients do not suffer from infarctive crises, and no specific therapy is required.

C. Hemoglobin SC disease

In this disease, some β-globin chains have the sickle cell mutation, whereas other β-globin chains carry the mutation found in HbC disease. [Note: Patients with HbSC disease are doubly heterozygous (**compound heterozygote**) because both of their β-globin genes are abnormal, although different from each other.] Hemoglobin levels tend to be higher in HbSC disease than in sickle cell disease, and may even be at the low end of the normal range. The clinical course of adults with HbSC disease differs from that of sickle cell disease, in which patients generally have painful crises beginning in childhood. It is common for patients with HbSC disease to remain well (and undiagnosed) until they suffer an infarctive crisis. This crisis often follows childbirth or surgery and may be fatal.

D. Methemoglobinemias

Oxidation of the heme component of hemoglobin to the ferric (Fe^{3+}) state forms methemoglobin, which cannot bind oxygen. This oxidation may be caused by the action of certain drugs, such as nitrates, or endogenous products, such as reactive oxygen intermediates (see p. 145). The oxidation may also result from inherited defects, for example, certain mutations in the α- or β-globin chain promote the formation of methemoglobin (HbM). Further, a deficiency of *NADH-cytochrome b_5 reductase* (also called *NADH-methemoglobin reductase*), the enzyme responsible for the conversion of methemoglobin (Fe^{3+}) to hemoglobin (Fe^{2+}), leads to the accumulation of methemoglobin. [Note: The erythrocytes of newborns have approximately one-half the capacity of those of adults to reduce methemoglobin. They are therefore particularly susceptible to the effects of methemoglobin-producing compounds]. The methemoglobinemias are characterized by "**chocolate cyanosis**" (a brownish-blue coloration of the skin and membranes) and chocolate colored-blood, as a result of the dark-colored methemoglobin. Symptoms are related to tissue hypoxia, and include anxiety, headache, and dyspnea. In rare cases, coma and death can occur.

E. Thalassemias

The thalassemias are hereditary hemolytic diseases in which an imbalance occurs in the synthesis of globin chains. As a group, they are the most common single gene disorders in humans. Normally, synthesis of the α- and β-globin chains are coordinated, so that each α-globin chain has a β-globin chain partner. This leads to the formation of $\alpha_2\beta_2$ (HbA). In the thalassemias, the synthesis of either the α- or β-globin chain is defective. A thalassemia can be caused by a variety of mutations, including entire gene deletions, or substitutions or deletions of one to many nucleotides in the DNA. [Note: Each thalassemia can be classified as either a disorder in which no globin chains are produced (α^o- or β^o-thalassemia), or one in which some chains are synthesized, but at a reduced rate (α^+- or β^+-thalassemia).]

1. **β-Thalassemias:** In these disorders, synthesis of β-globin chains is decreased or absent, whereas α-globin chain synthesis is normal. α-Globin chains cannot form stable tetramers and, therefore, precipitate, causing the premature death of cells initially destined to become mature red blood cells. Accumulation of $\alpha_2\gamma_2$ (HbF) and γ_4 (Hb Bart's) also occurs. There are only two copies of the β-globin gene in each cell (one on each chromosome 11). Therefore, individuals with β-globin gene defects have either **β-thalassemia trait** (**β-thalassemia minor**) if they have only one defective β-globin gene, or **β-thalassemia major** if both genes are defective (Figure 3.23). Because the β-globin gene is not expressed until late in fetal gestation, the physical manifestations of β-thalassemias appear only after birth. Those individuals with β-thalassemia minor make some β-chains, and usually do not require specific treatment. However, those infants born with β-thalassemia major have the sad fate of being seemingly healthy at birth, but becoming severely anemic, usually during the first or second year of life. These patients require regular transfusions of blood. [Note: Although this treatment is lifesaving, the cumulative effect of the transfusions is iron overload (a syndrome known as **hemosiderosis**), which typically causes death between the ages of 15 and 25 years.] The increasing use of bone marrow replacement therapy has been a boon to these patients.

2. **α-Thalassemias:** These are defects in which the synthesis of α-globin chains is decreased or absent. Because each individual's genome contains four copies of the α-globin gene (two on each chromosome 16), there are several levels of α-globin chain deficiencies (Figure 3.24). If one of the four genes is defective, the individual is termed a **silent carrier** of α-thalassemia, because no physical manifestations of the disease occur. If two α-globin genes are defective, the individual is designated as having **α-thalassemia trait**. If three α-globin genes are defective, the individual has **hemoglobin H (HbH)** disease—a mildly to moderately severe hemolytic anemia. [Note: The synthesis of unaffected γ- and then β-globin chains continues, resulting in the accumulation of γ tetramers in the newborn (**γ_4, Hb Bart's**) or β tetramers (**β_4, HbH**). Although these tetramers are fairly soluble, the subunits show no heme-heme interaction. Their oxygen dissociation curves are almost hyperbolic, indicating that these tetramers have very high oxygen affinities. This makes them essentially useless as oxygen deliverers to the tissues.] If all four α-globin genes are defective, **hydrops fetalis** and fetal death result, because α-globin chains are required for the synthesis of HbF.

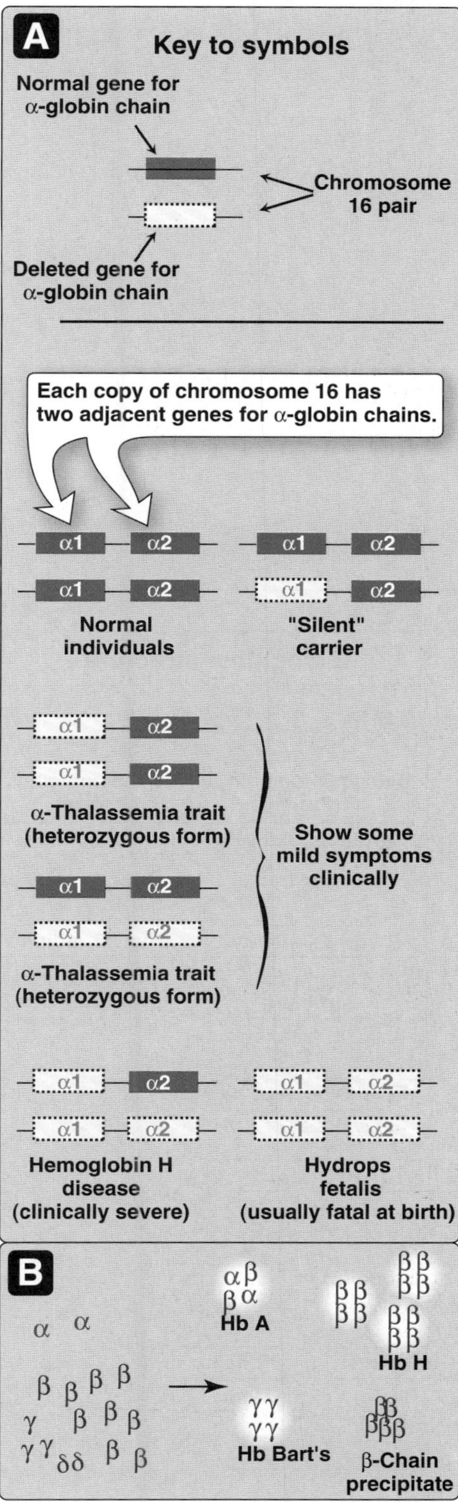

Figure 3.24
A. α-Globin gene deletions in the α-thalassemias. B. Hemoglobin tetramers formed in α-thalassemias.

V. CHAPTER SUMMARY

Hemoglobin A, the major hemoglobin in adults, is composed of four polypeptide chains (two α chains and two β chains, $\alpha_2\beta_2$) held together by noncovalent interactions. The subunits occupy different relative positions in deoxyhemoglobin compared with oxyhemoglobin. The **deoxy form** of hemoglobin is called the "**T**," or **taut** (**tense**) **form**. It has a constrained structure that limits the movement of the polypeptide chains. The T form is the **low oxygen-affinity form** of hemoglobin. The binding of oxygen to hemoglobin causes rupture of some of the ionic and hydrogen bonds. This leads to a structure called the "**R**," or **relaxed form**, in which the polypeptide chains have more freedom of movement. The R form is the **high oxygen affinity form** of hemoglobin. The **oxygen-dissociation curve** for hemoglobin is **sigmoidal** in shape (in contrast to that of **myoglobin**, which is **hyperbolic**), indicating that the subunits cooperate in binding oxygen. **Cooperative binding** of oxygen by the four subunits of hemoglobin means that the binding of an oxygen molecule at one heme group increases the oxygen affinity of the remaining heme

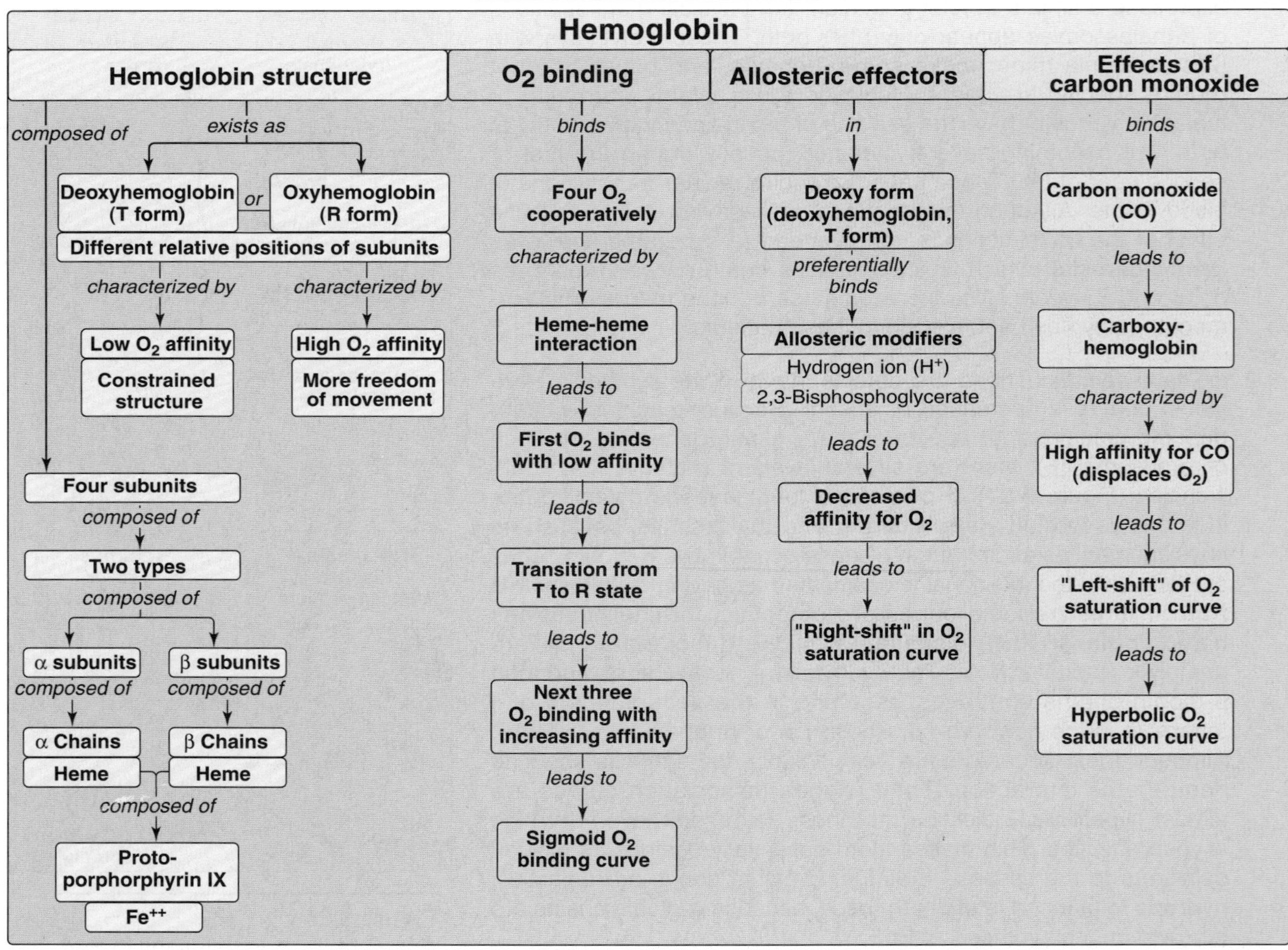

Figure 3.25
Key concept map for hemoglobin structure and function.

groups in the same hemoglobin molecule. Hemoglobin's **ability to bind oxygen reversibly** is affected by the pO_2 (through heme-heme interactions), the **pH** of the environment, the pCO_2, and the availability of **2,3-bisphosphoglycerate**. For example, the release of O_2 from Hb is enhanced when the pH is lowered or the pCO_2 is increased (the **Bohr effect**),such as in **exercising muscle**, and the oxygen-dissociation curve of Hb is shifted to the right. To cope long-term with the effects of **chronic hypoxia** or **anemia**, the concentration of **2,3-BPG** in **RBCs** increases. **2,3-BPG** binds to the Hb and decreases its oxygen affinity, and it, therefore, also shifts the oxygen-dissociation curve to the right. **Carbon monoxide (CO)** binds tightly (but reversibly) to the hemoglobin iron, forming **carbon monoxyhemoglobin**, HbCO. **Hemoglobinopathies** are disorders caused either by production of a **structurally abnormal hemoglobin** molecule, synthesis of **insufficient quantities** of normal hemoglobin subunits, or, rarely, both. **Sickle cell disease** (HbS disease), **hemoglobin C disease** (HbC disease), and the **thalassemia syndromes** are representative hemoglobinopathies that can have severe clinical consequences.

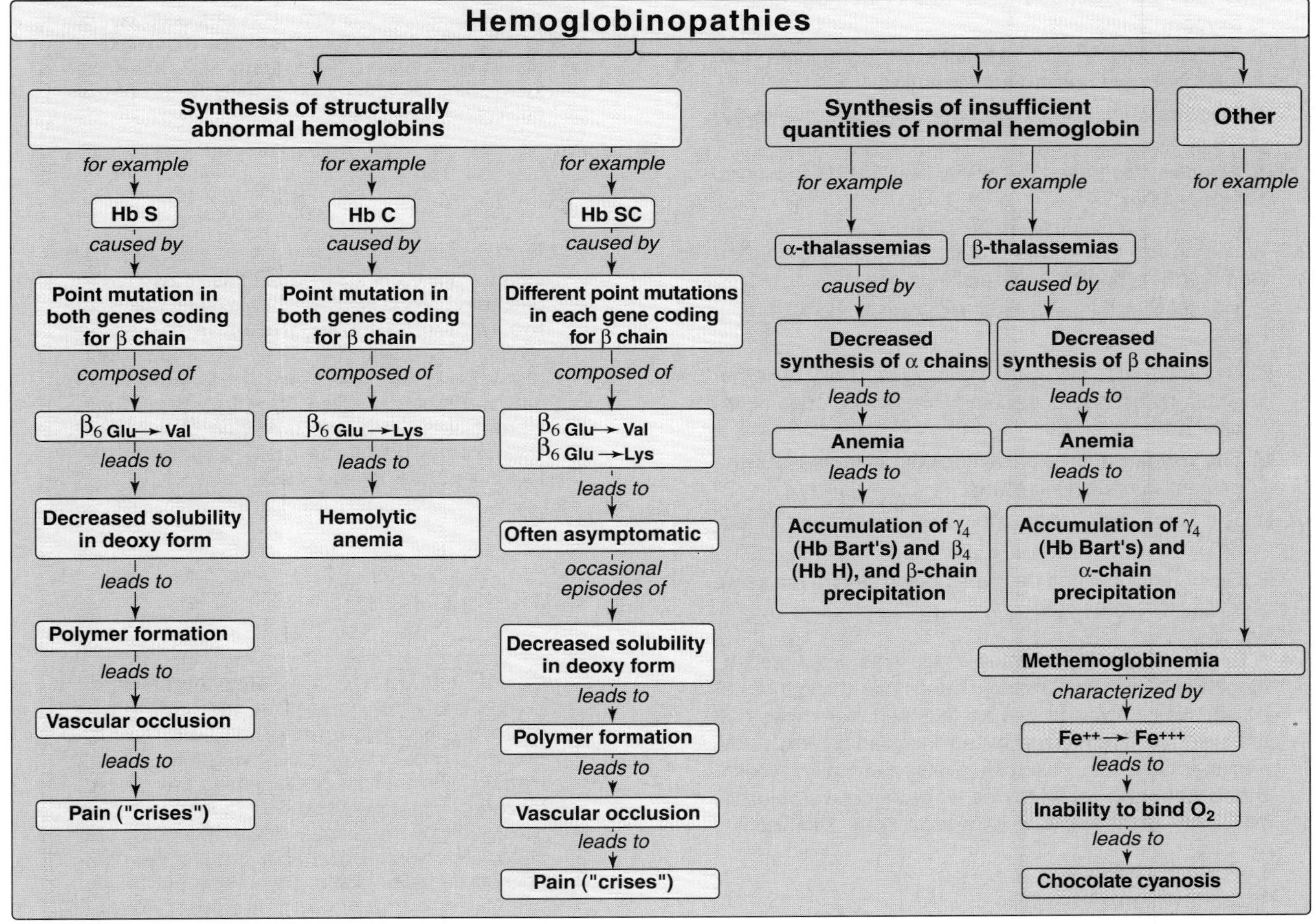

Figure 3.26
Key concept map for hemoglobinopathies.

Study Questions

Choose the ONE correct answer

3.1 Which one of the following statements concerning the hemoglobins is correct?

A. Fetal blood has a higher affinity for oxygen than does adult blood because HbF has a decreased affinity for 2,3-BPG.

B. Purified HbF (stripped of 2,3-BPG) has a higher affinity for oxygen than does purified HbA.

C. The chain composition of HbF is $\alpha_2\delta_2$.

D. HbA_{1c} differs from HbA by a single, genetically determined amino acid substitution.

E. HbA_2 appears early in fetal life.

Correct answer = A. Because 2,3-BPG reduces the affinity of hemoglobin for oxygen, the weaker interaction between 2,3-BPG and HbF results in a higher oxygen-affinity for HbF relative to HbA. In contrast, if both HbA and HbF are stripped of 2,3-BPG, they have a similar affinity for oxygen. HbF consists of $\alpha_2\gamma_2$. HbA_{1c} is a glycosylated form of HbA, formed nonenzymically in red cells. HbA_2 is a minor component of normal adult hemoglobin, first appearing about twelve weeks after birth.

3.2 Which one of the following statements concerning the ability of acidosis to precipitate a crisis in sickle cell disease is correct?

A. Acidosis decreases the solubility of HbS.

B. Acidosis increases the affinity of hemoglobin for oxygen.

C. Acidosis favors the conversion of hemoglobin from the taut to the relaxed conformation.

D. Acidosis shifts the oxygen-dissociation curve to the left.

E. Acidosis decreases the ability of 2,3-BPG to bind to hemoglobin.

Correct answer = A. HbS is significantly less soluble in the deoxygenated form, compared with oxy-HbS. A decrease in pH (acidosis) causes the oxygen-dissociation curve to shift to the right, indicating a decreased affinity for oxygen. This favors the formation of the deoxy, or taut, form of hemoglobin, and can precipitate a sickle cell crisis. The binding of 2,3-BPG is increased, because it binds only to the deoxy form of hemoglobins.

3.3 Which one of the following statements concerning the binding of oxygen by hemoglobin is correct?

A. The Bohr effect results in a lower affinity for oxygen at higher pH values.

B. Carbon dioxide increases the oxygen affinity of hemoglobin by binding to the amino terminal groups of the polypeptide chains.

C. The oxygen affinity of hemoglobin increases as the percent saturation increases.

D. The hemoglobin tetramer binds four molecules of 2,3-BPG.

E. Oxyhemoglobin and deoxyhemoglobin have the same affinity for protons (H^+).

Correct answer = C. The binding of oxygen at one heme group increases the oxygen affinity of the remaining heme groups in the same molecule. Carbon dioxide decreases oxygen affinity because it lowers the pH; also binding of carbon dioxide stabilizes the taut, deoxy form. Hemoglobin binds one molecule of 2,3-BPG. Deoxyhemoglobin has a greater affinity for protons and, therefore, is a weaker acid.

3.4 A 67-year-old man presented to the emergency department with a one week history of angina and shortness of breath. He complained that his face and extremities had a "blue color." His medical history included chronic stable angina treated with isosorbide dinitrate and nitroglycerin. Blood obtained for analysis was chocolate-colored. Which one of the following is the most likely diagnosis?

A. Sickle cell disease

B. Carboxyhemoglobinemia

C. Methemoglobinemia.

D. β-Thalassemia

E. Hemoglobin SC disease

Correct answer = C. Oxidation of the heme component of hemoglobin to the ferric (Fe^{3+}) state forms methemoglobin. This may be caused by the action of certain drugs, such as nitrates. The methemoglobinemias are characterized by chocolate cyanosis (a brownish-blue coloration of the skin and membranes), and chocolate-colored blood as a result of the dark-colored methemoglobin. Symptoms are related to tissue hypoxia, and include anxiety, headache, dyspnea, and, in rare cases, coma and death can occur.

Fibrous Proteins

4

I. OVERVIEW

Collagen and elastin are examples of common, well-characterized fibrous proteins that serve structural functions in the body. For example, collagen and elastin are found as components of skin, connective tissue, blood vessel walls, and sclera and cornea of the eye. Each fibrous protein exhibits special mechanical properties, resulting from its unique structure, which are obtained by combining specific amino acids into regular, secondary structural elements. This is in contrast to globular proteins, whose shapes are the result of complex interactions between secondary, tertiary, and, sometimes, quaternary structural elements.

II. COLLAGEN

Collagen is the most abundant protein in the human body. A typical collagen molecule is a long, rigid structure in which three polypeptides (referred to as "α-chains") are wound around one another in a rope-like triple-helix (Figure 4.1). Although these molecules are found throughout the body, their types and organization are dictated by the structural role collagen plays in a particular organ. In some tissues, collagen may be dispersed as a gel that give support to the structure, as in the extracellular matrix or the vitreous humor of the eye. In other tissues, collagen may be bundled in tight, parallel fibers that provide great strength, as in tendons. In the cornea of the eye, collagen is stacked so as to transmit light with a minimum of scattering. Collagen of bone occurs as fibers arranged at an angle to each other so as to resist mechanical shear from any direction.

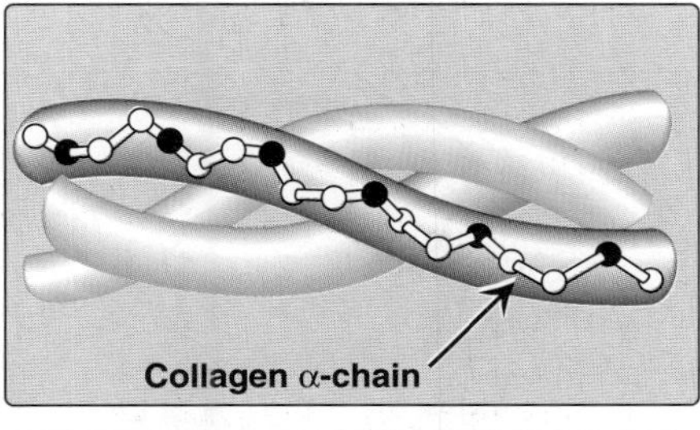

Figure 4.1
Triple-stranded helix of collagen.

A. Types of collagen

The collagen superfamily of proteins includes more than twenty collagen types, as well as additional proteins that have collagen-like domains. The three polypeptide α-chains are held together by hydrogen bonds between the chains. Variations in the amino acid sequence of the α-chains result in structural components that are about the same size (approximately 1000 amino acids long), but with slightly different properties. These α-chains are combined to form the various types of collagen found in the tissues. For example, the most common collagen, type I, contains two chains called α1 and one chain called

Lippincott's Illustrated Reviews: Biochemistry, Third Edition,
by Pamela C. Champe and Richard A. Harvey.

TYPE	TISSUE DISTRIBUTION
	Fibril-forming
I	Skin, bone, tendon, blood vessels, cornea
II	Cartilage, intervertebral disk, vitreous body
III	Blood vessels, fetal skin
	Network-forming
IV	Basement membrane
VII	Beneath stratified squamous epithelia
	Fibril-associated
IX	Cartilage
XII	Tendon, ligaments, some other tissues

Figure 4.2
The most abundant types of collagen.

$\alpha 2$ ($\alpha 1_2 \alpha 2$), whereas type II collagen contains three $\alpha 1$ chains ($\alpha 1_3$). The collagens can be organized into three groups, based on their location and functions in the body (Figure 4.2).

1. **Fibril-forming collagens:** Types I, II, and III are the fibrillar collagens, and have the rope-like structure described above for a typical collagen molecule. In the electron microscope, these linear polymers of fibrils have characteristic banding patterns, reflecting the regular staggered packing of the individual collagen molecules in the fibril (Figure 4.3). Type I collagen fibers are found in supporting elements of high tensile strength (for example, tendon and cornea), whereas fibers formed from type II collagen molecules are restricted to cartilaginous structures. The fibrils derived from type III collagen are prevalent in more distensible tissues, such as blood vessels.

2. **Network-forming collagens:** Types IV and VII form a three-dimensional mesh, rather than distinct fibrils (Figure 4.4). For example, type IV molecules assemble into a sheet or meshwork that constitutes a major part of basement membranes. [Note: **Basement membranes** are thin, sheet-like structures that provide mechanical support for adjacent cells, and function as a semipermeable filtration barrier for macromolecules in organs such as the kidney and the lung.]

3. **Fibril-associated collagens:** Types IX and XII bind to the surface of collagen fibrils, linking these fibrils to one another and to other components in the extracellular matrix (see Figure 4.2).

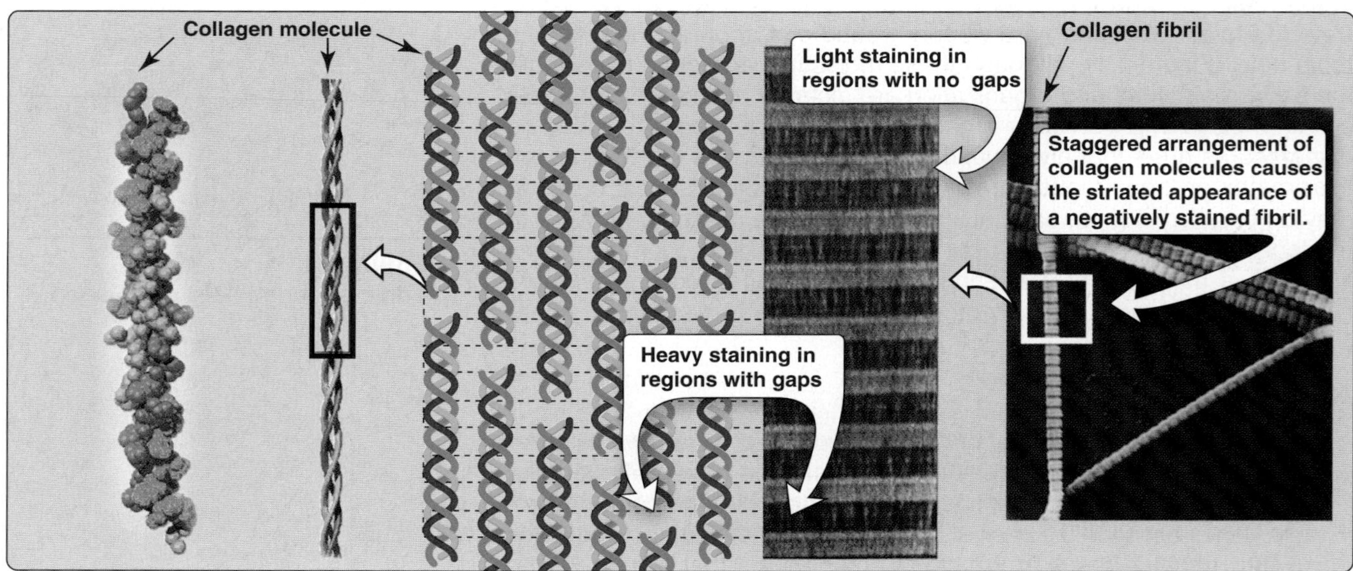

Figure 4.3
Collagen fibrils at right have a characteristic banding pattern, reflecting the regularly staggered packing of the individual collagen molecules in the fibril.

B. Structure of collagen

1. **Amino acid sequence:** Collagen is rich in proline and glycine, both of which are important in the formation of the triple-stranded helix. **Proline** facilitates the formation of the helical conformation of each α-chain because its ring structure causes "kinks" in the peptide chain. **Glycine**, the smallest amino acid, is found in every third position of the polypeptide chain. It fits into the restricted spaces where the three chains of the helix come together. The glycine residues are part of a repeating sequence, **–Gly–X–Y–**, where X is frequently proline and Y is often **hydroxyproline** or **hydroxylysine** (Figure 4.5). Thus, most of the α-chain can be regarded as a polytripeptide whose sequence can be represented as **(–Gly–X–Y–)$_{333}$**.

2. **Triple-helical structure:** Unlike most globular proteins that are folded into compact structures, collagen, a fibrous protein, has an elongated, triple-helical structure that places many of its amino acid side chains on the surface of the triple-helical molecule. [Note: This allows bond formation between the exposed R-groups of neighboring collagen monomers, resulting in their aggregation into long fibers.]

3. **Hydroxyproline and hydroxylysine:** Collagen contains hydroxyproline (hyp) and hydroxylysine (hyl), which are not present in most other proteins. These residues result from the hydroxylation of some of the proline and lysine residues after their incorporation into polypeptide chains (Figure 4.6). The hydroxylation is, thus, an example of **posttranslational modification** (see p. 440). Hydroxyproline is important in stabilizing the triple-helical structure of collagen because it maximizes interchain hydrogen bond formation.

4. **Glycosylation:** The hydroxyl group of the **hydroxylysine** residues of collagen may be enzymatically glycosylated. Most commonly, glucose and galactose are sequentially attached to the polypeptide chain prior to triple-helix formation (Figure 4.7).

C. Biosynthesis of collagen

The polypeptide precursors of the collagen molecule are formed in **fibroblasts** (or in the related **osteoblasts** of bone and **chondroblasts** of cartilage), and are secreted into the **extracellular matrix**. After enzymic modification, the mature collagen monomers aggregate and become cross-linked to form collagen fibrils.

1. **Formation of pro-α-chains:** Collagen is one of many proteins that normally function outside of cells. Like most proteins produced for export, the newly synthesized polypeptide precursors of α-chains contain a special amino acid sequence at their N-terminal ends. This acts as a signal that the polypeptide being synthesized is destined to leave the cell. The **signal sequence** facilitates the binding of ribosomes to the rough endoplasmic reticulum (RER), and directs the passage of the polypeptide chain into the cisternae of the RER. The signal sequence is rapidly cleaved in the endoplasmic reticulum to yield a precursor of collagen called a **pro-α-chain** (see Figure 4.7).

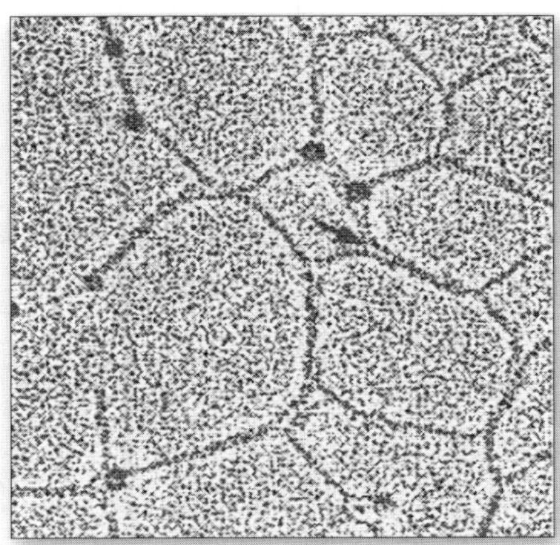

Figure 4.4
Electron micrograph of a polygonal network formed by association of collagen type IV monomers.

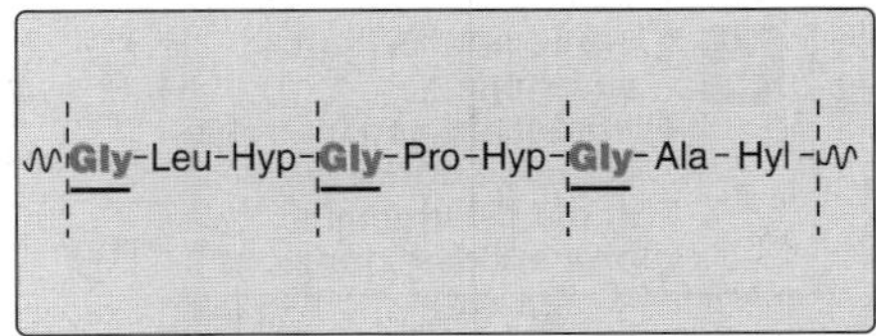

Figure 4.5
Amino acid sequence of a portion of the α1-chain of collagen. [Note: Hyp is hydroxyproline and Hyl is hydroxylysine.]

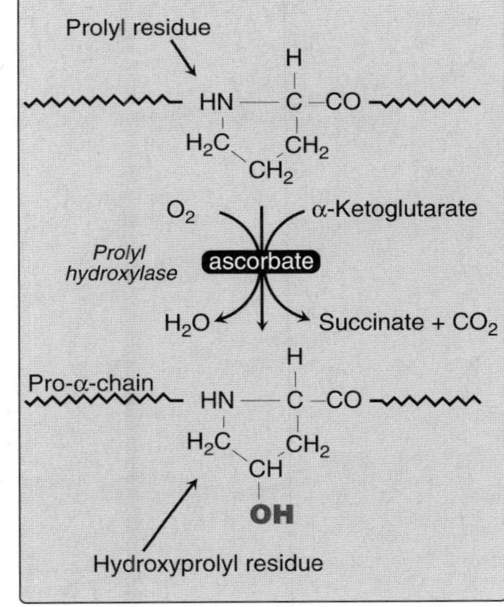

Figure 4.6
Hydroxylation of prolyl residues of pro-α-chains of collagen by *prolyl hydroxylase.*

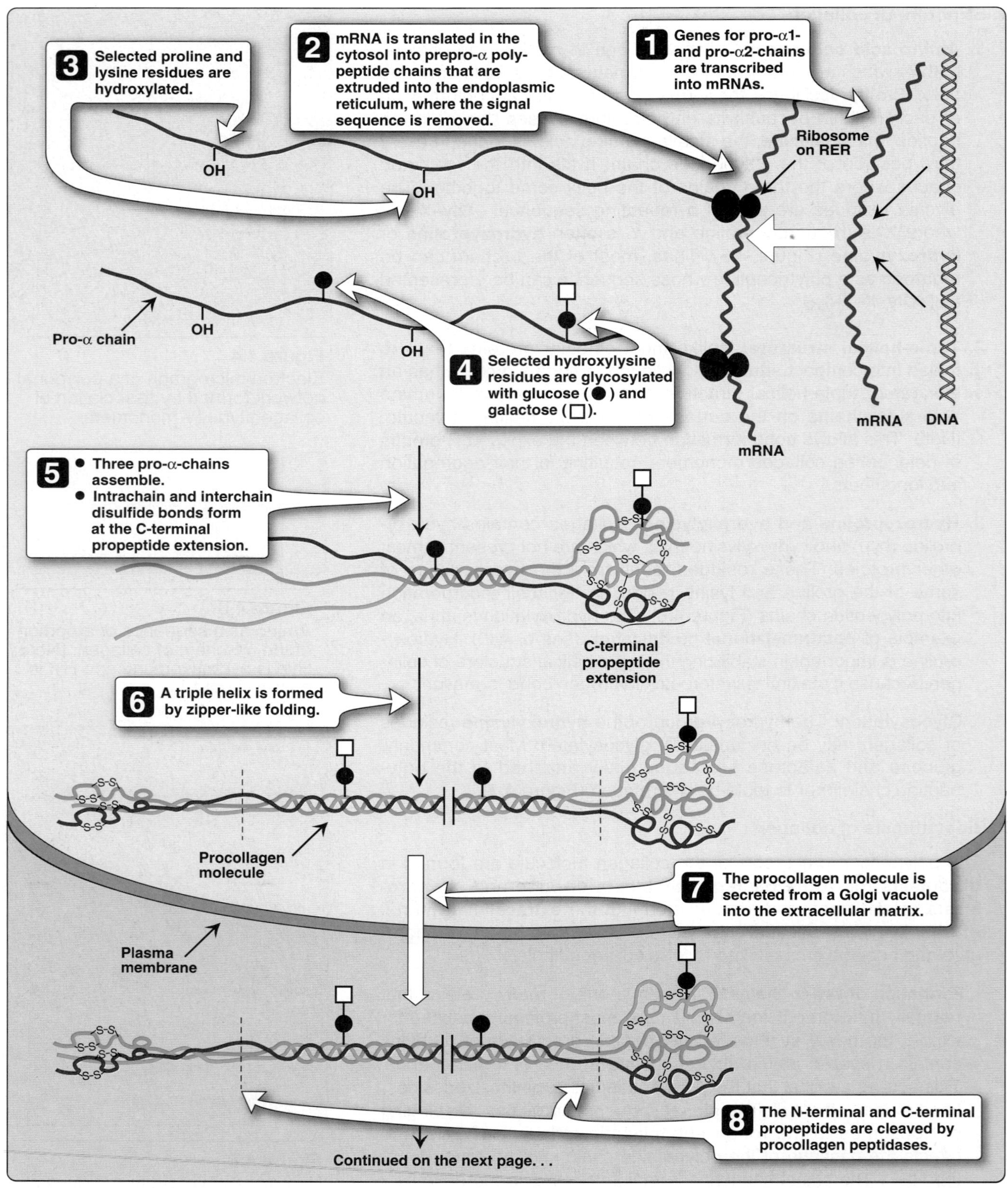

Figure 4.7
Formation of a collagen fibril. (*Continued on the next page*)

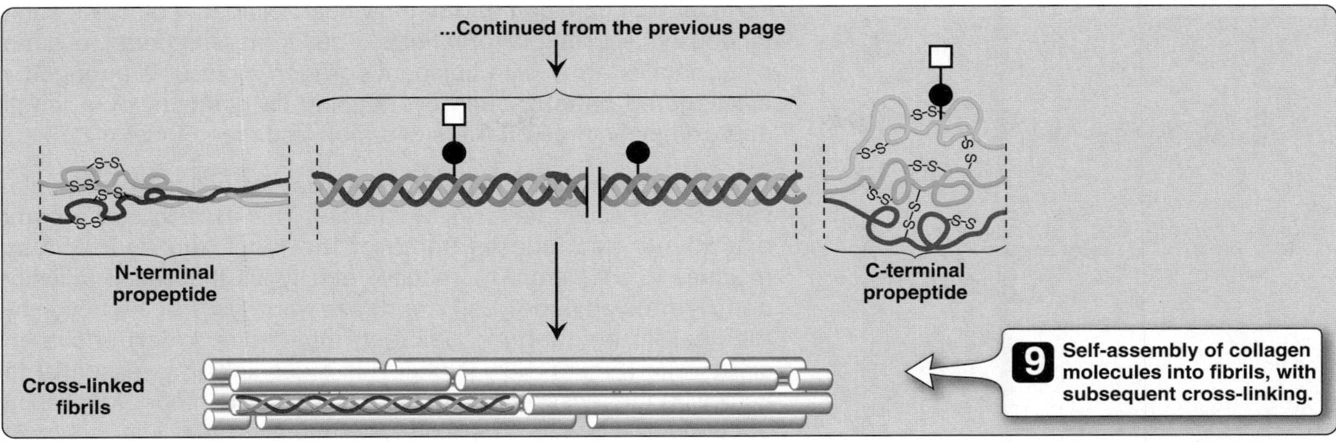

Figure 4.7
Formation of a collagen fibril. (*Continued from the previous page*)

2. **Hydroxylation:** The pro-α-chains are processed by a number of enzymic steps within the lumen of the RER while the polypeptides are still being synthesized (see Figure 4.7). Proline and lysine residues found in the Y-position of the –Gly–X–Y– sequence can be hydroxylated to form hydroxyproline and hydroxylysine residues. These hydroxylation reactions require molecular oxygen and the reducing agent vitamin C (ascorbic acid, see p. 375), without which the hydroxylating enzymes, *prolyl hydroxylase* and *lysyl hydroxylase*, are unable to function (see Figure 4.6). In the case of **ascorbic acid deficiency** (and, therefore, a lack of prolyl and lysyl hydroxylation), collagen fibers cannot be cross-linked (see below), greatly decreasing the tensile strength of the assembled fiber. One resulting deficiency disease is known as **scurvy**. Patients with ascorbic acid deficiency also often show bruises on the limbs as a result of subcutaneous extravascation of blood (**capillary fragility;** Figure 4.8).

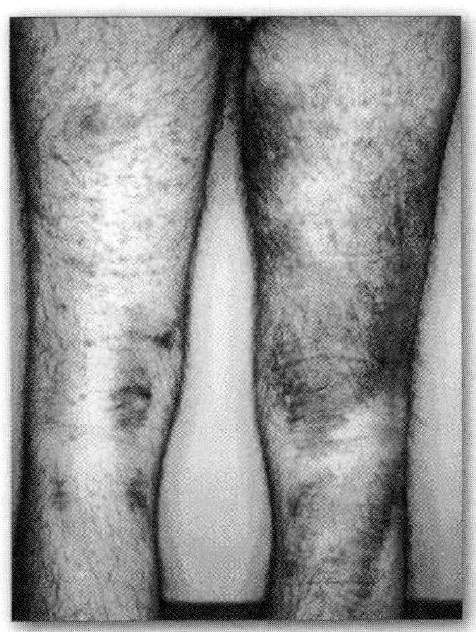

Figure 4.8
The legs of a 46-year-old man with scurvy.

3. **Glycosylation:** Some hydroxylysine residues are modified by glycosylation with glucose or glucosyl-galactose (see Figure 4.7).

4. **Assembly and secretion:** After hydroxylation and glycosylation, pro-α-chains form **procollagen**, a precursor of collagen that has a central region of triple helix flanked by the non-helical amino- and carboxyl-terminal extensions called **propeptides** (see Figure 4.7). The formation of procollagen begins with formation of interchain disulfide bonds between the C-terminal extensions of the pro-α-chains. This brings the three α-chains into an alignment favorable for helix formation. The procollagen molecules are translocated to the Golgi apparatus, where they are packaged in secretory vesicles. The vesicles fuse with the cell membrane, causing the release of procollagen molecules into the extracellular space.

5. **Extracellular cleavage of procollagen molecules:** After their release, the procollagen molecules are cleaved by *N-* and *C-procollagen peptidases,* which remove the terminal propeptides, releasing triple-helical collagen molecules.

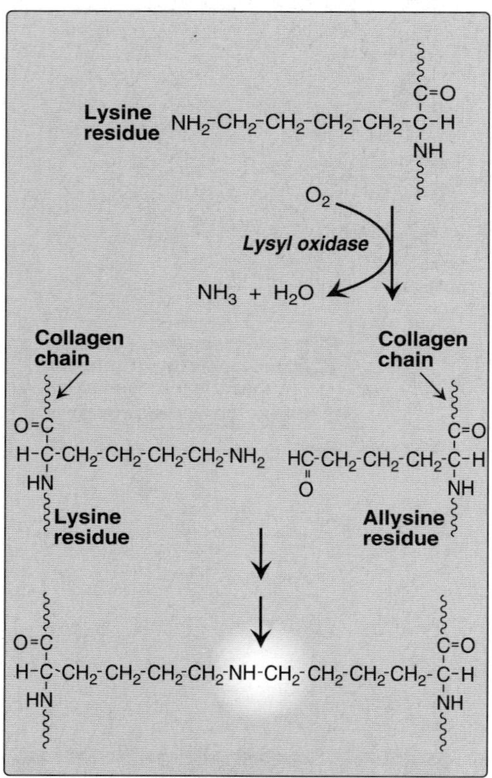

Figure 4.9
Formation of cross-links in collagen.

6. **Formation of collagen fibrils:** Individual collagen molecules spontaneously associate to form fibrils. They form an ordered, overlapping, parallel array, with adjacent collagen molecules arranged in a staggered pattern, each overlapping its neighbor by a length approximately three-quarters of a molecule (see Figure 4.7).

7. **Cross-link formation:** The fibrillar array of collagen molecules serves as a substrate for *lysyl oxidase*. This extracellular enzyme oxidatively deaminates some of the lysyl and hydroxylysyl residues in collagen. The reactive aldehydes that result (**allysine** and **hydroxyallysine**) can condense with lysyl or hydroxylysyl residues in neighboring collagen molecules to form covalent cross-links (Figure 4.9). [Note: This cross-linking is essential for achieving the tensile strength necessary for the proper functioning of connective tissue. Therefore, any mutation that interferes with the ability of collagen to form cross-linked fibrils almost certainly affects the stability of the collagen.]

D. Degradation of collagen

Normal collagens are highly stable molecules, having half-lives as long as several months. However, connective tissue is dynamic and is constantly being remodeled, often in response to growth or injury of the tissue. Breakdown of collagen fibrils is dependent on the proteolytic action of **collagenases**, which are part of a large family of matrix metalloproteinases. For type I collagen, the cleavage site is specific, generating three-quarter and one-quarter length fragments. These fragments are further degraded by other matrix proteinases to their constituent amino acids.

E. Collagen diseases

Defects in any one of the many steps in collagen fiber synthesis can result in a genetic disease involving an inability of collagen to form fibers properly and, thus, provide tissues with the needed tensile strength normally provided by collagen. More than 1000 mutations have been identified in 22 genes coding for twelve of the collagen types. The following are examples of diseases that are the result of defective collagen synthesis.

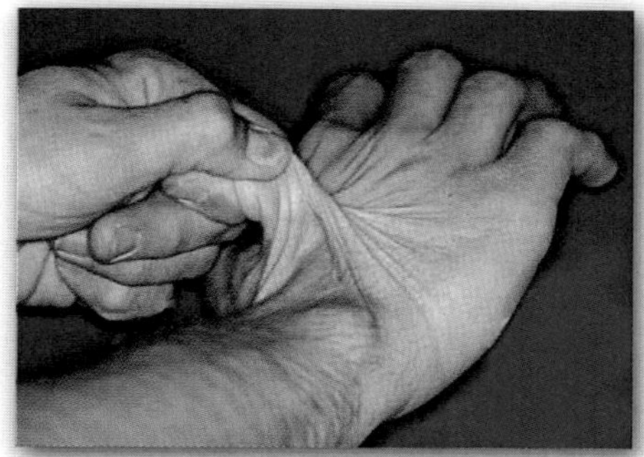

Figure 4.10
Stretchy skin of Ehlers-Danlos syndrome.

1. **Ehlers-Danlos syndrome (EDS):** This disorder is a heterogeneous group of generalized connective tissue disorders that result from inheritable defects in the metabolism of fibrillar collagen molecules. EDS can result from a deficiency of collagen-processing enzymes (for example, **lysyl-hydroxylase deficiency** or **procollagen peptidase deficiency**), or from mutations in the amino acid sequences of collagen types I, III, or V. The most clinically important mutations are found in the gene for type III collagen. Collagen containing mutant chains is not secreted, and is either degraded or accumulated to high levels in intracellular compartments. Because collagen type III is an important component of the arteries, potentially lethal vascular problems occur. [Note: Although collagen type III is only a minor component of the collagen fibrils in the skin, for unknown reasons, EDS patients also show defects in collagen type I fibrils. This results in stretchy skin and loose joints (Figure 4.10).]

2. **Osteogenesis imperfecta (OI):** This disease, known as **brittle bone syndrome**, is also a heterogeneous group of inherited disorders distinguished by bones that easily bend and fracture (Figure 4.11). Retarded wound healing and a rotated and twisted spine leading to a "humped-back" appearance are common features of the disease. Type I OI is called **osteogenesis imperfect tarda**. This disease presents in early infancy with fractures secondary to minor trauma, and may be suspected if prenatal ultrasound detects bowing or fractures of long bones. Type II OI, **osteogenesis imperfecta congenita**, is more severe, and patients die in utero or in the neonatal period of pulmonary hypoplasia. Most patients with severe OI have mutations in the gene for either the pro1– or pro2–α-chains of type I collagen. The most common mutations cause the substitution of single amino acids with bulky side chains for the glycine residues that appear as every third amino acid in the triple helix. The structurally abnormal pro–α-chains can prevent folding of the protein into a triple-helical conformation.

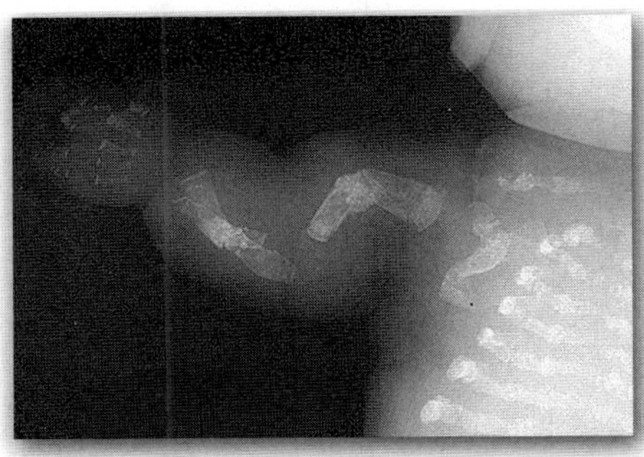

Figure 4.11
Lethal form of osteogenesis imperfecta in which the fractures appear in utero, as revealed by this radiograph of a stillborn fetus.

III. ELASTIN

In contrast to collagen, which forms fibers that are tough and have high tensile strength, elastin is a **connective tissue protein** with rubber-like properties. Elastic fibers composed of elastin and glycoprotein microfibrils are found in the lungs, the walls of large arteries, and elastic ligaments. They can be stretched to several times their normal length, but recoil to their original shape when the stretching force is relaxed.

A. Structure of elastin

Elastin is an insoluble protein polymer synthesized from a precursor, **tropoelastin**, which is a linear polypeptide composed of about 700 amino acids that are primarily small and nonpolar (for example, glycine, alanine, and valine, see p. 2). Elastin is also rich in proline and lysine, but contains only a little hydroxyproline and no hydroxylysine. Tropoelastin is secreted by the cell into the extracellular space. There it interacts with specific glycoprotein microfibrils, such as **fibrillin**, which function as a scaffold onto which tropoelastin is deposited. [Note: Mutations in the fibrillin gene are responsible for **Marfan's syndrome**.] Some of the lysyl side chains of the tropoelastin polypeptides are oxidatively deaminated by *lysyl oxidase*, forming allysine residues. Three of the allysyl side chains plus one unaltered lysyl side chain from the same or neighboring polypeptides form a desmosine cross-link (Figure 4.12). This produces **elastin**—an extensively interconnected, rubbery network that can stretch and bend in any direction when stressed, giving connective tissue elasticity (Figure 4.13).

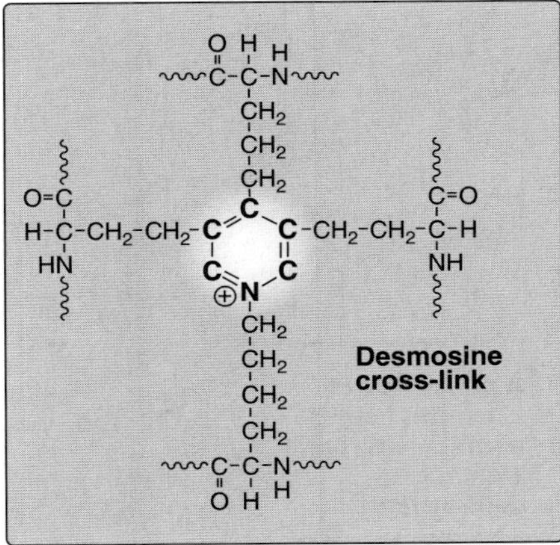

Figure 4.12
Desmosine cross-link in elastin.

B. Role of α_1-antitrypsin in elastin degradation

1. **α_1-Antitrypsin:** Blood and other body fluids contain a protein, α_1-antitrypsin (α_1-**AT**, currently also called α_1-**antiproteinase**), that inhibits a number of proteolytic enzymes (also called proteases or proteinases) that hydrolyze and destroy proteins. [Note: The inhibitor was originally named α_1-antitrypsin because it inhibits the activity of *trypsin* (a proteolytic enzyme synthesized as trypsinogen

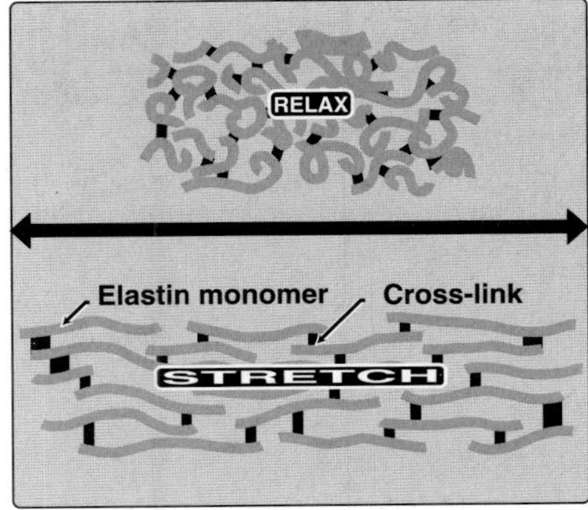

Figure 4.13
Elastin fibers in relaxed and stretched conformations.

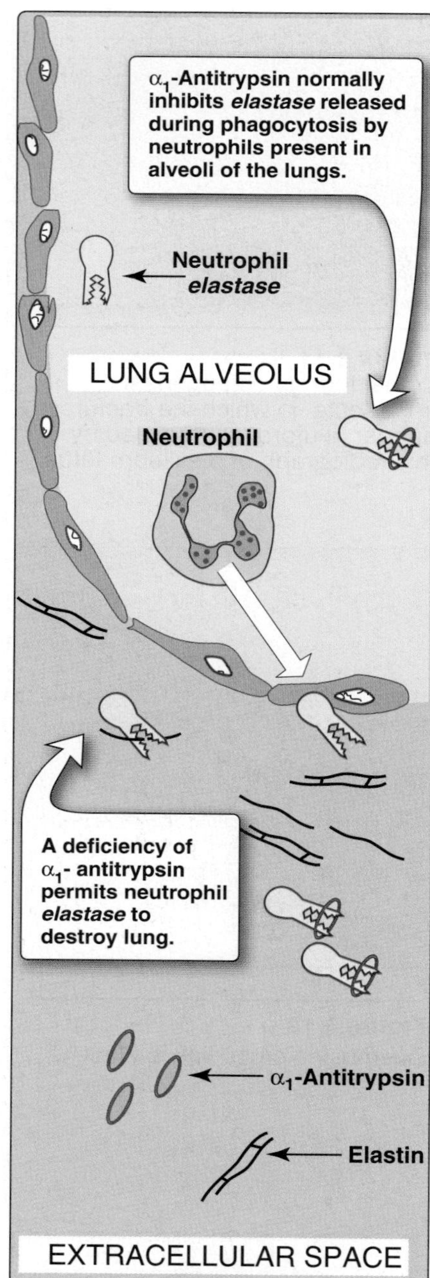

Figure 4.14
Destruction of alveolar tissue by *elastase* released from neutrophils.

by the pancreas, see p. 246).] α_1-AT comprises more than ninety percent of the α_1-globulin fraction of normal plasma. α_1-AT has the important physiologic role of inhibiting neutrophil *elastase*—a powerful protease that is released into the extracellular space, and degrades elastin of alveolar walls, as well as other structural proteins in a variety of tissues (Figure 4.14). Most of the α_1-AT found in plasma is synthesized and secreted by the liver. The remainder is synthesized by several tissues, including monocytes and alveolar macrophages, which may be important in the prevention of local tissue injury by *elastase*.

2. **Role of α_1-AT in the lungs:** In the normal lung, the alveoli are chronically exposed to low levels of neutrophil *elastase* released from activated and degenerating neutrophils. This proteolytic activity can destroy the elastin in alveolar walls if unopposed by the inhibitory action of α_1-AT, the most important inhibitor of neutrophil *elastase* (see Figure 4.14). Because lung tissue cannot regenerate, **emphysema** results from the destruction of the connective tissue of alveolar walls.

3. **Emphysema resulting from α_1-AT deficiency:** In the United States, approximately two to five percent of patients with emphysema are predisposed to the disease by inherited defects in α_1-AT. A number of different mutations in the α_1-AT gene are known to cause a deficiency of this protein, but one single purine base mutation (GAG $\rightarrow$ AAG, resulting in the substitution of lysine for glutamic acid at position 342 of the protein) is clinically the most widespread. An individual must inherit two abnormal α_1-AT alleles to be at risk for the development of emphysema. In a heterozygote, with one normal and one defective gene, the levels of α_1-AT are sufficient to protect the alveoli from damage. [Note: A specific α_1-AT methionine is required for the binding of the inhibitor to its target proteases. Smoking causes the oxidation and subsequent inactivation of that methionine residue, thereby rendering the inhibitor powerless to neutralize elastase. Smokers with α_1-AT deficiency, therefore, have a considerably elevated rate of lung destruction and a poorer survival rate than nonsmokers with the deficiency.] The deficiency of *elastase* inhibitor can be reversed by weekly intravenous administration of α_1-AT. The α_1-AT diffuses from the blood into the lung, where it reaches therapeutic levels in the fluid surrounding the lung epithelial cells.

IV. CHAPTER SUMMARY

Collagen molecules contain an abundance of **proline**, **lysine**, and **glycine**, the latter occurring at every third position in the primary structure. Collagen also contains **hydroxyproline**, **hydroxylysine**, **and glycosylated hydroxylysine**, each formed by posttranslational modification. Collagen molecules typically form **fibrils** containing a long, stiff, triple-stranded helical structure, in which three collagen polypeptide chains are wound around one another in a rope-like superhelix **triple helix**. Other types of collagen form mesh-like networks. **Elastin** is a connective tissue protein with rubber-like properties in tissues such as the lung. **α_1-Antitrypsin** (**α_1-AT**), produced primarily by the liver but also by tissues such as monocytes and alveolar macrophages, prevents elastin degradation in the alveolar walls. A deficiency of α_1-AT can cause **emphysema**.

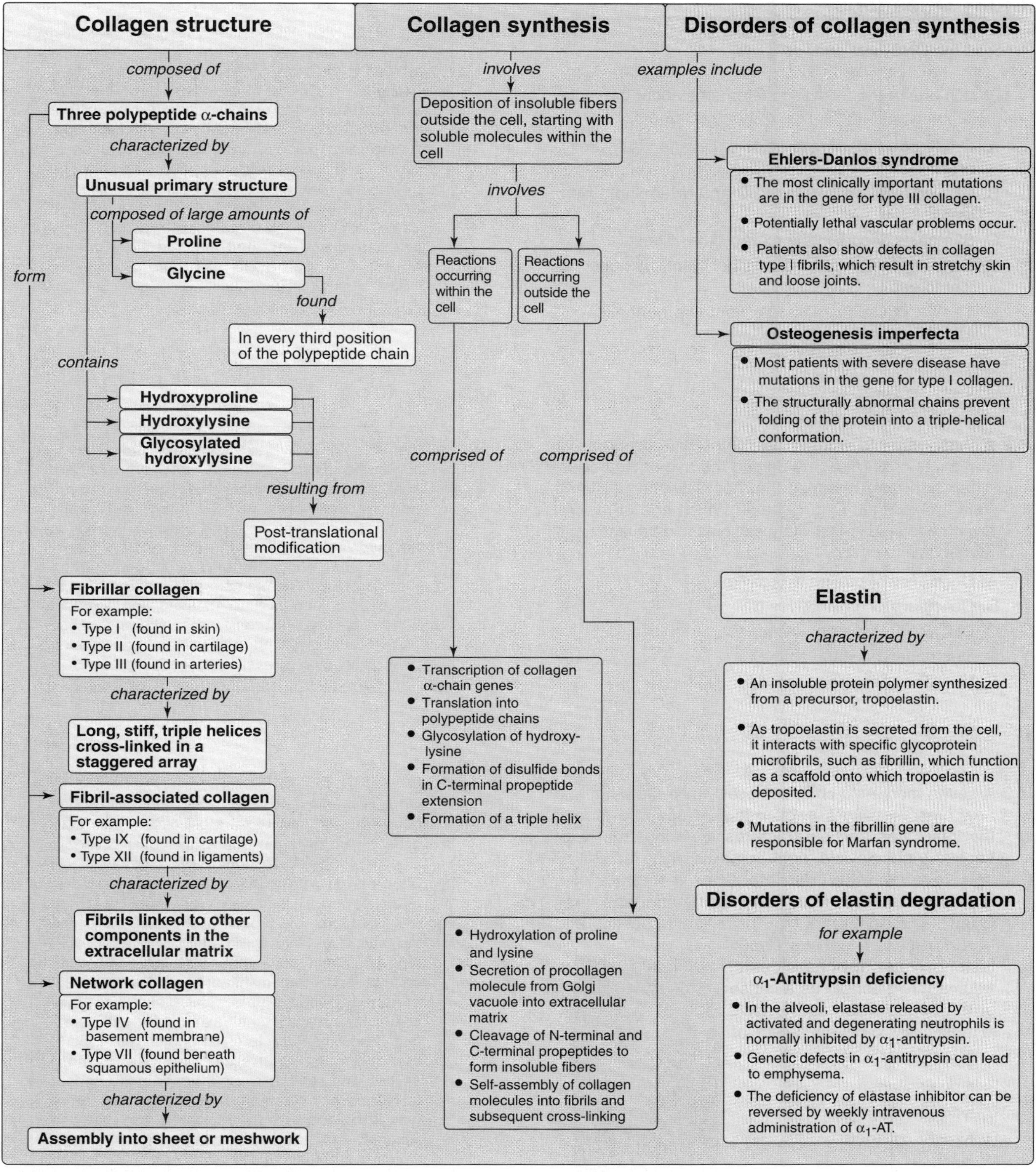

Figure 4.15
Key concept map for the fibrous proteins, collagen, and elastin.

Study Questions

Choose the ONE best answer

4.1 Which one of the following statements about the major collagen type found in skin or bond is correct?

A. One third of the amino acids of collagen is hydroxyproline.
B. Glycine is found only in the C-and N-terminal extensions.
C. Synthesis occurs in the extracellular matrix.
D. Collagen fibrils are held together solely by noncovalent forces.
E. The procollagen molecule contains nonhelical C- and N-terminal extensions.

Correct answer = E. The procollagen molecules contain a triple-helical center section with nonhelical C- and N-terminal extensions. One third of the amino acids of collagen is glycine, with a significant but lesser amount of hydroxyproline. Ascorbate is required for the hydroxylation of proline and lysine residues of pro-α-chains. The polypeptide precursors of the collagen molecule are formed in fibroblasts (or in the related osteoblasts of bone and chondroblasts of cartilage), and are secreted into the extracellular matrix. Collagen fibrils are held together by covalent cross-links.

4.2 A thirty-year-old woman presented with progressive shortness of breath. She denied the use of cigarettes. A family history revealed that her sister had suffered from unexplained lung disease. Which one of the following etiologies most likely explains this patient's pulmonary symptoms?

A. Deficiency of proline hydroxylase
B. Deficiency of α1-antitrypsin
C. Deficiency in dietary vitamin C
D Decreased elastase activity
E. Increase collaginase activity

Correct answer = B. α1-Antitrypsin deficiency is a genetic disorder that can cause pulmonary emphysema even in the absence of cigarette use. An deficiency of α1-antitrypsin permits increased elastase activity to destroy elastin in the alveolar walls, even in nonsmokers. α1-antitrypsin deficiency should be suspected when chronic obstructive pulmonary disease develops in a patient younger than 45 years who does not have a history of chronic bronchitis or tobacco use, or when multiple family members develop obstructive lung disease at an early age.

4.3 A seven-month-old child "fell over" while crawling, and now presents with a swollen leg. At age one month, the infant has multiple fractures in various states of healing (right clavicle, right humerus, right radius). At age seven months, the infant has a fracture of a bowed femur, secondary to minor trauma (see x-ray below). The bones are thin, have few trabecula, and thin cortices. A careful family history ruled out nonaccidental trauma (child abuse) as a cause of the bone fractures. The child is most likely to have a defect in:

A. fibrillin.
B. type I collagen.
C. type III collagen.
D. type IV collagen.
E. elastin.

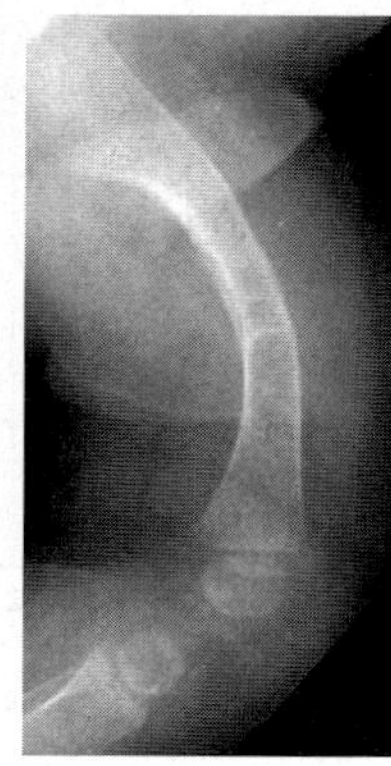

Correct answer = B. The child most likely has osteogenesis imperfecta. Most cases arise from a defect in the gene encoding type I collagen. Bones in affected patients are thin, osteoporotic, often bowed with a thin cortex and deficient trabeculae, and extremely prone to fracture. There are four types, but types I and II account for the majority of cases. This patient is affected with type I, osteogenesis imperfecta tarda. The disease presents in early infancy with fractures secondary to minor trauma. The disease may be suspected on prenatal ultrasound through detection of bowing or fractures of long bones. Type II, osteogenesis imperfecta congenita, is more severe, and patients die <u>in utero</u> or in the neonatal period of pulmonary hypoplasia. Defects in type III collagen are the most common cause of Ehlers-Danlos syndrome, characterized by lethal vascular problems and stretchy skin. Type IV collagen forms networks, not fibrils.

Enzymes

5

I. OVERVIEW

Virtually all reactions in the body are mediated by enzymes, which are protein catalysts that increase the rate of reactions without being changed in the overall process. Among the many biologic reactions that are energetically possible, enzymes selectively channel reactants (called substrates) into useful pathways. Enzymes thus direct all metabolic events. This chapter examines the nature of these catalytic molecules and their mechanism of action.

II. NOMENCLATURE

Each enzyme is assigned two names. The first is its short, **recommended name**, convenient for everyday use. The second is the more complete **systematic name,** which is used when an enzyme must be identified without ambiguity.

A. Recommended name

Most commonly used enzyme names have the suffix "**-ase**" attached to the **substrate** of the reaction (for example, *glucosidase, urease, sucrase*), or to a description of the **action performed** (for example, *lactate dehydrogenase* and *adenylyl cyclase*). [Note: Some enzymes retain their original trivial names, which give no hint of the associated enzymic reaction, for example, *trypsin* and *pepsin*.]

B. Systematic name

The International Union of Biochemistry and Molecular Biology (IUBMB) developed a system of nomenclature in which enzymes are divided into six major classes (Figure 5.1), each with numerous subgroups. The suffix -ase is attached to a fairly complete description of the chemical reaction catalyzed, for example *D-glyceraldehyde 3-phosphate:NAD oxidoreductase*. The IUBMB names are unambiguous and informative, but are sometimes too cumbersome to be of general use.

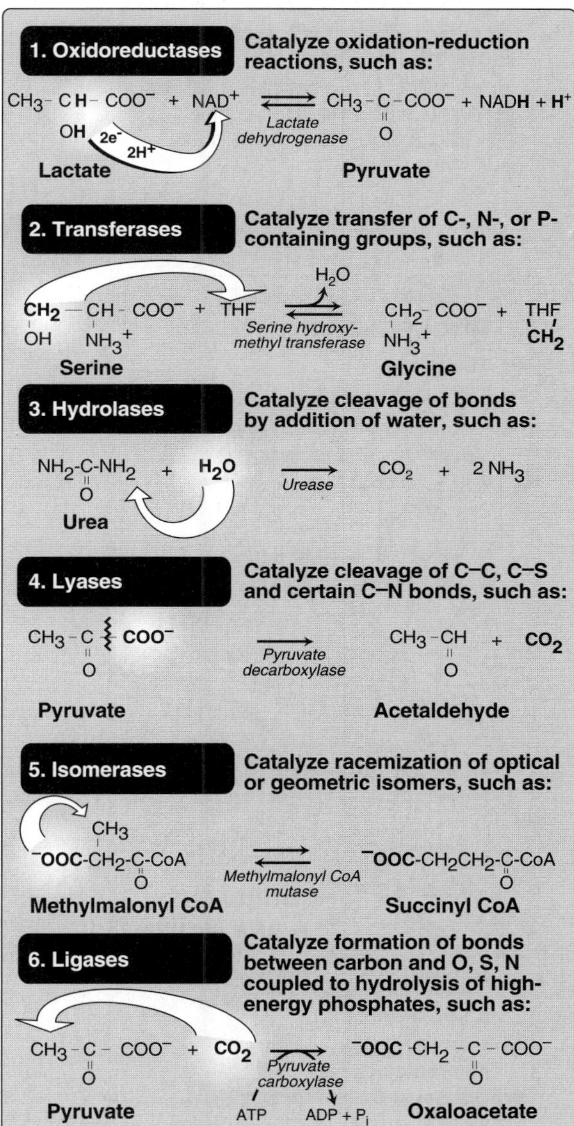

Figure 5.1
Examples of the six major classes of the international classification of enzymes (THF is tetrahydrofolate).

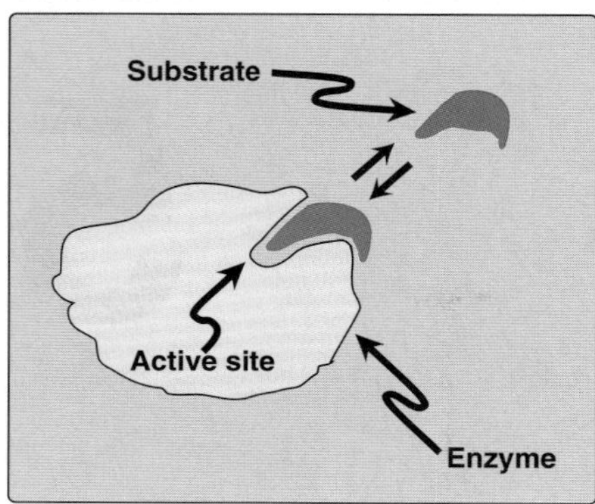

Figure 5.2
Schematic representation of an enzyme with one active site binding a substrate molecule.

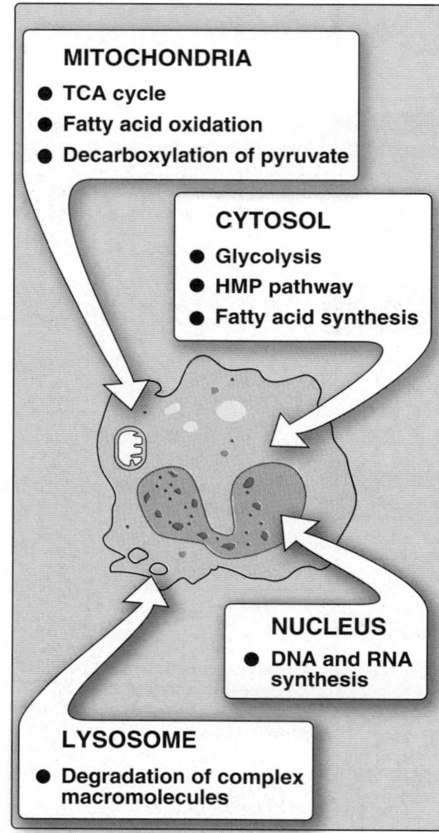

Figure 5.3
The intracellular location of some important biochemical pathways.

III. PROPERTIES OF ENZYMES

Enzymes are **protein catalysts** that increase the velocity of a chemical reaction, and are not consumed during the reaction they catalyze. [Note: Some types of RNA can act like enzymes, usually catalyzing the cleavage and synthesis of phosphodiester bonds. RNAs with catalytic activity are called **ribozymes** (see p. 436), and are much less commonly encountered than protein catalysts.]

A. Active sites

Enzyme molecules contain a special pocket or cleft called the active site. The active site contains amino acid side chains that create a three-dimensional surface complementary to the substrate (Figure 5.2). The active site binds the substrate, forming an enzyme-substrate (ES) complex. ES is converted to enzyme-product (EP), which subsequently dissociates to enzyme and product.

B. Catalytic efficiency

Most enzyme-catalyzed reactions are highly efficient, proceeding from 10^3 to 10^8 times faster than uncatalyzed reactions. Typically, each enzyme molecule is capable of transforming 100 to 1000 substrate molecules into product each second. The number of molecules of substrate converted to product per enzyme molecule per second is called the **turnover number**.

C. Specificity

Enzymes are highly specific, interacting with one or a few substrates and catalyzing only one type of chemical reaction.

D. Cofactors

Some enzymes associate with a **nonprotein cofactor** that is needed for enzymic activity. Commonly encountered cofactors include **metal ions** such as Zn^{2+} or Fe^{2+}, and **organic molecules**, known as **coenzymes**, that are often derivatives of vitamins. For example, the coenzyme NAD^+ contains niacin, FAD contains riboflavin, and coenzyme A contains pantothenic acid. (See pp. 371–379 for the role of vitamins as precursors of coenzymes.) **Holoenzyme** refers to the enzyme with its cofactor. **Apoenzyme** refers to the protein portion of the holoenzyme. In the absence of the appropriate cofactor, the apoenzyme typically does not show biologic activity. A **prosthetic group** is a tightly bound coenzyme that does not dissociate from the enzyme (for example, the biotin bound to carboxylases, see p. 379).

E. Regulation

Enzyme activity can be regulated, that is, enzymes can be **activated** or **inhibited**, so that the rate of product formation responds to the needs of the cell.

F. Location within the cell

Many enzymes are localized in specific organelles within the cell (Figure 5.3). Such **compartmentalization** serves to isolate the reaction substrate or product from other competing reactions. This provides a favorable environment for the reaction, and organizes the thousands of enzymes present in the cell into purposeful pathways.

IV. HOW ENZYMES WORK

The mechanism of enzyme action can be viewed from two different perspectives. The first treats catalysis in terms of energy changes that occur during the reaction, that is, enzymes provide an alternate, energetically favorable reaction pathway different from the uncatalyzed reaction. The second perspective describes how the active site chemically facilitates catalysis.

A. Energy changes occurring during the reaction

Virtually all chemical reactions have an **energy barrier** separating the reactants and the products. This barrier, called the **free energy of activation**, is the energy difference between that of the reactants and a high-energy intermediate that occurs during the formation of product. For example, Figure 5.4 shows the changes in energy during the conversion of a molecule of reactant A to product B as it proceeds through the **transition state (high-energy intermediate)**, **T***:

$$A \rightleftarrows T^* \rightleftarrows B$$

1. **Free energy of activation:** The peak of energy in Figure 5.4 is the difference in free energy between the reactant and T*, where the high-energy intermediate is formed during the conversion of reactant to product. Because of the high free energy of activation, the rates of uncatalyzed chemical reactions are often slow.

2. **Rate of reaction:** For molecules to react, they must contain sufficient energy to overcome the energy barrier of the transition state. In the absence of an enzyme, only a small proportion of a population of molecules may possess enough energy to achieve the transition state between reactant and product. The rate of reaction is determined by the number of such energized molecules. In general, the lower the free energy of activation, the more molecules have sufficient energy to pass through the transition state, and, thus, the faster the rate of the reaction.

3. **Alternate reaction pathway:** An enzyme allows a reaction to proceed rapidly under conditions prevailing in the cell by providing an alternate reaction pathway with a **lower free energy of activation** (see Figure 5.4). The enzyme does not change the free energies of the reactants or products and, therefore, does not change the equilibrium of the reaction.

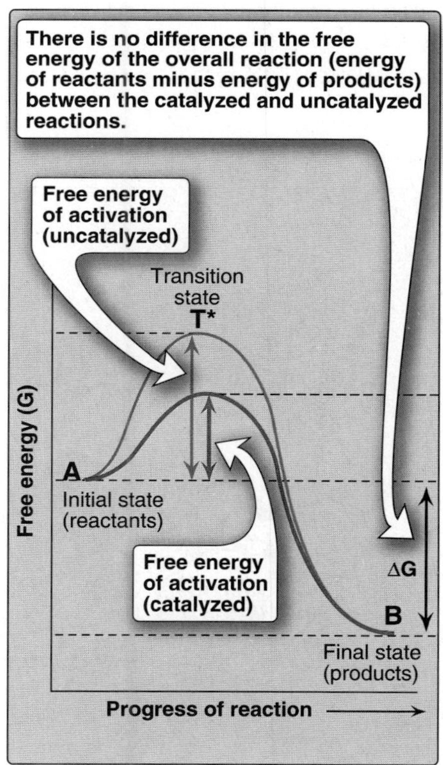

Figure 5.4
Effect of an enzyme on the activation energy of a reaction.

B. Chemistry of the active site

The active site is not a passive receptacle for binding the substrate, but rather is a complex molecular machine employing a diversity of chemical mechanisms to facilitate the conversion of substrate to product. A number of factors are responsible for the catalytic efficiency of enzymes, including the following:

1. **Transition-state stabilization:** The active site often acts as a flexible molecular template that binds the substrate in a geometric structure resembling the activated transition state of the molecule (see T^* at the top of the curve in Figure 5.4). By stabilizing the substrate in its transition state, the enzyme greatly increases the concentration of the reactive intermediate that can be converted to product and, thus, accelerates the reaction.

2. **Other mechanisms:** The active site can provide catalytic groups that enhance the probability that the transition state is formed. In some enzymes, these groups can participate in **general acid-base catalysis** in which amino acid residues provide or accept protons. In other enzymes, catalysis may involve the transient formation of a **covalent enzyme-substrate complex**.

3. **Visualization of the transition-state:** The enzyme-catalyzed conversion of substrate to product can be visualized as being similar to removing a sweater from an uncooperative infant (Figure 5.5). The process has a high energy of activation because the only reasonable strategy for removing the garment (short of ripping it off) requires that the random flailing of the baby results in both arms being fully extended over the head—an unlikely posture. However, we can envision a parent acting as an enzyme, first coming in contact with the baby (forming ES), then guiding the baby's arms into an extended, vertical position, analogous to the ES transition state. This posture (conformation) of the baby facilitates the removal of the sweater, forming the disrobed baby, which here represents product. [Note: The substrate bound to the enzyme (ES) is at a slightly lower energy than unbound substrate (S) and explains the small "dip" in the curve at ES.]

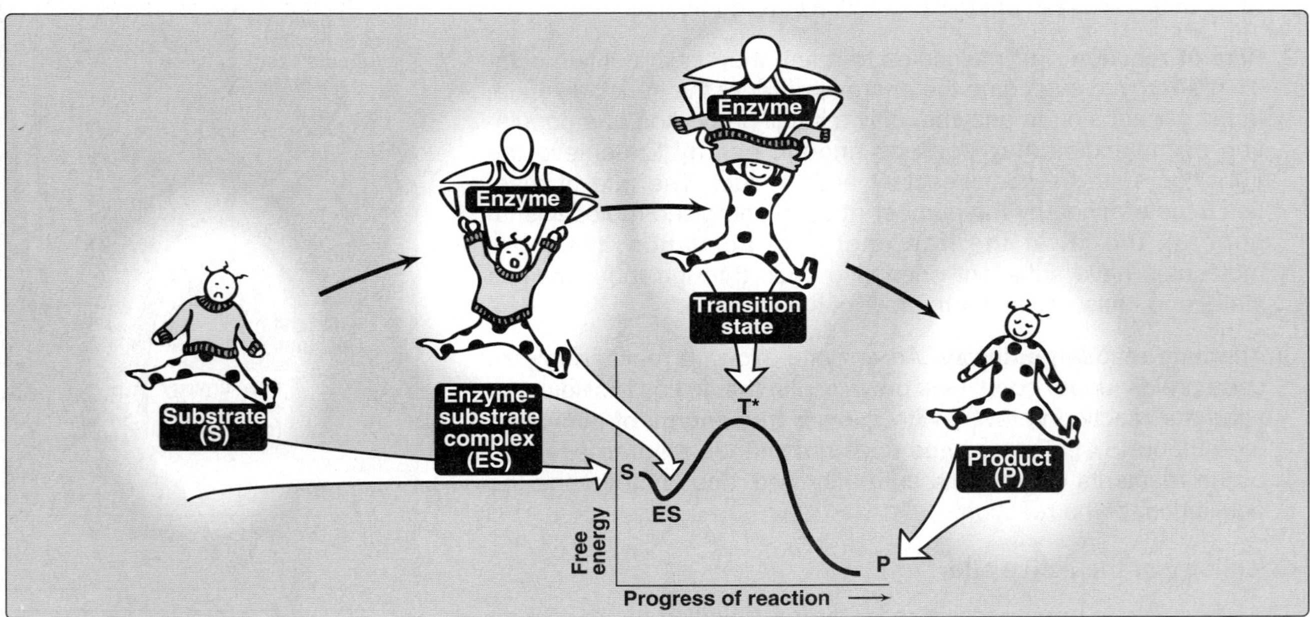

Figure 5.5
Schematic representation of energy changes accompanying formation of enzyme-substrate complex and subsequent formation of a transition-state complex.

V. FACTORS AFFECTING REACTION VELOCITY

Enzymes can be isolated from cells, and their properties studied in a test tube (that is, in vitro). Different enzymes show different responses to changes in substrate concentration, temperature, and pH. This section describes factors that influence the reaction velocity of enzymes. Enzymic responses to these factors give us valuable clues as to how enzymes function in living cells.

A. Substrate concentration

1. **Maximal velocity:** The **rate** or **velocity** of a reaction **(v)** is the number of substrate molecules converted to product per unit time; velocity is usually expressed as **µmol of product formed per minute**. The rate of an enzyme-catalyzed reaction increases with substrate concentration until a **maximal velocity** (V_{max}) is reached (Figure 5.6). The leveling off of the reaction rate at high substrate concentrations reflects the saturation with substrate of all available binding sites on the enzyme molecules present.

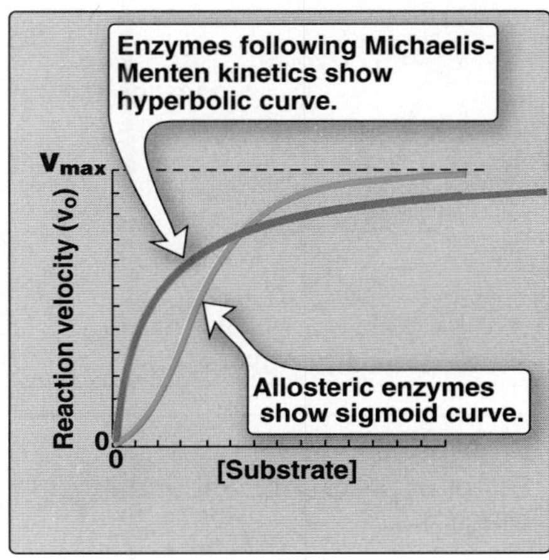

Figure 5.6
Effect of substrate concentration on reaction velocity.

2. **Hyperbolic shape of the enzyme kinetics curve:** Most enzymes show **Michaelis-Menten kinetics** (see p. 58), in which the plot of **initial reaction velocity**, v_o, against **substrate concentration [S]**, is **hyperbolic** (similar in shape to that of the oxygen-dissociation curve of myoglobin, see p. 29). In contrast, **allosteric enzymes** frequently show a sigmoidal curve (see p. 62) that is similar in shape to the oxygen-dissociation curve of hemoglobin (see p. 29).

B. Temperature

1. **Increase of velocity with temperature:** The reaction velocity increases with temperature until a peak velocity is reached (Figure 5.7). This increase is the result of the increased number of molecules having sufficient energy to pass over the energy barrier and form the products of the reaction.

2. **Decrease of velocity with higher temperature:** Further elevation of the temperature results in a decrease in reaction velocity as a result of temperature-induced **denaturation** of the enzyme (see Figure 5.7).

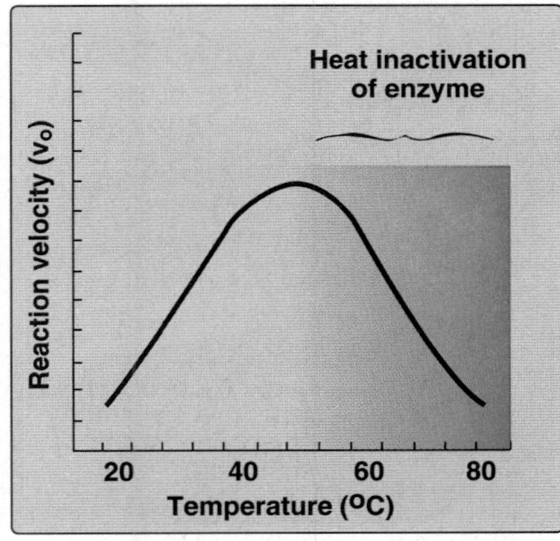

Figure 5.7
Effect of temperature on an enzyme-catalyzed reaction.

C. pH

1. **Effect of pH on the ionization of the active site:** The concentration of H^+ affects reaction velocity in several ways. First, the catalytic process usually requires that the enzyme and substrate have specific chemical groups in either an ionized or unionized state in order to interact. For example, catalytic activity may require that an amino group of the enzyme be in the protonated form ($-NH_3^+$). At alkaline pH, this group is deprotonated, and the rate of the reaction, therefore, declines.

2. **Effect of pH on enzyme denaturation:** Extremes of pH can also lead to denaturation of the enzyme, because the structure of the catalytically active protein molecule depends on the ionic character of the amino acid side chains.

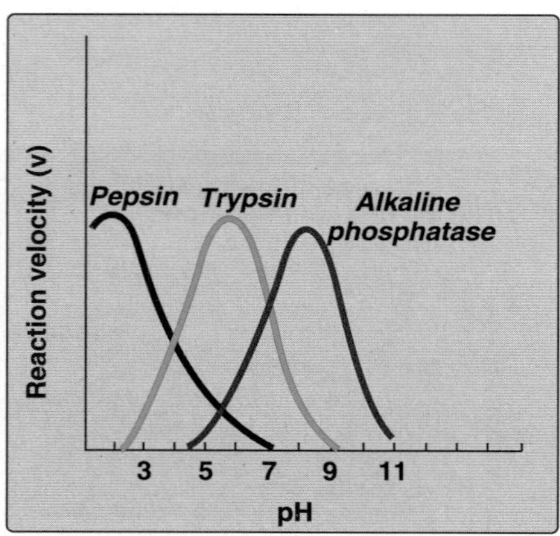

Figure 5.8
Effect of pH on enzyme-catalyzed reactions.

3. **The pH optimum varies for different enzymes:** The pH at which maximal enzyme activity is achieved is different for different enzymes, and often reflects the $[H^+]$ at which the enzyme functions in the body. For example, *pepsin*, a digestive enzyme in the stomach, is maximally active at pH 2, whereas other enzymes, designed to work at neutral pH, are denatured by such an acidic environment (Figure 5.8).

VI. MICHAELIS-MENTEN EQUATION

A. Reaction model

Michaelis and Menten proposed a simple model that accounts for most of the features of enzyme-catalyzed reactions. In this model, the enzyme reversibly combines with its substrate to form an ES complex that subsequently breaks down to product, regenerating the free enzyme. The model, involving one substrate molecule, is represented below:

$$E + S \underset{k_{-1}}{\overset{k_1}{\rightleftharpoons}} ES \xrightarrow{k_2} E + P$$

where S is the substrate
E is the enzyme
ES is the enzyme-substrate complex
P is the product
k_1, k_{-1}, and k_2 are rate constants

B. Michaelis-Menten equation

The Michaelis-Menten equation describes how reaction velocity varies with substrate concentration:

$$v_o = \frac{V_{max}[S]}{K_m + [S]}$$

where v_o = initial reaction velocity
V_{max} = maximal velocity
K_m = Michaelis constant $= (k_{-1} + k_2)/k_1$
[S] = substrate concentration

The following assumptions are made in deriving the Michaelis-Menten rate equation:

1. **Relative concentrations of E and S:** The concentration of substrate ([S]) is much greater than the concentration of enzyme ([E]), so that the percentage of total substrate bound by the enzyme at any one time is small.

2. **Steady-state assumption:** [ES] does not change with time (the steady-state assumption), that is, the rate of formation of ES is equal to that of the breakdown of ES (to E + S and to E + P). In general, an intermediate in a series of reactions is said to be in steady-state when its rate of synthesis is equal to its rate of degradation.

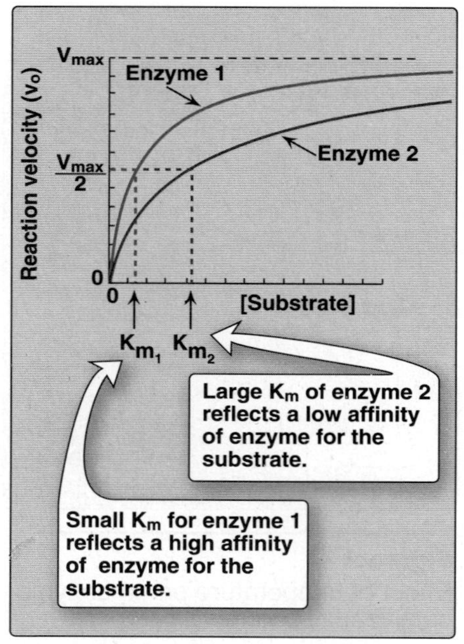

Figure 5.9
Effect of substrate concentration on reaction velocities for two enzymes: enzyme 1 with a small K_m, and enzyme 2 with a large K_m.

3. **Initial velocity:** Only initial reaction velocities (v_o) are used in the analysis of enzyme reactions. This means that the rate of the reaction is measured as soon as enzyme and substrate are mixed. At that time, the concentration of product is very small and, therefore, the rate of the back reaction from P to S can be ignored.

C. Important conclusions about Michaelis-Menten kinetics

1. **Characteristics of K_m:** K_m—the **Michaelis constant**—is characteristic of an enzyme and its particular substrate, and reflects the **affinity** of the enzyme for that substrate. K_m is numerically equal to the **substrate concentration** at which the reaction velocity is equal to $\frac{1}{2}$ V_{max}. K_m does not vary with the concentration of enzyme.

   a. **Small K_m:** A numerically small (low) K_m reflects a **high affinity** of the enzyme for substrate, because a low concentration of substrate is needed to half-saturate the enzyme—that is, reach a velocity that is $\frac{1}{2}$ V_{max} (Figure 5.9).

   b. **Large K_m:** A numerically large (high) K_m reflects a **low affinity** of enzyme for substrate because a high concentration of substrate is needed to half-saturate the enzyme.

2. **Relationship of velocity to enzyme concentration:** The rate of the reaction is directly proportional to the enzyme concentration at all substrate concentrations. For example, if the enzyme concentration is halved, the initial rate of the reaction (v_o), as well as that of V_{max}, are reduced to one half that of the original.

3. **Order of reaction:** When [S] is much less than K_m, the velocity of the reaction is approximately proportional to the substrate concentration (Figure 5.10). The rate of reaction is then said to be **first order** with respect to substrate. When [S] is much greater than K_m, the velocity is constant and equal to V_{max}. The rate of reaction is then independent of substrate concentration, and is said to be **zero order** with respect to substrate concentration (see Figure 5.10).

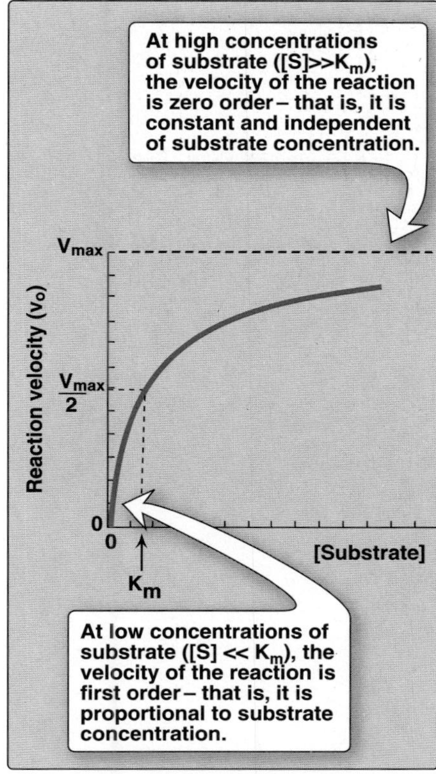

Figure 5.10
Effect of substrate concentration on reaction velocity for an enzyme-catalyzed reaction.

D. Lineweaver-Burke plot

When v_o is plotted against [S], it is not always possible to determine when V_{max} has been achieved, because of the gradual upward slope of the hyperbolic curve at high substrate concentrations. However, if $1/v_o$ is plotted versus 1/[S], a straight line is obtained (Figure 5.11). This plot, the **Lineweaver-Burke plot** (also called a **double-reciprocal plot**) can be used to calculate K_m and V_{max}, as well as to determine the mechanism of action of enzyme inhibitors.

1. The equation describing the Lineweaver-Burke plot is:

$$\frac{1}{v_o} = \frac{K_m}{V_{max}[S]} + \frac{1}{V_{max}}$$

where the intercept on the x axis is equal to $-1/K_m$, and the intercept on the y axis is equal to $1/V_{max}$.

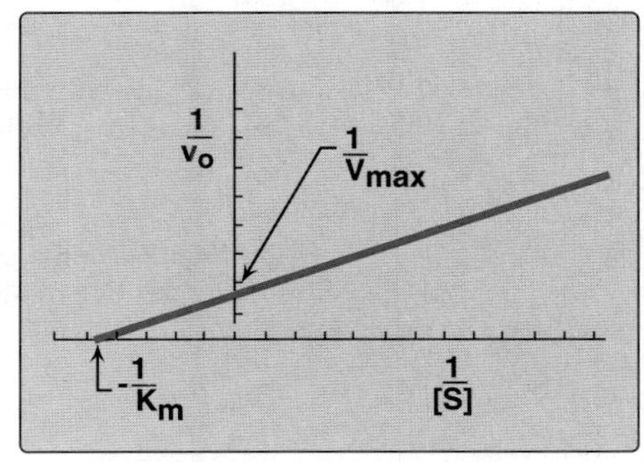

Figure 5.11
Lineweaver-Burke plot.

VII. INHIBITION OF ENZYME ACTIVITY

Any substance that can diminish the velocity of an enzyme-catalyzed reaction is called an **inhibitor**. **Reversible inhibitors** bind to enzymes through noncovalent bonds. Dilution of the enzyme-inhibitor complex results in dissociation of the reversibly bound inhibitor, and recovery of enzyme activity. **Irreversible inhibition** occurs when an inhibited enzyme does not regain activity on dilution of the enzyme-inhibitor complex. The two most commonly encountered types of inhibition are **competitive** and **noncompetitive**.

A. Competitive inhibition

This type of inhibition occurs when the inhibitor binds reversibly to the **same site** that the substrate would normally occupy and, therefore, competes with the substrate for that site.

1. **Effect on V_{max}:** The effect of a competitive inhibitor is reversed by increasing [S]. At a sufficiently high substrate concentration, the reaction velocity reaches the V_{max} observed in the absence of inhibitor (Figure 5.12).

2. **Effect on K_m:** A competitive inhibitor increases the **apparent K_m** for a given substrate. This means that, in the presence of a competitive inhibitor, more substrate is needed to achieve ½ V_{max}.

3. **Effect on Lineweaver-Burke plot:** Competitive inhibition shows a characteristic Lineweaver-Burke plot in which the plots of the inhibited and uninhibited reactions intersect on the y axis at $1/V_{max}$ (V_{max} is unchanged). The inhibited and uninhibited reactions show different x axis intercepts, indicating that the apparent K_m is increased in the presence of the competitive inhibitor (see Figure 5.12).

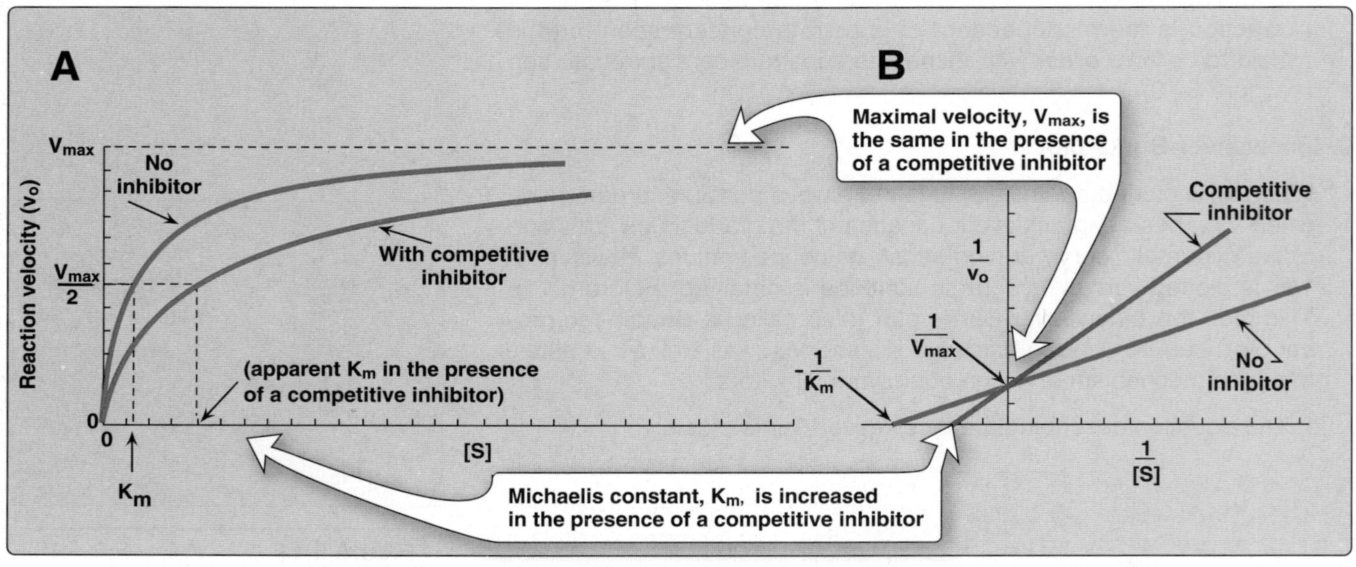

Figure 5.12
A. Effect of a competitive inhibitor on the reaction velocity (v_o) versus substrate [S] plot. B. Lineweaver-Burke plot of competitive inhibition of an enzyme.

4. **Statin drugs as examples of competitive inhibitors:** This group of **antihyperlipidemic agents** competitively inhibits the first committed step in cholesterol synthesis. This reaction is catalyzed by *hydroxymethylglutaryl CoA reductase* (*HMG CoA reductase*, see p. 219). Statin drugs, such as **atorvastatin** (**Lipitor**) and **simvastatin** (**Zocor**)[1] are **structural analogs** of the natural substrate for this enzyme, and compete effectively to inhibit *HMG CoA reductase*. By doing so, they inhibit de novo cholesterol synthesis, thereby lowering plasma cholesterol levels (Figure 5.13).

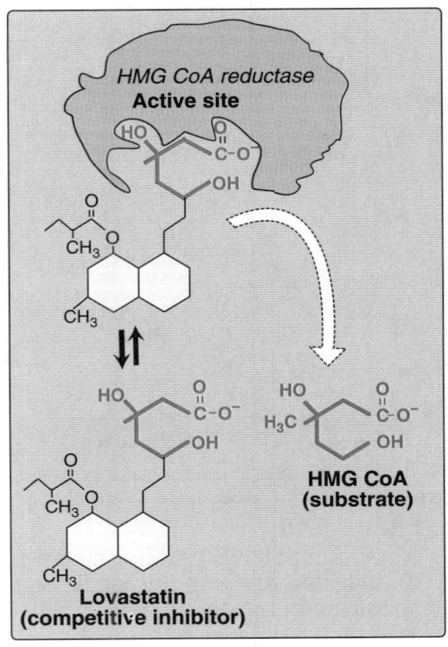

Figure 5.13
Lovastatin competes with HMG CoA for the active site of *HMG CoA reductase*.

B. Noncompetitive inhibition

This type of inhibition is recognized by its characteristic effect on V_{max} (Figure 5.14). Noncompetitive inhibition occurs when the inhibitor and substrate bind at **different sites** on the enzyme. The noncompetitive inhibitor can bind either free enzyme or the ES complex, thereby preventing the reaction from occurring (Figure 5.15).

1. **Effect on V_{max}:** Noncompetitive inhibition cannot be overcome by increasing the concentration of substrate. Thus, noncompetitive inhibitors decrease the V_{max} of the reaction.

2. **Effect on K_m:** Noncompetitive inhibitors do not interfere with the binding of substrate to enzyme. Thus, the enzyme shows the same K_m in the presence or absence of the noncompetitive inhibitor.

3. **Effect on Lineweaver-Burke plot:** Noncompetitive inhibition is readily differentiated from competitive inhibition by plotting $1/v_o$ versus $1/[S]$ and noting that V_{max} decreases in the presence of a noncompetitive inhibitor, whereas K_m is unchanged (see Figure 5.14).

4. **Examples of noncompetitive inhibitors:** Some inhibitors act by forming covalent bonds with specific groups of enzymes. For

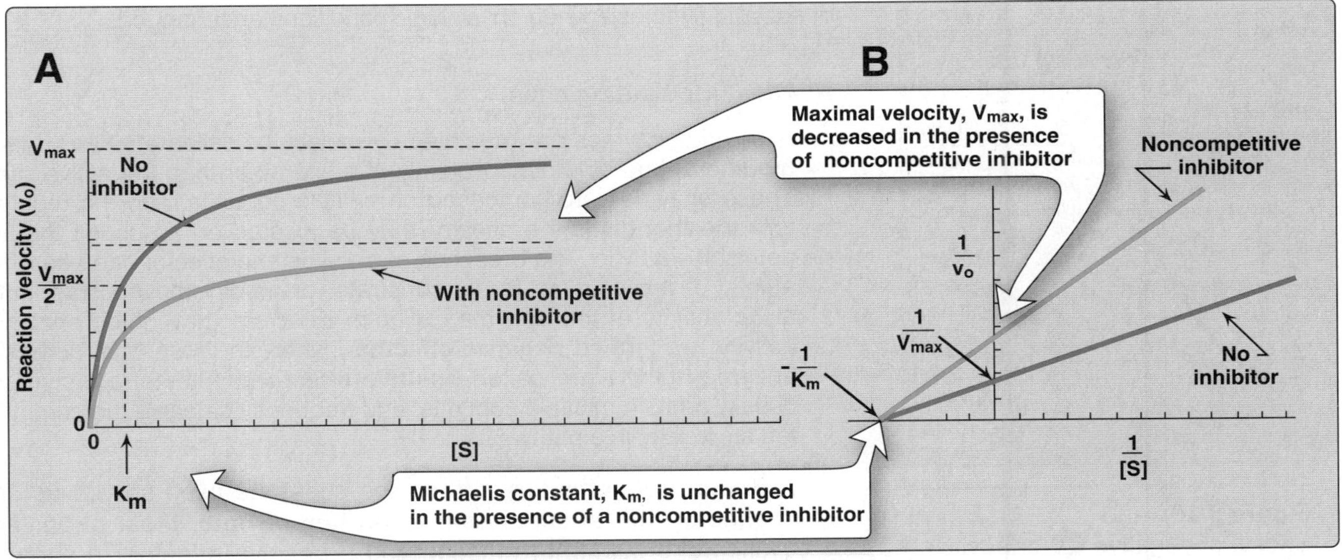

Figure 5.14
A. Effect of a noncompetitive inhibitor on the reaction velocity (v_o) versus substrate [S] plot. B. Lineweaver-Burke plot of noncompetitive inhibition of an enzyme.

INFO LINK [1]See Chapter 21 in ***Lippincott's Illustrated Reviews: Pharmacology*** (2nd and 3rd Eds.) for a more detailed discussion of drugs used to treat hyperlipidemia.

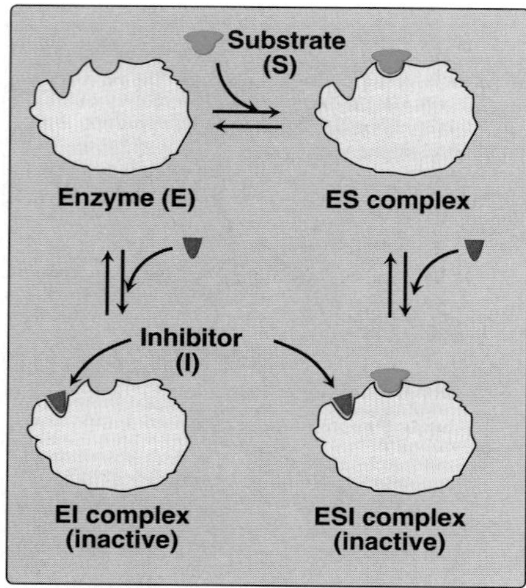

Figure 5.15
A noncompetitive inhibitor binding to both free enzyme and enzyme-substrate complex.

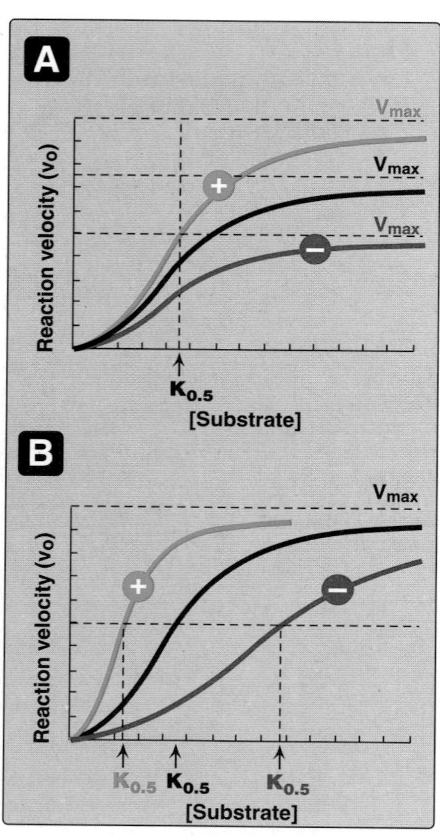

Figure 5.16
Effects of negative ⊖ or positive ⊕ effectors on an allosteric enyzme. A. V_{max} is altered. B. The substrate concentration that gives half-maximal velocity ($K_{0.5}$) is altered.

example, **lead** forms covalent bonds with the sulfhydryl side chains of **cysteine** in proteins. The binding of the heavy metal shows **noncompetitive inhibition**. *Ferrochelatase*, an enzyme that catalyzes the insertion of Fe^{2+} into protoporphyrin (a precursor of heme, see p. 277), is an example of an enzyme sensitive to inhibition by lead. Other examples of noncompetitive inhibition are certain insecticides, whose neurotoxic effects are a result of their irreversible binding at the catalytic site of the enzyme *acetylcholinesterase* (an enzyme that cleaves the neurotransmitter, acetylcholine.

C. Enzyme inhibitors as drugs

At least one half of the ten most commonly dispensed drugs in the United States act as enzyme inhibitors. For example, the widely prescribed β-lactam antibiotics, such as penicillin and amoxicillin,[2] act by inhibiting enzymes involved in bacterial cell wall synthesis. Drugs may also act by inhibiting extracellular reactions. This is illustrated by angiotensin-converting enzyme (ACE) inhibitors. They lower blood pressure by blocking the enzyme that cleaves angiotensin I to form the potent vasoconstrictor, angiotensin II. These drugs, which include captopril, enalapril, and lisinopril, cause vasodilation and a resultant reduction in blood pressure.[3]

VIII. REGULATION OF ENZYME ACTIVITY

The regulation of the reaction velocity of enzymes is essential if an organism is to coordinate its numerous metabolic processes. The rates of most enzymes are responsive to changes in substrate concentration, because the intracellular level of many substrates is in the range of the K_m. Thus, an increase in substrate concentration prompts an increase in reaction rate, which tends to return the concentration of substrate toward normal. In addition, some enzymes with specialized regulatory functions respond to allosteric effectors or covalent modification, or they show altered rates of enzyme synthesis when physiologic conditions are changed.

A. Allosteric binding sites

Allosteric enzymes are regulated by molecules called **effectors** (also **modifiers**) that bind noncovalently at a site other than the active site. These enzymes are composed of multiple subunits, and the regulatory site that binds the effector may be located on a subunit that is not itself catalytic. The presence of an allosteric effector can alter the affinity of the enzyme for its substrate, or modify the maximal catalytic activity of the enzyme, or both. Effectors that inhibit enzyme activity are termed **negative effectors**, whereas those that increase enzyme activity are called **positive effectors**. Allosteric enzymes usually contain multiple subunits, and frequently catalyze the committed step early in a pathway.

1. **Homotropic effectors:** When the substrate itself serves as an effector, the effect is said to be **homotropic**. Most often, an allosteric substrate functions as a positive effector. In such a case, the presence of a substrate molecule at one site on the enzyme enhances the catalytic properties of the other substrate-

[2]See Chapter 32 in ***Lippincott's Illustrated Reviews: Pharmacology*** (3rd Ed.) or Chapter 30 (2nd Ed.) for a discussion of inhibitors of bacterial cell wall synthesis.
[3]See Chapter 19 in ***Lippincott's Illustrated Reviews: Pharmacology*** (2nd and 3rd Eds.) for a discussion of angiotensin-converting enzyme inhibitors.

binding sites—that is, their binding sites exhibit **cooperativity**. These enzymes show a **sigmoidal curve** when reaction velocity (v_o) is plotted against substrate concentration [S] (Figure 5.16). This contrasts with the hyperbolic curve characteristic of enzymes following Michaelis-Menten kinetics, as previously discussed. [Note: The concept of cooperativity of substrate binding is analogous to the binding of oxygen to hemoglobin.] Positive and negative effectors of allosteric enzymes can affect either the Vmax or the Km, or both (see Figure 5.16).

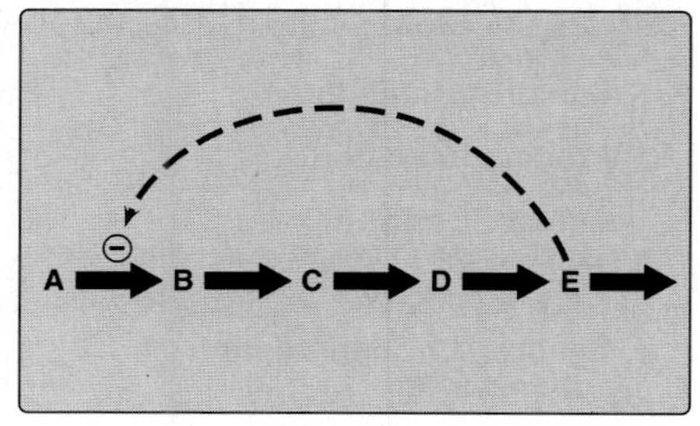

Figure 5.17
Feedback inhibition of a metabolic pathway.

2. **Heterotropic effectors:** The effector may be different from the substrate, in which case the effect is said to be **heterotropic**. For example, consider the **feedback inhibition** shown in Figure 5.17. The enzyme that converts A to B has an allosteric site that binds the end-product, E. If the concentration of E increases (for example, because it is not used as rapidly as it is synthesized), the initial enzyme in the pathway is inhibited. Feedback inhibition provides the cell with a product it needs by regulating the flow of substrate molecules through the pathway that synthesizes that product. [Note: Heterotropic effectors are commonly encountered, for example, the glycolytic enzyme *phosphofructokinase* is allosterically inhibited by citrate, which is not a substrate for the enzyme (see p. 97).]

B. Regulation of enzymes by covalent modification

Many enzymes may be regulated by covalent modification, most frequently by the addition or removal of phosphate groups from specific serine, threonine, or tyrosine residues of the enzyme. Protein phosphorylation is recognized as one of the primary ways in which cellular processes are regulated.

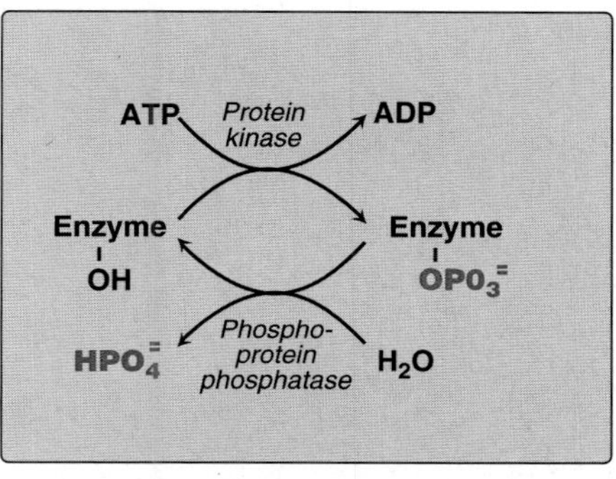

Figure 5.18
Covalent modification by the addition and removal of phosphate groups.

1. **Phosphorylation and dephosphorylation:** Phosphorylation reactions are catalyzed by a family of enzymes called *protein kinases* that use **adenosine triphosphate** (**ATP**) as a phosphate donor. Phosphate groups are cleaved from phosphorylated enzymes by the action of *phosphoprotein phosphatases* (Figure 5.18).

2. **Response of enzyme to phosphorylation:** Depending on the specific enzyme, the phosphorylated form may be more or less active than the unphosphorylated enzyme. For example, phosphorylation of *glycogen phosphorylase* (an enzyme that degrades glycogen) increases activity, whereas the addition of phosphate to *glycogen synthase* (an enzyme that synthesizes glycogen) decreases activity (see p. 132).

C. Induction and repression of enzyme synthesis

The regulatory mechanisms described above modify the activity of existing enzyme molecules. However, cells can also regulate the amount of enzyme present—usually by altering the rate of enzyme synthesis. The increased (induction) or decreased (repression) of enzyme synthesis leads to an alteration in the total population of active sites. [Note: The efficiency of existing enzyme molecules is not affected.] Enzymes subject to regulation of synthesis are often those that are needed at only one stage of development or under selected

REGULATOR EVENT	TYPICAL EFFECTOR	RESULTS	TIME REQUIRED FOR CHANGE
Substrate inhibition	Substrate	Change in velocity (v_o)	Immediate
Product inhibition	Product	Change in V_m and/or K_m	Immediate
Allosteric control	End product	Change in V_m and/or K_m	Immediate
Covalent modification	Another enzyme	Change in V_m and/or K_m	Immediate to minutes
Synthesis or degradation of enzyme	Hormone or metabolite	Change in the amount of enzyme	Hours to days

Figure 5.19
Mechanisms for regulating enzyme activity.

physiologic conditions. For example, elevated levels of insulin as a result of high blood glucose levels cause an increase in the synthesis of key enzymes involved in glucose metabolism (see p. 97). In contrast, enzymes that are in constant use are usually not regulated by altering the rate of enzyme synthesis. Alterations in enzyme levels as a result of induction or repression of protein synthesis are slow (hours to days), compared with allosterically regulated changes in enzyme activity, which occur in seconds to minutes. Figure 5.19 summarizes the common ways that enzyme activity is regulated.

IX. ENZYMES IN CLINICAL DIAGNOSIS

Plasma enzymes can be classified into two major groups. First, a relatively small group of enzymes are actively secreted into the blood by certain cell types. For example, the liver secretes **zymogens** (inactive precursors) of the enzymes involved in blood coagulation. Second, a large number of enzyme species are released from cells during normal cell turnover. These enzymes almost always function intracellularly, and have no physiologic use in the plasma. In healthy individuals, the levels of these enzymes are fairly constant, and represent a steady state in which the rate of release from damaged cells into the plasma is balanced by an equal rate of removal of the enzyme protein from the

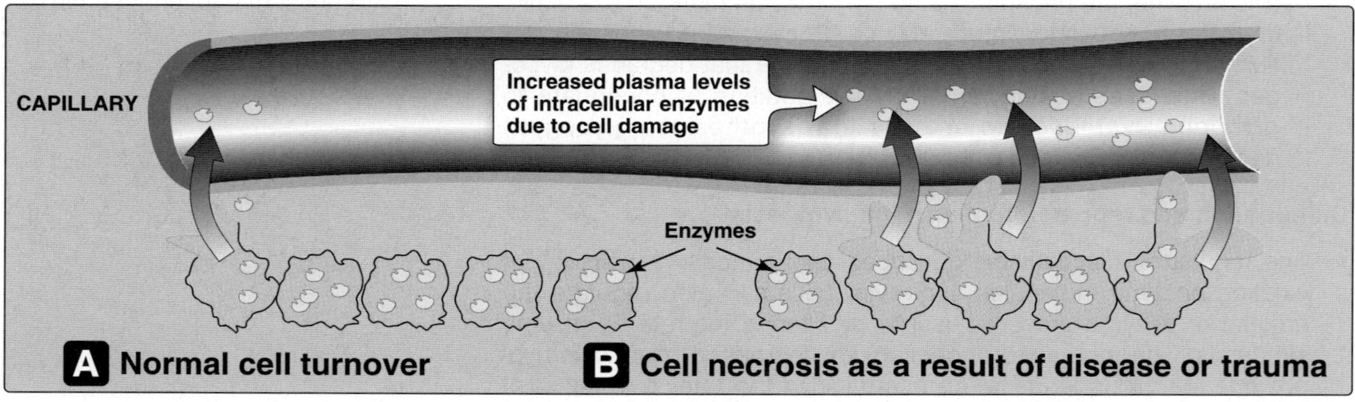

Figure 5.20
Release of enzymes from normal and diseased or traumatized cells.

plasma. The presence of elevated enzyme activity in the plasma may indicate tissue damage that is accompanied by increased release of intracellular enzymes (Figure 5.20). [Note: **Plasma** is the fluid, noncellular part of blood. Laboratory assays of enzyme activity most often use **serum,** which is obtained by centrifugation of whole blood after it has been allowed to coagulate. Plasma is a physiologic fluid, whereas serum is prepared in the laboratory.]

A. Alteration of plasma enzyme levels in disease states

Many diseases that cause tissue damage result in an increased release of intracellular enzymes into the plasma. The activities of many of these enzymes are routinely determined for diagnostic purposes in diseases of the heart, liver, skeletal muscle, and other tissues. The level of specific enzyme activity in the plasma frequently correlates with the extent of tissue damage. Thus, determining the degree of elevation of a particular enzyme activity in the plasma is often useful in evaluating the prognosis for the patient.

B. Plasma enzymes as diagnostic tools

Some enzymes show relatively high activity in only one or a few tissues. The presence of increased levels of these enzymes in plasma thus reflects damage to the corresponding tissue. For example, the enzyme *alanine aminotransferase* (*ALT*, see p. 248) is abundant in the liver. The appearance of elevated levels of *ALT* in plasma signals possible damage to hepatic tissue. Increases in plasma levels of enzymes with a wide tissue distribution provide a less specific indication of the site of cellular injury. This lack of tissue specificity limits the diagnostic value of many plasma enzymes.

C. Isoenzymes and diseases of the heart

Most isoenzymes (also called **isozymes**) are enzymes that catalyze the same reaction. However, they do not necessarily have the same physical properties because of genetically determined differences in amino acid sequence. For this reason, isoenzymes may contain different numbers of charged amino acids and may, therefore, be separated from each other by electrophoresis (Figure 5.21). Different organs frequently contain characteristic proportions of different isoenzymes. The pattern of isoenzymes found in the plasma may, therefore, serve as a means of identifying the site of tissue damage. For example, the plasma levels of *creatine kinase* (*CK*) and *lactate dehydrogenase* (*LDH*) are commonly determined in the diagnosis of myocardial infarction. They are particularly useful when the electrocardiogram is difficult to interpret, such as when there have been previous episodes of heart disease.

1. **Quaternary structure of isoenzymes:** Many isoenzymes contain different subunits in various combinations. For example, *creatine kinase* occurs as three isoenzymes. Each isoenzyme is a dimer composed of two polypeptides (called B and M subunits) associated in one of three combinations: *CK1* = BB, *CK2* = MB, and *CK3* = MM. Each *CK* isoenzyme shows a characteristic electrophoretic mobility (see Figure 5.21).

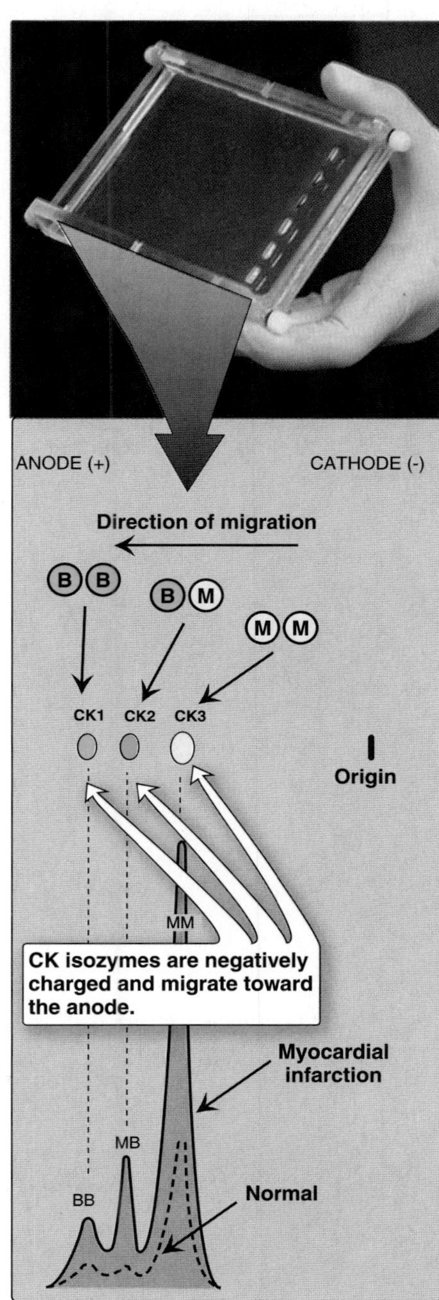

Figure 5.21
Subunit structure and electrophoretic mobility and enzyme activity of *creatine kinase* isoenzymes.

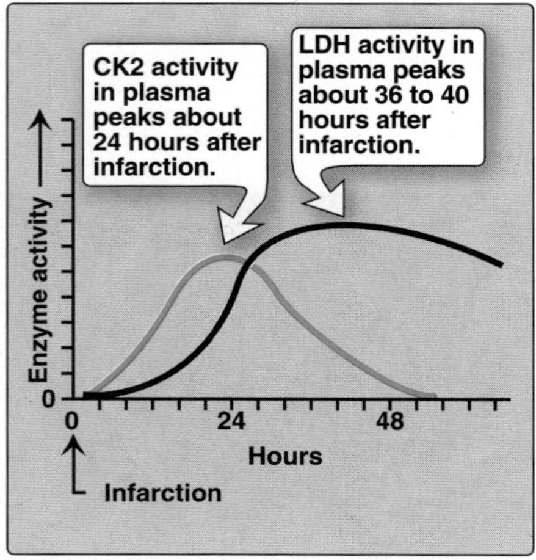

Figure 5.22
Appearance of *creatine kinase* (*CK*) and *lactate dehydrogenase* (*LDH*) inplasma after a myocardial infarction.

2. **Diagnosis of myocardial infarction:** Myocardial muscle is the only tissue that contains more than five percent of the total *CK* activity as the *CK2* (MB) isoenzyme. Appearance of this hybrid isoenzyme in plasma is virtually specific for infarction of the myocardium. Following an acute myocardial infarction, this isoenzyme appears approximately four to eight hours following onset of chest pain, and reaches a peak of activity at approximately 24 hours (Figure 5.22). [Note: *Lactate dehydrogenase* activity is also elevated in plasma following an infarction, peaking 36 to 40 hours after the onset of symptoms. *LDH* activity is, thus, of diagnostic value in patients admitted more than 48 hours after the infarction—a time when plasma *CK2* may provide equivocal results.]

3. **Newer markers for myocardial infarction: Troponin T** and **troponin I** are regulatory proteins involved in myocardial contractility. They are released into the plasma in response to cardiac damage. Elevated serum troponins are more predictive of adverse outcomes in unstable angina or myocardial infarction than the conventional assay of *CK2*.

X. CHAPTER SUMMARY

Enzymes are **protein catalysts** that increase the velocity of a chemical reaction by lowering the energy of the transition state. Enzymes are not consumed during the reaction they catalyze. Enzyme molecules contain a special pocket or cleft called the **active site**. The active site contains amino acid side chains that create a three-dimensional surface complementary to the substrate. The active site binds the substrate, forming an **enzyme-substrate (ES) complex**. ES is converted to enzyme-product (EP), which subsequently dissociates to enzyme and product. An enzyme allows a reaction to proceed rapidly under conditions prevailing in the cell by providing an **alternate reaction pathway** with a **lower free energy of activation**. The enzyme does not change the free energies of the reactants or products and, therefore, does not change the equilibrium of the reaction. Most enzymes show **Michaelis-Menten kinetics**, and a plot of the **initial reaction velocity, v_o**, against **substrate concentration, [S]**, has a **hyperbolic** shape similar to the oxygen dissociation curve of myoglobin. Any substance that can diminish the velocity of such enzyme-catalyzed reactions is called an **inhibitor**. The two most commonly encountered types of inhibition are **competitive** (which **increases** the **apparent K_m**) and **noncompetitive** (which **decreases** the **V_{max}**). In contrast, the **multi-subunit allosteric enzymes** frequently show a **sigmoidal curve** similar in shape to the oxygen dissociation curve of hemoglobin. They are frequently found catalyzing the **committed (rate-limiting) step(s)** of a pathway. Allosteric enzymes are regulated by molecules called **effectors** (also **modifiers**) that bind noncovalently at a site other than the active site. Effectors can be either **positive** (accelerate the enzyme-catalyzed reaction) or **negative** (slow down the reaction). An allosteric effector can alter the affinity of the enzyme for its substrate, or modify the maximal catalytic activity of the enzyme, or both.

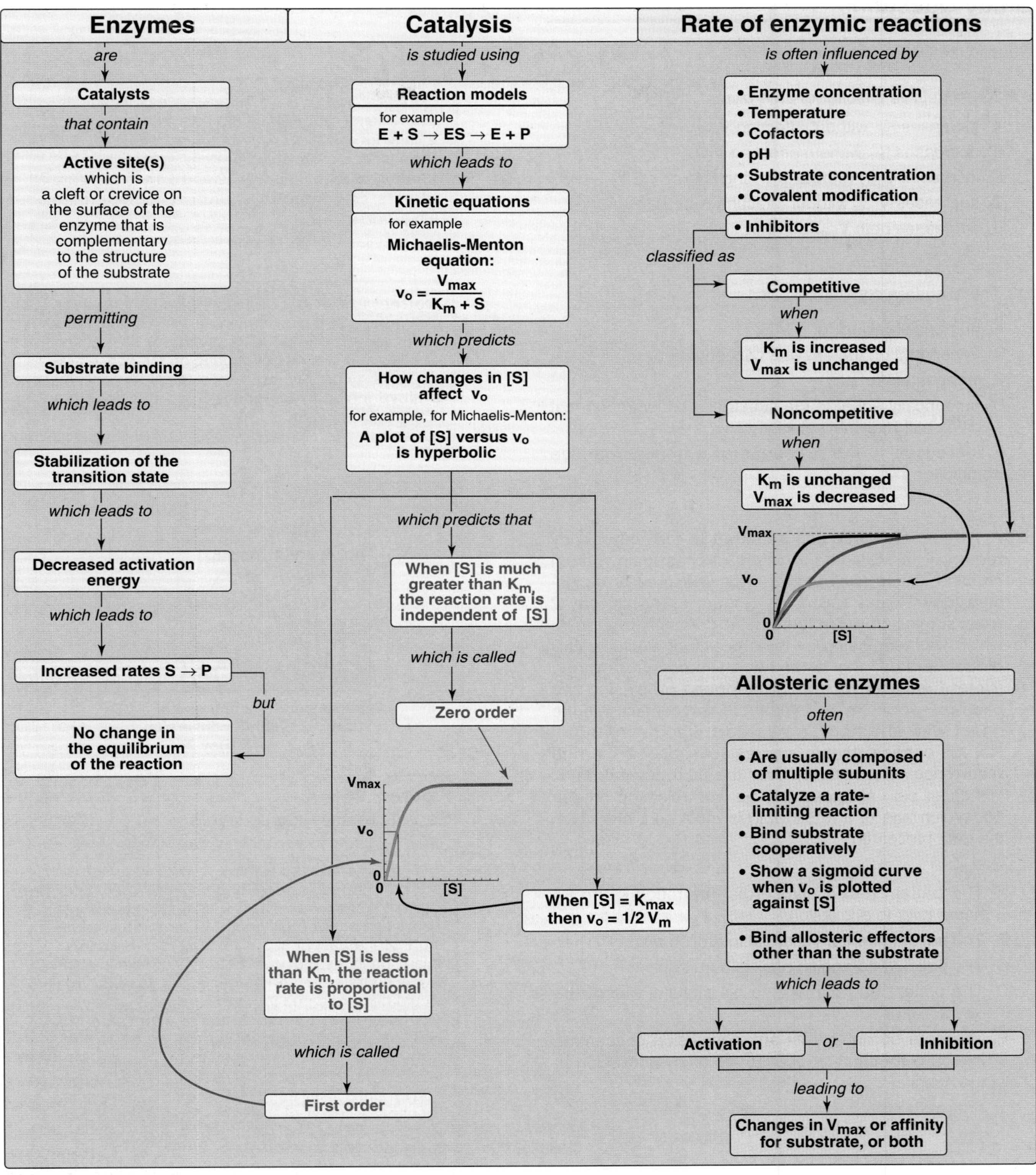

Figure 5.23
Key concept map for the enzymes. S = substrate, [S] = substrate concentration, P = product, E = enzyme, v_o = initial velocity, V_{max} = maximal velocity, K_m = Michaelis constant.

Study Questions

Choose the ONE correct answer

5.1 A competitive inhibitor of an enzyme:

A. increases K_m without affecting V_{max}.
B. decreases K_m without affecting V_{max}.
C. increases V_{max} without affecting K_m.
D. decreases V_{max} without affecting K_m.
E. decreases both V_{max} and K_m.

Correct answer = A. In the presence of a competitive inhibitor, an enzyme appears to have a lower affinity for substrate, but as the substrate level is increased, the observed velocity approaches V_m. (See panel B of Figures 5.12 and 5.14 to compare effects of competitive and noncompetitive inhibitors.)

5.2 The Michaelis constant, K_m, is:

A. numerically equal to $\frac{1}{2}$ V_{max}.
B. dependent on the enzyme concentration.
C. independent of pH.
D. numerically equal to the substrate concentration that gives half-maximal velocity.
E. increased in the presence of a noncompetitive inhibitor.

Correct answer = D. Remember K_m has the dimensions of concentration and is a characteristic of an enzyme under a given set of reaction conditions. K_m does not depend on the concentration of enzyme, but can vary with pH. A noncompetitive inhibitor decreases V_{max}, but does not alter K_m (see Figure 5.14).

5.3 A 70-year-old man was admitted to the emergency room with a twelve-hour history of chest pain. Serum creatine kinase (CK) activity was measured at admission (day 1) and once daily (Figure 5.24). On day 2 after admission, he experienced cardiac arrhythmia, which was terminated by three cycles of electrical cardioconversion, the latter two at maximum energy. [Note: Cardioconversion is performed by placing two paddles, twelve cm in diameter, in firm contact with the chest wall and applying a short electric voltage.] Normal cardiac rhythm was established. He had no recurrence of arrhythmia over the next several days. His chest pain subsided and he was released on day 10. Which one of the following is most consistent with the data presented?

A. The patient had a myocardial infarction 48 to 64 hours prior to admission.
B. The patient had a myocardial infarction on day 2.
C. The patient had angina prior to admission.
D. The patient had damage to his skeletal muscle on day 2.
E. The data do not permit any conclusion concerning myocardial infarction prior to, or after, admission to the hospital.

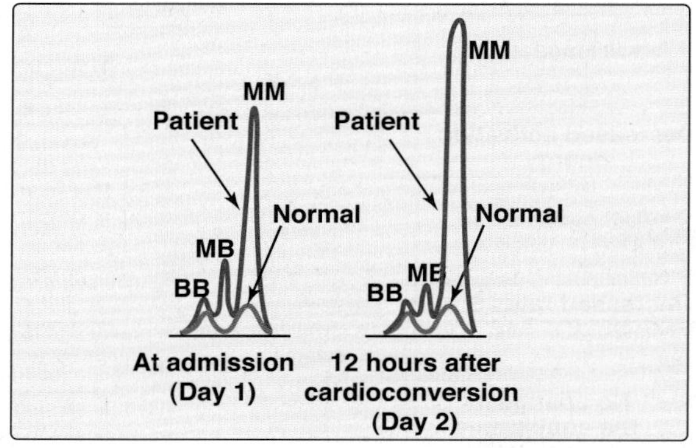

Figure 5.24
Serum creatine kinase levels.

Correct answer = D. The CK isoenzyme pattern at admission showed elevated MB isozyme, indicating that the patient had experienced a myocardial infarction in the previous 12 to 24 hours. [Note: 48 to 64 hours after an infarction, the MB isozyme would have returned to normal values.] On day 2, 12 hours after the cardioconversions, the MB isozyme had decreased, indicating no further damage to the heart. However, the patient showed an increased MM isozyme after cardoconversion. This suggests damage to muscle, probably a result of the convulsive muscle contractions caused by repeated cardioconversion. Angina is typically the result of transient spasms in the vasculature of the heart, and would not be expected to lead to tissue death that results in elevation in serum creatine kinase.

UNIT II: Intermediary Metabolism

Bioenergetics and Oxidative Phosphorylation

6

I. OVERVIEW

Bioenergetics describes the transfer and utilization of energy in biologic systems. It makes use of a few basic ideas from the field of thermodynamics, particularly the concept of free energy. Changes in free energy (ΔG) provide a measure of the energetic feasibility of a chemical reaction and can, therefore, allow prediction of whether a reaction or process can take place. Bioenergetics concerns only the **initial** and **final energy states** of reaction components, not the mechanism or how much time is needed for the chemical change to take place. In short, bioenergetics predicts if a process is possible, whereas kinetics measure how fast the reaction occurs (see p. 54).

II. FREE ENERGY

The direction and extent to which a chemical reaction proceeds is determined by the degree to which two factors change during the reaction. These are **enthalpy** (ΔH, a measure of the change in heat content of the reactants and products) and **entropy** (ΔS, a measure of the change in randomness or disorder of reactants and products, Figure 6.1). Neither of these thermodynamic quantities by itself is sufficient to determine whether a chemical reaction will proceed spontaneously in the direction it is written. However, when combined mathematically (see Figure 6.1), enthalpy and entropy can be used to define a third quantity, **free energy (G)**, which predicts the direction in which a reaction will spontaneously proceed.

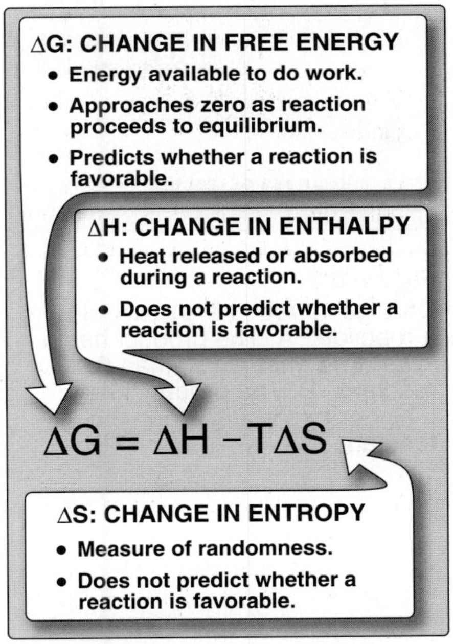

ΔG: CHANGE IN FREE ENERGY
- Energy available to do work.
- Approaches zero as reaction proceeds to equilibrium.
- Predicts whether a reaction is favorable.

ΔH: CHANGE IN ENTHALPY
- Heat released or absorbed during a reaction.
- Does not predict whether a reaction is favorable.

$$\Delta G = \Delta H - T\Delta S$$

ΔS: CHANGE IN ENTROPY
- Measure of randomness.
- Does not predict whether a reaction is favorable.

Figure 6.1
Relationship between changes in free energy (G), enthalpy (H), and entropy (S). T is the absolute temperature in degrees Kelvin (^oK): $^oK = {}^oC + 273$.

III. FREE ENERGY CHANGE

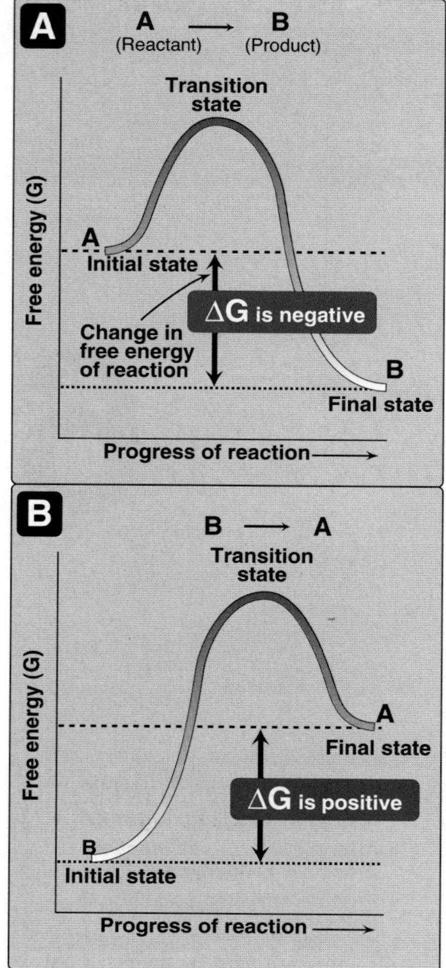

Figure 6.2
Change in free energy (ΔG) during a reaction. A. The product has a lower free energy (G) than the reactant. B. The product has a higher free energy than the reactant.

The change in free energy comes in two forms, ΔG and ΔG°. The first, ΔG (without the superscript "o"), is the more general because it predicts the change in free energy and, thus, the direction of a reaction at any specified concentration of products and reactants. This contrasts with the change in **standard free energy**, ΔG° (with the superscript "o"), which is the energy change when reactants and products are at a concentration of 1 mol/L. [Note: The concentration of protons is assumed to be 10^{-7} mol/L—that is, pH = 7.] Although ΔG° represents energy changes at these nonphysiologic concentrations of reactants and products, it is nonetheless useful in comparing the energy changes of different reactions. Further, ΔG° can readily be determined from measurement of the equilibrium constant (see p. 72). This section outlines the uses of ΔG; ΔG° is described on p. 71.

A. Sign of ΔG predicts the direction of a reaction

The change in free energy, ΔG, can be used to predict the direction of a reaction at constant temperature and pressure. Consider the reaction:

$$A \rightleftarrows B$$

1. **Negative ΔG:** If ΔG is a negative number, there is a net loss of energy, and the reaction goes spontaneously as written—that is, A is converted into B (Figure 6.2A). The reaction is said to be **exergonic**.
2. **Positive ΔG:** If ΔG is a positive number, there is a net gain of energy, and the reaction does not go spontaneously from B to A (see Figure 6.2B). The reaction is said to be **endergonic**, and energy must be added to the system to make the reaction go from B to A.
3. **ΔG is zero:** If ΔG = 0, the reactants are in equilibrium. [Note: When a reaction is proceeding spontaneously—that is, free energy is being lost—then the reaction continues until ΔG reaches zero and equilibrium is established.]

B. ΔG of the forward and back reactions

The free energy of the forward reaction (A → B) is equal in magnitude but opposite in sign to that of the back reaction (B → A). For example, if ΔG of the forward reaction is –5000 cal/mol, then that of the back reaction is +5000 cal/mol.

C. ΔG depends on the concentration of reactants and products

ΔG of the reaction A → B depends on the concentration of the reactant and product. At constant temperature and pressure, the following relationship can be derived:

$$\Delta G = \Delta G^{o} + RT \ln \frac{[B]}{[A]}$$

where ΔG° is the standard free energy change (see below).
R is the gas constant (1.987 cal/mol · degree).
T is the absolute temperature (°K).

[A] and [B] are the actual concentrations of the reactant and product.

ln represents the natural logarithm.

A reaction with a positive ΔG^o can proceed in the forward direction (have a negative overall ΔG) if the ratio of products to reactants ([B]/[A]) is sufficiently small (that is, the ratio of reactants to products is large). For example, consider the reaction:

$$\text{Glucose 6-phosphate} \rightleftarrows \text{fructose 6-phosphate}$$

Figure 6.3A shows reaction conditions in which the concentration of reactant, glucose 6-phosphate, is high compared with the concentration of product, fructose 6-phosphate. This means that the ratio of the product to reactant is small, and RT ln([fructose 6-phosphate]/[glucose 6-phosphate]) is large and negative, causing ΔG to be negative despite ΔG^o being positive. Thus, the reaction can proceed in the forward direction.

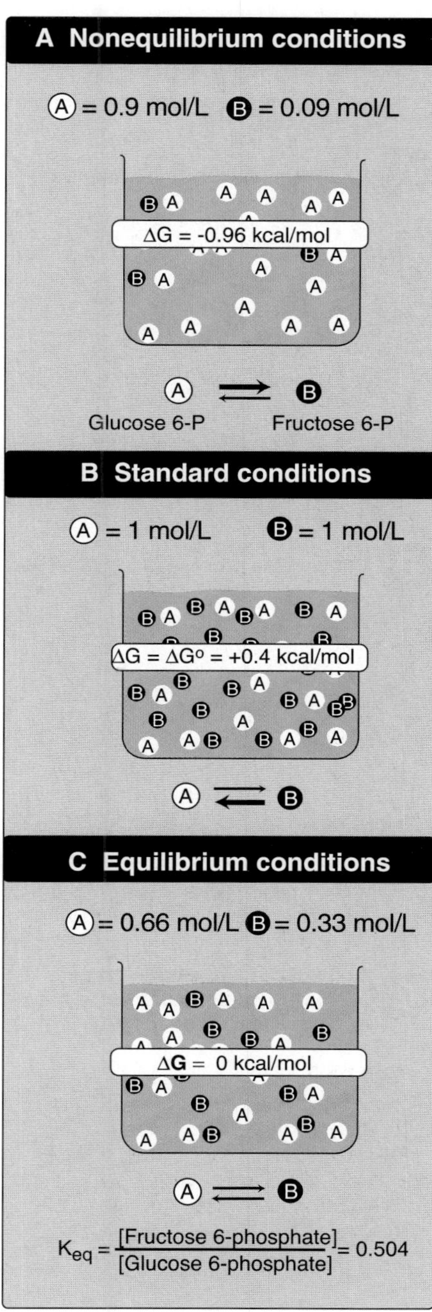

Figure 6.3
ΔG of a reaction depends on the concentration of reactant (A) and product (B). For the conversion of glucose 6-P to fructose 6-P, ΔG is negative when the ratio of reactant (A) to product (B) is large (top, panel A); is positive under standard conditions (middle, panel B); and is zero at equlibrium (bottom, panel C).

D. Standard free energy change, ΔG^o

ΔG^o is called the standard free energy change because it is equal to the free energy change, ΔG, under standard conditions—that is, when reactants and products are kept at 1 mol/L concentrations (see Figure 6.3B). Under these conditions, the natural logarithm (ln) of the ratio of products to reactants is zero (ln1 = 0) and, therefore, the equation shown at the bottom of p. 70 becomes:

$$\Delta G = \Delta G^o + 0$$

1. **ΔG^o is predictive only under standard conditions:** Under standard conditions, ΔG^o can be used to predict the direction a reaction proceeds because, under these conditions, ΔG^o is equal to ΔG. However, ΔG^o cannot predict the direction of a reaction under physiologic conditions, because it is composed solely of constants (R, T, and K_{eq}) and is, therefore, not altered by changes in product or substrate concentrations.

2. **Relationship between ΔG^o and K_{eq}:** In a reaction $A \rightarrow B$, a point of equilibrium is reached at which no further **net** chemical change takes place—that is, when A is being converted to B as fast as B is being converted to A. In this state, the ratio of [B] to [A] is constant, regardless of the actual concentrations of the two compounds:

$$K_{eq} = \frac{[B]_{eq}}{[A]_{eq}}$$

where K_{eq} is the equilibrium constant, and $[A]_{eq}$ and $[B]_{eq}$ are the concentrations of A and B at equilibrium. If the reaction $A \rightleftarrows B$ is allowed to go to equilibrium at constant temperature and pressure, then at equilibrium the overall free energy change (ΔG) is zero. Therefore,

$$\Delta G = 0 = \Delta G^o + RT \ln \frac{[B]_{eq}}{[A]_{eq}}$$

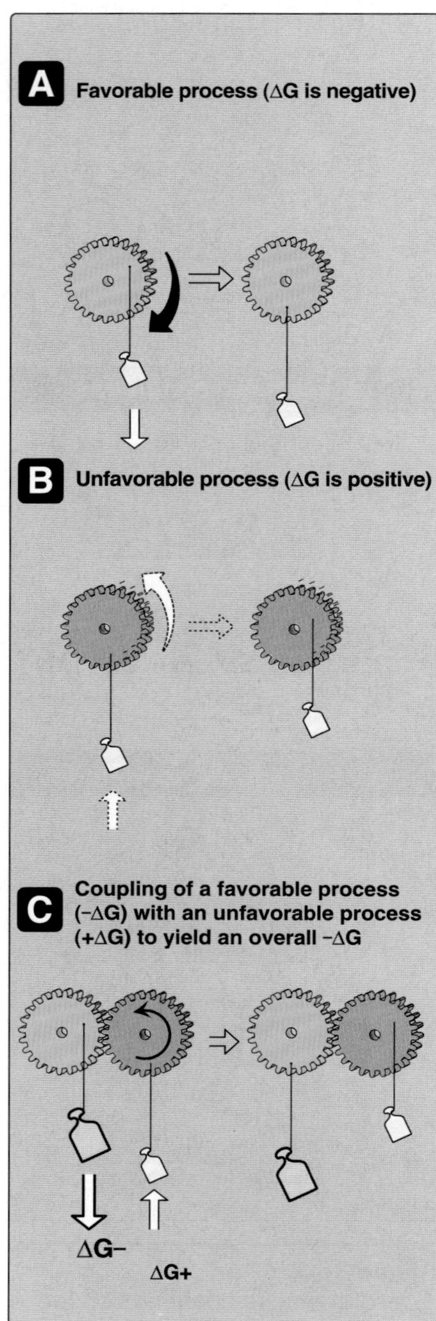

Figure 6.4
Mechanical model of coupling of favorable and unfavorable processes.

where the actual concentrations of A and B are equal to the equilibrium concentrations of reactant and product $[A]_{eq}$ and $[B]_{eq}$, and their ratio as shown above is equal to the K_{eq}. Thus,

$$\Delta G^o = -RT \ln K_{eq}$$

This equation allows some simple predictions:

If $K_{eq} = 1$, then $\Delta G^o = 0$	A $\rightleftarrows$ B
If $K_{eq} > 1$, then $\Delta G^o < 0$	A $\longrightarrow$ B
If $K_{eq} < 1$, then $\Delta G^o > 0$	A $\longleftarrow$ B

3. **ΔG^o of two consecutive reactions are additive:** The standard free energy changes (ΔG^o) are additive in any sequence of consecutive reactions, as are the free energy changes (ΔG).

   For example:

Glucose + ATP → glucose 6-P + ADP	ΔG^o = –4000 cal/mol
Glucose 6-P → fructose 6-P	ΔG^o = +400 cal/mol
Glucose + ATP → fructose 6-P + ADP	ΔG^o = –3600 cal/mol

4. **ΔGs of a pathway are additive:** This additive property of free energy changes is very important in biochemical pathways through which substrates must pass in a particular direction (for example, A → B → C → D →...). As long as the **sum** of the ΔGs of the individual reactions is **negative**, the pathway can potentially proceed as written, even if some of the individual component reactions of the pathway have a positive ΔG. The actual rate of the reactions does, of course, depend on the activity of the enzymes that catalyze the reactions.

IV. ATP AS AN ENERGY CARRIER

Reactions or processes that have a large positive ΔG, such as moving ions against a concentration gradient across a cell membrane, are made possible by coupling the endergonic movement of ions with a second spontaneous process with a large negative ΔG, such as the hydrolysis of adenosine triphosphate (ATP). Figure 6.4 shows a mechanical model of energy coupling. A gear with an attached weight spontaneously turns in the direction that achieves the lowest energy state, in this case the weight seeks its lowest position (see Figure 6.4A). The reverse motion (see Figure 6.4B) is energetically unfavored and does not occur spontaneously. Figure 6.4C shows that the energetically favored movement of one gear can be used to turn a second gear in a direction that it would not move spontaneously. The simplest example of energy coupling in biologic reactions occurs when the energy-requiring and the energy-yielding reactions share a **common intermediate**.

A. Reactions are coupled through common intermediates

Two chemical reactions have a common intermediate when they occur sequentially so that the product of the first reaction is a substrate for the second. For example, given the reactions

$$A + B \rightarrow C + D$$

$$D + X \rightarrow Y + Z$$

D is the common intermediate and can serve as a carrier of chemical energy between the two reactions. Many coupled reactions use ATP to generate a common intermediate. These reactions may involve ATP cleavage—that is, the transfer of a phosphate group from ATP to another molecule. Other reactions lead to ATP synthesis by transfer of phosphate from an energy-rich intermediate to ADP, forming ATP.

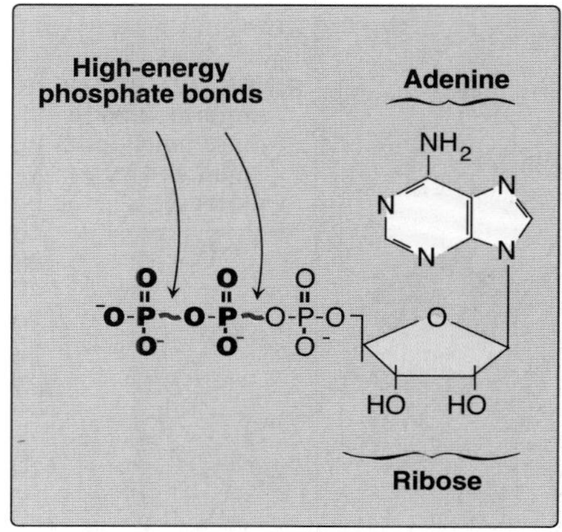

Figure 6.5
Adenosine triphosphate.

B. Energy carried by ATP

ATP consists of a molecule of adenosine (adenine + ribose) to which three phosphate groups are attached (Figure 6.5). If one phosphate is removed, adenosine diphosphate (ADP) is produced; if two phosphates are removed, adenosine monophosphate (AMP) results. The standard free energy of hydrolysis of ATP, ΔG^o, is approximately −7300 cal/mol for each of the two terminal phosphate groups. Because of this large, negative ΔG^o, ATP is called a high-energy phosphate compound.

V. ELECTRON TRANSPORT CHAIN

Energy-rich molecules, such as glucose, are metabolized by a series of oxidation reactions ultimately yielding CO_2 and water (Figure 6.6). The metabolic intermediates of these reactions donate electrons to specific coenzymes—nicotinamide adenine dinucleotide (NAD^+) and flavin adenine dinucleotide (FAD)—to form the energy-rich reduced coenzymes, NADH and $FADH_2$. These reduced coenzymes can, in turn, each donate a pair of electrons to a specialized set of electron carriers, collectively called the electron transport chain, described in this section. As electrons are passed down the electron transport chain, they lose much of their free energy. Part of this energy can be captured and stored by the production of ATP from ADP and inorganic phosphate (P_i). This process is called **oxidative phosphorylation** and is described on p. 77. The remainder of the free energy not trapped as ATP is released as heat.

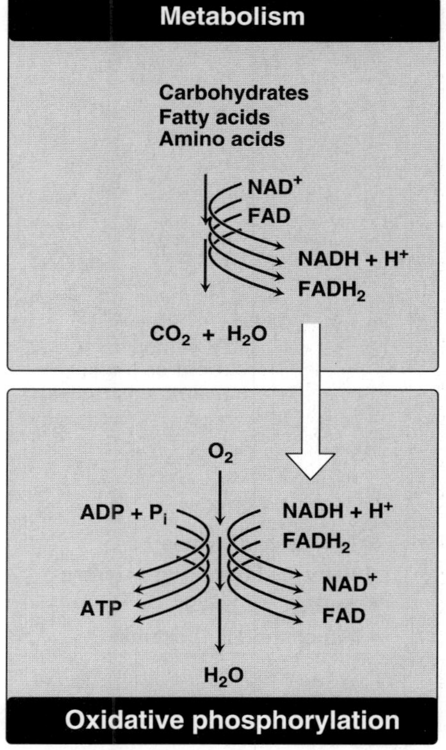

Figure 6.6
The metabolic breakdown of energy-yielding molecules.

A. Mitochondrion

The electron transport chain is present in the **inner mitochondrial membrane** and is the final common pathway by which electrons derived from different fuels of the body flow to oxygen. Electron transport and ATP synthesis by oxidative phosphorylation proceed continuously in all tissues that contain mitochondria.

1. **Structure of the mitochondrion:** The components of the electron transport chain are located in the inner membrane. Although the outer membrane contains special pores, making it freely perme-

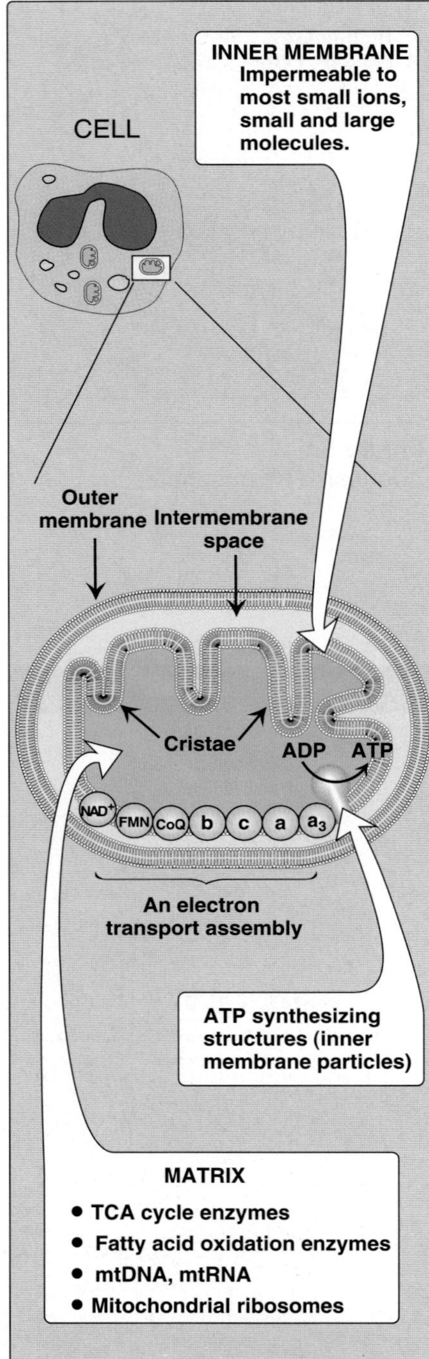

Figure 6.7
Structure of a mitochondrion showing schematic representation of the electron transport chain and ATP synthesizing structures on the inner membrane. mtDNA = mitochondrial DNA; mtRNA = mitochondrial RNA.

able to most ions and small molecules, the inner mitochondrial membrane is a specialized structure that is impermeable to most small ions, including H^+, Na^+, and K^+, small molecules such as ATP, ADP, pyruvate, and other metabolites important to mitochondrial function (Figure 6.7). Specialized carriers or transport systems are required to move ions or molecules across this membrane. The inner mitochondrial membrane is unusually rich in protein, half of which is directly involved in electron transport and oxidative phosphorylation. The inner mitochondrial membrane is highly convoluted. The convolutions, called **cristae**, serve to greatly increase the surface area of the membrane.

2. **ATP synthase complexes:** These complexes of proteins are referred to as **inner membrane particles** and are attached to the inner surface of the inner mitochondrial membrane. They appear as spheres that protrude into the mitochondrial matrix.

3. **Matrix of the mitochondrion:** This gel-like solution in the interior of mitochondria is fifty percent protein. These molecules include the enzymes responsible for the oxidation of pyruvate, amino acids, fatty acids (by β-oxidation), and those of the tricarboxylic acid (TCA) cycle. The synthesis of urea and heme occur partially in the matrix of mitochondria. In addition, the matrix contains NAD^+ and FAD (the oxidized forms of the two coenzymes that are required as hydrogen acceptors) and ADP and P_i, which are used to produce ATP. [Note: The matrix also contains mitochondrial RNA and DNA (mtRNA and mtDNA) and mitochondrial ribosomes.]

B. Organization of the chain

The inner mitochondrial membrane can be disrupted into five separate enzyme complexes, called complexes I, II, III, IV, and V. Complexes I to IV each contain part of the electron transport chain (Figure 6.8), whereas complex V catalyzes ATP synthesis (see p. 78). Each complex accepts or donates electrons to relatively mobile electron carriers, such as coenzyme Q and cytochrome c. Each carrier in the electron transport chain can receive electrons from an electron donor, and can subsequently donate electrons to the next carrier in the chain. The electrons ultimately combine with oxygen and protons to form water. This requirement for oxygen makes the electron transport process the **respiratory chain**, which accounts for the greatest portion of the body's use of oxygen.

C. Reactions of the electron transport chain

With the exception of coenzyme Q, all members of this chain are proteins. These may function as enzymes as is the case with the dehydrogenases, they may contain iron as part of an iron-sulfur center, they may be coordinated with a porphyrin ring as in the cytochromes, or they may contain copper, as does the cytochrome a + a_3 complex.

1. **Formation of NADH:** NAD^+ is reduced to NADH by dehydrogenases that remove two hydrogen atoms from their substrate. (For examples of these reactions, see the discussion of the dehydrogenases found in the TCA cycle, pp. 110-111.) Both electrons but

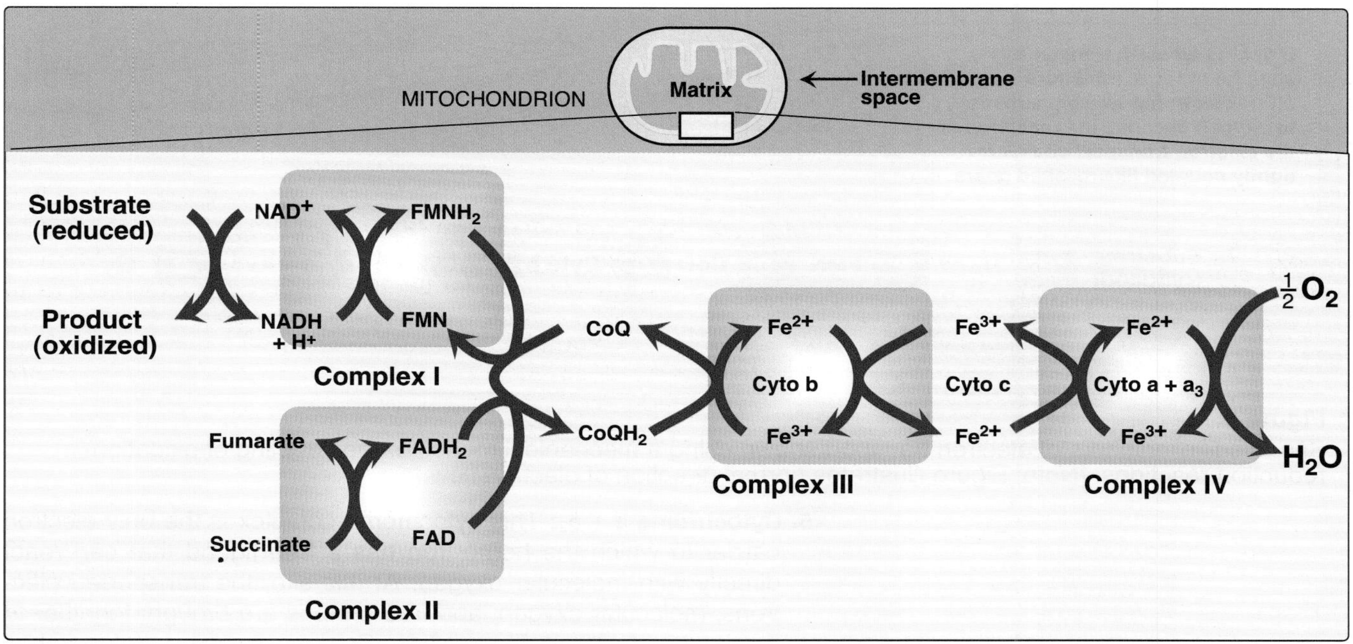

Figure 6.8
Electron transport chain. [Note: Complex V is not shown.]

only one proton (that is, a hydride ion, :H⁻) are transferred to the NAD^+, forming NADH plus a free proton, H^+.

2. **NADH dehydrogenase:** The free proton plus the hydride ion carried by NADH are next transferred to *NADH dehydrogenase*, an enzyme complex (Complex I) embedded in the inner mitochondrial membrane. This complex has a tightly bound molecule of flavin mononucleotide (FMN, a coenzyme structurally related to FAD, see Figure 28.5, p. 373) that accepts the two hydrogen atoms ($2\ e^- + 2H^+$), becoming $FMNH_2$. *NADH dehydrogenase* also contains several iron atoms paired with sulfur atoms to make **iron-sulfur centers** (Figure 6.9). These are necessary for the transfer of the hydrogen atoms to the next member of the chain, ubiquinone (known as coenzyme Q).

3. **Coenzyme Q:** Coenzyme Q is a quinone derivative with a long isoprenoid tail. It is also called **ubiquinone** because it is ubiquitous in biologic systems. Coenzyme Q can accept hydrogen atoms both from $FMNH_2$, produced by *NADH dehydrogenase,* and from $FADH_2$ (Complex II), which is produced by *succinate dehydrogenase* and *acyl CoA dehydrogenase* .

4. **Cytochromes:** The remaining members of the electron transport chain are cytochromes. Each contains a heme group made of a porphyrin ring containing an atom of iron (see p. 277). Unlike the heme groups of hemoglobin, the cytochrome iron atom is reversibly converted from its ferric (Fe^{3+}) to its ferrous (Fe^{2+}) form as a normal part of its function as a reversible carrier of electrons. Electrons are passed along the chain from coenzyme Q to cytochromes b and c (Complex III), and a + a_3 (Complex IV, see Figure 6.8).

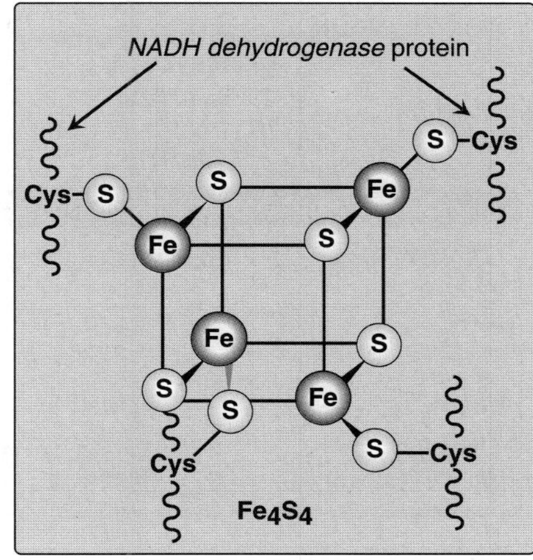

Figure 6.9
Iron-sulfur center of *NADH dehydro genase.*

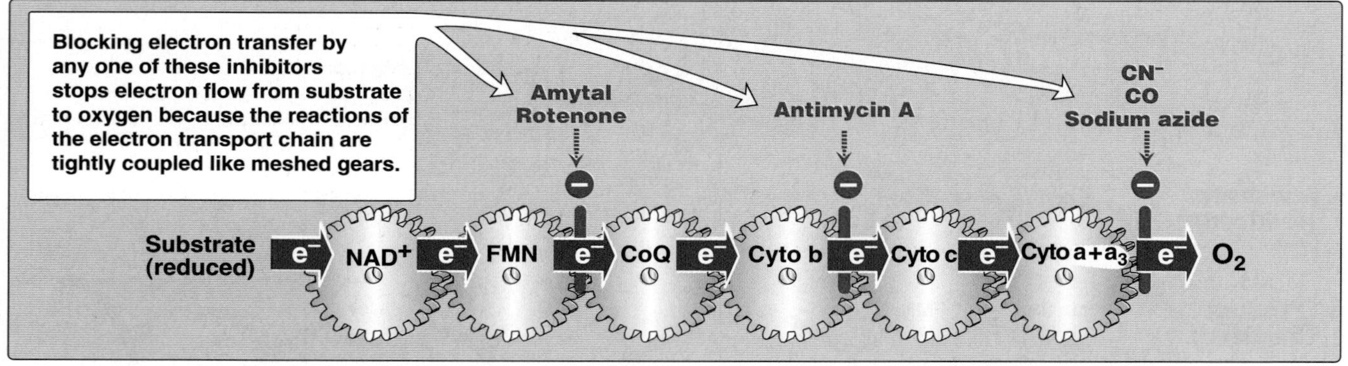

Figure 6.10
Site-specific inhibitors of electron transport shown using a mechanical model for the coupling of oxidation-reduction reactions. [Note: Figure illustrates normal direction of electron flow.]

5. **Cytochrome a + a_3:** This cytochrome complex is the only electron carrier in which the heme iron has a free ligand that can react directly with molecular oxygen. At this site, the transported electrons, molecular oxygen, and free protons are brought together to produce water (see Figure 6.8). Cytochrome a + a_3 (also called *cytochrome oxidase*) contains bound copper atoms that are required for this complex reaction to occur.

6. **Site-specific inhibitors:** Site-specific inhibitors of electron transport have been identified and are illustrated in Figure 6.10. These compounds prevent the passage of electrons by binding to a component of the chain, blocking the oxidation/reduction reaction. Therefore, all electron carriers before the block are fully reduced, whereas those located after the block are oxidized. [Note: Because electron transport and oxidative phosphorylation are **tightly coupled**, site-specific inhibition of the electron transport chain also inhibits ATP synthesis.]

C. Release of free energy during electron transport

Free energy is released as electrons are transferred along the electron transport chain from an electron donor (reducing agent or reductant) to an electron acceptor (oxidizing agent or oxidant). The electrons can be transferred in different forms, for example, as hydride ions ($:H^-$) to NAD^+, as hydrogen atoms ($\cdot H$) to FMN, coenzyme Q, and FAD, or as electrons ($\cdot e^-$) to cytochromes.

1. **Redox pairs:** Oxidation (loss of electrons) of one compound is always accompanied by reduction (gain of electrons) of a second substance. For example, Figure 6.11 shows the oxidation of NADH to NAD^+ accompanied by the reduction of FAD to $FADH_2$. Such oxidation-reduction reactions can be written as the sum of two half-reactions: an isolated oxidation reaction and a separate reduction reaction (see Figure 6.11). NAD^+ and NADH form a redox pair, as do FAD and $FADH_2$. Redox pairs differ in their tendency to lose electrons. This tendency is a characteristic of a particular redox pair, and can be quantitatively specified by a constant, $\mathbf{E_o}$ (the **standard reduction potential**), with units in volts.

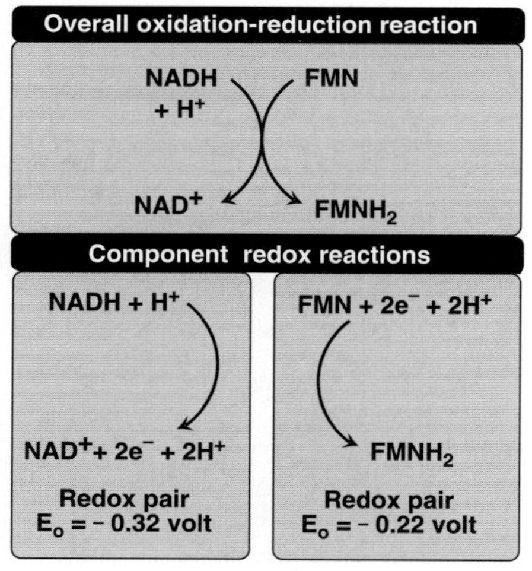

Figure 6.11
Oxidation of NADH by FMN, separated into two component redox pairs.

2. **Standard reduction potential (E_o):** The standard reduction potentials of various redox pairs can be listed to range from the most negative E_o to the most positive. The more negative the standard reduction potential of a redox pair, the greater the tendency of the reductant member of that pair to lose electrons. The more positive the E_o, the greater the tendency of the oxidant member of that pair to accept electrons. Therefore, electrons flow from the pair with the more negative E_o to that with the more positive E_o. The E_o values for some members of the electron transport chain are shown in Figure 6.12.

3. **ΔG^o is related to ΔE_o:** The change in free energy is related directly to the magnitude of the change in E_o:

$$\Delta G^o = -\,n\,F\,\Delta E_o \qquad \text{where}$$

n = number of electrons transferred (1 for a cytochrome, 2 for NADH, $FADH_2$, and coenzyme Q)

F = Faraday constant (23,062 cal/volt · mol)

ΔE_o = E_o of the electron-accepting pair minus the E_o of the electron-donating pair

ΔG^o = change in the standard free energy

4. **ΔG^o of ATP:** The standard free energy of hydrolysis of the terminal phosphate group of ATP is –7300 cal/mol. The transport of a pair of electrons from NADH to oxygen via the electron transport chain produces 52,580 cal and, therefore, more than sufficient energy is made available to produce 3 ATP from 3 ADP and 3 P_i (3 x 7300 = 21,900 cal). The remaining calories are released as heat. [Note: The transport of a pair of electrons from $FADH_2$ or $FMNH_2$ to oxygen via the electron transport chain produces more than sufficient energy to produce 2 ATP from 2 ADP and 2 P_i.]

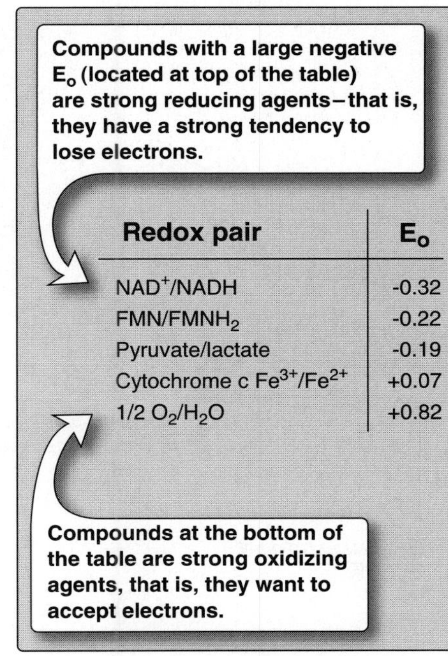

Redox pair	E_o
$NAD^+/NADH$	-0.32
$FMN/FMNH_2$	-0.22
Pyruvate/lactate	-0.19
Cytochrome c Fe^{3+}/Fe^{2+}	+0.07
1/2 O_2/H_2O	+0.82

Figure 6.12
Standard reduction potentials of some reactions.

VI. OXIDATIVE PHOSPHORYLATION

The transfer of electrons down the electron transport chain is energetically favored because NADH is a strong electron donor and molecular oxygen is an avid electron acceptor. However, the flow of electrons from NADH to oxygen does not directly result in ATP synthesis.

A. Chemiosmotic hypothesis

The chemiosmotic hypothesis (also known as the **Mitchell hypothesis**) explains how the free energy generated by the transport of electrons by the electron transport chain is used to produce ATP from ADP + P_i.

1. **Proton pump:** Electron transport is coupled to the phosphorylation of ADP by the transport of protons (H^+) across the inner mitochondrial membrane from the matrix to the intermembrane space. This process creates across the inner mitochondrial membrane an **electrical gradient** (with more positive charges on the outside of the membrane than on the inside) and a **pH gradient** (the outside of the

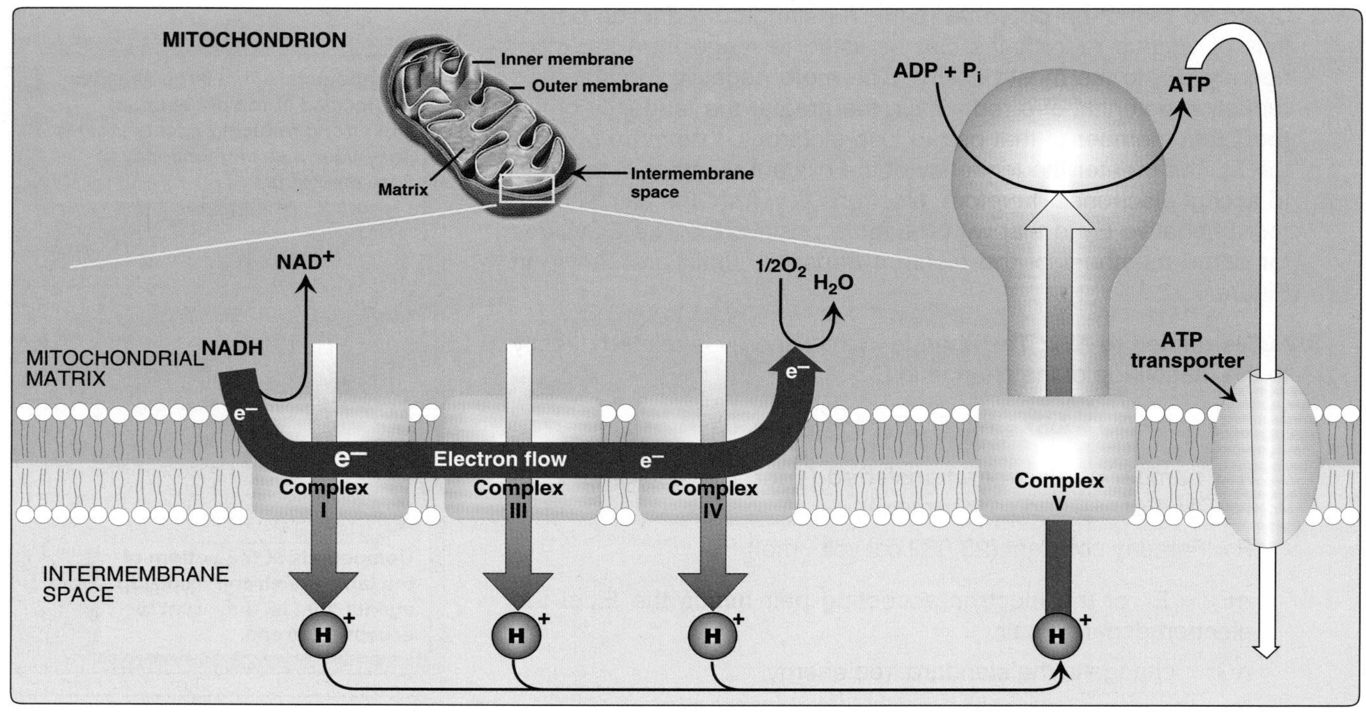

Figure 6.13
Electron transport chain shown coupled to the transport of protons. [Note: Complex II is not shown.]

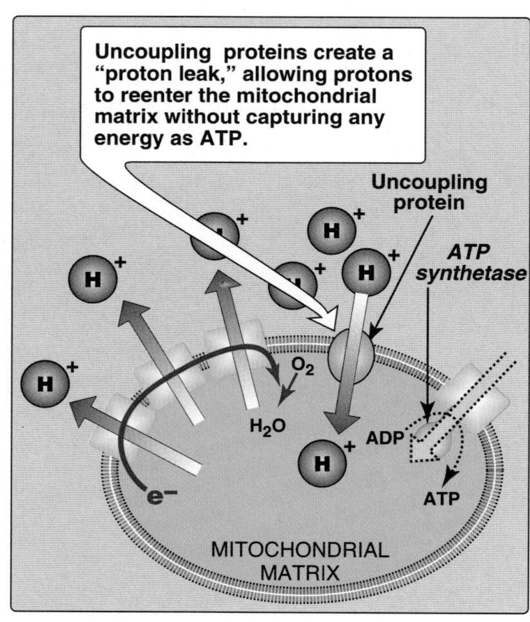

Figure 6.14
Transport of H^+ across mitochondrial membrane by 2,4-dinitrophenol.

membrane is at a lower pH than the inside; Figure 6.13). The energy generated by this proton gradient is sufficient to drive ATP synthesis. Thus, the proton gradient serves as the common intermediate that couples oxidation to phosphorylation.

2. **ATP synthase:** The enzyme complex *ATP synthase* (complex V, see Figure 6.13) synthesizes ATP, using the energy of the proton gradient generated by the electron transport chain. [Note: It is also called *ATPase,* because the isolated enzyme also catalyzes the hydrolysis of ATP to ADP and inorganic phosphate.] The chemiosmotic hypothesis proposes that after protons have been transferred to the cytosolic side of the inner mitochondrial membrane, they reenter the mitochondrial matrix by passing through a channel in the *ATP synthase complex*, resulting in the synthesis of ATP from ADP + P_i and, at the same time, dissipating the pH and electrical gradients.

   a. **Oligomycin:** This drug binds to the stalk of *ATP synthase,* closing the H^+ channel, and preventing reentry of protons into the mitochondrial matrix. Because the pH and electrical gradients cannot be dissipated in the presence of this drug, electron transport stops because of the difficulty of pumping any more protons against the steep gradients. Electron transport and phosphorylation are, therefore, again shown to be **tightly coupled** processes—inhibition of phosphorylation inhibits oxidation.

   b. **Uncoupling proteins (UCP):** UCPs occur in the inner mitochondrial membrane of mammals, including humans. These proteins create a "proton leak," that is, they allow protons to reenter the mitochondrial matrix without energy being captured as ATP (Figure 6.14). [Note: Energy is released in the form of heat.] **UCP1**, also called **thermogenin**, is responsible for the activation of fatty acid oxidation and heat production in the **brown adipocytes** of mammals. Brown fat, unlike the more abundant white fat, wastes amost ninety percent of its respiratory energy for thermogensis in response to cold, at birth, and during arousal in hibernating animals. However humans have little brown fat (except in the newborn), and UCP1 does not appear to play a major role in energy balance. Other uncoupling proteins (UCP2, UCP3) have been found in humans, but their significance remains controversial.

c. **Synthetic uncouplers:** Electron transport and phosphorylation can be uncoupled by compounds that increase the permeability of the inner mitochondrial membrane to protons. The classic example is **2,4-dinitrophenol**, a lipophilic proton carrier that readily diffuses through the mitochondial membrane. This uncoupler causes electron transport to proceed at a rapid rate without establishing a proton gradient, much as do the UCPs (see Figure 6.14). The energy produced by the transport of electrons is released as heat rather than being used to synthesize ATP. In high doses, the drug **aspirin** (as well as other salicylates) uncouples oxidative phosphorylation. This explains the fever that accompanies toxic overdoses of these drugs.

B. Membrane transport systems

The inner mitochondrial membrane is impermeable to most charged or hydrophilic substances. However, it contains numerous transport proteins that permit passage of specific molecules from the cytosol (or more correctly, the intermembrane space) to the mitochondrial matrix.

1. **ATP-ADP transport:** The inner mitochondrial membrane requires specialized carriers to transport ADP and P_i from the cytosol (where ATP is used and converted to ADP in many energy-requiring reactions) into mitochondria, where ATP can be resynthesized. An **adenine nucleotide carrier** transports one molecule of ADP from the cytosol into mitochondria, while exporting one ATP from the matrix back into the cytosol. This carrier is strongly inhibited by the plant toxin **atractyloside,** resulting in a depletion of the intramitochondrial ADP pool and cessation of ATP production. [Note: A phosphate carrier is responsible for transporting inorganic phosphate from the cytosol into mitochondria.]

2. **Transport of reducing equivalents:** The inner mitochondrial membrane lacks an NADH transport protein, and NADH produced in the cytosol cannot directly penetrate into mitochondria. However, two electrons of NADH (also called reducing equivalents) are transported from the cytosol into the mitochondria using shuttle mechanisms. In the **glycerophosphate shuttle** (Figure 6.15A), two electrons are transferred from NADH to *flavoprotein dehydrogenase* within the inner mitochondrial membrane. This enzyme then donates its electrons to the electron transport chain in a manner similar to that of *succinate dehydrogenase* (p. 111). The glycero-

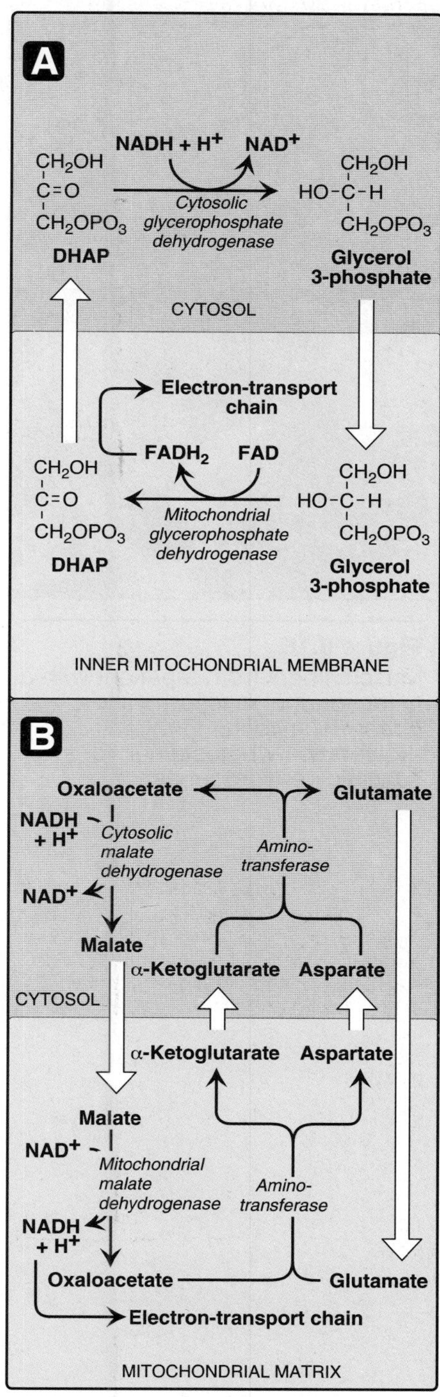

Figure 6.15
Shuttle pathways for the transport of electrons across the inner mitochondrial membrane. A. Glycerophosphate shuttle. B. Malate-aspartate shuttle.

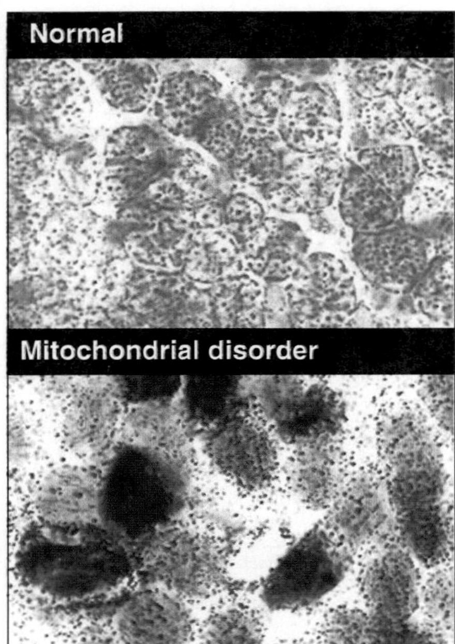

Figure 6.16
Muscle fibers from a patient with a mitochondrial myopathy show abnormal mitochondrial proliferation when stained for succinic dehydrogenase.

phosphate shuttle, therefore, results in the synthesis of two ATPs for each cytosolic NADH oxidized. This contrasts with the **malate-aspartate shuttle** (see Figure 6.15B), which produces NADH (rather than $FADH_2$) in the mitochondrial matrix and, therefore, yields three ATPs for each cytosolic NADH oxidized.

C. Inherited defects in oxidative phosphorylation

Thirteen of the approximately 100 polypeptides required for oxidative phosphorylation are coded for by mitochondrial DNA (mtDNA), whereas the remaining mitochondrial proteins are synthesized in the cytosol and transported into mitochondria. Defects in oxidative phosphorylation are more likely a result of alterations in mtDNA, which has a mutation rate about ten times greater than that of nuclear DNA. Tissues with the greatest ATP requirement (for example, CNS, skeletal and heart muscle, kidney, and liver) are most affected by defects in oxidative phosphorylation. Mutations in mtDNA are responsible for several diseases, including some cases of **mitochondrial myopathies** (Figure 6.16), and **Leber's hereditary optic neuropathy**, a disease in which bilateral loss of central vision occurs as a result of neuroretinal degeneration, including damage to the optic nerve. mtDNA is maternally inherited because mitochondria from the sperm cell do not enter the fertilized egg.

VI. CHAPTER SUMMARY

The change in **free energy** (**ΔG**) occuring during a reaction predicts the direction in which that reaction will spontaneously proceed. If **ΔG** is **negative** (that is, the product has a lower free energy than the substrate), the **reaction goes spontaneously**. If **ΔG** is **positive**, the reaction **does not go spontaneously**. If **ΔG = 0**, the reactants are in **equilibrium**. The change in free energy of the forward reaction ($A \rightarrow B$) is equal in magnitude but opposite in sign to that of the back reaction ($B \rightarrow A$). The **standard free energy changes** (**ΔG°s**) are **additive** in any sequence of consecutive reactions. Therefore, reactions or processes that have a large positive ΔG are made possible by **coupling** with hydrolysis of **adenosine triphosphate** (**ATP**), which has a large, negative ΔG°. The reduced coenzymes **NADH** and **$FADH_2$** each donate a pair of electrons to a specialized set of electron carriers, consisting of **FMN**, **coenzyme Q**, and a series of **cytochromes**, collectively called the **electron transport chain**. This pathway is present in the **inner mitochondrial membrane**, and is the final common pathway by which electrons derived from different fuels of the body flow to oxygen. The terminal cytochrome, **cytochrome a + a_3**, is the only cytochrome able to bind oxygen. **Electron transport** is coupled to the **transport of protons** (H^+) across the inner mitochondrial membrane from the matrix to the intermembrane space. This process creates an **electrical gradient** and a **pH gradient** across the inner mitochondrial membrane. After protons have been transferred to the cytosolic side of the inner mitochondrial membrane, they can reenter the mitochondrial matrix by passing through a channel in the **ATP synthase complex**, resulting in the synthesis of ATP from ADP + Pi, and at the same time dissipating the pH and electrical gradients. **Electron transport** and **phosphorylation** are thus said to be

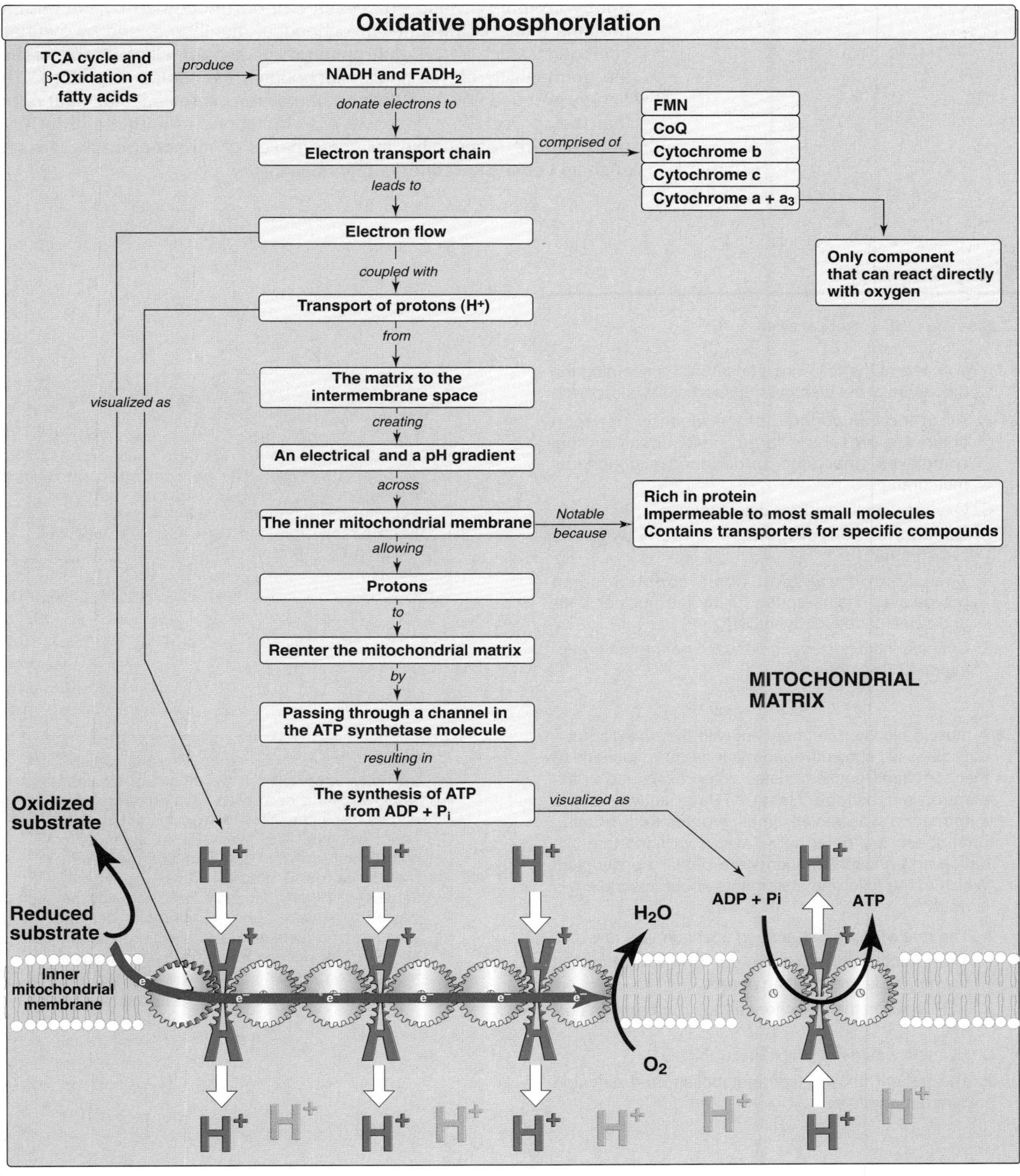

Figure 6.17
Summary of key concepts for oxidative phosphorylation. [Note: Electron flow and ATP synthesis are are envisioned as sets of interlocking gears to emphase the idea of coupling.]

tightly coupled. These processes can be **uncoupled** by **uncoupling proteins** found in the inner mitochondrial membrane, and by synthetic compounds such as **2,4-dinitrophenol** and **aspirin**, all of which increase the permeability of the inner mitochondrial membrane to protons. The energy produced by the transport of electrons is released as **heat** rather than being used to synthesize ATP. Mutations in **mitochondrial DNA (mtDNA)** are responsible for some cases of **mitochondrial diseases**, such as **Leber's hereditary optic neuropathy**.

Study Questions

Choose the ONE correct answer

6.1 Which one of the following statements concerning the components of the electron transport chain is correct?

A. All of the components of the electron transport chain are present in large, multisubunit protein complexes embedded in the inner mitochondrial membrane.
B. Oxygen directly oxidizes cytochrome c.
C. Succinate dehydrogenase directly reduces cytochrome c.
D. The electron transport chain contains some polypeptide chains coded for by the nuclear DNA and some coded for by mtDNA.
E. Cyanide inhibits electron flow, but not proton pumping or ATP synthesis.

Correct answer = D. Thirteen of the approximately 100 polypeptides required for oxidative phosphorylation are coded for by mitochondrial DNA, including the electron transport components cytochrome c and coenzyme Q. Oxygen directly oxidizes cytochrome oxidase. Succinate dehydrogenase directly reduces FAD. Cyanide inhibits electron flow, proton pumping, and ATP synthesis.

6.2 A muscle biopsy from a patient with the rare disorder, Luft disease, showed abnormally large mitochondria that contained packed cristae when examined in the electron microscope. Basal ATPase activity of the mitochodria was seven times greater than normal. From these and other data it was concluded that oxidation and phosphorylation were partially uncoupled. Which of the following statements about this patient is correct?

A. The rate of electron transport is abnormally low.
B. The proton gradient across the inner mitochondrial membrane is greater than normal.
C. ATP levels in the mitochondria are greater than normal.
D. Cyanide would not inhibit electron flow.
E. The patient shows hypermetabolism and elevated core temperature.

Corrrect answer = E. When phosphorylation is partially uncoupled from electron flow, one would expect a decrease in the proton gradient across the inner mitochondrial membrane and, hence, impaired ATP synthesis. In an attempt to compensate for this defect in energy capture, metabolism and electron flow to oxygen is increased. This hypermetabolism will be accompanied by elevated body temperature because the energy in fuels is largely wasted, appearing as heat. The electron transport chain will still be inhibited by cyanide.

Introduction to Carbohydrates

7

I. OVERVIEW

Carbohydrates are the most abundant organic molecules in nature. They have a wide range of functions, including providing a significant fraction of the energy in the diet of most organisms, acting as a storage form of energy in the body, and serving as cell membrane components that mediate some forms of intercellular communication. Carbohydrates also serve as a structural component of many organisms, including the cell walls of bacteria, the exoskeleton of many insects, and the fibrous cellulose of plants. The empiric formula for many of the simpler carbohydrates is $(CH_2O)_n$, hence the name "hydrate of carbon."

II. CLASSIFICATION AND STRUCTURE OF CARBOHYDRATES

Monosaccharides (simple sugars) can be classified according to the number of carbon atoms they contain. Examples of some monosaccharides commonly found in humans are listed in Figure 7.1. Carbohydrates with an aldehyde as their most oxidized functional group are called **aldoses**, whereas those with a keto group as their most oxidized functional group are called **ketoses** (Figure 7.2). For example, glyceraldehyde is an aldose, whereas dihydroxyacetone is a ketose. Carbohydrates that have a free carbonyl group have the suffix "-ose." [Note: Ketoses (with some exceptions, for example, fructose) have an additional two letters in their suffix; "-ulose," for example, xylulose.] Monosaccharides can be linked by **glycosidic bonds** to create larger structures (Figure 7.3). **Disaccharides** contain two monosaccharide units, **oligosaccharides** contain from three to about twelve monosaccharide units, whereas **polysaccharides** contain more than twelve monosaccharide units, and can be hundreds of sugar units in length.

Generic names		Examples
3 carbons:	trioses	Glyceraldehyde
4 carbons:	tetroses	Erythrose
5 carbons:	pentoses	Ribose
6 carbons:	hexoses	Glucose
7 carbons:	heptoses	Sedoheptulose
9 carbons:	nonoses	Neuraminic acid

Figure 7.1
Examples of monosaccharides found in humans, classified according to the number of carbons they contain.

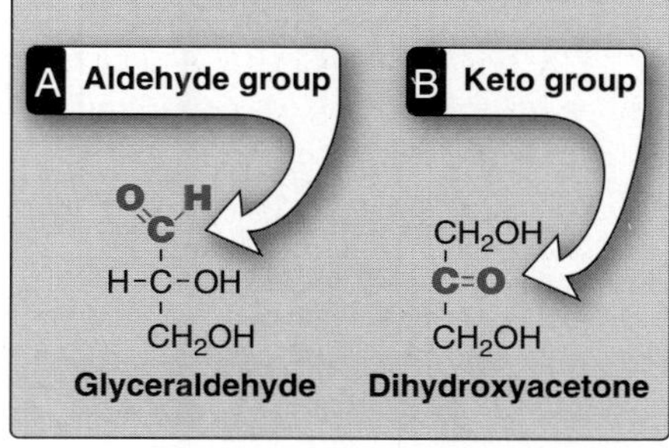

Figure 7.2
Examples of an aldose (A) and a ketose (B) sugar.

A. Isomers and epimers

Compounds that have the same chemical formula but have different structures are called **isomers**. For example, fructose, glucose, mannose, and galactose are all isomers of each other, having the same chemical formula, $C_6H_{12}O_6$. If two monosaccharides differ in configuration around only one specific carbon atom (with the exception of the carbonyl carbon, see "anomers" below), they are defined as **epimers** of each other. (Of course, they are also isomers!) For example, glucose

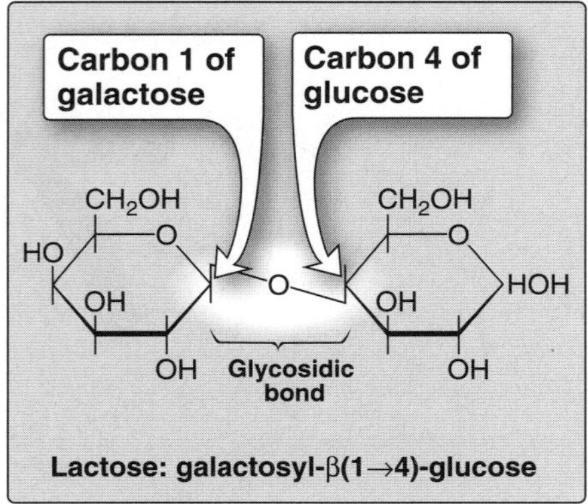

Figure 7.3
A glycosidic bond between two hexoses producing a disaccharide.

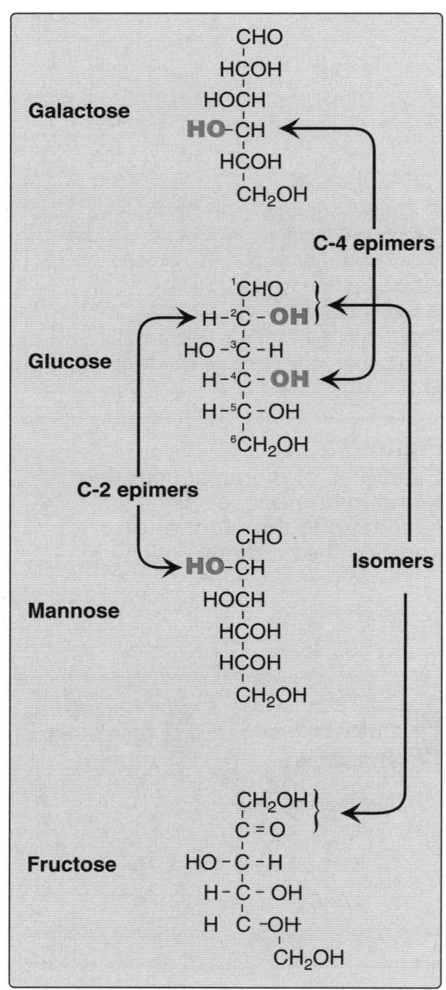

Figure 7.4
C-2 and C-4 epimers and an isomer of glucose.

and galactose are C-4 epimers—their structures differ only in the position of the –OH group at carbon 4. [Note: The carbons in sugars are numbered beginning at the end that contains the carbonyl carbon—that is, the aldehyde or keto group (Figure 7.4).] Glucose and mannose are C-2 epimers. However, galactose and mannose are NOT epimers—they differ in the position of –OH groups at two carbons (2 and 4) and are, therefore, defined only as isomers (see Figure 7.4).

B. Enantiomers

A special type of isomerism is found in the pairs of structures that are mirror images of each other. These mirror images are called **enantiomers**, and the two members of the pair are designated as a D- and an L-sugar (Figure 7.5). The vast majority of the sugars in humans are D-sugars.

C. Cyclization of monosaccharides

Less than one percent of each of the monosaccharides with five or more carbons exists in the open-chain (acyclic) form. Rather, they are predominantly found in a ring form, in which the aldehyde (or ketone) group has reacted with an alcohol group on the same sugar.

1. **Anomeric carbon:** Formation of a ring results in the creation of an anomeric carbon at carbon 1 of an aldose or at carbon 2 of a ketose. These structures are designated the **α or β configurations** of the sugar, for example, α-D-glucose and β-D-glucose (Figure 7.6). These two sugars are both glucose, but they are **anomers** of each other. Enzymes are able to distinguish between these two structures and use one or the other preferentially. For example, glycogen is synthesized from α-D-glucopyranose, whereas cellulose is synthesized from β-D-glucopyranose. The cyclic α and β anomers of a sugar in solution are in equilibrium with each other, and can be spontaneously interconverted (a process called **mutarotation**, see Figure 7.6).

2. **Reducing sugars:** If the oxygen on the anomeric carbon (the carbonyl group) of a sugar is not attached to any other structure, that sugar is a reducing sugar**.** A reducing sugar can react with chemical reagents (for example, Benedict's solution) and reduce the reactive component, with the **anomeric carbon** becoming oxidized. [Note: Only the state of the oxygen on the anomeric carbon determines if the sugar is reducing or nonreducing—the other hydroxyl groups on the molecule are not involved.]

D. Complex carbohydrates

Carbohydrates can be attached by glycosidic bonds to non-carbohydrate structures, including purines and pyrimidines (found in nucleic acids), aromatic rings (such as those found in steroids and bilirubin), proteins (found in glycoproteins and glycosaminoglycans), and lipids (found in glycolipids). The aldose, the carbon 1 of which (or ketose, the carbon 2 of which) participates in the glycosidic link, is called a **glycosyl residue**. For example, if the anomeric carbon of glucose participates in such a bond, that sugar is called a **glucosyl residue**; thus, the disaccharide lactose (see Figure 7.3) is galactosyl–glucose.

1. **O- and N-glycosides:** If the group on the non-carbohydrate molecule to which the sugar is attached is an –OH group, the structure is an **O-glycoside**. If the group is an $-NH_2$, the structure is an **N-glycoside** (Figure 7.7). [Note: All sugar–sugar glycosidic bonds are O-type linkages.]

2. **Naming glycosidic bonds:** Glycosidic bonds between sugars are named according to the numbers of the connected carbons, and also with regard to the position of the anomeric hydroxyl group of the sugar involved in the bond. If this anomeric hydroxyl group is in the α configuration, the linkage is an **α-bond**. If it is in the β configuration, the linkage is a **β-bond**. **Lactose**, for example, is synthesized by forming a glycosidic bond between carbon 1 of a β-galactose and carbon 4 of glucose. The linkage is, therefore, a β(1→4) glycosidic bond (see Figure 7.3). [Note: Because the anomeric end of the glucose residue is not involved in the glycosidic linkage it (and, therefore, lactose) remains a **reducing sugar**.]

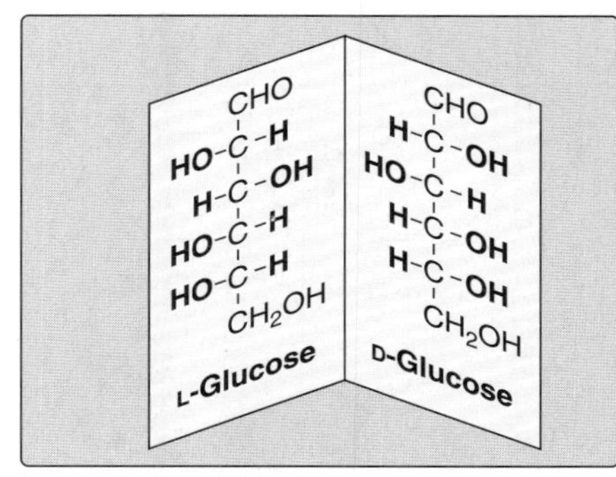

Figure 7.5
Enantiomers (mirror images) of glucose.

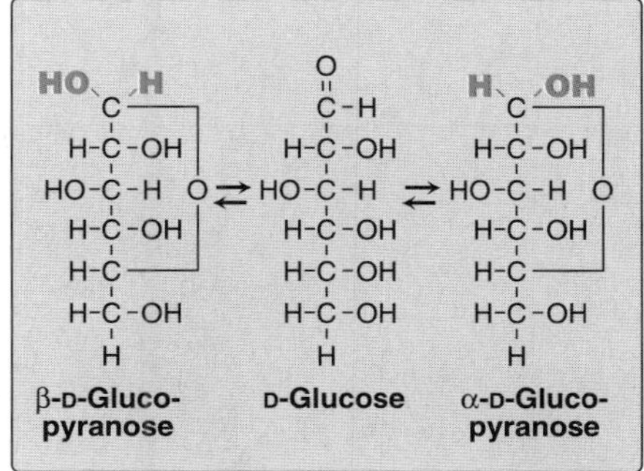

Figure 7.6
The interconversion of the α and β anomeric forms of glucose (mutarotation).

III. DIGESTION OF CARBOHYDRATES

The principal sites of dietary carbohydrate digestion are the mouth and intestinal lumen. This digestion is rapid and is generally completed by the time the stomach contents reach the junction of the duodenum and jejunum. There is little monosaccharide present in diets of mixed animal and plant origin. Therefore, the enzymes needed for degradation of most dietary carbohydrates are primarily disaccharidases and endoglycosidases (that break oligosaccharides and polysaccharides). Hydrolysis of glycosidic bonds is catalyzed by a family of glycosidases that degrade carbohydrates into their reducing sugar components (Figure 7.8). These enzymes are usually specific for the structure and configuration of the glycosyl residue to be removed, as well as for the type of bond to be broken.

A. Digestion of carbohydrates begins in the mouth

The major dietary polysaccharides are of animal (glycogen) and plant origin (starch, composed of amylose and amylopectin). During mastication, salivary *α-amylase* acts briefly on dietary starch in a random manner, breaking some α(1→4) bonds. [Note: There are both α(1→4)- and β(1→4)-endoglucosidases in nature, but humans do not produce and secrete the latter in digestive juices. Therefore, they are unable to digest cellulose—a carbohydrate of plant origin containing β(1→4) glycosidic bonds between glucose residues.] Because branched amylopectin and glycogen also contain α(1→6) bonds, the digest resulting from the action of *α-amylase* contains a mixture of smaller, branched oligosaccharide molecules (Figure 7.9). Carbohydrate digestion halts temporarily in the stomach, because the high acidity inactivates the salivary *α-amylase.*

B. Further digestion of carbohydrates by pancreatic enzymes occurs in the small intestine

When the acidic stomach contents reach the small intestine, they are neutralized by bicarbonate secreted by the pancreas, and pancreatic *α-amylase* continues the process of starch digestion.

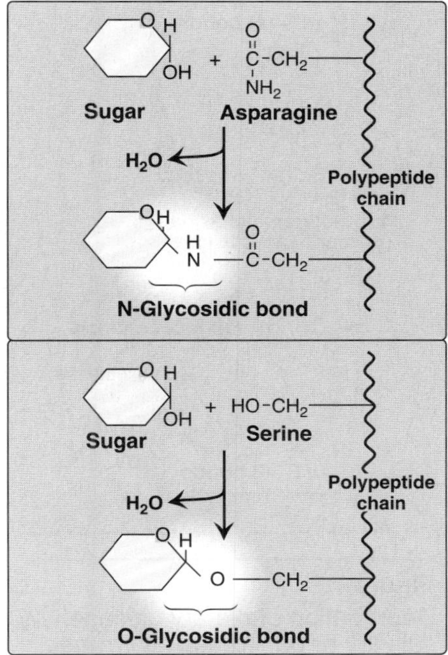

Figure 7.7
Glycosides: examples of N- and O-glycosidic bonds.

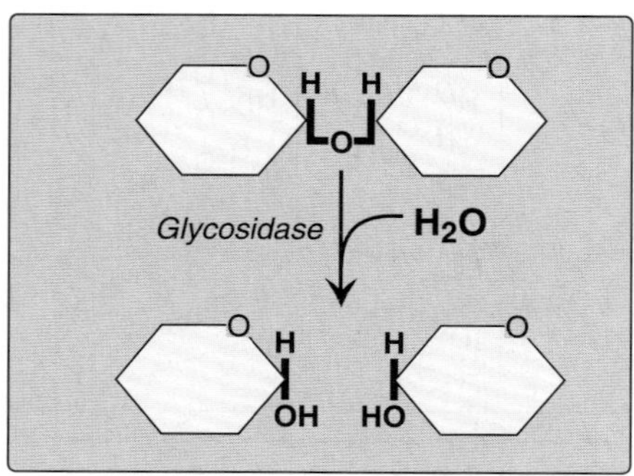

Figure 7.8
Hydrolysis of a glycosidic bond.

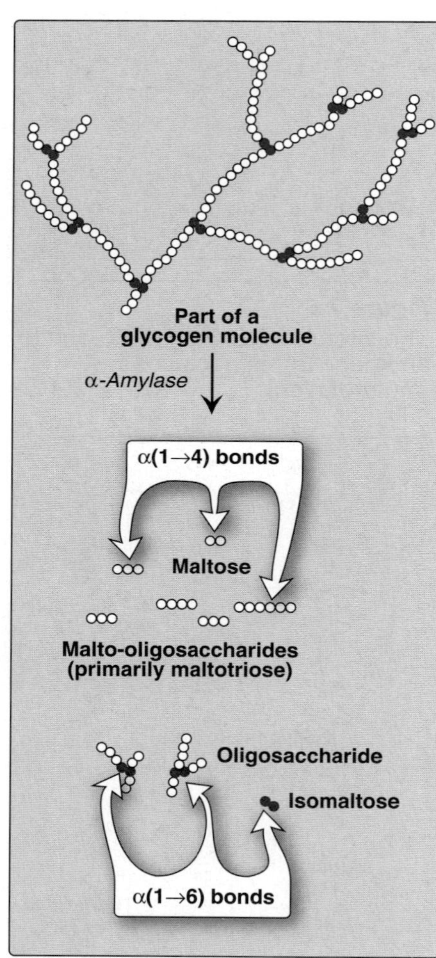

Figure 7.9
Degradation of dietary glycogen by salivary or pancreatic *α-amylase*.

C. Final carbohydrate digestion by enzymes synthesized by the intestinal mucosal cells

The final digestive processes occur at the mucosal lining of the upper jejunum, declining as they proceed down the small intestine, and include the action of several disaccharidases and oligosaccharidases (Figure 7.10). For example, *isomaltase* cleaves the α(1→6) bond in isomaltose and *maltase* cleaves maltose, both producing glucose, *sucrase* cleaves sucrose producing glucose and fructose, and *lactase* (*β-galactosidase*) cleaves lactose producing galactose and glucose. These enzymes are secreted through, and remain associated with, the luminal side of the brush border membranes of the intestinal mucosal cells.

D. Absorption of monosaccharides by intestinal mucosal cells

The duodenum and upper jejunum absorb the bulk of the dietary sugars. Insulin is not required for the uptake of glucose by intestinal cells. However, different sugars have different mechanisms of absorption. For example, galactose and glucose are transported into the mucosal cells by an active, energy-requiring process that involves a specific transport protein and requires a concurrent uptake of sodium ions. Fructose uptake requires a sodium-independent monosaccharide transporter (GLUT-5) for its absorption. All three monosaccharides are transported from the intestinal mucosal cell into the portal circulation by yet another transporter, GLUT-2. (See p. 95 for a discussion of these transporters.)

E. Abnormal degradation of disaccharides

The overall process of carbohydrate digestion and absorption is so efficient in healthy individuals that ordinarily all digestible dietary carbohydrate is absorbed by the time the ingested material reaches the lower jejunum. However, because predominantly monosaccharides are absorbed, any defect in a specific disaccharidase activity of the intestinal mucosa causes the passage of undigested carbohydrate into the large intestine. As a consequence of the presence of this osmotically active material, water is drawn from the mucosa into the large intestine, causing osmotic diarrhea. This is reinforced by the bacterial fermentation of the remaining carbohydrate to two- and three-carbon compounds (which are also osmotically active) plus large volumes of CO_2 and H_2 gas, causing abdominal cramps, diarrhea, and flatulence.

1. **Digestive enzyme deficiencies:** Hereditary deficiencies of the individual disaccharidases have been reported in infants and children with **disaccharide intolerance**. Alterations in disaccharide degradation can also be caused by a variety of intestinal diseases, malnutrition, or drugs that injure the mucosa of the small intestine. For example, brush border enzymes are rapidly lost in normal individuals with severe diarrhea, causing a temporary, acquired enzyme deficiency. Thus, patients suffering or recovering from such a disorder cannot drink or eat significant amounts of dairy products or sucrose without exacerbating the diarrhea.

2. **Lactose intolerance:** More than one half of the world's adults are lactose intolerant (Figure 7.11). This is particularly manifested in

certain races. For example, up to ninety percent of adults of African or Asian descent are lactase-deficient and, therefore, are less able to metabolize lactose than individuals of northern European origin. The mechanism by which the enzyme is lost is not clear, but it is determined genetically and represents a reduction in the amount of enzyme protein rather than a modified inactive enzyme. Treatment for this disorder is simply to remove lactose from the diet, or to take *lactase* in pill form prior to eating.

3. **Isomaltase-sucrase deficiency:** This enzyme deficiency results in an intolerance of ingested sucrose. This disorder is found in about ten percent of Greenland's Eskimos, whereas two percent of North Americans are heterozygous for the deficiency. Treatment is to withhold dietary sucrose.

4. **Diagnosis:** Identification of a specific enzyme deficiency can be obtained by performing oral tolerance tests with the individual disaccharides. Measurement of hydrogen gas in the breath is a reliable test for determining the amount of ingested carbohydrate not absorbed by the body, but which is metabolized instead by the intestinal flora (see Figure 7.11).

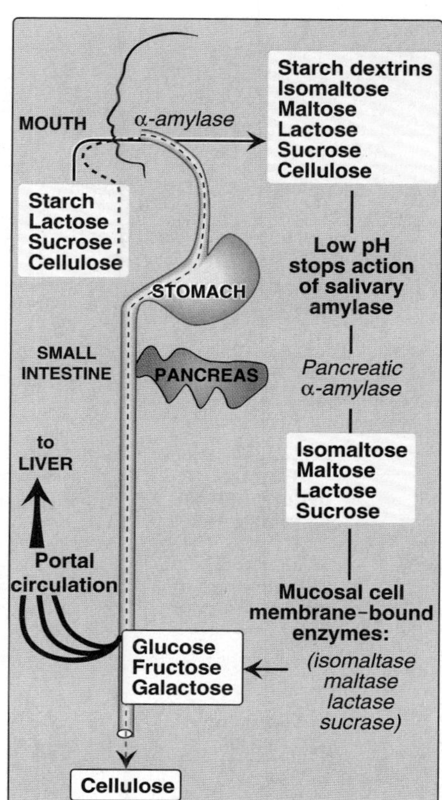

Figure 7.10
Digestion of carbohydrates.

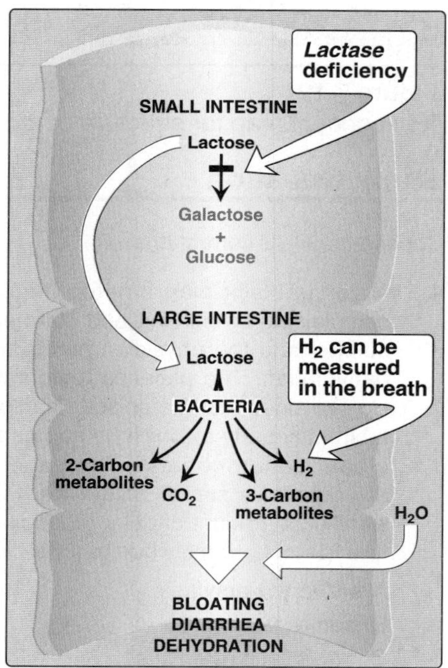

Figure 7.11
Abnormal lactose metabolism.

IV. CHAPTER SUMMARY

Monosaccharides (simple sugars) containing an aldehyde group are called **aldoses** and those with a keto group are called **ketoses**. **Disaccharides**, **oligosaccharides**, and **polysaccarides** consist of monosaccharides linked by **glycosidic bonds**. Compounds with the same chemical formula are called **isomers**. If two monosaccharide isomers differ in configuration around one specific carbon atom (with the exception of the carbonyl carbon), they are defined as **epimers** of each other. If a pair of sugars are mirror images of each other (**enantiomers**), the two members of the pair are designated as **D-** and **L-sugars**. When a sugar cyclizes, an **anomeric carbon** is created from the aldehyde group of an aldose or keto group of a ketose. This carbon can have two configurations, α or β. If the oxygen on the anomeric carbon is not attached to any other structure, that sugar is a **reducing sugar**. A sugar with its anomeric carbon linked to another structure is called a **glycosyl residue**. Sugars can be attached either to a $-NH_2$ or an –OH group, producing **N–** and **O–glycosides. Salivary α-amylase** acts on **dietary starch** (glycogen, amylose, amylopectin), producing **oligosaccharides. Pancreatic α-amylase** continues the process of starch digestion. The final digestive processes occur at the **mucosal lining** of the **small intestine**. Several **disaccharidases** [for example, **lactase** (**β-galactosidase**), **sucrase**, **maltase**, and **isomaltase**] produce monosaccharides (glucose, galactose, and fructose). These enzymes are secreted by and remain associated with the luminal side of the **brush border membranes** of **intestinal mucosal cells**. Absorption of the monosaccharides requires specific transporters. If carbohydrate degradation is deficient (as a result of heredity, intestinal disease, malnutrition, or drugs that injure the mucosa of the small intestine), undigested carbohydrate will pass into the large intestine, where it can cause **osmotic diarrhea**. Bacterial fermentation of the compounds produces large volumes of CO_2 and H_2 gas, causing abdominal cramps, diarrhea, and flatulence. **Lactose intolerance**, caused by a lack of **lactase**, is by far the most common of these deficiencies.

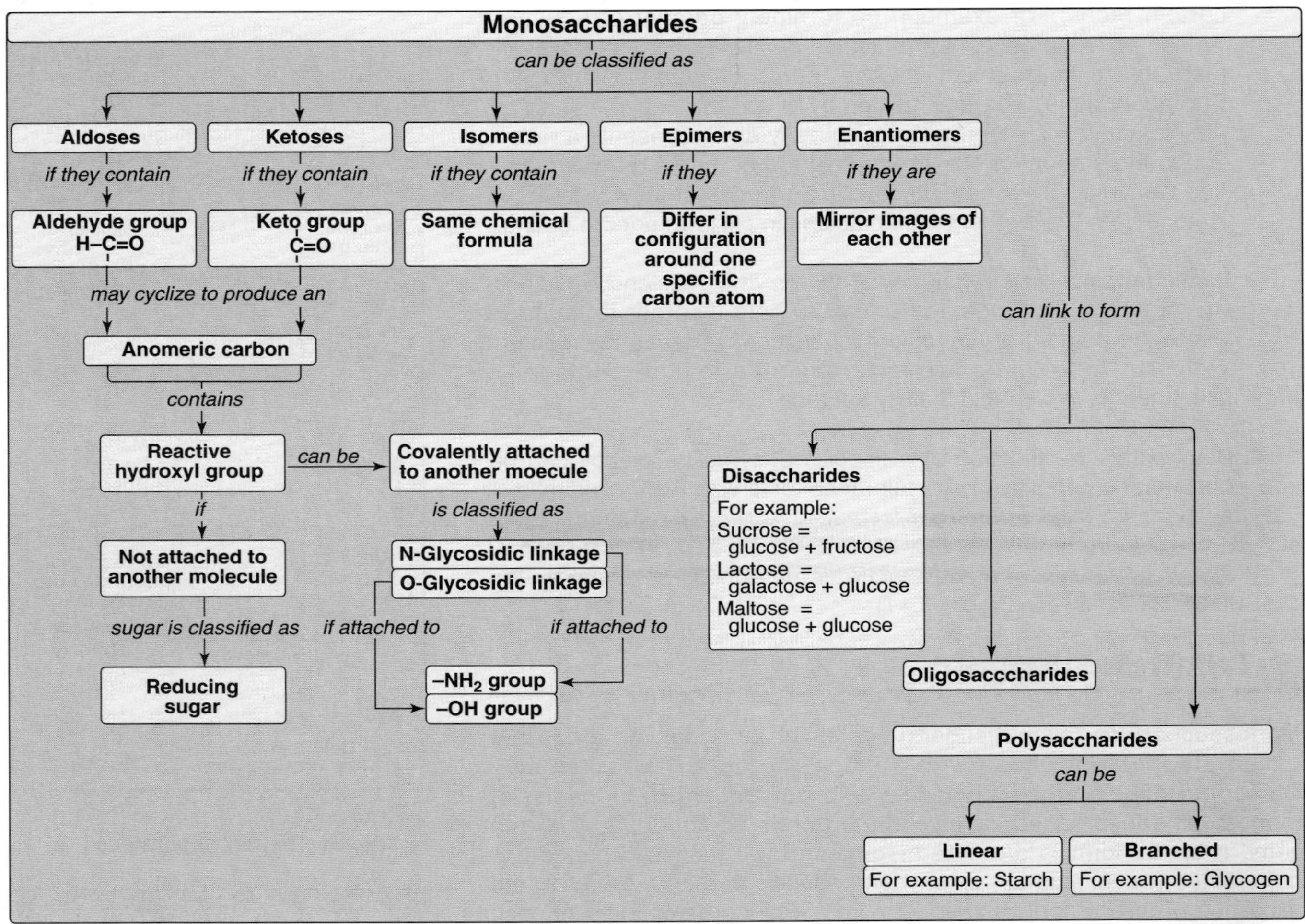

Figure 7.12
Key concept map for structure of monosaccharides.

Study Question

Choose the ONE correct answer

7.1 A young black man entered his physician's office complaining of bloating and diarrhea. His eyes were sunken and the physician noted additional signs of dehydration. The patient's temperature was normal. He explained that the episode had occurred following a birthday party at which he had participated in an ice cream eating contest. The patient reported prior episodes of a similar nature following ingestion of a significant amount of dairy products. This clinical picture is most probably due to a deficiency in:

A. salivary α-amylase.

B. isomaltase.

C. pancreatic α-amylase.

D. sucrase.

E. lactase.

Correct answer = E. The physical symptoms suggest a deficiency in an enzyme responsible for carbohydrate degradation. The symptoms observed following the ingestion of dairy products suggest that the patient is deficient in lactase.

Glycolysis

8

I. INTRODUCTION TO METABOLISM

In Chapter 5, individual enzymic reactions were analyzed in an effort to explain the mechanisms of catalysis. However, in cells, these reactions rarely occur in isolation, but rather are organized into multistep sequences called **pathways**, such as that of glycolysis (Figure 8.1). In a pathway, the product of one reaction serves as the substrate of the subsequent reaction. Different pathways can also intersect, forming an integrated and purposeful network of chemical reactions. These are collectively called **metabolism**, which is the sum of all the chemical changes occurring in a cell, a tissue, or the body. Most pathways can be classified as either **catabolic** (degradative) or **anabolic** (synthetic). Catabolic reactions break down complex molecules, such as proteins, polysaccharides, and lipids, to a few simple molecules, for example, CO_2, NH_3 (ammonia), and water. Anabolic pathways form complex end products from simple precursors, for example, the synthesis of the polysaccharide, gycogen, from glucose. In the following chapters, this text focuses on the central metabolic pathways that are involved in synthesizing and degrading carbohydrates, lipids, and amino acids.

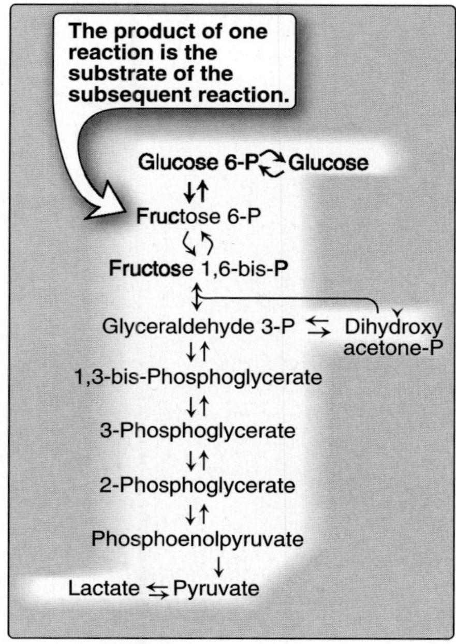

Figure 8.1
Glycolysis, an example of a metabolic pathway.

A. Metabolic map

It is convenient to investigate metabolism by examining its component pathways. Each pathway is composed of multienzyme sequences, and each enzyme, in turn, may exhibit important catalytic or regulatory features. To provide the reader with the "big picture," a metabolic map containing the important central pathways of energy metabolism is presented in Figure 8.2. This map is useful in tracing connections between pathways, visualizing the purposeful "movement" of metabolic intermediates, and picturing the effect on the flow of intermediates if a pathway is blocked, for example, by a drug or an inherited deficiency of an enzyme. Throughout the next three units of this book, each pathway under discussion will be repeatedly featured as part of the major metabolic picture shown in Figure 8.2.

B. Catabolic pathways

Catabolic reactions serve to capture chemical energy in the form of ATP from the **degradation** of energy-rich fuel molecules. Catabolism also allows molecules in the diet (or nutrient molecules stored in cells) to be converted into building blocks needed for the synthesis of complex molecules. Energy generation by degradation of complex molecules occurs in three stages as shown in Figure 8.3.

Lippincott's Illustrated Reviews: Biochemistry, Third Edition,
by Pamela C. Champe and Richard A. Harvey.
Lippincott Williams & Wilkins, Baltimore, MD © 2005.

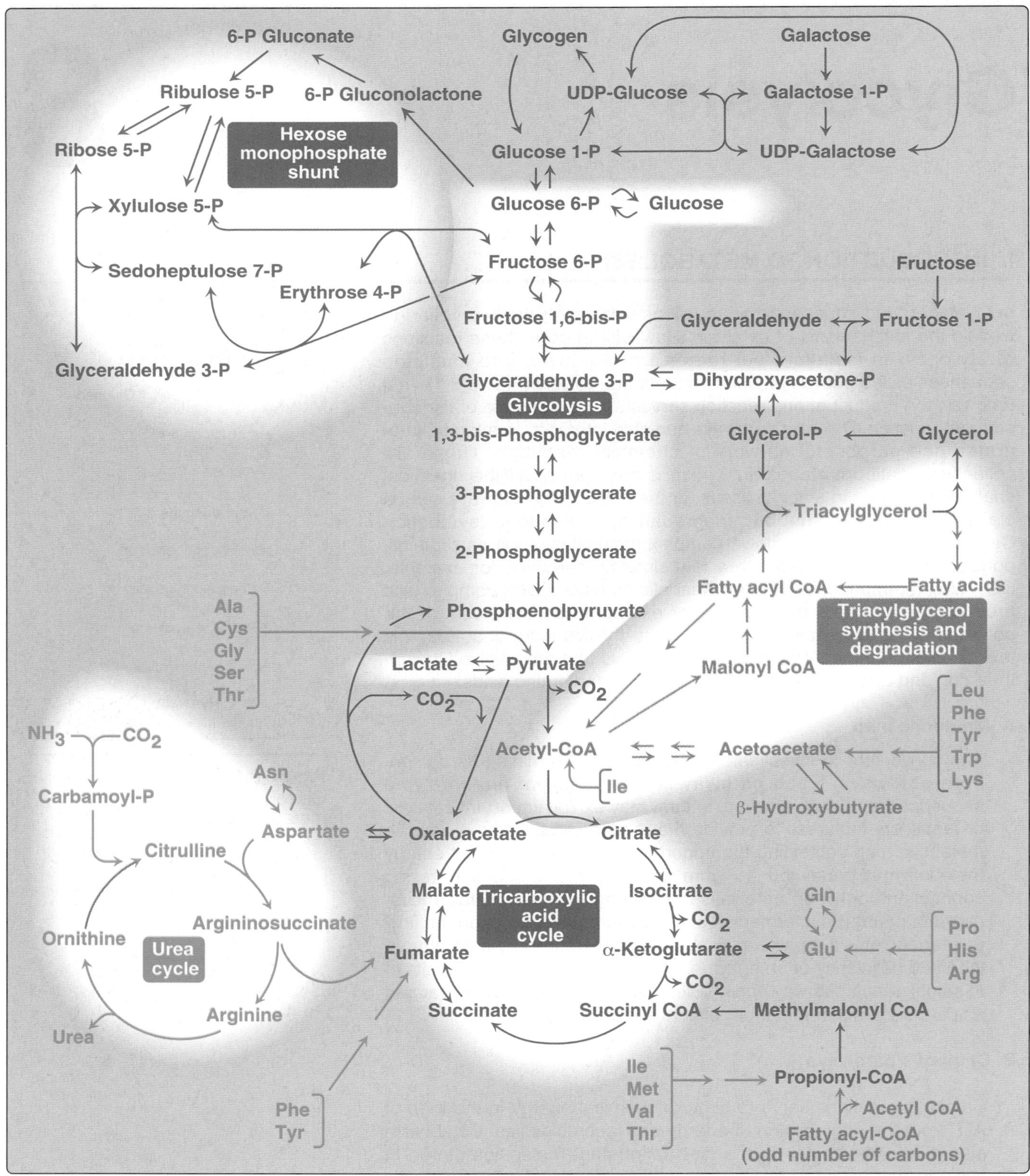

Figure 8.2
Important reactions of intermediary metabolism. Several important pathways to be discussed in later chapters are highlighted. Curved reaction arrows (⟲) indicate forward and reverse reactions that are catalyzed by different enzymes. The straight arrows (⇆) indicate forward and reverse reactions that are catalyzed by the same enzyme. Key: **Blue text** = intermediates of carbohydrate metabolism; **brown text** = intermediates of lipid metabolism; **green text** = intermediates of protein metabolism.

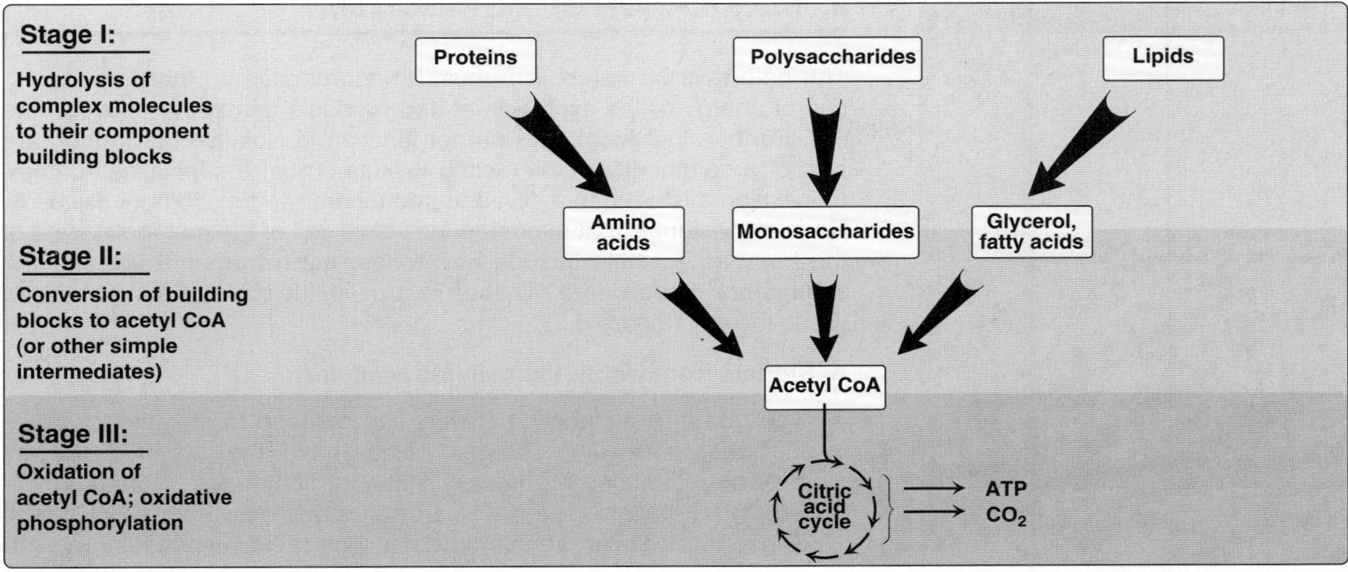

Figure 8.3
Three stages of catabolism.

1. **Hydrolysis of complex molecules:** In the first stage, complex molecules are broken down into their component building blocks. For example, proteins are degraded to amino acids, polysaccharides to monosaccharides, and triacylglycerols to free fatty acids and glycerol.

2. **Conversion of building blocks to simple intermediates:** In the second stage, these diverse building blocks are further degraded to **acetyl CoA** and a few other, simple molecules. Some energy is captured as **ATP**, but the amount is small compared with the energy produced during the third stage of catabolism.

3. **Oxidation of acetyl CoA:** The **tricarboxylic acid** (**TCA**) **cycle** (see p. 107) is the final common pathway in the oxidation of fuel molecules such as acetyl CoA. Large amounts of ATP are generated as electrons flow from **NADH** and **$FADH_2$** to **oxygen** via **oxidative phosphorylation** (see p. 77).

C. Anabolic pathways

Anabolic reactions combine small molecules, such as amino acids, to form complex molecules, such as proteins (Figure 8.4). Anabolic reactions require energy, which is generally provided by the breakdown of ATP to ADP and P_i. Anabolic reactions often involve chemical reductions in which the reducing power is most frequently provided by the electron donor NADPH (see p. 145). Note that **catabolism** is a **convergent process**—that is, a wide variety of molecules are transformed into a few common end products. By contrast, **anabolism** is a **divergent process** in which a few biosynthetic precursors form a wide variety of polymeric or complex products.

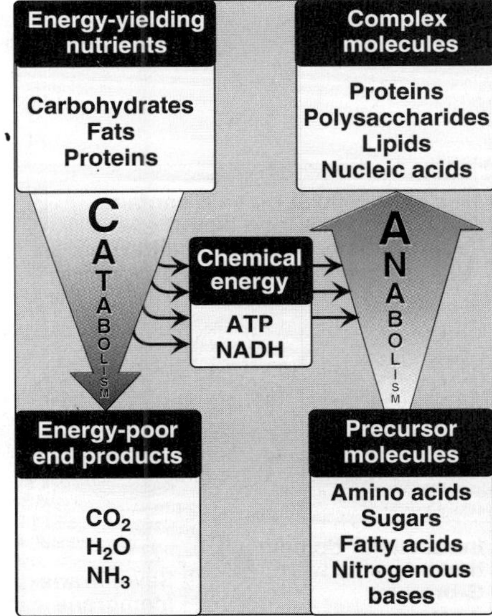

Figure 8.4
Comparison of catabolic and anabolic pathways.

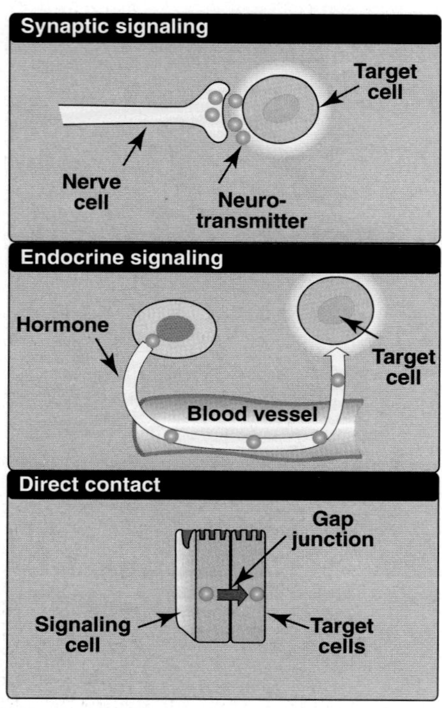

Figure 8.5
Some commonly used mechanisms for transmission of regulatory signals between cells.

II. REGULATION OF METABOLISM

The pathways of metabolism must be coordinated so that the production of energy or the synthesis of end products meets the needs of the cell. Further, individual cells do not function in isolation but, rather, are part of a community of interacting tissues. Thus, a sophisticated communication system has evolved to coordinate the functions of the body. Regulatory signals that inform an individual cell of the metabolic state of the body as a whole include **hormones**, **neurotransmitters**, and the **availability of nutrients**. These, in turn, influence signals generated within the cell (Figure 8.5).

A. Signals from within the cell (intracellular)

The rate of a metabolic pathway can respond to regulatory signals that arise from within the cell. For example, the rate of a pathway may be influenced by the availability of substrates, product inhibition, or alterations in the levels of allosteric activators or inhibitors. These intracellular signals typically elicit rapid responses, and are important for the moment-to-moment regulation of metabolism.

B. Communication between cells (intercellular)

The ability to respond to extracellular signals is essential for the survival and development of all organisms. Signaling between cells provides for long-range integration of metabolism, and usually results in a response that is slower than is seen with signals that originate within the cell. Communication between cells can be mediated by surface-to-surface contact and, in some tissues, by formation of gap junctions, allowing direct communication between the cytoplasms of adjacent cells. However, for energy metabolism, the most important route of communication is chemical signaling between cells, for example, by blood-borne hormones or by neurotransmitters.

C. Second messenger systems

Hormones or neurotransmitters can be thought of as signals, and a receptor as a signal detector. Each component serves as a link in the communication between extracellular events and chemical changes within the cell. Many receptors signal their recognition of a bound ligand by initiating a series of reactions that ultimately result in a specific intracellular response. "Second messenger" molecules—so named because they intervene between the original messenger (the neurotransmitter or hormone) and the ultimate effect on the cell—are part of the cascade of events that translates hormone or neurotransmitter binding into a cellular response. Two of the most widely recognized second messenger systems are the **calcium/phosphatidylinositol system** (see p. 203), and the **adenylyl cyclase system**, which is particularly important in regulating the pathways of intermediary metabolism.

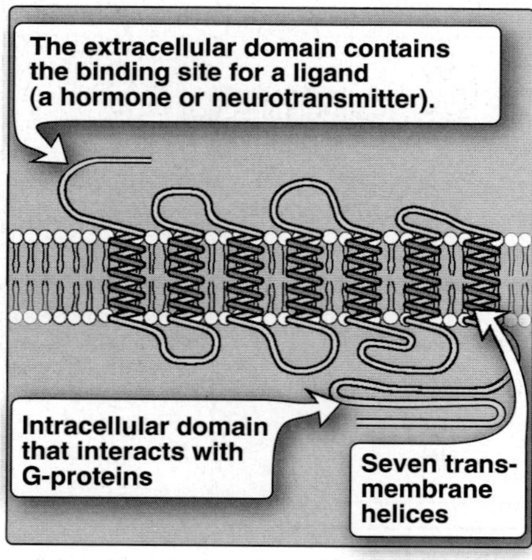

Figure 8.6
Structure of a typical membrane receptor.

D. Adenylyl cyclase

The recognition of a chemical signal by some membrane receptors, such as the β- and α_2-adrenergic receptors,[1] triggers either an increase or a decrease in the activity of *adenylyl cyclase*. This is a

[1]See Chapter 6 in ***Lippincott's Illustrated Reviews: Pharmacology*** (2nd and 3rd Eds.) for a discussion of adrenergic receptors.

membrane-bound enzyme that converts ATP to **3',5'-adenosine monophosphate** (also called **cyclic AMP** or **cAMP**). The chemical signals are most often hormones or neurotransmitters, each of which binds to a unique type of membrane receptor. Therefore, tissues that respond to more than one chemical signal must have several different receptors, each of which can be linked to *adenylyl cyclase*. [Note: Certain toxins, such as one produced by Vibrio cholerae, can also activate the *adenyl cyclase* cascade, with potentially disasterous consequences.[2]] These receptors are characterized by an extracellular ligand-binding region, seven transmembrane helices, and an intracellular domain that interacts with G-proteins (Figure 8.6).

1. **GTP-dependent regulatory proteins:** The effect of the activated, occupied receptor on second messenger formation is not direct but, rather, is mediated by specialized trimeric proteins in the cell membrane. These proteins, referred to as **G-proteins** because they bind guanosine nucleotides (GTP and GDP), form a link in the chain of communication between the receptor and *adenylyl cyclase*. The inactive form of a G-protein binds to GDP (Figure 8.7). The activated receptor interacts with G-proteins, triggering an exchange of GTP for GDP. The trimeric G-protein then dissociates into an α subunit and a $\beta\gamma$ dimer. The GTP-bound form of the α subunit moves from the receptor to *adenylyl cyclase*, which is thereby activated. Many molecules of active G-protein are formed by one activated receptor. [Note: The ability of a hormone or neurotransmitter to stimulate or inhibit *adenylyl cyclase* depends on the type of G-protein that is linked to the receptor. One family of G-proteins, designated G_s, is specific for stimulation of *adenylyl cyclase*; another family, designated G_i, causes inhibition of the enzyme (not shown in Figure 8.7).] The actions of the G-protein–GTP complex are short-lived because the G-protein has an inherent *GTPase* activity, resulting in the rapid hydrolysis of GTP to GDP. This causes the inactivation of G-protein.

2. **Protein kinases:** The next key link in the cAMP second-messenger system is the activation by cAMP of a family of enzymes called **cAMP-dependent protein kinases**, for example, *protein kinase A* (Figure 8.8). Cyclic AMP activates *protein kinase A* by binding to its two regulatory subunits, causing the release of active catalytic subunits. The active subunits catalyze the transfer of phosphate from ATP to specific serine or threonine residues of protein substrates. The phosphorylated proteins may act directly on the cell's ion channels, or may become activated or inhibited enzymes. *Protein kinase A* can also phosphorylate specific proteins that bind to promoter regions of DNA, causing increased expression of specific genes. [Note: Not all protein kinases respond to cAMP; there are several types of protein kinases that are not cAMP-dependent, for example, *protein kinase C* described on p. 203.]

3. **Dephosphorylation of proteins:** The phosphate groups added to proteins by protein kinases are removed by *protein phosphatases*—enzymes that hydrolytically cleave phosphate esters (see Figure 8.8). This ensures that changes in enzymic activity induced by protein phosphorylation are not permanent.

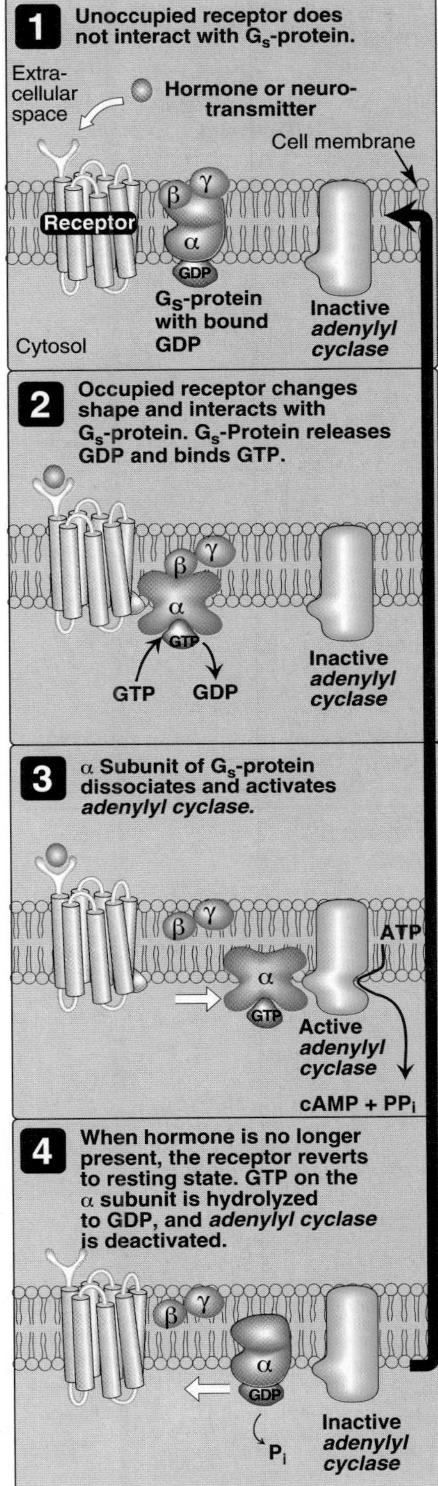

Figure 8.7
The recognition of chemical signals by certain membrane receptors triggers an increase (or, less often, a decrease) in the activity of *adenylyl cyclase*.

[2]See p. 185 in ***Lippincott's Illustrated Reviews: Microbiology*** for a discussion of cholera toxin.

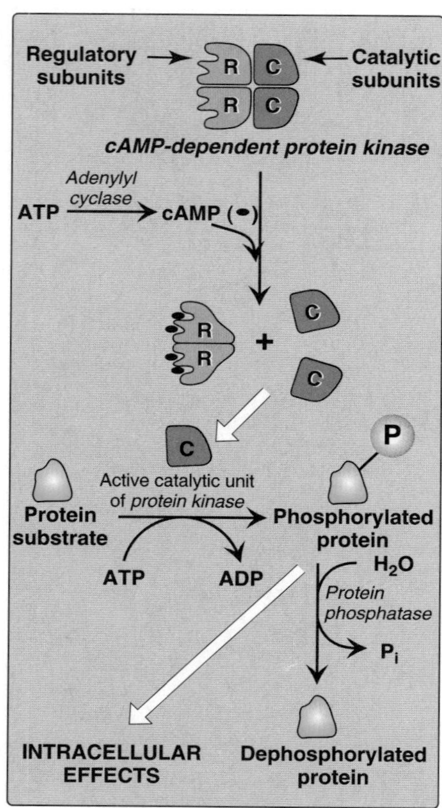

Figure 8.8
Actions of cAMP.

4. **Hydrolysis of cAMP:** cAMP is rapidly hydrolyzed to 5'-AMP by *cAMP phosphodiesterase*, one of a family of enzymes that cleave the cyclic 3',5'-phosphodiester bond. 5'-AMP is not an intracellular signalling molecule. Thus, the effects of neurotransmitter- or hormone-mediated increases of cAMP are rapidly terminated if the extracellular signal is removed. [Note: *Phosphodiesterase* is inhibited by methylxanthine derivatives, such as theophylline and caffeine.[3]]

III. OVERVIEW OF GLYCOLYSIS

The glycolytic pathway is employed by all tissues for the breakdown of glucose to provide energy (in the form of ATP) and intermediates for other metabolic pathways. Glycolysis is at the hub of carbohydrate metabolism because virtually all sugars—whether arising from the diet or from catabolic reactions in the body—can ultimately be converted to glucose (Figure 8.9A). Pyruvate is the end product of glycolysis in cells with mitochondria and an adequate supply of oxygen. This series of ten reactions is called **aerobic glycolysis** because oxygen is required to reoxidize the NADH formed during the oxidation of glyceraldehyde 3-phosphate (Figure 8.9B). Aerobic glycolysis sets the stage for the oxidative decarboxylation of pyruvate to acetyl CoA, a major fuel of the citric acid cycle. Alternatively, glucose can be converted to pyruvate, which is reduced by NADH to form lactate (Figure 8.9C). This conversion of glucose to lactate is called **anaerobic glycolysis** because it can occur without the participation of oxygen. Anaerobic glycolysis allows the continued production of ATP in tissues that lack mitochondria (for example, red blood cells) or in cells deprived of sufficient oxygen.

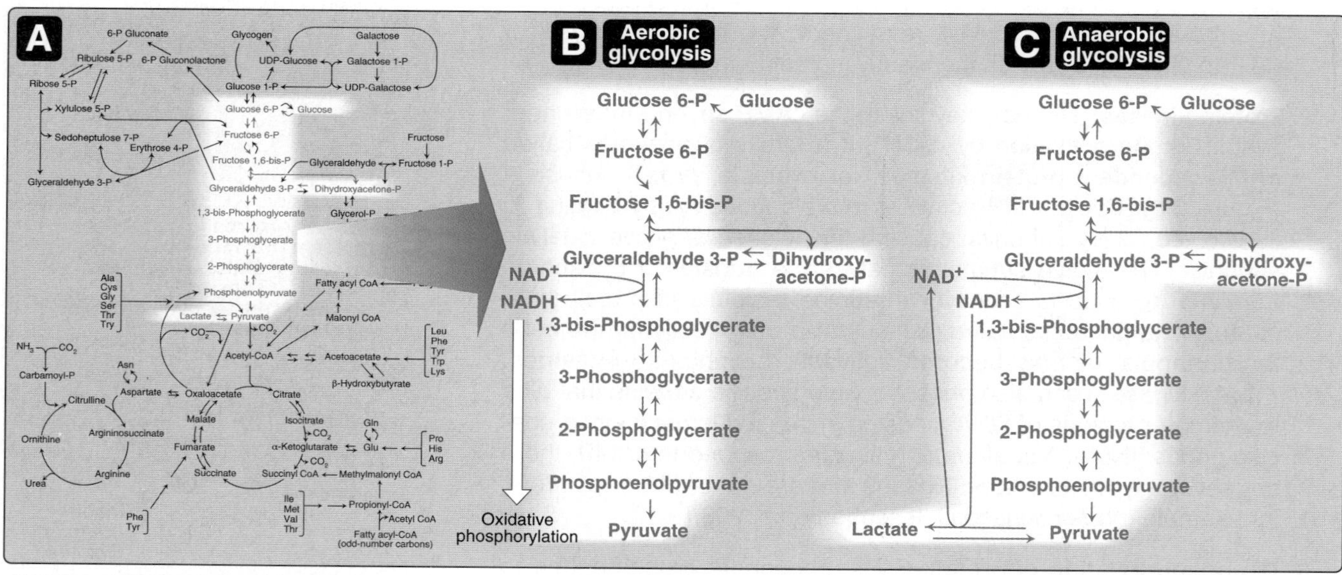

Figure 8.9
A. Glycolysis shown as one of the essential pathways of energy metabolism. B. Reactions of aerobic glycolysis. C. Reactions of anaerobic glycolysis.

[3]See Chapter 10 in ***Lippincott's Illustrated Reviews: Pharmacology*** (2nd and 3rd Eds.) for a discussion of methylxanthine derivates as drugs.

IV. TRANSPORT OF GLUCOSE INTO CELLS

Glucose cannot diffuse directly into cells, but enters by one of two transport mechanisms: a Na^+-independent, facilitated diffusion transport system or a Na^+–monosaccharide co-transporter system.

A. Na^+-independent facilitated diffusion transport

This system is mediated by a family of at least fourteen glucose transporters in cell membranes. They are designated **GLUT-1** to **GLUT-14** (**glu**cose **t**ransporter isoforms 1 to 14). These transporters exist in the membrane in two conformational states (Figure 8.10). Extracellular glucose binds to the transporter, which then alters its conformation, transporting glucose across the cell membrane.

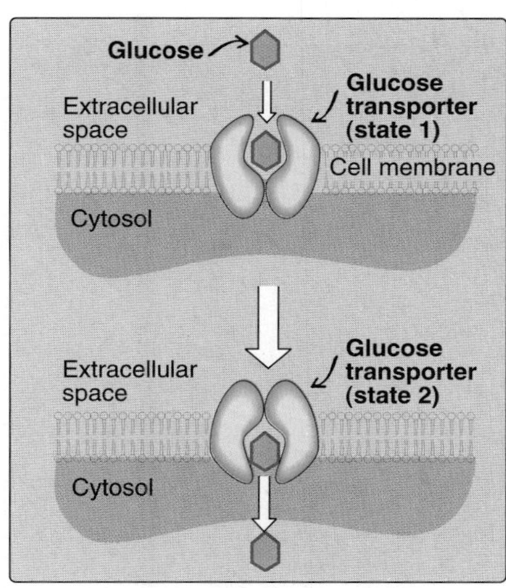

Figure 8.10
Schematic representation of the facilitated transport of glucose through a cell membrane.

1. **Tissue specificity of GLUT gene expression:** The glucose transporters display a tissue-specific pattern of expression. For example, GLUT-3 is the primary glucose transporter in neurons. GLUT-1 is abundant in erythrocytes and brain, but is low in adult muscle, whereas GLUT-4 is abundant in adipose tissue and skeletal muscle. [Note: The number of GLUT-4 transporters active in these tissues is increased by insulin. (See p. 310 for a discussion of insulin and glucose transport.)] The other GLUT isoforms also have tissue-specific distributions.

2. **Specialized functions of GLUT isoforms:** In facilitated diffusion, glucose movement follows a concentration gradient, that is, from a high glucose concentration to a lower one. For example, GLUT-1, GLUT-3, and GLUT-4 are primarily involved in glucose uptake from the blood. In contrast, GLUT-2, which is found in the liver, kidney, and β cells of the pancreas, can either transport glucose into these cells when blood glucose levels are high, or transport glucose from the cells to the blood when blood glucose levels are low (for example, during fasting). GLUT-5 is unusual in that it is the primary transporter for fructose (instead of glucose) in the small intestine and the testes. GLUT-7, which is expressed in the liver and other gluconeogenic tissues, mediates glucose flux across the endoplasmic reticular membrane.

B. Na^+-monosaccharide cotransporter system

This is an **energy-requiring** process that transports glucose "against" a concentration gradient—that is, from low glucose concentrations outside the cell to higher concentrations within the cell. This system is a carrier-mediated process in which the movement of glucose is coupled to the concentration gradient of Na^+, which is transported into the cell at the same time. This type of transport occurs in the epithelial cells of the intestine, renal tubules, and choroid plexus.

V. REACTIONS OF GLYCOLYSIS

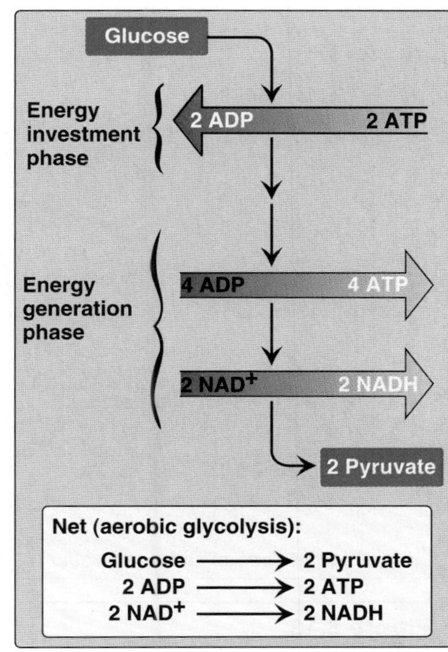

Figure 8.11
Two phases of aerobic glycolysis.

The conversion of glucose to pyruvate occurs in two stages (Figure 8.11). The first five reactions of glycolysis correspond to an energy investment phase in which the phosphorylated forms of intermediates are synthe-

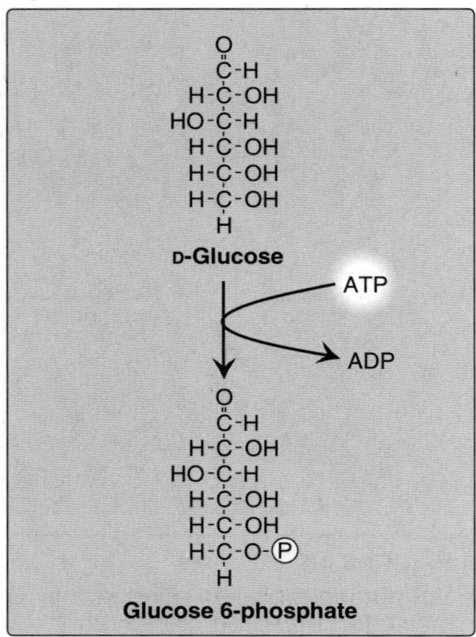

Figure 8.12
Energy investment phase: phosphorylation of glucose.

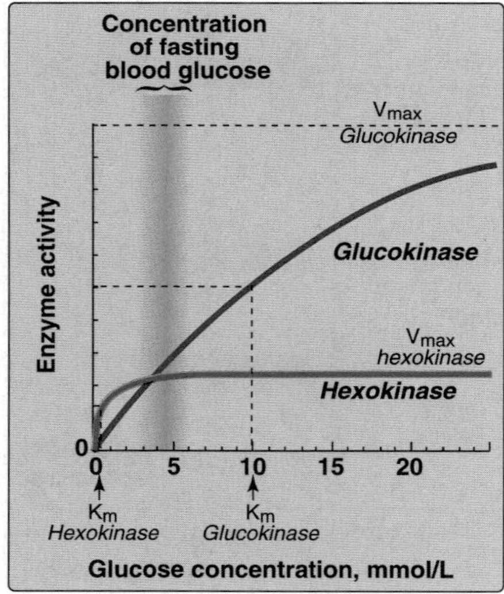

Figure 8.13
Effect of glucose concentration on the rate of phosphorylation catalyzed by *hexokinase* and *glucokinase*.

sized at the expense of ATP. The subsequent reactions of glycolysis constitute an energy generation phase in which a net of two molecules of ATP are formed by **substrate level phosphorylation** per glucose molecule metabolized. [Note: Two molecules of NADH are formed when pyruvate is produced (aerobic glycolysis), whereas NADH is reconverted to NAD^+ when lactate is the end product (anaerobic glycolysis).]

A. Phosphorylation of glucose

Phosphorylated sugar molecules do not readily penetrate cell membranes, because there are no specific transmembrane carriers for these compounds, and they are too polar to diffuse through the cell membrane. The irreversible phosphorylation of glucose (Figure 8.12), therefore, effectively traps the sugar as cytosolic glucose 6-phosphate, thus committing it to further metabolism in the cell. Mammals have several isozymes of the enzyme *hexokinase* that catalyze the phosphorylation of glucose to glucose 6-phosphate.

1. **Hexokinase:** In **most tissues**, the phosphorylation of glucose is catalyzed by *hexokinase*, one of three regulatory enzymes of glycolysis (see also *phosphofructokinase* and *pyruvate kinase*). *Hexokinase* has **broad substrate specificity** and is able to phosphorylate several hexoses in addition to glucose. *Hexokinase* is inhibited by the reaction product, glucose 6-phosphate, which accumulates when further metabolism of this hexose phosphate is reduced. *Hexokinase* has a **low K_m** (and, therefore, a **high affinity**, see p. 59) for glucose. This permits the efficient phosphorylation and subsequent metabolism of glucose even when tissue concentrations of glucose are low (Figure 8.13). *Hexokinase*, however, has a **low V_{max}** for glucose and, therefore, cannot sequester (trap) cellular phosphate in the form of phosphorylated hexoses, or phosphorylate more sugars than the cell can use.

2. **Glucokinase:** In **liver parenchymal cells** and **islet cells of the pancreas**, *glucokinase* (also called *hexokinase D*, or *type IV*) is the predominant enzyme responsible for the phosphorylation of glucose. In β cells, *glucokinase* functions as the glucose sensor, determining the threshold for insulin secretion. In the liver, the enzyme facilitates glucose phosphorylation during hyperglycemia. [Note: Despite the popular but misleading name "*glucokinase*," the sugar specificity of the enzyme is similar to that of other *hexokinase* isozymes.]

   a. **Kinetics:** *Glucokinase* differs from *hexokinase* in several important properties. For example, it has a **much higher K_m**, requiring a higher glucose concentration for half-saturation (see Figure 8.13). Thus, *glucokinase* functions only when the intracellular concentration of glucose in the hepatocyte is elevated, such as during the brief period following consumption of a carbohydrate-rich meal, when high levels of glucose are delivered to the liver via the portal vein. *Glucokinase* has a **high V_{max}**, allowing the liver to effectively remove the flood of glucose delivered by the portal blood. This prevents large amounts of glucose from entering the systemic circulation following a carbohydrate-rich meal, and thus minimizes hyperglycemia during the absorptive period. [Note: GLUT-2 insures that blood glucose equilibrates rapidly across the membrane of the hepatocyte.]

b. **Regulation by fructose 6-phosphate and glucose:** *Glucokinase* activity is not allosterically inhibited by glucose 6-phosphate as are the other *hexokinases*, but rather is **indirectly inhibited** by **fructose 6-phosphate** (which is in equilibrium with glucose 6-phosphate), and is **stimulated indirectly** by **glucose** via the following mechanism. A **glucokinase regulatory protein** exists in the **nucleus** of hepatocytes. In the presence of fructose 6-phosphate, *glucokinase* is translocated into the nucleus and binds tightly to the regulatory protein, thus rendering the enzyme inactive (Figure 8.14). When glucose levels in the blood (and also in the hepatocyte, as a result of GLUT-2) increase, the glucose causes the release of *glucokinase* from the regulatory protein, and the enzyme enters the cytosol where it phosphorylates glucose to glucose 6-phosphate. As free glucose levels fall, fructose 6-phosphate causes *glucokinase* to translocate back into the nucleus and bind to the regulatory protein, thus inhibiting the enzyme's activity.

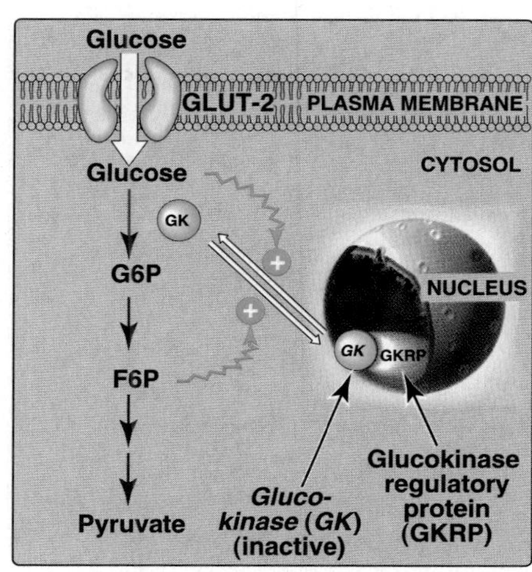

Figure 8.14
Regulation of *glucokinase* activity by glucokinase regulatory protein.

c. **Regulation by insulin:** *Glucokinase* activity in hepatocytes is also increased by insulin. As blood glucose levels rise following a meal, the β cells of the pancreas are stimulated to release insulin into the portal circulation. [Note: Approximately one half of the newly secreted insulin is extracted by the liver during the first pass through that organ. Therefore, the liver is exposed to twice as much insulin as is found in the systemic circulation.] Insulin also promotes transcription of the *glucokinase* gene, resulting in an increase in liver enzyme protein and, therefore, of total *glucokinase* activity. [Note: The absence of insulin in patients with diabetes causes a deficiency in hepatic *glucokinase*. This contributes to an inability of the patient to efficiently decrease blood glucose levels.]

B. Isomerization of glucose 6-phosphate

The isomerization of glucose 6-phosphate to fructose 6-phosphate is catalyzed by *phosphoglucose isomerase* (Figure 8.15). The reaction is readily reversible and is not a rate-limiting or regulated step.

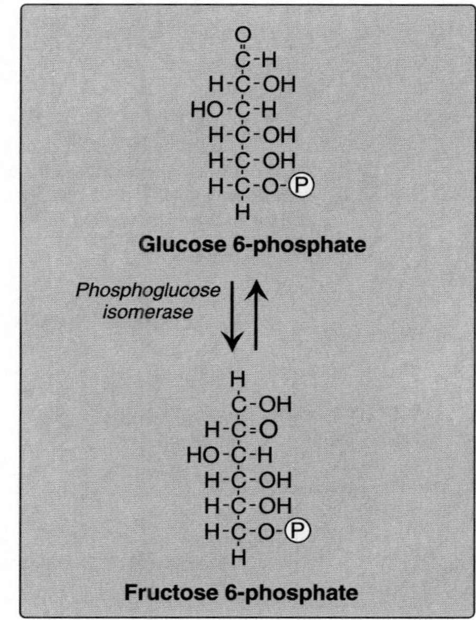

Figure 8.15
Isomerization of glucose 6-phosphate to fructose 6-phosphate.

C. Phosphorylation of fructose 6-phosphate

The irreversible phosphorylation reaction catalyzed by *phosphofructokinase-1* (*PFK-1*) is the most important control point and the rate-limiting step of glycolysis (Figure 8.16). *PFK-1* is controlled by the available concentrations of the substrates ATP and fructose 6-phosphate, and by regulatory substances described below.

1. **Regulation by energy levels within the cell:** *PFK-1* is **inhibited** allosterically by elevated levels of **ATP**, which act as an "energy-rich" signal indicating an abundance of high-energy compounds. Elevated levels of **citrate**, an intermediate in the tricarboxylic acid cycle (see p. 109), also inhibit *PFK-1*. Conversely, *PFK-1* is **activated** allosterically by high concentrations of **AMP**, which signal that the cell's energy stores are depleted.

2. **Regulation by fructose 2,6-bisphosphate:** Fructose 2,6-bisphosphate is the most potent activator of *PFK-1* (see Figure 8.16). This compound also acts as an inhibitor of *fructose 1,6-bisphosphatase*

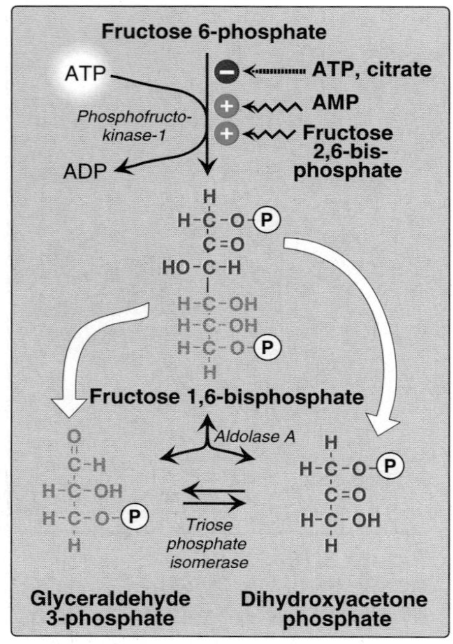

Figure 8.16
Energy investment phase (continued): Conversion of fructose 6-phosphate to triose phosphates.

(see p. 119 for a discussion of the regulation of gluconeogenesis). The reciprocal actions of fructose 2,6-bisphosphate on glycolysis and gluconeogenesis ensure that both pathways are not fully active at the same time. [Note: This would result in a "futile cycle" in which glucose would be converted to pyruvate followed by resynthesis of glucose from pyruvate.] Fructose 2,6-bisphosphate is formed by *phosphofructokinase-2* (*PFK-2*), an enzyme different than *phosphofructokinase-1*. Fructose 2,6-bisphosphate is converted back to fructose 6-phosphate by *fructose bisphosphatase-2* (Figure 8.17). [Note: The *kinase* and *phosphatase* activities are different domains of one bifunctional polypeptide molecule.]

a. **During the well-fed state:** Decreased levels of glucagon and elevated levels of insulin, such as occur following a carbohydrate-rich meal, cause an increase in fructose 2,6-bisphosphate and thus in the rate of glycolysis in the liver (see Figure 8.17). Fructose 2,6-bisphosphate, therefore, acts as an intracellular signal, indicating that glucose is abundant.

b. **During starvation:** Elevated levels of glucagon and low levels of insulin, such as occur during fasting (see p. 327), decrease the intracellular concentration of hepatic fructose 2,6-bisphosphate. This results in a decrease in the overall rate of glycolysis and an increase in gluconeogenesis.

D. Cleavage of fructose 1,6-bisphosphate

Aldolase A cleaves fructose 1,6-bisphosphate to dihydroxyacetone phosphate and glyceraldehyde 3-phosphate (see Figure 8.16). The

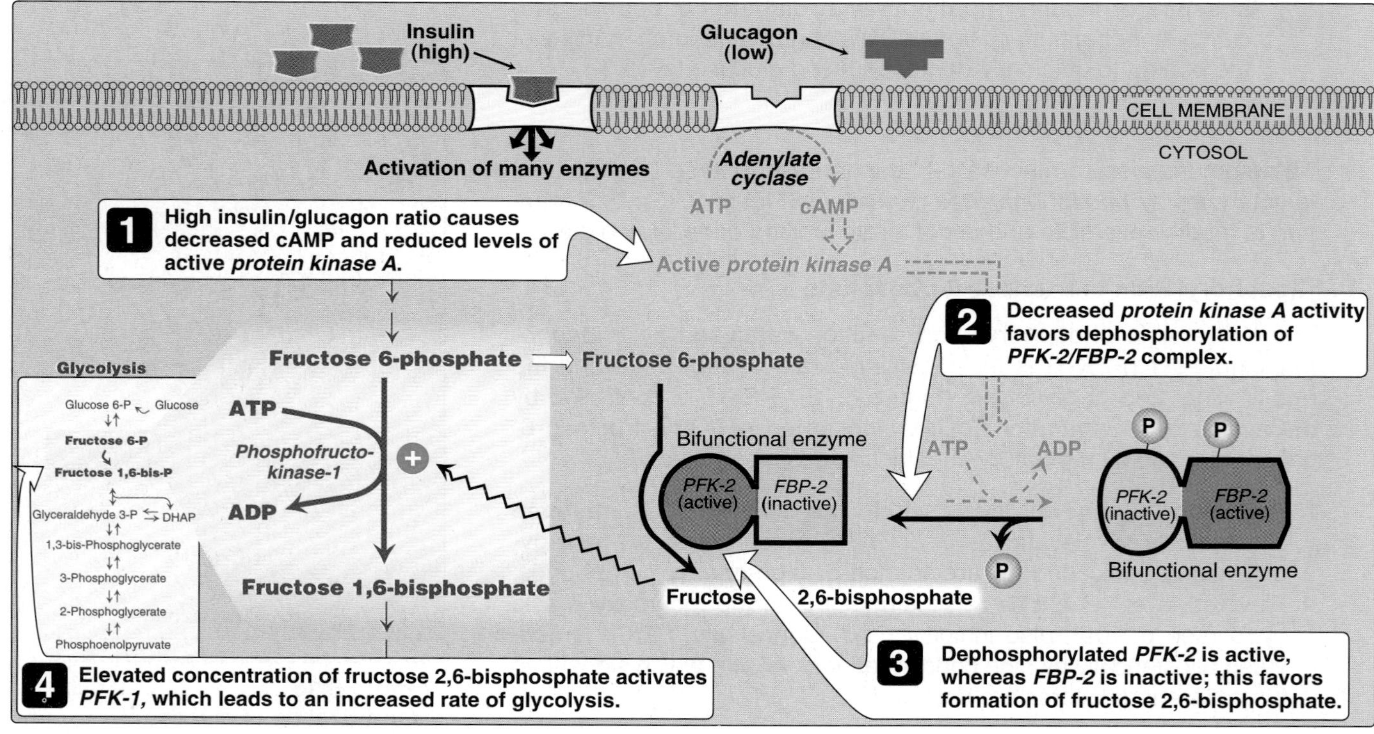

Figure 8.17
Effect of elevated insulin concentration on the intracellular concentration of fructose 2,6-bisphosphate in liver. *PFK-2 = phosphofructokinase-2; FBP-2 = Fructose bisphospate phosphatase-2.*

reaction is reversible and not regulated. [Note: *Aldolase B* in the liver and kidney also cleaves fructose 1,6-bisphosphate, and functions in the metabolism of dietary fructose (see p. 136).]

E. Isomerization of dihydroxyacetone phosphate

Triose phosphate isomerase interconverts dihydroxyacetone phosphate and glyceraldehyde 3-phosphate (see Figure 8.16). Dihydroxyacetone phosphate must be isomerized to glyceraldehyde 3-phosphate for further metabolism by the glycolytic pathway. This isomerization results in the net production of two molecules of glyceraldehyde 3-phosphate from the cleavage products of fructose 1,6-bisphosphate.

F. Oxidation of glyceraldehyde 3-phosphate

The conversion of glyceraldehyde 3-phosphate to 1,3-bisphosphoglycerate by *glyceraldehyde 3-phosphate dehydrogenase* is the first oxidation-reduction reaction of glycolysis (Figure 8.18). [Note: Because there is only a limited amount of NAD^+ in the cell, the NADH formed by this reaction must be reoxidized to NAD^+ for glycolysis to continue. Two major mechanisms for oxidizing NADH are: 1) the NADH-linked conversion of pyruvate to lactate (see p. 101), and 2) oxidation of NADH via the respiratory chain (see p. 75).]

1. **Synthesis of 1,3-bisphosphoglycerate:** The oxidation of the aldehyde group of glyceraldehyde 3-phosphate to a carboxyl group is coupled to the attachment of P_i to the carboxyl group. The high-energy phosphate group at carbon 1 of 1,3-bisphosphoglycerate (1,3-BPG) conserves much of the free energy produced by the oxidation of glyceraldehyde 3-phosphate. The energy of this high-energy phosphate drives the synthesis of ATP in the next reaction of glycolysis.

2. **Mechanism of arsenic poisoning:** The toxicity of arsenic is explained primarily by the inhibition of enzymes such as *pyruvate dehydrogenase*, which require lipoic acid as a cofactor (see p. 107). However, pentevalent arsenic (arsenate) also prevents net ATP and NADH production by glycolysis, without inhibiting the pathway itself. The poison does so by competing with inorganic phosphate as a substrate for *glyceraldehyde 3-phosphate dehydrogenase*, forming a complex that spontaneously hydrolyzes to form 3-phosphoglycerate (see Figure 8.18). By bypassing the synthesis and dephosphorylation of 1,3-BPG, the cell is deprived of energy usually obtained from the glycolytic pathway.

3. **Synthesis of 2,3-bisphosphoglycerate in red blood cells:** Some of the 1,3-bisphosphoglycerate is converted to 2,3-bisphosphoglycerate (2,3-BPG) by the action of *bisphosphoglycerate mutase* (see Figure 8.18). 2,3-BPG, which is found in only trace amounts in most cells, is present at high concentration in red blood cells (see p. 31). 2,3-BPG is hydrolyzed by a *phosphatase* to 3-phosphoglycerate, which is also an intermediate in glycolysis (see Figure 8.18). In the red blood cell, glycolysis is modified by inclusion of these "shunt" reactions.

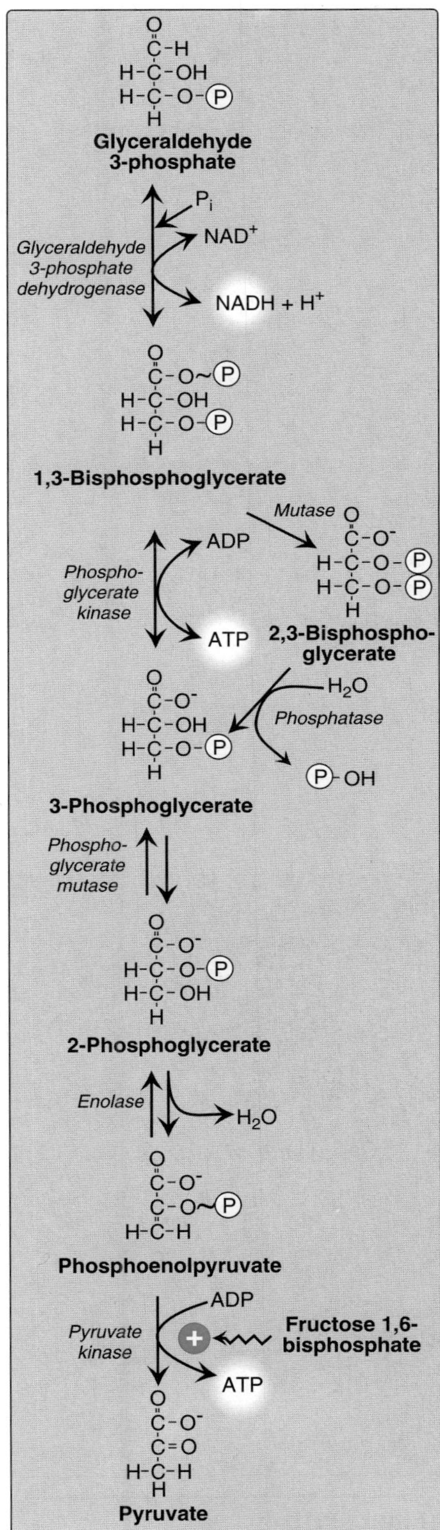

Figure 8.18
Energy-generating phase: conversion of glyceraldehyde 3-phosphate to pyruvate.

G. Synthesis of 3-phosphoglycerate producing ATP

When 1,3-BPG is converted to 3-phosphoglycerate, the high-energy phosphate group of 1,3-BPG is used to synthesize ATP from ADP (see Figure 8.18). This reaction is catalyzed by *phosphoglycerate kinase*, which, unlike most other kinases, is physiologically reversible. Because two molecules of 1,3-BPG are formed from each glucose molecule, this kinase reaction replaces the two ATP molecules consumed by the earlier formation of glucose 6-phosphate and fructose 1,6-bisphosphate. [Note: This is an example of substrate-level phosphorylation, in which the production of a high-energy phosphate is coupled directly to the oxidation of a substrate, instead of resulting from oxidative phosphorylation via the electron transport chain.]

H. Shift of the phosphate group from carbon 3 to carbon 2

The shift of the phosphate group from carbon 3 to carbon 2 of phosphoglycerate by *phosphoglycerate mutase* is freely reversible (see Figure 8.18).

I. Dehydration of 2-phosphoglycerate

The dehydration of 2-phosphoglycerate by *enolase* redistributes the energy within the 2-phosphoglycerate molecule, resulting in the formation of phosphoenolpyruvate (PEP), which contains a high-energy enol phosphate (see Figure 8.18). The reaction is reversible despite the high-energy nature of the product.

J. Formation of pyruvate producing ATP

The conversion of PEP to pyruvate is catalyzed by *pyruvate kinase*, the third irreversible reaction of glycolysis. The equilibrium of the *pyruvate kinase* reaction favors the formation of ATP (see Figure 8.18). [Note: This is another example of **substrate-level phosphorylation**.]

1. **Feed-forward regulation:** In liver, *pyruvate kinase* is **activated** by **fructose 1,6-bisphosphate**, the product of the *phosphofructokinase* reaction. This feed-forward (instead of the more usual feedback) regulation has the effect of linking the two kinase activities: increased *phosphofructokinase* activity results in elevated levels of fructose 1,6-bisphosphate, which activates *pyruvate kinase*.

2. **Covalent modulation of pyruvate kinase:** Phosphorylation by a *cAMP-dependent protein kinase* leads to inactivation of *pyruvate kinase* in the liver (Figure 8.19). When blood glucose levels are low, elevated glucagon increases the intracellular level of cAMP, which causes the phosphorylation and inactivation of *pyruvate kinase*. Therefore, phosphoenolpyruvate is unable to continue in glycolysis, but instead enters the gluconeogenesis pathway. This, in part, explains the observed inhibition of hepatic glycolysis and stimulation of gluconeogenesis by glucagon. Dephosphorylation of *pyruvate kinase* by a *phosphoprotein phosphatase* results in reactivation of the enzyme.

3. **Pyruvate kinase deficiency:** The normal, mature erythrocyte lacks mitochondria and is, therefore, completely dependent on glycoly-

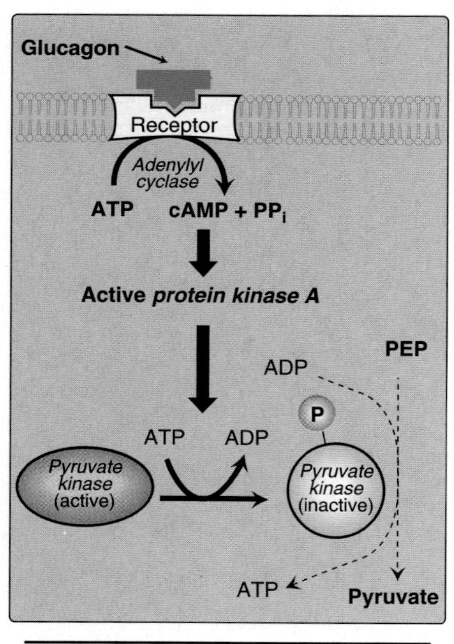

Figure 8.19
Covalent modification of *pyruvate kinase* results in inactivation of enzyme.

sis for production of ATP. This high-energy compound is required to meet the metabolic needs of the red blood cell, and also to fuel the pumps necessary for the maintenance of the bi-concave, flexible shape of the cell, which allows it to squeeze through narrow capillaries. The anemia observed in glycolytic enzyme deficiencies is a consequence of the reduced rate of glycolysis, leading to decreased ATP production. The resulting alterations in the red blood cell membrane lead to changes in the shape of the cell and, ultimately, to phagocytosis by the cells of the reticuloendothelial system, particularly macrophages of the spleen. The premature death and lysis of the red blood cell result in hemolytic anemia. Among patients exhibiting genetic defects of glycolytic enzymes, about 95 percent show a deficiency in *pyruvate kinase*, and four percent exhibit *phosphoglucose isomerase deficiency.* **Pyruvate kinase (PK) deficiency** is the second most common cause (after glucose-6-phosphatase dehydrogenase deficiency) of enzymatic-related hemolytic anemia. PK deficiency is restricted to the erythrocytes, and produces mild to severe **chronic hemolytic anemia** (erythrocyte destruction), with the severe form requiring regular cell transfusions. The severity of the disease depends both on the degree of enzyme deficiency (generally 5 to 25 percent of normal levels), and on the extent to which the individual's red blood cells compensate by synthesizing increased levels of 2,3-BPG (see p. 31). Almost all individuals with *PK* deficiency have a mutant enzyme that shows abnormal properties—most often altered kinetics (Figure 8.20).

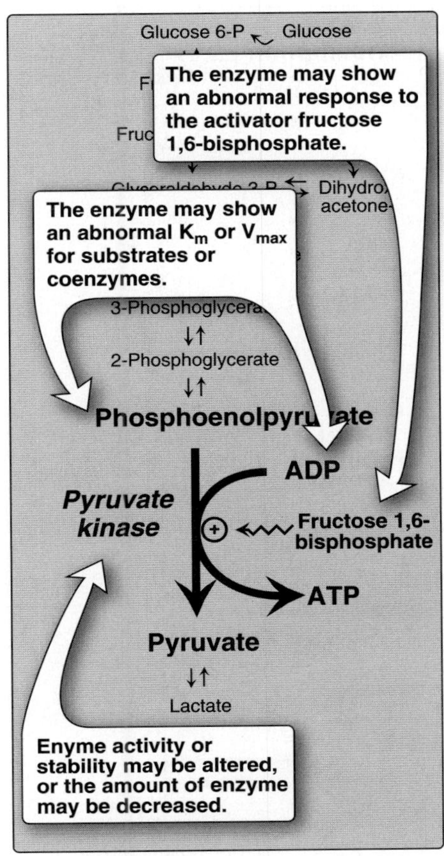

Figure 8.20
Alterations observed with various mutant forms of *pyruvate kinase*.

K. Reduction of pyruvate to lactate

Lactate, formed by the action of *lactate dehydrogenase*, is the final product of anaerobic glycolysis in eukaryotic cells (Figure 8.21). The formation of lactate is the major fate for pyruvate in red blood cells, lens and cornea of the eye, kidney medulla, testes, and leukocytes.

1. **Lactate formation in muscle:** In exercising skeletal muscle, NADH production (by *glyceraldehyde 3-phosphate dehydrogenase* and by the three NAD^+-linked dehydrogenases of the citric acid cycle) exceeds the oxidative capacity of the respiratory chain. This results in an elevated $NADH/NAD^+$ ratio, favoring reduction of pyruvate to lactate. Therefore, during intense exercise, lactate accumulates in muscle, causing a drop in the intracellular pH, potentially resulting in cramps. Much of this lactate eventually diffuses into the bloodstream, and can be used by the liver to make glucose (see p. 116).

2. **Lactate consumption:** The direction of the *lactate dehydrogenase* reaction depends on the relative intracellular concentrations of pyruvate and lactate, and on the ratio of $NADH/NAD^+$ in the cell. For example, in liver and heart, the ratio of $NADH/NAD^+$ is lower than in exercising muscle. These tissues oxidize lactate (obtained from the blood) to pyruvate. In the liver, pyruvate is either converted to glucose by gluconeogenesis or oxidized in the TCA cycle. Heart muscle exclusively oxidizes lactate to CO_2 and H_2O via the citric acid cycle.

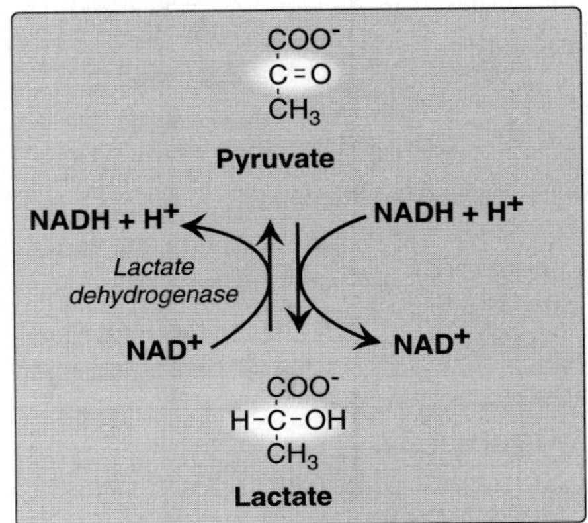

Figure 8.21
Interconversion of pyruvate and lactate.

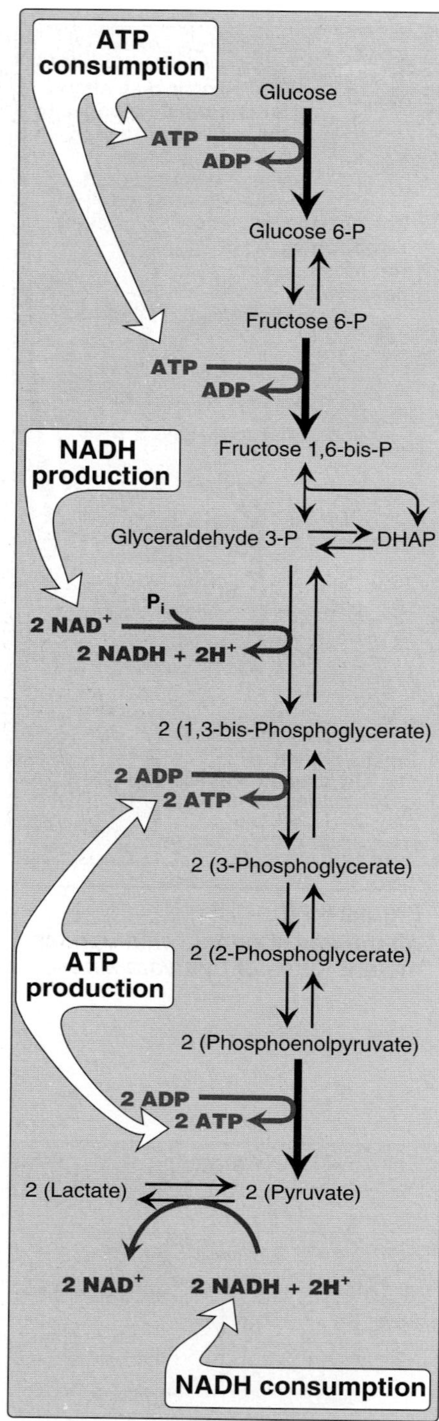

Figure 8.22
Summary of anaerobic glycolysis. Reactions involving the production or consumption of ATP or NADH are indicated. The irreversible reactions of glycolysis are shown with thick arrows. DHAP = dihydroxyacetone phosphate.

3. **Lactic acidosis:** Elevated concentrations of lactate in the plasma, termed lactic acidosis, occur when there is a collapse of the circulatory system, such as in myocardial infarction, pulmonary embolism, and uncontrolled hemorrhage, or when an individual is in shock. The failure to bring adequate amounts of oxygen to the tissues results in impaired oxidative phosphorylation and decreased ATP synthesis. To survive, the cells use anaerobic glycolysis as a backup system for generating ATP, producing lactic acid as the end-product. [Note: Production of even meager amounts of ATP may be life-saving during the period required to reestablish adequate blood flow to the tissues.] The excess oxygen required to recover from a period when the availability of oxygen has been inadequate is termed the **oxygen debt**. The oxygen debt is often related to patient morbidity or mortality. In many clinical situations, measuring the blood levels of lactic acid provides for the rapid, early detection of oxygen debt in patients. For example, blood lactic acid levels can be used to measure the presence and severity of shock, and to monitor the patient's recovery.

L. Energy yield from glycolysis

Despite the production of some ATP during glycolysis, the end products, pyruvate or lactate, still contain most of the energy originally contained in glucose. The TCA cycle is required to release that energy completely (see p. 107).

1. **Anaerobic glycolysis:** Two molecules of ATP are generated for each molecule of glucose converted to two molecules of lactate (Figure 8.22). There is no net production or consumption of NADH. Anaerobic glycolysis, although releasing only a small fraction of the energy contained in the glucose molecule, is a valuable source of energy under several conditions, including 1) when the oxygen supply is limited, as in muscle during intensive exercise; and 2) for tissues with few or no mitochondria, such as the medulla of the kidney, mature erythrocytes, leukocytes, and cells of the lens, cornea, and testes.

2. **Aerobic glycolysis:** The direct formation and consumption of ATP is the same as in anaerobic glycolysis—that is, a net gain of two ATP per molecule of glucose. Two molecules of NADH are also produced per molecule of glucose. Ongoing aerobic glycolysis requires the oxidation of most of this NADH by the electron transport chain, producing approximately three ATP for each NADH molecule entering the chain (see p. 77).

VI. HORMONAL REGULATION OF GLYCOLYSIS

The regulation of glycolysis by allosteric activation or inhibition, or the phosphorylation/dephosphorylation of rate-limiting enzymes, is short-term—that is, they influence glucose consumption over periods of minutes or hours. Superimposed on these moment-to-moment effects are slower, often more profound, hormonal influences on the amount of enzyme protein synthesized. These effects can result in ten-fold to twenty-fold increases in enzyme activity that typically occur over hours to days. Although the current focus is on glycolysis, reciprocal changes occur in the rate-limiting enzymes of gluconeogenesis, which are described in Chapter 10 (see p. 115). Regular consumption of meals rich in carbohydrate or administration of insulin initiates an increase in the amount of *glucokinase*, *phosphofructokinase*, and *pyruvate kinase* in liver (Figure 8.23). These changes reflect an increase in gene transcription, resulting in increased enzyme synthesis. High activity of these three enzymes favors the conversion of glucose to pyruvate, a characteristic of the well-fed state (see p. 319). Conversely, gene transcription and synthesis of *glucokinase*, *phosphofructokinase,* and *pyruvate kinase* are decreased when plasma glucagon is high and insulin is low, for example, as seen in fasting or diabetes.

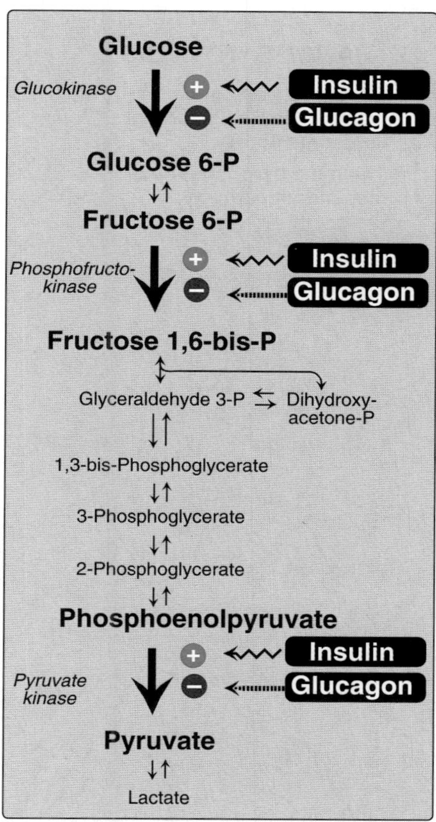

Figure 8.23
Effect of insulin and glucagon on the synthesis of key enzymes of glycolysis in liver.

VII. ALTERNATE FATES OF PYRUVATE

A. Oxidative decarboxylation of pyruvate

Oxidative decarboxylation of pyruvate by *pyruvate dehydrogenase complex* is an important pathway in tissues with a high oxidative capacity, such as cardiac muscle (Figure 8.24). *Pyruvate dehydrogenase* irreversibly converts pyruvate, the end product of glycolysis, into acetyl CoA, a major fuel for the tricarboxylic acid cycle (see p. 107) and the building block for fatty acid synthesis (see p. 181).

B. Carboxylation of pyruvate to oxaloacetate

Carboxylation of pyruvate to oxaloacetate (OAA) by *pyruvate carboxylase* is a biotin-dependent reaction (see Figure 8.24). This reaction is important because it replenishes the citric acid cycle intermediates, and provides substrate for gluconeogenesis (see p. 116).

C. Reduction of pyruvate to ethanol (microorganisms)

The conversion of pyruvate to ethanol occurs by the two reactions summarized in Figure 8.24. The decarboxylation of pyruvate by *pyruvate decarboxylase* occurs in yeast and certain microorganisms, but not in humans. The enzyme requires thiamine pyrophosphate as a coenzyme, and catalyzes a reaction similar to that described for *pyruvate dehydrogenase* (see p. 108).

VIII. CHAPTER SUMMARY

Most pathways can be classified as either **catabolic** (**degrade** complex molecules to a few simple products) or **anabolic** (**synthesize** complex end products from simple precursors). **Catabolic reactions** also **capture chemical energy** in the form of **ATP** from the degradation of energy-rich molecules. **Anabolic reactions require energy**, which is generally provided by the breakdown of ATP. The rate of a metabolic pathway can respond to **regulatory signals**, for example **allosteric activators** or **inhibitors**, that arise from **within the cell**. Signaling **between cells** provides for the integration of metabolism. The most important route of this communication is **chemical signaling** between cells, for example, by **hormones** or **neurotransmitters**. **Second messenger molecules** convey the intent of a chemical signal (hormone or neurotransmitter) to appropri-

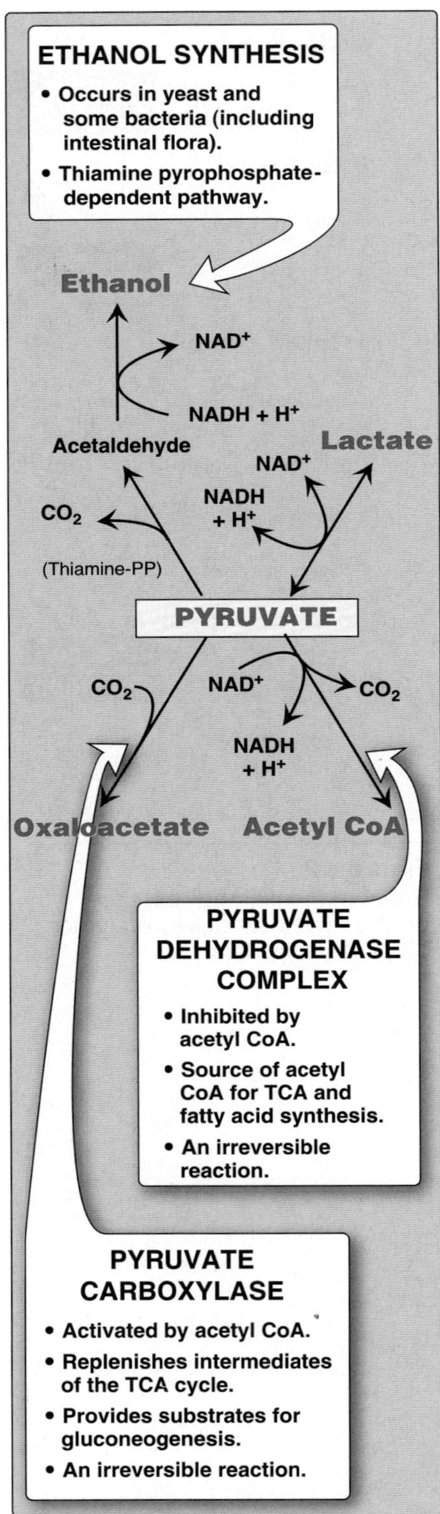

Figure 8.24
Summary of the metabolic fates of pyruvate.

ate intracellular responders. **Adenylyl cyclase** is a membrane-bound enzyme that synthesizes **cyclic AMP** (**cAMP**) in response to chemical signals, such as the hormones **glucagon** and **epinephrine**. Following binding of a hormone to its **cell-surface receptor**, a GTP-dependent regulatory protein (**G-protein**) is activated that, in turn, activates adenylyl cyclase. The cAMP activates a **protein kinase**, which phosphorylates a cadre of enzymes, causing their activation or deactivation. Phosphorylation is reversed by **protein phosphatases. Aerobic glycolysis**, in which pyruvate is the end product, occurs in cells with mitochondria and an adequate supply of oxygen. **Anaerobic glycolysis**, in which lactic acid is the end product, occurs in cells that lack mitochondria, or in cells deprived of sufficient oxygen. Glucose is transported across membranes by one of at least fourteen **glucose transporter isoforms** (**GLUTs**). **GLUT-1** is abundant in **erythrocytes** and **brain**, **GLUT-4** (which is **insulin-dependent**) is found in **muscle** and **adipose tissue**, and **GLUT-2** is found in **liver** and the **β cells** of the pancreas. The conversion of glucose to pyruvate (**glycolysis**) occurs in two stages: an **energy investment phase** in which phosphorylated intermediates are synthesized at the expense of ATP, and an **energy generation phase**, in which ATP is produced. In the energy investment phase, glucose is phosphorylated by **hexokinase** (found in **most tissues**) or **glucokinase** (a hexokinase found in **liver cells** and the **β cells** of the pancreas). **Hexokinase** has a **high affinity** (**low K_m**) and a **small V_{max}** for glucose, and is **inhibited** by **glucose 6-phosphate**. **Glucokinase** has a **large K_m** and a **large V_{max}** for glucose. It is indirectly **inhibited** by **fructose 6-phosphate** and **activated** by **glucose**, and the **transcription** of the glucokinase gene is **enhanced by insulin**. Glucose 6-phosphate is isomerized to fructose 6-phosphate, which is phosphorylated to **fructose 1,6-bisphosphate** by **phosphofructokinase**. This enzyme is **allosterically inhibited** by **ATP** and **citrate**, and **activated** by **AMP**. **Fructose 2,6-bisphosphate**, whose synthesis is **activated** by **insulin**, is the most potent allosteric activator of this enzyme. A total of **two ATP are used** during this phase of glycolysis. Fructose 1,6-bisphosphate is cleaved to form two trioses that are further metabolized by the glycolytic pathway, forming pyruvate. During these reactions, **four ATP** and **two NADH are produced** from ADP and NAD^+. The final step in pyruvate synthesis from phosphoenolpyruvate is catalyzed by **pyruvate kinase**. This enzyme is **allosterically activated** by **fructose 1,6-bisphosphate**, and **hormonally activated** by **insulin** and **inhibited** by **glucagon** via the **cAMP pathway. Pyruvate kinase deficiency** accounts for 95 percent of all inherited defects in glycolytic enzymes. It is restricted to **erythrocytes**, and causes mild to severe **chronic hemolytic anemia**. In **anaerobic glycolysis**, NADH is reoxidized to NAD^+ by the **conversion of pyruvate to lactic acid**. This occurs in cells, such as **erythrocytes**, that have few or no mitochondria, and in tissues, such as **exercising muscle**, where production of NADH exceeds the oxidative capacity of the respiratory chain. Elevated concentrations of lactate in the plasma (**lactic acidosis**) occur when there is a **collapse of the circulatory system**, or when an individual is in **shock**. Pyruvate can be: 1) **oxidatively decarboxylated** by **pyruvate dehydrogenase**, producing **acetyl CoA**; 2) **carboxylated** to **oxaloacetate** (a TCA cycle intermediate) by **pyruvate carboxylase**; or 3) **reduced** by microorganisms to **ethanol** by **pyruvate decarboxylase**.

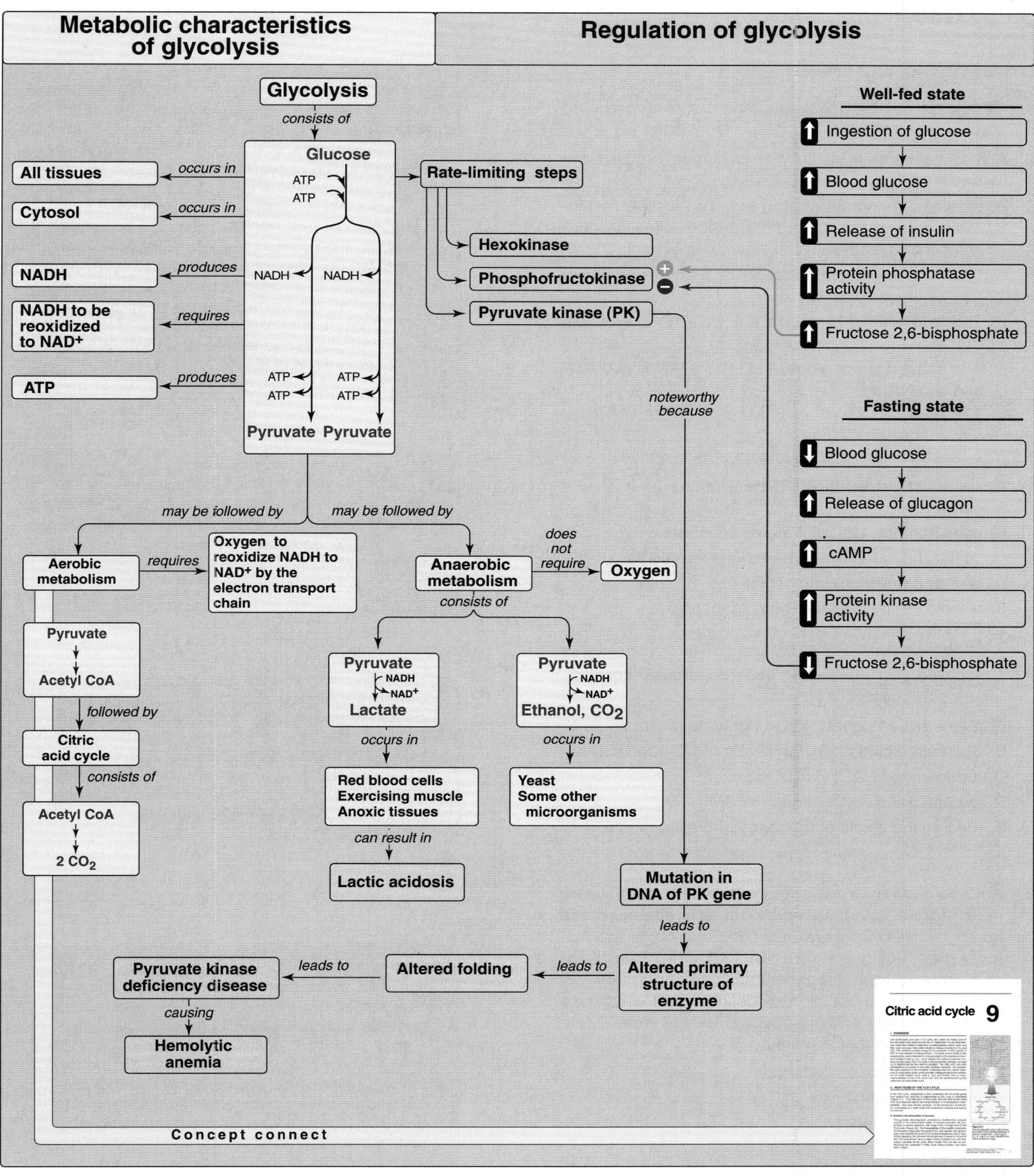

Figure 8.25
Key concept map for glycolysis.

Study Questions

Choose the ONE best answer

8.1 Which one of the following statements concerning glycolysis is correct?

A. The conversion of glucose to lactate requires the presence of oxygen.
B. Hexokinase is important in hepatic glucose metabolism only in the absorptive period following consumption of a carbohydrate-containing meal.
C. Fructose 2,6-bisphosphate is a potent inhibitor of phosphofructokinase.
D. The rate-limiting reactions are also the irreversible reactions.
E. The conversion of glucose to lactate yields two ATP and two NADH.

Correct answer = D. Hexokinase, phosphofructokinase and pyruvate kinase are all irreversible and are the regulated steps in glycolysis. The conversion of glucose to lactate (anaerobic glycolysis) is a process that does not involve a net oxidation or reduction and, thus, oxygen is not required. Glucokinase (not hexokinase) is important in hepatic glucose metabolism only in the absorptive period following consumption of a carbohydrate-containing meal. Fructose 2,6-bisphosphate is a potent activator (not inhibitor) of phosphofructokinase. The conversion of glucose to lactate yields two ATP but no net production of NADH.

8.2 The reaction catalyzed by phosphofructokinase:

A. is activated by high concentrations of ATP and citrate.
B. uses fructose 1-phosphate as substrate.
C. is the regulated reaction of the glycolytic pathway.
D. is near equilibrium in most tissues.
E. is inhibited by fructose 2,6-bisphosphate.

Correct answer = C. Phosphofructokinase is the pace-setting enzyme of glycolysis. It is inhibited by ATP and citrate, uses fructose 6-phosphate as substrate, and catalyzes a reaction that is far from equilibrium. The reaction is activated by fructose 2,6-bisphosphate.

8.3 Compared with the resting state, vigorously contracting muscle shows:

A. an increased conversion of pyruvate to lactate.
B. decreased oxidation of pyruvate to CO_2 and water.
C. a decreased NADH/NAD^+ ratio.
D. a decreased concentration of AMP.
E. decreased levels of fructose 2,6-bisphosphate.

Correct answer = A. Vigorously contracting muscle shows an increased formation of lactate and an increased rate of pyruvate oxidation compared with resting skeletal muscle. The levels of AMP and NADH increase, whereas change in the concentration of fructose 2,6-bisphosphate is not a key regulatory factor in muscle.

8.4 A 43-year-old man presented with symptoms of weakness, fatigue, shortness of breath, and dizziness. His hemoglobin levels were between 5 to 7 g/dl (normal for a male being greater than 13.5 g/dl). Red blood cells isolated from the patient showed abnormally low level of lactate production. A deficiency of which one of the following enzymes would be the most likely cause of this patient's anemia.

A. Phosphoglucose isomerase
B. Phosphofructokinase
C. Pyruvate kinase
D. Hexokinase
E. Lactate dehydrogenase

Correct answer = C. Decreased lactate production in the erythrocyte indicates a defect in glycolysis. Among patients exhibiting genetic defects of glycolytic enzymes, about 95 percent show a deficiency in pyruvate kinase, and four percent exhibit phosphoglucose isomerase deficiency. Pyruvate kinase deficiency is the second most common cause (after glucose 6-phosphate dehydrogenase deficiency) of enzyme deficiency-related hemolytic anemia.

Tricarboxylic Acid Cycle

9

I. OVERVIEW

The tricarboxylic acid cycle (**TCA cycle**, also called the **Krebs cycle** or the **citric acid cycle**) plays several roles in metabolism. It is the final pathway where the oxidative metabolism of carbohydrates, amino acids, and fatty acids converge, their carbon skeletons being converted to CO_2 and H_2O. This oxidation provides energy for the production of the majority of ATP in most animals, including humans. The cycle occurs totally in the mitochondria and is, therefore, in close proximity to the reactions of electron transport (see p. 73), which oxidize the reduced coenzymes produced by the cycle. The TCA cycle is thus an aerobic pathway, because O_2 is required as the final electron acceptor. The citric acid cycle also participates in a number of important synthetic reactions. For example, the cycle functions in the formation of glucose from the carbon skeletons of some amino acids, and it provides building blocks for the synthesis of some amino acids (see p. 265) and heme (see p. 276). Intermediates of the TCA cycle can also be synthesized by the catabolism of some amino acids. Therefore, this cycle should not be viewed as a closed circle, but instead as a traffic circle with compounds entering and leaving as required.

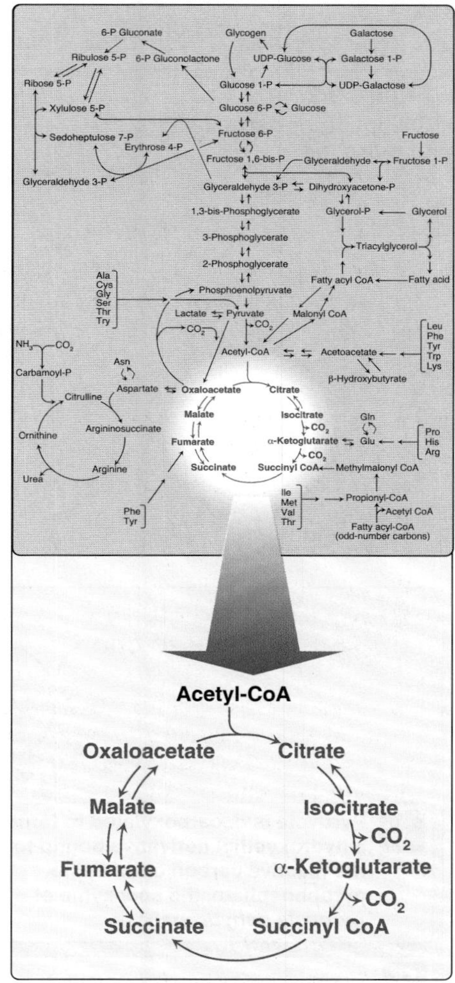

Figure 9.1
The tricarboxylic acid cycle shown as a part of the central pathways of energy metabolism. (See Figure 8.2, p. 90 for a more detailed view of the metabolic map.)

II. REACTIONS OF THE TCA CYCLE

In the TCA cycle, oxaloacetate is first condensed with an acetyl group from acetyl CoA, and then is regenerated as the cycle is completed (Figure 9.1). Thus, the entry of one acetyl CoA into one round of the TCA cycle does not lead to the net production or consumption of intermediates.

A. Oxidative decarboxylation of pyruvate

Pyruvate, the end-product of aerobic glycolysis, must be transported into the mitochondrion before it can enter the TCA cycle. This is accomplished by a specific pyruvate transporter that helps pyruvate cross the inner mitochondrial membrane. Once in the matrix, pyruvate is converted to acetyl CoA by the *pyruvate dehydrogenase complex*, which is a multienzyme complex (Figure 9.2). [Note: The **irreversibility** of the reaction precludes the formation of pyruvate from acetyl CoA, and explains why glucose cannot be formed from acetyl CoA via gluconeogenesis.] Strictly speaking, the *pyruvate dehydrogenase complex* is not part of the TCA cycle proper, but is a major source of acetyl CoA—the two-carbon substrate for the cycle.

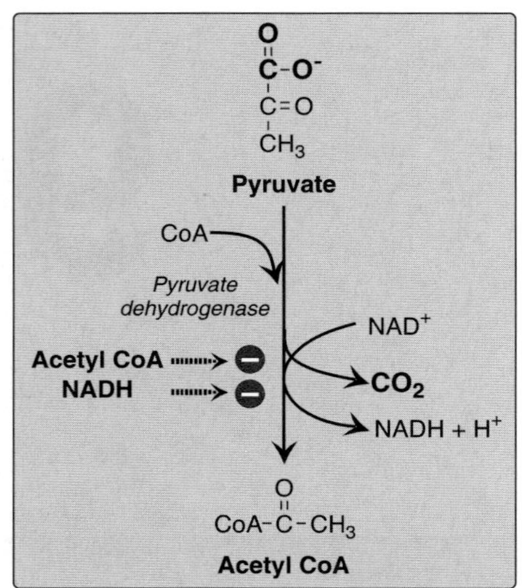

Figure 9.2
Oxidative decarboxylation of pyruvate.

1. **Component enzymes:** The *pyruvate dehydrogenase complex* is a multimolecular aggregate of three enzymes, *pyruvate dehydrogenase* (*E_1*, also called a *decarboxylase*), *dihydrolipoyl transacetylase* (*E_2*), and *dihydrolipoyl dehydrogenase* (*E_3*). Each is present in multiple copies, and each catalyzes a part of the overall reaction (Figure 9.3). Their physical association links the reactions in proper sequence without the release of intermediates. In addition to the enzymes participating in the conversion of pyruvate to acetyl CoA, the complex also contains two tightly bound regulatory enzymes, *protein kinase* and *phosphoprotein phosphatase*.

2. **Coenzymes:** The *pyruvate dehydrogenase complex* contains five coenzymes that act as carriers or oxidants for the intermediates of the reactions shown in Figure 9.3. *E_1* requires **thiamine pyrophosphate**, *E_2* requires **lipoic acid** and **coenzyme A**, and *E_3* requires **FAD** and **NAD^+**. [Note: Deficiencies of thiamine or niacin can cause serious central nervous system problems. This is because brain cells are unable to produce sufficient ATP (via the TCA cycle) for proper function if *pyruvate dehydrogenase* is inactive.]

3. **Regulation of the pyruvate dehydrogenase complex:** The two regulatory enzymes that are part of the complex alternately activate and inactivate *E_1*: the *cyclic AMP-independent protein kinase* phosphorylates and, thereby, inhibits *E_1*, whereas *phosphoprotein phosphatase* activates *E_1* (Figure 9.4). The *kinase* is allosterically activated by ATP, acetyl CoA, and NADH. Therefore, in the presence of these high-energy signals, the *pyruvate dehydrogenase complex* is turned off. Acetyl CoA and NADH also allosterically inhibit the dephosphorylated (active) form of *E_1*. *Protein kinase* is allosterically inactivated by NAD^+ and coenzyme A—low-energy signals that thus turn *pyruvate dehydrogenase* on (see Figure 9.2). Pyruvate is also a potent inhibitor of *protein kinase*. Therefore, if pyruvate concentrations are elevated, *E_1* will be maximally active. Calcium is a strong activator of *protein*

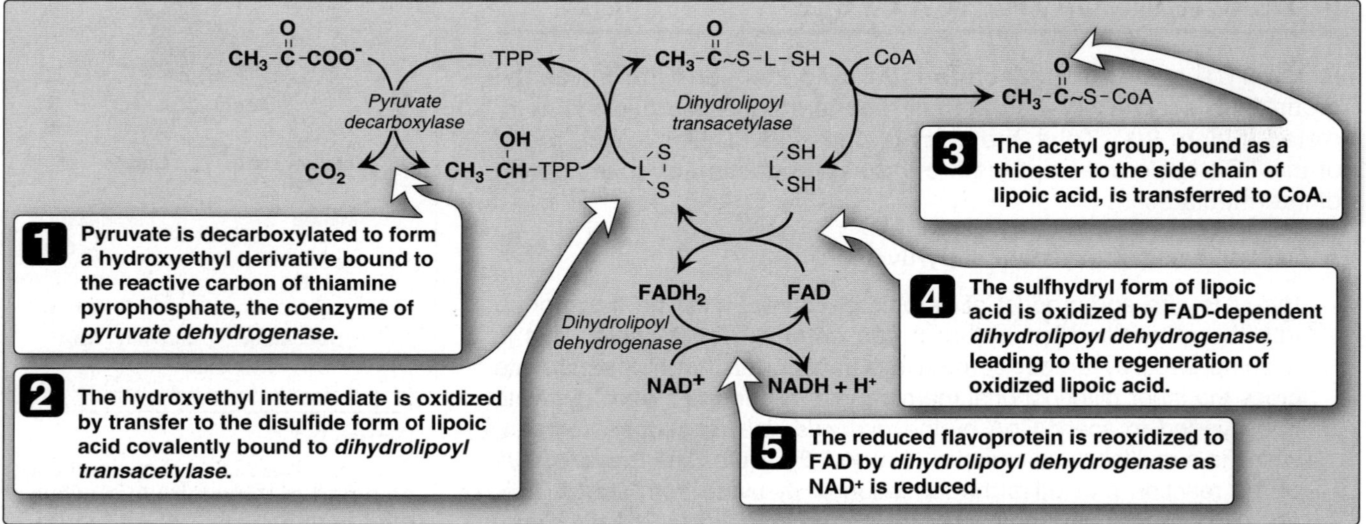

Figure 9.3
Mechanism of action of the *pyruvate dehydrogenase complex.* TPP = thiamine pyrophosphate; L = lipoic acid.

phosphatase, stimulating E_1 activity. [Note: This is particularly important in skeletal muscle, where release of Ca^{++} during contraction stimulates the *pyruvate dehydrogenase complex*, and thereby energy production.]

4. **Pyruvate dehydrogenase deficiency:** A deficiency in the *pyruvate dehydrogenase complex* is the most common biochemical cause of **congenital lactic acidosis**. This enzyme deficiency results in an inability to convert pyruvate to acetyl CoA, causing pyruvate to be shunted to lactic acid via *lactate dehydrogenase* (see p. 101). This causes particular problems for the brain, which relies on the TCA cycle for most of its energy, and is particularly sensitive to acidosis. The most severe form of this deficiency causes overwhelming lactic acidosis with neonatal death. A second form produces moderate lactic acidosis, but causes profound psychomotor retardation, with damage to the cerebral cortex, basal ganglia, and brain stem, leading to death in infancy. A third form of the deficiency causes episodic ataxia (an inability to coordinate voluntary muscles) that is induced by a carbohydrate-rich meal. The E_1 defect is X-linked, but because of the importance of the enzyme in the brain, it affects both males and females. Therefore, the defect is classified as **X-linked dominant**. There is no proven treatment for *pyruvate dehydrogenase complex* deficiency, although a **ketogenic diet** (one low in carbohydrate and enriched in fats) has been shown in some cases to be of benefit. Such a diet provides an alternate fuel supply in the form of ketone bodies (see p. 193) that can be used by most tissues including the brain, but not the liver (see p. 194).

5. **Mechanism of arsenic poisoning:** As previously described (see p. 99), arsenic can interfere with glycolysis at the *glyceraldehyde 3-phosphate* step, thereby decreasing ATP production. "Arsenic poisoning" is, however, due primarily to inhibition of enzymes that require **lipoic acid** as a cofactor, including *pyruvate dehydrogenase*, *α-ketoglutarate dehydrogenase* (see below), *and branched-chain α-keto acid dehydrogenase* (see p. 264). Arsenite (the trivalent form of arsenic) forms a stable complex with the thiol (–SH) groups of lipoic acid, making that compound unavailable to serve as a coenzyme. When it binds to lipoic acid in the *pyruvate dehydrogenase complex*, pyruvate (and consequently lactate) accumulate. Like *pyruvate dehydrogenase complex* deficiency, this particularly affects the brain, causing neurologic disturbances and death.

B. Synthesis of citrate from acetyl CoA and oxaloacetate

The condensation of acetyl CoA and oxaloacetate to form citrate is catalyzed by *citrate synthase* (Figure 9.5). This aldol condensation has an equilibrium far in the direction of citrate synthesis. *Citrate synthase* is allosterically activated by Ca^{2+} and ADP, and inhibited by ATP, NADH, succinyl CoA, and fatty acyl CoA derivatives (see Figure 9.9). However, the primary mode of regulation is also determined by the availability of its substrates, acetyl CoA and oxaloacetate. [Note: Citrate, in addition to being an intermediate in the TCA cycle, provides a source of acetyl CoA for the cytosolic synthesis of

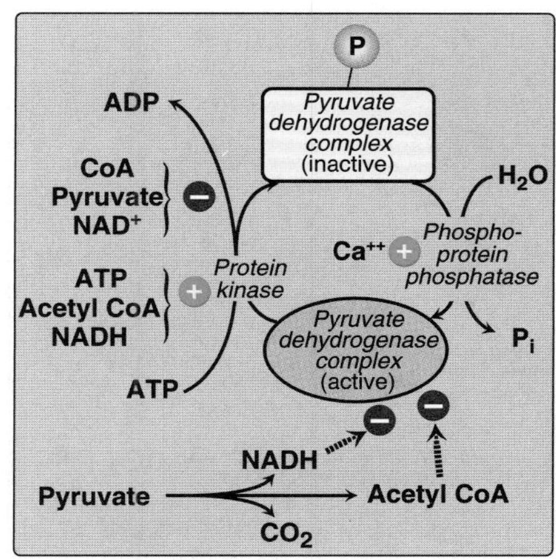

Figure 9.4
Regulation of *pyruvate dehydrogenase complex.*

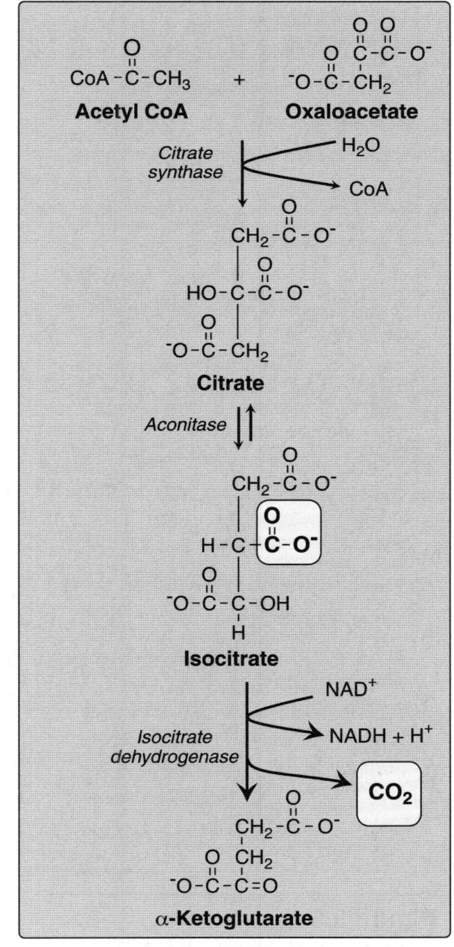

Figure 9.5
Formation of α-ketoglutarate from acetyl CoA and oxaloacetate.

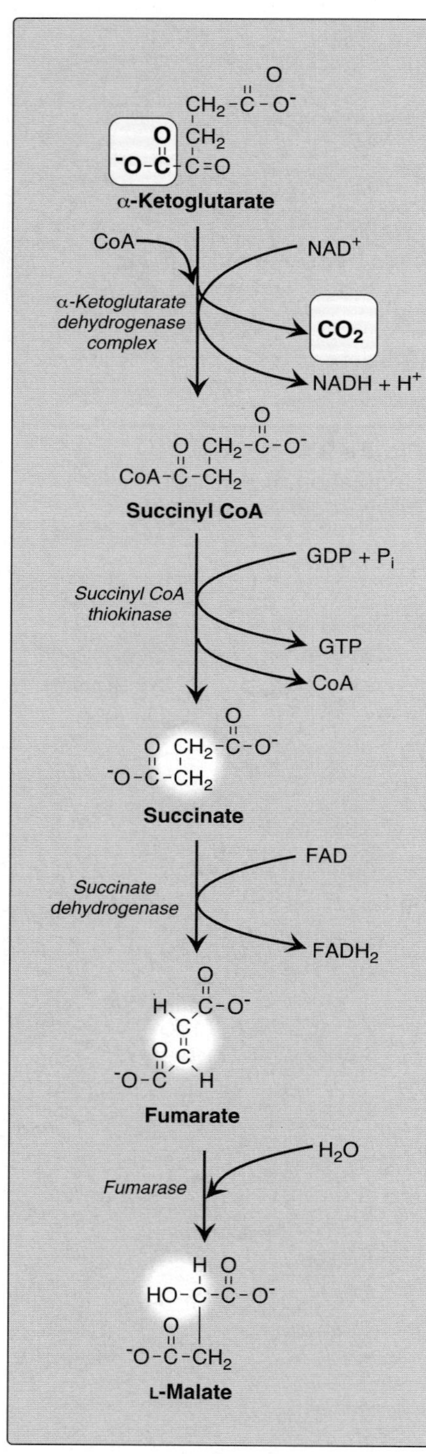

Figure 9.6
Formation of malate from α-ketoglutarate.

fatty acids (see p. 181). Citrate also inhibits *phosphofructokinase*, the rate-setting enzyme of glycolysis (see p. 97), and activates *acetyl CoA carboxylase* (the rate-limiting enzyme of fatty acid synthesis; see p. 181).]

C. Isomerization of citrate

Citrate is isomerized to isocitrate by *aconitase* (see Figure 9.5). [Note: *Aconitase* is inhibited by **fluoroacetate**, a compound that is used as a rat poison. Fluoroacetate is converted to fluoroacetyl CoA, which condenses with oxaloacetate to form fluorocitrate—a potent inhibitor of *aconitase*—resulting in citrate accumulation.]

D. Oxidation and decarboxylation of isocitrate

Isocitrate dehydrogenase catalyzes the irreversible oxidative decarboxylation of isocitrate, yielding the first of three NADH molecules produced by the cycle, and the first release of CO_2 (see Figure 9.5). This is one of the rate-limiting steps of the TCA cycle. The enzyme is allosterically activated by ADP (a low-energy signal) and Ca^{++}, and is inhibited by ATP and NADH, whose levels are elevated when the cell has abundant energy stores.

E. Oxidative decarboxylation of α-ketoglutarate

The conversion of α-ketoglutarate to succinyl CoA is catalyzed by the *α-ketoglutarate dehydrogenase complex*, which consists of three enzymatic activities (Figure 9.6). The mechanism of this oxidative decarboxylation is very similar to that used for the conversion of pyruvate to acetyl CoA. The reaction releases the second CO_2 and produces the second NADH of the cycle. The coenzymes required are **thiamine pyrophosphate**, **lipoic acid**, **FAD**, **NAD⁺**, and **coenzyme A**. Each functions as part of the catalytic mechanism in a way analogous to that described for *pyruvate dehydrogenase complex* (see p. 108). The equilibrium of the reaction is far in the direction of succinyl CoA—a high-energy thioester similar to acetyl CoA. *α-Ketoglutarate dehydrogenase complex* is inhibited by ATP, GTP, NADH, and succinyl CoA, and activated by Ca^{++}. However, it is not regulated by phosphorylation/dephosphorylation reactions as described for *pyruvate dehydrogenase complex*. [Note: α-Ketoglutarate is also produced by the oxidative deamination or transamination of the amino acid, glutamate.]

F. Cleavage of succinyl CoA

Succinate thiokinase (also called *succinyl CoA synthetase*) cleaves the high-energy thioester bond of succinyl CoA (see Figure 9.6). This reaction is coupled to phosphorylation of GDP to GTP. GTP and ATP are energetically interconvertible by the *nucleoside diphosphate kinase* reaction:

$$\text{GTP} + \text{ADP} \rightleftarrows \text{GDP} + \text{ATP}$$

The generation of GTP by *succinate thiokinase* is another example of **substrate-level phosphorylation** (see p. 100). [Note: Succinyl CoA is also produced from propionyl CoA derived from the metabolism of fatty acids with an odd number of carbon atoms (see p. 191), and from metabolism of several amino acids (see p. 264).

G. Oxidation of succinate

Succinate is oxidized to fumarate by *succinate dehydrogenase*, producing the reduced coenzyme $FADH_2$ (see Figure 9.6). [Note: FAD, rather than NAD^+, is the electron acceptor because the reducing power of succinate is not sufficient to reduce NAD^+.] *Succinate dehydrogenase* is inhibited by oxaloacetate.

H. Hydration of fumarate

Fumarate is hydrated to malate in a freely reversible reaction catalyzed by *fumarase* (also called *fumarate hydratase*, see Figure 9.6). [Note: Fumarate is also produced by the urea cycle (see p. 251), in purine synthesis (see p. 293), and during catabolism of the amino acids, phenylalanine and tyrosine (see p. 261).]

I. Oxidation of malate

Malate is oxidized to oxaloacetate by *malate dehydrogenase* (Figure 9.7). This reaction produces the third and final NADH of the cycle. [Note: Oxaloacetate is also produced by the transamination of the amino acid, aspartic acid.]

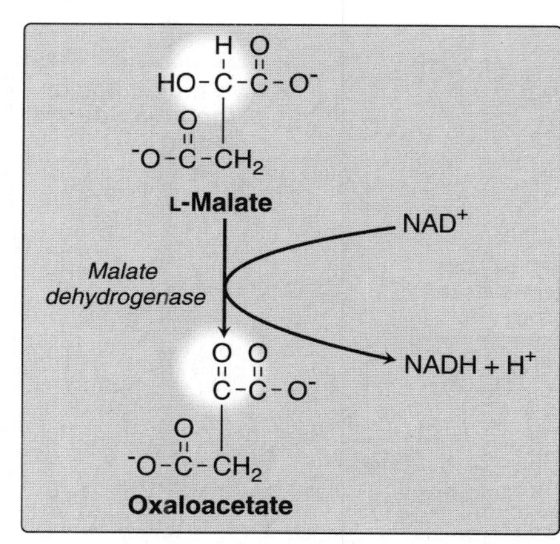

Figure 9.7
Formation of oxaloacetate from malate.

III. ENERGY PRODUCED BY THE TCA CYCLE

Two carbon atoms enter the cycle as acetyl CoA and leave as CO_2. The cycle does not involve net consumption or production of oxaloacetate or of any other intermediate. Four pairs of electrons are transferred during one turn of the cycle: three pairs of electrons reducing NAD^+ to NADH and one pair reducing FAD to $FADH_2$. Oxidation of one NADH by the electron transport chain (see p. 73) leads to formation of approximately three ATP, whereas oxidation of $FADH_2$ yields approximately two ATP. The total yield of ATP from the oxidation of one acetyl CoA is shown in Figure 9.8. Figure 9.9 summarizes the reactions of the TCA cycle.

Energy-producing reaction	Number of ATP produced
3 NADH ⟶ 3 NAD^+	9
$FADH_2$ ⟶ FAD	2
GDP + P_i ⟶ GTP	1
	12 ATP/acetyl CoA oxidized

Figure 9.8
Number of ATP molecules produced from the oxidation of one molecule ofacetyl CoA (using both substrate-level and oxidative phosphorylation).

IV. REGULATION OF THE TCA CYCLE

A. Regulation by activation and inhibition of enzyme activities

In contrast to glycolysis, which is regulated primarily by *phosphofructokinase,* the TCA cycle is controlled by the regulation of several enzyme activities (see Figure 9.9). The most important of these regulated enzymes are *citrate synthase*, *isocitrate dehydrogenase*, and *α-ketoglutarate dehydrogenase complex*.

B. Regulation by the availability of ADP

1. **Effects of elevated ADP:** Energy consumption as a result of muscular contraction, biosynthetic reactions, or other processes results in the hydrolysis of ATP to ADP and P_i. The resulting increase in the concentration of ADP accelerates the rate of reactions that use ADP to generate ATP, most important of which is oxidative phosphorylation (see p. 77). Production of ATP increases until it matches the rate of ATP consumption by energy-requiring reactions.

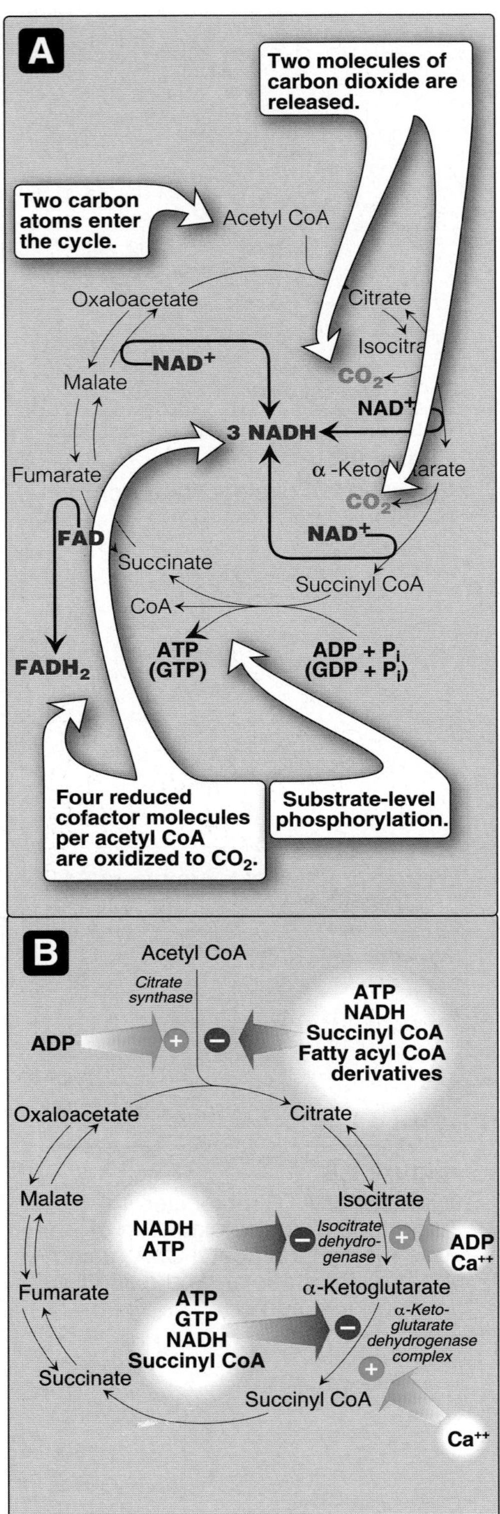

Figure 9.9
A. Production of reduced coenzymes, ATP, and CO_2 in the citric acid cycle.
B. Inhibitors and activators of the cycle.

2. **Effects of low ADP:** If ADP (or P_i) is present in limiting concentration, the formation of ATP by oxidative phosphorylation decreases as a result of the lack of phosphate acceptor (ADP) or inorganic phosphate (P_i). The rate of oxidative phosphorylation is proportional to [ADP][Pi]/[ATP]; this is known as **respiratory control** of energy production. The oxidation of NADH and $FADH_2$ by the electron transport chain also stops if ADP is limiting. This is because the processes of oxidation and phosphorylation are tightly coupled and occur simultaneously (see p. 78). As NADH and $FADH_2$ accumulate, their oxidized forms become depleted, causing the oxidation of acetyl CoA by the TCA cycle to be inhibited as a result of a lack of oxidized coenzymes.

V. CHAPTER SUMMARY

Pyruvate is **oxidatively decarboxylated** by **pyruvate dehydrogenase complex**, producing **acetyl CoA**, which is the major fuel for the tricarboxylic acid cycle (TCA cycle). This enzyme complex requires five coenzymes: **thiamine pyrophosphate**, **lipoic acid**, **FAD**, **NAD^+**, and **coenzyme A** (which contains the vitamin pantothenic acid). The reaction is activated by NAD^+, coenzyme A, and pyruvate, and inhibited by ATP, acetyl CoA, NADH, and calcium. **Pyruvate dehydrogenase deficiency** is the most common biochemical cause of **congenital lactic acidosis**. Because the deficiency deprives the brain of acetyl CoA, the central nervous system is particularly affected, with profound **psychomotor retardation** and **death** occurring in most patients. The deficiency is **X-linked dominant**. **Arsenic poisoning** causes inactivation of pyruvate dehydrogenase by binding to lipoic acid. **Citrate** is synthesized from **oxaloacetate** (OAA) and **acetyl CoA** by **citrate synthase**. This enzyme is allosterically activated by ADP, and inhibited by ATP, NADH, succinyl CoA, and fatty acyl CoA derivatives. Citrate is isomerized to **isocitrate** by **aconitase**. **Isocitrate** is oxidized and decarboxylated by **isocitrate dehydrogenase** to **α-ketoglutarate**, producing **CO_2** and **NADH**. The enzyme is inhibited by ATP and NADH, and activated by ADP and Ca^{++}. **α-Ketoglutarate** is oxidatively decarboxylated to **succinyl CoA** by the **α-ketoglutarate dehydrogenase complex**, producing **CO_2** and **NADH**. The enzyme is very similar to pyruvate dehydrogenase and uses the same coenzymes. α-Ketoglutarate dehydrogenase complex is activated by calcium and inhibited by ATP, GTP, NADH, and succinyl CoA. **Succinyl CoA** is cleaved by **succinate thiokinase** (also called **succinyl CoA synthetase**), producing **succinate** and **GTP**. This is an example of **substrate-level phosphorylation**. **Succinate** is oxidized to **fumarate** by **succinate dehydrogenase**, producing **$FADH_2$**. This enzyme is inhibited by oxaloacetate. **Fumarate** is hydrated to **malate** by **fumarase** (**fumarate hydratase**), and **malate** is oxidized to **oxaloacetate** by **malate dehydrogenase**, producing **NADH**. **Three NADH**, **one $FADH_2$**, and **one GTP** (whose terminal phosphate can be transferred to ADP by nucleoside diphosphate kinase, producing ATP) are produced by one round of the TCA cycle. Oxidation of the NADHs and $FADH_2$ by the electron transport chain yields approximately eleven ATPs, making **twelve** the total number of ATPs produced.

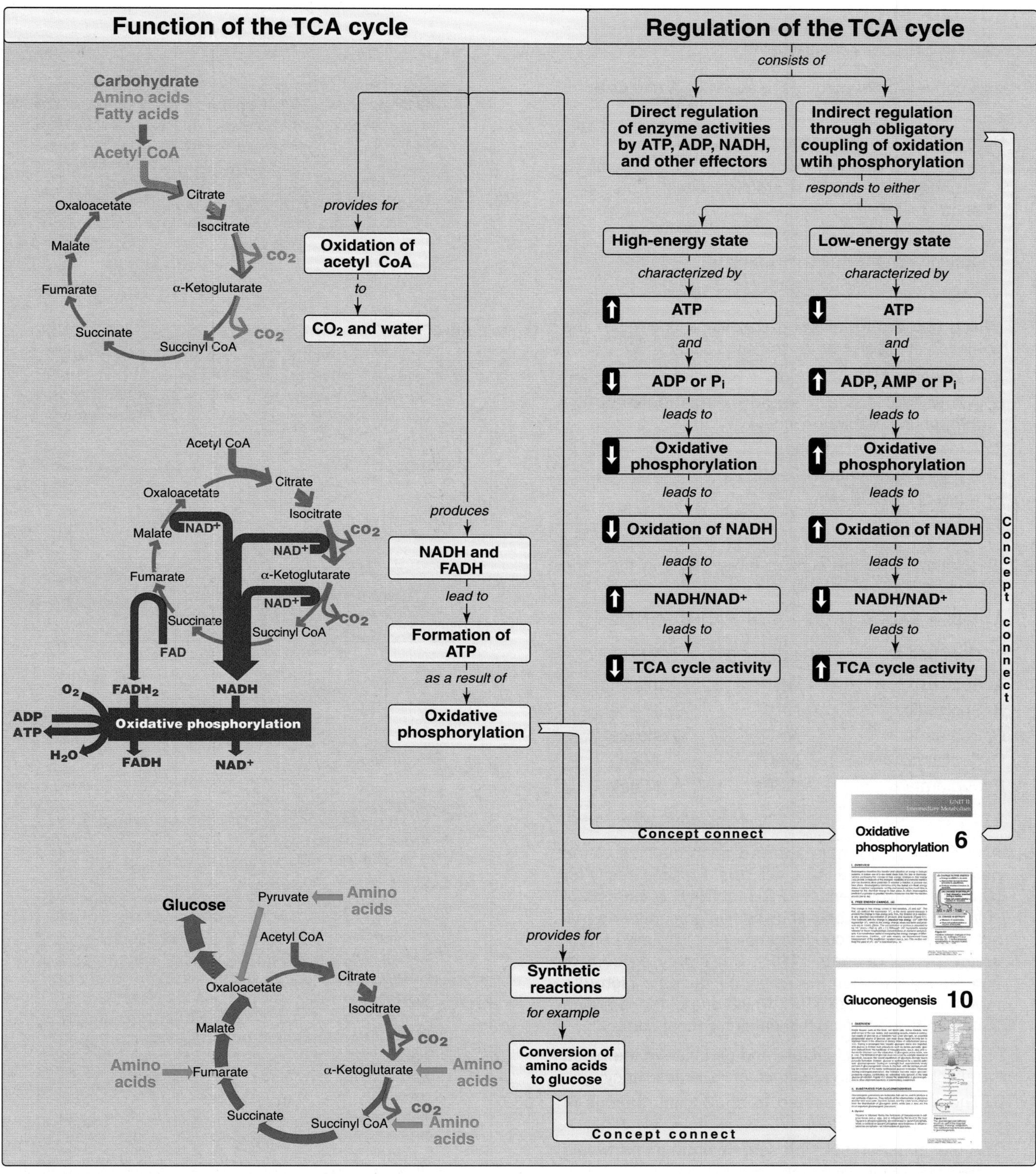

Figure 9.10
Key concept map for tricarboxylic acid cycle.

Study Questions

Choose the ONE correct answer

9.1 The conversion of pyruvate to acetyl CoA and CO_2:

A. is reversible.
B. involves the participation of lipoic acid.
C. is activated when pyruvate dehydrogenase complex is phosphorylated by a protein kinase in the presence of ATP.
D. occurs in the cytosol.
E. depends on the coenzyme biotin.

Correct answer = B. Lipoic acid is an intermediate acceptor of the acetyl group formed in the reaction. Pyruvate dehydrogenase complex catalyzes an irreversible reaction that is inhibited when the enzyme is phosphorylated. The enzyme is located in the mitochondrial matrix.

9.2 Which one of the following conditions decreases the oxidation of acetyl CoA by the citric acid cycle?

A. A low ATP/ADP ratio
B. A low NADH due to rapid oxidization to NAD+ through the respiratory chain
C. A low NAD+/NADH ratio
D. A high concentration of AMP
E. A low GTP/GDP ratio

Correct answer = C. A low NAD^+/NADH ratio limits the rates of the NAD^+-requiring dehydrogenases. A low ATP/ADP or GTP/GDP ratio stimulates the cycle. AMP does not directly affect the cycle.

9.3 The following is the sum of three steps in the citric acid cycle:

A + B + FAD + H_2O → C + $FADH_2$ + NADH

	Reactant A	Reactant B	Reactant C
A.	Succinyl CoA	GDP	Succinate
B.	Succinate	NAD^+	Oxaloacetate
C.	Fumarate	NAD^+	Oxaloacetate
D.	Succinate	NAD^+	Malate
E.	Fumarate	GTP	Malate

Correct answer = B. Succinate + NAD^+ + FAD → oxaloacetate + NADH + $FADH_2$

9.4 A one-month-old male showed abnormalities of the nervous system and lactic acidosis. Enzyme assay for pyruvate dehydrogenase (PDH) activity on extracts of cultured skin fibroblasts showed five percent of normal activity, with a low concentration (1 x 10^{-4} mM) of thiamine pyrophosphate (TPP), but eighty percent of normal activity when the assay contained a high (0.4 mM) concentration of TPP. Which one of the following statements concerning this patient is most correct?

A. Elevated levels of lactate and pyruvate in the blood reliably predict the presence of PDH deficiency.
B. The patient is expected to show disturbances in fatty acid degradation.
C. A diet consisting of high carbohydrate intake would be expected to be beneficial in this patient.
D. Alanine concentration in the blood is expected to be less than normal.
E. Administration of thiamine is expected to reduce his serum lactate concentration and improve his clinical symptoms.

Correct answer = E. The patient appears to have a thiamine-responsive PDH deficiency. The enzyme fails to bind thiamine pyrophosphate at low concentration, but shows significant activity at a high concentration of the cofactor. This mutation, which affects the K_m of the enzyme for the cofactor, is present in some, but not all cases of PDH deficiency. All inborn errors of PDH are associated with elevated levels of lactate, pyruvate, and alanine (the transamination product of pyruvate). Patients routinely show neuroanatomic defects, developmental delay, and often early death. Elevated lactate and pyruvate are also observed in pyruvate carboxylase deficiency, another rare defect in pyruvate metabolism. Because PDH is an integral part of carbohydrate metabolism, a diet low in carbohydrates (not high) would be expected to blunt the effects of the enzyme deficiency. By contrast, fatty acid degradation occurs via conversion to acetyl CoA by β-oxidation, a process that does not involve pyruvate as an intermediate. Thus, fatty acid metabolism is not disturbed in this enzyme deficiency.

Gluconeogenesis

10

I. OVERVIEW

Some tissues, such as the brain, red blood cells, kidney medulla, lens and cornea of the eye, testes, and exercising muscle, require a continuous supply of glucose as a metabolic fuel. Liver glycogen, an essential postprandial source of glucose, can meet these needs for only ten to eighteen hours in the absence of dietary intake of carbohydrate (see p. 328). During a prolonged fast, hepatic glycogen stores are depleted, and glucose is formed from precursors such as lactate, pyruvate, glycerol (derived from the backbone of triacylglycerols, see p. 188), and α-ketoacids (derived from the catabolism of glucogenic amino acids, see p. 260). The formation of glucose does not occur by a simple reversal of glycolysis, because the overall equilibrium of glycolysis strongly favors pyruvate formation. Instead, glucose is synthesized by a special pathway, gluconeogenesis. During an overnight fast, approximately ninety percent of gluconeogenesis occurs in the liver, with the kidneys providing ten percent of the newly synthesized glucose molecules. However, during prolonged fasting, the kidneys become major glucose-producing organs, contributing an estimated forty percent of the total glucose production. Figure 10.1 shows the relationship of gluconeogenesis to other important reactions of intermediary metabolism

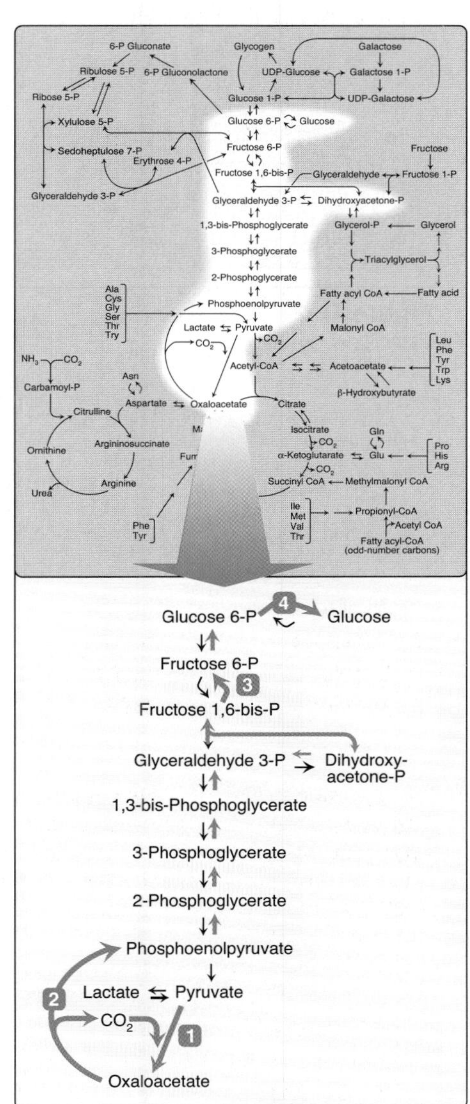

Figure 10.1
The gluconeogenesis pathway shown as part of the essential pathways of energy metabolism. The numbered reactions are unique to gluconeogenesis. (See Figure 8.2, p. 90 for a more detailed view of the metabolic map.)

II. SUBSTRATES FOR GLUCONEOGENESIS

Gluconeogenic precursors are molecules that can be used to produce a net synthesis of glucose. They include all the intermediates of glycolysis and the citric acid cycle. Glycerol, lactate, and the α-keto acids obtained from the deamination of glucogenic amino acids are the most important gluconeogenic precursors.

A. Glycerol

Glycerol is released during the hydrolysis of triacylglycerols in adipose tissue (see p. 188), and is delivered by the blood to the liver. Glycerol is phosphorylated by *glycerol kinase* to glycerol phosphate, which is oxidized by *glycerol phosphate dehydrogenase* to dihydroxyacetone phosphate—an intermediate of glycolysis. [Note: Adipocytes cannot phosphorylate glycerol because they lack *glycerol kinase*.]

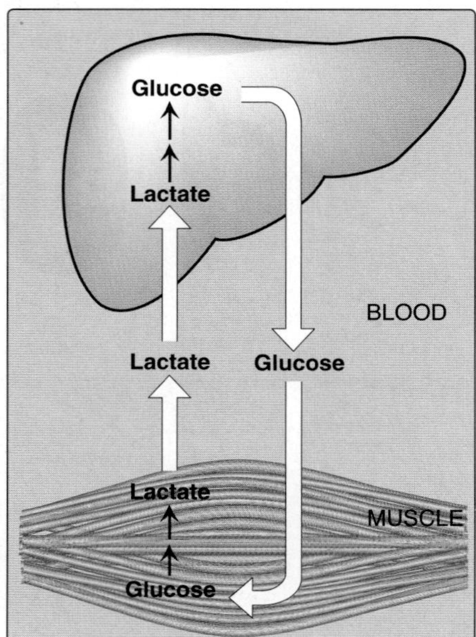

Figure 10.2
The Cori cycle.

B. Lactate

Lactate is released into the blood by exercising skeletal muscle, and by cells that lack mitochondria, such as red blood cells. In the **Cori cycle**, blood-borne glucose is converted by exercising muscle to lactate, which diffuses into the blood. This lactate is taken up by the liver and reconverted to glucose, which is released back into the circulation (Figure 10.2).

C. Amino acids

Amino acids derived from hydrolysis of tissue proteins are the major sources of glucose during a fast. α-Ketoacids, such as oxaloacetate and α-ketoglutarate, are derived from the metabolism of **glucogenic amino acids** (see p. 259). These substances can enter the citric acid cycle and form oxaloacetate—a direct precursor of phosphoenolpyruvate. [Note: Acetyl CoA and compounds that give rise to acetyl CoA (for example, acetoacetate and amino acids such as lysine and leucine) cannot give rise to a net synthesis of glucose. This is due to the irreversible nature of the *pyruvate dehydrogenase* reaction, which converts pyruvate to acetyl CoA (see p. 107). These compounds give rise instead to ketone bodies (see p. 193) and are therefore termed **ketogenic**.]

III. REACTIONS UNIQUE TO GLUCONEOGENESIS

Seven glycolytic reactions are reversible and are used in the synthesis of glucose from lactate or pyruvate. However, three of the reactions are irreversible and must be circumvented by four alternate reactions that energetically favor the synthesis of glucose. These reactions, unique to gluconeogenesis, are described below.

A. Carboxylation of pyruvate

The first "roadblock" to overcome in the synthesis of glucose from pyruvate is the irreversible conversion in glycolysis of pyruvate to phosphoenolpyruvate (PEP) by *pyruvate kinase.* In gluconeogenesis, pyruvate is first carboxylated by *pyruvate carboxylase* to oxaloacetate (OAA), which is then converted to PEP by the action of *PEP-carboxykinase* (Figure 10.3).

1. **Biotin is a coenzyme:** *Pyruvate carboxylase* contains biotin (see p. 379), which is covalently bound to the enzyme protein through the ε-amino group of lysine, forming an active enzyme (see Figure 10.3). This covalently bound form of biotin is called **biocytin**. Cleavage of a high-energy phosphate of ATP drives the formation of an enzyme–biotin–CO_2 intermediate. This high-energy complex subsequently carboxylates pyruvate to form oxaloacetate. [Note: This reaction occurs in the mitochondria of liver and kidney cells, and has two purposes: to provide an important substrate for gluconeogenesis, and to provide OAA that can replenish TCA cycle intermediates that may become depleted, depending on the synthetic needs of the cell. Muscle cells also contain *pyruvate carboxylase*, but use the OAA produced only for the latter purpose—they do not synthesize glucose.]

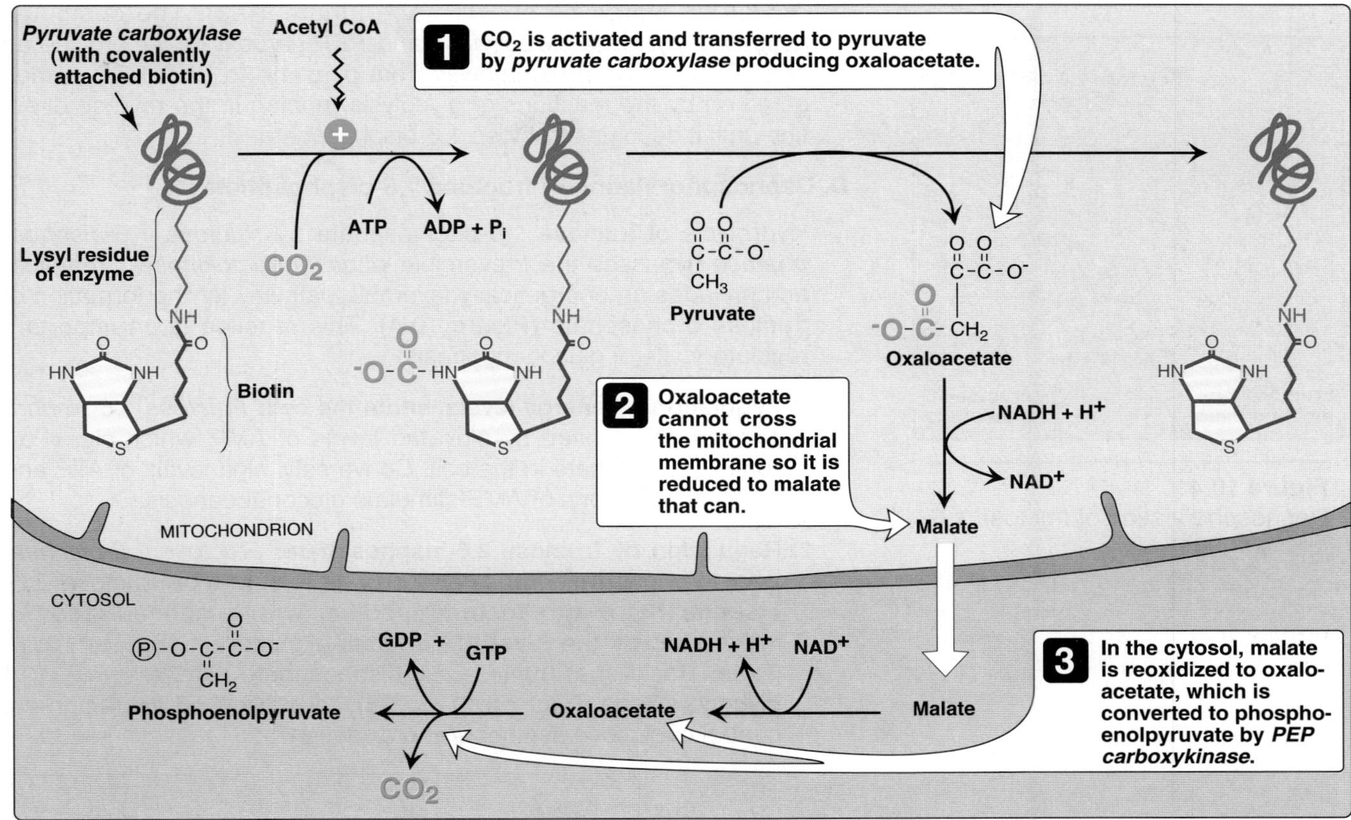

Figure 10.3
Activation and transfer of CO_2 to pyruvate, followed by transport of oxaloacetate to the cytosol and subsequent decarboxylation.

2. **Allosteric regulation:** *Pyruvate carboxylase* is allosterically activated by acetyl CoA. Elevated levels of acetyl CoA may signal one of several metabolic states in which the increased synthesis of oxaloacetate is required. For example, this may occur during fasting in which OAA is used for the synthesis of glucose by gluconeogenesis in the liver and kidney. Conversely, at low levels of acetyl CoA, *pyruvate carboxylase* is largely inactive, and pyruvate is primarily oxidized by *pyruvate dehydrogenase* to produce acetyl CoA that can be further oxidized by the TCA cycle (see p. 107).

B. Transport of oxaloacetate to the cytosol

Oxaloacetate, formed in the mitochondria, must enter the cytosol where the other enzymes of gluconeogenesis are located. However, OAA is unable to directly cross the inner mitochondrial membrane; it must first be reduced to malate by mitochondrial *malate dehydrogenase*. Malate can be transported from the mitochondria to the cytosol, where it is reoxidized to oxaloacetate by cytosolic *malate dehydrogenase* (see Figure 10.3).

C. Decarboxylation of cytosolic oxaloacetate

Oxaloacetate is decarboxylated and phosphorylated in the cytosol by *PEP-carboxykinase* (also referred to as *PEPCK*). The reaction is

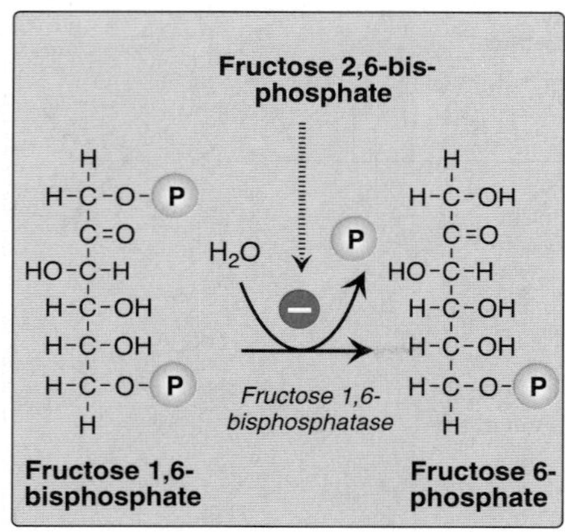

Figure 10.4
Dephosphorylation of fructose 1,6-bisphosphate.

driven by hydrolysis of GTP (see Figure 10.3). The combined actions of *pyruvate carboxylase* and *PEP-carboxykinase* provide an energetically favorable pathway from pyruvate to PEP. PEP is then acted on by the reactions of glycolysis running in the reverse direction until it becomes fructose 1,6-bisphosphate.

D. Dephosphorylation of fructose 1,6-bisphosphate

Hydrolysis of fructose 1,6-bisphosphate by *fructose 1,6-bisphosphatase* bypasses the irreversible *phosphofructokinase-1* reaction, and provides an energetically favorable pathway for the formation of fructose 6-phosphate (Figure 10.4). This reaction is an important regulatory site of gluconeogenesis.

1. **Regulation by energy levels within the cell:** *Fructose 1,6-bisphosphatase* is inhibited by elevated levels of AMP, which signal an "energy-poor" state in the cell. Conversely, high levels of ATP and low concentrations of AMP stimulate gluconeogenesis.

2. **Regulation by fructose 2,6-bisphosphate:** *Fructose 1,6-bisphosphatase*, found in liver and kidney, is inhibited by fructose 2,6-bisphosphate, an allosteric modifier whose concentration is influenced by the level of circulating glucagon (Figure 10.5). [Note: Recall that fructose 2,6-bisphosphate activates *PFK-1* of glycolysis (see Figure 8.17, p. 98), thus allowing for reciprocal control of glucose synthesis and oxidation.]

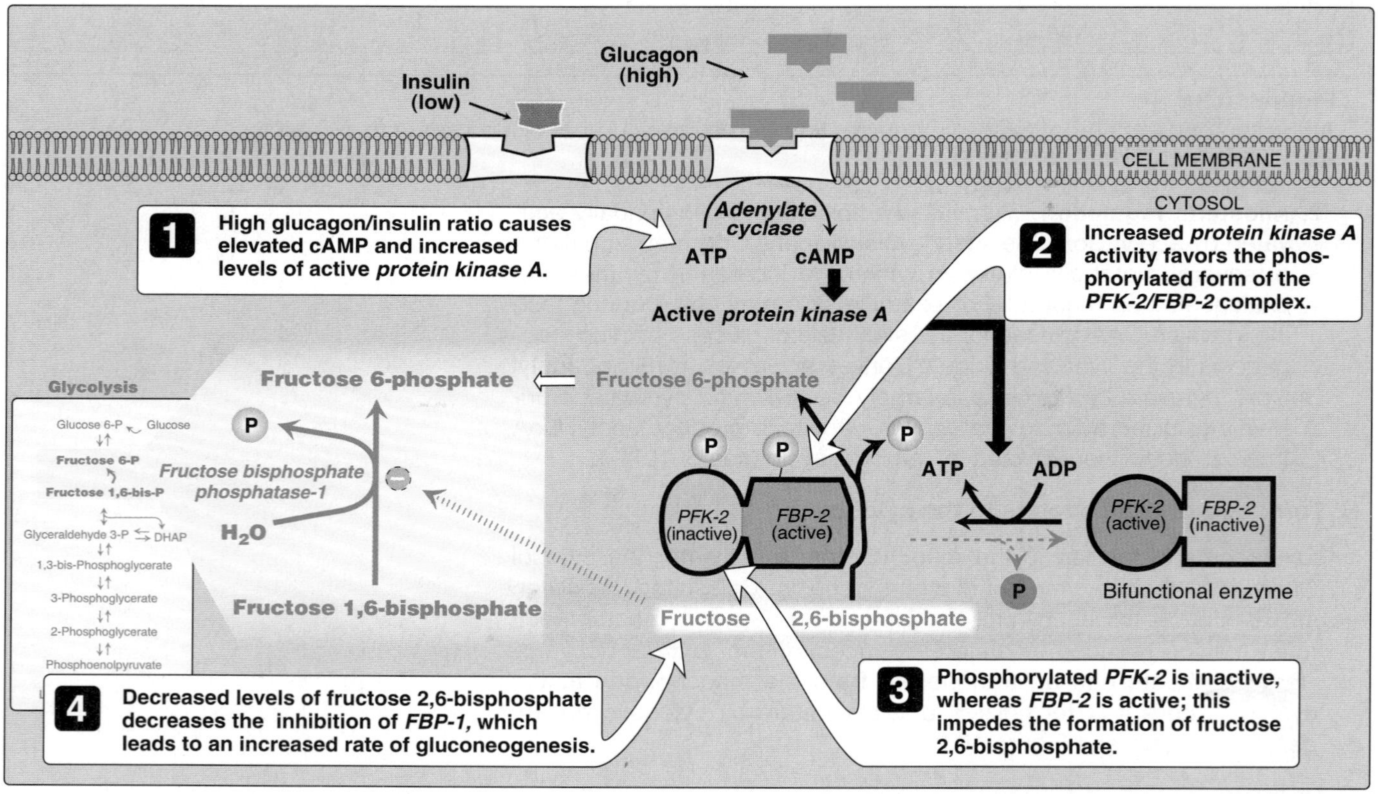

Figure 10.5
Effect of elevated glucagon on the intracellular concentration of fructose 2,6-bisphosphate in the liver. *PFK-2 = phosphofructokinase-2; FBP-2 = Fructose bisphospate phosphatase-2.*

E. Dephosphorylation of glucose 6-phosphate

Hydrolysis of glucose 6-phosphate by *glucose 6-phosphatase* bypasses the irreversible *hexokinase* reaction, and provides an energetically favorable pathway for the formation of free glucose (Figure 10.6). Liver and kidney are the only organs that release free glucose from glucose 6-phosphate. This process actually requires two enzymes: *glucose 6-phosphate translocase*, which transports glucose 6-phosphate across the endoplasmic reticular (ER) membrane, and a second ER enzyme, *glucose 6-phosphatase* (found only in gluconeogenic cells), which removes the phosphate, producing free glucose (Figure 10.6). [Note: These enzymes are required for the final step in glycogenolysis (see p. 127), as well as gluconeogenesis. **Type Ia glycogen storage disease** (see p. 128) results from an inherited deficiency of one of these enzymes.] Specific transporters are responsible for releasing free glucose and phosphate back into the cytosol, and in hepatocytes, into the blood. [Note: Muscle lacks *glucose 6-phosphatase* and, therefore, cannot provide blood glucose by gluconeogenesis. Also, glucose 6-phosphate derived from muscle glycogen cannot be dephosphorylated to yield free glucose.]

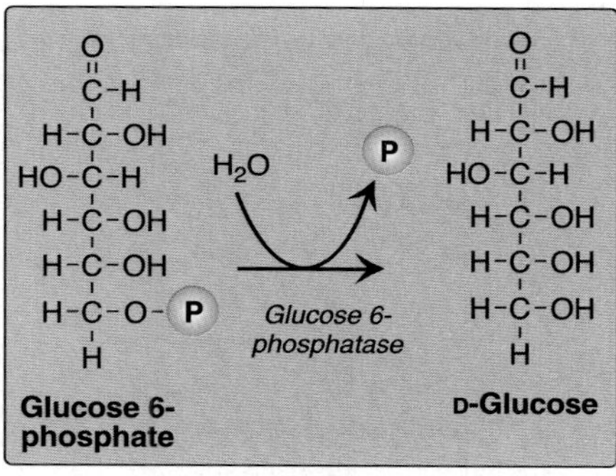

Figure 10.6
Dephosphorylation of glucose 6-phosphate.

F. Summary of the reactions of glycolysis and gluconeogenesis

Of the eleven reactions required to convert pyruvate to free glucose, seven are catalyzed by reversible glycolytic enzymes (Figure 10.7). The irreversible reactions of glycolysis catalyzed by *hexokinase*, *phophofructokinase*, and *pyruvate kinase* are circumvented by *glucose 6-phosphatase*, *fructose 1,6-bisphosphatase*, and *pyruvate carboxylase/PEP carboxykinase*. In gluconeogenesis, the equilibria of the seven reversible reactions of glycolysis are pushed in favor of glucose synthesis as a result of the essentially irreversible formation of PEP, fructose 6-phosphate, and glucose catalyzed by the gluconeogenic enzymes. [Note: The stoichiometry of gluconeogenesis from pyruvate couples the cleavage of six high-energy phosphate bonds and the oxidation of two NADH with the formation of each molecule of glucose (see Figure 10.7).]

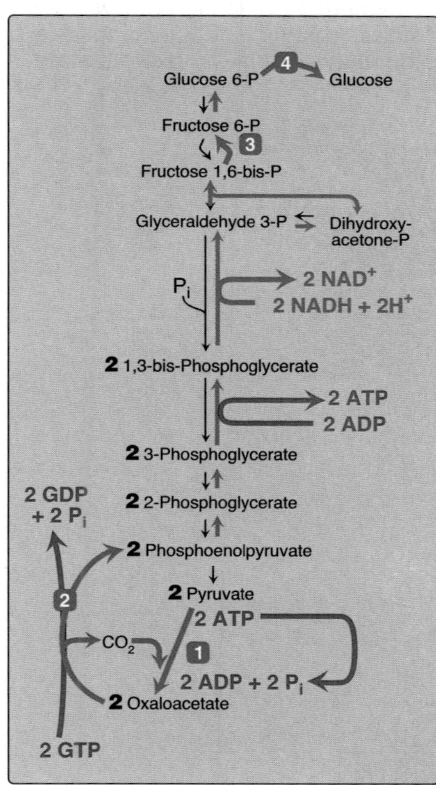

Figure 10.7
Summary of the reactions of glycolysis and gluconeogenesis, showing the energy requirements of gluconeogenesis.

IV. REGULATION OF GLUCONEOGENESIS

The moment-to-moment regulation of gluconeogenesis is determined primarily by the circulating level of glucagon, and by the availability of gluconeogenic substrates. In addition, slow adaptive changes in enzyme activity result from an alteration in the rate of enzyme synthesis or degradation, or both. [Note: Hormonal control of the glucoregulatory system is presented in Chapter 23, p. 305.]

A. Glucagon

This pancreatic islet hormone (see p. 311) stimulates gluconeogenesis by three mechanisms.

1. **Changes in allosteric effectors:** Glucagon lowers the level of fructose 2,6-bisphosphate, resulting in activation of *fructose 1,6-bisphosphatase* and inhibition of *phosphofructokinase* (see Figure

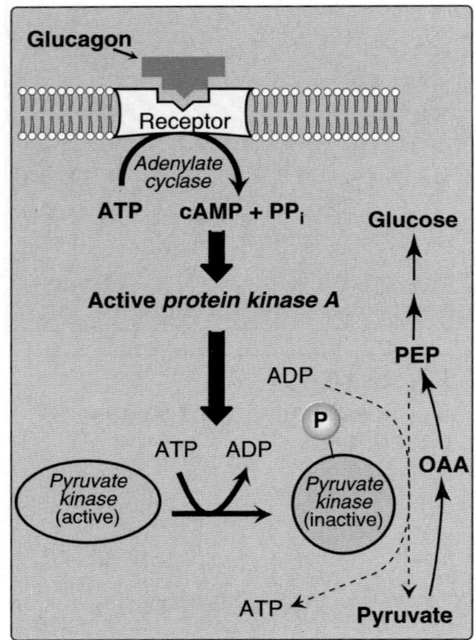

Figure 10.8
Covalent modification of *pyruvate kinase* results in inactivation of the enzyme. OAA = oxaloacetate.

10.5). [Note: See p. 98 for the role of fructose 2,6-bisphosphate in the regulation of glycolysis.]

2. **Covalent modification of enzyme activity:** Glucagon, via an elevation in cAMP level and *cAMP-dependent protein kinase* activity, stimulates the conversion of *pyruvate kinase* to its inactive (phosphorylated) form. This decreases the conversion of PEP to pyruvate, which has the effect of diverting PEP to the synthesis of glucose (Figure 10.8).

3. **Induction of enzyme synthesis:** Glucagon increases the transcription of the *PEP carboxykinase* gene, thereby increasing the availability of this enzyme's activity as levels of its substrate rise during fasting. [Note: Insulin causes decreased transcription of the mRNA for this enzyme.]

B. Substrate availability

The availability of gluconeogenic precursors, particularly glucogenic amino acids, significantly influences the rate of hepatic glucose synthesis. Decreased levels of insulin favor mobilization of amino acids from muscle protein, and provide the carbon skeletons for gluconeogenesis.

C. Allosteric activation by acetyl CoA

Allosteric activation of hepatic *pyruvate carboxylase* by acetyl CoA occurs during fasting. As a result of excessive lipolysis in adipose tissue, the liver is flooded with fatty acids (see p. 328). The rate of formation of acetyl CoA by β-oxidation of these fatty acids exceeds the capacity of the liver to oxidize it to CO_2 and H_2O. As a result, acetyl CoA accumulates and leads to activation of *pyruvate carboxylase*. [Note: Acetyl CoA inhibits *pyruvate dehydrogenase* (see p. 108). Thus, this single compound can divert pyruvate toward gluconeogenesis and away from the TCA cycle.]

D. Allosteric inhibition by AMP

Fructose 1,6-bisphosphatase is inhibited by AMP—a compound that activates *phosphofructokinase*. Elevated AMP thus stimulates pathways that oxidize nutrients to provide energy for the cell. [Note: ATP and NADH, produced in large quantities during fasts by catalytic pathways, such as fatty acid oxidation, are required for gluconeogenesis.]

V. CHAPTER SUMMARY

Gluconeogenic precursors include all the **intermediates of glycolysis** and the **citric acid cycle**, **glycerol** released during the hydrolysis of triacylglycerols in adipose tissue, **lactate** released into the blood by cells that lack mitochondria and by exercising skeletal muscle, and **α-ketoacids** derived from the metabolism of glucogenic amino acids. Seven of the reactions of glycolysis are reversible and are used for gluconeogenesis in the liver and kidneys. Three reactions are **physiologically irreversible** and must be circumvented. These reactions are catalyzed by the glycolytic enzymes **pyruvate kinase**, **phosphofructoki-**

nase, and **hexokinase**. **Pyruvate** is converted to **phosphoenolpyruvate** (PEP) by **pyruvate carboxylase** and **PEP carboxykinase**. The carboxylase requires **biotin** and **ATP**, and is allosterically activated by acetyl CoA. PEP carboxykinase requires **GTP**. The transcription of its mRNA is increased by glucagon and decreased by insulin. **Fructose 1,6-bisphosphate** is converted to **fructose 1-phosphate** by **fructose 1,6-bisphosphatase**. This enzyme is **inhibited** by elevated levels of **AMP** and **activated** by elevated levels of **ATP**. The enzyme is also **inhibited** by **fructose 2,6-bisphosphate**, the primary allosteric activator of glycolysis. **Glucose 6-phosphate** is converted to **glucose** by **glucose 6-phosphatase**. This enzyme activity is required for the final step in glycogen degradation, as well as gluconeogenesis. A deficiency of this enzyme results in **type Ia glycogen storage disease**.

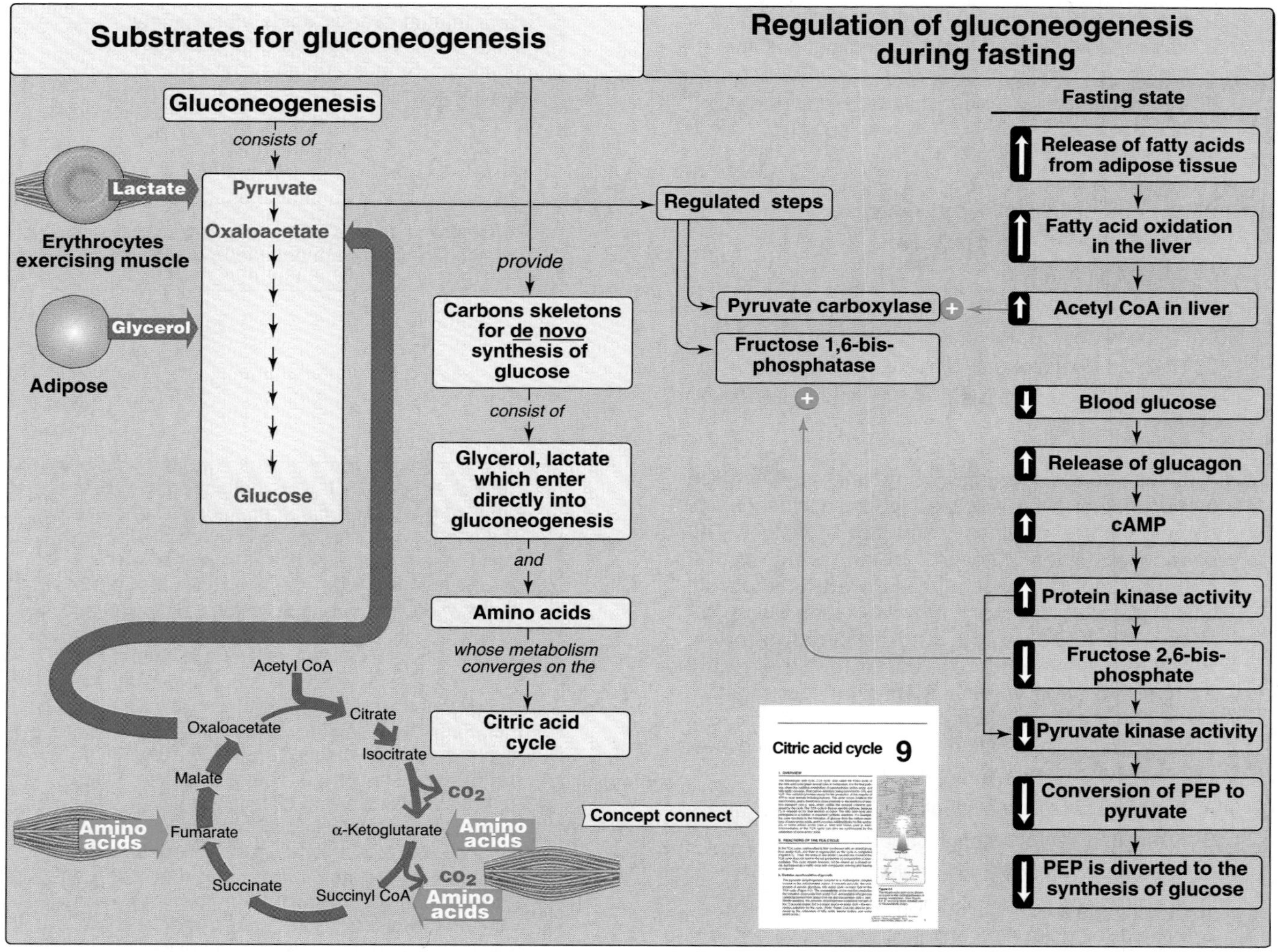

Figure 10.9
Key concept map for gluconeogenesis.

Study Questions

Choose the ONE correct answer

10.1 The synthesis of glucose from pyruvate by gluconeogenesis:

A. occurs exclusively in the cytosol.
B. is inhibited by an elevated level of glucagon.
C. requires the participation of biotin.
D. involves lactate as an intermediate.
E. requires the oxidation/reduction of FAD.

Correct answer = C. Biotin is the coenzyme-/prosthetic group of pyruvate carboxylase. The carboxylation of pyruvate occurs in the mitochondria, whereas the other reactions of gluconeogenesis occur in the cytosol. Glucagon stimulates gluconeogenesis. Lactate is not an intermediate in the conversion of pyruvate to glucose.

10.2 Which one of the following statements concerning gluconeogenesis is correct?

A. It occurs in muscle.
B. It is stimulated by fructose 2,6-bisphosphate.
C. It is inhibited by elevated levels of acetyl CoA.
D. It is important in maintaining blood glucose during the normal overnight fast.
E. It uses carbon skeletons provided by degradation of fatty acids.

Correct answer = D. During the overnight fast, glycogen is partially depleted and gluconeogenesis provides blood glucose. Gluconeogenesis is inhibited by fructose 2,6-bisphosphate and stimulated by elevated levels of acetyl CoA. Degradation of fatty acids yields acetyl CoA, which cannot be converted to glucose. Carbon skeletons of most amino acids are, however, gluconeogenic.

10.3 Which one of the following reactions is unique to gluconeogenesis?

A. Lactate → pyruvate
B. Phosphoenolpyruvate → pyruvate
C. Oxaloacetate → phosphoenolpyruvate
D. Glucose 6-phosphate → fructose 6-phosphate
E. 1,3-Bisphosphoglycerate → 3-phosphoglycerate

Correct answer = C. The other reactions are common to both gluconeogenesis and glycolysis.

10.4 A thirteen-year-old female was brought to your office by her mother, who was troubled by her daughter's chronic fatigue, dizziness, and loss of weight. The patient was 5 feet 5 inches tall and weighed 103 pounds. Laboratory results showed leukopenia, blood glucose = 50 mg/dl (normal fasting blood glucose = 70 to 90 mg/dl), and elevated ketones. Persistent questioning revealed that the young woman had been virtually fasting for four months, hoping to obtain a "skinny face" as a prelude to a career in modeling. Which one of the following best explains the patient hypoglycemia?

A. Impaired secretion of insulin
B. Enhanced secretion of glucagon
C. Impaired hydrolysis of liver glycogen
D. Impaired conversion of amino acids to glucose
E. Impaired mobilization of triglycerides

Correct answer = D. This patient shows many signs of the eating disorder anorexia nervosa. The condition is characterized by an aversion to food that leads to a state of fasting and emaciation. Patients often have a distorted image of their own weight or shape and are unconcerned by the serious health consequences of their low weight. After several months of near starvation, the blood glucose in this patient is maintained by gluconeogenesis, primarily from amino acids mobilized from tissue proteins. This process involves several vitamins (pyridoxine, thiamine, biotin) and although not determined, the patient potentially has vitamin deficiencies. The alternate explanations are untenable. Liver glycogen is exhausted in the first days of fasting. Impaired secretion of insulin or enhanced secretion of glucagon would lead to hyperglycemia. Decreased hydrolysis of triacylglycerols in adipose tissue would not lead to hypoglycemia.

Glycogen Metabolism

11

I. OVERVIEW

A constant source of blood glucose is an absolute requirement for human life. Glucose is the greatly preferred energy source for the brain, and the required energy source for cells with few or no mitochondria, such as mature erythrocytes. Glucose is also essential as an energy source for exercising muscle, where it is the substrate for anaerobic glycolysis. Blood glucose can be obtained from three primary sources: the diet, degradation of glycogen, and gluconeogenesis. Dietary intake of glucose and glucose precursors, such as starch, monosaccharides, and disaccharides, is sporadic and, depending on the diet, is not always a reliable source of blood glucose. In contrast, gluconeogenesis (see p. 115) can provide sustained synthesis of glucose, but it is somewhat slow in responding to a falling blood glucose level. Therefore, the body has developed mechanisms for storing a supply of glucose in a rapidly mobilizable form, namely, glycogen. In the absence of a dietary source of glucose, this compound is rapidly released from liver and kidney glycogen. Similarly, muscle glycogen is extensively degraded in exercising muscle to provide that tissue with an important energy source. When glycogen stores are depleted, specific tissues synthesize glucose <u>de novo</u>, using amino acids from the body's proteins as a primary source of carbons for the gluconeogenic pathway. Figure 11.1 shows the reactions of glycogen synthesis and degradation as part of the essential pathways of energy metabolism.

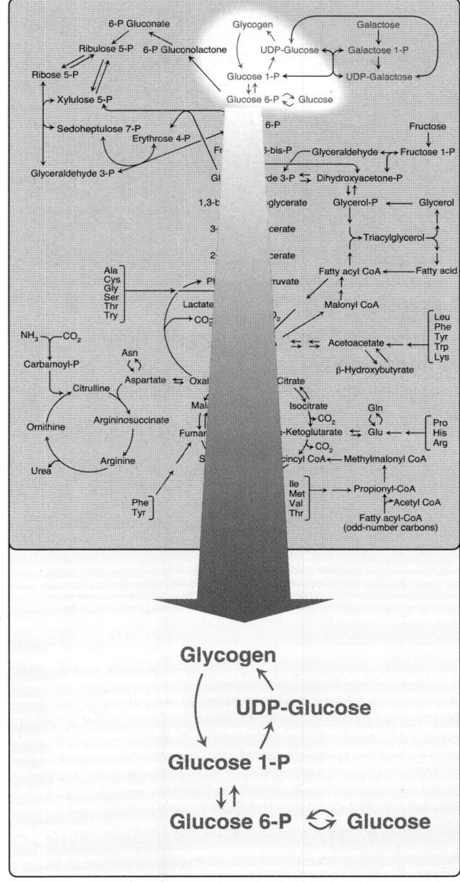

Figure 11.1
Glycogen synthesis and degradation shown as a part of the essential reactions of energy metabolism (see Figure 8.2, p. 90, for a more detailed view of the overall reactions of metabolism).

II. STRUCTURE AND FUNCTION OF GLYCOGEN

The main stores of glycogen in the body are found in skeletal muscle and liver, although most other cells store small amounts of glycogen for their own use. The function of muscle glycogen is to serve as a fuel reserve for the synthesis of ATP during muscle contraction. That of liver glycogen is to maintain the blood glucose concentration, particularly during the early stages of a fast (Figure 11.2, and see p. 328).

A. Amounts of liver and muscle glycogen

Approximately 400 g of glycogen make up one to two percent of the fresh weight of resting muscle, and approximately 100 g of glycogen make up to ten percent of the fresh weight of a well-fed adult liver. What limits the production of glycogen at these levels is not clear.

Lippincott's Illustrated Reviews: Biochemistry, Third Edition,
by Pamela C. Champe and Richard A. Harvey.
Lippincott Williams & Wilkins, Baltimore, MD © 2005.

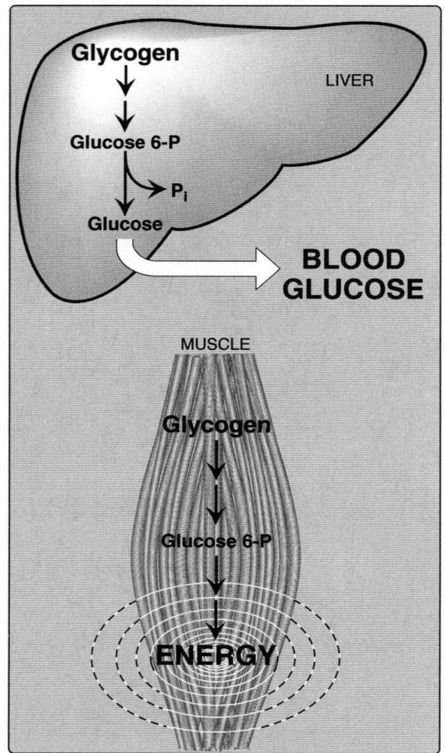

Figure 11.2
Functions of muscle and liver glycogen.

However, in some glycogen storage diseases (see p. 128), the amount of glycogen in the liver and/or muscle can be significantly higher.

B. Structure of glycogen

Glycogen is a branched-chain homopolysaccharide made exclusively from **α-D-glucose**. The primary glycosidic bond is an α(1→4) linkage. After an average of eight to ten glucosyl residues, there is a branch containing an α(1→6) linkage (Figure 11.3). A single molecule of glycogen can have a molecular mass of up to 10^8 daltons. These molecules exist in discrete cytoplasmic granules that contain most of the enzymes necessary for glycogen synthesis and degradation.

C. Fluctuation of glycogen stores

Liver glycogen stores increase during the well-fed state (see p. 321), and are depleted during a fast (see p. 328). Muscle glycogen is not affected by short periods of fasting (a few days) and is only moderately decreased in prolonged fasting (weeks). Muscle glycogen is synthesized to replenish muscle stores after they have been depleted, for example, following strenuous exercise. [Note: Synthesis and degradation of glycogen are processes that go on continuously. The differences between the rates of these two processes determine the levels of stored glycogen during specific physiologic states.]

III. SYNTHESIS OF GLYCOGEN (GLYCOGENESIS)

Glycogen is synthesized from molecules of α-D-glucose. The process occurs in the cytosol, and requires energy supplied by ATP (for the phosphorylation of glucose) and uridine triphosphate (UTP).

A. Synthesis of UDP-glucose

α-D-Glucose attached to uridine diphosphate (UDP) is the source of all of the glucosyl residues that are added to the growing glycogen molecule. UDP-glucose (Figure 11.4) is synthesized from glucose 1-phosphate and UTP by *UDP-glucose pyrophosphorylase* (Figure

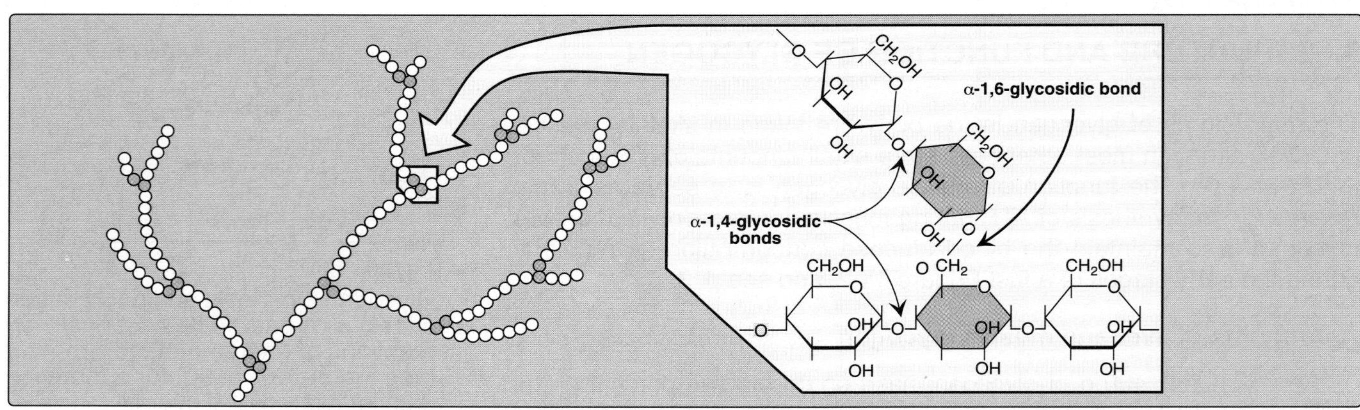

Figure 11.3
Branched structure of glycogen, showing α-1,4 and α-1,6 linkages.

11.5). The high-energy bond in pyrophosphate (PP_i), the second product of the reaction, is hydrolyzed to two inorganic phosphates (P_i) by *pyrophosphatase*, which ensures that synthesis of UDP-glucose proceeds in the direction of UDP-glucose production. [Note: Glucose 6-phosphate is converted to glucose 1-phosphate by *phosphoglucomutase.* Glucose 1,6-bisphosphate is an obligatory intermediate in this reaction (Figure 11.6).]

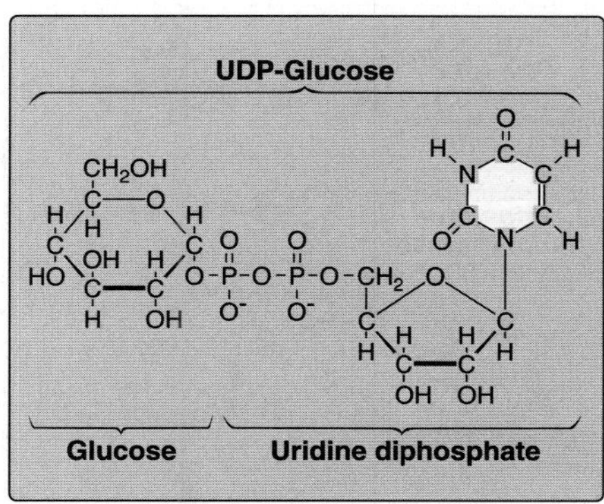

Figure 11.4
The structure of UDP-glucose.

B. Synthesis of a primer to initiate glycogen synthesis

Glycogen synthase is responsible for making the α(1→4) linkages in glycogen. This enzyme cannot initiate chain synthesis using free glucose as an acceptor of a molecule of glucose from UDP-glucose. Instead, it can only elongate already existing chains of glucose. Therefore, a fragment of glycogen can serve as a primer in cells whose glycogen stores are not totally depleted. In the absence of a glycogen fragment, a protein, called **glycogenin**, can serve as an acceptor of glucose residues (see Figure 11.5). The side chain hydroxyl group of a specific tyrosine serves as the site at which the initial glucosyl unit is attached. Transfer of the first few molecules of glucose from UDP-glucose to glycogenin is catalyzed by glycogenin itself, which can then transfer additional glucosyl units to the growing α(1→4)–linked glucosyl chain. This short chain serves as an acceptor of future glucose residues as described below. [Note: Glycogenin stays associated with and is found in the center of the completed glycogen molecule.]

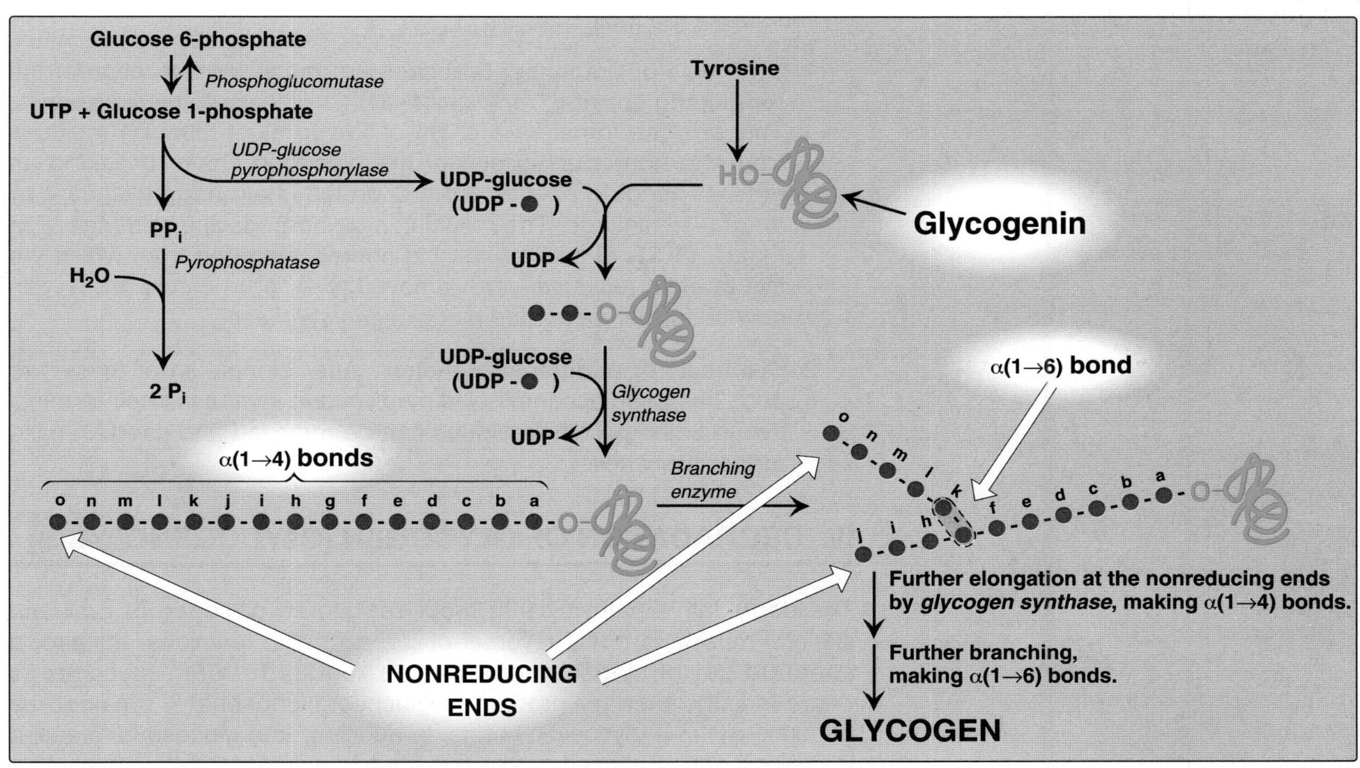

Figure 11.5
Glycogen synthesis.

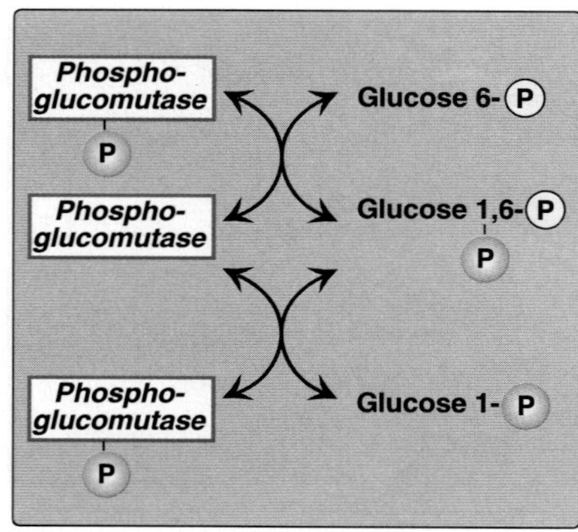

Figure 11.6
Interconversion of glucose 6-phosphate and glucose 1-phosphate by *phosphoglucomutase* .

C. Elongation of glycogen chains by glycogen synthase

Elongation of a glycogen chain involves the transfer of glucose from UDP-glucose to the nonreducing end of the growing chain, forming a new glycosidic bond between the anomeric hydroxyl of carbon 1 of the activated glucose and carbon 4 of the accepting glucosyl residue (see Figure 11.5). [Note: The "nonreducing end" of a carbohydrate chain is one in which the anomeric carbon of the terminal sugar is linked by a glycosidic bond to another compound, making the terminal sugar "nonreducing" (see p. 84).] The enzyme responsible for making the $\alpha(1\rightarrow4)$ linkages in glycogen is *glycogen synthase.* [Note: The UDP released when the new $\alpha(1\rightarrow4)$ glycosidic bond is made can be converted back to UTP by *nucleoside diphosphate kinase* (UDP + ATP $\rightleftarrows$ UTP + ADP, see p. 294).]

D. Formation of branches in glycogen

If no other synthetic enzyme acted on the chain, the resulting structure would be a linear (unbranched) molecule of glucosyl residues attached by $\alpha(1\rightarrow4)$ linkages. Such a compound is found in plant tissues, and is called **amylose**. In contrast, glycogen has branches located, on average, eight glucosyl residues apart, resulting in a highly branched, treelike structure (see Figure 11.3) that is far more soluble than the unbranched amylose. Branching also increases the number of nonreducing ends to which new glucosyl residues can be added (and also, as described later, from which these residues can be removed), thereby greatly accelerating the rate at which glycogen synthesis and degradation can occur, and dramatically increasing the size of the molecule.

1. **Synthesis of branches:** Branches are made by the action of the "**branching enzyme**," *amylo*-$\alpha(1\rightarrow4) \rightarrow \alpha(1\rightarrow6)$-*transglucosidase*. This enzyme transfers a chain of five to eight glucosyl residues from the nonreducing end of the glycogen chain [breaking an $\alpha(1\rightarrow4)$ bond] to another residue on the chain and attaches it by an $\alpha(1\rightarrow6)$ linkage. The resulting new, nonreducing end (see "j" in Figure 11.5), as well as the old nonreducing end from which the five to eight residues were removed (see "o" in Figure 11.5), can now be further elongated by *glycogen synthase*.

2. **Synthesis of additional branches:** After elongation of these two ends has been accomplished by *glycogen synthase*, their terminal five to eight glucosyl residues can be removed and used to make further branches.

IV. DEGRADATION OF GLYCOGEN (GLYCOGENOLYSIS)

The degradative pathway that mobilizes stored glycogen in liver and skeletal muscle is not a reversal of the synthetic reactions. Instead, a separate set of cytosolic enzymes is required. When glycogen is degraded, the primary product is glucose 1-phosphate, obtained by breaking $\alpha(1\rightarrow4)$ glycosidic bonds. In addition, free glucose is released from each $\alpha(1\rightarrow6)$-linked glucosyl residue.

A. Shortening of chains

Glycogen phosphorylase sequentially cleaves the $\alpha(1\rightarrow4)$ glycosidic bonds between the glucosyl residues at the nonreducing ends of the glycogen chains by simple phosphorolysis until four glucosyl units remain on each chain before a branch point (Figure 11.7). [Note: This enzyme contains a molecule of covalently bound **pyridoxal phosphate** that is required as a coenzyme.] The resulting structure is called a **limit dextrin**, and *phosphorylase* cannot degrade it any further (Figure 11.8).

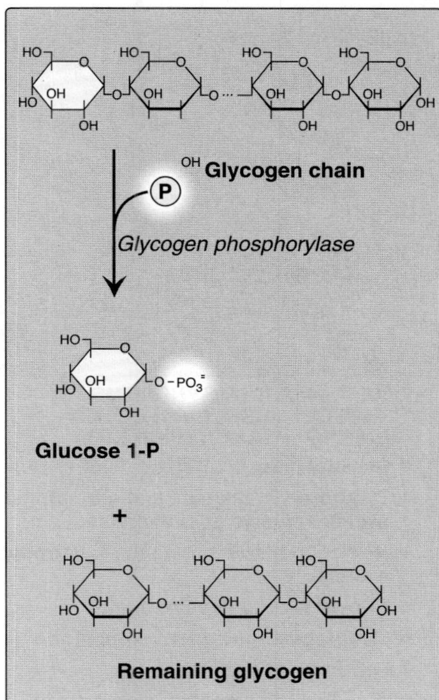

Figure 11.7
Cleavage of an $\alpha(1\rightarrow4)$-glycosidic bond.

B. Removal of branches

Branches are removed by two enzymic activities (see Figure 11.8). First, *oligo-α(1→4)→α(1→4)-glucan transferase* removes the outer three of the four glucosyl residues attached at a branch. It next transfers them to the non-reducing end of another chain, lengthening it accordingly. Thus, an $\alpha(1\rightarrow4)$ bond is broken and an $\alpha(1\rightarrow4)$ bond is made. Next, the remaining single glucose residue attached in an $\alpha(1\rightarrow6)$ linkage is removed hydrolytically by *amylo-α(1→6)-glucosidase* activity, releasing free glucose. [Note: Both the *transferase* and *glucosidase* are domains of a single polypeptide molecule, the "**debranching enzyme.**"] The glucosyl chain is now available again for degradation by *glycogen phosphorylase* until four glucosyl units from the next branch are reached.

C. Conversion of glucose 1-phosphate to glucose 6-phosphate

Glucose 1-phosphate, produced by *glycogen phosphorylase*, is converted in the cytosol to glucose 6-phosphate by *phosphoglucomutase*—a reaction that produces glucose 1,6-bisphosphate as a temporary but essential intermediate (see Figure 11.6). In the liver, glucose 6-phosphate is translocated into the endoplasmic reticulum (ER) by *glucose 6-phosphate translocase*. There it is converted to glucose by *glucose 6-phosphatase*—the same enzyme used in the last step of gluconeogenesis (see p. 119). The resulting glucose is then transported out of the ER to the cytosol. Hepatocytes release glycogen-derived glucose into the blood to help maintain blood glucose levels until the gluconeogenic pathway is actively producing glucose. [Note: In the muscle, glucose 6-phosphate cannot be dephosphorylated because of a lack of *glucose 6-phosphatase*. Instead, it enters glycolysis, providing energy neded for muscle contraction.]

D. Lysosomal degradation of glycogen

A small amount of glycogen is continuously degraded by the lysosomal enzyme, $\alpha(1\rightarrow4)$-*glucosidase* (*acid maltase*). The purpose of this pathway is unknown. However, a deficiency of this enzyme causes accumulation of glycogen in vacuoles in the cytosol, resulting in the serious **glycogen storage disease type II** (**Pompe disease**, see Figure 11.8).

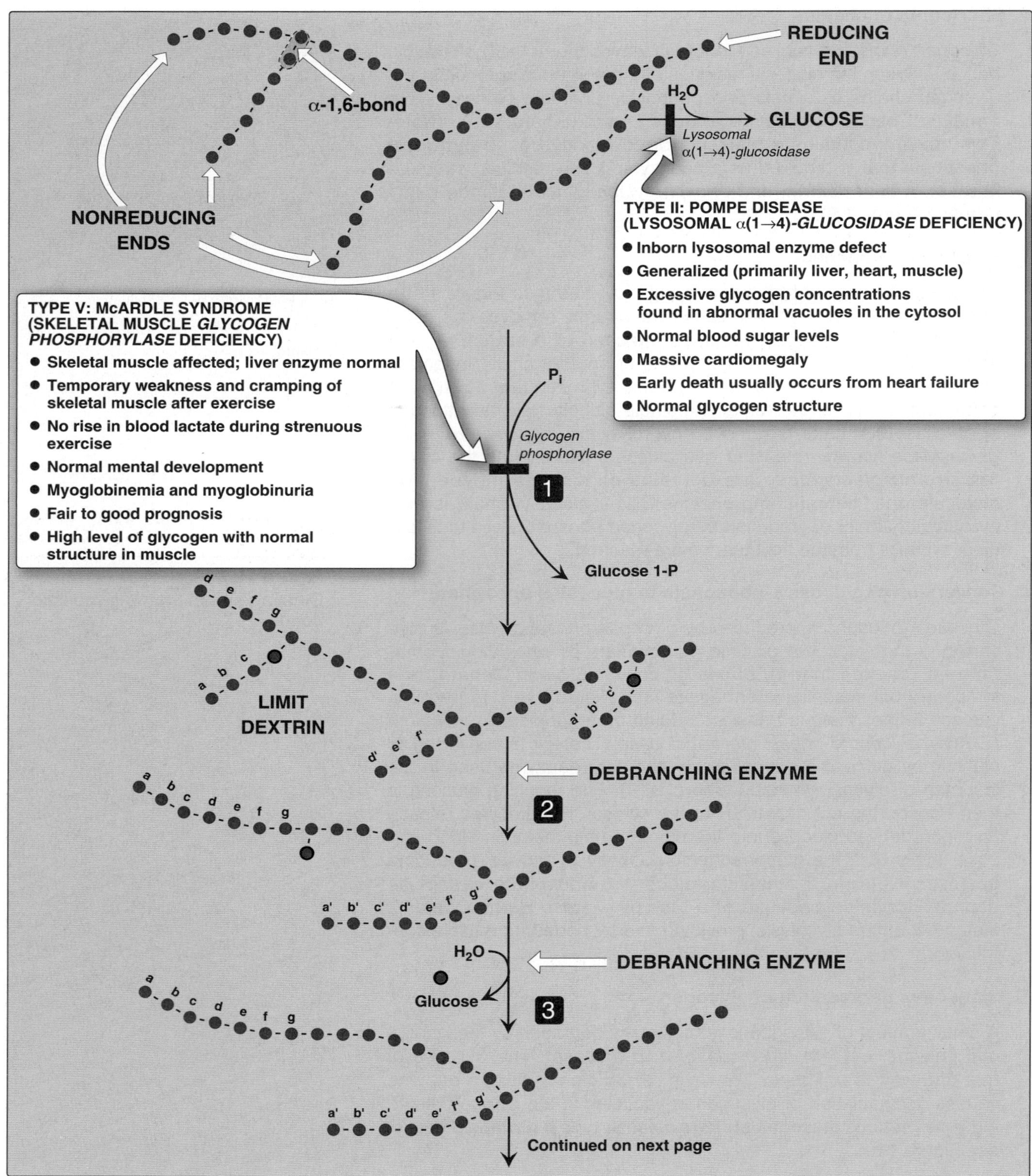

Figure 11.8
Glycogen degradation, showing some of the glycogen storage diseases. (*Continued on next page.*)

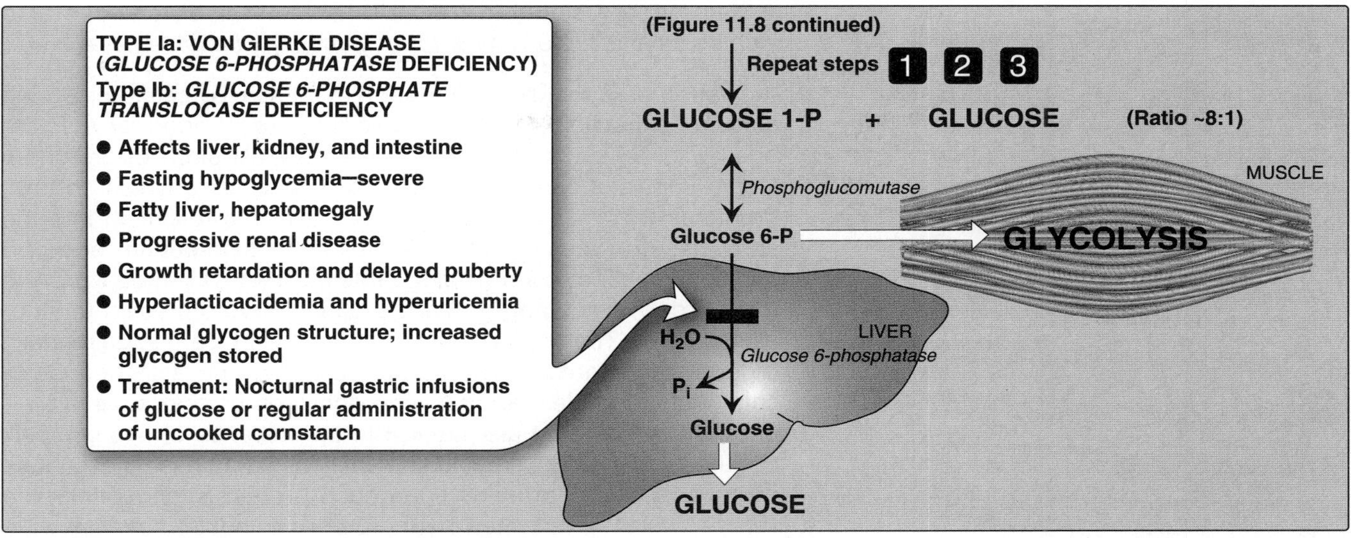

Figure 11.8 (*Continued*)

V. REGULATION OF GLYCOGEN SYNTHESIS AND DEGRADATION

Because of the importance of maintaining blood glucose levels, the synthesis and degradation of its glycogen storage form are tightly regulated. In the liver, glycogen synthesis accelerates during periods when the body has been well fed, whereas glycogen degradation accelerates during periods of fasting. In skeletal muscle, glycogen degradation occurs during active exercise, and synthesis begins as soon as the muscle is again at rest. Regulation of glycogen synthesis and degradation is accomplished on two levels. First, *glycogen synthase* and *glycogen phosphorylase* are **allosterically controlled**. Second, the pathways of glycogen synthesis and degradation are **hormonally regulated**. [Note: The regulation of glycogen synthesis and degradation is extremely complex, involving many enzymes (for example, protein kinases and phosphatases), calcium, and enzyme inhibitors, among others. A complete discussion of these pathways is beyond the scope of a review book. This section, therefore, presents an overview of the basic mechanisms of regulation of glycogen synthesis and degradation.]

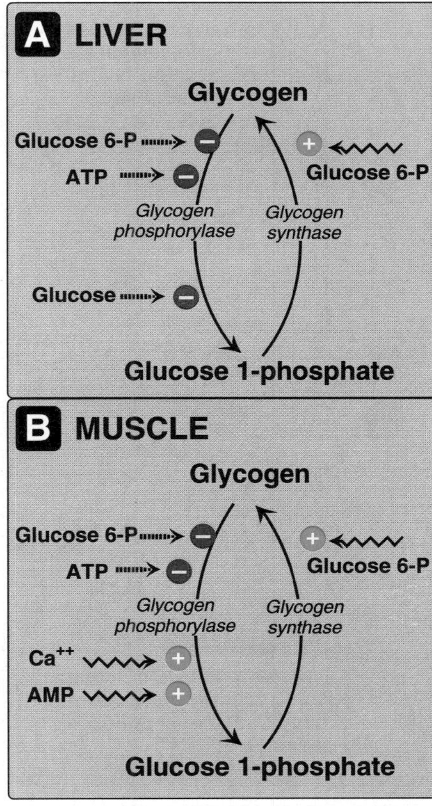

Figure 11.9
Allosteric regulation of glycogen synthesis and degradation in A. Liver, and B. Muscle.

A. Allosteric regulation of glycogen synthesis and degradation

Glycogen synthase and *glycogen phosphorylase* respond to the levels of metabolites and energy needs of the cell. It is logical, therefore, that glycogen synthesis is stimulated when substrate availability and energy levels are high, whereas glycogen degradation is increased when energy levels and available glucose supplies are low.

1. Regulation of glycogen synthesis and degradation in the well-fed state:

In the well-fed state, *glycogen synthase* is allosterically activated by glucose 6-phosphate when it is present in elevated concentrations (Figure 11.9). In contrast, *glycogen phosphorylase* is allosterically inhibited by glucose 6-phosphate, as well as by ATP,

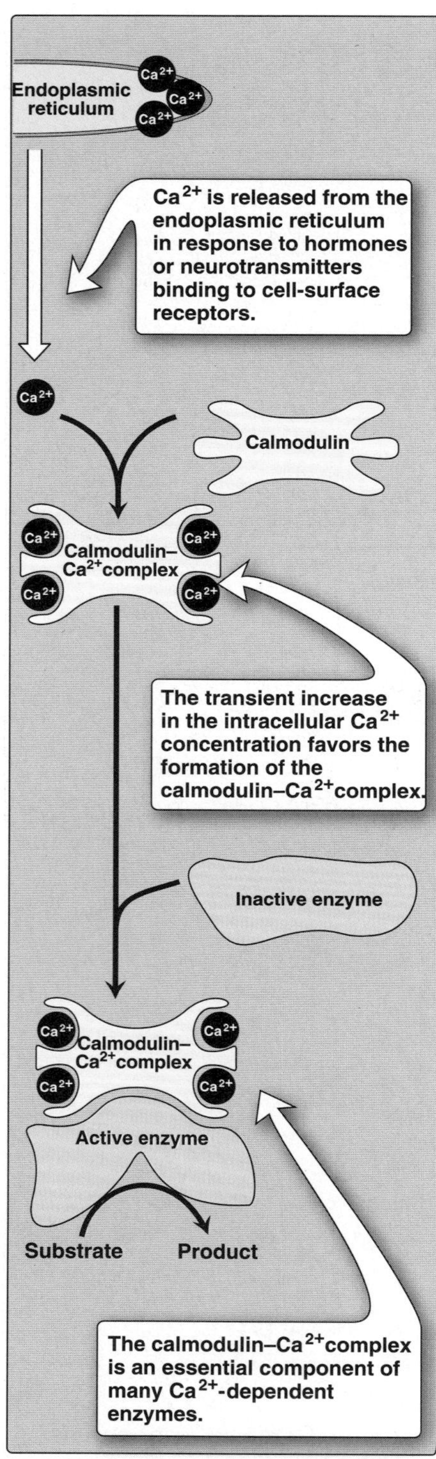

Figure 11.10
Calmodulin mediates many effects of intracellular calcium.

a high-energy signal in the cell. [Note: In liver, glucose also serves as an allosteric inhibitor of *glycogen phosphorylase* .]

2. **Activation of glycogen degradation in muscle by calcium:** During muscle contraction, there is a rapid and urgent need for ATP, the energy for which is supplied by the muscle's store of glycogen. Nerve impulses cause membrane depolarization, which, in turn, promotes **Ca^{2+}** release from the sarcoplasmic reticulum into the sarcoplasm of muscle cells. Ca^{2+} binds to **calmodulin**, one of a family of small, calcium-binding proteins. [Note: Calmodulin is the most widely distributed of these proteins, and is present in virtually all cells.] The binding of four molecules of Ca^{2+} to calmodulin triggers a conformational change such that the activated Ca^{2+}-calmodulin complex binds to and activates protein molecules—often enzymes—that are inactive in the absence of this complex (Figure 11.10). Thus, calmodulin functions as an essential subunit of many complex proteins. One such protein is *phosphorylase kinase*, which is activated by the Ca^{2+}-calmodulin complex without the need for the *kinase* to be phosphorylated by *cAMP-dependent protein kinase* (see below and Figure 11.11). When the muscle relaxes, Ca^{2+} returns to the sarcoplasmic reticulum, and *phosphorylase kinase* becomes inactive. [Note: *Phosphorylase kinase* is maximally active in exercising muscle when it is both phosphorylated and bound to Ca^{2+}.]

3. **Activation of glycogen degradation in muscle by AMP:** Muscle *glycogen phosphorylase* is active in the presence of the high AMP concentrations that occur in the muscle under extreme conditions of anoxia and ATP depletion. AMP binds to the inactive form of *glycogen phosphorylase*, causing its activation without phosphorylation (see Figure 11.11).

B. Activation of glycogen degradation by cAMP-directed pathway

The binding of hormones, such as glucagon or epinephrine, to membrane receptors signals the need for glycogen to be degraded—either to elevate blood glucose levels or to provide energy for exercising muscle.

1. **Activation of protein kinase:** Binding of glucagon or epinephrine to their specific cell-membrane receptors results in the cAMP-mediated activation of *cAMP-dependent protein kinase*. This enzyme is a tetramer, having two regulatory subunits (R) and two catalytic subunits (C). cAMP binds to the regulatory subunit dimer, releasing individual catalytic subunits that are active (see Figure 11.11). [Note: When cAMP is removed, the inactive tetramer, R_2C_2, is again formed.]

2. **Activation of phosphorylase kinase:** *Phosphorylase kinase* exists in two forms: an **inactive "b" form** and an **active "a" form**. Active *cAMP-dependent protein kinase* phosphorylates the inactive form of *phosphorylase kinase*, resulting in its activation (see Figure 11.11). [Note: The phosphorylated enzyme can be inactivated by the hydrolytic removal of its phosphate by *protein phosphatase 1*. This enzyme is activated by a kinase-mediated signal cascade initiated by insulin (see p. 309).]

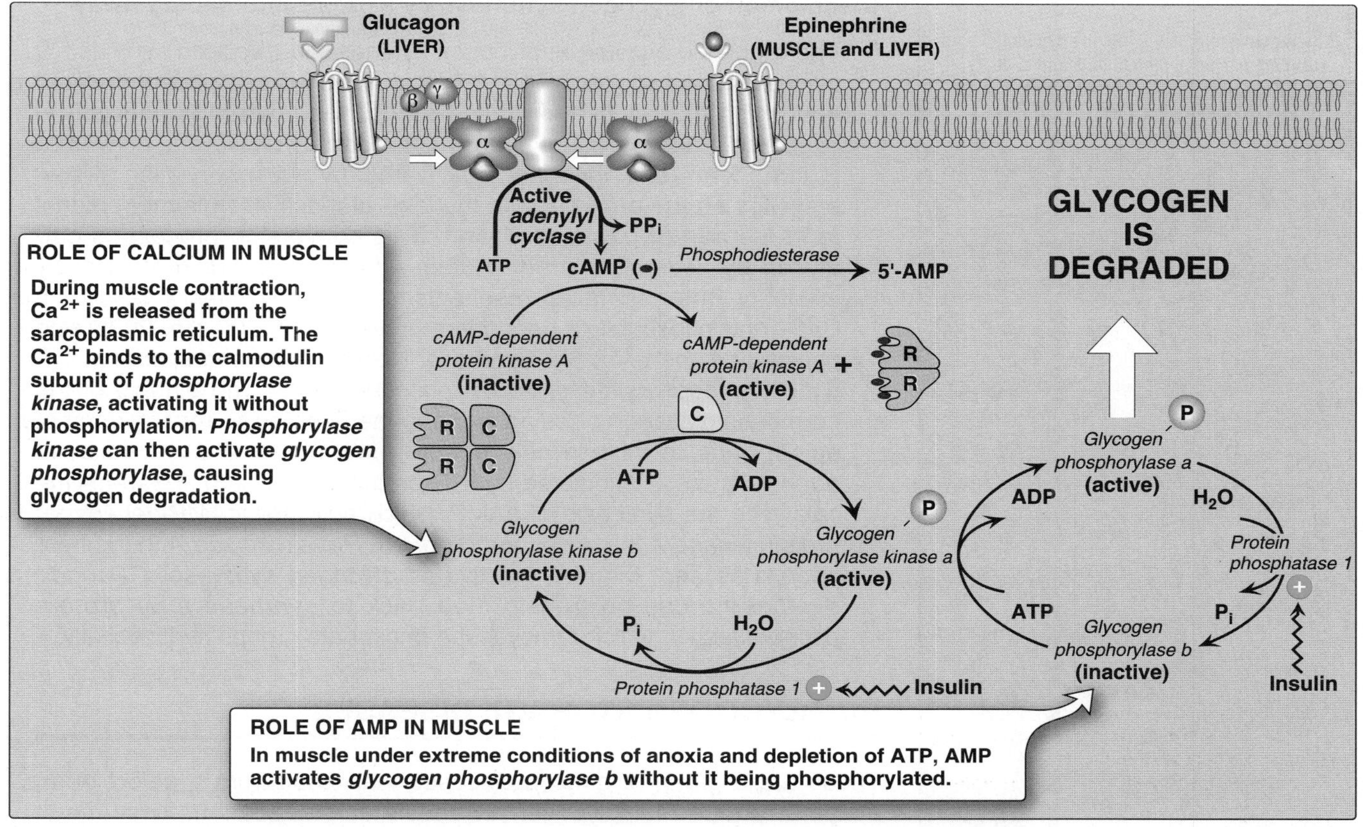

Figure 11.11
Stimulation and inhibition of glycogen degradation.

3. **Activation of glycogen phosphorylase:** *Glycogen phosphorylase* also exists in two forms: the dephosphorylated, **inactive b form** and the phosphorylated, **active a form**. Active *phosphorylase kinase* phosphorylates *glycogen phosphorylase a,* which then begins glycogen breakdown. *Phosphorylase a* is reconverted to *phosphorylase b* by the hydrolysis of its phosphate by *protein phosphatase 1*. [Note: When glucose is bound to *glycogen phosphorylase a*, thus signalling that glycogen degradation is no longer required, the complex becomes a better substrate for *protein phosphatase 1*. In addition, when muscle *glycogen phosphorylase b* is bound to glucose, it cannot be allosterically activated by AMP. In the muscle, insulin indirectly inhibits the enzyme by increasing the uptake of glucose (see p. 308), leading to an increased level of glucose 6-phosphate—a potent allosteric inhibitor of *glycogen phosphorylase* (see Figure 11.9.]

4. **Summary of the regulation of glycogen degradation:** The cascade of reactions listed above result in glycogen degradation. The large number of sequential steps serves to amplify the effect of the hormonal signal, that is, a few hormone molecules binding to their receptors result in a number of *protein kinase* molecules being activated that can each activate many *phosphorylase kinase* molecules. This causes the production of many active *glycogen phosphorylase a* molecules that can degrade glycogen.

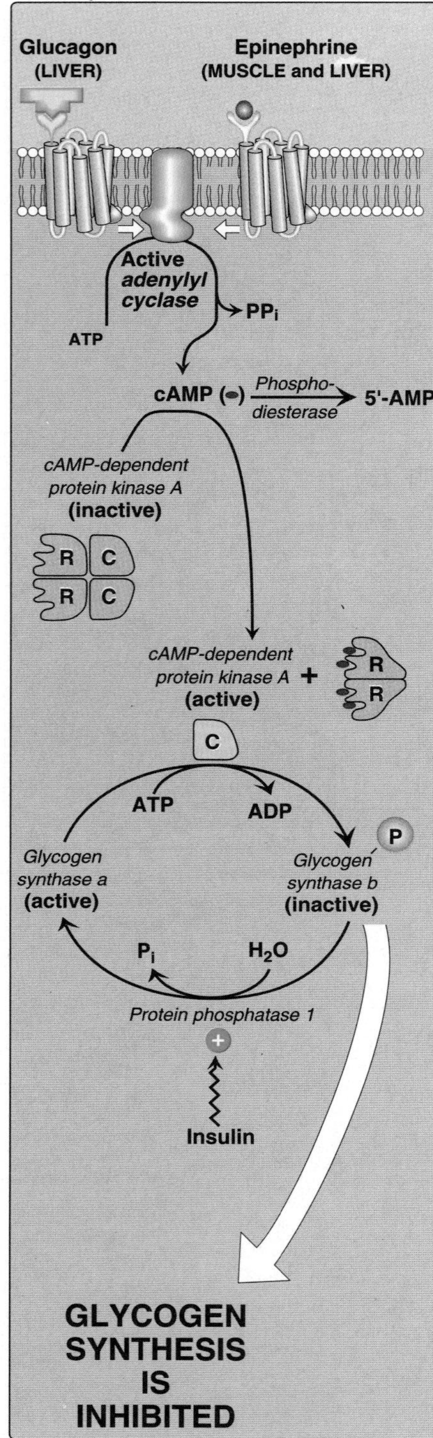

Figure 11.12
Hormonal regulation of glycogen synthesis. [Note: In contrast to *glycogen phosphorylase*, *glycogen synthase* is inactive if phosphorylated.]

C. Inhibition of glycogen synthesis by a cAMP-directed pathway

The regulated enzyme in glycogen synthesis is *glycogen synthase*. It also exists in two forms, the "a" form, which is not phosphorylated and is the most active form, and the "b" form, which is phosphorylated and inactive (Figure 11.12). *Glycogen synthase a* is converted to the *b* form (and, therefore, inactivated) by phosphorylation at several sites on the enzyme, with the level of inactivation is proportional to its degree of phosphorylation. This conversion process is catalyzed by several different protein kinases that are regulated by cAMP or other signaling mechanisms. [Note: *Protein kinase C*, a Ca^{2+}- and phospholipid-dependent protein kinase (see p. 203), also phosphorylates *glycogen synthase*. Neither *protein kinase A* nor *C* directly phosphorylate *glycogen phosphorylase*.] The binding of the hormones glucagon or epinephrine to the hepatocyte receptors, or of epinephrine to muscle cell receptors, results in the activation of *adenylyl cyclase*, mediated by a G-protein (see p. 93). This enzyme catalyzes the synthesis of cAMP, which activates *cAMP-dependent protein kinase A*, as described on p. 93. *Protein kinase A* then phosphorylates and, thereby, inactivates *glycogen synthase*. *Glycogen synthase b* can be transformed back to *synthase a* by *protein phosphatase 1,* which removes the phosphate groups hydrolytically.

VI. GLYCOGEN STORAGE DISEASES

These are a group of genetic diseases that result from a defect in an enzyme required for glycogen synthesis or degradation. They result either in formation of glycogen that has an abnormal structure, or in the accumulation of excessive amounts of normal glycogen in specific tissues as a result of impaired degradation. A particular enzyme may be defective in a single tissue, such as the liver, or the defect may be more generalized, affecting liver, muscle, kidney, intestine, and myocardium. The severity of the glycogen storage diseases (GSDs) ranges from fatal in infancy to mild disorders that are not life-threatening. Some of the more prevalent GSDs are illustrated in Figure 11.8.

VII. CHAPTER SUMMARY

The **main stores** of glycogen in the body are found in **skeletal muscle**, where they serve as a **fuel reserve** for the synthesis of ATP during **muscle contraction**, and in the **liver**, where glycogen is used to **maintain the blood glucose** concentration, particularly during the **early stages of a fast**. Glycogen is a **highly branched** polymer of **α-D-glucose**. The primary glycosidic bond is an **α(1→4) linkage**. After about eight to ten glucosyl residues, there is a **branch** containing an **α(1→6) linkage**. **UDP-glucose**, the **building block** of glycogen, is synthesized from **glucose 1-phosphate** and **UTP** by **UDP-glucose pyrophosphorylase**. Glucose from UDP-glucose is transferred to the non-reducing ends of glycogen chains by **glycogen synthase**, which makes **α(1→4) linkages**. **Branches** are formed by **amylo-α(1→4) → α(1→6)-transglucosidase**, which transfers a chain of five to eight glucosyl residues from the nonreducing end of the glycogen chain (**breaking** an **α(1→4) linkage**), and attaches it with an **α(1→4) linkage** to another residue in the chain.

Glycogen phosphorylase cleaves the **α(1→4) bonds** between glucosyl residues at the **nonreducing ends** of the glycogen chains, producing **glucose 1-phosphate**. It requires **pyridoxyl phosphate** as a coenzyme. This sequential degradation continues until four glucosyl units remain on each chain before a branch point. The resulting structure is called a **limit dextrin**. **Oligo-α(1→4)→α(1→4)-glucan transferase** [common name, **glucosyl (4:4) transferase**] removes the outer three of the four glucosyl residues attached at a branch, and transfers them to the non-reducing end of another chain where they can be converted to glucose 1-phosphate by glycogen phosphorylase. Next, the remaining single glucose residue attached in an **α-1,6 linkage** is removed hydrolytically by the **amylo-α-1,6-glucosidase** activity, releasing **free glucose**. **Glucose 1-phosphate** is converted to **glucose 6-phosphate** by **phosphoglucomutase**. In the **muscle**, glucose 6-phosphate enters glycolysis. In the **liver**, the phosphate is removed by **glucose 6-phosphatase**, releasing **free glucose** that can be used to maintain blood glucose levels at the beginning of a fast. A **deficiency** of the **phosphatase** causes **glycogen storage disease type 1** (**Von Gierke disease**). This disease results in an inability of the liver to provide free glucose to the body during a fast. It affects both glycogen degradation and the last step in gluconeogenesis. **Glycogen synthase** and **glycogen phosphorylase** are **allosterically regulated**. In the well-fed state, **glycogen synthase** is **activated** by **glucose 6-phosphate**, but **glycogen phosphorylase** is **inhibited** by **glucose 6-phosphate**, as well as by **ATP**. In the **liver**, **glucose** also serves an an allosteric **inhibitor** of **glycogen phosphorylase**. Ca^{2+} is released from the **sarcoplasmic reticulum** during exercise. It activates **phosphorylase kinase** in the muscle by binding to the enzyme's

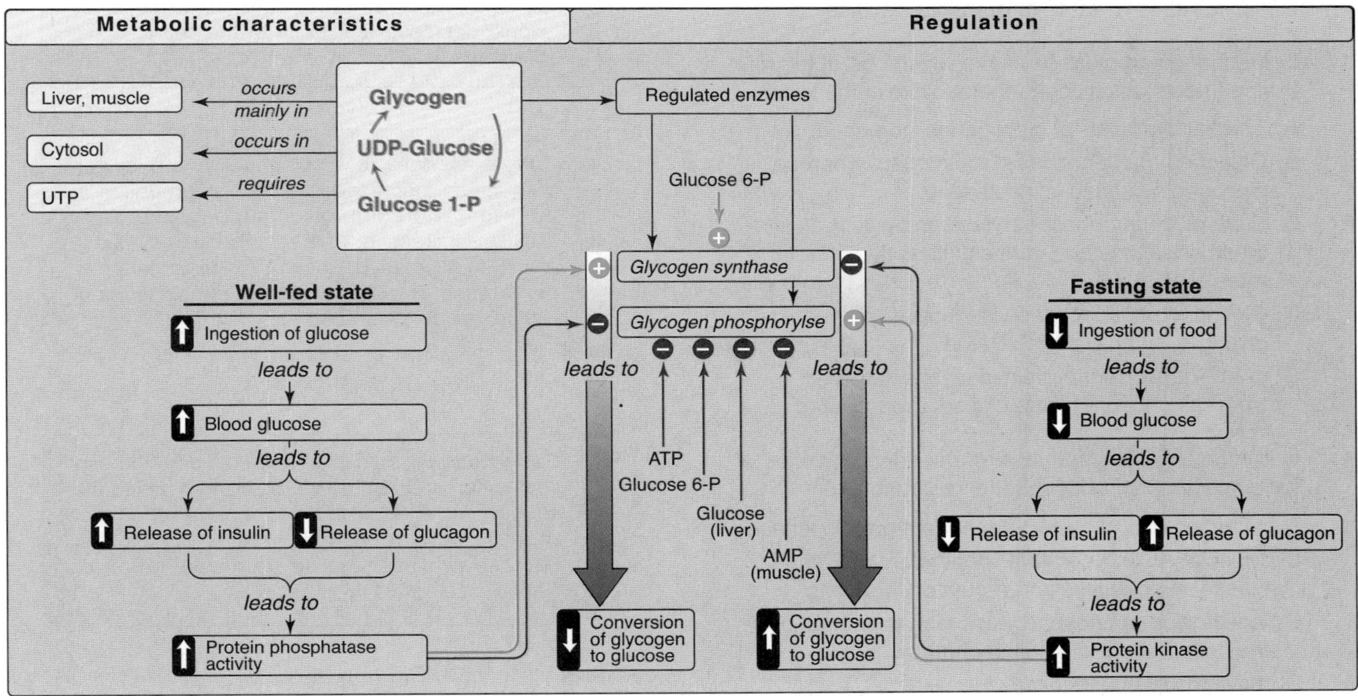

Figure 11.13
Key concept map for glycogen metabolism in liver.

calmodulin subunit. This allows the enzyme to activate **glycogen phosphorylase**, thereby causing glycogen degradation. Glycogen synthesis and degradation are reciprocally regulated by the same hormonal signals, namely, an **elevated insulin** level results in overall **increased glycogen synthesis** and **decreased glycogen degradation**, whereas an **elevated glucagon** (or **epinephrine**) level causes **increased glycogen degradation** and **decreased glycogen synthesis**. Key enzymes are phosphorylated by a family of **protein kinases**, some of which are **cAMP-dependent** (a compound increased by **glucagon** and **epinephrine**). Phosphate groups are removed by **protein phosphatase 1** (activated when **insulin** levels are elevated).

Study Questions

Choose the ONE correct answer

11.1 A two year-old boy was brought into the emergency room, suffering from severe fasting hypoglycemia. On physical examination, he was found to have hepatomegaly. Laboratory tests indicated that he also had hyperlacticacidemia and hyperuricemia. A liver biopsy indicated that hepatocytes contained greater than normal amounts of glycogen that was of normal structure. The child was subsequently found to be missing which of the following enzymes?

A. Glycogen synthase
B. Glycogen phosphorylase
C. Glucose 6-phosphatase
D. Amylo-α(1→6)-glucosidase
E. Amylo-α(1→4)→α(1→6)-transglucosylase

Correct answer = C. A deficiency of glucose 6-phosphatase (Von Gierke disease) prevents the liver from releasing free glucose into the blood, causing severe fasting hypoglycemia, hyperlacticacidemia, and hyperuricemia. A deficiency of glycogen phosphorylase would result in a decrease in glycogen degradation, causing fasting hypoglycemia, but not the other symptoms. A deficiency of glycogen synthase would result in lower amounts of stored glycogen. Amylo-α(1→6)-glucosidase removes single glucosyl residues attached to the glycogen chain through an α(1→6)-glycosidic bond. A deficiency in this enzyme would result in a decreased ability of the cell to degrade glycogen branches, producing limit dextrins. Amylo-α(1→4)→α(1→6)-transglucosidase deficiency would decrease the ability of the cell to make branches.

11.2 Epinephrine and glucagon have which one of the following effects on glycogen metabolism in the liver?

A. The net synthesis of glycogen is increased.
B. Glycogen phosphorylase is activated, whereas glycogen synthase is inactivated.
C. Both glycogen phosphorylase and glycogen synthase are activated but at significantly different rates.
D. Glycogen phosphorylase is inactivated, whereas glycogen synthase is activated.
E. cAMP-dependent protein kinase is activated, whereas phosphorylase kinase is inactivated.

Correct answer = B. Epinephrine and glucagon both cause increased glycogen degradation in the liver. Therefore, glycogen phosphorylase activity is increased, whereas glycogen synthase activity is decreased. Both cAMP-dependent protein kinase and its substrate, phosphorylase kinase, are also activated.

11.3 In contracting skeletal muscle, a sudden elevation of the cytosolic Ca^{2+} concentration will result in:

A. activation of cyclic AMP-dependent protein kinase.
B. dissociation of cyclic AMP-dependent protein kinase into catalytic and regulatory subunits.
C. inactivation of phosphorylase kinase caused by the action of a protein phosphatase.
D. activation of phosphorylase kinase.
E. conversion of cAMP to AMP by phosphodiesterase.

Correct answer = D. Ca^{2+} released from the sarcoplasmic reticulum during exercise binds to the calmodulin subunit of phosphorylase kinase, thereby activating this enzyme. The other choices are not caused by an elevation of cytosolic calcium.

Metabolism of Monosaccharides and Disaccharides

12

I. OVERVIEW

Although many monosaccharides have been identified in nature, only a few sugars appear as metabolic intermediates or as structural components in mammals. Glucose is the most common monosaccharide consumed by humans, and its metabolism has been discussed extensively. However, two other monosaccharides—fructose and galactose—occur in significant amounts in the diet, and make important contributions to energy metabolism. In addition, galactose is an important component of cell structural carbohydrates. Figure 12.1 shows the metabolism of fructose and galactose as part of the essential pathways of energy metabolism in the body.

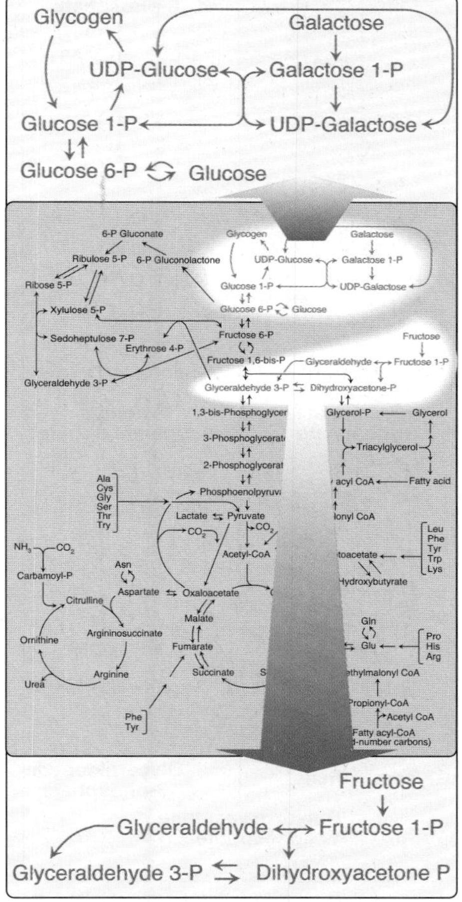

Figure 12.1
Galactose and fructose metabolism as part of the essential pathways of energy metabolism (see Figure 8.2, p. 90, for a more detailed view of the overall reactions of metabolism).

II. FRUCTOSE METABOLISM

About ten percent of the calories contained in the Western diet are supplied by fructose (approximately fifty g/day). The major source of fructose is the disaccharide **sucrose**, which, when cleaved in the intestine, releases equimolar amounts of fructose and glucose (see p. 86). Fructose is also found as a free monosaccharide in high-fructose corn syrup (55 percent fructose/45 percent glucose, which is used to sweeten most cola drinks), in many fruits, and in honey. Entry of fructose into cells is not insulin-dependent (unlike that of glucose into certain tissues, see p. 95), and, in contrast to glucose, fructose does not promote the secretion of insulin.

A. Phosphorylation of fructose

For fructose to enter the pathways of intermediary metabolism, it must first be phosphorylated (Figure 12.2). This can be accomplished by either *hexokinase* or *fructokinase* (also called *ketohexokinase*). *Hexokinase* phosphorylates glucose in all cells of the body (see p. 96), and several additional hexoses can serve as substrates for this enzyme. However, it has a low affinity (that is, a high K_m, see p. 59) for fructose. Therefore, unless the intracellular concentration of fructose becomes unusually high, the normal presence of saturating concentrations of glucose means that little fructose is converted

Lippincott's Illustrated Reviews: Biochemistry, Third Edition,
by Pamela C. Champe and Richard A. Harvey.
Lippincott Williams & Wilkins, Baltimore, MD © 2005.

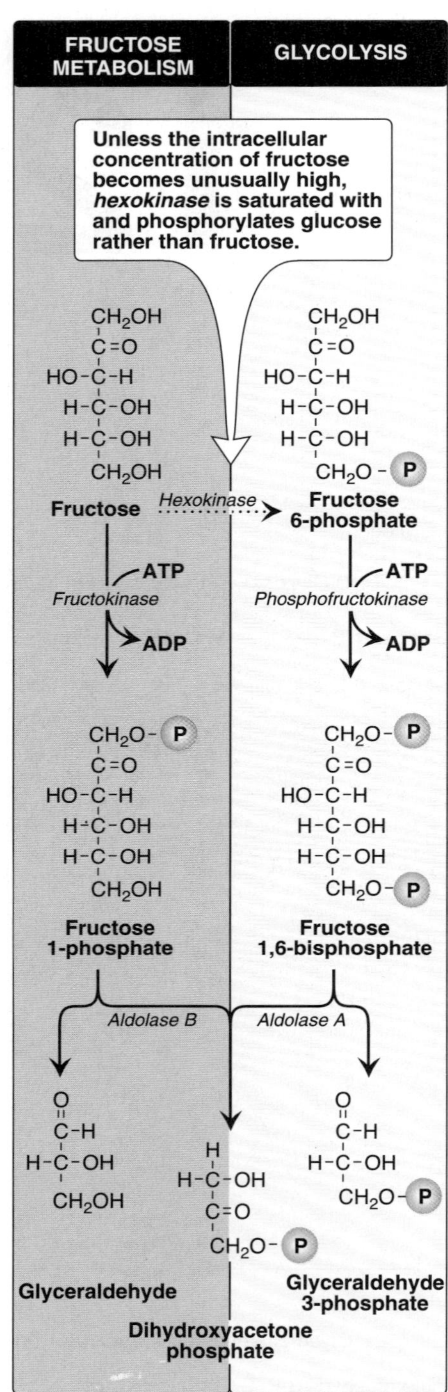

Figure 12.2
Phosphorylation products of fructose and their cleavage.

to fructose 6-phosphate by *hexokinase*. *Fructokinase* provides the primary mechanism for fructose phosphorylation (see Figure 12.2). It is found in the **liver** (which processes most of the dietary fructose), **kidney**, and the **small intestinal mucosa**, and converts fructose to **fructose 1-phosphate**, using ATP as the phosphate donor. [Note: These three tissues also contain *aldolase B*, discussed below.]

B. Cleavage of fructose 1-phosphate

Fructose 1-phosphate is not converted to fructose 1,6-bisphosphate as is fructose 6-phosphate (see p. 97), but is cleaved by *aldolase B* (also called *fructose 1-phosphate aldolase*) to **dihydroxyacetone phosphate** (**DHAP**) and **glyceraldehyde**. [Note: Both *aldolase A* (found in all tissues) and *aldolase B* cleave fructose 1,6-bisphosphate produced during glycolysis to DHAP and glyceraldehyde 3-phosphate (see p. 99).] DHAP can directly enter glycolysis or gluconeogenesis, whereas glyceraldehyde can be metabolized by a number of pathways, as illustrated in Figure 12.3.

C. Kinetics of fructose metabolism

The rate of fructose metabolism is more rapid than that of glucose because the trioses formed from fructose 1-phosphate bypass *phosphofructokinase*—the major rate-limiting step in glycolysis (see p. 97). [Note: Loading the liver with fructose, for example, by intravenous infusion, can significantly elevate the rate of lipogenesis, caused by the enhanced production of acetyl CoA.]

D. Disorders of fructose metabolism

A deficiency of one of the key enzymes required for the entry of fructose into intermediary metabolic pathways can result in either a benign condition (**fructokinase deficiency**), or a severe disturbance of liver and kidney metabolism as a result of *aldolase B* deficiency (**hereditary fructose intolerance**, **HFI**), which is estimated to occur in 1:20,000 live births. The first symptoms appear when a baby is weaned and begins to be fed food containing sucrose or fructose. Fructose 1-phosphate accumulates, and ATP and inorganic phosphate levels fall significantly, with adenine being converted to uric acid, causing hyperuricemia. The decreased availability of hepatic ATP affects gluconeogenesis (causing hypoglycemia with vomiting), and protein synthesis (causing a decrease in blood clotting factors and other essential proteins). If fructose (and, therefore, sucrose) is not removed from the diet, liver failure and death can occur. Diagnosis of HFI can be made on the basis of fructose in the urine, or by a restriction fragment length polymorphism test (see p. 454).

E. Conversion of mannose to fructose 6-phosphate

Mannose, the C-2 epimer of glucose (see p. 84), is an important component of glycoproteins (see p. 164). *Hexokinase* phosphorylates mannose, producing mannose 6-phosphate, which, in turn, is (reversibly) isomerized to fructose 6-phosphate by *phosphomannose isomerase*. [Note: There is little mannose in dietary carbohydrates. Most intracellular mannose is synthesized from fructose, or is preexisting mannose produced by the degradation of structural carbohydrates and salvaged by *hexokinase*.]

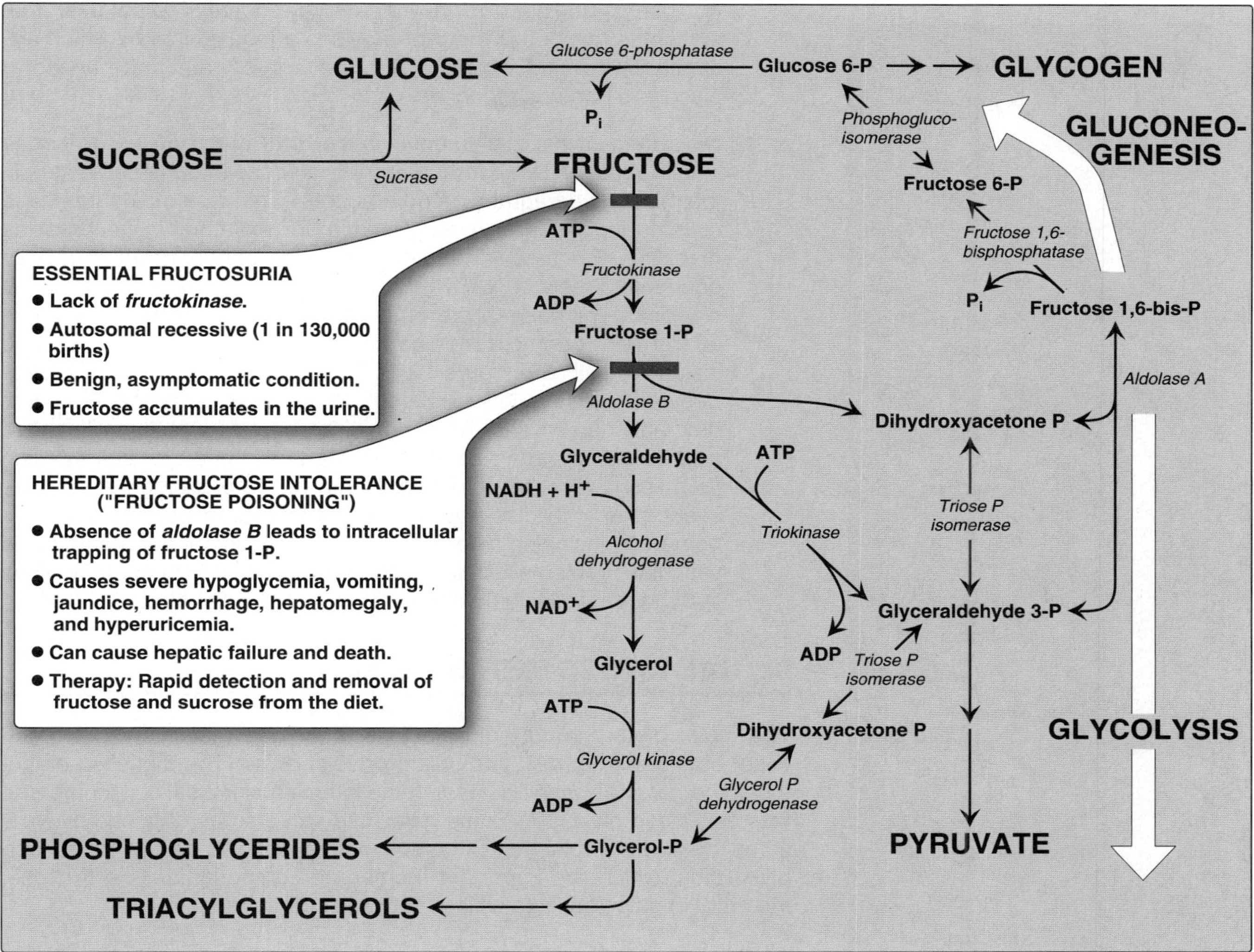

Figure 12.3
Summary of fructose metabolism.

F. Conversion of glucose to fructose via sorbitol

Most sugars are rapidly phosphorylated following their entry into cells. They are thereby trapped within the cells, because organic phosphates cannot freely cross membranes without specific transporters. An alternate mechanism for metabolizing a monosaccharide is to convert it to a **polyol** by the reduction of an aldehyde group, thereby producing an additional hydroxyl group.

1. **Synthesis of sorbitol:** *Aldose reductase* reduces glucose, producing sorbitol (glucitol, Figure 12.4). This enzyme is found in many tissues, including the lens, retina, Schwann cells of peripheral nerves, liver, kidney, placenta, red blood cells, and in cells of the ovaries and seminal vesicles. In cells of the liver, ovaries, sperm, and seminal vesicles, there is a second enzyme, *sorbitol dehydrogenase,* that can oxidize the sorbitol to produce fructose (see Figure 12.4). The two-reaction pathway from glucose to fructose in the seminal vesicles is for the benefit of sperm cells, which use

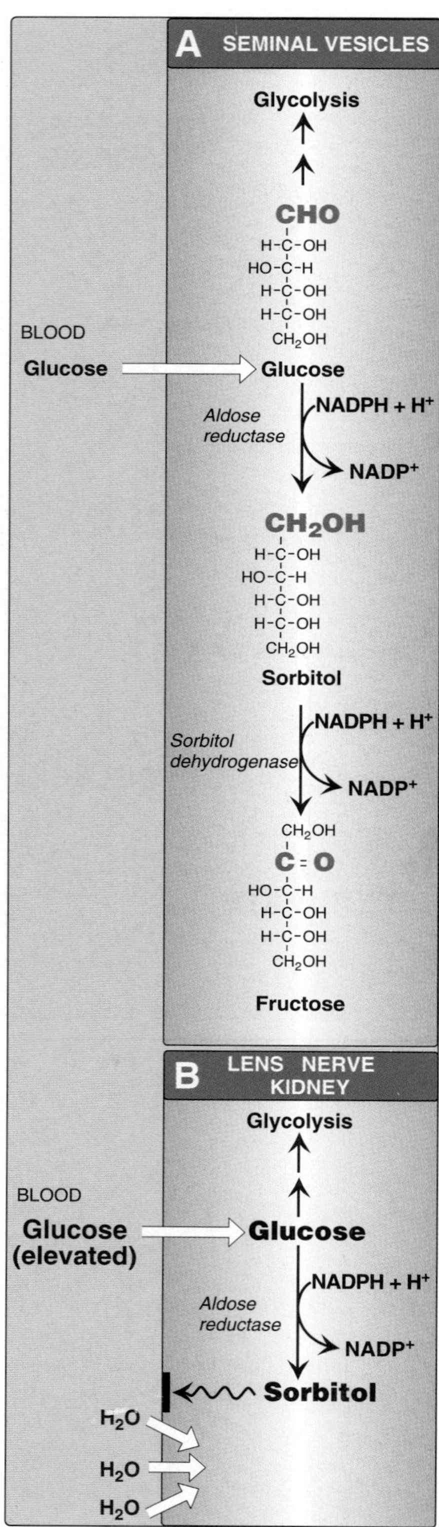

Figure 12.4
Sorbitol metabolism.

fructose as a major carbohydrate energy source. The pathway from sorbitol to fructose in the liver provides a mechanism by which any available sorbitol is converted into a substrate that can enter glycolysis or gluconeogenesis.

2. **The effect of hyperglycemia on sorbitol metabolism:** Because insulin is not required for the entry of glucose into the cells listed in the previous paragraph, large amounts of glucose may enter these cells during times of hyperglycemia, for example, in uncontrolled diabetes. Elevated intracellular glucose concentrations and an adequate supply of NADPH cause *aldose reductase* to produce a significant increase in the amount of sorbitol, which cannot pass efficiently through cell membranes and, therefore, remains trapped inside the cell (see Figure 12.4). This is exacerbated when *sorbitol dehydrogenase* is low or absent, for example, in retina, lens, kidney, and nerve cells. As a result, sorbitol accumulates in these cells, causing strong osmotic effects and, therefore, cell swelling as a result of water retention. Some of the pathologic alterations associated with diabetes can be attributed, in part, to this phenomenon, including cataract formation, peripheral neuropathy, and vascular problems leading to nephropathy and retinopathy. (See p. 343 for a disussion of the complications of diabetes.)

III. GALACTOSE METABOLISM

The major dietary source of galactose is **lactose** (galactosyl β-1,4-glucose) obtained from milk and milk products. [Note: The digestion of lactose by *β-galactosidase* (*lactase*) of the intestinal mucosal cell membrane was discussed on p. 86.] Some galactose can also be obtained by lysosomal degradation of complex carbohydrates, such as glycoproteins, and glycolipids, which are important membrane components. Like fructose, the entry of galactose into cells is not insulin-dependent.

A. Phosphorylation of galactose

Like fructose, galactose must be phosphorylated before it can be further metabolized. Most tissues have a specific enzyme for this purpose, *galactokinase*, which produces galactose 1-phosphate (Figure 12.5). ATP is the phosphate donor.

B. Formation of UDP-galactose

Galactose 1-phosphate cannot enter the glycolytic pathway unless it is first converted to UDP-galactose (see Figure 12.5). This occurs in an exchange reaction, in which UMP is removed from UDP-glucose (leaving behind glucose 1-phosphate), and is then transferred to the galactose 1-phosphate, producing UDP-galactose (Figure 12.6). The enzyme that catalyzes this reaction is *galactose 1-phosphate uridyltransferase.*

C. Use of UDP-galactose as a carbon source for glycolysis or gluconeogenesis

In order for UDP-galactose to enter the mainstream of glucose metabolism, it must first be converted to its C-4 epimer, UDP-glucose, by *UDP-hexose 4-epimerase.* This "new" UDP-glucose

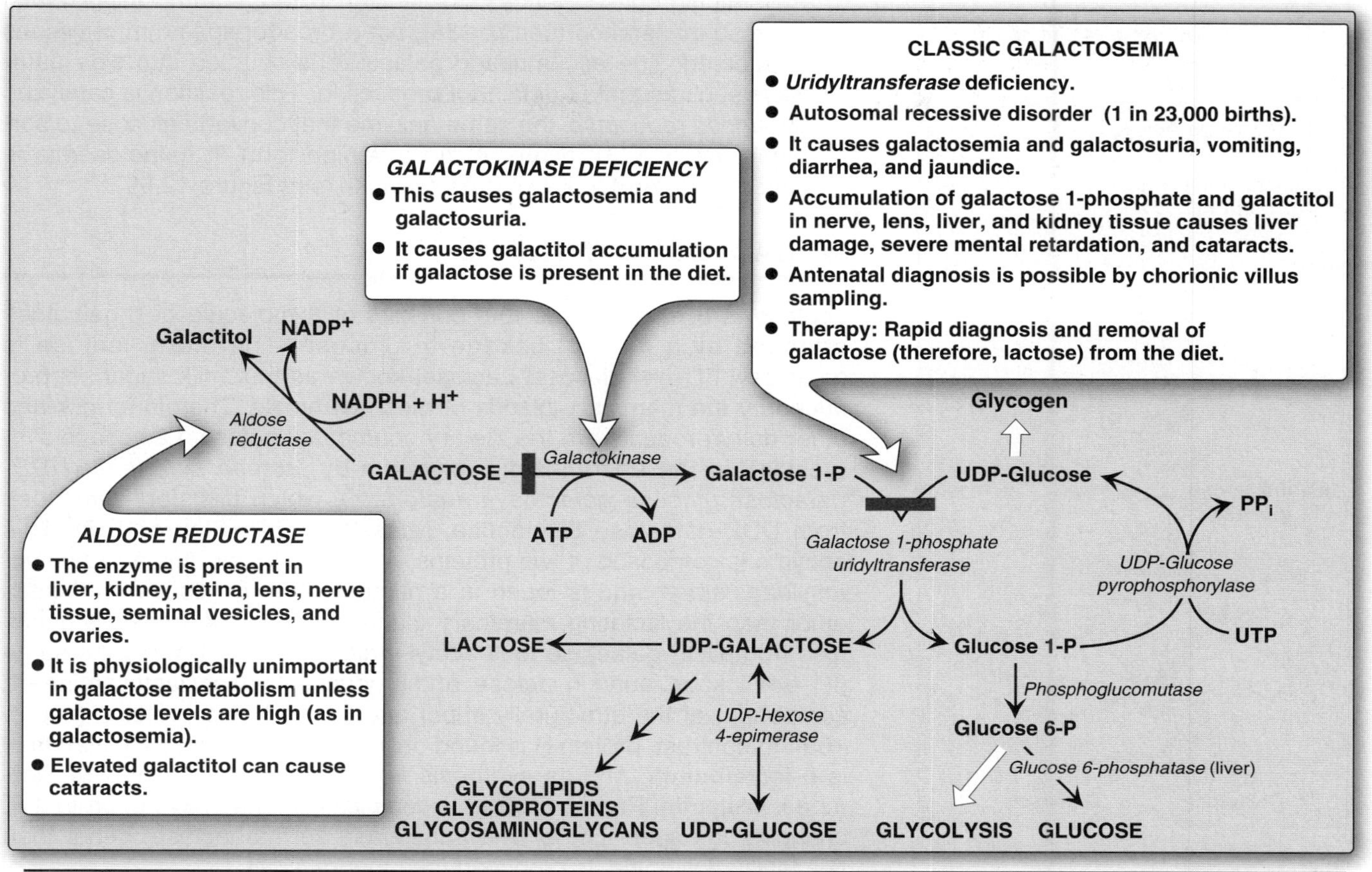

CLASSIC GALACTOSEMIA

- ***Uridyltransferase*** **deficiency.**
- **Autosomal recessive disorder (1 in 23,000 births).**
- **It causes galactosemia and galactosuria, vomiting, diarrhea, and jaundice.**
- **Accumulation of galactose 1-phosphate and galactitol in nerve, lens, liver, and kidney tissue causes liver damage, severe mental retardation, and cataracts.**
- **Antenatal diagnosis is possible by chorionic villus sampling.**
- **Therapy: Rapid diagnosis and removal of galactose (therefore, lactose) from the diet.**

GALACTOKINASE DEFICIENCY

- **This causes galactosemia and galactosuria.**
- **It causes galactitol accumulation if galactose is present in the diet.**

ALDOSE REDUCTASE

- **The enzyme is present in liver, kidney, retina, lens, nerve tissue, seminal vesicles, and ovaries.**
- **It is physiologically unimportant in galactose metabolism unless galactose levels are high (as in galactosemia).**
- **Elevated galactitol can cause cataracts.**

Galactitol
$NADP^+$
$NADPH + H^+$
Aldose reductase
GALACTOSE
Galactokinase
ATP
ADP
Galactose 1-P
Galactose 1-phosphate uridyltransferase
Glycogen
UDP-Glucose
PP_i
UDP-Glucose pyrophosphorylase
UTP
LACTOSE
UDP-GALACTOSE
Glucose 1-P
Phosphoglucomutase
Glucose 6-P
UDP-Hexose 4-epimerase
Glucose 6-phosphatase (liver)
GLYCOLIPIDS
GLYCOPROTEINS
GLYCOSAMINOGLYCANS
UDP-GLUCOSE
GLYCOLYSIS
GLUCOSE

Figure 12.5
Metabolism of galactose.

(produced from the original UDP-galactose) can then participate in many biosynthetic reactions, as well as being used in the *uridyltransferase* reaction described above, converting another galactose 1-phosphate into UDP-galactose, and releasing glucose 1-phosphate, whose carbons are those of the original galactose. (See Figure 12.5 for a summary of this interconversion.)

D. Role of UDP-galactose in biosynthetic reactions

UDP-galactose can serve as the donor of galactose units in a number of synthetic pathways, including synthesis of lactose (see below), glycoproteins (see p. 164), glycolipids (see p. 207), and glycosaminoglycans (see p. 155). [Note: If galactose is not provided by the diet (for example, when it cannot be released from lactose as a result of a lack of *β-galactosidase* in people who are lactose-intolerant), all tissue requirements for UDP-galactose can be met by the action of *UDP-hexose 4-epimerase* on UDP-glucose, which is efficiently produced from glucose 1-phosphate (see Figure 12.5).]

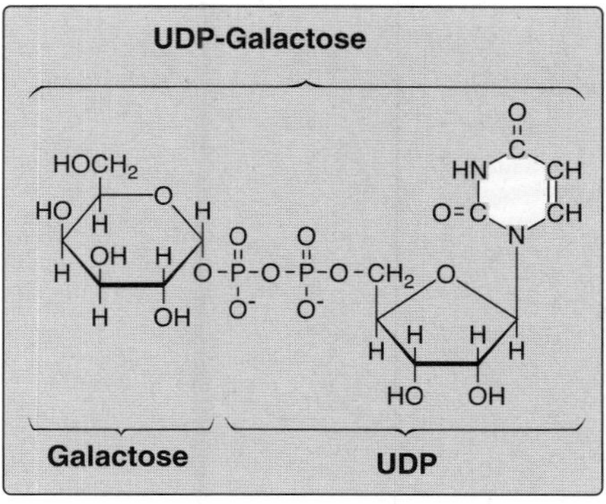

Figure 12.6
Structure of UDP-galactose.

E. Disorders of galactose metabolism

Galactose 1-phosphate uridyltransferase is missing in individuals with **classic galactosemia** (see Figure 12.5). In this disorder, galactose 1-phosphate and, therefore, galactose, accumulate in cells.

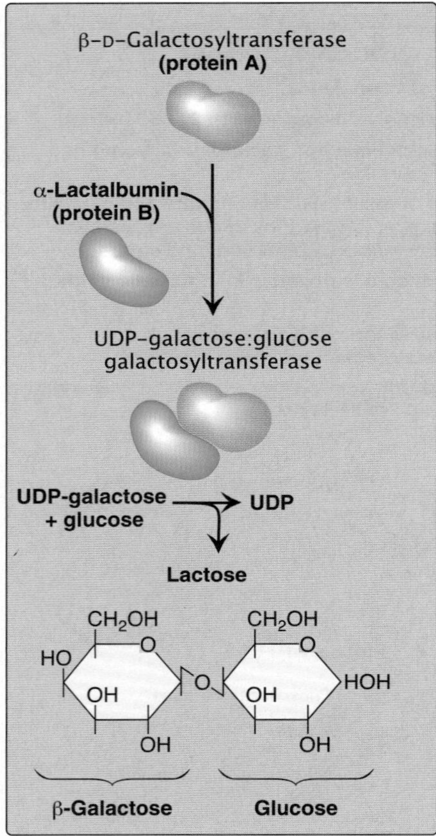

Figure 12.7
Lactose synthesis.

Physiologic consequences are similar to those found in essential fructose intolerance (see p. 136), but a broader spectrum of tissues is affected. The accumulated galactose is shunted into side pathways such as that of **galactitol** production. This reaction is catalyzed by *aldose reductase,* the same enzyme that converts glucose to sorbitol (see p. 137). [Note: A more benign form of galactosemia is caused by a deficiency of *galactokinase* (see Figure 12.5).]

IV. LACTOSE SYNTHESIS

Lactose is a disaccharide that consists of a molecule of β-galactose attached by a β(1→4) linkage to glucose. Therefore, lactose is galactosyl β(1→4)-glucose. Lactose, known as the "milk sugar," is produced by the mammary glands of most mammals. Therefore, milk and other dairy products are the dietary sources of lactose. Lactose is synthesized in the endoplasmic reticulum by *lactose synthase (UDP-galactose:glucose galactosyltransferase),* which transfers galactose from UDP-galactose to glucose, releasing UDP (Figure 12.7). This enzyme is composed of two proteins, A and B. Protein A is a *β-D-galactosyltransferase,* and is found in a number of body tissues. In tissues other than the lactating mammary gland, this enzyme transfers galactose from UDP-galactose to N-acetyl-D-glucosamine, forming the same β(1→4) linkage found in lactose, and producing N-acetyllactosamine—a component of the structurally important N-linked glycoproteins (see p. 165). In contrast, protein B is found only in lactating mammary glands. It is **α-lactalbumin**, and its synthesis is stimulated by the peptide hormone, **prolactin**. Protein B forms a complex with the enzyme, protein A, changing the specificity of that transferase so that lactose, rather than N-acetyllactosamine, is produced (see Figure 12.7).

V. CHAPTER SUMMARY

The major source of fructose is **sucrose**, which when cleaved releases equimolar amounts of fructose and glucose. Entry of fructose into cells is **insulin-independent**. Fructose is first phosphorylated to **fructose 1-phosphate** by **fructokinase**, and then cleaved by **aldolase B** to **dihydroxyacetone phosphate** and **glyceraldehyde**. These enzymes are found in the **liver**, **kidney**, and **small intestinal mucosa**. A **deficiency** of **fructokinase** causes a benign condition (**fructosuria**), but a **deficiency** of **aldolase B** causes **hereditary fructose intolerance**, in which **severe hypoglycemia** and **liver failure** lead to **death** if the amount of fructose (and, therefore, sucrose) in the diet is not severely limited. **Mannose**, an important component of **glycoproteins**, is phosphorylated by **hexokinase** to **mannose 6-phosphate**, which is reversibly isomerized to **fructose 6-phosphate** by **phosphomannose isomerase**. **Glucose** can be reduced to **sorbitol** (**glucitol**) by **aldose reductase** in many tissues, including the **lens**, **retina**, **Schwann cells**, **liver**, **kidney**, **ovaries**, and **seminal vesicles**. In cells of the **liver**, **ovaries**, and **seminal vesicles**, a second enzyme, **sorbitol dehydrogenase**, can oxidize sorbitol to produce **fructose**. **Hyperglycemia** results in the accumulation of sorbitol in those cells lacking *sorbitol dehydrogenase.* The resulting **osmotic events** cause cell swelling, and may contribute to the **cataract formation**, **peripheral neuropathy**, **nephropathy**, and **retinopathy** seen in **diabetes**. The major dietary source of galactose is **lactose**. The entry of galactose into cells is not insulin-dependent. Galactose is first phosphorylated by **galactokinase** which produces **galactose 1-phosphate**. This compound is converted to **UDP-galactose** by **galactose 1-phosphate uridyltransferase**, with the nucleotide supplied by UDP-glucose. A **deficiency** of this enzyme causes **classic galactosemia**. Galactose 1-phosphate accumulates, and excess galactose is converted to **galactitol** by **aldose reductase**. This causes **liver damage**, **severe mental retardation**, and **cataracts**. Treatment requires removal of galactose (and, therefore, lactose) from the diet. In order for UDP-galactose to enter the mainstream of glucose metabolism, it must first be converted to UDP-glucose by **UDP-hexose 4-epimerase**. This enzyme can also be used to produce UDP-galactose from UDP-glucose when the former is required for the synthesis of structural carbohydrates. **Lactose** is a disaccharide that consists of **galactose** and **glucose**. Milk and other dairy products are the dietary sources of lactose. Lactose is synthesized by **lactase synthase** from **UDP-galactose** and **glucose** in the **lactating mammary gland**. The enzyme has two subunits, **protein A** (which is a **galactosyl transferase** found in most cells where it synthesizes **N-acetyllactosamine**) and **protein B** (**α-lactalbumin**, which is found only in the lactating mammary glands, and whose synthesis is stimulated by the peptide hormone, **prolactin**). When both subunits are present, the transferase produces lactose.

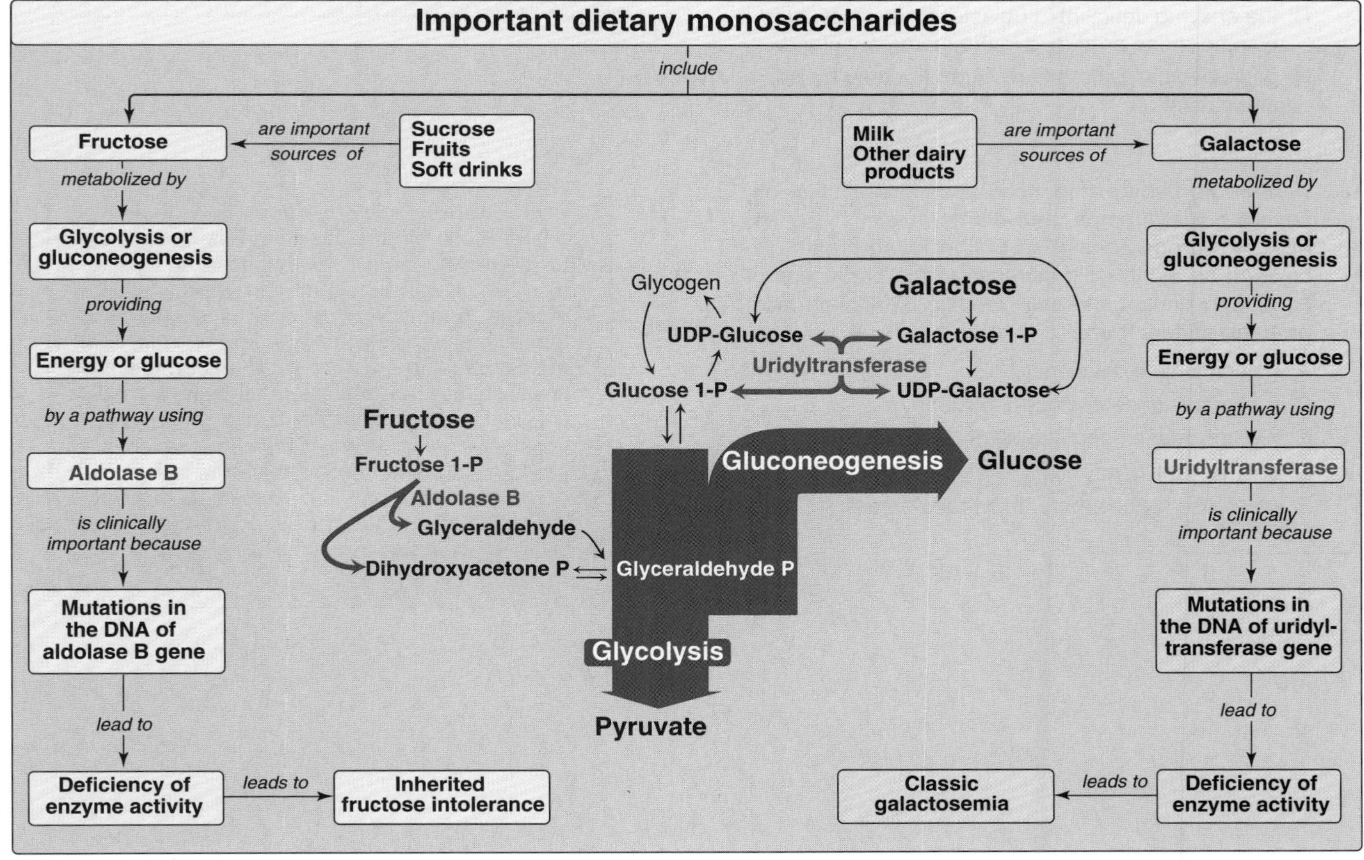

Figure 12.8
Key concept map for metabolism of fructose and galactose.

Study Questions

Choose the ONE correct answer

12.1 Following the intravenous injection of lactose into a rat, none of the lactose is metabolized. However, ingestion of lactose leads to rapid metabolism of this disaccharide. The difference in these observations is a result of:

A. the presence of lactase in the serum.
B. the absence of hepatic galactokinase.
C. the absence of maltase in the serum.
D. the presence of lactase in the intestine.

Correct answer = D. Lactase and maltase are intestinal enzymes not found in the serum. Therefore, ingested lactose is degraded, injected lactose is not. If hepatic galactokinase is absent, the galactose segment of the lactose is not metabolized, but the glucose segment of the lactose can still be metabolized.

12.2 A galactosemic female is able to produce lactose because:

A. free (nonphosphorylated) galactose is the acceptor of glucose transferred by lactose synthase in the synthesis of lactose.
B. galactose can be produced from a glucose metabolite by epimerization.
C. hexokinase can efficiently phosphorylate dietary galactose to galactose 1-phosphate.
D. the enzyme deficient in galactosemia is activated by a hormone produced in the mammary gland.
E. galactose can be produced from fructose by isomerization.

Correct answer = B. UDP-hexose 4-epimerase converts UDP-glucose to UDP-galactose, thus providing the appropriate form of galactose for lactose synthesis. UDP-galactose, not free galactose, is the source of the galactose portion of lactose. Galactose is not converted to galactose 1-phosphate by hexokinase. Isomerization of fructose to galactose does not occur in the human body.

12.3 A newborn baby experienced abdominal distension, severe bowel cramps, and diarrhea after being fed milk. A hydrogen analysis of his exhaled breath discovered an eighty–fold increase in the production of H_2 ninety minutes after milk feeding. The infant most probably suffers from:

A. galactokinase deficiency.
B. β-galactosidase (lactase) deficiency.
C. β-glucosidase (isomaltase) deficiency.
D. sucrase deficiency.
E. galactose 1-phosphate uridyltransferase.

Correct answer = B. Lactase is a digestive enzyme found in the intestine. It is responsible for degrading lactose into glucose and galactose, which can be absorbed by the intestinal mucosa. If lactase is deficient, intestinal flora ferment the disaccharide, producing H_2 and CO_2 gases, causing abdominal distension, severe bowel cramps, and diarrhea after being fed milk. Galactokinase or galactose 1-phosphate uridyltransferase deficiencies affect many tissues, but galactosemia does not produce the symptoms described. Isomaltase and sucrase are digestive enzymes that degrade isomaltose and sucrose—compounds not found in milk.

Pentose Phosphate Pathway and NADPH

13

I. OVERVIEW

The pentose phosphate pathway (also called the **hexose monophosphate shunt**, or **6–phosphogluconate pathway**) occurs in the cytosol of the cell. It consists of two, irreversible oxidative reactions, followed by a series of reversible sugar–phosphate interconversions (Figure 13.1). No ATP is directly consumed or produced in the cycle. Carbon one of glucose 6-phosphate is released as CO_2, and two NADPH are produced for each glucose 6-phosphate molecule entering the oxidative part of the pathway. The rate and direction of the reversible reactions of the pentose phosphate pathway are determined by the supply of and demand for intermediates of the cycle. The pathway provides a major portion of the body's NADPH, which functions as a biochemical reductant. It also produces ribose 5-phosphate required for the biosynthesis of nucleotides (see p. 290), and provides a mechanism for the metabolic use of five-carbon sugars obtained from the diet or the degradation of structural carbohydrates in the body.

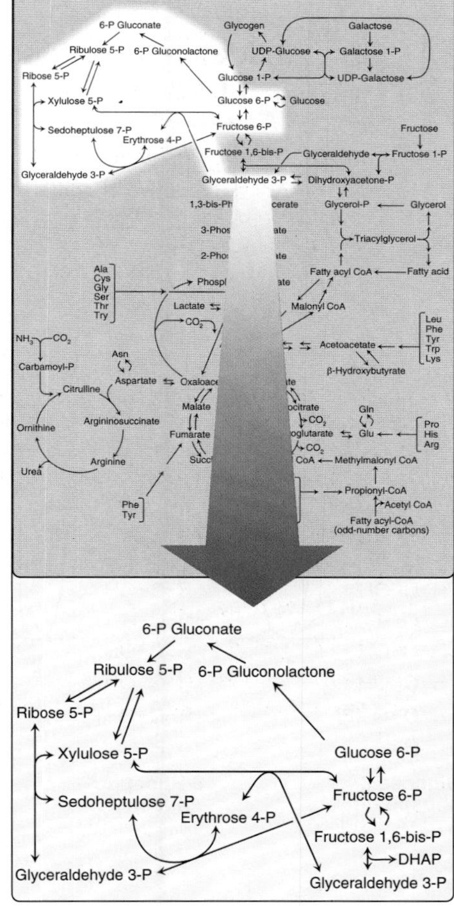

Figure 13.1
Hexose monophosphate pathway shown as a component of the metabolic map (see Figure 8.2, p. 90 for a more detailed view of the metabolic pathway).

II. IRREVERSIBLE OXIDATIVE REACTIONS

The oxidative portion of the pentose phosphate pathway consists of three reactions that lead to the formation of ribulose 5-phosphate, CO_2, and two molecules of NADPH for each molecule of glucose 6-phosphate oxidized (Figure 13.2). This portion of the pathway is particularly important in the liver and lactating mammary glands, which are active in the biosynthesis of fatty acids, in the adrenal cortex, which is active in the NADPH-dependent synthesis of steroids, and in erythrocytes, which require NADPH to keep glutathione reduced.

A. Dehydrogenation of glucose 6-phosphate

Glucose 6-phosphate dehydrogenase (*G6PD*) catalyzes an irreversible oxidation of glucose 6-phosphate to 6-phosphogluconolactone in a reaction that is specific for $NADP^+$ as its coenzyme. The pentose phosphate pathway is regulated primarily at the *glucose 6-phosphate dehydrogenase* reaction. **NADPH** is a potent **competitive**

Lippincott's Illustrated Reviews: Biochemistry, Third Edition,
by Pamela C. Champe and Richard A. Harvey.

inhibitor of the enzyme, and, under most metabolic conditions, the ratio of NADPH/NADP+ is sufficiently high to substantially inhibit enzyme activity. However, with increased demand for NADPH, the ratio of NADPH/NADP+ ratio decreases and flux through the cycle increases in response to the enhanced activity of *glucose 6-phosphate dehydrogenase*. **Insulin** enhances G6PD gene expression, and flux through the pathway increases in the well-fed state.

B. Formation of ribulose 5-phosphate

6-Phosphogluconolactone is hydrolyzed by *6-phosphogluconolactone hydrolase*. The reaction is irreversible and not rate-limiting. The subsequent oxidative decarboxylation of 6-phosphogluconate is catalyzed by *6-phosphogluconate dehydrogenase*. This irreversible reaction produces a pentose sugar–phosphate (ribulose 5-phosphate), CO_2 (from carbon 1 of glucose), and a second molecule of NADPH (see Figure 13.2).

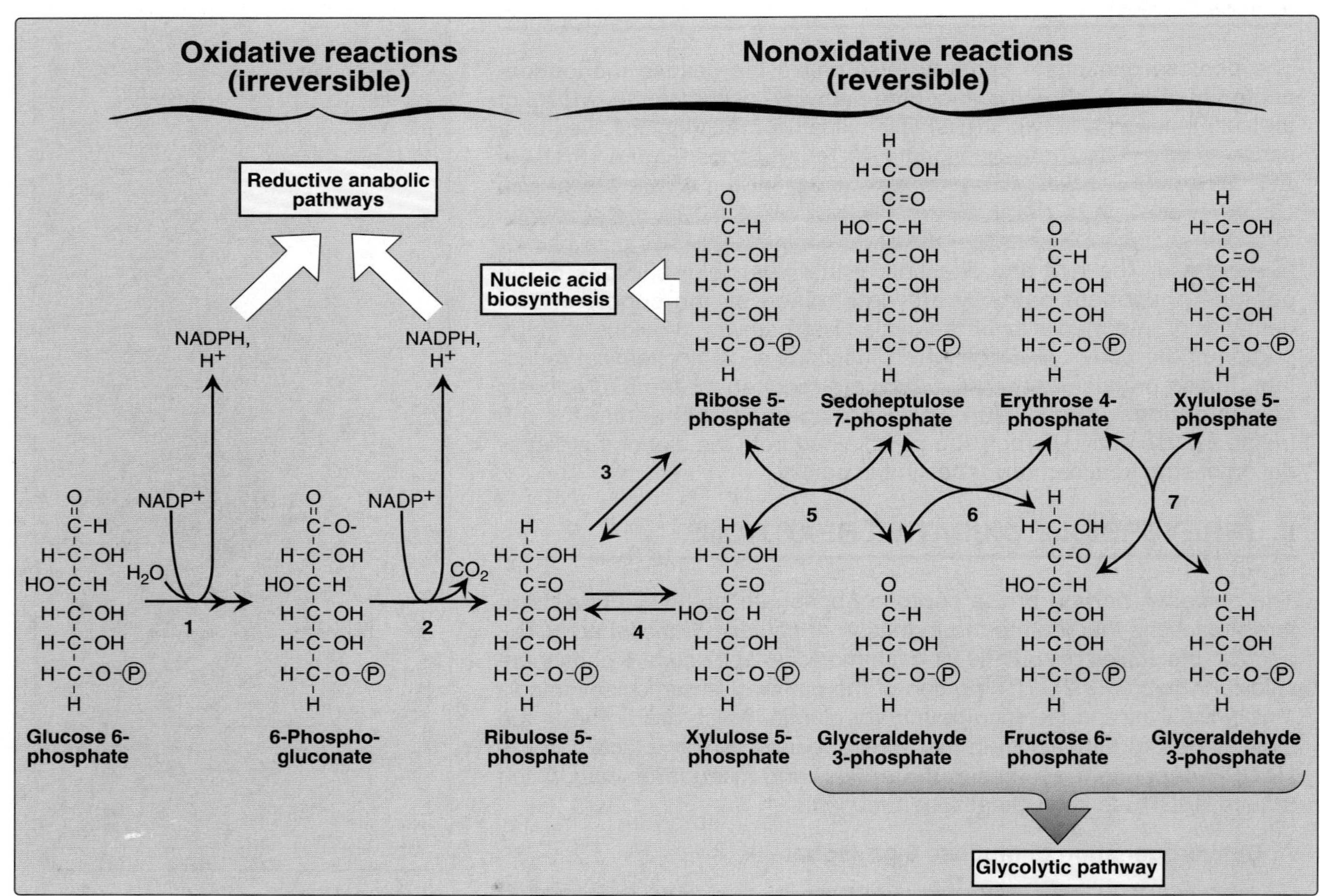

Figure 13.2
Reactions of the hexose monophosphate pathway. Enzymes numbered above are 1) *glucose 6-phosphate dehydrogenase* and *6-phosphogluconolactone hydrolase*, 2) *6-phosphogluconate dehydrogenase*, 3) *ribose 5-phosphate isomerase*, 4) *phosphopentose epimerase*, 5) and 7) *transketolase* (coenzyme: thiamine pyrophosphate), and 6) *transaldolase*.

III. REVERSIBLE NONOXIDATIVE REACTIONS

The nonoxidative reactions of the pentose phosphate pathway occur in all cell types synthesizing nucleotides and nucleic acids. These reactions catalyze the interconversion of three-, four-, five-, six-, and seven-carbon sugars (see Figure 13.2). These reversible reactions permit ribulose 5-phosphate (produced by the oxidative portion of the pathway) to be converted either to ribose 5-phosphate (needed for nucleotide synthesis, see p. 291) or to intermediates of glycolysis—fructose 6-phosphate and glyceraldehyde 3-phosphate. For example, many cells that carry out reductive biosynthetic reactions have a greater need for NADPH than for ribose 5-phosphate. In this case, *transketolase* (which transfers two-carbon units) and *transaldolase* (which transfers three-carbon units) convert the ribulose 5-phosphate produced as an end-product of the oxidative reactions to glyceraldehyde 3-phosphate and fructose 6-phosphate, which are intermediates of glycolysis. In contrast, under conditions in which the demand for ribose for incorporation into nucleotides and nucleic acids is greater than the need for NADPH, the nonoxidative reactions can provide the biosynthesis of ribose 5-phosphate from glyceraldehyde 3-phosphate and fructose 6-phosphate in the absence of the oxidative steps (Figure 13.3).

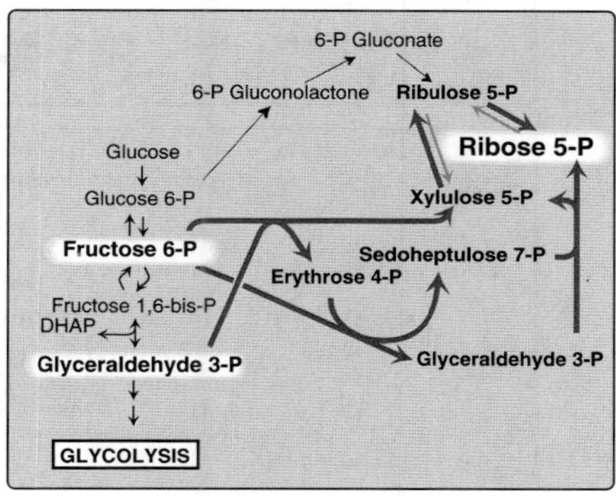

Figure 13.3
Formation of ribose 5-phosphate from intermediates of glycolysis.

IV. USES OF NADPH

The coenzyme $NADP^+$ differs from NAD^+ only by the presence of a phosphate group ($-PO_4^=$) on one of the ribose units (Figure 13.4). This seemingly small change in structure allows $NADP^+$ to interact with $NADP^+$-specific enzymes that have unique roles in the cell. For example, the steady-state ratio of $NADP^+$/NADPH in the cytosol of hepatocytes is approximately 0.1, which favors the use of NADPH in reductive biosynthetic reactions. This contrasts with the high ratio of NAD^+/NADH (approximately 1000 in the cytosol of hepatocytes), which favors an oxidative role for NAD^+. This section summarizes some important $NADP^+$ or NADPH-specific functions.

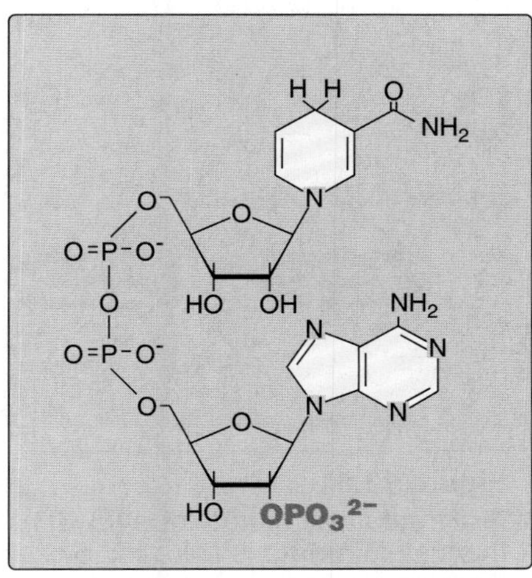

Figure 13.4
Structure of NADPH.

A. Reductive biosynthesis

NADPH can be thought of as a high-energy molecule, much in the same way as NADH. However, the electrons of NADPH are destined for use in reductive biosynthesis, rather than for transfer to oxygen as is the case with NADH (see p. 73). Thus, in the metabolic transformations of the pentose phosphate pathway, part of the energy of glucose 6-phosphate is conserved in NADPH—a molecule that can be used in reactions requiring a high electron-potential electron donor.

B. Reduction of hydrogen peroxide

Hydrogen peroxide is one of a family of **reactive oxygen species** that are formed from the partial reduction of molecular oxygen (Figure 13.5A). These compounds are formed continuously as by-products of aerobic metabolism, through reactions with drugs and environmental toxins, or when the level of antioxidants is diminished, all creating the condition of **oxidative stress**. The highly reactive oxygen intermediates can cause serious chemical damage to DNA, proteins, and

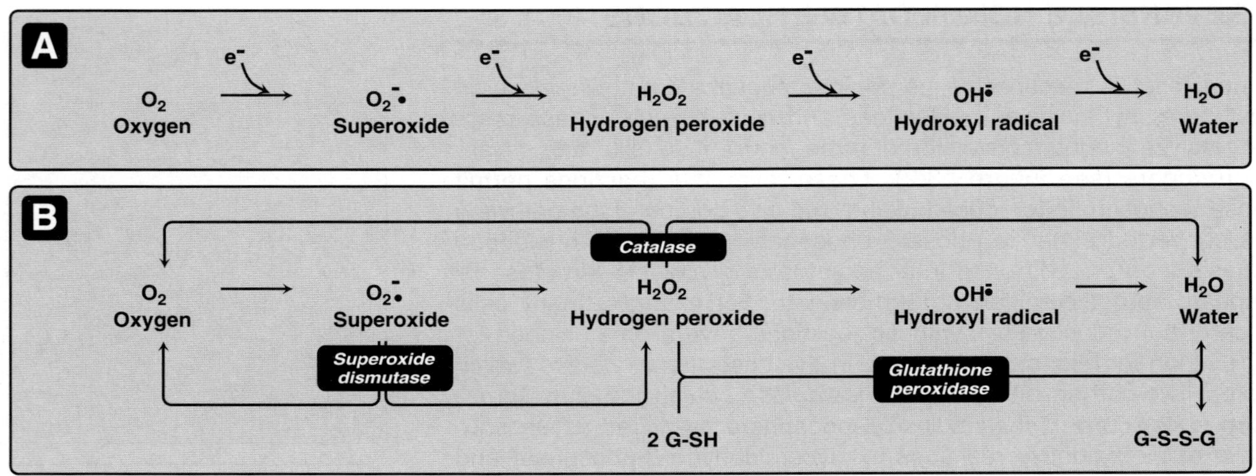

Figure 13.5
A. Formation of reactive intermediates from molecular oxygen. B. Actions of antioxidant enzymes. G-SH = reduced glutathione; G-S-S-G = oxidized glutathione.

unsaturated lipids, and can lead to cell death. These reactive oxygen species have been implicated in a number of pathologic processes, including reperfusion injury, cancer, inflammatory disease, and aging. The cell has several protective mechanisms that minimize the toxic potential of these compounds.

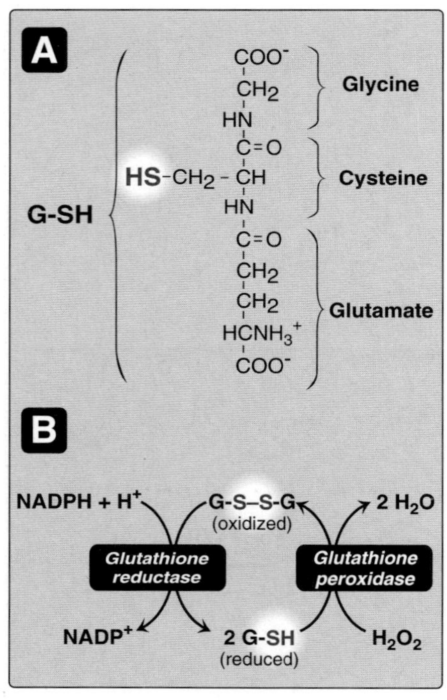

Figure 13.6
A. Structure of glutathione (G-SH). [Note: Glutamate is linked to cysteine through a γ-carboxyl, rather than an α-carboxyl.] B. Glutathione-mediated reduction of hydrogen peroxide by NADPH.

1. **Enzymes that catalyze antioxidant reactions:** Reduced **glutathione**, a tripeptide-thiol (γ-glutamylcysteinylglycine) present in most cells, can chemically detoxify hydrogen peroxide (Figure 13.5B). This reaction, catalyzed by the selenium-requiring *glutathione peroxidase*, forms oxidized glutathione, which no longer has protective properties. The cell regenerates reduced glutathione in a reaction catalyzed by *glutathione reductase,* using **NADPH** as a source of reducing electrons. Thus, NADPH indirectly provides electrons for the reduction of hydrogen peroxide (Figure 13.6). [Note: Erythrocytes are totally dependent on the pentose phosphate pathway for their supply of NADPH because, unlike other cell types, erythrocytes do not have an alternate source for this essential coenzyme. If *glucose 6-phosphate dehydrogenase* is compromised in some way, NADPH levels will fall, and oxidized glutathione cannot be reduced. As a result, hydrogen peroxide will accumulate, threatening membrane stability and causing red cell lysis.] Additional enzymes, such as *superoxide dismutase* and *catalase*, catalyze the conversion of other toxic oxygen intermediates to harmless products (see Figure 13.5B). As a group, these enzymes serve as a defense system to guard against the toxic effects of reactive oxygen species.

2. **Antioxidant chemicals:** A number of intracellular reducing agents, such as **ascorbate** (see p. 375), **vitamin E** (see p. 389), and **β-carotene** (see p. 380), are able to reduce and, thus, detoxify oxygen intermediates in the laboratory. Consumption of foods rich in these antioxidant compounds has been correlated with a reduced risk for certain types of cancers, as well as decreased

frequency of certain other chronic health problems. Thus, it is tempting to speculate that the effects of these compounds are, in part, an expression of their ability to quench the toxic effect of oxygen intermediates. However, clinical trials with antioxidants as dietary supplements have failed to show clear beneficial effects. In the case of dietary supplemention with β-carotene, the rate of lung cancer in smokers increased rather than decreased. Thus, the health-promoting effects of dietary fruits and vegetables probably reflects a complex interaction among many naturally occuring compounds, which has not been duplicated by consumption of isolated antioxidant compounds.

C. Cytochrome P450 monooxygenase system

Monooxygenases (*mixed function oxidases*) incorporate one atom from molecular oxygen into a substrate (creating a hydroxyl group), with the other atom being reduced to water. In the *cytochrome P450 monooxygenase* system, **NADPH** provides the reducing equivalents required by this series of reactions (Figure 13.7). This system performs different functions in two separate locations in cells. The overall reaction catalyzed by a cytochrome P450 enzyme is:

$$R\text{-}H + O_2 + NADPH + H^+ \rightarrow R\text{-}OH + H_2O + NADP^+$$

where R may be a steroid, drug, or other chemical. [Note: Cytochrome P450s (CYPs) are actually a superfamily comprised of hundreds of genes, coding for related, heme-containing enzymes that participate in a broad variety of reactions.]

1. **Mitochondrial system:** The function of the mitochondrial *cytochrome P450 monooxygenase* system is to participate in the **hydroxylation of steroids**, a process that makes these hydrophobic compounds more water soluble. For example, in the steroid hormone-producing tissues, such as the placenta, ovaries, testes, and adrenal cortex, it is used to hydroxylate intermediates in the conversion of cholesterol to steroid hormones. The liver uses this system in bile acid synthesis (see p. 222), and the kidney uses it to hydroxylate vitamin 25-hydroxycholecalciferol (vitamin D, see p. 384) to its biologically active 1,25-hydroxylated form.

2. **Microsomal system:** An extremely important function of the microsomal *cytochrome P450 monooxygenase* system found associated with the membranes of the **smooth endoplasmic reticulum** (particularly in the **liver**) is the **detoxification** of foreign compounds (**xenobiotics**). These include numerous drugs and such varied pollutants as petroleum products, carcinogens, and pesticides. The *cytochrome P450 monooxygenase* system can be used to hydroxylate these toxins, again using NADPH as the source of reducing equivalents. The purpose of these modifications is two-fold. First, it may itself activate or inactivate a drug or second, make a toxic compound more soluble, thus facilitating its excretion in the urine or feces. Frequently, however, the new hydroxyl group will serve as a site for conjugation with a polar compound, such as glucuronic acid (see p. 159), which will significantly increase the compound's solubility.

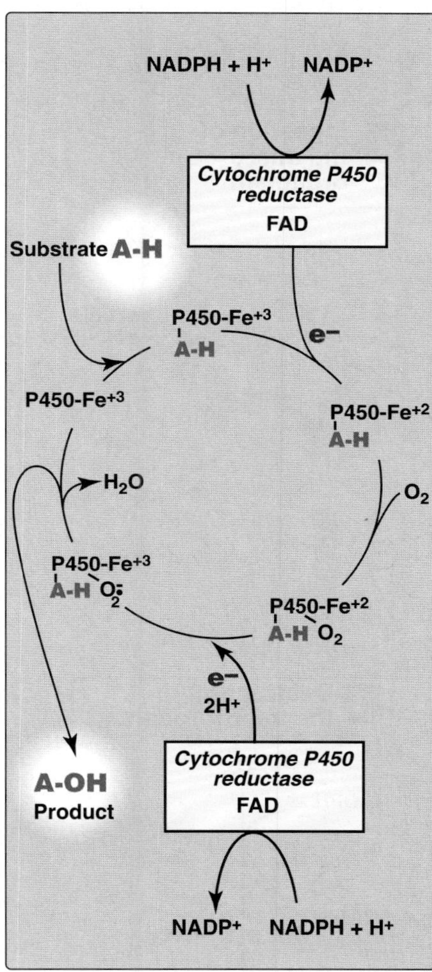

Figure 13.7
Cytochrome P450 monooxygenase cycle.

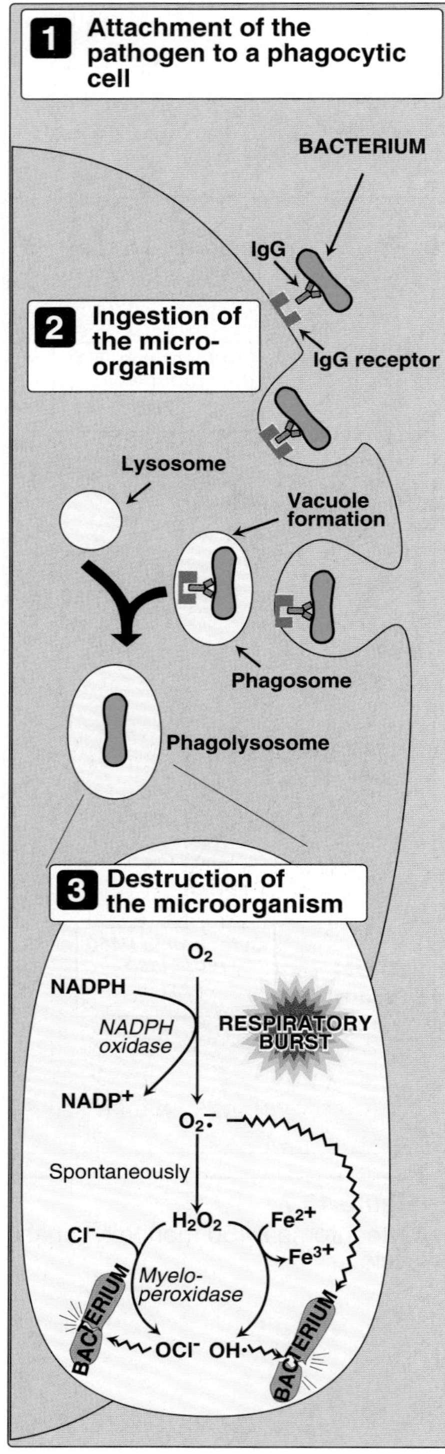

Figure 13.8
Phagocytosis and the oxygen dependent pathway of microbial killing. IgG = the antibody immunoglobulin G.

D. Phagocytosis by white blood cells

Phagocytosis is the ingestion by receptor-mediated endocytosis of microorganisms, foreign particles, and cellular debris by cells such as neutrophils and macrophages (monocytes). It is an important body defense mechanism, particularly in bacterial infections. Neutrophils and monocytes are armed with both oxygen-independent and oxygen-dependent mechanisms for killing bacteria. The **oxygen-dependent mechanisms** include the *myeloperoxidase* (*MPO*) system and a system that generates oxygen-derived free radicals (Figure 13.8). **Oxygen-independent systems** use pH changes in phagolysosomes and lysosomal enzymes to destroy pathogens. Overall, the *MPO* system is the most potent of the bactericidal mechanisms. An invading bacterium is recognized by the immune system and attacked by antibodies that bind it to a receptor on a phagocytic cell. After internalization of the microorganism has occurred, *NADPH oxidase*, located in the leukocyte cell membrane, converts molecular oxygen from the surrounding tissue into **superoxide**. The rapid consumption of molecular oxygen that accompanies formation of superoxide is referred to as the **respiratory burst**. [Note: *NADPH oxidase* is a complex enzyme, with subunits containing a cytochrome and a flavin coenzyme group. Genetic deficiencies in this enzyme cause **chronic granulomatosis**, a disease characterized by severe, persistant, chronic pyogenic infections.] Next, superoxide is spontaneously converted into hydrogen peroxide. Any superoxide that escapes the phagolysosome is converted to **hydrogen peroxide** by *superoxide dismutase* (*SOD*). This product is then neutralized by *catalase* or *glutathione peroxidase* (see Figure 13.5). In the presence of *MPO*, a lysosomal enzyme present within the phagolysosome, peroxide plus chloride ions are converted into **hypochlorous acid** (HOCl, the major component of household bleach), which kills the bacteria. Excess peroxide is either neutralized by *catalase* or by *glutathione peroxidase*.

E. Synthesis of nitric oxide

Nitric oxide (NO) is recognized as a mediator in a broad array of biologic systems. NO is the **endothelium-derived relaxing factor**, which causes vasodilation by relaxing vascular smooth muscle. NO also acts as a neurotransmitter, prevents platelet aggregation, and plays an essential role in macrophage function. [Note: NO is a free radical gas that is often confused with nitrous oxide (N_2O), the "laughing gas" that is used as an anesthetic[1] and is chemically stable.] NO has a very short half-life in tissues (three to ten seconds) because it reacts with oxygen and superoxide, and then is converted into nitrates and nitrites.

1. **Synthesis of NO:** Arginine, O_2, and NADPH are substrates for cytosolic *NO synthase* (Figure 13.9). Flavin mononucleotide (FMN), flavin adenine dinucleotide (FAD), heme, and tetrahydrobiopterin are coenzymes for the enzyme, and NO and citrulline are products of the reaction. Three *NO synthases* have been identified. Two are constitutive (synthesized at a constant rate regardless of physiologic demand), Ca^{2+}–calmodulin-dependent enzymes. They are found primarily in endothelium (*eNOS*), and neural tissue

[1]See Chapter 11 in ***Lippincott's Illustrated Reviews: Pharmacology*** (2nd and 3rd Eds.) for a more detailed discussion of nitrous oxide.

(*nNOS*), and constantly produce low levels of NO. An inducible, Ca^{2+}-independent enzyme (*iNOS*) can be expressed in many cells, including hepatocytes, macrophages, monocytes, and neutrophils. The specific inducers for *NO synthase* vary with cell type, and include tumor necrosis factor-α, bacterial endotoxins, and inflammatory cytokines. These compounds have been shown to promote synthesis of *iNOS*, which can result in large amounts of NO being produced over hours or even days.

2. **Actions of NO on vascular endothelium:** NO is an important mediator in the control of vascular smooth muscle tone. NO is synthesized by *eNOS* in endothelial cells and diffuses to vascular smooth muscle, where it activates the cytosolic form of *guanylate cyclase*. [Note: This reaction is analogous to the formation of cAMP by *adenylate cyclase* (see p. 92), except that this *guanylate cyclase* is not membrane-associated.] The resultant rise in cGMP causes muscle relaxation through activation of *protein kinase G,* which phosphorylates *myosin light-chain kinase* and renders it inactive, thereby decreasing smooth muscle contraction. [Note: Vasodilator nitrates, such as nitroglycerin and nitroprusside,[2] are metabolized to nitric oxide, which causes relaxation of vascular smooth muscle and, therefore, lowers blood pressure. Thus, NO can be envisioned as an endogenous nitrovasodilator.]

3. **Role of NO in mediating macrophage bactericidal activity:** In macrophages, *iNOS* activity is normally low, but synthesis of the enzyme is significantly stimulated by bacterial lipopolysaccharide and γ-interferon release in response to infection. Activated macrophages form superoxide radicals (see p. 148) that combine with NO to form intermediates that decompose, forming the highly bactericidal $OH\bullet^-$ radical. [Note: NO production in macrophages is also effective against viral, fungal, helmintic, and protozoan infections.]

4. **Other functions of NO:** NO is a potent inhibitor of platelet aggregation (by activating the cGMP pathway). It is also characterized as a neurotransmitter in the brain.

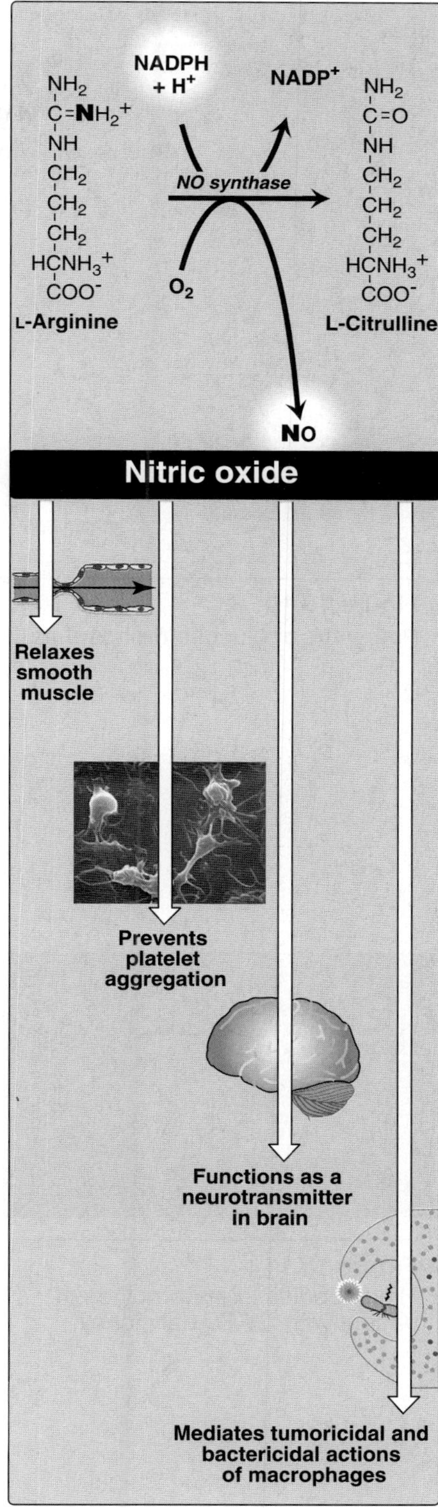

Figure 13.9
Synthesis and some of the actions of nitric oxide.

V. GLUCOSE 6-P DEHYDROGENASE DEFICIENCY

Glucose 6-phosphate dehydrogenase (*G6PD*) deficiency is an inherited disease characterized by hemolytic anemia caused by the inability to detoxify oxidizing agents. *G6PD* deficiency is the most common disease-producing enzyme abnormality in humans, affecting more than 200 million individuals worldwide. This deficiency has the highest prevelance in the Middle East, tropical Africa and Asia, and parts of the Mediterranean. *G6PD* deficiency is **X-linked**, and is, in fact, a family of deficiencies caused by more than 400 different mutations in the gene coding for *G6PD*. Only some of these mutations cause clinical symptoms. The life span of many individuals with *G6PD* deficiency is somewhat shortened as a result of complications arising from chronic hemolysis. This slightly negative effect of *G6PD* deficiency has been balanced in evolution by an advantage in survival—an increased resistance to falciparum malaria shown by female carriers of the mutation. [Note: Sickle cell trait and β-thallasemia minor also confer resistance.]

INFO LINK [2]See Chapter 19 in ***Lippincott's Illustrated Reviews: Pharmacology*** (2nd and 3rd Eds.) for a more detailed discussion of vasodilators.

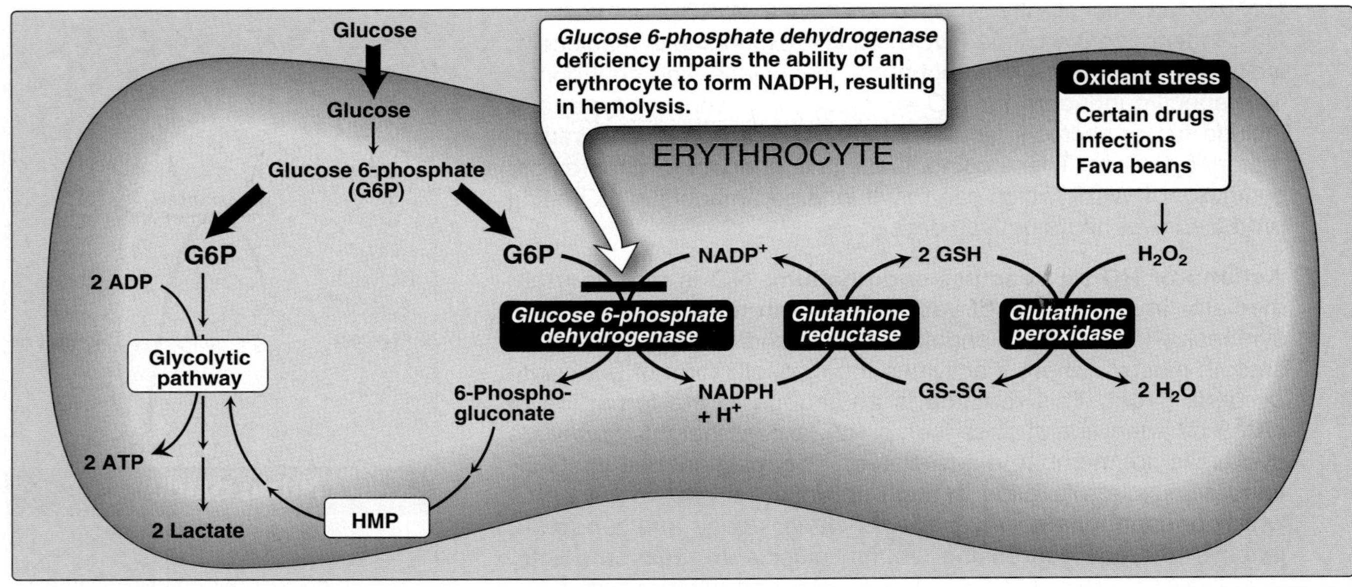

Figure 13.10
Pathways of glucose 6-phosphate metabolism in the erythrocyte.

A. Role of G6PD in red blood cells

Diminished *G6PD* activity impairs the ability of the cell to form the NADPH that is essential for the maintenance of the reduced glutathione pool. This results in a decrease in the cellular detoxification of free radicals and peroxides formed within the cell (Figure 13.10). Glutathione also helps maintain the reduced states of sulfhydryl groups in proteins, including hemoglobin. Oxidation of those sulfhydral groups leads to the formation of denatured proteins that form insoluble masses (called Heinz bodies) that attach to the red cell membranes (Figure 13.11). Additional oxidation of membrane proteins causes the red cells to be rigid and nondeformable, and they are removed from the circulation by macrophages in the spleen and liver. Although *G6PD* deficiency occurs in all cells of the affected individual, it is most severe in erythrocytes, where the pentose phosphate pathway provides the only means of generating NADPH. Other tissues have alternative sources for NADPH production (such as *NADP+-dependent malate dehydrogenases*, see Figure 16.11, p. 184) that can keep glutathione reduced. The erythrocyte has no nucleus or ribosomes and cannot renew its supply of the enzyme. Thus, red blood cells are particularly vulnerable to enzyme variants with diminished stability.

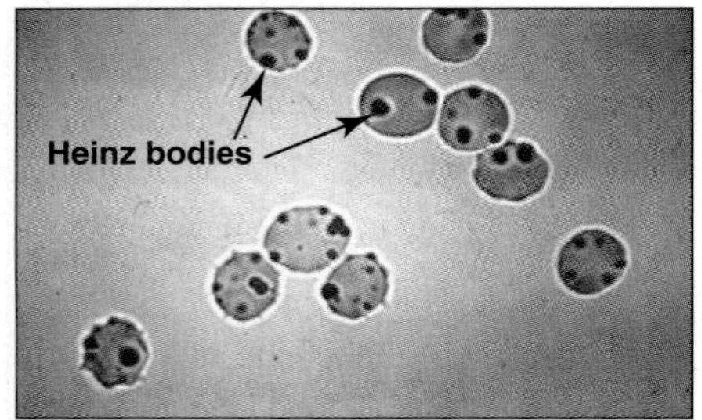

Figure 13.11
Heinz bodies in erythrocytes of patient with *G6PD* deficiency.

B. Precipitating factors in G6PD deficiency

Most individuals who have inherited one of the many *G6PD* mutations do not show clinical manifestations (Figure 13.12. However, some patients with *G6PD* deficiency develop hemolytic anemia if they are treated with an oxidant drug, ingest fava beans, or contract a severe infection.

1. **Oxidant drugs:** Commonly used drugs that produce hemolytic anemia in patients with *G6PD* deficiency are best remembered from the mnemonic **AAA** = **A**ntibiotics (for example, sulfa-

methoxazole and chloramphenicol), **A**ntimalarials (for example, primaquine but not quinine), and **A**ntipyretics (for example, acetanilid but not acetaminophen).

2. **Favism:** Some forms of *G6PD* deficiency, for example the Mediterranean variant, are particularly susceptible to the hemolytic effect of the fava bean, a dietary staple in the Mediterranean region. Favism, the hemolytic effect of ingesting fava beans, is not observed in all individuals with *G6PD* deficiency, but all patients with favism have *G6PD* deficiency.

3. **Infection:** Infection is the most common precipitating factor of hemolysis in *G6PD* deficiency. The inflammatory response to infection results in the generation of free radicals in macrophages, which can diffuse into the red blood cells and cause oxidative damage.

4. **Neonatal jaundice:** Babies with *G6PD* deficiency may experience neonatal jaundice appearing one to four days after birth. The jaundice, which may be severe, results from impaired hepatic catabolism of heme or increased production of bilirubin.

Class	Clinical symptoms	Residual enzyme activity
I	Very severe	<2%
II	Severe	<10%
III	Moderate	10–50%
IV	None	60–150%

Figure 13.12
Classification of G6PD deficiency variants.

C. Properties of the variant enzymes

Almost all *G6PD* variants are caused by point mutations in the *G6PD* gene. Some mutations do not disrupt the structure of the enzyme's active site and, hence, do not affect enzymic activity. However, many mutant enzymes show altered kinetic properties. For example, variant enzymes may show decreased catalytic activity, decreased stability, or an alteration of binding affinity for $NADP^+$, NADPH, or glucose 6-phosphate. The severity of the disease usually correlates with the amount of residual enzyme activity in the patient's red blood cells. For example, variants can be classified as shown in Figure 13.13. ***G6PD* A$^-$** is the prototype of the moderate (class III) form of the disease. The red cells contain an unstable, but kinetically normal *G6PD*, with most of the enzyme activity present in the reticulocytes and younger erythrocytes (see Figure 13.13). The oldest cells, therefore, have the lowest level of enzyme activity, and are preferentially removed in a hemolytic episode. ***G6PD* Mediterranean** is the prototype of a more severe (class II) deficiency in which the enzyme shows normal stability but scarcely detectable activity in all red blood cells. Class I mutations are often associated with **chronic nonspherocytic anemia**, which occurs even in the absence of oxidative stress.

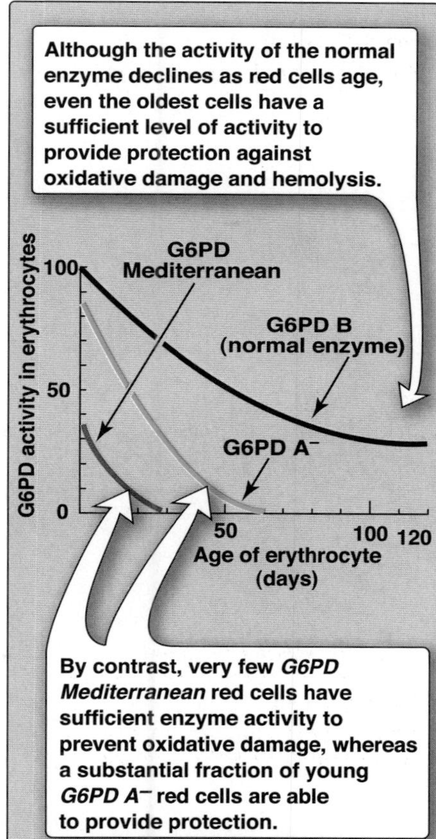

Figure 13.13
Decline of erythrocyte *G6PD* activity with cell age for the three most commonly ecountered forms of the enzyme.

D. Molecular biology of G6PD

The cloning of the *G6PD* gene and the sequencing of its complementary DNA (see p. 448 for a discussion of cDNA) have permitted the identification of mutations that cause *G6PD* deficiency. More than 300 different mutations or mutation combinations have been identified in this gene, a finding that explains the numerous biochemical variants that have been described. Most of these DNA changes are missense, point mutations (see p. 431). Both ***G6PD* A$^-$** and ***G6PD* Mediterranean** represent mutant enzymes that differ from the respective normal variants by a single amino acid. Large deletions or frameshift mutations have not been identified, suggesting that complete absence of G6PD activity is probably lethal.

VI. CHAPTER SUMMARY

The pentose phosphate pathway consists of two irreversible oxidative reactions followed by a series of reversible sugar–phosphate interconversions. No ATP is directly consumed or produced in the cycle. The **oxidative portion** of the pentose phosphate pathway is particularly important in liver and mammary glands, which are active in the biosynthesis of fatty acids, in the adrenal cortex, which is active in the NADPH-dependent synthesis of steroids, and in erythrocytes, which require NADPH to keep glutathione reduced. **Glucose 6-phosphate** is irreversibly converted to **ribulose 5-phosphate**, and **two NADPH** are produced. The regulated step is **glucose 6-phosphate dehydrogenase** (G6PD), which is strongly **inhibited by NADPH**. **Reversible nonoxidate reactions** interconvert sugars. This part of the pathway is the **source of ribose 5-phosphate** required for nucleotide and nucleic acid synthesis. Because the reactions are reversible, they can be entered from fructose 6-phosphate and glyceraldehyde 3-phosphate (glycolytic intermediates) if ribose is needed and glucose 6-phosphate dehydrogenase is inhibited. **NADPH** is a source of reducing equivalents in **reductive biosynthesis**, such as the production of fatty acids and steroids. It is also required for the **reduction of hydrogen peroxide**, providing the reducing equivalents required by **glutathione** (GSH). GSH is used by **glutathione peroxidase** to reduce peroxide to water. The oxidized glutathione is reduced by **glutathione reductase**, using NADPH as the source of electrons. NADPH provides reducing equivalents for the **cytochrome P450 monooxygenase system**, which is used in the **hydroxylation of steroids** to produce steroid hormones, **bile acid synthesis** by the liver, and **activation of vitamin D**. The system also **detoxifies** foreign compounds, such as drugs and varied pollutants, including carcinogens, pesticides, and petroleum products. NADPH provides the reducing equivalents for phagocytes in the process of eliminating invading microorganisms. **NADPH oxidase** uses molecular oxygen and NADPH electrons to produce **superoxide radicals**, which, in turn, can be converted to peroxide, hypochlorous acid, and hydroxyl radicals. **Myeloperoxidase** is an important enzyme in this pathway. A genetic defect in NADPH oxidase causes **chronic granulomatosis**, a disease characterized by severe, persistant, chronic pyogenic infections. NADPH is required for the synthesis of **nitric oxide** (**NO**), an important molecule that causes **vasodilation** by relaxing vascular smooth muscle, acts as a kind of **neurotransmitter**, **prevents platelet aggregation**, and helps mediate **macrophage bactericidal activity**. This deficiency is a **genetic disease** characterized by **hemolytic anemia**. **G6PD deficiency** impairs the ability of the cell to form the NADPH that is essential for the maintenance of the reduced glutathione pool. The cells most affected are the **red blood cells** because they do not have additional sources of NADPH. **Free radicals** and **peroxides** formed within the cells cannot be neutralized, causing denaturation of protein (for example, hemoglobin, forming Heinz bodies) and membrane proteins. The cells become rigid, and they are removed by the reticuloendothelial system of the spleen and liver. Hemolytic anemia can be caused by the production of free radicals and peroxides following the taking of **oxidant drugs**, ingestion of **fava beans**, or severe **infections**. Babies with G6PD deficiency may experience **neonatal jaundice** appearing one to four days after birth. The degree of severity of the anemia depends on the location of the mutation in the G6PD gene. Class I mutations are the most severe (for example, **G6PD Mediterranean**). They are often associated with **chronic nonspherocytic anemia**. Class III mutations (for example, **G6PD A$^-$**) cause a more moderate form of the disease.

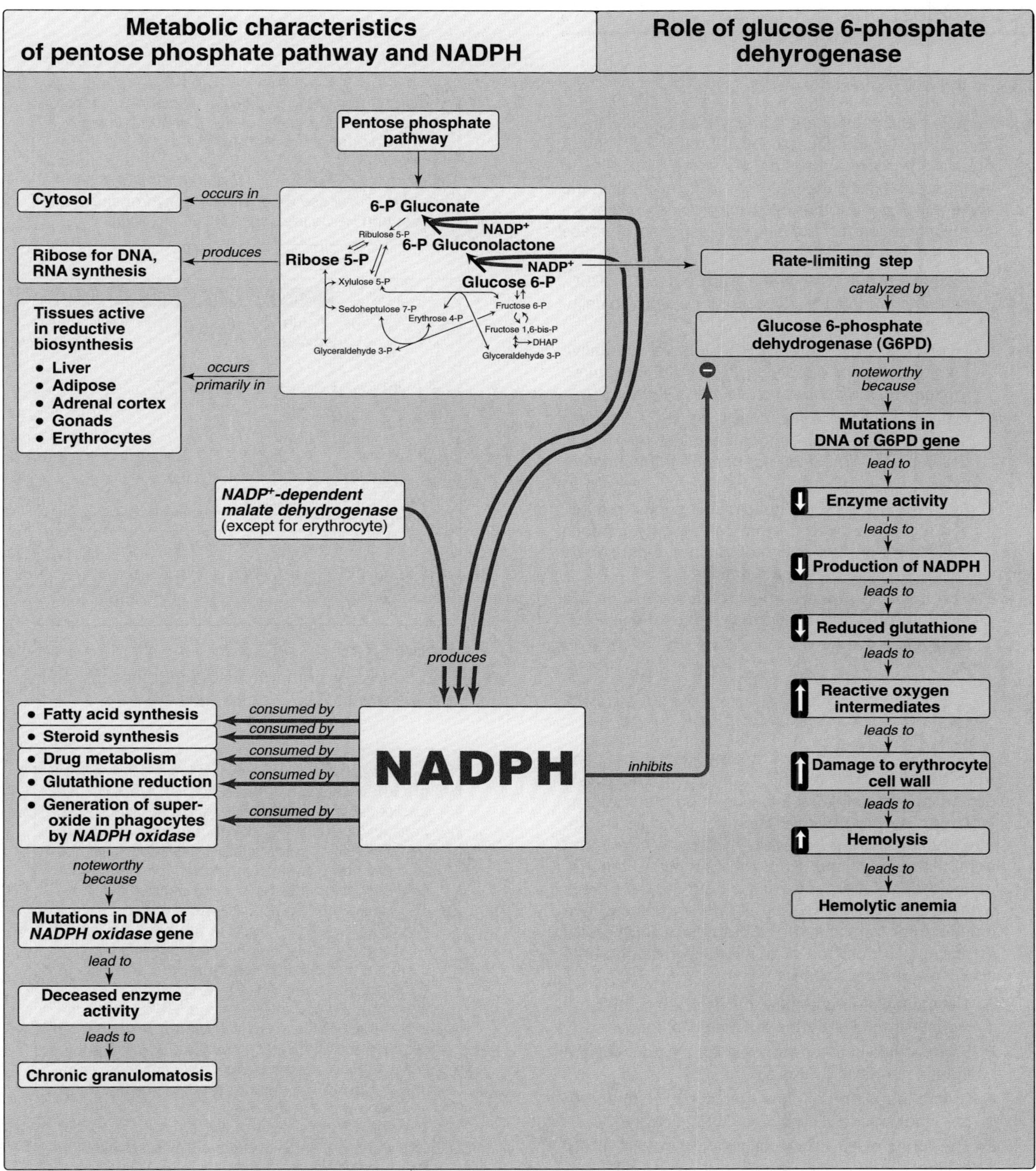

Figure 13.14
Key concept map for the pentose phosphate pathway and NADPH.

Study Questions

Choose the ONE correct answer

13.1 In male patients who are homozygous for glucose 6-phosphate dehydrogenase (G6PD) deficiency, pathophysiologic consequences are more apparent in erythrocytes (RBC) than in other cells, such as in the liver. Which one of the following provides a reasonable explanation for this different response by these individual tissue types?

A. Excess glucose 6-P in the liver, but not in RBC, can be channeled to glycogen, thus averting cellular damage.
B. Liver cells, in contrast to RBCs, have alternative mechanisms for supplying the NADPH required for keeping metabolic and cellular integrity.
C. Glucose 6-phosphatase activity in RBCs removes the excess glucose 6-phosphate, thus resulting in cell damage. This does not happen in the hepatocyte.
D. Because RBCs do not have mitochondria, production of ATP required to keep cell integrity depends exclusively on the routing of glucose 6-phosphate via the pentose phosphate pathway.
E. The catalytic properties of the liver enzyme are significantly different than those of the RBC enzyme.

Correct answer = B. Cellular damage is directly related to decreased ability of the cell to regenerate reduced glutathione, for which large amounts of NADPH are needed. RBCs have plenty of glutathione peroxidase. Catalytic properties of glucose 6-phosphate dehydrogenase in liver and RBCs are very similar.

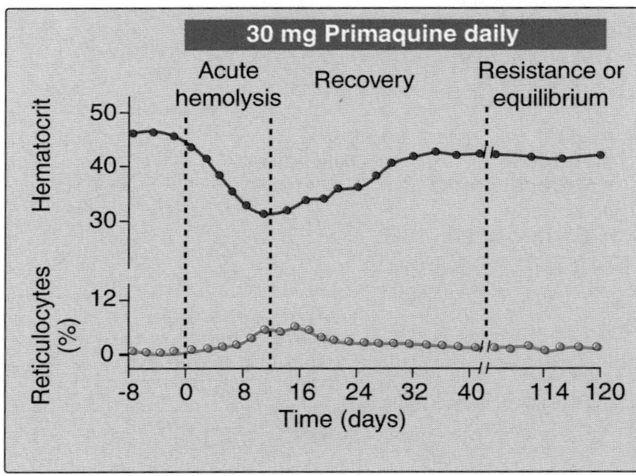

Figure 13.15
The course of primaquine-induced hemolysis in a patient with G6PD deficiency.

13.2 A G6PD A^- individual was treated with primaquine from day 0 to day 120 (Figure 13.15). Hemolysis occurs immediately after initiation of drug therapy, as indicated by progressive anemia, hemoglobinuria, and reticulocytosis. However, despite continued administration of the drug, the hemolysis spontaneously decreases and red cell survival improves with time. Red cell G6PD activity measured two months after termination of therapy was ten percent of normal. Which one of the following statements about this patient is correct?

A. The patient will continue to be resistant to drug-induced hemolysis after six months or longer.
B. Erythrocytes in this patient exhibit a longer lifetime than in normal individuals.
C. During the period of peak hemolysis, erythrocytes from the patient will show no G6PD activity.
D. The intracellular concentration of NADPH in the patient's erythrocytes is greater than normal.
E. The patient shows increased bone marrow erythropoiesis.

Correct answer = E. As red cells age, the activity of G6PD declines (Figure 13.13). Despite this loss of enzyme activity, normal old red blood cells contain sufficient G6PD activity to generate NADPH and thereby sustain GSH levels in the face of oxidant stress. In contrast, the G6PD variants with hemolysis have much shorter half-lives. The clinical correlate of this age-related enzyme instability is that hemolysis in patients with G6PD A^- generally is mild and limited to the older deficient erythrocytes. The anemia is self-limited because the older, vulnerable population of erythrocytes is replaced by younger RBCs with sufficient G6PD activity to withstand an oxidative assault. Although red cell survival remains shortened as long as use of the drug continues, compensation by the erythroid marrow effectively abolishes the anemia in subjects with G6PD A^-. The individual's continuing sensitivity to the effects of the drug is revealed by discontinuing the drug for several months to allow the rate of red cell production by the bone marrow to normalize; during this phase, the older red cells are able to survive, and the red cell population is rendered sensitive to drug-induced hemolysis.

Glycosaminoglycans and Glycoproteins

14

I. OVERVIEW OF GLYCOSAMINOGLYCANS

Glycosaminoglycans (GAGs) are large complexes of negatively charged heteropolysaccharide chains. They are generally associated with a small amount of protein, forming **proteoglycans**, which typically consist of over 95 percent carbohydrate. [Note: This is in comparison to the glycoproteins, which consist primarily of protein with a small amount of carbohydrate. Glycoproteins are discussed in this chapter, beginning on p. 163.] Glycosaminoglycans have the special ability to bind large amounts of water, thereby producing the gel-like matrix that forms the basis of the body's **ground substance**. The viscous, lubricating properties of mucous secretions also result from the presence of glycosaminoglycans, which led to the original naming of these compounds as **mucopolysaccharides.**

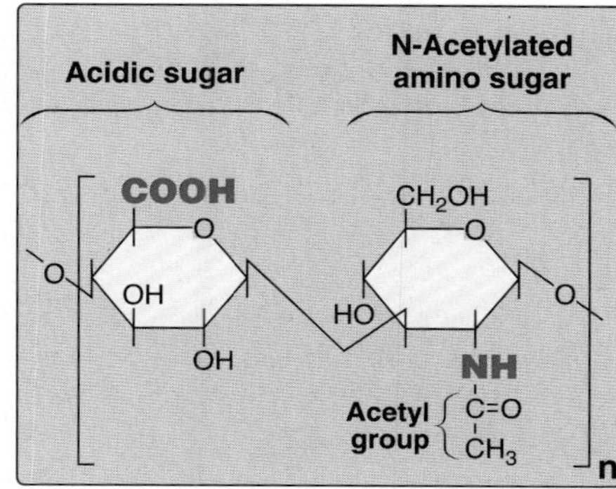

Figure 14.1
Repeating disaccharide unit.

II. STRUCTURE OF GLYCOSAMINOGLYCANS

Glycosaminoglycans are long, unbranched, heteropolysaccharide chains generally composed of a **repeating disaccharide unit** [acidic sugar–amino sugar]$_n$ (Figure 14.1). The **amino sugar** is either D-glucosamine or D-galactosamine, in which the amino group is usually acetylated, thus eliminating its positive charge. The amino sugar may also be sulfated on carbon 4 or 6 or on a nonacetylated nitrogen. The **acidic sugar** is either D-glucuronic acid or its carbon-5 epimer, L-iduronic acid (Figure 14.2). [Note: A single exception is keratan sulfate, in which galactose rather than an acidic sugar is present.] These acidic sugars contain carboxyl groups that are negatively charged at physiologic pH and, together with the sulfate groups, give glycosaminoglycans their strongly negative nature.

A. Relationship between glycosaminoglycan structure and function

Because of their large number of negative charges, these heteropolysaccharide chains tend to be extended in solution. They repel each other and are surrounded by a shell of water molecules. When brought together, they "slip" past each other, much as two magnets with the same polarity seem to slip past each other. This produces the "slippery" consistency of mucous secretions and synovial fluid. When a solution of glycosaminoglycans is compressed,

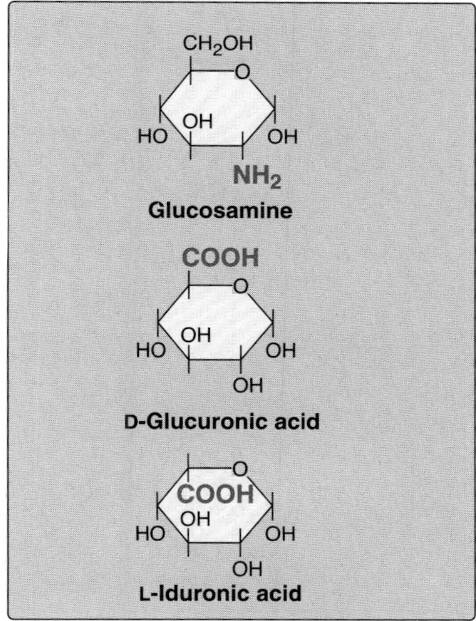

Figure 14.2
Some monsaccharide units found in glycosaminoglycans.

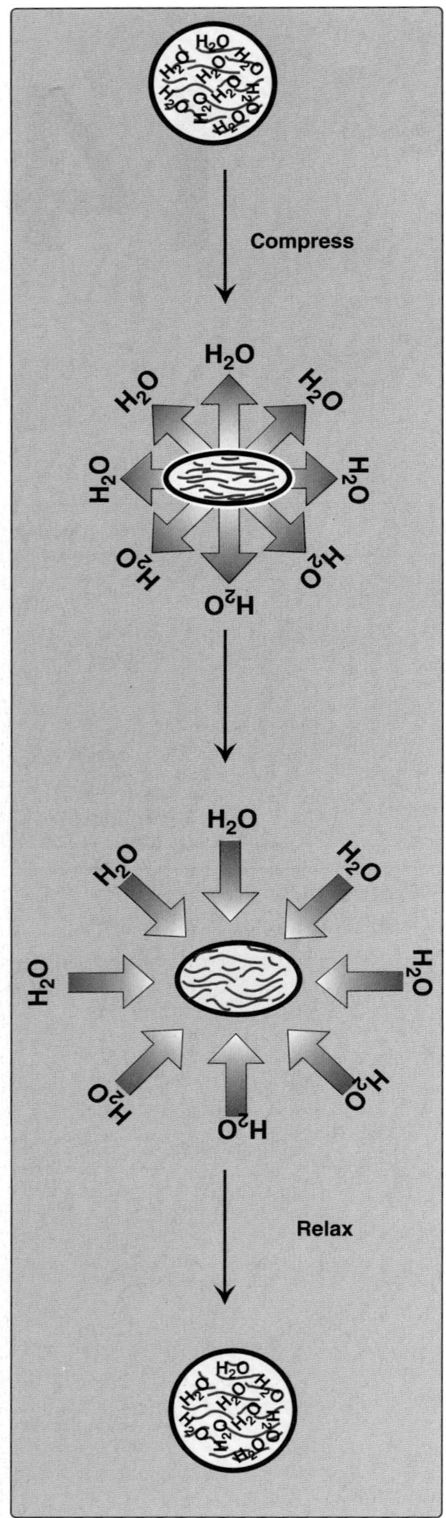

Figure 14.3
Resilience of glycosaminoglycans.

the water is "squeezed out" and the glycosaminoglycans are forced to occupy a smaller volume. When the compression is released, the glycosaminoglycans spring back to their original, hydrated volume because of the repulsion of their negative charges. This property contributes to the resilience of synovial fluid and the vitreous humor of the eye (Figure 14.3).

B. Classification of the glycosaminoglycans

The six major classes of glycosaminoglycans are divided according to monomeric composition, type of glycosidic linkages, and degree and location of sulfate units. The structure of the glycosaminoglycans and their distribution in the body is illustrated in Figure 14.4.

C. Structure of proteoglycans

All of the glycosaminoglycans, except hyaluronic acid, are found covalently attached to protein, forming **proteoglycan monomers**.

1. **Structure of proteoglycan monomers:** A proteoglycan monomer found in cartilage consists of a **core protein** to which the linear glycosaminoglycan chains are covalently attached. These chains, which may each be composed of more than 100 monosaccharides, extend out from the core protein and remain separated from each other because of charge repulsion. The resulting structure resembles a "bottle brush" (Figure 14.5). In cartilage proteoglycan, the species of glycosaminoglycans include chondroitin sulfate and keratan sulfate. [Note: A number of proteoglycans have been characterized and named based on their structure and functional location. For example, syndecan is an integral membrane proteoglycan, versican and aggrecan are the predominant extracelular proteoglycans, and neurocan and cerebrocan are found primarily in the nervous system.]

2. **Linkage between the carbohydrate chain and the protein:** This linkage is most commonly through a **trihexoside** (galactose-galactose-xylose) and a serine residue, respectively. An O-glycosidic bond is formed between the xylose and the hydroxyl group of the serine (Figure 14.6).

3. **Proteoglycan aggregates:** The proteoglycan monomers associate with a molecule of **hyaluronic acid** to form **proteoglycan aggregates**. The association is not covalent, but occurs primarily through ionic interactions between the core protein and the hyaluronic acid. The association is stabilized by additional small proteins called **link proteins** (Figure 14.7).

III. SYNTHESIS OF GLYCOSAMINOGLYCANS

The polysaccharide chains are elongated by the sequential addition of alternating acidic and amino sugars, donated by their UDP-derivatives. The reactions are catalyzed by a family of specific transferases. The synthesis of the glycosaminoglycans is analogous to that of glycogen (p. 124) except that the glycosaminoglycans are produced for export from the cell. Their synthesis occurs, therefore, in the **endoplasmic reticulum** and the **Golgi**, rather than in the cytosol.

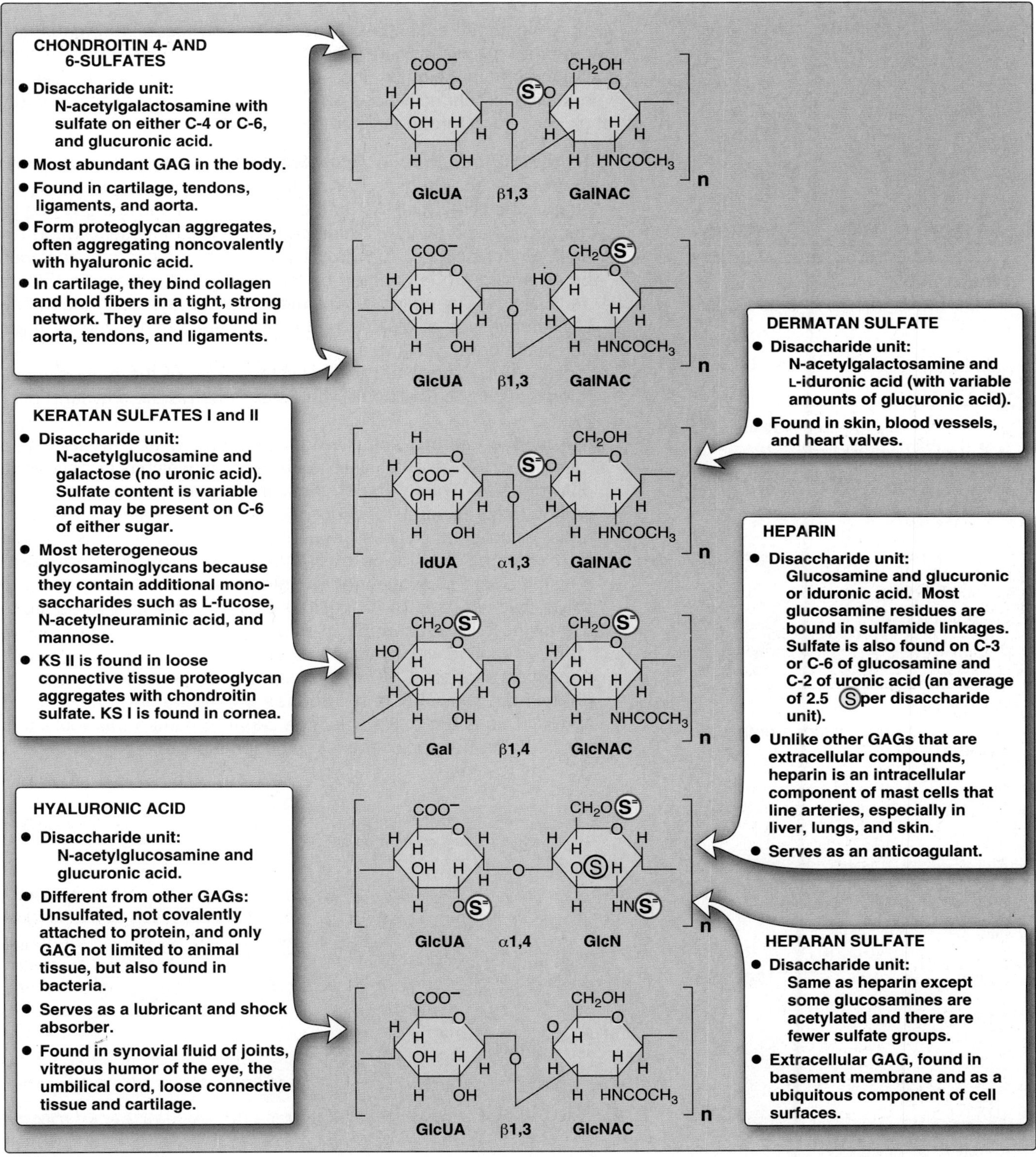

Figure 14.4

Structure and distribution of glycosaminoglycans (GAGs). Sulfate groups (Ⓢ) are shown in all possible positions. GlcUA = glucuronic acid; IdUA = iduronic acid; GalNAC = N-acetylgalactosamine; GlcNAC = N-acetylglucosamine; GlcN = glucosamine; Gal = galactose.

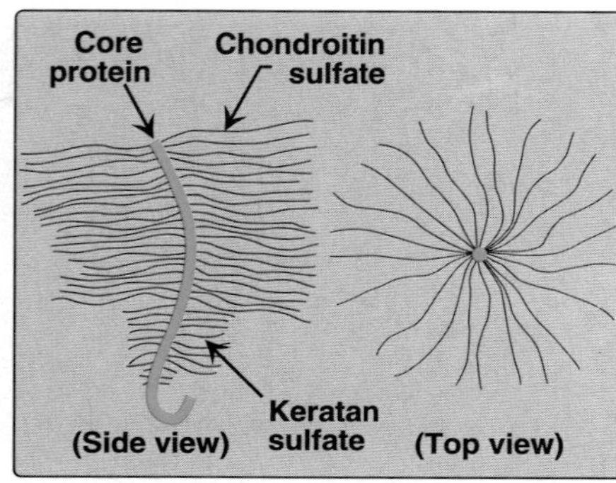

Figure 14.5
"Bottle-brush" model of a cartilage proteoglycan monomer.

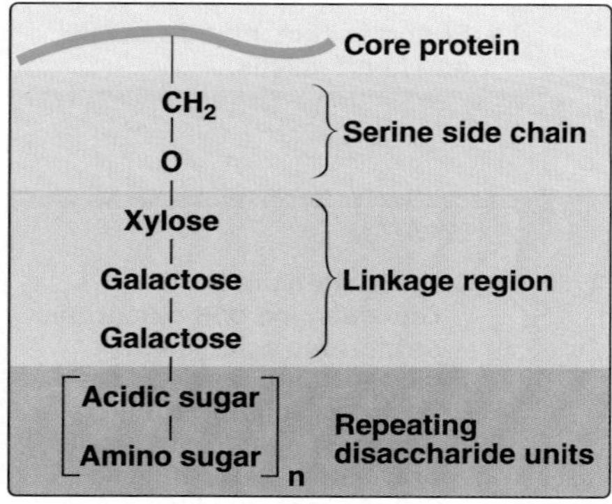

Figure 14.6
Linkage region of glycosamino-glycans.

A. Synthesis of amino sugars

Amino sugars are essential components of glycosaminoglycans, glycoproteins, glycolipids, and certain oligosaccharides, and are also found in some antibiotics. The synthetic pathway of amino sugars is very active in connective tissues, where as much as twenty percent of glucose flows through this pathway.

1. **N-Acetylglucosamine (glcNAc) and N-acetylgalactosamine (galNAc):** The monosaccharide **fructose 6-phosphate** is the precursor of gluNAc, galNAc, and the sialic acids, including N-acetylneuraminic acid (NANA, a nine-carbon, acidic monosaccharide). In each of these sugars, a hydroxyl group of the precursor is replaced by an amino group donated by the amino acid, glutamine (Figure 14.8). [Note: The amino groups are almost always acetylated.] The UDP-derivatives of gluNAc and galNAc are synthesized by reactions analogous to those described for UDP-glucose synthesis (see p. 124). These are the activated forms of the monosaccharides that can be used to elongate the carbohydrate chains.

2. **N-Acetylneuraminic acid:** N-Acetylneuraminic acid (NANA) is a member of the family of sialic acids, each of which is acylated at a different site. These compounds are usually found as terminal carbohydrate residues of oligosaccharide side chains of glycoproteins, glycolipids, or, less frequently, of glycosaminoglycans. The carbons and nitrogens in NANA come from N-acetylmannosamine and phosphoenolpyruvate (an intermediate in the glycolytic pathway, see p. 100). [Note: Before NANA can be added to a growing oligosaccharide, it must be converted into its active form by reacting with cytidine triphosphate (CTP). The enzyme *N-acetylneuraminate-CMP-pyrophosphorylase* removes pyrophosphate from the CTP and attaches the remaining CMP to the NANA. This is the only nucleotide sugar in human metabolism in which the carrier nucleotide is a monophosphate.]

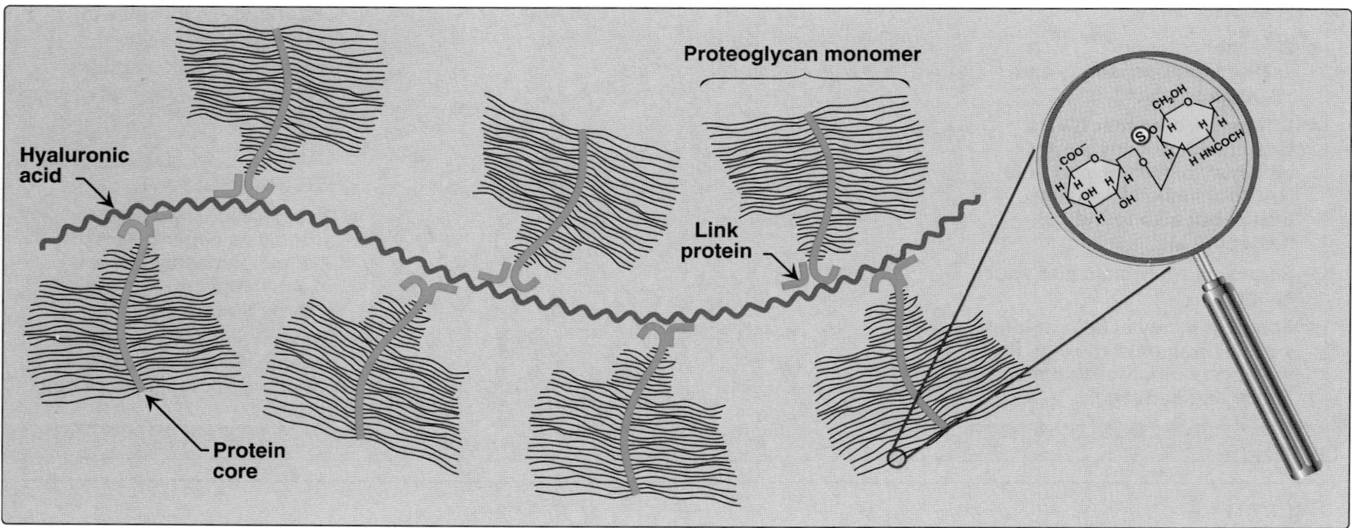

Figure 14.7
Proteoglycan aggregate.

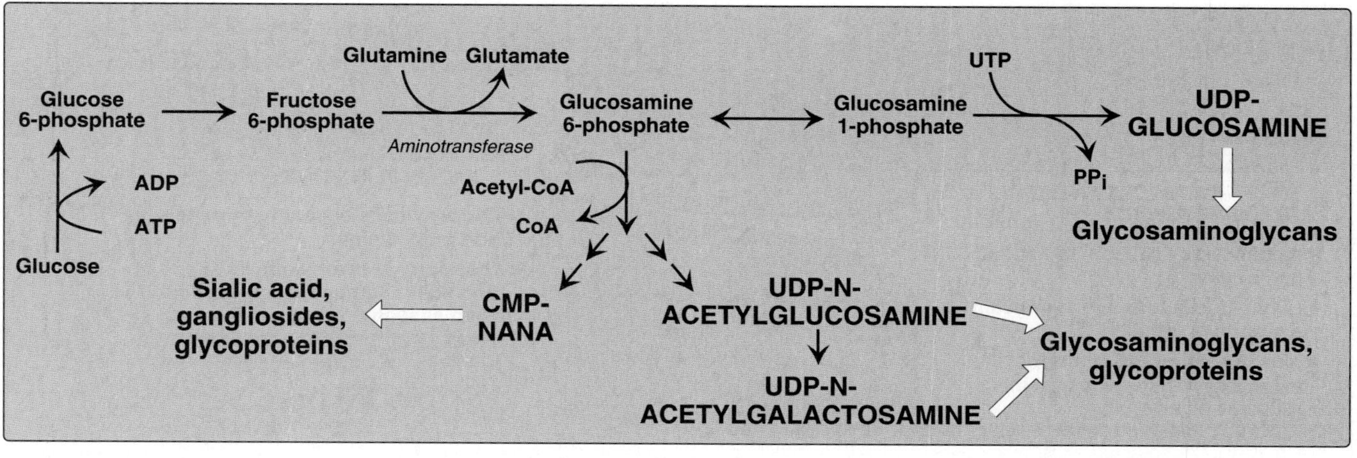

Figure 14.8
Synthesis of the amino sugars.

B. Synthesis of acidic sugars

D-Glucuronic acid, whose structure is that of glucose with an oxidized carbon 6 ($-CH_2OH \rightarrow -COOH$), and its C-5 epimer, L-iduronic acid, are essential components of glycosaminoglycans. Glucuronic acid is also required in detoxification reactions of a number of insoluble compounds, such as bilirubin (see p. 280), steroids, and several drugs. In plants and mammals (other than guinea pigs and primates, including man), glucuronic acid serves as a precursor of ascorbic acid (vitamin C). The uronic acid pathway also provides a mechanism by which dietary D-xylulose can enter the central metabolic pathways.

1. **Glucuronic acid:** Glucuronic acid can be obtained in small amounts from the diet. It can also be obtained from the intracellular lysosomal degradation of glycosaminoglycans, or via the uronic acid pathway. The end-product of glucuronic acid metabolism in humans is D-xylulose 5-phosphate, which can enter the hexose monophosphate pathway and produce the glycolytic intermediates glyceraldehyde 3-phosphate and fructose 6-phophate (Figure 14.9; see also Figure 13.2, p. 144). The active form of glucuronic acid that donates the sugar in glycosaminoglycan synthesis and other glucuronylating reactions is UDP-glucuronic acid, which is produced by oxidation of UDP-glucose (Figure 14.10).

2. **L-Iduronic acid synthesis:** Synthesis of L-iduronic acid residues occurs after D-glucuronic acid has been incorporated into the carbohydrate chain. *Uronosyl 5-epimerase* causes epimerization of the D- to the L-sugar.

C. Synthesis of the core protein

The core protein is synthesized on and enters the rough endoplasmic reticulum (RER). The protein is then glycosylated by membrane-bound transferases as it moves through the ER.

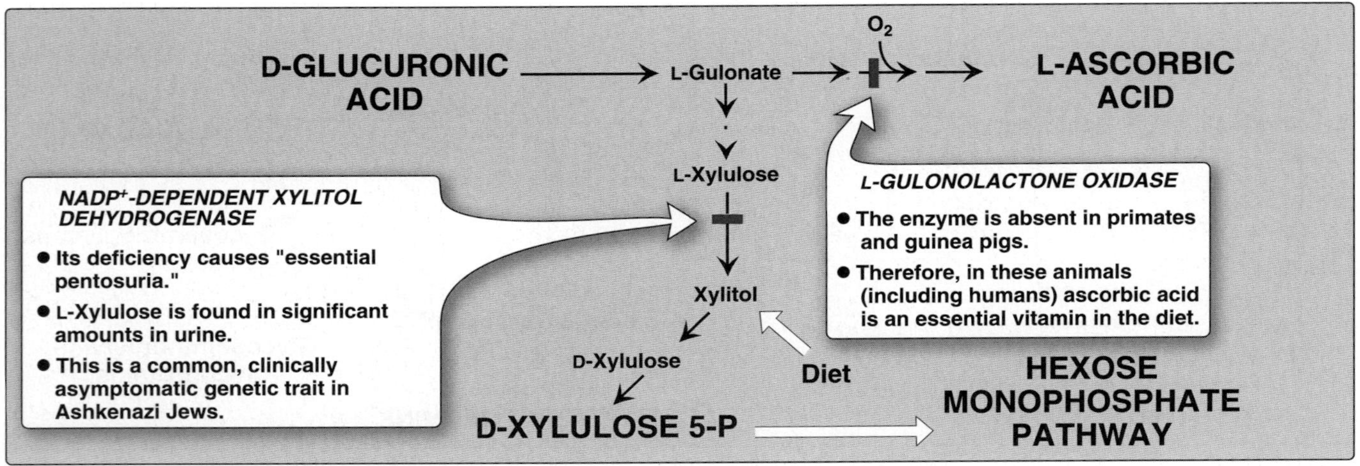

Figure 14.9
Uronic acid pathway.

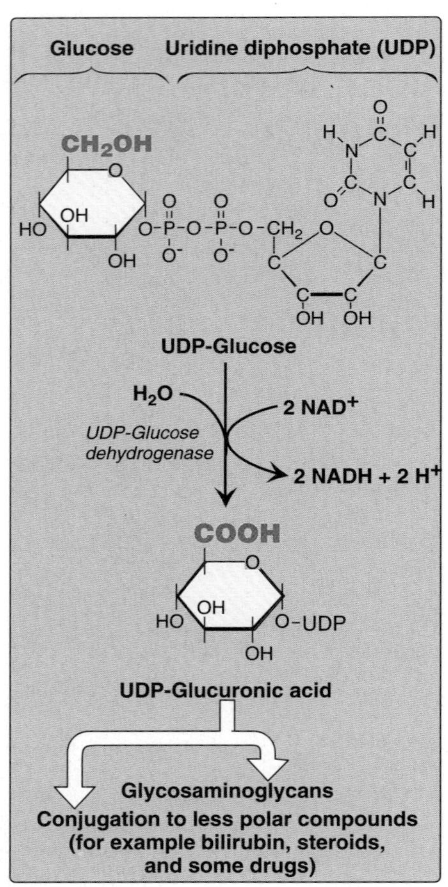

Figure 14.10
Oxidation of UDP-glucose to UDP-glucuronic acid.

D. Synthesis of the carbohydrate chain

Carbohydrate chain formation begins by synthesis of a short **linkage region** on the core protein on which carbohydrate chain synthesis will be initiated. The most common linkage region is formed by the transfer of a xylose from UDP-xylose to the hydroxyl group of a serine (or threonine) catalyzed by *xylosyltransferase*. Two galactose molecules are then added, completing the trihexoside. This is followed by sequential addition of alternating acidic and amino sugars (Figure 14.11), and conversion of some D-glucuronyl to L-iduronyl residues.

E. Addition of sulfate groups

Sulfation of the carbohydrate chain occurs after the monosaccharide to be sulfated has been incorporated into the growing carbohydrate chain. The source of the sulfate is **3'-phosphoadenosyl-5'-phosphosulfate** (PAPS, a molecule of AMP with a sulfate group attached to the 5'–phosphate). *Sulfotransferases* cause the sulfation of the carbohydrate chain at specific sites. [Note: An example of the synthesis of a sulfated glycosaminoglycan, chondroitin sulfate, is shown in Figure 14.11.] PAPS is also the sulfur donor in glycosphingolipid synthesis. [Note: A defect in the sulfation process results in one of several autosomal recessive disorders that affect the proper development and maintenance of the skeletal system. This illustrates the importance of the sulfation step.]

IV. DEGRADATION OF GLYCOSAMINOGLYCANS

Glycosaminoglycans are degraded in **lysosomes**, which contain hydrolytic enzymes that are most active at a pH of approximately 5. [Note: Therefore, as a group, these enzymes are called **acid hydrolases**.] The low pH optimum is a protective mechanism that prevents the enzymes from destroying the cell should leakage occur into the cytosol where the pH is neutral. With the exception of keratan sulfate, which has a half-life of greater than 120 days, the glycosaminoglycans have a relatively short half-life, ranging from about three days for hyaluronic acid to ten days for chondroitin and dermatan sulfate.

A. Phagocytosis of extracellular glycosaminoglycans

Because glycosaminoglycans are **extracellular** or **cell-surface** compounds, they must be engulfed by an invagination of the cell membrane (phagocytosis), forming a vesicle inside of which the glycosaminoglycans are to be degraded. This vesicle then fuses with a lysosome, forming a single digestive vesicle in which the glycosaminoglycans are efficiently degraded (see p. 148 for a discussion of phagocytosis).

B. Lysosomal degradation of glycosaminoglycans

The lysosomal degradation of glycosaminoglycans requires a large number of acid hydrolases for complete digestion. First, the polysaccharide chains are cleaved by endoglycosidases, producing oligosaccharides. Further degradation of the oligosaccharides occurs sequentially from the non-reducing end of each chain (see p. 127), the last group (sulfate or sugar) added during synthesis being the first group removed. Examples of some of these enzymes and the bonds they hydrolyze are shown in Figure 14.12.

V. MUCOPOLYSACCHARIDOSES

The mucopolysaccharidoses are hereditary disorders that are clinically progressive. They are characterized by accumulation of glycosaminoglycans in various tissues, causing varied symptoms, such as skeletal and extracellular matrix deformities, and mental retardation. Mucopolysaccharidoses are caused by a deficiency of one of the lysosomal hydrolases normally involved in the degradation of heparan sulfate and/or dermatan sulfate (see Figure 14.12). This results in the presence of oligosaccharides in the urine, because of incomplete lysosomal degradation of glycosaminoglycans. These fragments can be used to diagnose the specific mucopolysaccharidosis, namely by identifying the structure present on the nonreducing end of the oligosaccharide. That residue would have been the substrate for the missing enzyme. Diagnosis is confirmed by measuring the patient's cellular level of lysosomal hydrolases. Children who are homozygous for one of these diseases are apparently normal at birth, then gradually deteriorate. In severe cases, death occurs in childhood. All of the deficiencies are autosomal and recessively inherited except Hunter syndrome, which is X-linked. Bone marrow transplants are currently being used successfully to treat Hunter syndrome; the transplanted macrophages produce the sulfatase needed to degrade glycosaminoglycans in the extracellular space. [Note: Some of the lysosomal enzymes required for the degradation of glycosaminoglycans also participate in the degradation of glycolipids and glycoproteins. Therefore, an individual suffering from a specific mucopolysaccharidosis may also have a lipidosis or glycoprotein-oligosaccharidosis.]

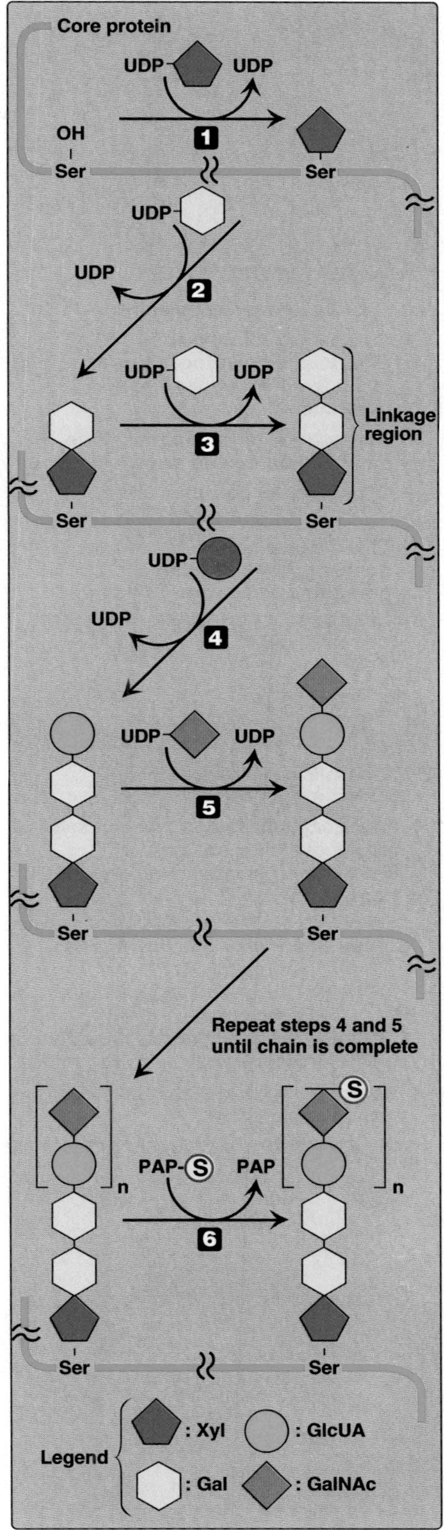

Figure 14.11
Synthesis of chondroitin sulfate.

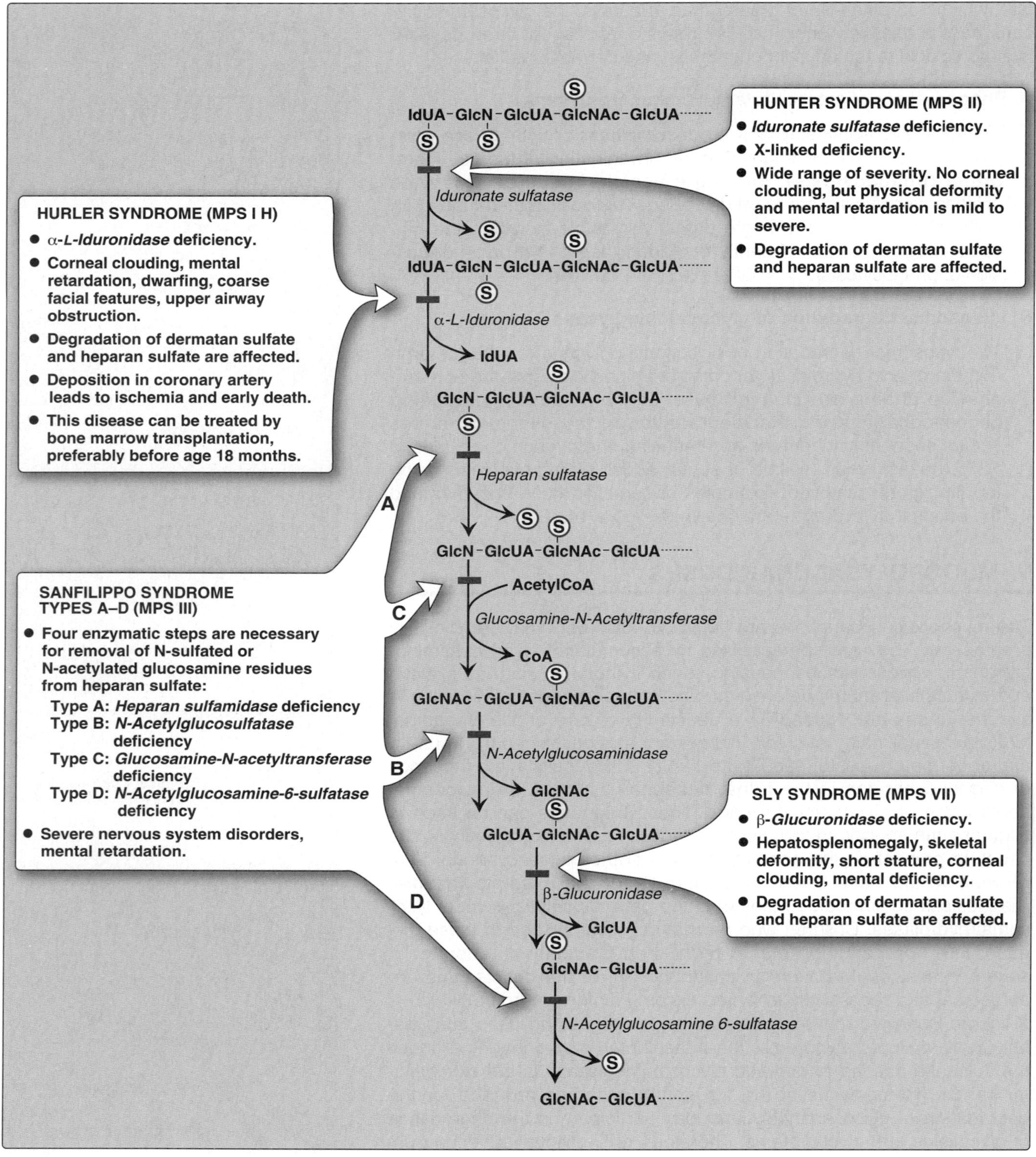

Figure 14.12
Degradation of the glycosaminoglycan heparan sulfate by lysosomal enzymes, indicating sites of enzyme deficiencies in some representative mucopolysaccharidoses.

VI. OVERVIEW OF GLYCOPROTEINS

Glycoproteins are proteins to which oligosaccharides are covalently attached. They differ from the proteoglycans (which might be considered a special case of glycoproteins) in that the length of the glycoprotein's carbohydrate chain is relatively short (usually two to ten sugar residues in length, although they can be longer), whereas it can be very long in the glycosaminoglycans (see p. 155). In addition, whereas glycosaminoglycans have diglucosyl repeat units, the carbohydrates of glycoproteins do not have serial repeats. The glycoprotein carbohydrate chains are often branched instead of linear, and may or may not be negatively charged. Glycoproteins contain highly variable amounts of carbohydrate. For example, the immunoglobulin IgG, contains less than four percent of its mass as carbohydrate, whereas human gastric glycoprotein **(mucin)** contains more than eighty percent carbohydrate. Membrane-bound glycoproteins participate in a broad range of cellular phenomena, including cell surface recognition (by other cells, hormones, viruses), cell surface antigenicity (such as the blood group antigens), and as components of the extracellular matrix and of the mucins of the gastrointestinal and urogenital tracts, where they act as protective biologic lubricants. In addition, almost all of the globular proteins present in human plasma are glycoproteins. (See Figure 14.13 for a summary of some of the functions of glycoproteins.)

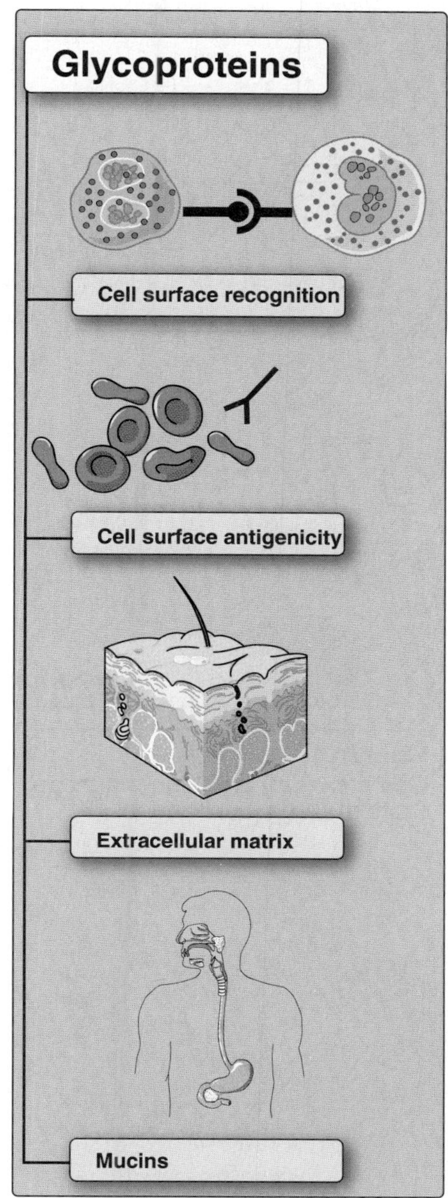

Figure 14.13
Functions of glycoproteins.

VII. STRUCTURE OF GLYCOPROTEIN OLIGOSACCHARIDES

The oligosaccharide components of glycoproteins are generally branched heteropolymers composed primarily of D-hexoses, with the addition in some cases of neuraminic acid, and of L-fucose—a 6-deoxyhexose.

A. Structure of the linkage between carbohydrate and protein

The oligosaccharide may be attached to the protein through an N- or an O-glycosidic link (see p. 85). In the former case, the sugar chain is attached to the amide group of an asparagine side chain, and in the latter case, to a hydroxyl group of either a serine or threonine R-group. [Note: In the case of collagen, there is an O-glycosidic linkage between galactose or glucose and the hydroxyl group of hydroxylysine (see p. 45).]

B. N- and O-linked oligosaccharides

A glycoprotein may contain only one type of glycosidic linkage (N- or O-linked), or may have both O- and N-linked oligosaccharides within the same molecule.

1. **O-Linked oligosaccharides:** The O-linked oligosaccharides may have one or more of a wide variety of sugars arranged in either a linear or a branched pattern. Many O-linked oligosaccharides are found as membrane glycoprotein components or in extracellular glycoproteins. For example, O-linked oligosaccharides help provide the ABO blood group determinants.

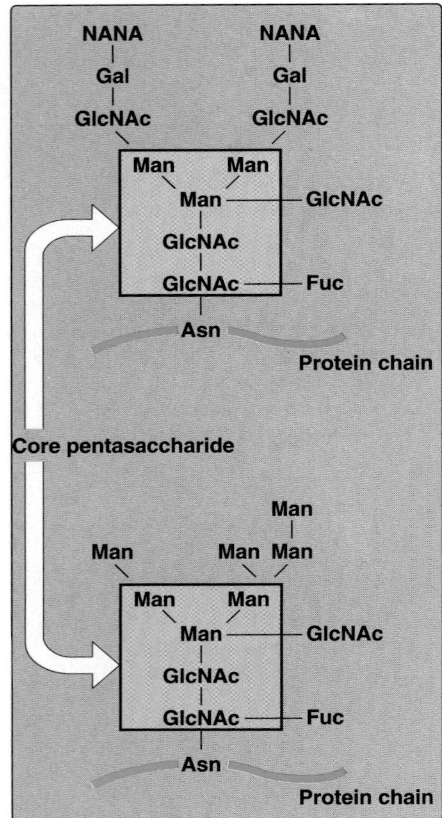

Figure 14.14
Complex (top) and high-mannose (bottom) oligosaccharides.

2. **N-linked oligosaccharides:** The N-linked oligosaccharides fall into two broad classes: complex oligosaccharides and high-mannose oligosaccharides. Both contain the same **core pentasaccharide** shown in Figure 14.14, but the complex oligosaccharides contain a diverse group of additional sugars, for example, N-acetylglucosamine (GlcNAc), L-fucose (Fuc), N-acetylneuraminic acid (NANA), whereas the high-mannose oligosaccharides contain primarily mannose (Man).

VIII. SYNTHESIS OF GLYCOPROTEINS

Most proteins are destined for the cytoplasm and are synthesized on free ribosomes in the cytosol. However, proteins, including many glycoproteins that are destined for cellular membranes, lysosome, or to be exported from the cell, are synthesized on ribosomes attached to the RER. These proteins contain specific signal sequences at their N-terminal end that act as molecular "address labels," which direct the proteins to their proper destinations. These signal sequences allow the growing polypeptide to be extruded into the lumen of the RER. The proteins are then transported via secretory vesicles to the Golgi complex, which acts as a sorting center (Figure 14.15). In the Golgi those glycoproteins that are to be secreted from the cell (or are targeted for lysosomes) remain free in the lumen, whereas those that are to become components of the cell membrane become integrated into the Golgi membrane, their carbohydrate portions oriented toward the lumen. Vesicles bud off from the Golgi and fuse with the cell membrane, either releasing the free glycoproteins, or adding the membrane-bound proteins of the vesicle to the cell membrane. The membrane glycoproteins are thus oriented with the carbohydrate portion on the outside of the cell (Figure 14.15).

A. Carbohydrate components of glycoproteins

The precursors of the carbohydrate components of glycoproteins are **sugar nucleotides**, which include UDP-glucose, UDP-galactose, UDP-N-acetylglucosamine, and UDP-N-acetylgalactosamine. In addition, GDP-mannose, GDP-L-fucose (which is synthesized from GDP-mannose), and CMP-N-acetylneuraminic acid may donate sugars to the growing chain. [Note: When NANA is present, the oligosaccharide has a negative charge at physiologic pH.] The oligosaccharides are covalently attached to specific amino acid R-groups of the protein, where the three-dimensional structure of the protein determines whether or not a specific amino acid R-group is glycosylated.

B. Synthesis of O-linked glycosides

The synthesis of the O-linked glycosides is very similar to that of the glycosaminoglycans (see p. 156). First, the protein to which the oligosaccharides are to be attached is synthesized on the RER, and extruded into its lumen. Glycosylation begins immediately, with the transfer of an N-acetylgalactosamine (from UDP-N-acetylgalactosamine) onto a specific seryl or threonyl R-group.

1. **Role of glycosyltransferases:** The glycosyltransferases responsible for the stepwise synthesis of the oligosaccharides are bound to the membranes of the ER or the Golgi apparatus. They act in a

specific order, without using a template as is required for DNA, RNA, and protein synthesis (see Unit VI of this text), but rather by recognizing the actual structure of the growing oligosaccharide as the appropriate substrate.

C. Synthesis of the N-linked glycosides

The synthesis of N-linked glycosides also occurs in the lumen of the ER and in the Golgi. However, these structures undergo additional processing steps, and require the participation of a lipid (**dolichol**) and its phosphorylated derivative, dolichol pyrophosphate (Figure 14.16).

1. **Synthesis of dolichol-linked oligosaccharide:** First, as with the O-linked glycosides, protein is synthesized on the RER and enters its lumen. The protein itself does not become glycosylated with individual sugars at this stage of glycoprotein synthesis, but rather a lipid-linked oligosaccharide is first constructed. This consists of dolichol (an ER membrane lipid 80 to 100 carbons long) attached through a pyrophosphate linkage to an oligosaccharide containing N-acetylglucosamine, mannose, and glucose. The sugars to be added to the dolichol by the membrane-bound *glycosyltransferases* are first N-acetylglucosamine, followed by mannose and glucose (see Figure 14.16). The oligosaccharide is transferred from the dolichol to an asparagine side group of the protein by a *protein-oligosaccharide transferase* present in the endoplasmic reticulum.

2. **Final processing of N-linked oligosaccharides:** After incorporation into the protein, the N-linked oligosaccharide is processed by the removal of specific mannosyl and glucosyl residues as the

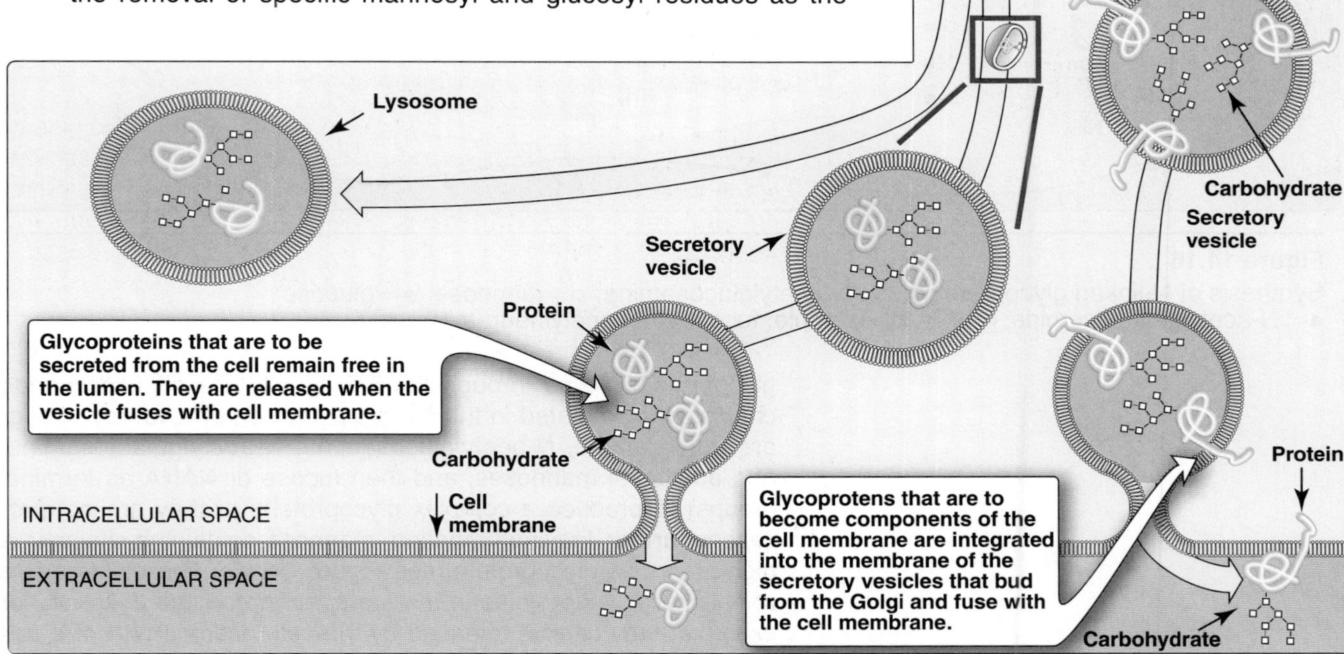

Figure 14.15
Transport of glycoproteins through the Golgi apparatus and their subsequent release or incorporation into a lysosome or the cell membrane.

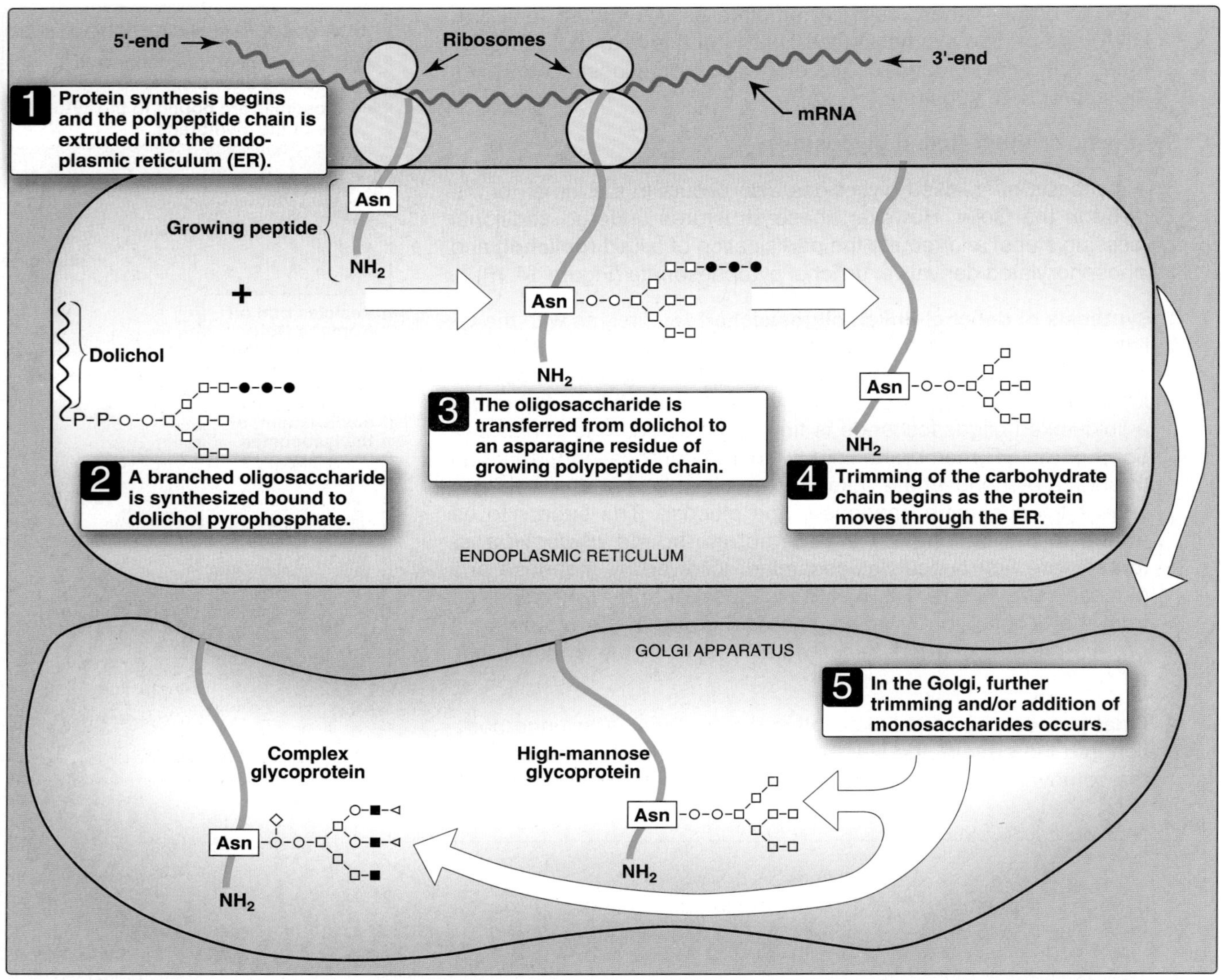

Figure 14.16
Synthesis of N-linked glycoproteins. ○ = N-acetylglucosamine; □ = mannose; ● = glucose; ■ = N-acetylgalactosamine; ◇ or ◁ for example, fucose or N-acetylneuraminic acid.

glycoprotein moves through the ER. Finally, the oligosaccharide chains are completed in the Golgi by addition of a variety of sugars (for example, N-acetylglucosamine, N-acetylgalactosamine, and additional mannoses, and then fucose or NANA as terminal groups) to produce a complex glycoprotein, or they are not processed further, leaving branched, mannose-containing chains in a high-mannose glycoprotein (see Figure 14.16). The ultimate fate of N-linked glycoproteins is the same as that of the O-linked, for example, they can be released by the cell, become part of a cell membrane, or alternatively, N-linked glycoproteins can be translocated to the lysosomes.

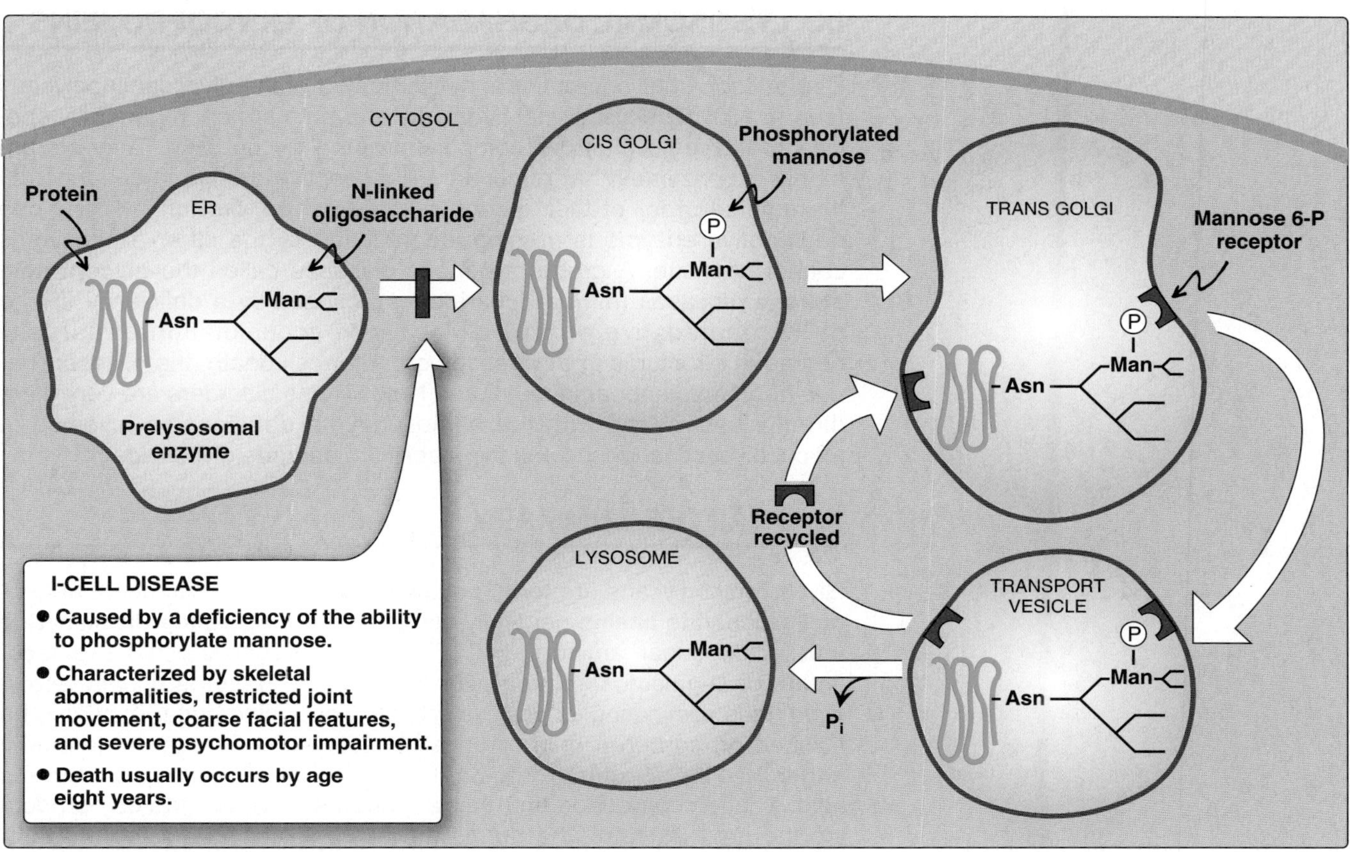

Figure 14.17
Mechanism for transport of N-linked glycoproteins to the lysosomes.

3. **Enzymes destined for lysosomes:** N-linked glycoproteins being processed through the Golgi can be phosphorylated at one or more specific mannosyl residues. Mannose 6-P receptors, located in the Golgi apparatus, bind the mannose 6-P residues of these targeted enzymes, resulting in their translocation to the lysosomes. **I-cell disease** is a rare syndrome in which the acid hydrolase enzymes normally found in lysosomes are absent, resulting in an accumulation of substrates normally degraded by lysosomal enzymes within these vesicles. [Note: I-cell disease is so-named because of the large inclusion bodies seen in cells of patients with this disease.] In addition, high amounts of lysosomal enzymes are found in the patient's plasma, suggesting that the targeting process to lysosomes (rather than the synthetic pathway of these enzymes) is deficient. It has been determined that individuals with I-cell disease are lacking the enzymic ability to phosphorylate the mannose residues of potential lysosomal enzymes, causing an incorrect targeting of these proteins to extracellular sites, rather than lysosomal vesicles (Figure 14.17). I-cell disease is characterized by skeletal abnormalities, restricted joint movement, coarse facial features, and severe psychomotor impairment. Death usually occurs by age eight years.

IX. LYSOSOMAL DEGRADATION OF GLYCOPROTEINS

Degradation of glycoproteins is similar to that of the glycosaminoglycans (see p. 160). The lysosomal hydrolytic enzymes are each generally specific for the removal of one component of the glycoprotein. They are primarily exoenzymes that remove their respective groups in sequence in the reverse order of their incorporation ("last on, first off"). If any one degradative enzyme is missing, degradation by the other exoenzymes cannot continue. A group of genetic diseases called the **glycoprotein storage diseases** (**oligosaccharidoses**), caused by a deficiency of one of the degradative enzymes, results in accumulation of partially degraded structures in the lysosomes. After cell death, the oligosaccharide fragments appear in the urine. [Note: These disorders are very often directly associated with the same enzyme deficiencies involved in mucopolysaccharidoses and the inability to degrade glycolipids.]

X. CHAPTER SUMMARY

Glycosaminoglycans are **long**, **negatively charged**, **unbranched**, **heteropolysaccharide chains** generally composed of a **repeating disaccharide unit** [acidic sugar–amino sugar]$_n$. The **amino sugar** is either **D-glucosamine** or **D-galactosamine** in which the amino group is usually acetylated, thus eliminating its positive charge. The amino sugar may also be sulfated on carbon 4 or 6 or on a nonacetylated nitrogen. The **acidic sugar** is either **D-glucuronic acid** or its carbon-5 epimer, **L-iduronic acid**. These compounds **bind large amounts of water**, thereby producing the gel-like matrix that forms the basis of the body's **ground substance.** The viscous, lubricating properties of **mucous secretions** are also caused by the presence of glycosaminoglycans, which led to the original naming of these compounds as **mucopolysaccharides.** As essential components of cell surfaces, glycosaminoglycans play an important role in mediating **cell–cell signaling** and **adhesion**. There are **six major classes** of glycosaminoglycans, including **chondroitin 4- and 6-sulfates**, **keratan sulfate**, **dermatan sulfate**, **heprin**, **heparan sulfate**, and **hyaluronic acid**. All of the glycosaminoglycans, except hyaluronic acid, are found covalently attached to protein, forming **proteoglycan monomers**, which consist of a **core protein** to which the linear glycosaminoglycan chains are covalently attached. The proteoglycan monomers associate with a molecule of **hyaluronic acid** to form **proteoglycan aggregates**. Glycosaminoglycans are synthesized in the **endoplasmic reticulum** and the **Golgi**. The polysaccharide chains are elongated by the sequential addition of alternating acidic and amino sugars, donated by their **UDP-derivatives**. The last step in synthesis is sulfation of some of the amino sugars. The source of the sulfate is **3'-phosphoadenosyl-5'-phosphosulfate**. Glycosaminoglycans are **degraded** by **lysosomal hydrolases**. They are first broken down to oligosaccharides, which are degraded sequentially from the non-reducing end of each chain. A **deficiency** of one of the hydrolases results in a **mucopolysaccharidosis**. These are hereditary disorders in which glycosaminoglycans accumulate in tissues, causing symptoms such as **skeletal** and **extracellular matrix deformities**, and **mental retardation**. Examples of these genetic diseases include **Hunter** and **Hurler syndromes**. **Glycoproteins** are proteins to which **oligosaccharides** are

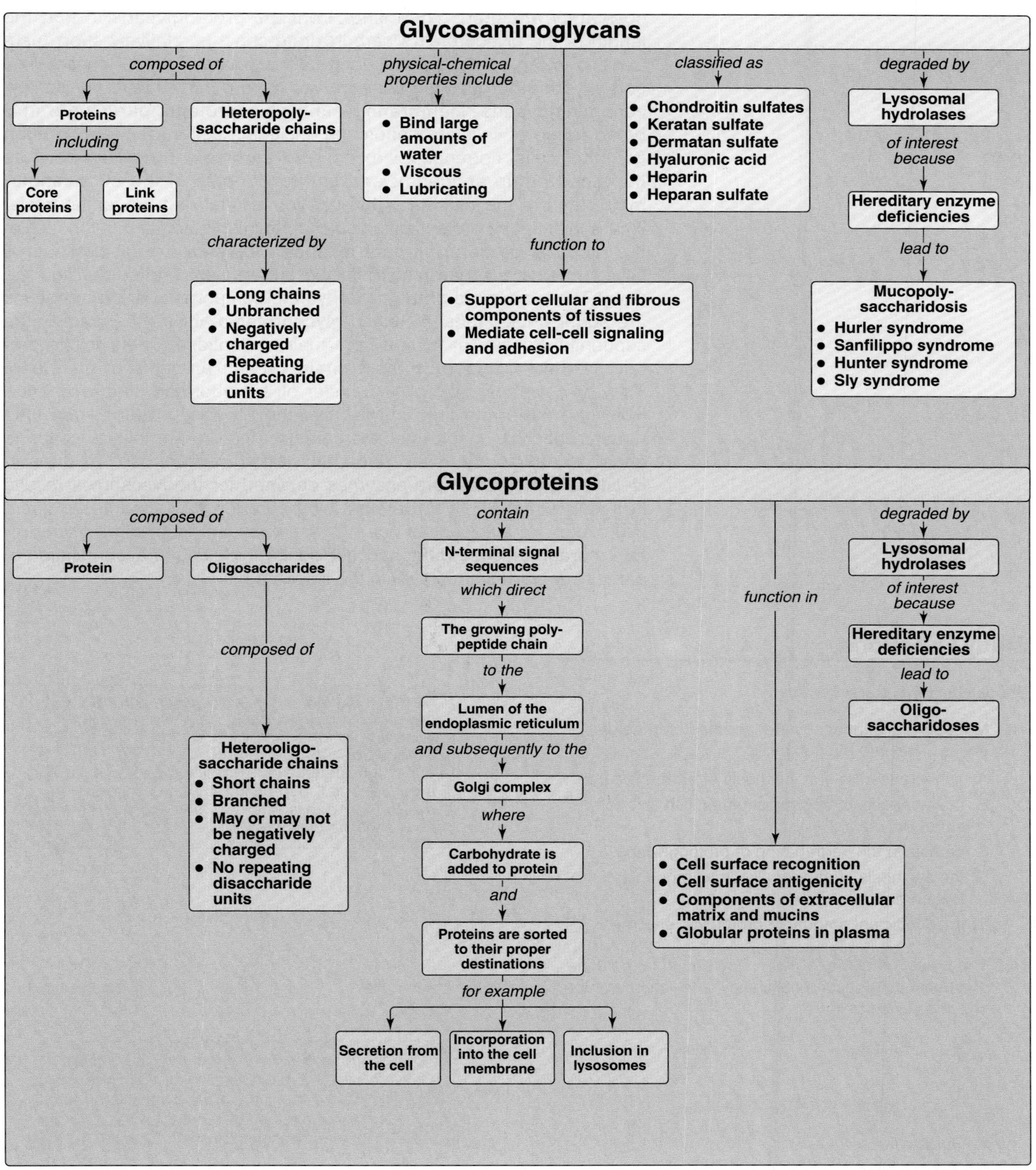

Figure 14.18
Key concept map for glycosaminoglycans and glycoproteins.

covalently attached. They differ from the proteoglycans in that the length of the glycoprotein's carbohydrate chain is relatively short (usually two to ten sugar residues long, although they can be longer). The carbohydrates of glycoproteins do not have serial repeats as do glycosaminoglycans. **Membrane-bound** glycoproteins participate in a broad range of cellular phenomena, including **cell surface recognition** (by other cells, hormones, viruses), **cell surface antigenicity** (such as the blood group antigens), and as components of the **extracellular matrix** and of the **mucins** of the gastrointestinal and urogenital tracts, where they act as protective biologic lubricants. In addition, almost all of the globular proteins present in human plasma are glycoproteins. Glycoproteins are **synthesized** in the **endoplasmic reticulum** and the **Golgi**.The precursors of the carbohydrate components of glycoproteins are **sugar nucleotides**. **O–linked glycoproteins** are synthesized by the sequential transfer of sugars from their nucleotide carriers to the protein. **N–linked glycoproteins** contain varying amounts of **mannose**. They are synthesized by the transfer of a pre-formed oligosaccharide from its membrane lipid carrier, dolichol, to the protein. They also require **dolichol**, an intermediate carrier of the growing oligosaccharide chain. A deficiency in the phosphorylation of mannose residues in N–linked glycoprotein pre-enzymes destined for the lysosomes results in **I–cell disease**. Glycoproteins are degraded in lysosomes by acid hydrolases. A deficiency of one of these enzymes results in a **glycoprotein storage disease** (**oligosaccharidosis**), resulting in accumulation of partially degraded structures in the lysosome.

Study Questions

Choose the ONE correct answer

14.1 Mucopolysaccharidoses are inherited storage diseases. They are caused by:

A. an increased rate of synthesis of proteoglycans.
B. the synthesis of polysaccharides with an altered structure.
C. defects in the degradation of proteoglycans.
D. the synthesis of abnormally small amounts of protein cores.
E. an insufficient amount of proteolytic enzymes.

Correct answer = C. In mucopolysaccharidoses, synthesis of proteoglycans is unaffected, both in terms of the structure and the amount of material synthesized. The diseases are caused by a deficiency of one of the lysosomal, hydrolytic enzymes responsible for the degradation of glycosaminoglycans (not the core protein).

14.2 The presence of the following compound in the urine of a patient suggests a deficiency in which one of the enzymes listed below?

```
Sulfate              Sulfate
   |                    |
GalNac—GlcUA—GalNAc—
```

A. Galactosidase
B. Glucosidase
C. Glucuronidase
D. Mannosidase
E. Sulfatase

Correct answer = E. Degradation of glycoproteins follows the rule "last on, first off." Because sulfation is the last step in the synthesis of this sequence, a sulfatase is required for the next step in the degradation of the above compound.

UNIT III:
Lipid Metabolism

Metabolism of Dietary Lipids

15

I. OVERVIEW

Lipids are a heterogeneous group of water-insoluble (hydrophobic) organic molecules that can be extracted from tissues by nonpolar solvents (Figure 15.1). Because of their insolubility in aqueous solutions, body lipids are generally found compartmentalized, as in the case of membrane-associated lipids or droplets of triacylglycerol in adipocytes, or transported in plasma in association with protein, as in lipoprotein particles (see p. 175). Lipids are a major source of energy for the body, and they also provide the hydrophobic barrier that permits partitioning of the aqueous contents of cells and subcellular structures. Lipids serve additional functions in the body, for example, some fat-soluble vitamins have regulatory or coenzyme functions, and the prostaglandins and steroid hormones play major roles in the control of the body's homeostasis. Not surprisingly, deficiencies or imbalances of lipid metabolism can lead to some of the major clinical problems encountered by physicians, such as atherosclerosis and obesity.

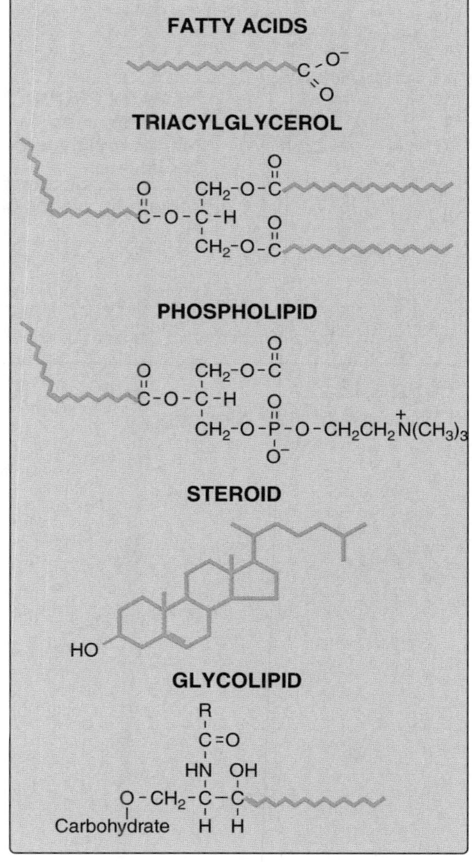

Figure 15.1
Structures of some common classes of lipids. Hydrophobic portions of the molecules are shown in orange.

II. DIGESTION, ABSORPTION, SECRETION, AND UTILIZATION OF DIETARY LIPIDS

An adult ingests about 60 to 150 g of lipids per day, of which more than ninety percent is normally triacylglycerol (formerly called triglyceride). The remainder of the dietary lipids consists primarily of cholesterol, cholesteryl esters, phospholipids, and unesterified ("free") fatty acids. The digestion of dietary lipids is summarized in Figure 15.2.

A. Processing of dietary lipid in the stomach

The digestion of lipids begins in the stomach, catalyzed by an acid-stable lipase that originates from glands at the back of the tongue (*lingual lipase*). Triacylglycerol molecules, particularly those containing

Lippincott's Illustrated Reviews: Biochemistry, 3rd Edition
by Pamela C. Champe and Richard A. Harvey.
Lippincott Williams & Wilkins, Baltimore, MD © 2005.

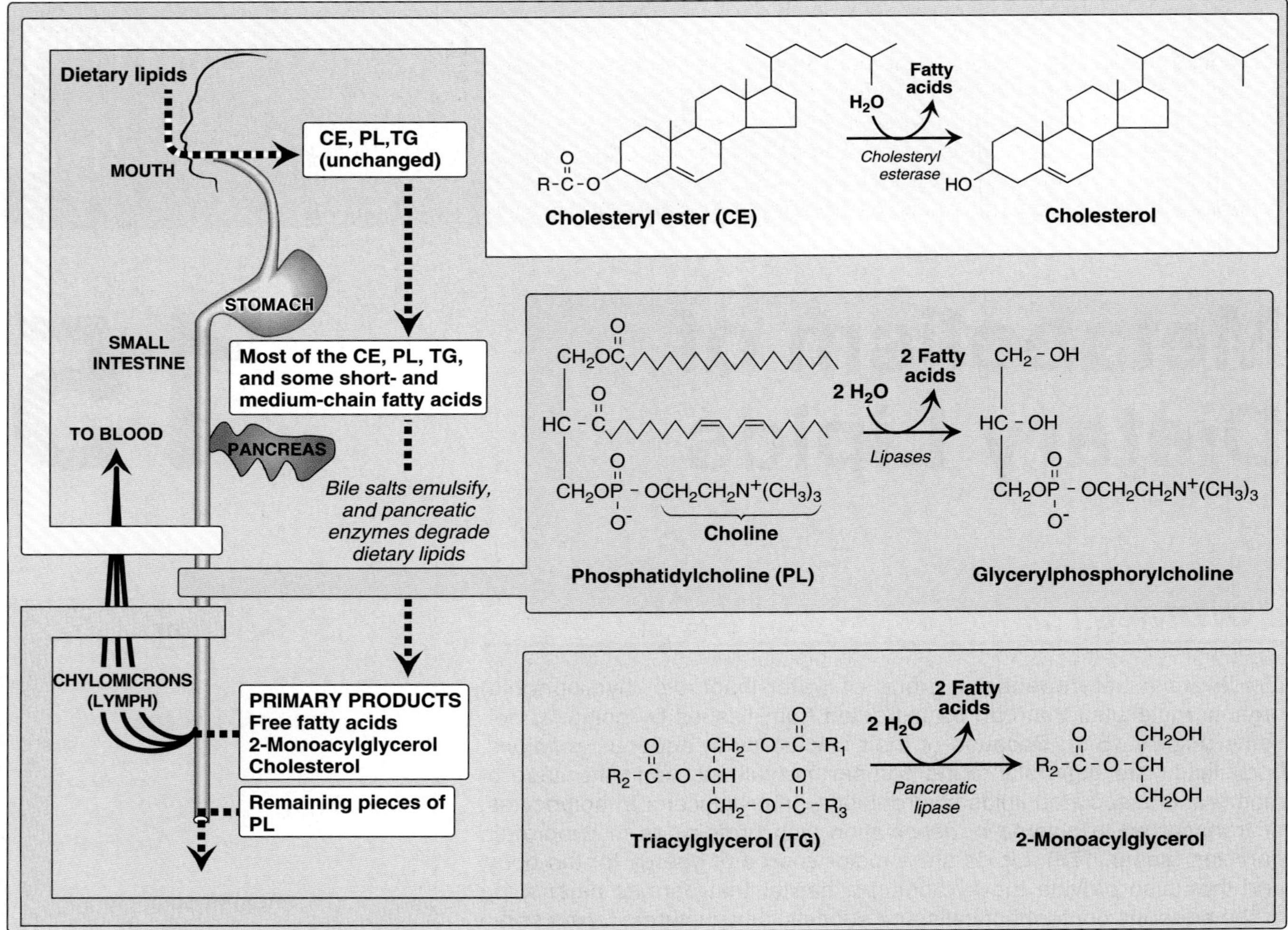

Figure 15.2
Overview of lipid digestion.

fatty acids of short- or medium-chain length (less than twelve carbons, such as are found in milk fat), are the primary target of this enzyme. These same triacylglycerols are also degraded by a separate *gastric lipase*, secreted by the gastric mucosa. Both enzymes are relatively acid-stable, with pH optimums of pH 4 to pH 6. These "acid lipases" play a particularly important role in lipid digestion in **neonates**, for whom milk fat is the primary source of calories. They are also important digestive enzymes in individuals with pancreatic insufficiency, such as those with **cystic fibrosis**. *Lingual* and *gastric lipases* aid these patients in degrading triacylglycerol molecules (especially those with short- to medium-chain length fatty acids) despite a near or complete absence of *pancreatic lipase* (see below).

B. Emulsification of dietary lipid in the small intestine

The critical process of emulsification of dietary lipids occurs in the duodenum. Emulsification increases the surface area of the hydrophobic lipid droplets so that the digestive enzymes, which work at the interface of the droplet and the surrounding aqueous solution, can act effectively. Emulsification is accomplished by two complementary mechanisms, namely, use of the detergent properties of the **bile salts**, and mechanical mixing due to **peristalsis**. Bile salts, made in the liver and stored in the gallbladder, are derivatives of cholesterol (see p. 222). They consist of a sterol ring structure with a side chain to which a molecule of glycine or taurine is covalently attached by an amide linkage (Figure 15.3). These emulsifying agents interact with the dietary lipid particles and the aqueous duodenal contents, thereby stabilizing the particles as they become smaller, and preventing them from coalescing. A more complete discussion of bile salt metabolism is given on p. 223.

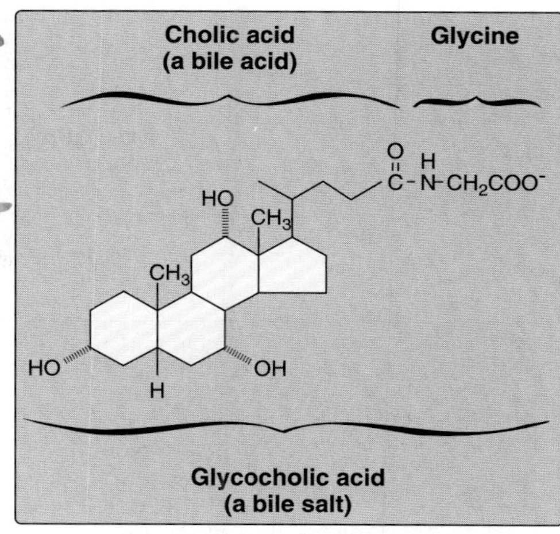

Figure 15.3
Structure of glycocholic acid.

C. Degradation of dietary lipids by pancreatic enzymes

The dietary triacylglycerol, cholesteryl esters, and phospholipids are enzymically degraded ("digested") by pancreatic enzymes, whose secretion is hormonally controlled.

1. **Triacylglycerol degradation:** Triacylglycerol molecules are too large to be taken up efficiently by the mucosal cells of the intestinal villi. They are, therefore, acted upon by an esterase, *pancreatic lipase*, which preferentially removes the fatty acids at carbons 1 and 3. The primary products of hydrolysis are thus a mixture of **2-monoacylglycerol** and **free fatty acids** (see Figure 15.2). [Note: This enzyme is found in high concentrations in pancreatic secretions (two to three percent of the total protein present), and it is highly efficient catalytically, thus insuring that only severe pancreatic deficiency, such as that seen in **cystic fibrosis**, results in significant malabsorption of fat.] A second protein, **colipase**, also secreted by the pancreas, binds the *lipase* at a ratio of one to one, and anchors it at the lipid-aqueous interface. [Note: Colipase is secreted as the zymogen, procolipase, which is activated in the intestine by *trypsin*.] **Orlistat**, an antiobesity drug, inhibits *gastric* and *pancreatic lipases,* thereby decreasing fat absorption, resulting in loss of weight.[1]

2. **Cholesteryl ester degradation:** Most dietary cholesterol is present in the free (nonesterified) form, with ten to fifteen percent present in the esterified form. Cholesteryl esters are hydrolyzed by pancreatic *cholesterol ester hydrolase* (*cholesterol esterase*), which produces cholesterol plus free fatty acids (see Figure 15.2). *Cholesteryl ester hydrolase* activity is greatly increased in the presence of bile salts.

3. **Phospholipid degradation:** Pancreatic juice is rich in the proenzyme of *phospholipase A_2* that, like procolipase, is activated by *trypsin* and, like *cholesterol ester hydrolase*, requires bile salts for optimum activity. *Phospholipase A_2* removes one fatty acid from carbon 2 of a phospholipid, leaving a **lysophospholipid**. For example, phosphatidylcholine (the predominant phospholipid during digestion) becomes lysophosphatidylcholine. The remaining fatty acid at carbon 1 can be removed by *lysophospholipase*, leaving a **glycerylphosphoryl base** (for example, glycerylphosphorylcholine, see Figure 15.2) that may be excreted in the feces, further degraded, or absorbed.

INFO LINK [1]See Chapter 28 in ***Lippincott's Illustrated Reviews: Pharmacology*** (3rd Ed.) and Chapter 42 (2nd Ed.) for a discussion of orlistat.

4. **Control of lipid digestion:** Pancreatic secretion of the hydrolytic enzymes that degrade dietary lipids in the small intestine is **hormonally controlled** (Figure 15.4). Cells in the mucosa of the jejunum and lower duodenum produce a small peptide hormone, **cholecystokinin** (**CCK**, formerly called **pancreozymin**), in response to the presence of lipids and partially digested proteins entering these regions of the upper small intestine. CCK acts on the gallbladder (causing it to contract and release bile), and on the exocrine cells of the pancreas (causing them to release digestive enzymes). It also decreases gastric motility, resulting in a slower release of gastric contents into the small intestine. Other intestinal cells produce another small peptide hormone, **secretin**, in response to the low pH of the chyme entering the intestine. Secretin causes the pancreas and the liver to release a watery solution rich in bicarbonate that helps neutralize the pH of the intestinal contents, bringing them to the appropriate pH for enzymic digestive activity by pancreatic enzymes.

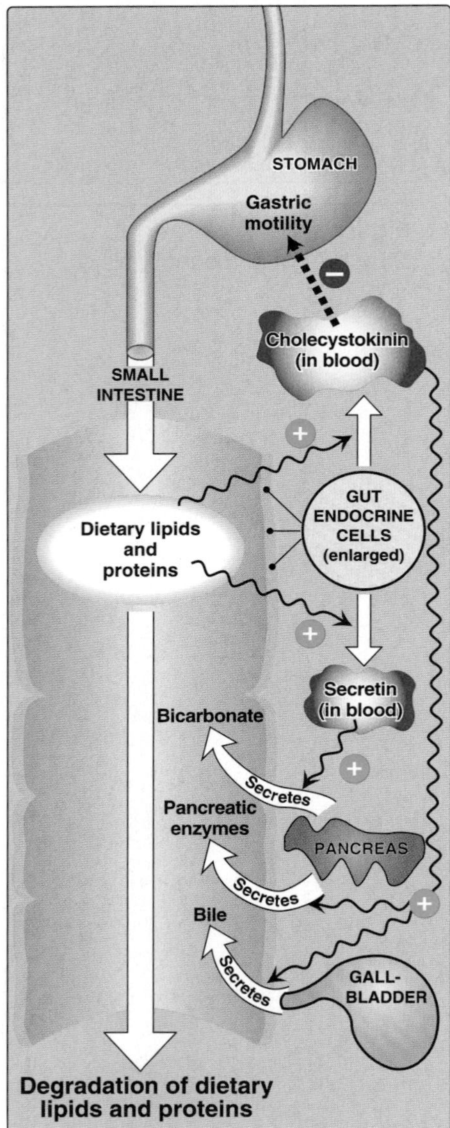

Figure 15.4
Hormonal control of lipid metabolism in the small intestine.

D. Absorption of lipids by intestinal mucosal cells (enterocytes)

Free fatty acids, free cholesterol, and 2-monoacylglycerol are the primary products of dietary lipid degradation in the jejunum. These, together with bile salts, form **mixed micelles**—disk-shaped clusters of amphipathic lipids that coalesce with their hydrophobic groups on the inside and their hydrophilic groups on the outside of the cluster. Mixed micelles are, therefore, soluble in the aqueous environment of the intestinal lumen (Figure 15.5). These particles approach the primary site of lipid absorption, the **brush border membrane** of the enterocytes (mucosal cell). This membrane is separated from the liquid contents of the intestinal lumen by an **unstirred water layer** that mixes poorly with the bulk fluid. The hydrophilic surface of the micelles facilitates the transport of the hydrophobic lipids through the unstirred water layer to the brush border membrane where they are absorbed. [Note: Short- and medium-chain length fatty acids do not require the assistance of mixed micelles for absorption by the intestinal mucosa. This is an important consideration in the dietary therapy for individuals with malabsorption of other lipids.]

E. Resynthesis of triacylglycerol and cholesteryl esters

The mixture of lipids absorbed by the enterocytes migrates to the endoplasmic reticulum where biosynthesis of complex lipids takes place. Fatty acids are first converted into their activated form by *fatty acyl CoA synthetase* (*thiokinase*) (Figure 15.6). Using the fatty acyl CoA derivatives, the 2-monoacylglycerols absorbed by the enterocytes are converted to triacylglycerols by the enzyme complex, *triacylglycerol synthase*. This complex synthesizes triacylglycerol by the consecutive actions of two enzyme activities— *monoacylglycerolacyltransferase* and *diacylglycerolacyltransferase*. Lysophospholipids are reacylated to form phospholipids by a family of *acyltransferases*, and cholesterol is esterified to a fatty acid primarily by *acyl CoA:cholesterolacyltransferase* (see 232). [Note: Virtually all long-chain fatty acids entering the enterocytes are used in this fashion to form triacylglycerols, phospholipids, and cholesteryl esters.

Short- and medium-chain length fatty acids are not converted to their CoA derivatives, and are not reesterified to 2-monoacylglycerol. Instead, they are released into the portal circulation, where they are carried by serum albumin to the liver.]

F. Lipid malabsorption

Lipid malabsorption, resulting in increased lipid (including the fat-soluble vitamins A, D, E, and K, and essential fatty acids) in the feces (that is, **steatorrhea**), can be caused by disturbances in lipid digestion and/or absorption (Figure 15.7). Such disturbances can result from several conditions, including **cystic fibrosis** (causing poor digestion) and **shortened bowel** (causing decreased absorption).

G. Secretion of lipids from enterocytes

The newly synthesized triacylglycerols and cholesteryl esters are very hydrophobic, and aggregate in an aqueous environment. It is, therefore, necessary that they be packaged as lipid droplets surrounded by a thin layer composed of phospholipids, unesterified cholesterol, and a single protein molecule (**apolipoprotein B-48**, see p. 226). This layer stabilizes the particle and increases its solubility, thereby preventing multiple particles from coalescing. [Note: The presence of these particles in the lymph after a lipid-rich meal gives the lymph a milky appearance. This lymph is called **chyle** (as opposed to **chyme**, the name given to the semifluid mass of partially digested food that passes from the stomach to the duodenum). The small particles are named **chylomicrons**.] Chylomicrons are released by exocytosis from enterocytes into the **lacteals** (lymphatic vessels originating in the villi of the small intestine). They follow the lymphatic system to the thoracic duct, and are then conveyed to the left subclavian vein, where they enter the blood. The steps in the production of chylomicrons are summarized in Figure 15.6. (For a more detailed description of chylomicron structure and metabolism, see p. 226.)

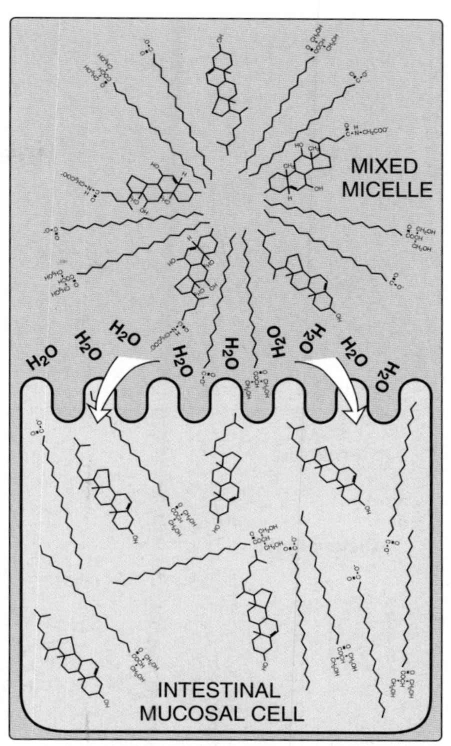

Figure 15.5
Absorption of lipids contained in a mixed micelle by an intestinal mucosal cell.

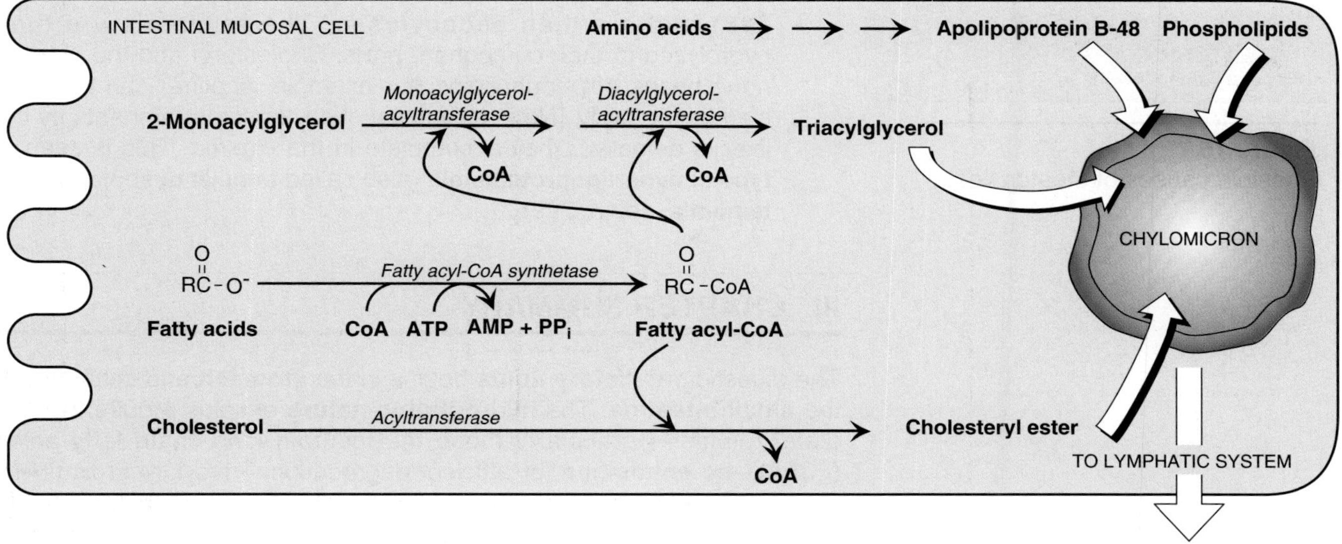

Figure 15.6
Assembly and secretion of chylomicrons by intestinal mucosal cells.

H. Use of dietary lipids by the tissues

Triacylglycerol contained in chylomicrons is broken down primarily in the capillaries of skeletal muscle and adipose tissues, but also those of the heart, lung, kidney, and liver. Triacylglycerol in chylomicrons is degraded to free fatty acids and glycerol by *lipoprotein lipase*. This enzyme is synthesized primarily by adipocytes and muscle cells. It is secreted and becomes associated with the luminal surface of endothelial cells of the capillary beds of the peripheral tissues. [Note: **Familial lipoprotein lipase deficiency** (**type I hyperlipoproteinemia**) is a rare, autosomal recessive disorder that results from a deficiency of *lipoprotein lipase* or its coenzyme, apo C-II (see p. 227). The result is massive chylomicronemia.]

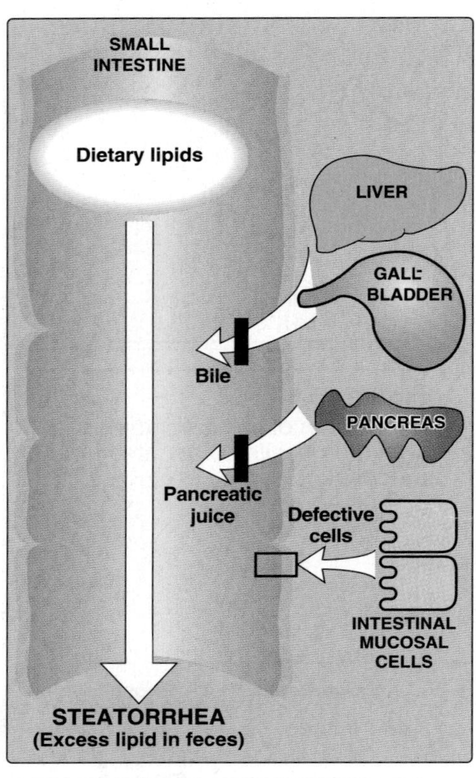

Figure 15.7
Possible causes of steatorrhea.

1. **Fate of free fatty acids:** The free fatty acids derived from the hydrolysis of triacylglycerol may directly enter adjacent muscle cells or adipocytes. Alternatively, the free fatty acids may be transported in the blood in association with **serum albumin** until they are taken up by cells. [Note: Serum albumin is a large protein secreted by the liver. It transports a number of primarily hydrophobic compounds in the circulation, including free fatty acids and some drugs.[2]] Most cells can oxidize fatty acids to produce energy (see p. 188). Adipocytes can also reesterify free fatty acids to produce triacylglycerol molecules, which are stored until the fatty acids are needed by the body (see p. 185).

2. **Fate of glycerol:** Glycerol that is released from triacylglycerol is used almost exclusively by the liver to produce glycerol 3-phosphate, which can enter either glycolysis or gluconeogenesis by oxidation to dihydroxyacetone phosphate (see p. 188).

3. **Fate of the remaining chylomicron components:** After most of the triacylglycerol has been removed, the **chylomicron remnants** (which contain cholesteryl esters, phospholipids, apolipoproteins, and some triacylglycerol) bind to receptors on the **liver** (see p. 228) and are then endocytosed. The remnants are then hydrolyzed to their component parts. Cholesterol and the nitrogenous bases of phopholipids (for example, choline) can be recycled by the body. [Note: If removal of chylomicron remnants by the liver is defective, they accumulate in the plasma. This is seen in **type III hyperlipoproteinemia** (also called **familial dysbetalipoproteinemia**, see p. 229).

III. CHAPTER SUMMARY

The digestion of **dietary lipids** begins in the **stomach** and continues in the **small intestine**. The **hydrophobic nature** of lipids require that the dietary lipids—particularly those that contain long-chain fatty acids (LCFA)—be **emulsified** for efficient degradation. Triacylglycerols (TAG) obtained from milk contain **short-** to **medium-chain length fatty acids** that can be degraded in the **stomach** by the acid lipases (**lingual lipase** and **gastric lipase**). Cholesteryl esters (CE), phospholipids (PL), and TAG containing LCFAs are degraded in the **small intestine**

INFO LINK [2]See Chapter 1 in ***Lippincott's Illustrated Reviews: Pharmacology*** (2nd and 3rd Eds.) for a discussion of the interaction of drugs with serum albumin.

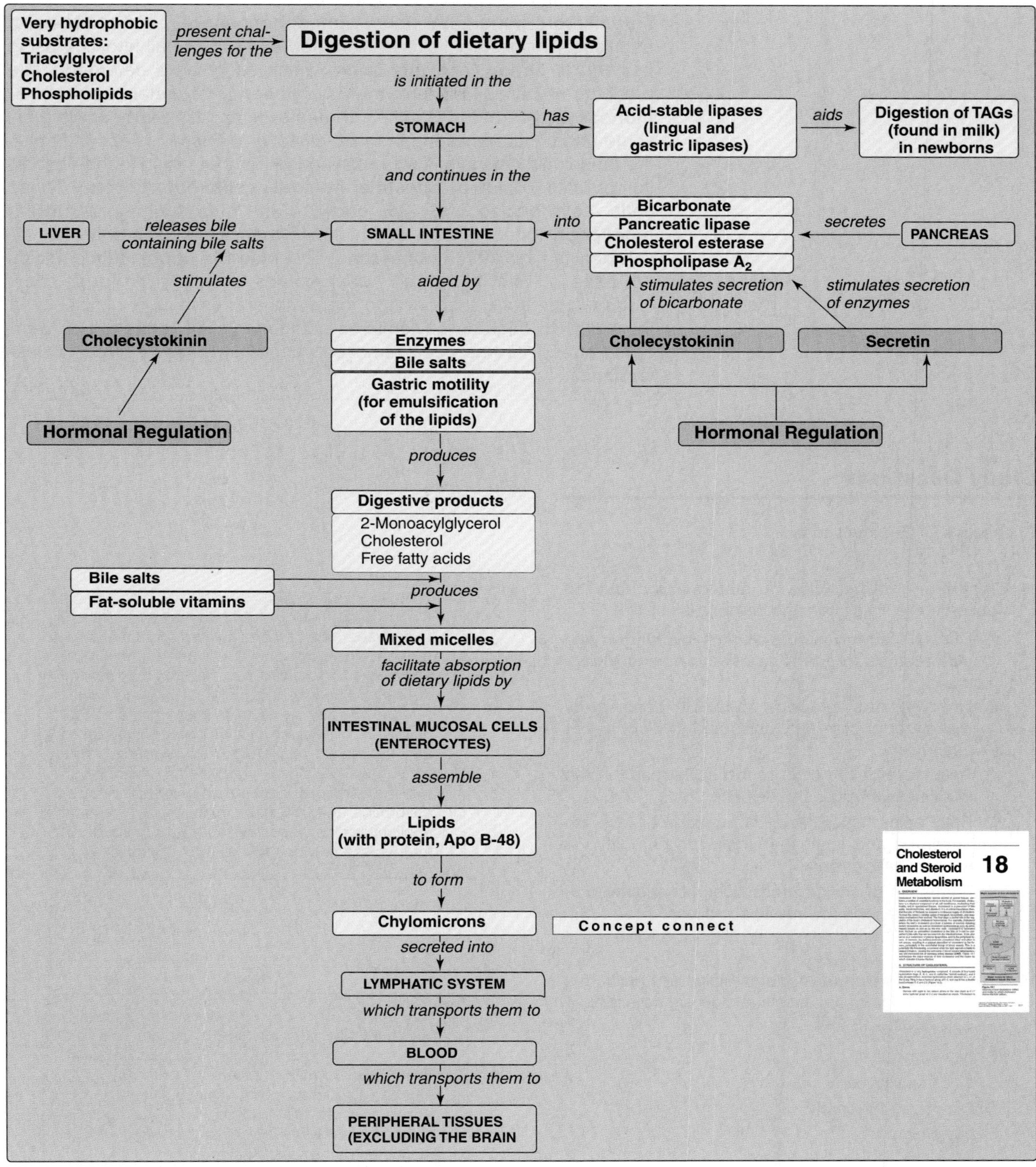

Figure 15.8
Key concept map for metabolism of dietary lipids.

by enzymes secreted by the **pancreas**. The most important of these enzymes are **pancreatic lipase**, **phospholipase A_2**, and **cholesterol esterase**. The dietary lipids are **emulsified** in the small intestine using **peristaltic action**, and **bile salts**, which serve as a detergent. The products resulting from enzymatic degradation of dietary lipid are **2-monoacylglycerol**, unesterified **cholesterol**, and **free fatty acids** (plus some fragments remaining from PL digestion). These, compounds plus the fat-soluble vitamins, form **mixed micelles** that facilitate the absorption of dietary lipids by **intestinal mucosal cells** (**enterocytes**). These cells resynthesize TAG, CE, and PL, and also synthesize protein (**apolipoprotein B-48**), all of which are then assembled with the fat-soluble vitamins into **chylomicrons**. These **serum lipoprotein** particles are released into the **lymph**, which carries them to the **blood**. Thus, dietary lipids are transported to the peripheral tissues. A deficiency in the ability to degrade chylomicron components, or remove their remnants after TAG has been removed, results in **massive hypercholesterolemia**.

Study Questions

Choose the ONE correct answer

15.1 Which one of the following statements about the absorption of lipids from the intestine is correct?

A. Dietary triacylglycerol is partially hydrolyzed and absorbed as free fatty acids and monoacylglycerol.
B. Dietary triacylglycerol must be completely hydrolyzed to free fatty acids and glycerol before absorption.
C. Release of fatty acids from triacylglycerol in the intestine is inhibited by bile salts.
D. Fatty acids that contain ten carbons or less are absorbed and enter the circulation primarily via the lymphatic system.
E. Formation of chylomicrons does not require protein synthesis in the intestinal mucosa.

Correct answer = A. Pancreatic lipase hydrolyzes dietary triacylglycerol primarily to 2-monoacylglycerol plus two fatty acids. These products of hydrolysis can be absorbed by the intestinal mucosal cells. Bile salts do not inhibit release of fatty acids from triacylglycerol, but rather are necessary for the proper solubilization and hydrolysis of dietary triacylglycerol in the small intestine. Short- and medium-chain length fatty acids enter the portal circulation after absorption from the small intestine. Synthesis of apolipoproteins, especially apo B-48, is essential for the assembly and secretion of chylomicrons.

15.2 The form in which most dietary lipids are packaged and exported from the intestinal mucosal cells is as:

A. free fatty acids.
B. mixed micelles.
C. free triacylglycerol.
D. 2-monoacylglycerol.
E. chylomicrons.

Correct answer = E. Chylomicrons contain a lipid core that is composed of dietary lipid and lipid synthesized in the intestinal mucosal cells. Free fatty acids are esterified primarily to 2-monoacylglycerol, forming triacylglycerol, prior to export from the intestinal mucosal cells in chylomicrons. Mixed micelles are found only in the lumen of the small intestine.

Fatty Acid and Triacylglycerol Metabolism

16

I. OVERVIEW

Fatty acids exist free in the body (that is, they are unesterified), and, also are found as fatty acyl esters in more complex molecules, such as triacylglycerols. Low levels of free fatty acids occur in all tissues, but substantial amounts sometimes can be found in the plasma, particularly during fasting. Plasma free fatty acids (transported by serum albumin) are in route from their point of origin (triacylglycerol of adipose tissue or circulating lipoproteins) to their site of consumption (most tissues). Free fatty acids can be oxidized by many tissues—particularly liver and muscle—to provide energy. Fatty acids are also structural components of membrane lipids, such as phospholipids and glycolipids (see Chapter 17, p. 199). Fatty acids are attached to certain intracellular proteins to enhance the ability of those proteins to associate with membranes. Fatty acids are also precursors of the hormone-like prostaglandins (see p. 211). Esterified fatty acids, in the form of triacylglycerols stored in adipose cells, serve as the major energy reserve of the body. Figure 16.1 illustrates the metabolic pathways of fatty acid synthesis and degradation, and their relationship to carbohydrate metabolism.

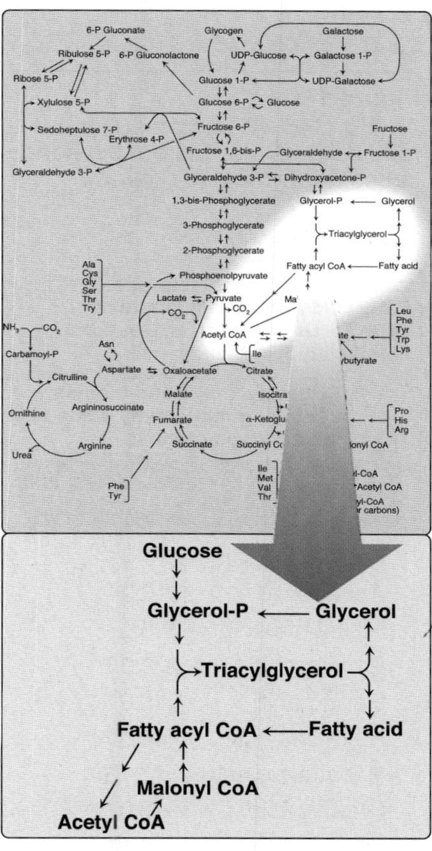

Figure 16.1
Triacylglycerol synthesis and degradation.

II. STRUCTURE OF FATTY ACIDS

A fatty acid consists of a hydrophobic hydrocarbon chain with a terminal carboxyl group that has a pK_a of about 4.8 (Figure 16.2). At physiologic pH, the terminal carboxyl group (–COOH) ionizes, becoming $–COO^-$. This anionic group has an affinity for water, giving the fatty acid its **amphipathic nature** (having both a hydrophilic and a hydrophobic region). However, for long-chain fatty acids (LCFA), the hydrophobic portion is predominant. These molecules are highly water-insoluble, and must be transported in the circulation in association with protein. More than ninety percent of the fatty acids found in plasma are in the form of fatty acid esters (primarily triacylglycerol, cholesteryl esters, and phospholipids) contained in circulating **lipoprotein particles** (see p. 225). Unesterified fatty acids are transported in the circulation in association with **albumin**.

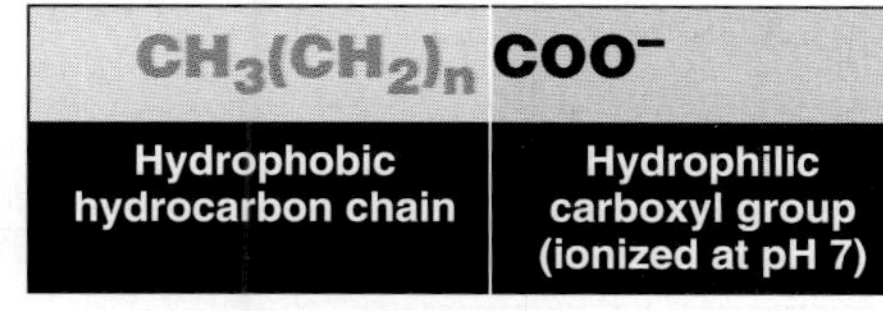

Figure 16.2
Structure of a fatty acid.

Lippincott's Illustrated Reviews: Biochemistry, by Pamela C. Champe and Richard A. Harvey.
Lippincott Williams & Wilkins, Baltimore, MD © 2005.

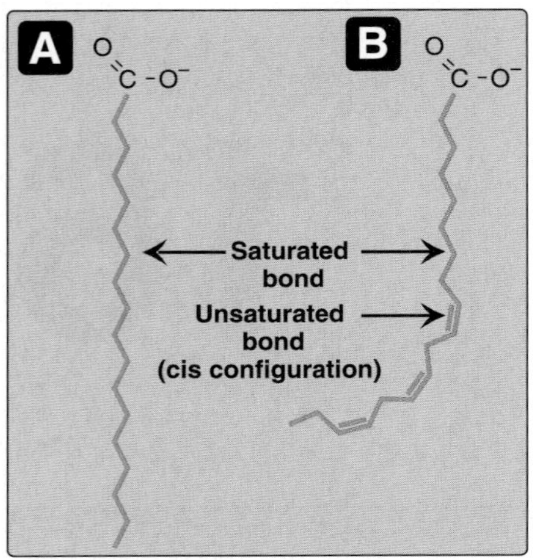

Figure 16.3
A saturated (A) and an unsaturated (B) fatty acid. [Note: Cis double bonds cause a fatty acid to "kink."]

A. Saturation of fatty acids

Fatty acid chains may contain no double bonds—that is, be **saturated,** or contain one or more double bonds—that is, be **mono-** or **polyunsaturated**. When double bonds are present, they are nearly always in the **cis** rather than in the trans configuration. (See p. 362 for a discussion of the dietary occurrence of cis and trans unsaturated fatty acids.) The introduction of a cis double bond causes the fatty acid to bend or "kink" at that position (Figure 16.3). If the fatty acid has two or more double bonds, they are always spaced at three carbon intervals. [Note: In general, addition of double bonds decreases the **melting temperature** (**T_m**) of a fatty acid, whereas increasing the chain length increases the T_m. Because membrane lipids typically contain LCFA, the presence of double bonds in some fatty acids helps maintain the fluid nature of those lipids.]

B. Chain lengths of fatty acids

The common names and structures of some fatty acids of physiologic importance are listed in Figure 16.4. In this figure, the carbon atoms are numbered, beginning with the carboxyl carbon as carbon 1. The number before the colon indicates the number of carbons in the chain, and those after the colon indicate the numbers and positions of double bonds. For example, as shown in Figure 16.5A, arachidonic acid, 20:4(5, 8, 11, 14), is 20 carbons long and has 4 double bonds (between carbons 5–6, 8–9, 11–12, and 14–15). [Note: The carbon to which the carboxyl group is attached (carbon 2) is also called the α–carbon, carbon 3 is the β–carbon, and carbon 4 is the γ–carbon. The carbon of the terminal methyl group is called the ω–carbon regardless of the chain length.] The carbons in a fatty acid can also be counted beginning at the ω- (or methyl-terminal) end of the chain. Arachidonic acid is referred to as an **ω-6** (also called an n–6, see p. 360) **fatty acid** because the closest double bond to the ω end begins six carbons from that end (Figure 16.5B). Another ω-6 fatty acid is the essential **linoleic acid**, 18:2(9,12). In contrast, **linolenic acid**, 18:3(9,12,15), is an **ω-3 fatty acid**. (See p. 360 for a discussion of the nutritional significance of ω-3 and ω-6 fatty acids.)

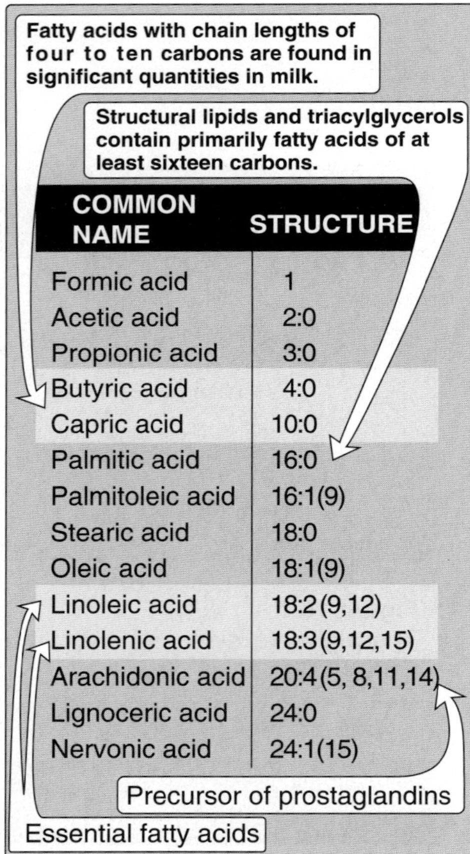

COMMON NAME	STRUCTURE
Formic acid	1
Acetic acid	2:0
Propionic acid	3:0
Butyric acid	4:0
Capric acid	10:0
Palmitic acid	16:0
Palmitoleic acid	16:1(9)
Stearic acid	18:0
Oleic acid	18:1(9)
Linoleic acid	18:2(9,12)
Linolenic acid	18:3(9,12,15)
Arachidonic acid	20:4(5, 8,11,14)
Lignoceric acid	24:0
Nervonic acid	24:1(15)

Figure 16.4
Some fatty acids of physiologic importance.

C. Essential fatty acids

Two fatty acids are dietary essentials in humans (see p. 361): **linoleic acid**, which is the precursor of arachidonic acid, the substrate for prostaglandin synthesis (see p. 211), and **linolenic acid**, the precursor of other ω-3 fatty acids important for growth and development. [Note: A deficiency of linolenic acid results in decreased vision and altered learning behaviors.] Arachidonic acid becomes essential if linoleic acid is deficient in the diet.

III. DE NOVO SYNTHESIS OF FATTY ACIDS

A large proportion of the fatty acids used by the body is supplied by the diet. Carbohydrates, protein, and other molecules obtained from the diet in excess of the body's needs for these compounds can be converted to fatty acids, which are stored as triacylglycerols. (See Chapter 24, p. 319, for a discussion of the metabolism of dietary nutrients in the well-fed state.) In humans, fatty acid synthesis occurs primarily in the **liver** and **lactating mammary glands** and, to a lesser extent, in adipose tissue. The process incorporates carbons from acetyl CoA into the growing fatty acid chain, using ATP and reduced nicotinamide adenine dinucleotide phosphate (NADPH).

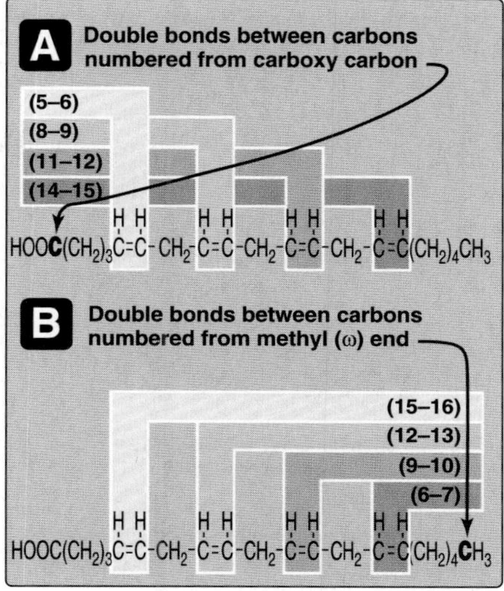

Figure 16.5
Arachidonic acid, illustrating position of double bonds.

A. Production of cytosolic acetyl CoA

The first step in de novo fatty acid synthesis is the transfer of acetate units from mitochondrial acetyl CoA to the cytosol. Mitochondrial acetyl CoA is produced by the oxidation of pyruvate (see p. 107), and by the catabolism of fatty acids (see p. 188), ketone bodies (see p. 194), and certain amino acids (see p. 263). The coenzyme A portion of acetyl CoA, however, cannot cross the mitochondrial membrane; only the acetyl portion is transported to the cytosol. It does so in the form of **citrate** produced by the condensation of oxaloacetate (OAA) and acetyl CoA (Figure 16.6). [Note: This process of **translocation** of citrate from the mitochondrion to the cytosol, where it is cleaved by *ATP-citrate lyase* to produce cytosolic acetyl CoA and OAA, occurs when the mitochondrial citrate concentration is high. This is observed when ***isocitrate dehydrogenase*** is inhibited by the presence of large amounts of ATP, causing citrate and isocitrate to accumulate (see p. 110). Therefore, cytosolic citrate may be viewed as a high-energy signal.] Because a large amount of ATP is needed for fatty acid synthesis, the increase in both ATP and citrate enhances this pathway.

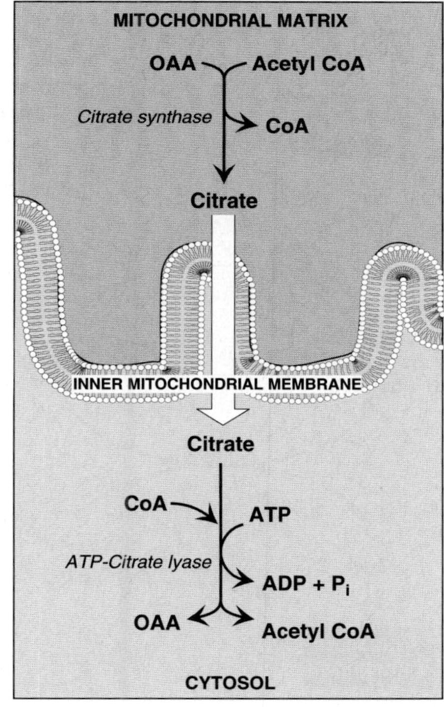

Figure 16.6
Production of cytosolic acetyl CoA.

B. Carboxylation of acetyl CoA to form malonyl CoA

The energy for the carbon-to-carbon condensations in fatty acid synthesis is supplied by the process of carboxylation and then decarboxylation of acetyl groups in the cytosol. The carboxylation of acetyl CoA to form malonyl CoA is catalyzed by *acetyl CoA carboxylase* (Figure 16.7), and requires HCO_3^- and ATP. The coenzyme is the vitamin, **biotin**, which is covalently bound to a lysyl residue of the *carboxylase*.

1. **Short-term regulation of acetyl CoA carboxylase:** This carboxylation is both the **rate-limiting** and the regulated step in fatty acid synthesis (see Figure 16.7). The inactive form of *acetyl CoA carboxylase* is a protomer (dimer). The enzyme undergoes **allosteric activation** by **citrate**, which causes dimers to polymerize. The enzyme can be **allosterically inactivated** by **long-chain fatty acyl CoA** (the end product of the pathway), which causes its depolymerization. A second mechanism of short-term regulation is by **reversible phosphorylation**. In the presence of counterregulatory hormones, such as **epinephrine** and **glucagon**, *acetyl CoA carboxylase* is **phosphorylated** and, thereby, **inactivated** (Figure 16.8). [Note: This is analogous to the mechanism of inactivation of *glycogen synthase*, p. 129.] In the presence of **insulin**, *acetyl CoA carboxylase* is **dephosphorylated** and, thereby, **activated**.

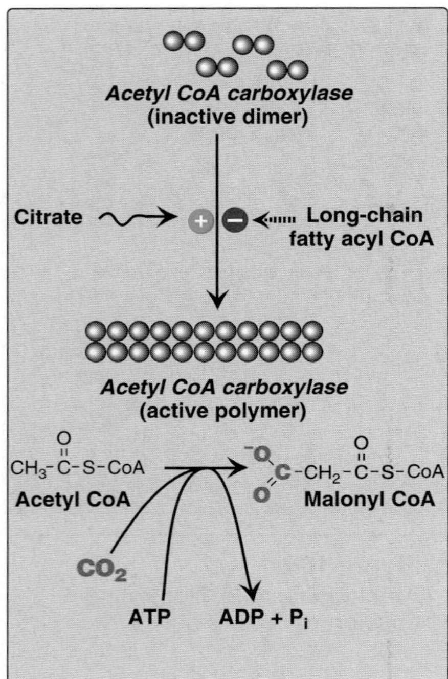

Figure 16.7
Allosteric regulation of malonyl CoA synthesis by *acetyl CoA carboxylase.* The carboxyl group contributed by dissolved CO_2 is shown in blue.

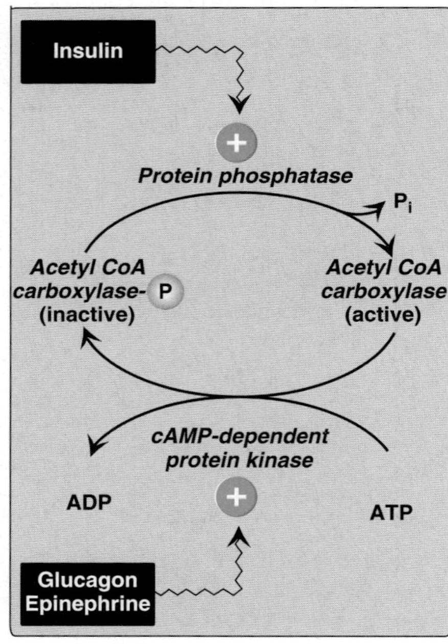

Figure 16.8
Hormone-mediated, covalent regulation of *acetyl CoA carboxylase.*

2. **Long-term regulation of acetyl CoA carboxylase:** Prolonged consumption of a diet containing excess calories (particularly, high-calorie, high-carbohydrate diets) causes an increase in *acetyl CoA carboxylase* synthesis, thus increasing fatty acid synthesis. Conversely, a low-calorie diet or fasting causes a reduction in fatty acid synthesis by decreasing the synthesis of *acetyl CoA carboxylase.* [Note: *Fatty acid synthase* (see below) is similarly regulated by this type of dietary manipulation.]

C. Fatty acid synthase: a multifunctional enzyme in eukaryotes

The remaining series of reactions of fatty acid synthesis in eukaryotes is catalyzed by the multifunctional, dimeric enzyme, *fatty acid synthase.* Each *fatty acid synthase* monomer is a multicatalytic polypeptide with seven different enzymic activities plus a domain that covalently binds a molecule of **4'-phosphopantetheine**. [Note: 4'-Phosphopantetheine, a derivative of the vitamin **pantothenic acid** (see p. 379), carries acetyl and acyl units on its terminal thiol (–SH) group during fatty acid synthesis. It also is a component of coenzyme A.] In prokaryotes, *fatty acid synthase* is a multienzyme complex, and the 4'-phosphopantetheine domain is a separate protein, referred to as the **acyl carrier protein** (**ACP**). ACP is used below to refer to the phosphopantetheine-binding domain of the eukaryotic *fatty acid synthase* molecule. The reaction numbers in brackets below refer to Figure 16.9. [Note: The enzyme activities listed are actually separate catalytic domains present in each multicatalytic *fatty acid synthase* monomer.]

[1] A molecule of acetate is transferred from acetyl CoA to the –SH group of the ACP. Domain: *Acetyl CoA-ACP acetyltransacylase.*

[2] Next, this two-carbon fragment is transferred to a temporary holding site, the thiol group of a cysteine residue on the enzyme.

[3] The now-vacant ACP accepts a three-carbon malonate unit from malonyl CoA. Domain: *Malonyl CoA-ACP-transacylase.*

[4] The malonyl group loses the HCO_3^- originally added by *acetyl CoA carboxlyase*, facilitating its nucleophilic attack on the thioester bond linking the acetyl group to the cysteine residue. The result is a four-carbon unit attached to the ACP domain. The loss of free energy from the decarboxylation drives the reaction. Domain: *3-Ketoacyl-ACP synthase.*

The next three reactions convert the 3-ketoacyl group to the corresponding saturated acyl group by a pair of reductions requiring NADPH and a dehydration step.

[5] The keto group is reduced to an alcohol. Domain: *3-Ketoacyl-ACP reductase.*

[6] A molecule of water is removed to introduce a double bond. Domain: *3-Hydroxyacyl-ACP dehydratase.*

[7] A second reduction step occurs. Domain: *Enoyl-ACP reductase.*

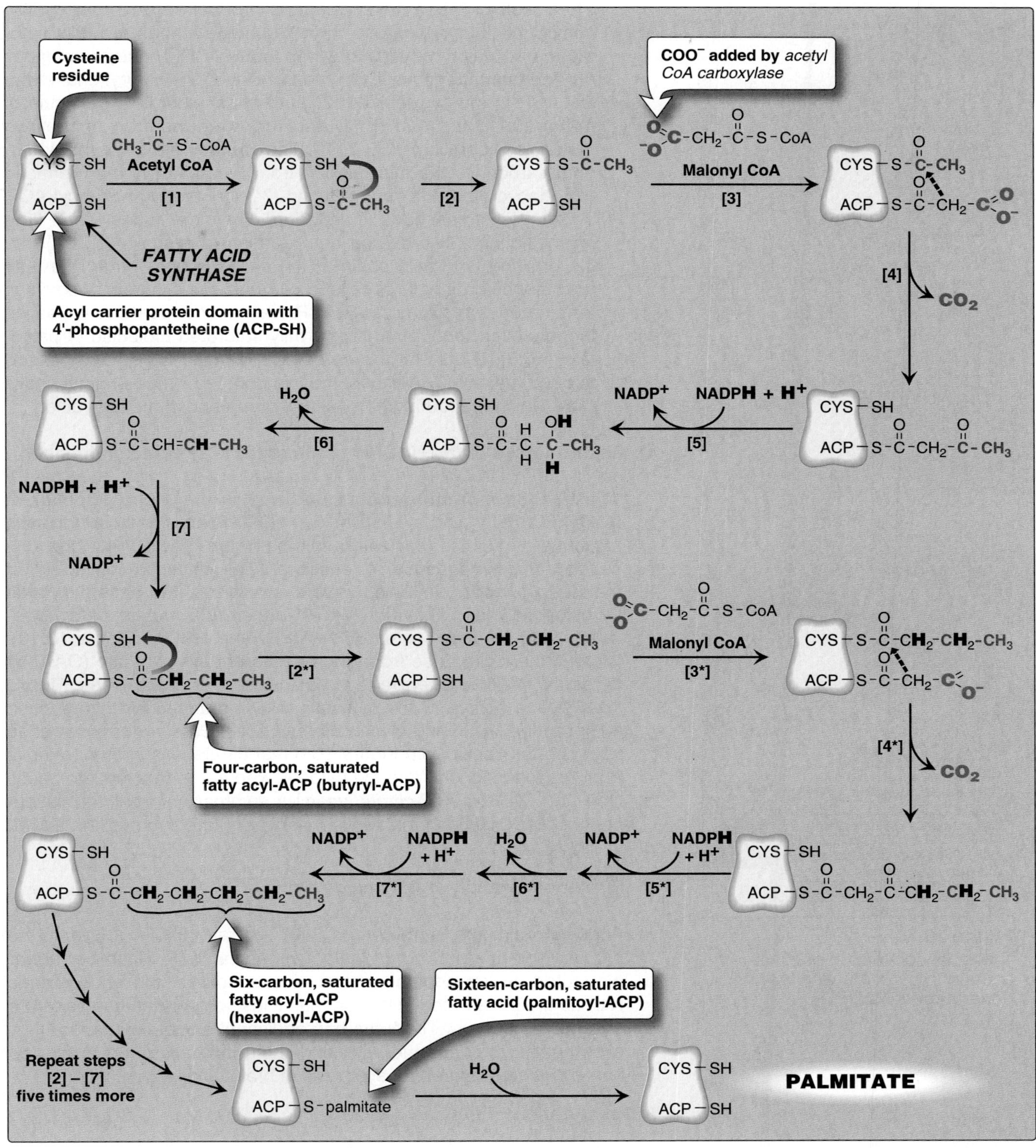

Figure 16.9
Synthesis of palmitate (16:0) by the *fatty acid synthase* complex. [Note: Numbers in brackets correspond to bracketed numbers in the text. A second repetition of the steps is indicated by numbers with an asterisk (*). Carbons provided directly by acetyl CoA are shown in red.]

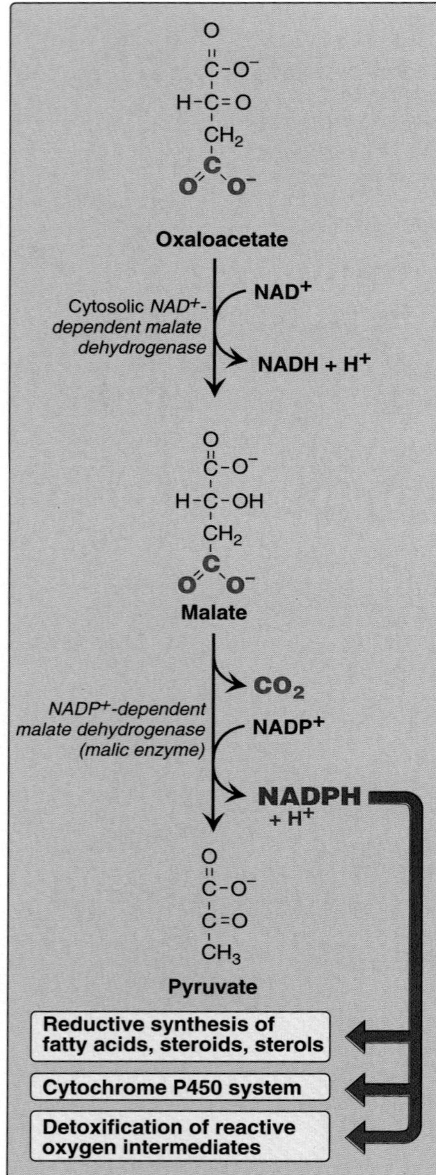

Figure 16.10
Cytosolic conversion of oxaloacetate to pyruvate with the generation of NADPH.

The result of these seven steps is production of a four-carbon compound (butyryl) whose three terminal carbons are fully saturated, and which remains attached to the ACP. These seven steps are repeated, beginning with the transfer of the butyryl chain from the ACP to the cys residue [2*], the attachment of a molecule of malonate to the ACP [3*], and the condensation of the two molecules liberating HCO_3^- [4*]. The carbonyl group at the β–carbon (carbon 3—the third carbon from the sulfur) is then reduced (5*), dehydrated (6*), and reduced (7*), generating hexanoyl-ACP. This cycle of reactions is repeated five more times, each time incorporating a two-carbon unit (derived from malonyl CoA) into the growing fatty acid chain at the carboxyl end. When the fatty acid reaches a length of sixteen carbons, the synthetic process is terminated with palmitoyl-S-ACP. *Palmitoyl thioesterase* cleaves the thioester bond, producing a fully saturated molecule of palmitate (16:0). [Note: All the carbons in palmitic acid have passed through malonyl CoA except the two donated by the original acetyl CoA, which are found at the methyl-group end of the fatty acid.]

D. Major sources of the NADPH required for fatty acid synthesis

The **hexose monophosphate pathway** is the major supplier of NADPH for fatty acid synthesis. Two NADPH are produced for each molecule of glucose that enters this pathway. (See p. 143 for a discussion of this sequence of reactions.) The cytosolic conversion of malate to pyruvate, in which malate is oxidized and decarboxylated by cytosolic *malic enzyme* (*NADP⁺-dependent malate dehydrogenase*), also produces cytosolic NADPH (and HCO_3^-, Figure 16.10). [Note: Malate can arise from the reduction of oxaloacetate (OAA) by cytosolic *NAD⁺-dependent malate dehydrogenase* (see Figure 16.10). One source of the cytosolic NADH required for this reaction is that produced during glycolysis (see 102). OAA, in turn, can arise from citrate. Recall from Figure 16.6 that citrate was shown to move from the mitochondria into the cytosol, where it is cleaved into acetyl CoA and OAA by *ATP-citrate lyase*.] A summary of the interrelationship between glucose metabolism and palmitate synthesis is shown in Figure 16.11.

E. Further elongation of fatty acid chains

Although palmitate, a 16-carbon, fully saturated LCFA (16:0), is the primary end-product of *fatty acid synthase* activity, it can be further elongated by the addition of two-carbon units in the endoplasmic reticulum (ER) and the mitochondria. These organelles use separate enzymic processes. The brain has additional elongation capabilities, allowing it to produce the very-long-chain fatty acids (up to 24 carbons) that are required for synthesis of brain lipids.

F. Desaturation of fatty acid chains

Enzymes present in the ER are responsible for desaturating fatty acids (that is, adding cis double bonds). Termed **mixed-function oxidases**, the desaturation reactions require NADPH and O_2. A variety of polyunsaturated fatty acids (PUFA) can be made through additional desaturation combined with elongation. [Note: Humans lack the ability to introduce double bonds between carbon 9 and the ω-

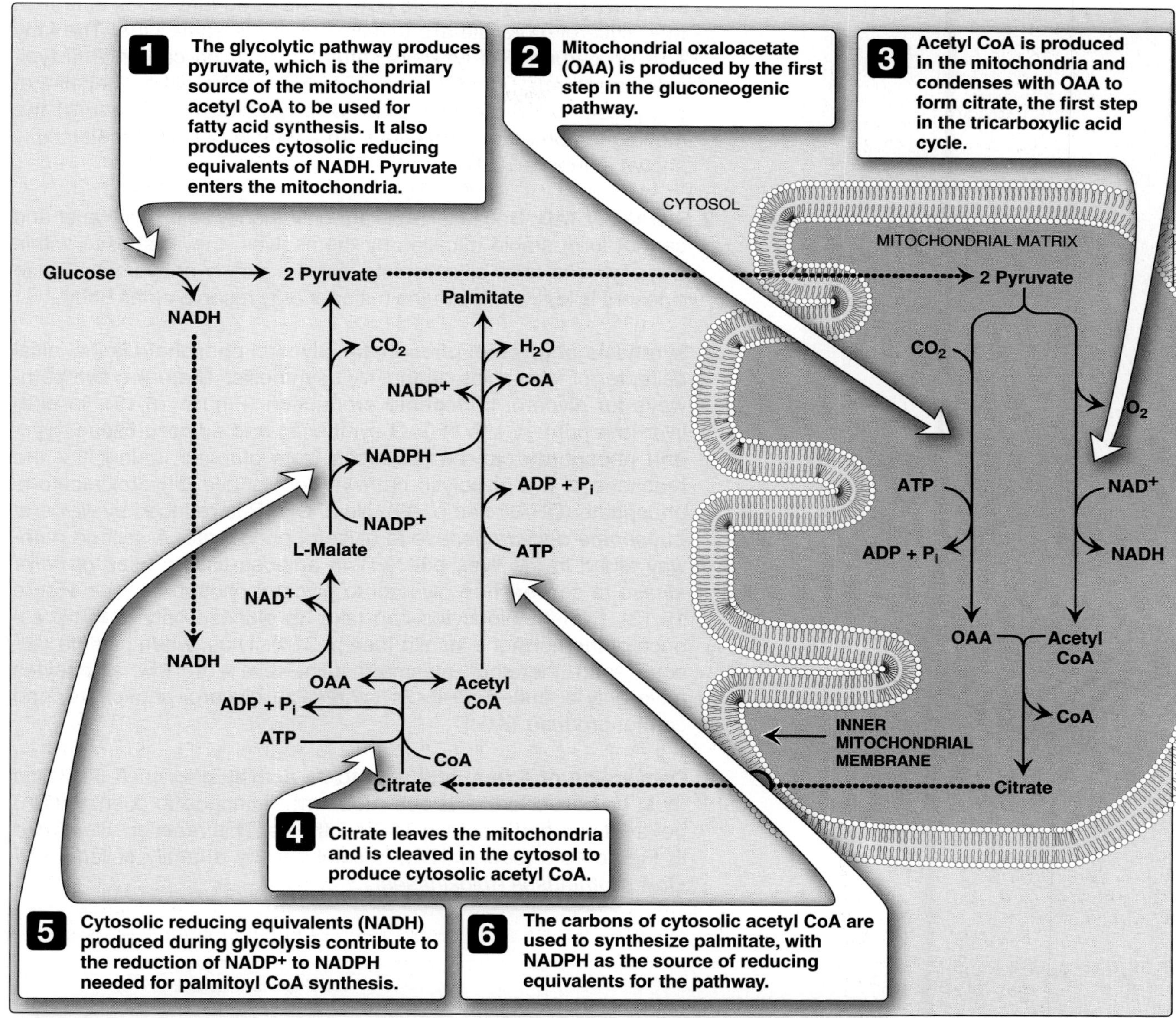

Figure 16.11
Interrelationship between glucose metabolism and palmitate synthesis.

end of the chain and, therefore, must have the polyunsaturated linoleic and linolenic acids provided in the diet (see Figure 16.4 for their structures).]

G. Storage of fatty acids as components of triacylglycerols

Mono-, di-, and triacylglycerols consist of one, two, or three molecules of fatty acid esterified to a molecule of glycerol. Fatty acids are esterified through their carboxyl groups, resulting in a loss of negative charge and formation of "**neutral fat**." [Note: If a species of acylglycerol is **solid** at room temperature, it is called a "**fat**"; if **liquid**, it is called an "**oil**."]

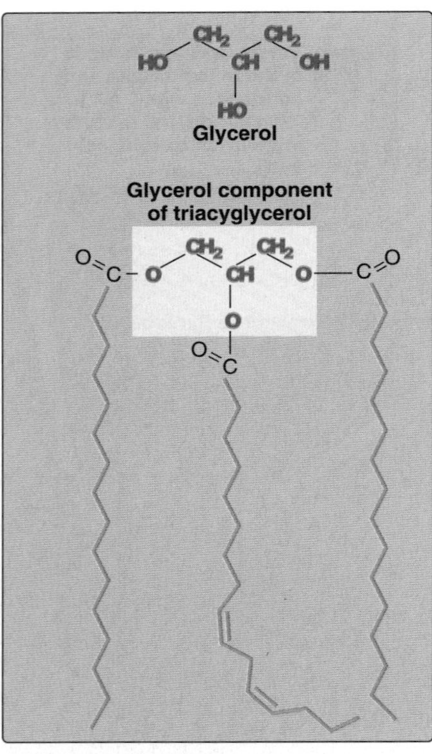

Figure 16.12
A triacylglycerol.

1. **Structure of triacylglycerols (TAG):** The three fatty acids esterified to a glycerol molecule are usually not of the same type. The fatty acid on carbon 1 is typically saturated, that on carbon 2 is typically unsaturated, and that on carbon 3 can be either. Recall that the presence of the unsaturated fatty acid(s) decrease(s) the melting temperature of the lipid. An example of a TAG molecule is shown in Figure 16.12.

2. **Storage of TAG:** Because TAGs are only slightly soluble in water and cannot form stable micelles by themselves, they coalesce within adipocytes to form oily droplets that are nearly anhydrous. These cytosolic lipid droplets are the major energy reserve of the body.

3. **Synthesis of glycerol phosphate:** Glycerol phosphate is the initial acceptor of fatty acids during TAG synthesis. There are two pathways for glycerol phosphate production (Figure 16.13). In both liver (the primary site of TAG synthesis) and adipose tissue, glycerol phosphate can be produced from glucose, using first the reactions of the glycolytic pathway to produce dihydroxyacetone phosphate (DHAP, see p. 99). Next, DHAP is reduced by *glycerol phosphate dehydrogenase* to glycerol phosphate. A second pathway found in the liver, but NOT in adipose tissue, uses *glycerol kinase* to convert free glycerol to glycerol phosphate (see Figure 16.13). [Note: Adipocytes can take up glucose only in the presence of the hormone insulin (see p. 310). Thus, when plasma glucose—and, therefore, plasma insulin—levels are low, adipocytes have only a limited ability to synthesize glycerol phosphate, and cannot produce TAG.]

4. **Conversion of a free fatty acid to its activated form:** A fatty acid must be converted to its activated form (attached to coenzyme A) before it can participate in TAG synthesis. This reaction, illustrated in Figure 15.6 (see p. 175), is catalyzed by a family of *fatty acyl CoA synthetases* (*thiokinases*).

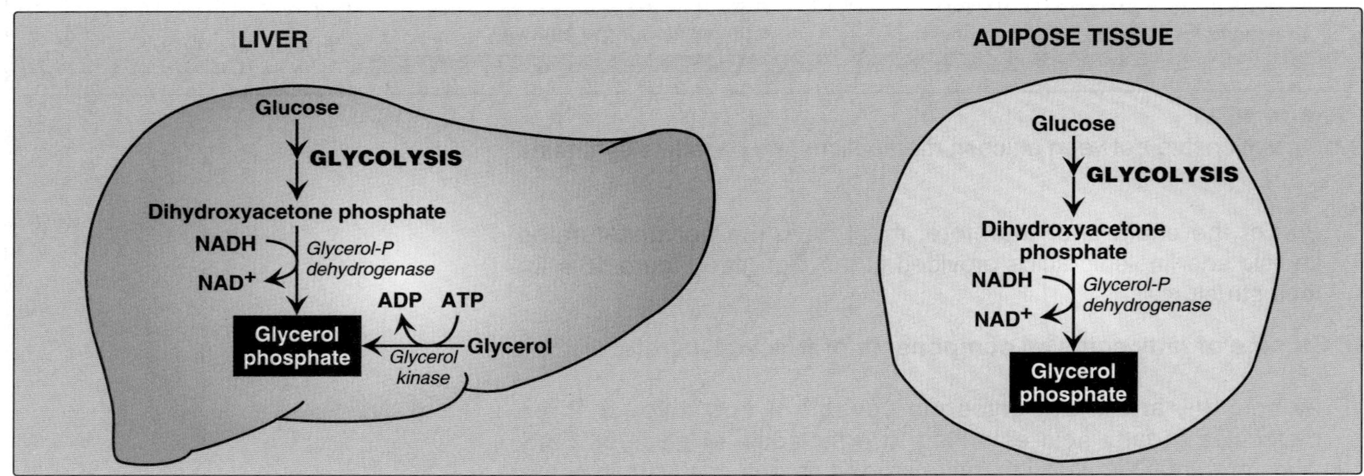

Figure 16.13
Pathways for production of glycerol phosphate in liver and adipose tissue.

5. **Synthesis of a molecule of TAG from glycerol phosphate and fatty acyl CoA:** This pathway involves four reactions, shown in Figure 16.14. These include the sequential addition of two fatty acids from fatty acyl CoA, the removal of phosphate, and the addition of the third fatty acid.

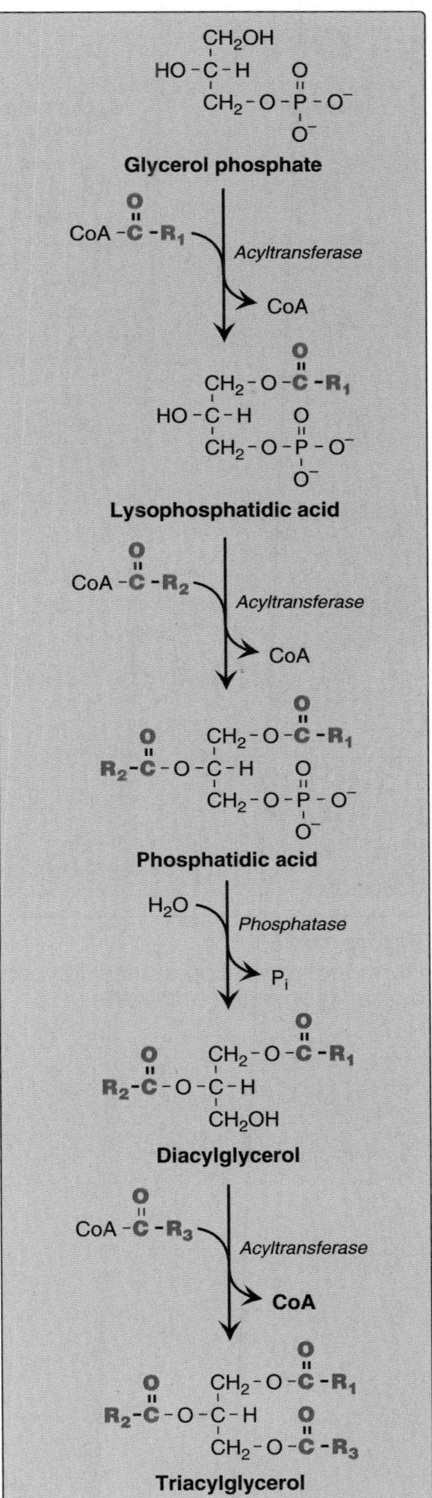

Figure 16.14
Synthesis of triacylglycerol.

H. Different fates of TAG in the liver and adipose tissue

In adipose tissue, TAG is stored in the cytosol of the cells in a nearly anhydrous form. It serves as "**depot fat**," ready for mobilization when the body requires it for fuel. Little TAG is stored in the liver. Instead, most is exported, packaged with cholesteryl esters, cholesterol, phospholipid, and protein (apolipoprotein B-100, see p. 229) to form lipoprotein particles called **very low density lipoproteins** (**VLDL**). Nascent VLDL are secreted into the blood where they mature and function to deliver the endogenously-derived lipids to the peripheral tissues. [Note: Recall that chylomicrons deliver primarily dietary (exogenously-derived) lipids.] Plasma lipoproteins are discussed in Chapter 18, p. 225.

IV. MOBILIZATION OF STORED FATS AND OXIDATION OF FATTY ACIDS

Fatty acids stored in adipose tissue, in the form of neutral TAG, serve as the body's major fuel storage reserve. TAGs provide concentrated stores of metabolic energy because they are highly reduced and largely anhydrous. The yield from complete oxidation of fatty acids to CO_2 and H_2O is nine kcal/g of fat (as compared to four kcal/g of protein or carbohydrate, see Figure 27.5, p. 357).

A. Release of fatty acids from TAG

The mobilization of stored fat requires the hydrolytic release of fatty acids and glycerol from their TAG form. This process is initiated by *hormone-sensitive lipase*, which removes a fatty acid from carbon 1 and/or carbon 3 of the TAG. Additional lipases specific for diacylglycerol or monoacylglycerol remove the remaining fatty acids.

1. **Activation of hormone-sensitive lipase (HSL):** This enzyme is activated when phosphorylated by a *3′,5′-cyclic AMP-dependent protein kinase*. **3′,5′-Cyclic AMP** is produced in the adipocyte when one of several hormones (primarily epinephrine) binds to receptors on the cell membrane, and activates *adenylate cyclase* (Figure 16.15). The process is similar to that of the activation of *glycogen phosphorylase* (see Figure 11.12, p. 131). [Note: Because *acetyl CoA carboxylase* is inhibited by hormone-directed phosphorylation when the cAMP-mediated cascade is activated (see Figure 16.8), fatty acid synthesis is turned off when TAG degradation is turned on.] In the presence of high plasma levels of insulin and glucose, *HSL* is dephosphorylated, and becomes inactive.

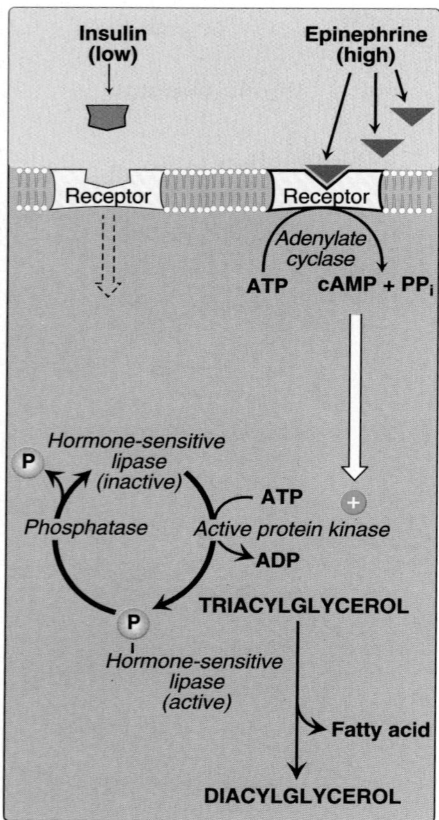

Figure 16.15
Hormonal regulation of triacylglycerol degradation in the adipocyte.

2. **Fate of glycerol:** The glycerol released during TAG degradation cannot be metabolized by adipocytes because they lack *glycerol kinase*. Rather, glycerol is transported through the blood to the liver, where it can be phosphorylated. The resulting glycerol phosphate can be used to form TAG in the liver, or can be converted to DHAP by reversal of the *glycerol phosphate dehydrogenase* reaction illustrated in Figure 16.13. DHAP can participate in glycolysis or gluconeogenesis.

3. **Fate of fatty acids:** The free (unesterified) fatty acids move through the cell membrane of the adipocyte, and immediately bind to albumin in the plasma. They are transported to the tissues, where the fatty acids enter cells, get activated to their CoA derivatives, and are oxidized for energy. [Note: Active transport of fatty acids across membranes is mediated by a membrane **fatty acid binding protein**.] Regardless of their blood levels, plasma free fatty acids cannot be used for fuel by erythrocytes, which have no mitochondria, or by the brain because of the impermeable blood-brain barrier.

B. β–Oxidation of fatty acids

The major pathway for catabolism of saturated fatty acids is a **mitochondrial pathway** called β-oxidation, in which two-carbon fragments are successively removed from the carboxyl end of the fatty acyl CoA, producing acetyl CoA, NADH, and $FADH_2$.

1. **Transport of long-chain fatty acids (LCFA) into the mitochondria:** After a LCFA enters a cell, it is converted to the CoA derivative by *long-chain fatty acyl CoA synthetase* (*thiokinase*) in the cytosol (see p. 174). Because β-oxidation occurs in the mitochondrial matrix, the fatty acid must be transported across the mitochondrial inner membrane. Therefore, a specialized carrier transports the long-chain acyl group from the cytosol into the mitochondrial matrix. This carrier is **carnitine**, and the transport process is called the **carnitine shuttle** (Figure 16.16).

   a. **Steps in LCFA translocation:** First, an acyl group is transferred from the cytosolic coenzyme A to carnitine by *carnitine palmitoyltransferase I* (*CPT-I*)—an enzyme associated with the outer mitochondrial membrane. [Note: *CPT-I* is also known as *CAT-I* for *carnitine acyltransferase I*.] This reaction forms acylcarnitine, and regenerates free coenzyme A. Second, the acylcarnitine is transported into the mitochondrion in exchange for free carnitine by *carnitine–acylcarnitine translocase*. *Carnitine palmitoyltransferase II* (*CPT-II*, or *CAT-II*)—an enzyme of the inner mitochondrial membrane—catalyzes the transfer of the acyl group from carnitine to coenzyme A in the mitochondrial matrix, thus regenerating free carnitine.

   b. **Inhibitor of the carnitine shuttle: Malonyl CoA** inhibits *CPT-I*, thus preventing the entry of long-chain acyl groups into the mitochondrial matrix. Therefore, when fatty acid synthesis is occurring in the cytosol (as indicated by the presence of

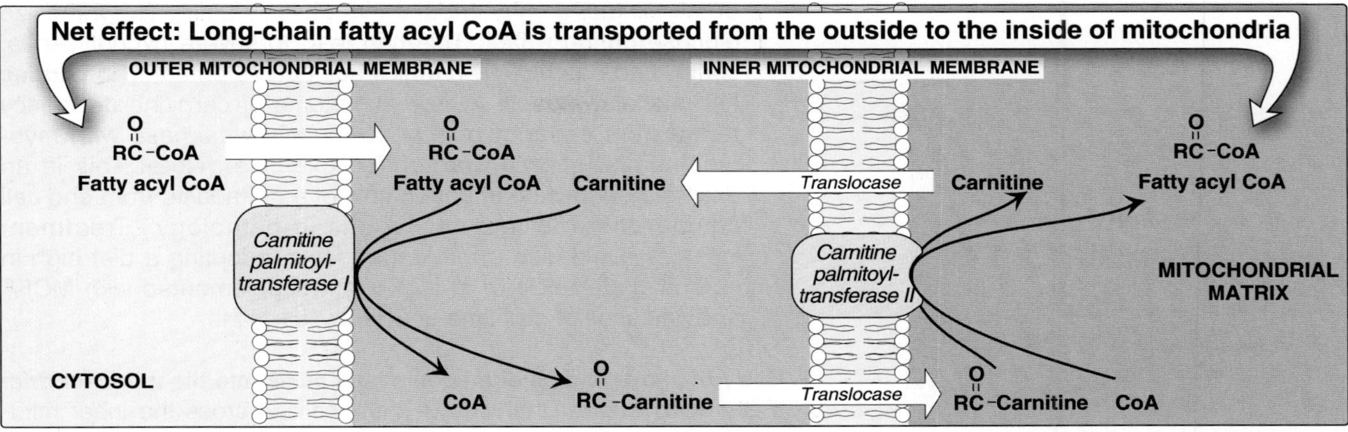

Figure 16.16
Carnitine shuttle.

malonyl CoA), the newly made palmitate cannot be transferred into the mitochondria and degraded.

c. **Sources of carnitine:** Carnitine can be obtained from the diet, where it is found primarily in meat products. Carnitine can also be synthesized from the amino acids lysine and methionine by an enzymatic pathway found in the liver and kidney but not in skeletal or heart muscle. Therefore, these tissues are totally dependent on carnitine provided by hepatocytes or the diet, and distributed by the blood. [Note: Skeletal muscle contains about 97 percent of all carnitine in the body.]

d. **Additional functions of carnitine:** The carnitine system also allows the export from the mitochondria of branched-chain acyl groups (such as those produced during the catabolism of the branched-chain amino acids). In addition, the carnitine system is involved in the trapping and excretion via the kidney of acyl groups that cannot be metabolized by the body.

e. **Carnitine deficiencies:** Such deficiencies result in a decreased ability of tissues to use LCFA as a metabolic fuel, and it can also cause the accumulation of toxic amounts of free fatty acids and branched-chain acyl groups in cells. Secondary carnitine deficiency occurs for many reasons, including 1) in patients with liver disease causing decreased synthesis of carnitine, 2) in individuals suffering from malnutrition or those on strictly vegetarian diets, 3) in those with an increased requirement for carnitine as a result of, for example, to pregnancy, severe infections, burns, or trauma, or 4) in those undergoing hemodialysis, which removes carnitine from the blood (Figure 16.17). **Congenital deficiencies** in one of the components of the *carnitine palmatoyltransferase* system, in tubular reabsorption of carnitine, or a deficiency in carnitine uptake by cells, can also cause carnitine deficiency. Genetic *CPT-I* deficiency affects the liver, where an inability to use

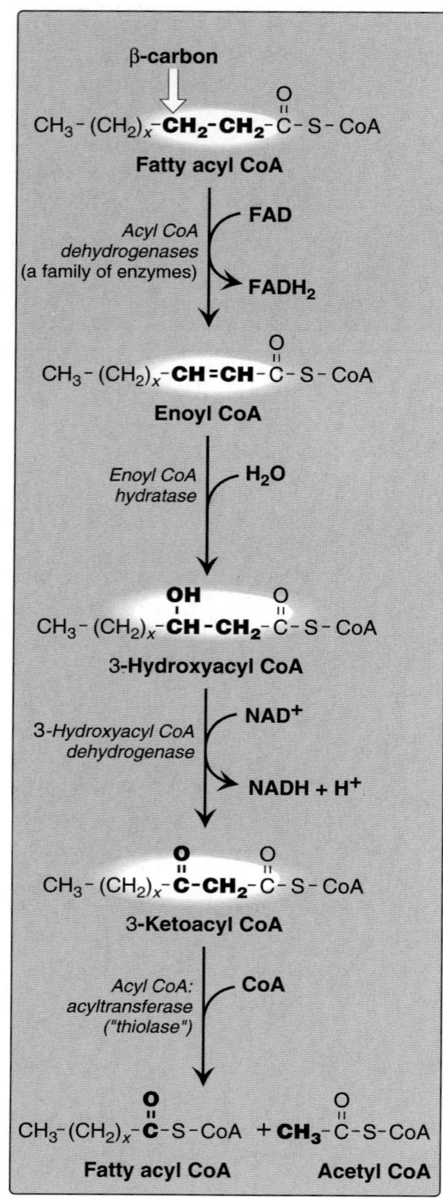

Figure 16.17
Enzymes involved in the β-oxidation of fatty acyl CoA.

LCFA for fuel greatly impairs that tissue's ability to synthesize glucose during a fast. This can lead to severe hypoglycemia, coma, and death. *CPT-II* deficiency occurs primarily in **cardiac** and **skeletal muscle**, where symptoms of carnitine deficiency range from cardiomyopathy, to muscle weakness with myoglobinemia following prolonged exercise. [Note: This is an example of how the impaired flow of a metabolite from one cell compartment to another results in pathology.] Treatment includes avoidance of prolonged fasts, adopting a diet high in carbohydrate and low in LCFA, but supplemented with MCFA and, in cases of carnitine deficiency, carnitine.

2. **Entry of short- and medium-chain fatty acids into the mitochondria:** Fatty acids shorter than twelve carbons can cross the inner mitochondrial membrane without the aid of carnitine or the *CPT* system. Once inside the mitochondria, they are activated to their coenzyme A derivatives by matrix enzymes, and are oxidized. [Note: MCFAs are plentiful in human milk. Because their oxidation is not dependent on *CPT-I*, it is not subject to inhibition by malonyl CoA.]

3. **Reactions of β-oxidation:** The first cycle of β-oxidation is shown in Figure 16.18. It consists of a sequence of four reactions that result in shortening the fatty acid chain by two carbons. The steps include an oxidation that produces $FADH_2$, a hydration step, a second oxidation that produces NADH, and a thiolytic cleavage that releases a molecule of acetyl CoA. These four steps are repeated for saturated fatty acids of even-numbered carbon chains (n/2)–1 times (where n is the number of carbons), each cycle producing an acetyl group plus one NADH and one $FADH_2$. The final thiolytic cleavage produces two acetyl groups. [Note: Acetyl CoA is a positive allosteric effector of *pyruvate carboxylase* (see p. 116), thus linking fatty acid oxidation and gluconeogenesis.]

4. **Energy yield from fatty acid oxidation:** The energy yield from the β-oxidation pathway is high. For example, the oxidation of a molecule of palmitoyl CoA to CO_2 and H_2O yields 131 ATPs (Figure 16.19). A comparison of the processes of synthesis and degradation of saturated fatty acids with an even number of carbon atoms is provided in Figure 16.20.

5. **Medium-chain fatty acyl CoA dehydrogenase (MCAD) deficiency:** In mitochondria, there are four *fatty acyl CoA dehydrogenase* species, each of which has a specificity for either short-, medium-, long-, or very-long-chain fatty acids. MCAD deficiency, an autosomal, recessive disorder, is one of the most common inborn errors of metabolism, and the most common inborn error of fatty acid oxidation, being found in 1 in 12,000 births in the west, and 1 in 40,000 worldwide. It causes a decrease in fatty acid oxidation and severe hypoglycemia (because the tissues cannot obtain full energetic benefit from fatty acids and, therefore, must now rely on glucose). Treatment includes a carbohydrate-rich diet. [Note: Infants are particularly affected by MCAD deficiency, because they rely for their nourishment on milk, which contains primarily MCADs.

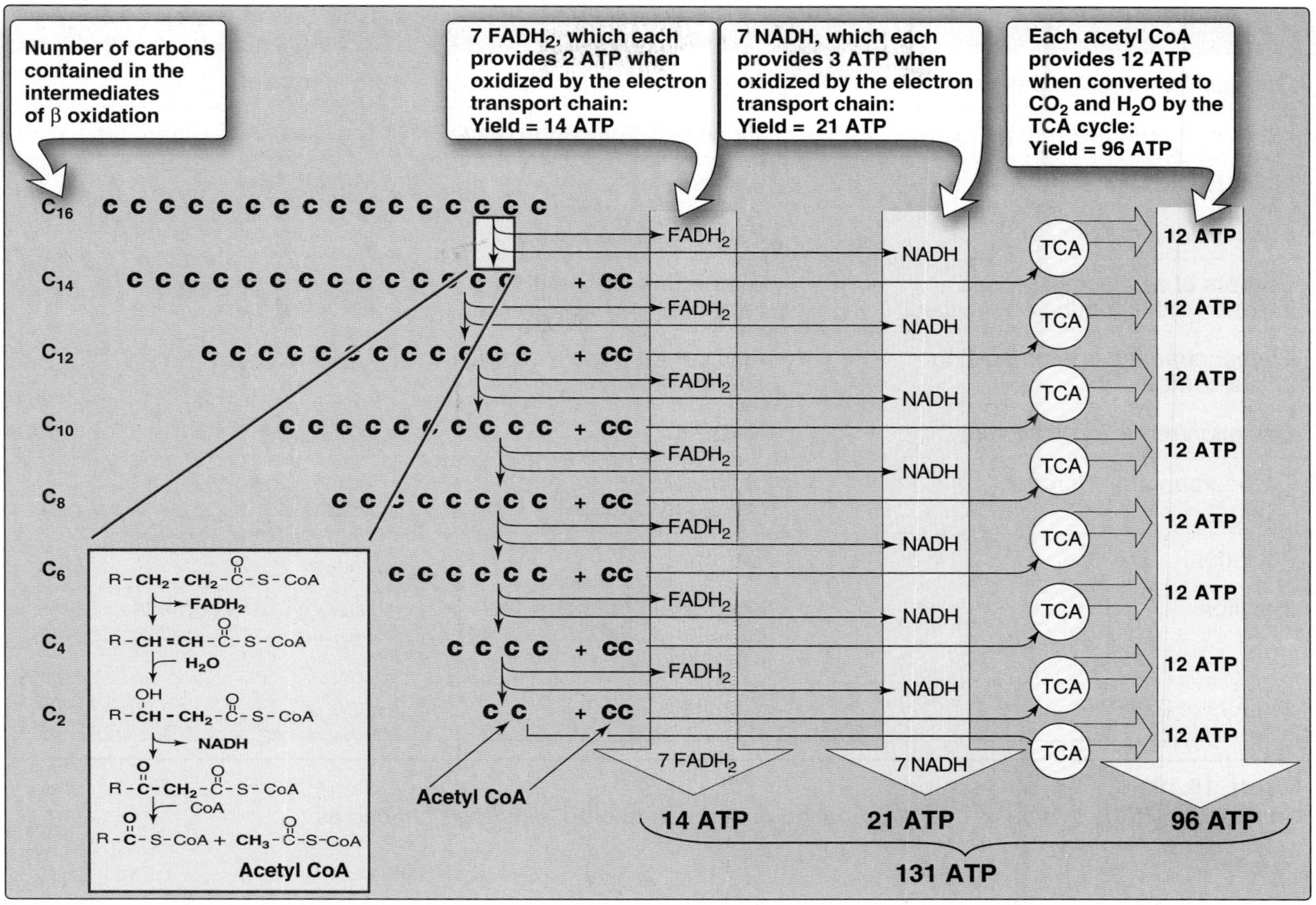

Figure 16.18
Summary of the energy yield from the oxidation of palmityl CoA (16 carbons). CC = acetyl CoA.

MCAD dehydrogenase deficiency has been identified as the cause of some cases originally reported as sudden infant death syndrome (SIDS) or Reye's syndrome.]

6. **Oxidation of fatty acids with an odd number of carbons:** The β-oxidation of a saturated fatty acid with an odd number of carbon atoms proceeds by the same reaction steps as that of fatty acids with an even number, until the final three carbons are reached. This compound, **propionyl CoA**, is metabolized by a three-step pathway (Figure 16.21). [Note: Propionyl CoA is also produced during the metabolism of certain amino acids (see Figure 20.10, p. 264).]

   a. **Synthesis of D-methylmalonyl CoA:** First, propionyl CoA is carboxylated, forming **D**-methylmalonyl CoA. The enzyme *propionyl CoA carboxylase* has an absolute requirement for the coenzyme **biotin**, as do other carboxylases (see p. 379).

   b. **Formation of L-methylmalonyl CoA:** Next, the D-isomer is converted to the L-form by the enzyme, *methylmalonyl CoA racemase*.

	SYNTHESIS	DEGRADATION
Greatest flux through pathway	After carbohydrate-rich meal	In starvation
Hormonal state favoring pathway	High insulin/glucagon ratio	Low insulin/glucagon ratio
Major tissue site	Primarily liver	Muscle, liver
Subcellular location	Primarily cytosol	Primarily mitochondria
Carriers of acyl/acetyl groups between mitochondria and cytosol	Citrate (mitochondria to cytosol)	Carnitine (cytosol to mitochondria)
Phosphopantetheine-containing active carriers	Acyl carrier protein domain, coenzyme A	Coenzyme A
Oxidation/reduction cofactors	NADPH	NAD^+, FAD
Two-carbon donor/product	Malonyl CoA: donor of one acetyl group	Acetyl CoA: product of β-oxidation
Activator	Citrate	
Inhibitor	Long-chain fatty acyl CoA (inhibits *acetyl CoA carboxylase*)	Malonyl CoA (inhibits *carnitine palmitoyltransferase*)
Product of pathway	Palmitate	Acetyl CoA

Figure 16.19
Comparison of the synthesis and degradation of even-numbered, saturated fatty acids.

c. **Synthesis of succinyl CoA:** Finally, the carbons of L-methylmalonyl CoA are rearranged, forming succinyl CoA, which can enter the TCA cycle (see p. 110). The enzyme, *methylmalonyl CoA mutase*, requires a coenzyme form of **vitamin B_{12}** (**deoxyadenosylcobalamin**) for its action. The *mutase* reaction is one of only two reactions in the body that require vitamin B_{12} (see p. 373). [Note: In patients with **vitamin B_{12} deficiency**, both propionate and methylmalonate are excreted in the urine. Two types of inheritable **methylmalonic acidemia** and **aciduria** have been described: one in which the *mutase* is missing or deficient (or has reduced affinity for the coenzyme), and one in which the patient is unable to convert vitamin B_{12} into its coenzyme form. Either type results in metabolic acidosis, with developmental retardation seen in some patients.

7. **Oxidation of unsaturated fatty acids:** The oxidation of unsaturated fatty acids provides less energy than that of saturated fatty acids because they are less highly reduced and, therefore, fewer reducing equivalents can be produced from these structures. Oxidation of monounsaturated fatty acids, such as 18:1(9) (oleic acid) requires one additional enzyme, *3,2-enoyl CoA isomerase*, which converts the 3-cis derivative obtained after three rounds of β-oxidation to the 2-trans derivative that can serve as a substrate for the *hydratase*. Oxidation of polyunsaturated fatty acids, such

as 18:2(9,12) (linoleic acid). requires an *NADPH-dependent reductase* in addition to the *isomerase*.

8. **β-Oxidation in the peroxisome:** Very-long-chain fatty acids (VLCFA), twenty carbons long or longer, undergo a preliminary β-oxidation in peroxisomes. The shortened fatty acid is then transferred to a mitochondrion for further oxidation. In contrast to mitochondrial β-oxidation, the initial dehydrogenation in peroxisomes is catalyzed by an **FAD**-containing *acyl CoA oxidase*. The $FADH_2$ produced is oxidized by molecular oxygen, which is reduced to H_2O_2. The H_2O_2 is reduced to H_2O by *catalase* (see 146). [Note: The genetic defects **Zellweger (cerebrohepatorenal) syndrome** (a defect in peroxisomal biogenesis in all tissues) and **X-linked adrenoleukodystrophy** (a defect in peroxisomal activation of VLCFA) lead to accumulation of VLCFA in the blood and tissues.]

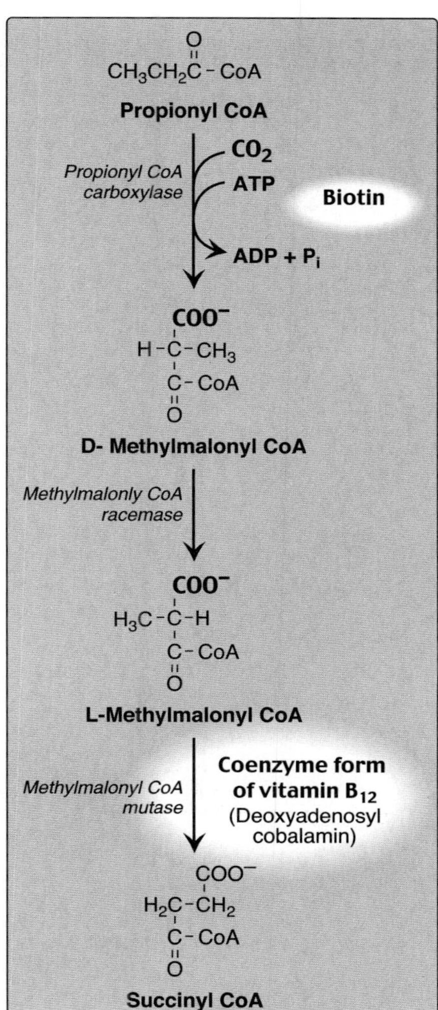

Figure 16.20
Metabolism of propionyl CoA.

C. α-Oxidation of fatty acids

The **branched-chain fatty acid**, **phytanic acid**, is not a substrate for *acyl CoA dehydrogenase* due to the methyl group on its third (β) carbon (Figure 16.22). Instead, it is hydroxylated at the α-carbon by *fatty acid α-hydroxylase*. The product is decarboxylated and then activated to its CoA derivative, which is a substrate for the enzymes of β-oxidation. [Note: **Refsum disease** is a rare, autosomal recessive disorder caused by a deficiency of *α-hydroxylase*. This results in the accumulation of phytanic acid in the plasma and tissues. The symptoms are primarily neurologic, and the treatment involves dietary restriction to halt disease progression.]

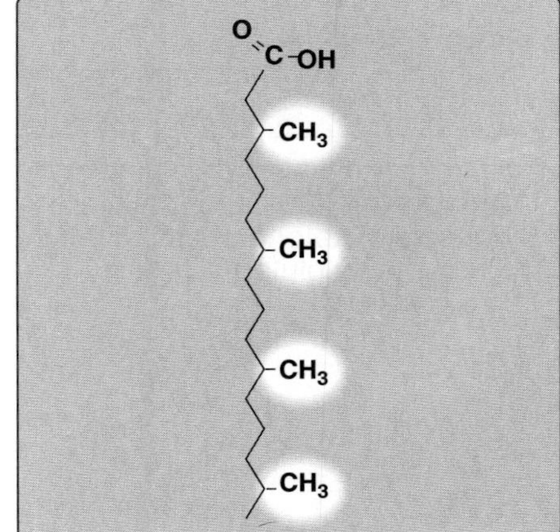

Figure 16.21
Phytanic acid–a branched-chain fatty acid.

V. KETONE BODIES: AN ALTERNATE FUEL FOR CELLS

Liver mitochondria have the capacity to convert acetyl CoA derived from fatty acid oxidation into ketone bodies. The compounds categorized as ketone bodies are **acetoacetate**, **3-hydroxybutyrate** (formerly called β-hydroxybutyrate), and **acetone** (a nonmetabolizable side product, Figure 16.23). [Note: The two functional ketone bodies are actually organic acids.] Acetoacetate and 3-hydroxybutyrate are transported in the blood to the peripheral tissues. There they can be reconverted to acetyl CoA, which can be oxidized by the TCA cycle. Ketone bodies are important sources of energy for the peripheral tissues because 1) they are soluble in aqueous solution and, therefore, do not need to be incorporated into lipoproteins or carried by albumin as do the other lipids; 2) they are produced in the liver during periods when the amount of acetyl CoA present exceeds the oxidative capacity of the liver; and 3) they are used in proportion to their concentration in the blood by extrahepatic tissues, such as the skeletal and cardiac muscle and renal cortex. Even the brain can use ketone bodies to help meet its energy needs if the blood levels rise sufficiently. [Note: This is important during prolonged periods of fasting, see p. 330.]

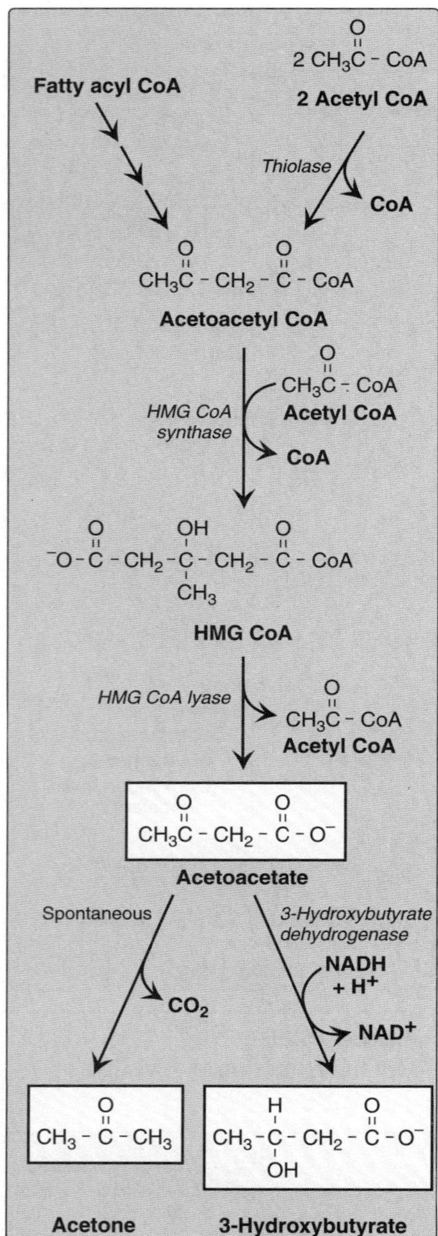

Figure 16.22
Synthesis of ketone bodies. HMG = hydroxymethylglutaryl CoA.

A. Synthesis of ketone bodies by the liver

During a fast, the liver is flooded with fatty acids mobilized from adipose tissue. The resulting elevated hepatic acetyl CoA produced primarily by fatty acid degradation inhibits *pyruvate dehydrogenase* (see p. 108), and activates *pyruvate carboxylase* (see p. 117). The oxaloacetate thus produced is used by the liver for gluconeogenesis rather than for the TCA cycle. Therefore, acetyl Co A is channeled into ketone body synthesis.

1. **Synthesis of 3-hydroxy-3-methylglutaryl CoA (HMG CoA):** The first synthetic step, formation of acetoacetyl CoA, occurs by reversal of the *thiolase* reaction of fatty acid oxidation (see Figure 16.18). Mitochondrial *HMG CoA synthase* combines a third molecule of acetyl CoA with acetoacetyl CoA to produce HMG CoA. [Note: HMG CoA is also a precursor of cholesterol (see p. 218). These pathways are separated by location in, and conditions of, the cell (see p. 218).] *HMG CoA synthase* is the rate-limiting step in the synthesis of ketone bodies, and is present in significant quantities only in the liver.

2. **Synthesis of the ketone bodies:** HMG CoA is cleaved to produce acetoacetate and acetyl CoA, as shown in Figure 16.23. Acetoacetate can be reduced to form 3-hydroxybutyrate with NADH as the hydrogen donor. Acetoacetate can also spontaneously decarboxylate in the blood to form acetone—a volatile, biologically non-metabolized compound that can be released in the breath. [Note: The equilibrium between acetoacetate and 3-hydroxybutyrate is determined by the NAD^+/NADH ratio. Because this ratio is high during fatty acid oxidation, 3-hydroxybutyrate synthesis is favored.]

B. Use of ketone bodies by the peripheral tissues

Although the liver constantly synthesizes low levels of ketone bodies, their production becomes much more significant during fasting when ketone bodies are needed to provide energy to the peripheral tissues. 3-Hydroxybutyrate is oxidized to acetoacetate by *3-hydroxybutyrate dehydrogenase*, producing NADH (Figure 16.24). Acetoacetate is then provided with a coenzyme A molecule taken from succinyl CoA by *succinyl CoA:acetoacetate CoA transferase* (*thiophorase*). This reaction is reversible, but the product, acetoacetyl CoA, is actively removed by its conversion to two acetyl CoAs. Extrahepatic tissues, including the brain but excluding cells lacking mitochondria (for example, red blood cells), efficiently oxidize acetoacetate and 3-hydroxybutyrate in this manner. In contrast, although the liver actively produces ketone bodies, it lacks *thiophorase* and, therefore, is unable to use ketone bodies as fuel.

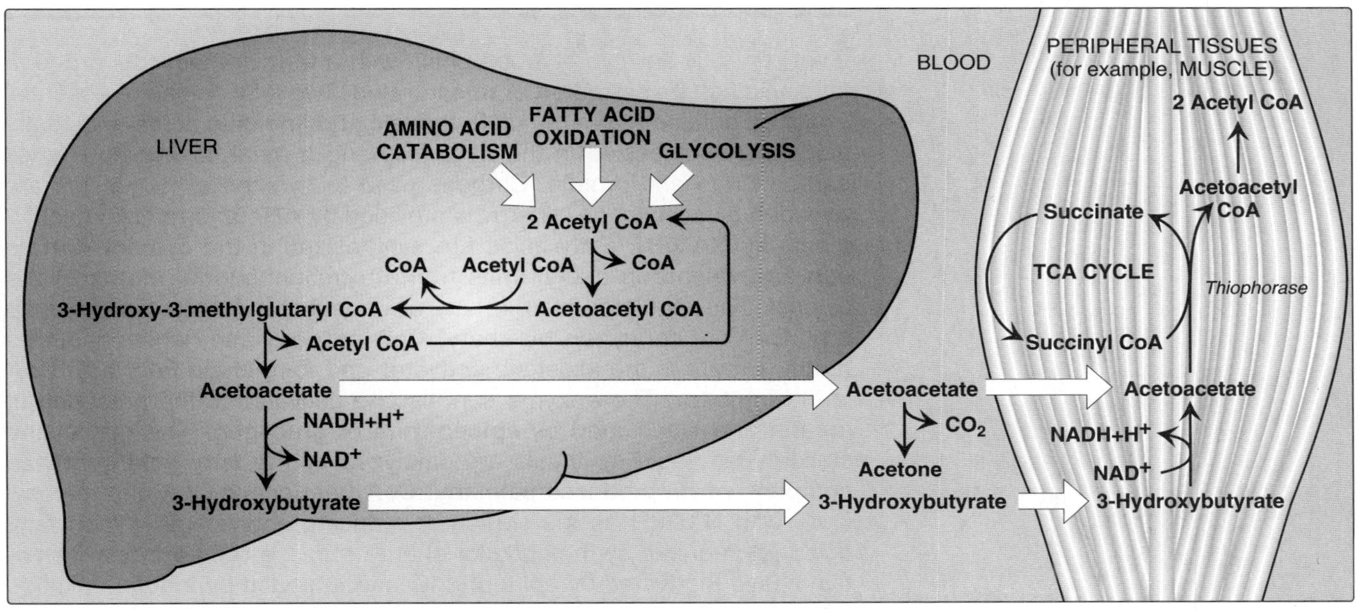

Figure 16.23
Ketone body synthesis in the liver and use in peripheral tissues.

C. Excessive production of ketone bodies in diabetes mellitus

When the rate of formation of ketone bodies is greater than the rate of their use, their levels begin to rise in the blood (**ketonemia**) and eventually in the urine (**ketonuria**). These two conditions are seen most often in cases of uncontrolled, type 1 (insulin-dependent) diabetes mellitus. In such individuals, high fatty acid degradation produces excessive amounts of acetyl CoA. It also depletes the NAD^+ pool and increases the NADH pool, which slows the TCA cycle (see p. 112). This forces the excess acetyl CoA into the ketone body pathway. In diabetic individuals with severe ketosis, urinary excretion of the ketone bodies may be as high as 5000 mg/24 hr, and the blood concentration may reach 90 mg/dl (versus less than 3 mg/dL in normal individuals). A frequent symptom of diabetic ketoacidosis is a fruity odor on the breath which result from increased production of acetone. An elevation of the ketone body concentration in the blood results in **acidemia**. [Note: The carboxyl group of a ketone body has a pK_a about 4. Therefore, each ketone body loses a proton (H^+) as it circulates in the blood, which lowers the pH of the body. Also, excretion of glucose and ketone bodies in the urine results in dehydration of the body. Therefore, the increased number of H^+, circulating in a decreased volume of plasma, can cause severe acidosis (ketoacidosis)]. **Ketoacidosis** may also be seen in cases of fasting (see p. 327).

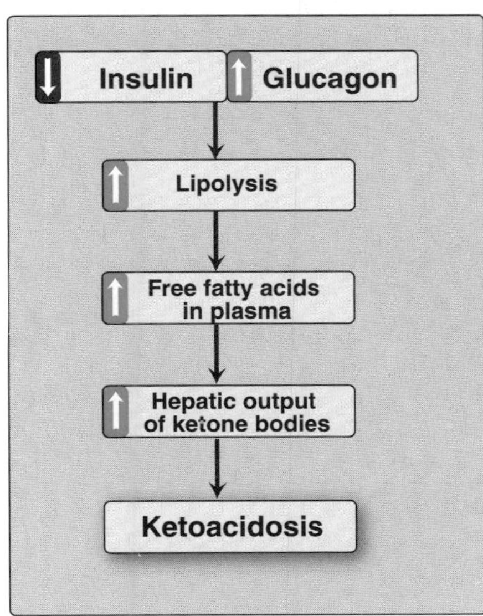

Figure 16.24
Mechanism of diabetic ketoacidosis seen in type 1 diabetes.

VI. CHAPTER SUMMARY

Generally a linear hydrocarbon chain with a terminal carboxyl group, a fatty acid can be **saturated** or **unsaturated**. Two fatty acids are essential (must be obtained from the diet): **linoleic** and **linolenic acids**. Most fatty acids are synthesized in the **liver** following a meal containing excess carbohydrate and protein. Carbons used to synthesize fatty acids are provided by **acetyl CoA**, energy is provided by **ATP**, and reducing equivalents by **NADPH**. Fatty acids are synthesized in the **cytosol**. **Citrate** carries two-carbon acetyl units from the mitochondrial matrix to the cytosol. The regulated step in fatty acid synthesis (acetyl CoA → malonyl CoA) is catalyzed by **acetyl CoA carboxylase**, which requires **biotin**. **Citrate** is the allosteric **activator** and **long-chain fatty acyl CoA** is the **inhibitor**. The enzyme can also be activated in the presence of **insulin** and inactivated by **epinephrine** or **glucagon**. The rest of the steps in fatty acid synthesis are catalyzed by the **fatty acid synthase complex**, which produces **palmitoyl CoA** from acetyl CoA and malonyl CoA, with NADPH as the source of reducing equivalents. When fatty acids are required by the body for energy, adipose cell **hormone- sensitive lipase** (**activated** by **epinephrine**, and **inhibited** by **insulin**) initiates degradation of stored triacylglycerol. Fatty acids are carried by **serum albumin** to the liver and peripheral tissues, where oxidation of the fatty acids provides energy. The **glycerol** backbone of the degraded triacylglycerol is carried by the blood to the **liver**, where it serves as an important **gluconeogenic precursor**. Fatty acid degradation (**β-oxidation**) occurs in **mitochondria**. The **carnitine shuttle** is required to transport fatty acids from the cytosol to the mitochondria. Enzymes required are **carnitine palmitoyltransferases I** and **II**. CPT I is **inhibited** by **malonyl CoA**. This prevents fatty acids being synthesized in the cytosol from malonyl CoA from being transported into the mitochondria where they would be degraded. Once in the mitochondria, fatty acids are oxidized, producing acetyl CoA, NADH, and $FADH_2$. The first step in the β-oxidation pathway is catalyzed by one of a family of four acyl CoA dehydrogenases, each of which has a specificity for either short-, medium-, long-, or very-long-chain fatty acids. **Medium-chain fatty acyl CoA dehydrogenase** (**MCAD**) **deficiency** is one of the most common in-born errors of metabolism. It causes a decrease in fatty acid oxidation, resulting in severe hypoglycemia. Treatment includes a carbohydrate-rich diet. Oxidation of fatty acids with an odd number of carbons proceeds two carbons at a time (producing acetyl CoA) until the last three carbons (**propionyl CoA**). This compound is converted to **methylmalonyl CoA** (a reaction requiring **biotin**), which is then converted to **succinyl CoA** by methylmalonyl CoA mutase (requiring **vitamin B_{12}**). A genetic error in the mutase or vitamin B_{12} deficiency causes **methylmalonic acidemia** and **aciduria**. Liver mitochondria can convert acetyl CoA derived from fatty acid oxidation into the ketone bodies, **acetoacetate** and **3-hydroxybutyrate**. Peripheral tissues possessing mitochondria can oxidize 3-hydroxybutyrate to acetoacetate, which can be reconverted to acetyl CoA, thus producing energy for the cell. Unlike fatty acids, ketone bodies can be utilized by the **brain** and, therefore, are important fuels during a fast. The liver lacks the ability to degrade ketone bodies, and so synthesizes them specifically for the peripheral tissues. **Ketoacidosis** occurs when the rate of formation of ketone bodies is greater than their rate of use, as is seen in cases of uncontrolled, **type 1 (insulin-dependent) diabetes mellitus**.

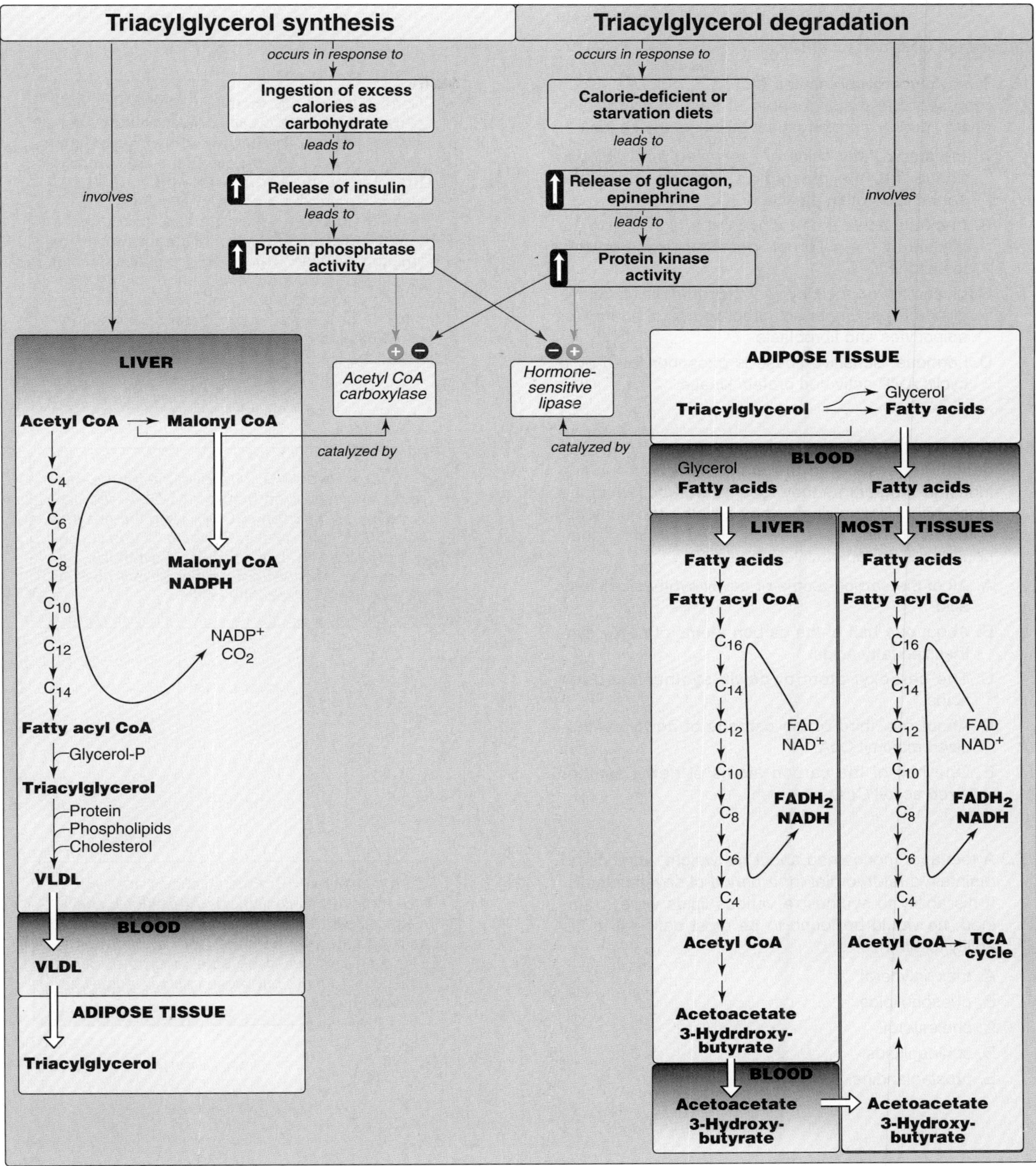

Figure 16.25
Key concept map for fatty acid and triacylglycerol metabolism.

Study Questions

Choose the ONE correct answer

16.1 Triacylglycerol molecules stored in adipose tissue represent the major reserve of substrate providing energy during a prolonged fast. During such a fast:

A. the stored fatty acids are released from adipose tissue into the plasma as components of the serum lipoprotein particle, VLDL.

B. free fatty acids are produced at a high rate in the plasma by the action of lipoprotein lipase on chylomicrons.

C. glycerol produced by the degradation of triacylglycerol is an important direct source of energy for adipocytes and fibroblasts.

D. hormone- sensitive lipase is phosphorylated by a cyclic AMP-activated protein kinase.

Correct answer = D. Hormone sensitive lipase is phosphorylated by cyclic AMP-activated protein kinase, which is itself activated by insulin. Fatty acids released from adipose tissue are carried in the plasma by serum albumin, not VLDL. During a fast, the amount of circulating triacylglycerol (found in chylomicra and VLDL) will be low. Therefore, there is little substrate for lipoprotein lipase. The glycerol produced during triacylglycerol degradation cannot be metabolized by adipocytes or fibroblasts, but rather must go to the liver where it can be phosphorylated (by glycerol kinase).

16.2 A low level of carbon dioxide labeled with ^{14}C is accidentally released into the atmosphere surrounding industrial workers as they resume work following the lunch hour. Unknowingly, they breathe the contaminated air for one hour. Which of the following compounds will be radioactively labeled?

A. All of the carbon atoms of newly synthesized fatty acid.

B. About one half of the carbon atoms of newly synthesized fatty acids.

C. The carboxyl atom of newly synthesized fatty acids.

D. About one third of the carbons of newly synthesized malonyl CoA.

E. One half of the carbon atoms of newly synthesized acetyl CoA.

Correct answer = D. Malonyl CoA (three carbons) is synthesized from acetyl CoA (two carbons) by the addition of CO_2, using the enzyme acetyl CoA carboxylase. Because CO_2 is subsequently removed during fatty acid synthesis, the radioactive label will not appear at any position in newly synthesized fatty acids.

16.3 A teenager, concerned about his weight, attempts to maintain a fat-free diet for a period of several weeks. If his ability to synthesize various lipids were examined, he would be found to be most deficient in his ability to synthesize:

A. triacylglycerol.

B. phospholipids.

C. cholesterol.

D. sphingolipids.

E. prostaglandins.

Correct answer = E. Prostaglandins are synthesized from arachidonic acid. Arachidonic acid is synthesized from linoleic acid, an essential fatty acid obtained by humans from dietary lipids. The teenager would be able to synthesize all other compounds, but presumably in somewhat depressed amounts.

Complex Lipid Metabolism

17

I. OVERVIEW OF PHOSPHOLIPIDS

Phospholipids are polar, ionic compounds composed of an alcohol that is attached by a phosphodiester bridge to either diacylglycerol or sphingosine. Like fatty acids, phospholipids are amphipathic in nature, that is, each has a hydrophilic head (the phosphate group plus whatever alcohol is attached to it, for example, serine, ethanolamine, and choline, highlighted in blue in Figure 17.1A), and a long, hydrophobic tail (containing fatty acids or fatty acid-derived hydrocarbons, shown in orange in Figure 17.1A). Phospholipids are the predominant lipids of cell membranes. In membranes, the hydrophobic portion of a phospholipid molecule is associated with the nonpolar portions of other membrane constituents, such as glycolipids, proteins, and cholesterol. The hydrophilic (polar) head of the phospholipid extends outward, facing the intracellular or extracellular aqueous environment (see Figure 17.1A). Membrane phospholipids also function as a reservoir for intracellular messengers, and, for some proteins, phospholipids serve as anchors to cell membranes. Non–membrane-bound phospholipids serve additional functions in the body, for example, as components of lung surfactant and essential components of bile, where their detergent properties aid in the solubilization of cholesterol.

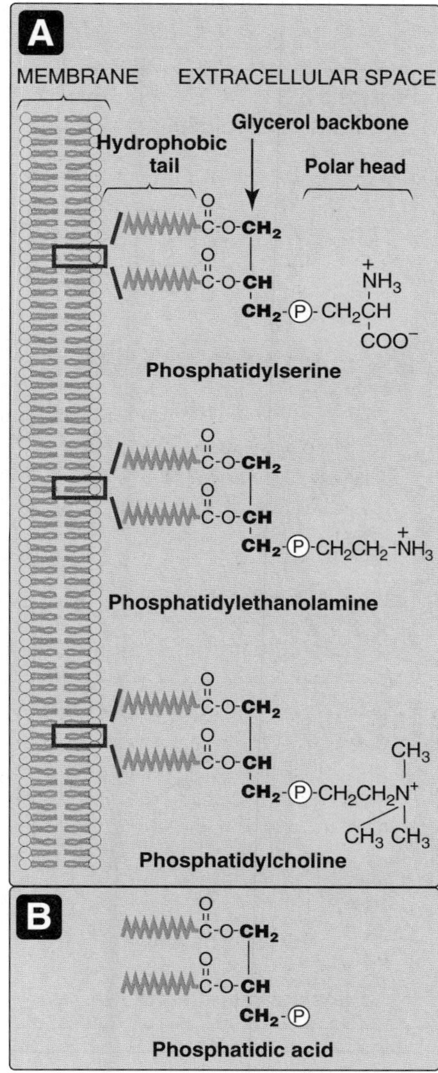

Figure 17.1
A. Structures of some glycerophospholipids. B. Phosphatidic acid. Ⓟ = phosphate, PO_4^{-2}.

II. STRUCTURE OF PHOSPHOLIPIDS

There are two classes of phospholipids: those that have glycerol as a backbone and those that contain sphingosine. Both classes are found as structural components of membranes, and both play a role in the generation of lipid-signaling molecules.

A. Glycerophospholipids

Phospholipids that contain **glycerol** are called **glycerophospholipids** (or **phosphoglycerides**). Glycerophospholipids constitute the major class of phospholipids. All contain (or are derivatives of) phosphatidic acid (diacylglycerol with a phosphate group on the third carbon, Figure 17.1B). **Phosphatidic acid** is the simplest phosphoglyceride, and is the precursor of the other members of this group.

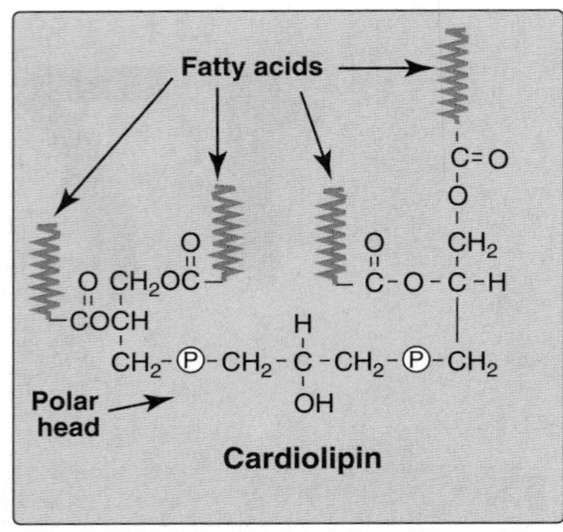

Figure 17.2
Structure of cardiolipin.

1. **Glycerophospholipids are formed from phosphatidic acid and an alcohol:** The phosphate group on phosphatidic acid (PA) can be esterified to another compound containing an alcohol group (see Figure 17.1). For example:

Serine	+	PA	→	phosphatidylserine
Ethanolamine	+	PA	→	phosphatidylethanolamine (cephalin)
Choline	+	PA	→	phosphatidylcholine (lecithin)
Inositol	+	PA	→	phosphatidylinositol
Glycerol	+	PA	→	phosphatidylglycerol

2. **Cardiolipin:** Two molecules of phosphatidic acid esterified through their phosphate groups to an additional molecule of glycerol are called cardiolipin (diphosphatidylglycerol, Figure 17.2). This is the only human glycerophospholipid that is **antigenic**. For example, cardiolipin is recognized by antibodies raised against Treponema pallidum, the bacterium that causes syphylis. [Note: Cardiolipin is an important component of the **inner mitochondrial membrane** and bacterial membranes.]

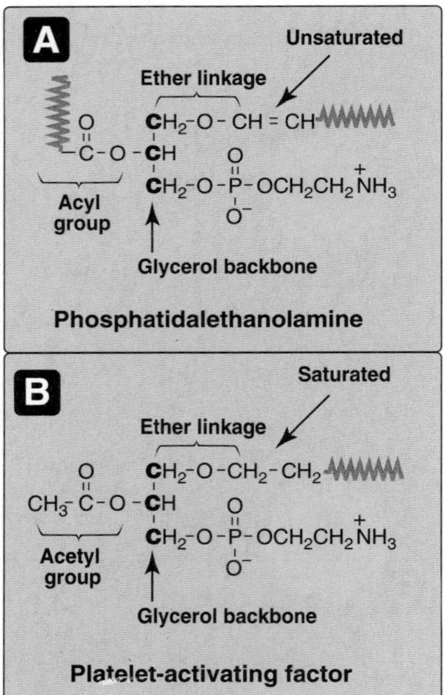

Figure 17.3
A. The plasmalogen phosphatidal-ethanolamine. B. Platelet-activating factor.

3. **Plasmalogens:** When the fatty acid at carbon 1 of a glycerolphospholipid is replaced by an **unsaturated alkyl group** attached by an **ether** (rather than by an ester) linkage to the core glycerol molecule, a plasmalogen is produced. For example, **phosphatidalethanolamine** (abundant in nerve tissue, Figure 17.3A) is the plasmalogen that is similar in structure to phosphatidylethanolamine. **Phosphatidalcholine** (abundant in heart muscle) is the other quantitatively significant ether lipid in mammals.

4. **Platelet-activating factor (PAF)** is an unusual ether glycerophospholipid, with a **saturated alkyl group** in an **ether** link to carbon 1 and an **acetyl residue** (rather than a fatty acid) at carbon 2 of the glycerol backbone (Figure 17.3B). PAF is synthesized and released by a variety of cell types. It binds to surface receptors, triggering potent thrombotic and acute inflammatory events. For example, PAF activates inflammatory cells and mediates hypersensitivity, acute inflammatory, and anaphylactic reactions. It causes platelets to aggregate and degranulate, and neutrophils and alveolar macrophages to generate superoxide radicals (see p. 148 for a discussion of the role of superoxides in killing bacteria). [Note: PAF is one of the most potent bioactive molecules known, causing effects at concentrations as low as 10^{-12} mol/L.]

B. Sphingophospholipids: sphingomyelin

The backbone of sphingomyelin is the amino alcohol **sphingosine**, rather than glycerol (Figure 17.4). A long-chain fatty acid is attached to the amino group of sphingosine through an amide linkage, producing a **ceramide**, which can also serve as a precursor of glycolipids (see p. 207). The alcohol group at carbon 1 of sphingosine is esterified to phosphorylcholine, producing **sphingomyelin**, the only significant sphingophospholipid in humans. Sphingomyelin is an important constituent of the **myelin** of nerve fibers. [Note: The myelin sheath is a layered, membranous structure that insulates and protects neuronal fibers of the central nervous system.]

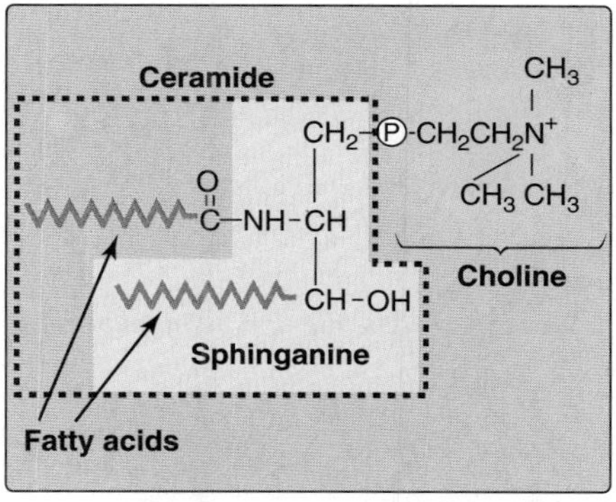

Figure 17.4
Structure of sphingomyelin, showing sphinganine (in green box) and ceramide components (in dashed box).

III. PHOSPHOLIPID SYNTHESIS

Glycerophospholipid synthesis involves either the donation of phosphatidic acid from CDP-diacylglycerol to an alcohol, or the donation of the phosphomonoester of the alcohol from CDP-alcohol to 1,2-diacylglycerol (Figure 17.5). [Note: CDP is the nucleotide cytidine diphosphate, (see p. 289).] In both cases, the CDP-bound structure is considered an "activated intermediate," and CMP is released as a side product of glycerophospholipid synthesis. A key concept in phosphoglyceride synthesis, therefore, is activation—either of diacylglycerol or the alcohol to be added—by linkage with CDP. [Note: This is similar in principle to the activation of sugars by their attachment to UDP (see p. 124).] The fatty acids esterified to the glycerol alcohol groups can vary widely, contributing to the heterogeneity of this group of compounds. Phospholipids are synthesized in the smooth endoplasmic reticulum. From there, they are transported to the Golgi apparatus and then to membranes of organelles or the plasma membrane, or are secreted from the cell by exocytosis.

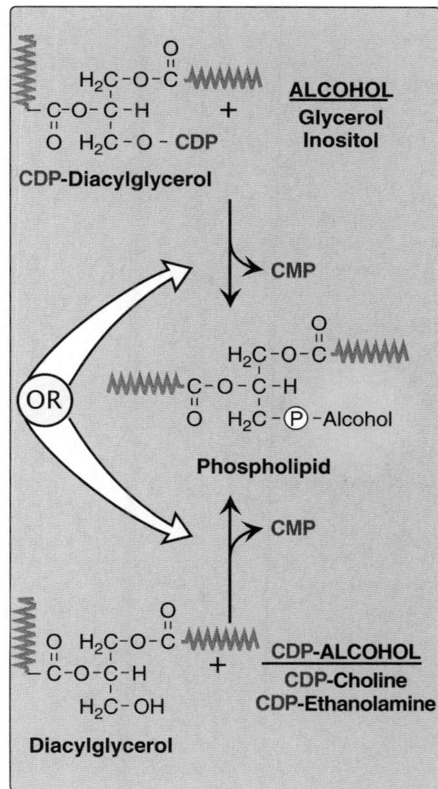

Figure 17.5
Activation of either diacylglycerol OR an alcohol by linkage to a nucleoside diphosphate (CDP) promotes phospholipid synthesis.

A. Synthesis of phosphatidic acid (PA)

PA is the precursor of many other phosphoglycerides. The steps in its synthesis from glycerol phosphate and two fatty acyl CoAs were illustrated in Figure 16.14, p. 187, in which PA is shown as a precursor of triacylglycerol. [Note: Essentially all cells except mature erythrocytes can synthesize phospholipids, whereas triacylglycerol synthesis occurs essentially only in liver, adipose tissue, lactating mammary glands, and intestinal mucosal cells.]

B. Synthesis of phosphatidylethanolamine (PE) and phosphatidylcholine (PC)

PC and PE are the most abundant phospholipids in most eukaryotic cells. The primary route of their synthesis uses choline and ethanolamine obtained either from the diet or from the turnover of the body's phospholipids. [Note: In the liver, PC also can be synthesized from phosphatidylserine (PS) and PE (see below).]

1. **Synthesis of PE and PC from preexisting choline and ethanolamine:** These synthetic pathways involve the phosphorylation of choline or ethanolamine by kinases, followed by conversion to the activated form, CDP-choline or CDP-ethanolamine. Finally, choline-phosphate or ethanolamine-phosphate is transferred from the nucleotide (leaving CMP) to a molecule of diacylglycerol (see Figure 17.5).

   a. **Significance of choline reutilization:** The reutilization of choline is important because, whereas humans can synthesize choline de novo, the amount made is insufficient for our needs. Thus, choline is an **essential dietary nutrient** with an Adequate Intake (p. 356) of 550 mg for men and 420 mg for women.

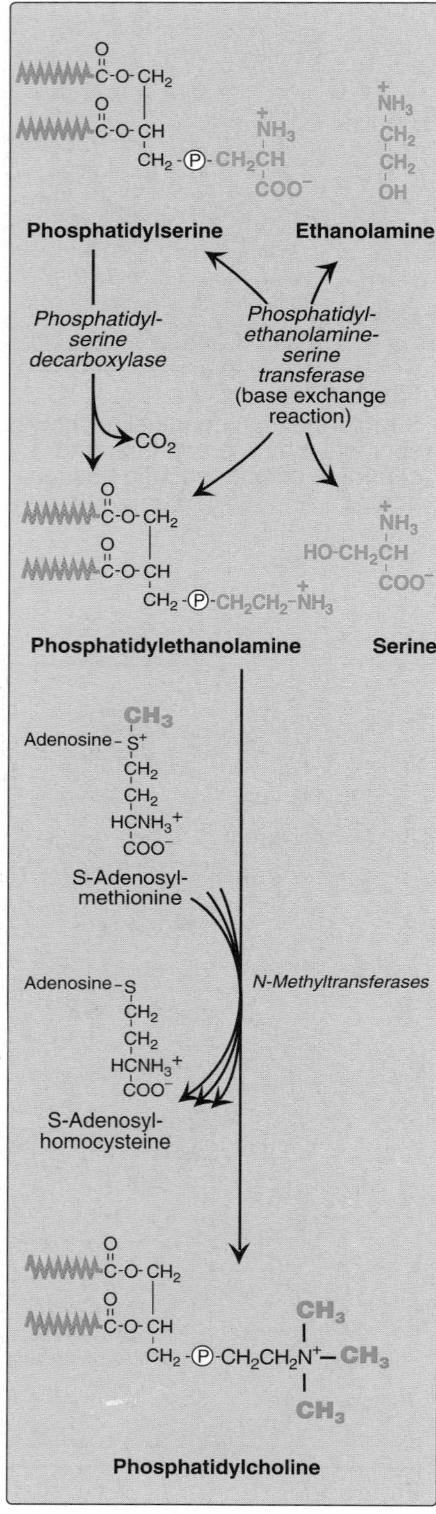

Figure 17.6
Synthesis of phosphatidylcholine from phosphatidylserine in the liver.

b. **Role of PC in lung surfactant:** The pathway described above is the principal pathway for the synthesis of **dipalmitoyl-phosphatidylcholine** (**DPPC**, or **dipalmitoylecithin**). In DPPC, positions 1 and 2 on the glycerol are occupied by palmitate. DPPC, made and secreted by granular pneumocytes, is the major lipid component of lung surfactant—the extracellular fluid layer lining the alveoli. Surfactant serves to decrease the surface tension of this fluid layer, reducing the pressure needed to reinflate alveoli, thereby preventing alveolar collapse (atelectasis). **Respiratory distress syndrome** (RDS) in pre-term infants is associated with insufficient surfactant production, and is a significant cause of all neonatal deaths in western countries. [Note: Lung maturity of the fetus can be gauged by determining the ratio of DPPC to sphingomyelin, usually written as the L (for lecithin)/S ratio, in amniotic fluid. A ratio of two or above is evidence of maturity, because it reflects the major shift from sphingomyelin to DPPC synthesis that occurs in the pneumocytes at about 32 weeks of gestation. Lung maturation can be accelerated by giving the mother glucocorticoids shortly before delivery. Administration of natural or synthetic surfactant (by intratracheal instillation) is also used in the prevention and treatment of infant RDS.] Respiratory distress syndrome due to an insufficient amount of surfactant can also occur in adults whose surfactant-producing pneumocytes have been damaged or destroyed, for example, as an adverse side effect of immunosuppressive medication or chemotherapeutic drug use.

2. **Synthesis of PC from phosphatidylserine (PS) in the liver:** The liver requires a mechanism for producing PC, even when free choline levels are low, because it exports significant amounts of PC in the bile and as a component of serum lipoproteins. To provide the needed PC, PS is decarboxylated to phosphatidylethanolamine (PE) by *PS decarboxylase*, an enzyme requiring pyridoxal phosphate as a cofactor. PE then undergoes three methylation steps to produce PC, as illustrated in Figure 17.6. S-adenosylmethionine (SAM) is the methyl group donor (see p. 262).

C. Phosphatidylserine (PS)

The primary pathway for synthesis of PS in mammalian tissues is provided by the **base exchange reaction**, in which the ethanolamine of PE is exchanged for free serine (see Figure 17.6). This reaction, although reversible, is used primarily to produce the PS required for membrane synthesis.

D. Phosphatidylinositol (PI)

PI is synthesized from free inositol and CDP-diacylglycerol as shown in Figure 17.5. PI is an unusual phospholipid in that it often contains stearic acid on carbon 1 and arachidonic acid on carbon 2 of the glycerol. PI, therefore, serves as a reservoir of **arachidonic acid** in membranes and, thus, provides the substrate for prostaglandin synthesis when required (see p. 211 for a discussion of these compounds).

1. **Role of PI in signal transmission across membranes:** The phosphorylation of membrane-bound phosphatidylinositol produces polyphosphoinositides, for example, **phosphatidylinositol 4,5-bisphosphate** (**PIP_2**, Figure 17.7). The degradation of PIP_2 by *phospholipase C* occurs in response to the binding of a variety of neurotransmitters, hormones, and growth factors to receptors on the cell membrane (Figure 17.8). The products of this degradation, **inositol 1,4,5-trisphosphate** (**IP_3**) and **diacylglycerol** (**DAG**), mediate the mobilization of intracellular calcium and the activation of *protein kinase C*, respectively, which act synergistically to evoke specific cellular responses. Signal transmission across the membrane is thus accomplished.

2. **Role of PI in membrane protein anchoring:** Specific proteins can be covalently attached via a carbohydrate bridge to membrane-bound PI (Figure 17.9). [Note: Examples of such proteins include *alkaline phosphatase* (a digestive enzyme found on the surface of the small intestine that attacks organic phosphates), and *acetylcholine esterase* (an enzyme of the postsynaptic membrane that

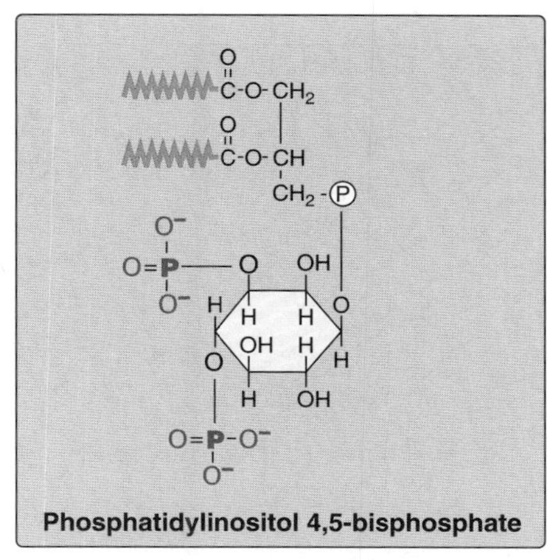

Figure 17.7
Structure of phosphatidylinositol 4,5-bisphosphate.

1. Hormone binds to a specific receptor.
2. Occupied receptor interacts with G_s protein.
3. G_s protein releases GDP and binds GTP.
4. α-Subunit of G_s protein dissociates and activates *phospholipase C.*
5. Active *phospholipase C* cleaves phosphatidylinositol 4,5-bisphospate to inositol trisphosphate (IP_3) and diacylglycerol.
6. IP_3 binds to a specific receptor on the endoplasmic reticulum, causing release of sequestered Ca^{++}.
7. Calcium and diacylglycerol activate *protein kinase C.*
8. *Protein kinase C* catalyzes phosphorylation of cellular proteins that mediate cellular response to the hormone.

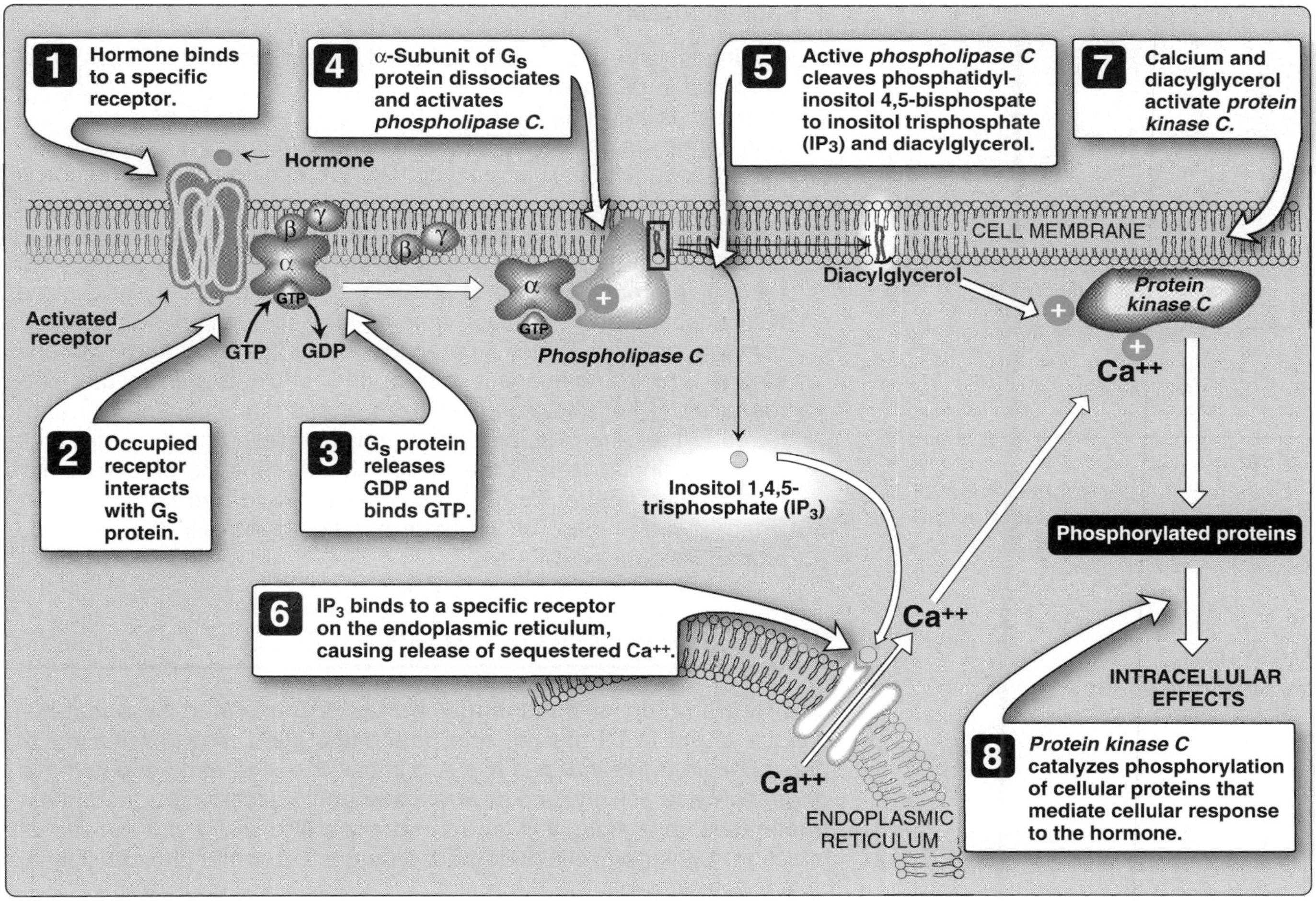

Figure 17.8
Role of inositol trisphosphate in intracellular signaling.

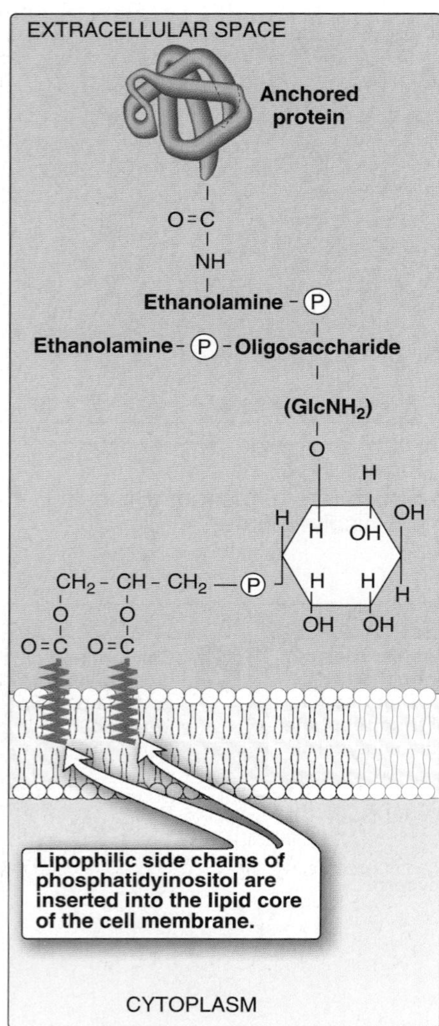

Figure 17.9
Example of a phosphatidylinositol glycan membrane protein anchor. $GlcNH_2$ = glucosamine.

degrades the neurotransmitter acetylcholine). Cell surface proteins bound to **glycosyl phosphatidylinositol** (**GPI**) are also found in a variety of parasitic protozoans (for example, trypanosomes and leishmania).] Being attached to a membrane lipid (rather than being an integral part of the membrane) allows GPI-anchored proteins rapid lateral mobility on the surface of the plasma membrane. The protein can be cleaved from its anchor by the action of *phospholipase C* (see Figure 17.8), releasing diacylglycerol. [Note: A deficiency in the synthesis of GPI in hematopoietic cells results in a hemolytic disease, **paroxysmal nocturnal hemoglobinuria**.]

E. Phosphatidylglycerol (PG) and cardiolipin

PG occurs in relatively large amounts in mitochondrial membranes and is a precursor of cardiolipin. It is synthesized by a two-step reaction from CDP-diacylglycerol and glycerol 3-phosphate. Cardiolipin (diphosphatidylglycerol, see Figure 17.2) is composed of two molecules of phosphatidic acid connected by a molecule of glycerol. It is synthesized by the transfer of diacylglycerophosphate from CDP-diacylglycerol to a preexisting molecule of phosphatidylglycerol.

F. Sphingomyelin

Sphingomyelin, a sphingosine-based phospholipid, is a major structural lipid in the membranes of nerve tissue. The synthesis of sphingomyelin is shown in Figure 17.10. Briefly, palmitoyl CoA condenses with serine, as coenzyme A and the carboxyl group (as CO_2) of serine are lost. [Note: This reaction, like the decarboxylation reactions involving amino acids, requires **pyridoxal phosphate** (a derivative of vitamin B_6) as a coenzyme (see p. 376).] The product is reduced in an **NADPH-requiring** reaction to **sphinganine**, which is acylated at the amino group with one of a variety of long-chain fatty acids, and then desaturated to produce a ceramide—the immediate precursor of sphingomyelin. [Note: A ceramide with a fatty acid thirty carbons long is a major component of skin, and regulates skin's water permeability.] Phosphorylcholine from phosphatidylcholine is transferred to the ceramide, producing sphingomyelin and diacylglycerol. [Note: Sphingomyelin of the myelin sheath contains predominantly longer-chain fatty acids such as lignoceric acid and nervonic acid, whereas gray matter of the brain has sphingomyelin that contains primarily stearic acid.]

IV. DEGRADATION OF PHOSPHOLIPIDS

The degradation of phosphoglycerides is performed by phospholipases found in all tissues and pancreatic juice (see discussion of phospholipid digestion, p. 173). A number of toxins and venoms have phospholipase activity, and several pathogenic bacteria produce phospholipases that dissolve cell membranes and allow the spread of infection. Sphingomyelin is degraded by the lysosomal phospholipase, *sphingomyelinase*.

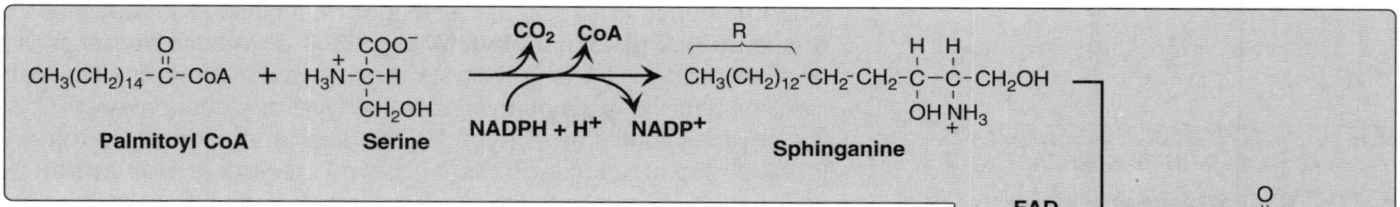

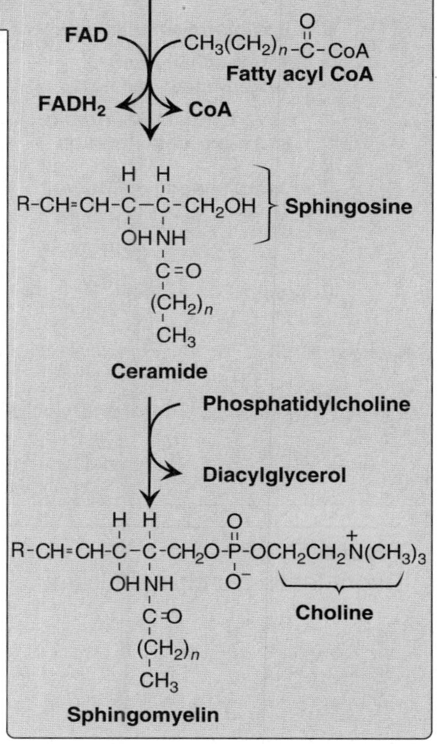

$CH_3(CH_2)_{14}$-C(=O)-CoA + $H_3\overset{+}{N}$-CH(COO^-)-CH_2OH → (CO_2, CoA released; NADPH + H^+ → $NADP^+$) → $CH_3(CH_2)_{12}$-CH_2-CH_2-CH(OH)-CH($\overset{+}{N}H_3$)-CH_2OH

Palmitoyl CoA Serine Sphinganine

FAD, $CH_3(CH_2)_n$-C(=O)-CoA (Fatty acyl CoA) → $FADH_2$, CoA

R-CH=CH-CH(OH)-CH(NH-C(=O)-$(CH_2)_n$-CH_3)-CH_2OH (Sphingosine portion)

Ceramide

Phosphatidylcholine → Diacylglycerol

R-CH=CH-CH(OH)-CH(NH-C(=O)-$(CH_2)_n$-CH_3)-CH_2O-P(=O)(O^-)-$OCH_2CH_2\overset{+}{N}(CH_3)_3$ (Choline)

Sphingomyelin

Figure 17.10
Synthesis of sphingomyelin.

A. Degradation of phosphoglycerides

Phospholipases hydrolyze the phosphodiester bonds of phosphoglycerides, with each enzyme cleaving the phospholipid at a specific site. The major enzymes responsible for degrading phosphoglycerides are shown in Figure 17.11. [Note: Removal of the fatty acid from carbon 1 or 2 of a phosphoglyceride produces a **lysophosphoglyceride**, which is the substrate for lysophospholipases.] Phospholipases release molecules that can serve as messengers (for example, DAG and IP_3), or that are the substrates for synthesis of messengers (for example, arachidonic acid). [Note: Phospholipases are responsible not only for degrading phospholipids, but also for "remodeling" them. For example, *phospholipases A_1* and *A_2* remove specific fatty acids from membrane-bound phospholipids; these can be replaced with alternative fatty acids using *fatty acyl CoA transferase*. This mechanism is used as one way to create the unique lung surfactant, dipalmitoylphosphatidylcholine (see p. 202), and to insure that carbon 2 of PI (and sometimes of PC) is bound to arachidonic acid.]

B. Degradation of sphingomyelin

Sphingomyelin is degraded by *sphingomyelinase*, a lysosomal enzyme that hydrolytically removes phosphorylcholine, leaving a ceramide. The ceramide is, in turn, cleaved by *ceramidase* into sphingosine and a free fatty acid (Figure 17.12). [Note: The ceramide and sphingosine released by the degradation of sphingomyelin play a role as intracellular messengers. Ceramides appear to be involved

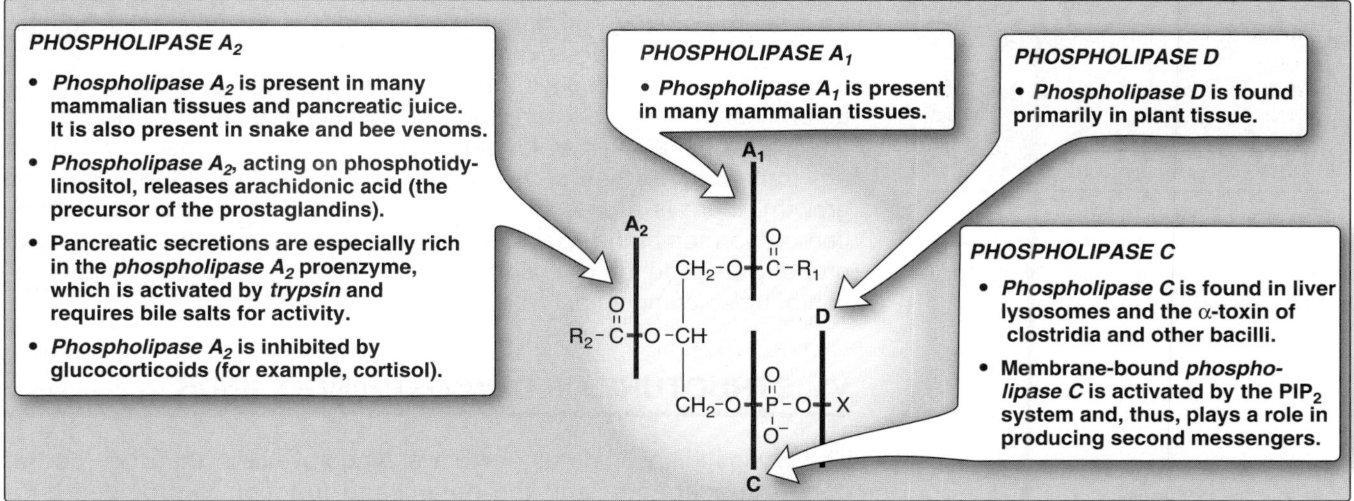

Figure 17.11
Degradation of glycerophospholipids by phospholipases.

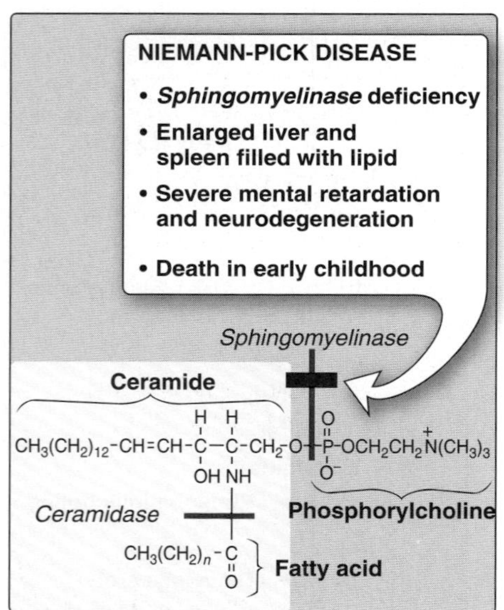

Figure 17.12
Degradation of sphingomyelin.

in the response to stress, and sphingosine inhibits *protein kinase C*.] **Niemann-Pick disease** (**Types A** and **B**) is an autosomal recessive disease caused by the inability to degrade sphingomyelin. The deficient enzyme is *sphingomyelinase*—a type of *phospholipase C*. In the severe infantile form (type A), the liver and spleen are the primary sites of lipid deposits and are, therefore, tremendously enlarged. The lipid consists primarily of the sphingomyelin that cannot be degraded (Figure 17.13). Infants with this disease experience rapid and progressive **neurodegeneration** as a result of deposition of sphingomyelin in the CNS, and they die in early childhood. A less severe variant (type B) causes little to no damage to neural tissue, but lungs, spleen, liver, and bone marrow are affected, resulting in a chronic form of the disease, with a life expectancy only to early adulthood. Although Niemann-Pick disease occurs in all ethnic groups, both type A and B occur with greater frequency in the Ashkenazi Jewish population than in the general population. [Note: In the Ashkenazi Jewish population, the incidence of type A is 1:40,000 live births, and that of type B is 1:80,000. The incidence of Niemann-Pick disease in the general population is less than 1:100,000.]

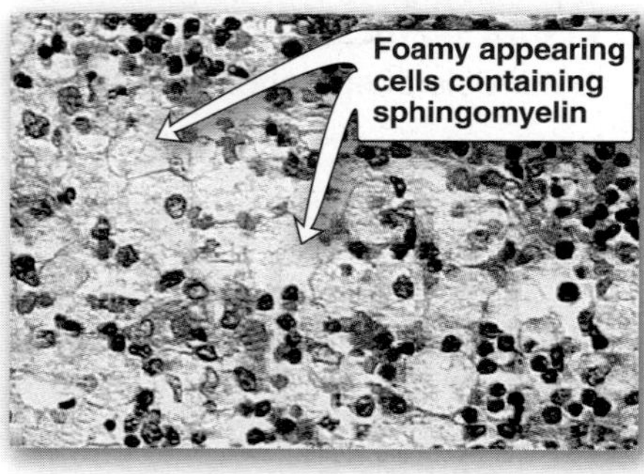

Figure 17.13
Accumulation of lipids in spleen cells from a patient with Niemann-Pick disease.

V. OVERVIEW OF GLYCOLIPIDS

Glycolipids are molecules that contain both carbohydrate and lipid components. Like the phospholipid sphingomyelin, almost all glycolipids are derivatives of ceramides in which a long-chain fatty acid is attached to the amino alcohol sphingosine. They are, therefore, more precisely called **glycosphingolipids**. [Note: Ceramides, then, are the precursors of both phosphorylated and glycosylated sphingolipids.] Like the phospholipids, glycosphingolipids are essential components of all membranes in the body, but they are found in greatest amounts in nerve tissue. They are located in the outer leaflet of the plasma membrane, where they interact with the extracellular environment. As such, they play a role in the regulation of cellular interactions, growth, and development. Glycosphingolipids are antigenic, and they have been identified as a source of blood group antigens, various embryonic antigens specific for particular stages of fetal development, and some tumor antigens. [Note: The carbohydrate portion of a glycolipid is the antigenic determinant.] They also serve as cell surface receptors for cholera and tetanus toxins, as well as for certain viruses and microbes. When cells are transformed (that is, when they lose control of cell division and growth), there is a dramatic change in the glycosphingolipid composition of the membrane. Genetic disorders associated with an inability to properly degrade the glycosphingolipids result in intracellular accumulation of these compounds.

VI. STRUCTURE OF GLYCOSPHINGOLIPIDS

The glycosphingolipids differ from sphingomyelin in that they do not contain phosphate, and the polar head function is provided by a monosaccharide or oligosaccharide attached directly to the ceramide by an O-glycosidic bond (Figure 17.14). The number and type of carbohydrate moieties present help determine the type of glycosphingolipid.

A. Neutral glycosphingolipids

The simplest neutral (uncharged) glycosphingolipids are the **cerebrosides**. These are ceramide monosaccharides that contain either a molecule of galactose (**galactocerebroside**—the most common cerebroside found in membranes, see Figure 17.14) or glucose (**glucocerebroside**, which serves primarily as an intermediate in the synthesis and degradation of the more complex glycosphingolipids). [Note: Members of a group of galactocerebrosides (or glucocerebrosides) may also differ from each other in the type of fatty acid attached to the sphingosine.] As their name implies, cerebrosides are found predominantly in the brain and peripheral nervous tissue, with high concentrations in the myelin sheath. Ceramide oligosaccharides (or **globosides**) are produced by attaching additional monosaccharides (including GalNAc) to a glucocerebroside. Examples of these compounds include:

Cerebroside (glucocerebroside):	Cer-Glc
Globoside (lactosylceramide):	Cer-Glc-Gal
Globoside (Forssman antigen):	Cer-Glc-Gal-Gal-GalNac-GalNac

(Cer = ceramide, Glc = glucose, Gal = galactose, GalNac = N-acetylgalactosamine)

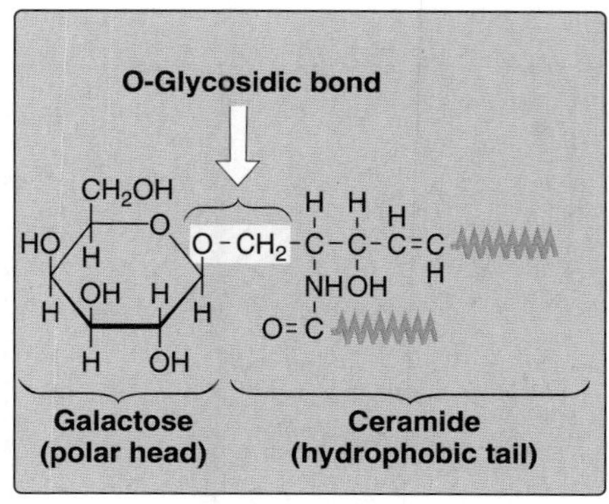

Figure 17.14
Structure of a neutral glycosphingolipid, galactocerebroside (R is a fatty acid hydrocarbon chain).

B. Acidic glycosphingolipids

Acidic glycosphingolipids are negatively charged at physiologic pH. The negative charge is provided by **N-acetylneuraminic acid** (**NANA**, Figure 17.15) in **gangliosides**, or by **sulfate groups** in **sulfatides**. [Note: NANA is also referred to as **sialic acid**.]

1. **Gangliosides:** These are the most complex glycosphingolipids, and are found primarily in the ganglion cells of the central nervous system, particularly at the nerve endings. They are derivatives of ceramide oligosaccharides, and contain one or more molecules of NANA. The notation for these compounds is G (for ganglioside), plus a subscript M, D, T, or Q to indicate whether there is one (mono), two, three, or four (quatro) molecules of NANA in the ganglioside, respectively. Additional numbers and letters in the subscript designate the sequence of the carbohydrate attached to the ceramide. (See Figure 17.15 for the structure of G_{M2}.) Gangliosides are of medical interest because several lipid storage disorders involve the accumulation of NANA-containing glycosphingolipids in cells (see Figure 17.20, p. 210).

2. **Sulfatides:** Sulfoglycosphingolipids (sulfatides) are cerebrosides that contain sulfated galactosyl residues, and are therefore negatively charged at physiologic pH. Sulfatides are found predominantly in nerve tissue and kidney.

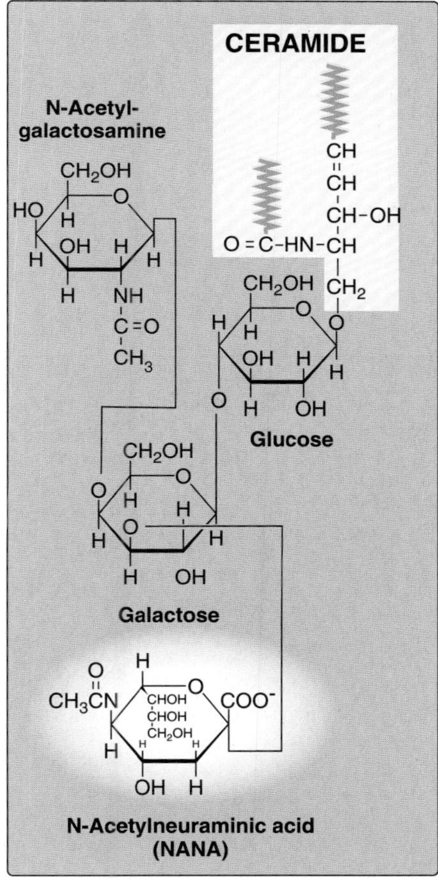

Figure 17.15
Structure of the ganglioside G_{M2}.

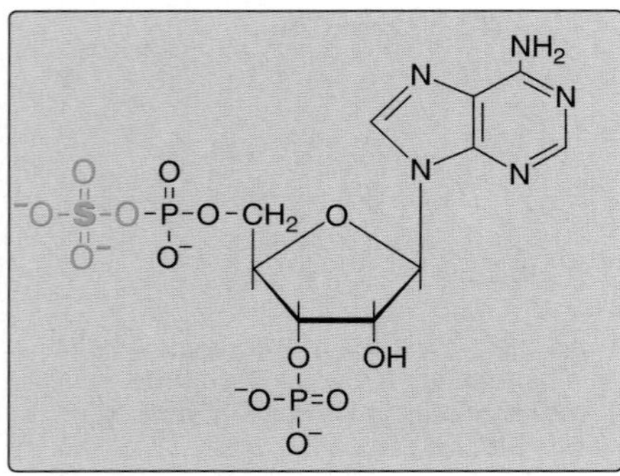

Figure 17.16
Structure of 3'-phospho-adenosine-5'-phosphosulfate.

VII. SYNTHESIS AND DEGRADATION OF GLYCOSPHINGOLIPIDS

Synthesis of glycosphingolipids occurs in the **endoplasmic reticulum** and **Golgi** by sequential addition of glycosyl monomers transferred from sugar-nucleotide donors to the acceptor molecule. The mechanism is similar to that used in glycoprotein synthesis (see p. 164).

A. Enzymes involved in synthesis

The enzymes involved in the synthesis of glycosphingolipids are glycosyl transferases, each specific for a particular sugar-nucleotide and acceptor. [Note: These enzymes may recognize both glycosphingolipids and glycoproteins as substrates.]

B. Addition of sulfate groups

A sulfate group from the sulfate carrier, **3'-phosphoadenosine-5'-phosphosulfate** (**PAPS,** Figure 17.16), is added by a *sulfotransferase* to the 3'-hydroxyl group of the galactose in a galactocerebroside. **Galactocerebroside 3-sulfate** is the major sulfatide in the brain (Figure 17.17). [Note: PAPS is also the sulfur donor in glycosaminoglycan synthesis (see p. 160), and steroid hormone catabolism (see p. 238).] An overview of the synthesis of sphingolipids is shown in Figure 17.18.

C. Degradation of glycosphingolipids

Glycosphingolipids are internalized by **endocytosis** as described for the glycosaminoglycans. All of the enzymes required for the degradative process are present in the **lysosomes**, which fuse with the endocytotic vesicles. The lysosomal enzymes hydrolytically and irreversibly cleave specific bonds in the glycosphingolipid. As seen with the glycosaminoglycans (see p. 161) and glycoproteins (see p. 168), degradation is a sequential process following the rule "last on, first off," in which the last group added during synthesis is the first group removed in degradation.

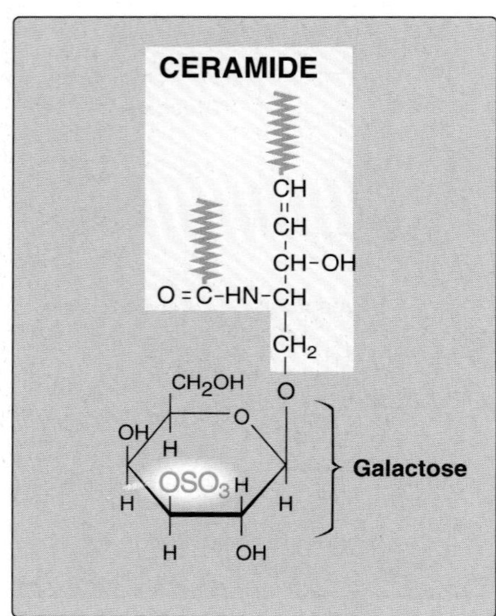

Figure 17.17
Structure of galactocerebroside 3-sulfate.

D. Sphingolipidoses

In a normal individual, synthesis and degradation of sphingolipids are balanced, so that the amount of these compounds present in membranes is constant. If a specific hydrolase required for the degradation process is partially or totally missing, a sphingolipid accumulates in the lysosomes. Lipid storage diseases caused by these deficiencies are called sphingolipidoses. The result of a specific hydrolase deficiency may be seen dramatically in nerve tissue, where neurologic deterioration can lead to early death. [Note: Ganglioside turnover in the central nervous system is extensive during neonatal development.] (See Figure 17.20 for an outline of the pathway of sphingolipid degradation and descriptions of some sphingolipidoses.)

1. **Common properties:** A specific lysosomal hydrolytic enzyme is deficient in each disorder. Therefore, usually only a single sphin-

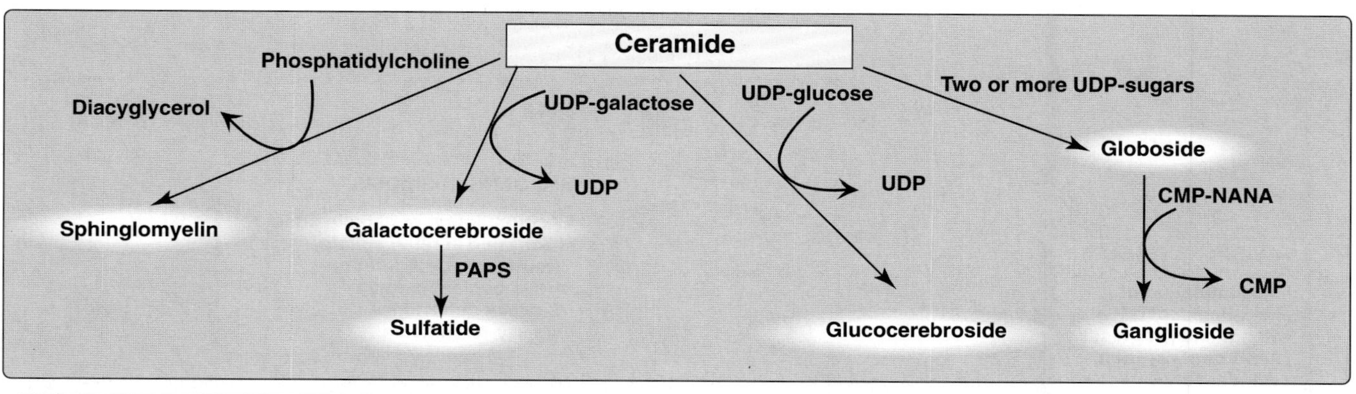

Figure 17.18
Overview of sphingolipid synthesis.

golipid (the substrate for the deficient enzyme) accumulates in the involved organs in each disease. [Note: The rate of biosynthesis of the accumulating lipid is normal.] The disorders are progressive and, although many are fatal in childhood, extensive phenotypic variability is seen leading to the designation of different clinical types, such as A and B in **Niemann-Pick disease**. Genetic variability is also seen, because a given disorder can be caused by any one of a variety of mutations within a single gene. The sphingolipidoses are autosomal recessive diseases, except for Fabry disease, which is X-linked. The incidence of the sphingolipidoses is low in most populations, except for **Gaucher** and **Tay-Sachs diseases**, which, like Niemann-Pick disease, show a high frequency in the Ashkenazi Jewish population.

2. **Diagnosis and treatment:** A specific sphingolipidosis can be diagnosed by measuring enzyme activity in cultured fibroblasts or peripheral leukocytes, or by analysis of DNA (see Chapter 32, p. 445). Histologic examination of the affected tissue is also useful. [Note: Shell-like inclusion bodies are seen in Tay-Sachs, and a wrinkled tissue paper appearance of the cytosol is seen in Gaucher (Figure 17.19).] Prenatal diagnosis, using cultured amniocytes or chorionic villi, is available. The sphingolipid that accumulates in the lysosomes in each disease is the structure that cannot be further degraded as a result of the specific enzyme deficiency. Gaucher disease, in which macrophages become engorged with glucocerebroside, and **Fabry disease**, in which globosides accumulate in the vascular endothelial lysosomes of the brain, heart, kidneys, and skin, have been successfully treated by recombinant human enzyme replacement therapy, but the cost is extremely high. Gaucher is also treated by bone marrow transplantation (because macrophages are derived from hematopoietic stem cells).

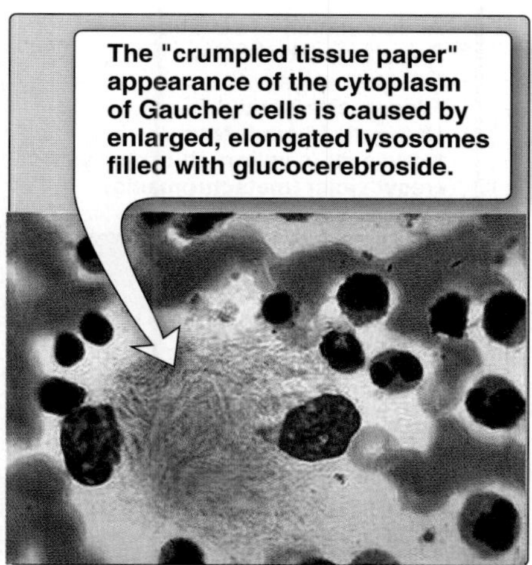

Figure 17.19
Aspirated bone marrow cells from patient with Gaucher disease.

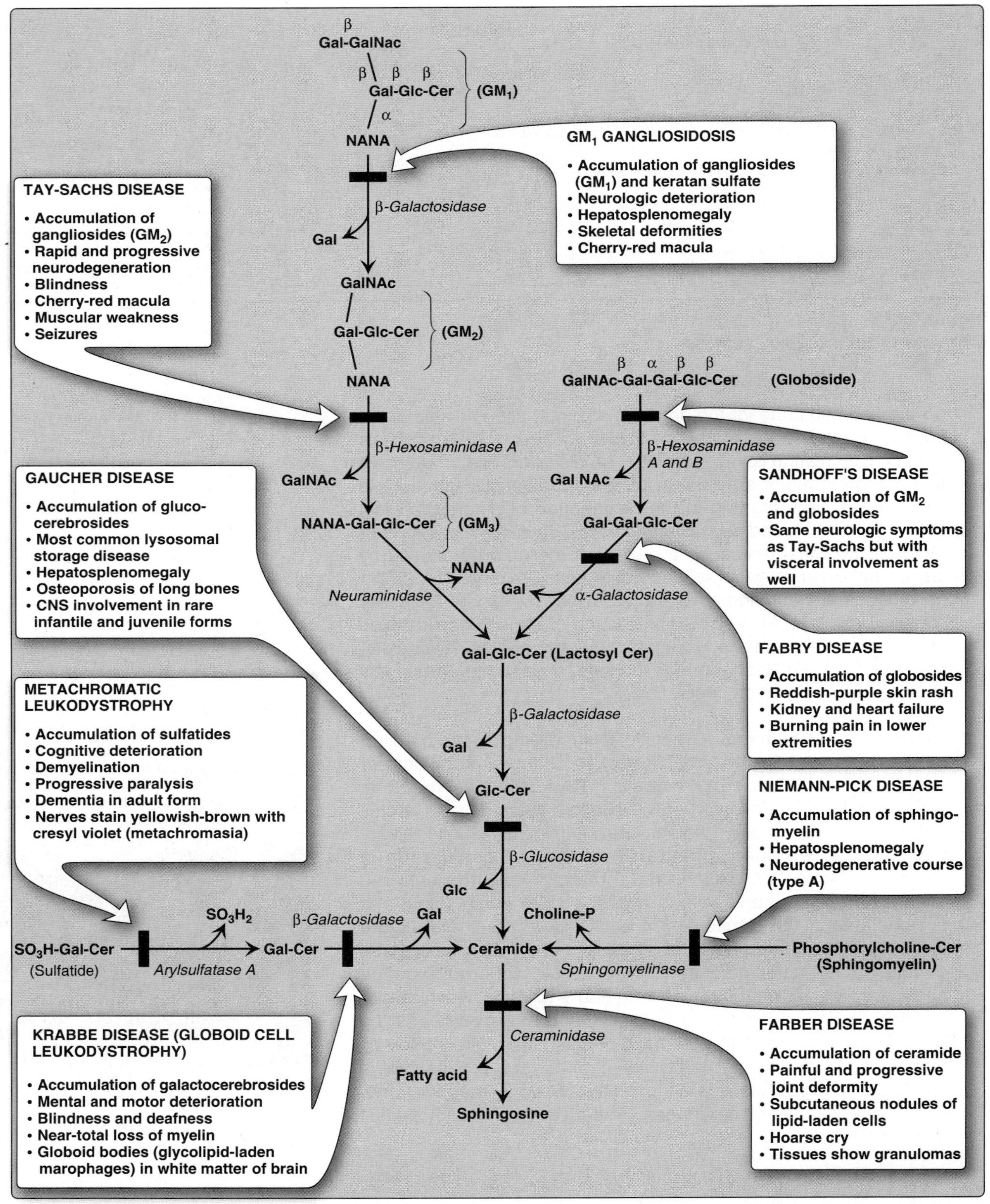

Figure 17.20
Degradation of sphingolipids showing the enzymes missing in related genetic diseases, the sphingolipidoses. All of the diseases are autosomal recessive except Fabry disease which is X-linked, and all can be fatal in early life. (Cer = ceramide).

VIII. PROSTAGLANDINS AND RELATED COMPOUNDS

Prostaglandins (**PG**), and the related compounds **thromboxanes** (**TX**) and **leukotrienes** (**LT**), are collectively known as **eicosanoids** to reflect their origin from polyunsaturated fatty acids with twenty carbons. They are extremely potent compounds that elicit a wide range of responses, both physiologic and pathologic. Although they have been compared to hormones in terms of their actions, eicosanoids differ from the true hormones in that they are produced in very small amounts in almost all tissues rather than in specialized glands. They also act locally rather than after transport in the blood to distant sites, as occurs with true hormones such as insulin. Eicosanoids are not stored, and they have an extremely short half-life, being rapidly metabolized to inactive products at their site of synthesis. Their biologic actions are mediated by plasma and nuclear membrane receptors, which are different in different organ systems. Examples of prostaglandins and related structures are shown in Figure 17.21.

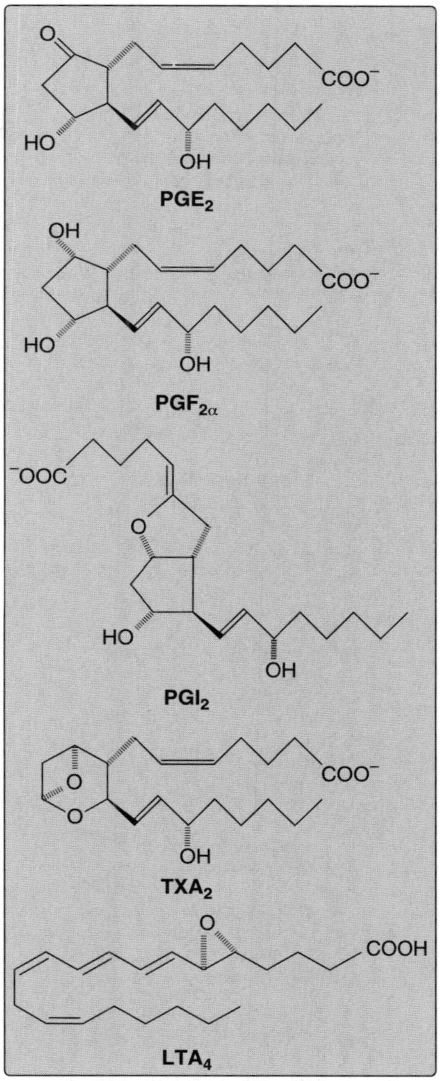

A. Synthesis of prostaglandins and thromboxanes

The dietary precursor of the prostaglandins is the essential fatty acid, **linoleic acid**. It is elongated and desaturated to **arachidonic acid**, the immediate precursor of the predominant class of prostaglandins (those with two double bonds) in humans (Figure 17.22). [Note: Arachidonic acid is released from membrane-bound phospholipids by *phospholipase A_2* in response to a variety of signals (Figure 17.23).]

1. **Synthesis of PGH_2:** The first step in the synthesis of prostaglandins is the oxidative cyclization of free arachidonic acid to yield PGH_2 by *prostaglandin endoperoxide synthase*. This enzyme is a microsomal protein that has two catalytic activities: *fatty acid cyclooxygenase* (*COX*), which requires two molecules of O_2, and *peroxidase*, which is dependent on reduced glutathione (see p. 146). PGH_2 is converted to a variety of prostaglandins and thromboxanes, as shown in Figure 17.23, by cell-specific synthases.

   a. **Isozymes of prostaglandin endoperoxide synthase:** Two isozymes, usually denoted as *COX-1* and *COX-2*, of the *synthase* are known. COX-1 is made **constitutively** in most tissues, and is required for maintenance of healthy gastric tissue, renal homeostasis, and platelet aggregation. COX-2 is **inducible** in a limited number of tissues in response to products of activated immune and inflammatory cells. [Note: The increase in prostaglandin synthesis subsequent to the induction of *COX-2* mediates the pain, heat, redness, and swelling of inflammation, and the fever of infection.]

2. **Inhibition of prostaglandin synthesis:** The synthesis of prostaglandins can be inhibited by a number of unrelated compounds. For example, **cortisol** (a steroidal anti-inflammatory agent) inhibits *phospholipase A_2* activity (see Figure 17.23) and, therefore, the precursor of the prostaglandins, arachidonic acid, is

Figure 17.21
Examples of prostaglandin structures. Prostaglandins are named as follows: PG plus a third letter (for example, A,D,E,F), which designates the type and arrangement of functional groups in the molecule. PGI_2 is known as prostacyclin. The subscript number indicates the number of double bonds in the molecule. Thromboxanes are designated by TX and leukotrienes by LT.

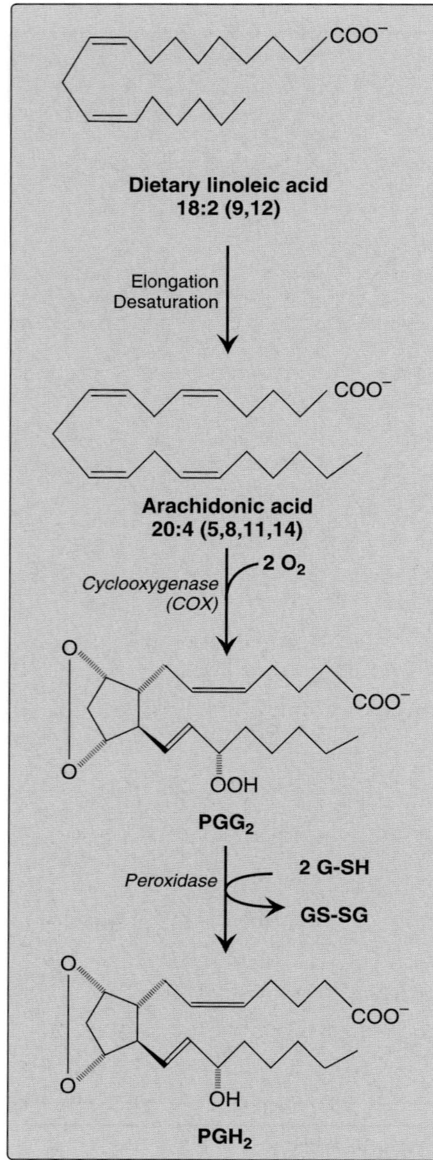

Figure 17.22
Oxidation and cyclization of arachidonic acid by the two catalytic activities of prostaglandin endoperoxide synthase. G-SH = reduced glutathione; GS-SG = oxidized gluathione.

not available. Cortisol also inhibits *COX-2*, but not *COX-1*. **Aspirin**, **indomethacin**, and **phenylbutazone** (all nonsteroidal anti-inflammatory agents or NSAIDS) inhibit both *COX-1* and *COX-2* and, therefore, prevent the synthesis of the parent prostaglandin, PGH_2. [Note: Systemic inhibition of *COX-1*, with subsequent damage to the stomach and the kidneys, and impaired clotting of blood, is the basis of aspirin's toxicity.] Inhibitors specific for *COX-2* (for example, **celecoxib**[1]) are designed to reduce pathologic inflammatory processes while maintaining the physiologic functions of *COX-1*.

B. Synthesis of leukotrienes

Arachidonic acid is converted to a variety of linear hydroperoxy acids by a separate pathway involving a family of lipoxygenases. For example, neutrophils contain *5-lipoxygenase*, which converts arachidonic acid to 5-hydroxy-6,8,11,14 eicosatetraenoic acid (5-HPETE; see Figure 17.23). 5-HPETE is converted to a series of leukotrienes, the nature of the final products varying according to the tissue. Lipoxygenases are not affected by NSAIDS. Leukotrienes are mediators of allergic response and inflammation. [Note: Inhibitors of *5-lipoxygenase* and leukotriene receptor antagonists are used in the treatment of **asthma.**[2]]

C. Role of prostaglandins in platelet homeostasis

In addition to their roles in mediating inflammation, fever, and allergic response, and ensuring gastric integrity and renal function, eicosanoids are involved in a diverse group of physiologic functions, including ovarian and uterine function, bone metabolism, nerve and brain function, smooth muscle regulation, and platelet homeostasis. **Thromboxane A_2** (**TXA_2**) is produced by **activated platelets**. It promotes adherence and aggregation of circulating platelets, and contraction of vascular smooth muscle, thus promoting formation of **blood clots** (thrombi). **Prostacyclin** (**PGI_2**), produced by **vascular endothelial cells**, inhibits platelet aggregation and stimulates vasodilation, and so **impedes thrombogenesis**. The opposing effects of TXA_2 and PGI_2 limit thrombi formation to sites of vascular injury. [Note: Aspirin has an antithrombogenic effect. It inhibits thromboxane A_2 synthesis from arachidonic acid in platelets by irreversible acetylation and inhibition of *COX-1* (Figure 17.24). This irreversible inhibition of *COX-1* cannot be overcome in anucleate platelets, but can be overcome in endothelial cells, because they have a nucleus and, therefore, can generate more of the enzyme. This difference is the basis of **low-dose aspirin therapy** used to lower the risk of stroke and heart attacks by decreasing formation of thrombi.]

[1]See Ch. 43 in ***Lippincott's Illustrated Reviews: Pharmacology*** (3rd Ed.) and Ch. 39 (2nd Ed.) for a discussion of anti-inflammatory drugs.
[2]See Ch. 26 in ***Lippincott's Illustrated Reviews: Pharmacology*** (3rd Ed.) and Ch. 22 (2nd Ed.) for a discussion of the treatment of asthma.

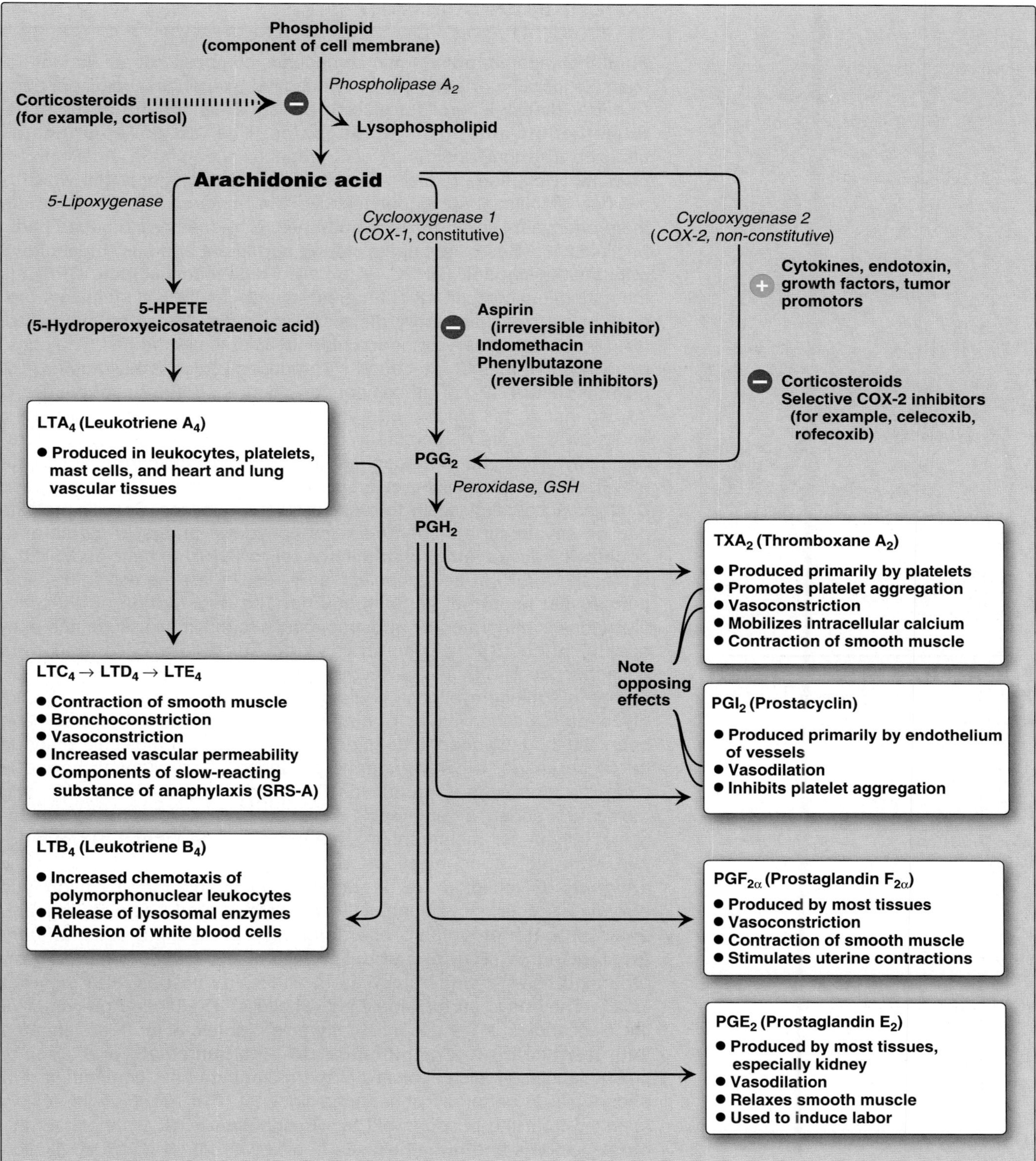

Figure 17.23
Overview of the biosynthesis and function of some important prostaglandins, leukotrienes, and a thromboxane from arachidonic acid.

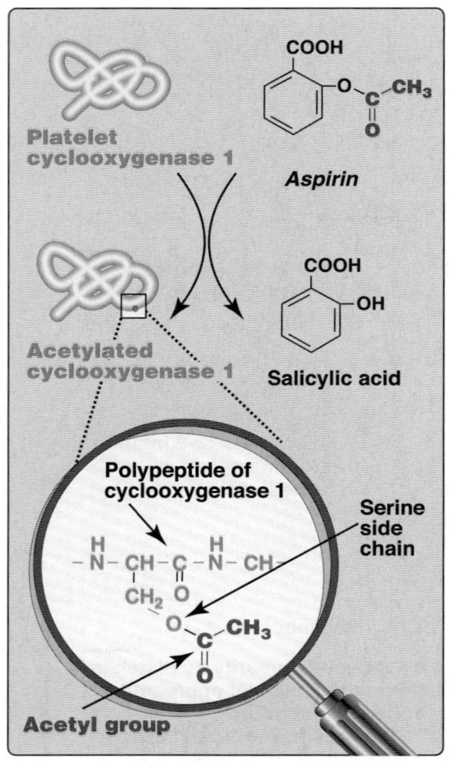

Figure 17.24
Acetylation of cyclooxygenase 1 by *aspirin*.

IX. CHAPTER SUMMARY

Phospholipids are **polar**, **ionic** compounds composed of an **alcohol** (for example, **choline** or **ethanolamine**) attached by a phosphodiester bridge to either **diacylglycerol** (producing **phosphatidylcholine** or **phosphatidylethanolamine)** or to **sphingosine**. The alcohol **sphingosine** attached to a long-chain fatty acid produces a **ceramide**. Addition of a **phosphorylcholine** produces the phospholipid **sphingomyelin**, which is the only significant sphingophospholipid in humans. Phospholipids are the predominant lipids of **cell membranes**. Non–membrane-bound phospholipids serve as components of **lung surfactant** and **bile. Dipalmitoylphosphatidylcholine** (DPPC, also called **dipalmitoyllecithin**, DPPL) is the major lipid component of **lung surfactant**. Insufficient surfactant production causes **respiratory distress syndrome. Phosphatidylinositol (PI)** serves as a reservoir for **arachidonic acid** in membranes. The phosphorylation of membrane-bound PI produces **phosphatidylinositol 4,5-bisphosphate** (**PIP_2**). This compound is degraded by **phospholipase C** in response to the binding of a variety of neurotransmitters, hormones, and growth factors to membrane receptors. The products of this degradation, **inositol 1,4,5-trisphosphate** (**IP_3**) and **diacylglycerol** mediate the mobilization of intracellular **calcium** and the activation of **protein kinase C**, which act synergistically to evoke cellular responses. Specific proteins can be covalently attached via a carbohydrate bridge to membrane-bound PI (**glycosylphosphatidylinositol**, or **GPI**). A deficiency in the synthesis of GPI in hematopoietic cells results in a hemolytic disease, **paroxysmal nocturnal hemoglobinuria**. The **degradation** of phosphoglycerides is performed by **phospholipases** found in all tissues and pancreatic juice. **Sphingomyelin** is degraded to a ceramide plus phosphorylcholine by the lysosomal enzyme **sphingomyelinase.** A deficiency in sphingomyelinase causes **Niemann-Pick disease**. Almost all **glycolipids** are derivatives of **ceramides** to which carbohydrates have been attached (**glycosphingolipids**). When one sugar molecule is added to the ceramide, a **cerebroside** is produced. If an oligosaccharide is added, a **globoside** is produced. If an acidic N-acetylneuraminic acid molecule is added, a **ganglioside** is produced. Glycolipids are found predominantly in cell membranes of the **brain** and **peripheral nervous tissue**, with high concentrations in the **myelin sheath**. They are very **antigenic**. Glycolipids are degraded in the **lysosomes** by hydrolytic enzymes. A deficiency of one of these enzymes produces a **sphingolipidosis**, in each of which a characteristic sphingolipid accumulates. **Prostaglandins** (**PG**), **thromboxanes** (**TX**), and **leukotrienes** (**LT**) are produced in very small amounts in almost all tissues, and they act locally. They have an extremely short half-life. The dietary precursor of the eicosanoids is the essential fatty acid, **linoleic acid**. It is elongated and desaturated to **arachidonic acid**—the immediate precursor of prostaglandins—which is stored in the membrane as a component of a phospholipid, generally phosphatidylinositol (PI). Arachidonic acid is released from the phospholipid by **phospholipase A_2**. Synthesis of the **prostaglandins** and **thromboxanes** begins with the oxidative cyclization of free arachidonic acid to yield PGH_2 by **prostaglandin endoperoxide synthase**—a microsomal protein that has two catalytic activities: **fatty acid cyclooxygenase** (**COX**) and **peroxidase.** There are two isozymes of the synthase: **COX-1** and **COX-2**. **Leukotrienes** are produced by the **5-lipoxygenase** pathway.

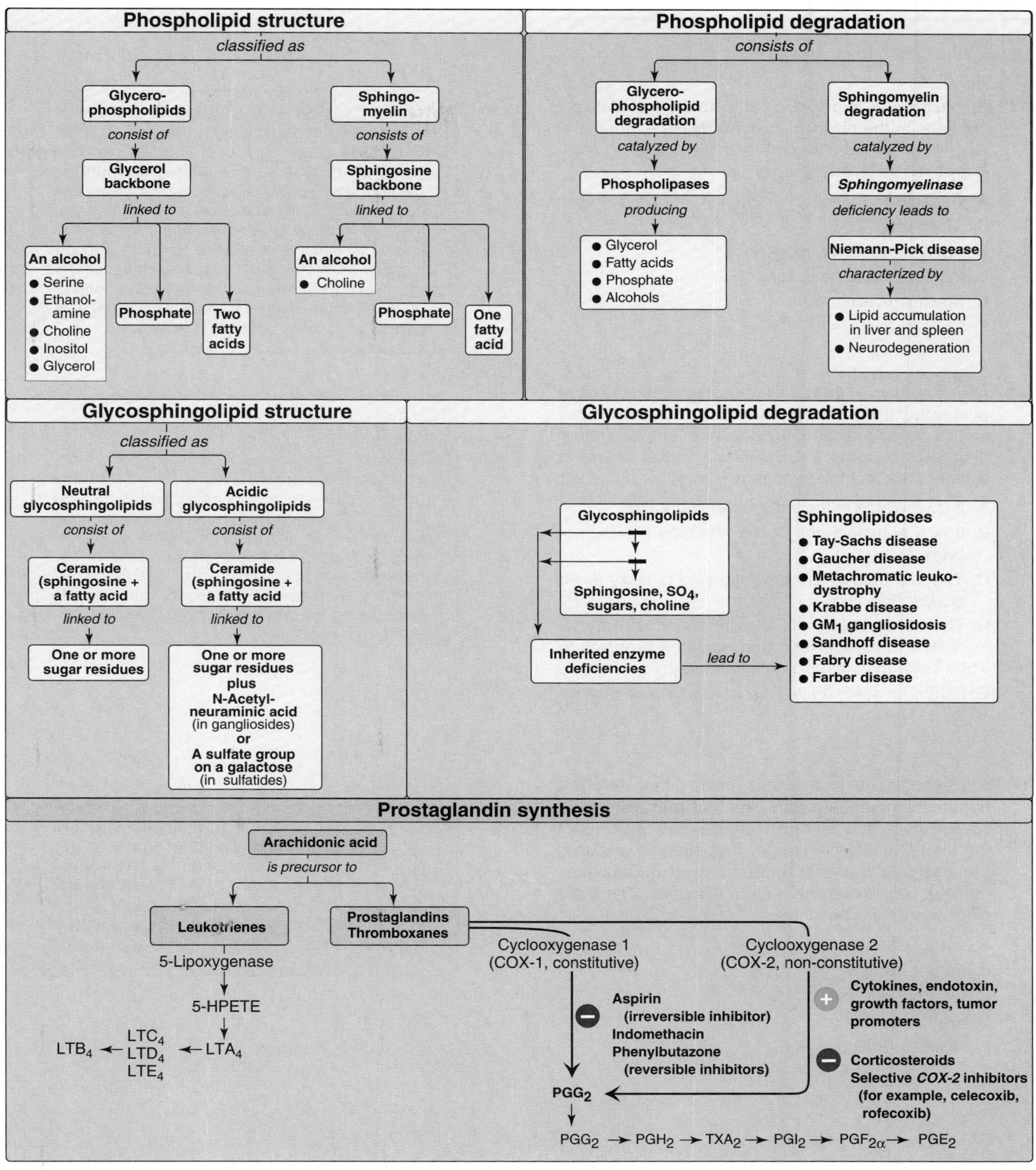

Figure 17.25
Key concept map for complex lipids.

Study Questions

Choose the ONE correct answer

17.1 Autoantibodies to a lipid in the membrane of platelets are seen in the disease, systemic lupus erythematosus. Which of the following membrane lipids is most likely to be involved?

A. Cardiolipin
B. Ceramide
C. Dipalmitoylphosphatidylcholine
D. Platelet-activating factor
E. Sphingomyelin

Correct answer = A. Cardiolipin is the only human glycerophospholipid that is antigenic. Ceramides are precursors of phospholipids and glycolipids, but are not themselves found in membranes. Dipalmitoylphosphatidylcholine is a component of lung surfactant, not membranes. Platelet-activating factor is not a membrane lipid, but it binds to membrane receptors, triggering potent thrombotic and acute inflammatory events, for example, it causes platelets to aggregate and degranulate. Sphingomyelin is not antigenic, and is found primarily in myelin.

17.2 An infant, born at 28 weeks of gestation, rapidly gave evidence of respiratory distress. Lab and x-ray results supported the diagnosis of infant respiratory distress syndrome (RDS). Which of the following statements about this syndrome is true?

A. It is unrelated to the baby's premature birth.
B. It is a consequence of too few type II pneumocytes.
C. The L/S ratio in the amniotic fluid is likely to be greater than two.
D. The concentration of dipalmitoylphosphatidylcholine in the amniotic fluid would be expected to be lower than that of a full-term baby.
E. RDS is an easily treated disorde,r with low mortality.

Correct answer = D. Dipalmitoylphosphatidylcholine (DPPC, or dipalmitoyllecithin) is the lung surfactant found in mature, healthy lungs. RDS can occur in lungs that make too little of this compound. If the lecithin/sphingomyelin ratio in amniotic is greater than two, a newborn's lungs are considered sufficiently mature—premature lungs would be expected to have a ratio lower than two. The RDS would not be due to too few type II pneumocytes—these cells would simply be secreting sphingomyelin rather than DPPC.

17.3 A 25-year-old woman with a history that included hepatosplenomegaly with eventual removal of the spleen, bone and joint pain with several fractures of the femur, and a liver biopsy that showed wrinkled-looking cells with accumulations of glucosylceramides was presented at Grand Rounds. The likely diagnosis for this patient is:

A. Fabry disease.
B. Farber disease.
C. Gaucher disease.
D. Krabbe disease.
E. Niemann-Pick disease.

Correct answer = C. The adult form of Gaucher disease causes hepatosplenomegaly, osteoporosis of the long bones, and the characteristicly wrinkled appearance of the cytosol of cells. This is also the sphingolipidosis in which glucosylceramides accumulate. The deficient enzyme is β-glucosidase.

Cholesterol and Steroid Metabolism

18

I. OVERVIEW

Cholesterol, the characteristic steroid alcohol of animal tissues, performs a number of essential functions in the body. For example, cholesterol is a structural component of all cell membranes, modulating their fluidity, and, in specialized tissues, cholesterol is a precursor of bile acids, steroid hormones, and vitamin D. It is therefore of critical importance that the cells of the body be assured a continuous supply of cholesterol. To meet this need, a complex series of transport, biosynthetic, and regulatory mechanisms has evolved. The liver plays a central role in the regulation of the body's cholesterol homeostasis. For example, cholesterol enters the liver's cholesterol pool from a number of sources including dietary cholesterol, as well as cholesterol synthesized de novo by extrahepatic tissues as well as by the liver itself. Cholesterol is eliminated from the liver as unmodified cholesterol in the bile, or it can be converted to bile salts that are secreted into the intestinal lumen. It can also serve as a component of plasma lipoproteins sent to the peripheral tissues. In humans, the balance between cholesterol influx and efflux is not precise, resulting in a gradual deposition of cholesterol in the tissues, particularly in the endothelial linings of blood vessels. This is a potentially life-threatening occurrence when the lipid deposition leads to plaque formation, causing the narrowing of blood vessels (**atherosclerosis**) and increased risk of **coronary artery disease (CAD)**. Figure 18.1 summarizes the major sources of liver cholesterol and the routes by which cholesterol leaves the liver.

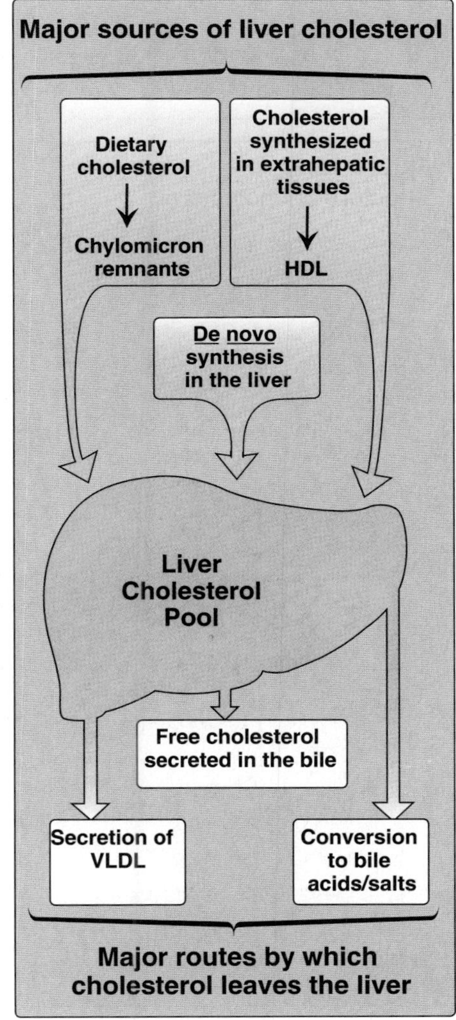

Figure 18.1
Sources of liver cholesterol (influx) and routes by which cholesterol leaves the liver (efflux).

II. STRUCTURE OF CHOLESTEROL

Cholesterol is a very **hydrophobic** compound. It consists of four fused hydrocarbon rings (A, B, C, and D, called the "steroid nucleus"), and it has an eight-carbon, branched hydrocarbon chain attached to C-17 of the D ring. Ring A has a hydroxyl group at C-3, and ring B has a double bond between C-5 and C-6 (Figure 18.2).

A. Sterols

Steroids with eight to ten carbon atoms in the side chain at C-17 and a hydroxyl group at C-3 are classified as sterols. Cholesterol is

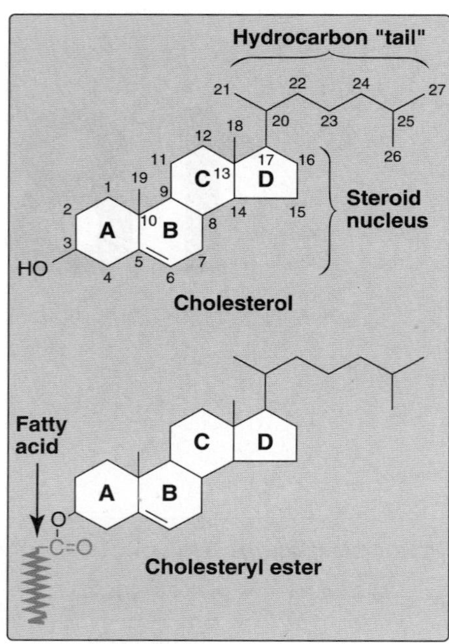

Figure 18.2
Structure of cholesterol.

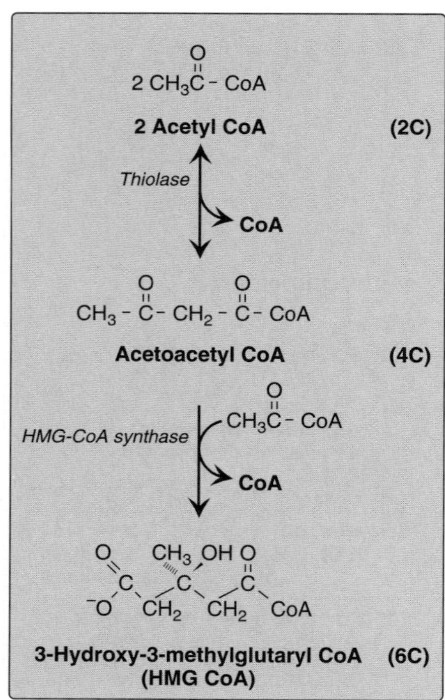

Figure 18.3
Synthesis of 3-hydroxy-3-methylglutaryl CoA (HMG CoA).

the major sterol in animal tissues. [Note: Plant sterols, such as **β–sitosterol** are poorly absorbed by humans. After entering the enterocytes, they are actively transported back into the intestinal lumen. Because some cholesterol is transported as well, plant sterols appear to block the absorption of dietary cholesterol. This has led to clinically useful dietary treatment for hypercholesteremia. Daily ingestion of plant steroid esters (in the form of commercially available trans fatty acid-free margarine) is one of a number of dietary strategies leading to the reduction of plasma cholesterol levels (see 362).]

B. Cholesteryl esters (CE)

Most **plasma cholesterol** is in an esterified form (with a fatty acid attached at C-3, see Figure 18.2), which makes the structure even more hydrophobic than free cholesterol. Cholesteryl esters are not found in membranes, and are normally present only in low levels in most cells. Because of their hydrophobicity, cholesterol and its esters must be transported in association with protein as a component of a lipoprotein particle (see p. 225) or be solubilized by phospholipids and bile salts in the bile (see p. 223).

III. SYNTHESIS OF CHOLESTEROL

Cholesterol is synthesized by virtually all tissues in humans, although **liver**, **intestine**, **adrenal cortex**, and **reproductive tissues**, including ovaries, testes, and placenta, make the largest contributions to the body's cholesterol pool. As with fatty acids, all the carbon atoms in cholesterol are provided by **acetate**, and **NADPH** provides the reducing equivalents. The pathway is driven by hydrolysis of the high-energy thioester bond of acetyl CoA and the terminal phosphate bond of ATP. Synthesis occurs in the cytoplasm, with enzymes in both the **cytosol** and the membrane of the **endoplasmic reticulum**. The pathway is responsive to changes in cholesterol concentration, and regulatory mechanisms exist to balance the rate of cholesterol synthesis within the body against the rate of cholesterol excretion. An imbalance in this regulation can lead to an elevation in circulating levels of plasma cholesterol, with the potential for CAD.

A. Synthesis of 3-hydroxy-3-methylglutaryl CoA (HMG CoA)

The first two reactions in the cholesterol synthetic pathway are similar to those in the pathway that produces ketone bodies (see Figure 16.22, p. 194). They result in the production of 3-hydroxy-3-methylglutaryl CoA (HMG CoA, Figure 18.3). First, two acetyl CoA molecules condense to form acetoacetyl CoA. Next, a third molecule of acetyl CoA is added, producing HMG CoA, a six-carbon compound. [Note: Liver parenchymal cells contain two isoenzymes of *HMG CoA synthase*. The **cytosolic enzyme** participates in **cholesterol synthesis**, whereas the **mitochondrial enzyme** functions in the pathway for **ketone body synthesis**.]

B. Synthesis of mevalonic acid (mevalonate)

The next step, the reduction of HMG CoA to mevalonic acid, is catalyzed by *HMG CoA reductase*, and is the rate-limiting step in cholesterol synthesis. It occurs in the cytosol, uses two molecules of NADPH as the reducing agent, and releases CoA, making the reaction irreversible (Figure 18.4). [Note: *HMG CoA reductase* is an intrinsic membrane protein of the endoplasmic reticulum, with the enzyme's catalytic domain projecting into the cytosol. Regulation of *HMG CoA reductase* activity is discussed below.]

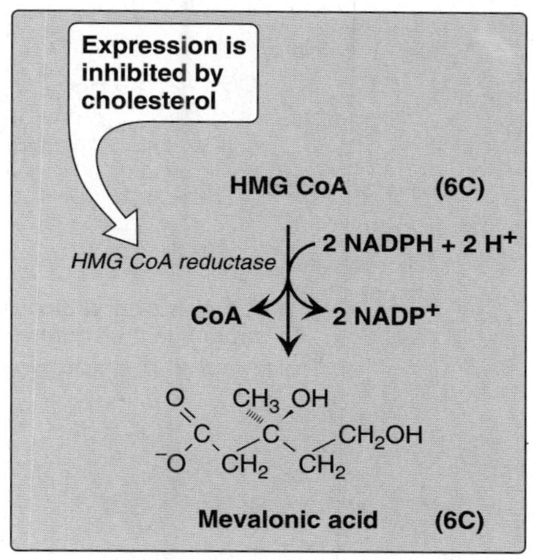

Figure 18.4
Synthesis of mevalonic acid.

C. Synthesis of cholesterol

The reactions and enzymes involved in the synthesis of cholesterol from mevalonate are illustrated in Figure 18.5. [Note: The numbers shown in brackets below correspond to numbered reactions shown in this figure.]

[1] Mevalonic acid is converted to **5-pyrophosphomevalonate** in two steps, each of which transfers a phosphate group from ATP.

[2] A five-carbon isoprene unit—**isopentenyl pyrophosphate** (**IPP**)—is formed by the decarboxylation of 5-pyrophosphomevalonate. The reaction requires ATP. [Note: IPP is the precursor of a family of molecules with diverse functions, the **isoprenoids**. Cholesterol is a sterol isoprenoid. Non-sterol isoprenoids include dolichol (see p. 165) and ubiquinone (see p. 75).]

[3] IPP is isomerized to **3,3-dimethylallyl pyrophosphate** (**DPP**).

[4] IPP and DPP condense to form ten-carbon **geranyl pyrophosphate** (**GPP**).

[5] A second molecule of IPP then condenses with GPP to form 15-carbon **farnesyl pyrophosphate** (**FPP**). [Note: Covalent attachment of farnesyl to proteins, a process known as "**prenylation**," is one mechanism for anchoring proteins to plasma membranes.

[6] Two molecules of farnesyl pyrophosphate combine, releasing pyrophosphate, and are reduced, forming the 30-carbon compound **squalene**. [Note: Squalene is formed from six isoprenoid units. Because three ATP are hydrolysed per mevalonic acid residue converted to IPP, a total of eighteen ATP are required to make the polyisoprenoid squalene.]

[7] Squalene is converted to the sterol **lanosterol** by a sequence of reactions that use molecular oxygen and NADPH. The hydroxylation of squalene triggers the cyclization of the structure to lanosterol.

[8] The conversion of lanosterol to cholesterol is a multistep process, resulting in the shortening of the carbon chain from 30 to 27, removal of the two methyl groups at C-4, migration of the double bond from C-8 to C-5, and reduction of the double bond between C-24 and C-25. [Note: This pathway has been proposed to include more than 18 different enzymatic reactions, but it has not yet been completely solved. **Smith-Lemli-Opitz syndrome** (**SLOS**), a relatively common autosomal recessive disorder of cholesterol biosynthesis, is caused by a partial deficiency in *7-dehydroholesterol-7-reductase*, an enzyme involved in the migration of the double bond. SLOS is characterized by multisystem anomalies, reflecting the importance of cholesterol in embryonic development.

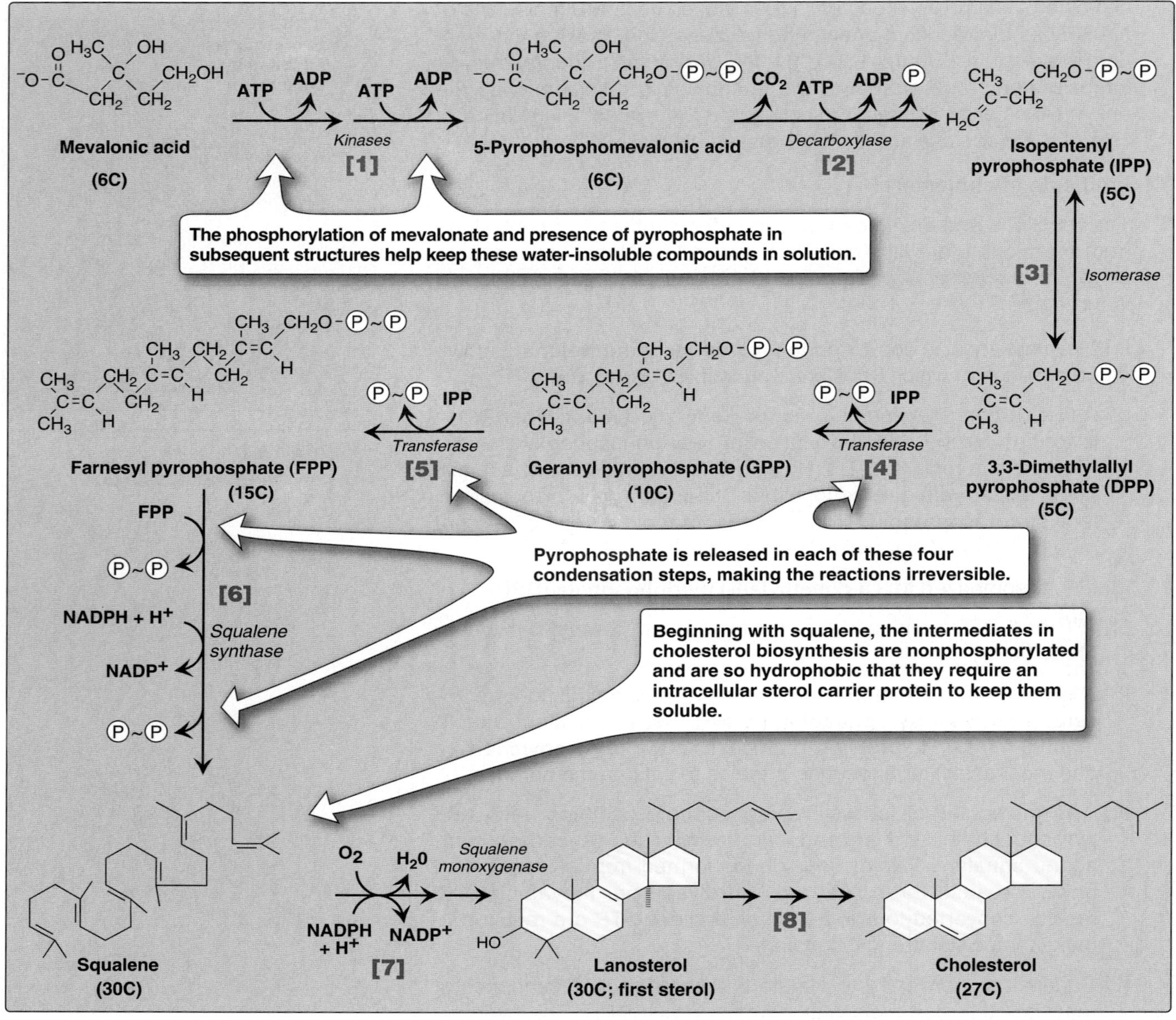

Figure 18.5
Synthesis of cholesterol from mevalonic acid.

D. Regulation of cholesterol synthesis

HMG CoA reductase, the rate-limiting enzyme, is the major control point for cholesterol biosynthesis, and is subject to different kinds of metabolic control.

1. **Sterol-dependent regulation of gene expression:** Expression of the *HMG CoA reductase* gene is controlled by a transcription factor (**sterol regulatory element-binding protein**, or **SREBP**) that binds to DNA at the **sterol regulatory element** (**SRE**) located upstream of the *reductase* gene. The SREBP is initially associated with the ER membrane, but proteolytic cleavage liberates the active form, which travels to the nucleus. When the SREBP binds,

expression of the *reductase* gene increases. When cholesterol levels are low, activation of SREBP occurs, resulting in increased *HMG CoA reductase* and, therefore, more cholesterol synthesis. Conversely, high levels of cholesterol prevent activation of the transcription factor. (This regulatory mechanism is summarized in Figure 18.6.) Cholesterol content also affects the stability of the *HMG CoA reductase* protein and its mRNA, with increased cholesterol leading to decreased stability (and therefore increased degradation) of both.

2. **Sterol-independent phosphorylation/dephosphorylation:** *HMG CoA reductase* activity is controlled covalently through the actions of a *protein kinase* and a *phosphoprotein phosphatase* (see Figure 18.6). The phosphorylated form of the enzyme is inactive, whereas the dephosphorylated form is active. [Note: *Protein kinase* is activated by AMP, so cholesterol synthesis is decreased when ATP availability is decreased.]

3. **Hormonal regulation:** The amount (and, therefore, the activity) of *HMG CoA reductase* is controlled hormonally. An increase in insulin favors upregulation of the expression of the *HMG CoA reductase* gene. Glucagon has the opposite effect.

4. **Inhibition by drugs:** The statin drugs, including **simvastatin**, **lovastatin**, and **mevastatin**, are structural analogs of HMG CoA, and are reversible, competitive inhibitors of *HMG CoA reductase* (Figure 18.7). They are used to decrease plasma cholesterol levels in patients with **hypercholesterolemia**.[1]

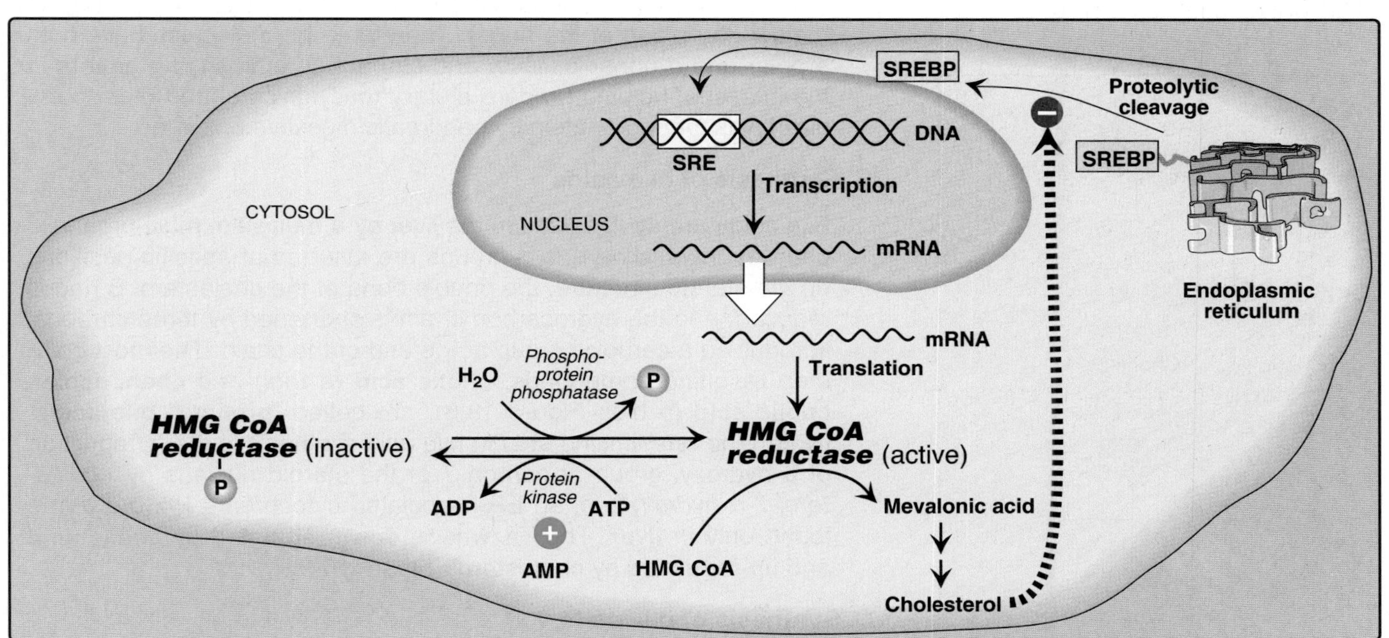

Figure 18.6
Regulation of *HMG CoA reductase*. SRE = sterol regulatory element; SREBP = sterol regulatory element-binding protein.

[1]See Ch. 21 in ***Lippincott's Illustrated Reviews: Pharmacology*** (2nd and 3rd Eds.) for a discussion of antihyperlipidemic drugs.

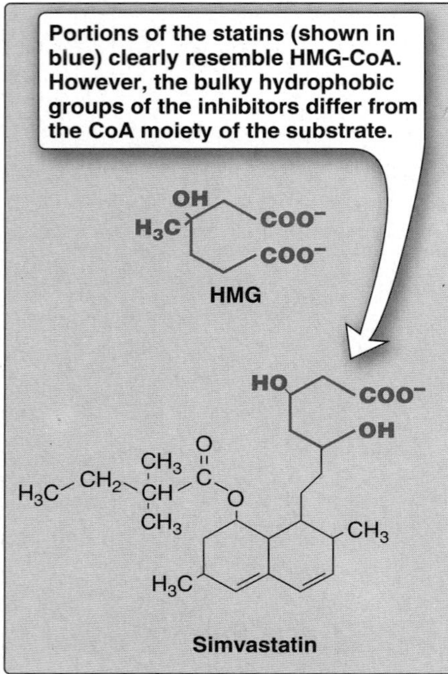

Figure 18.7
Structural similarity of HMG and simvastatin, a clinically useful cholesterol-lowering drug of the "statin" family.

IV. DEGRADATION OF CHOLESTEROL

The ring structure of cholesterol cannot be metabolized to CO_2 and H_2O in humans. Rather, the intact sterol nucleus is eliminated from the body by conversion to **bile acids** and **bile salts**, which are excreted in the feces, and by secretion of cholesterol into the bile, which transports it to the intestine for elimination. Some of the cholesterol in the intestine is modified by **bacteria** before excretion. The primary compounds made are the isomers **coprostanol** and **cholestanol**, which are reduced derivatives of cholesterol. Together with cholesterol, these compounds make up the bulk of **neutral fecal sterols**.

V. BILE ACIDS AND BILE SALTS

Bile consists of a watery mixture of organic and inorganic compounds. **Phosphatidylcholine** (**lecithin**, see p. 201) and **bile salts** (conjugated bile acids) are quantitatively the most important organic components of bile. Bile can either pass directly from the liver where it is synthesized into the duodenum through the common bile duct, or be stored in the gallbladder when not immediately needed for digestion.

A. Structure of the bile acids

The bile acids contain 24 carbons, with two or three hydroxyl groups and a side chain that terminates in a carboxyl group. The carboxyl group has a pK_a of about 6 and, therefore, is not fully ionized at physiologic pH—hence, the term "bile acid." The bile acids are **amphipathic** in that the hydroxyl groups are α in orientation (they lie "above" the plane of the rings) and the methyl groups are β (they lie "below" the plane of the rings). Therefore, the molecules have both a polar and a nonpolar face, and can act as **emulsifying agents** in the intestine, helping prepare dietary triacylglycerol and other complex lipids for degradation by pancreatic digestive enzymes.

B. Synthesis of bile acids

Bile acids are synthesized in the liver by a multistep, multi-organelle pathway in which hydroxyl groups are inserted at specific positions on the steroid structure, the double bond of the cholesterol B ring is reduced, and the hydrocarbon chain is shortened by three carbons, introducing a carboxyl group at the end of the chain. The most common resulting compounds, **cholic acid** (a triol) and **chenodeoxycholic acid** (a diol, Figure 18.8), are called **"primary" bile acids**. [Note: The rate-limiting step in bile acid synthesis is the introduction of a hydroxyl group at carbon 7 of the steroid nucleus by *cholesterol-7-α-hydroxylase*, an ER-associated cytochrome P450 enzyme found only in **liver**. The enzyme is **down-regulated** by **cholic acid** and **up-regulated** by **cholesterol** (Figure 18.9).]

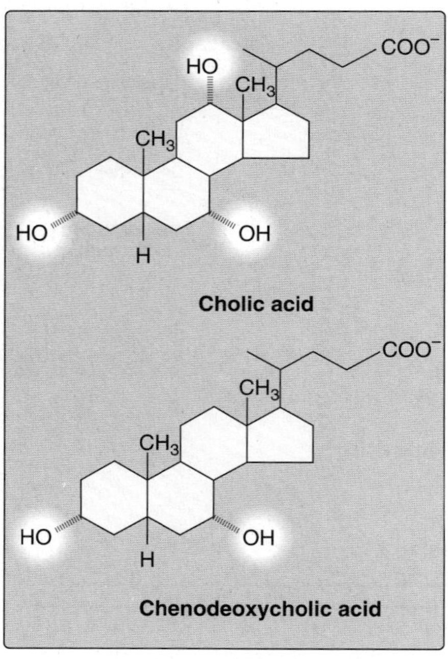

Figure 18.8
Bile acids.

C. Synthesis of bile salts

Before the bile acids leave the liver, they are conjugated to a molecule of either **glycine** or **taurine** (an end-product of cysteine metabolism) by an amide bond between the carboxyl group of the bile acid and the amino group of the added compound. These new structures are called **bile salts** and include **glycocholic** and **glycochenodeoxycholic acids**, and **taurocholic** and **taurochenodeoxycholic acids** (Figure 18.10). The ratio of glycine to taurine forms in the bile is approximately 3:1. Addition of glycine or taurine results in the presence of a carboxyl group with a lower pK_a (from glycine) or a sulfate group (from taurine), both of which are fully ionized (negatively charged) at physiologic pH. Bile salts are more effective detergents than bile acids because of their **enhanced amphipathic nature**. Therefore, only the conjugated forms—that is, the bile salts—are found in the bile. [Note: Bile salts provide the only significant mechanism for cholesterol excretion, both as a metabolic product of cholesterol and as an essential solubilizer for cholesterol excretion in bile. Individuals with genetic deficiencies in the conversion of cholesterol to bile acids are treated with exogenously supplied chenodeoxycholic acid.]

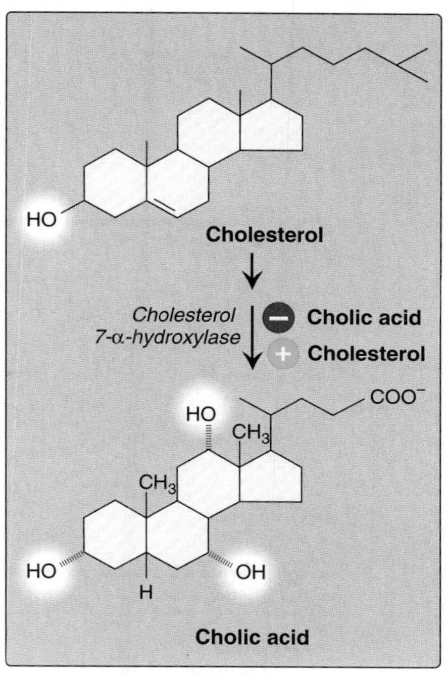

Figure 18.9
Synthesis of cholic acid.

D. Action of intestinal flora on bile salts

Bacteria in the intestine can remove glycine and taurine from bile salts, regenerating bile acids. They can also convert some of the primary bile acids into "**secondary**" **bile acids** by removing a hydroxyl group, producing **deoxycholic acid** from cholic acid and **lithocholic acid** from chenodeoxycholic acid (Figure 18.11).

E. Enterohepatic circulation

Bile salts secreted into the intestine are efficiently reabsorbed (greater than 95 percent) and reused. The mixture of primary and secondary bile acids and bile salts is absorbed primarily in the ileum. They are actively transported from the intestinal mucosal cells into the portal blood, and are efficiently removed by the liver parenchymal cells. [Note: Bile acids are hydrophobic and require a carrier in the portal blood. Albumin carries them in a noncovalent complex, just as it transports fatty acids in blood (see p. 179).] The liver converts both primary and secondary bile acids into bile salts by conjugation with glycine or taurine, and secretes them into the bile. The continuous process of secretion of bile salts into the bile, their passage through the duodenum where some are converted to bile acids, and their subsequent return to the liver as a mixture of bile acids and salts is termed the **enterohepatic circulation** (see Figure 18.11). Between 15 and 30 g of bile salts are secreted from the liver into the duodenum each day, yet only about 0.5 g is lost daily in the feces. Approximately 0.5 g per day is synthesized from cholesterol in the liver to replace the lost bile acids. Bile acid sequestrants, such as **cholestyramine**,[2] bind bile acids in the gut, prevent their reabsorption, and so promote their excretion. They are used in the treatment of hypercholesterolemia because the removal of bile acids relieves the inhibition on bile acid synthesis in the liver, thereby diverting additional cholesterol into that pathway. [Note: **Dietary fiber** also binds bile acids and increases their excretion.]

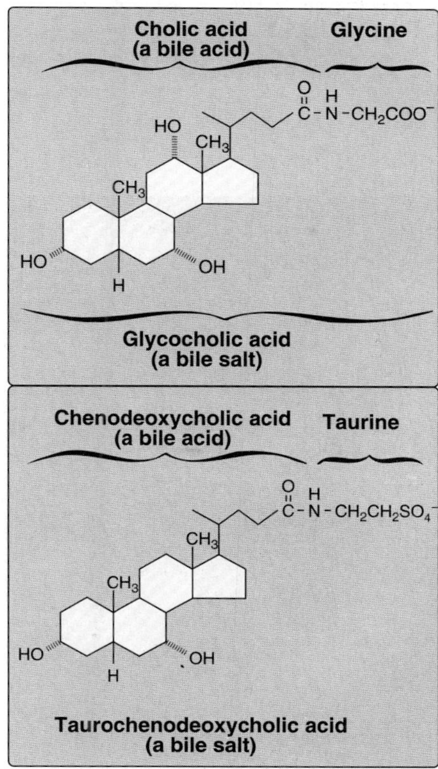

Figure 18.10
Bile salts.

F. Bile salt deficiency: cholelithiasis

The movement of cholesterol from the liver into the bile must be accompanied by the simultaneous secretion of phospholipid and

[2]See Chapter 21 in ***Lippincott's Illustrated Reviews: Pharmacology*** (2nd and 3rd Eds.) for a more detailed discussion of drugs used to treat hyperlipidemia.

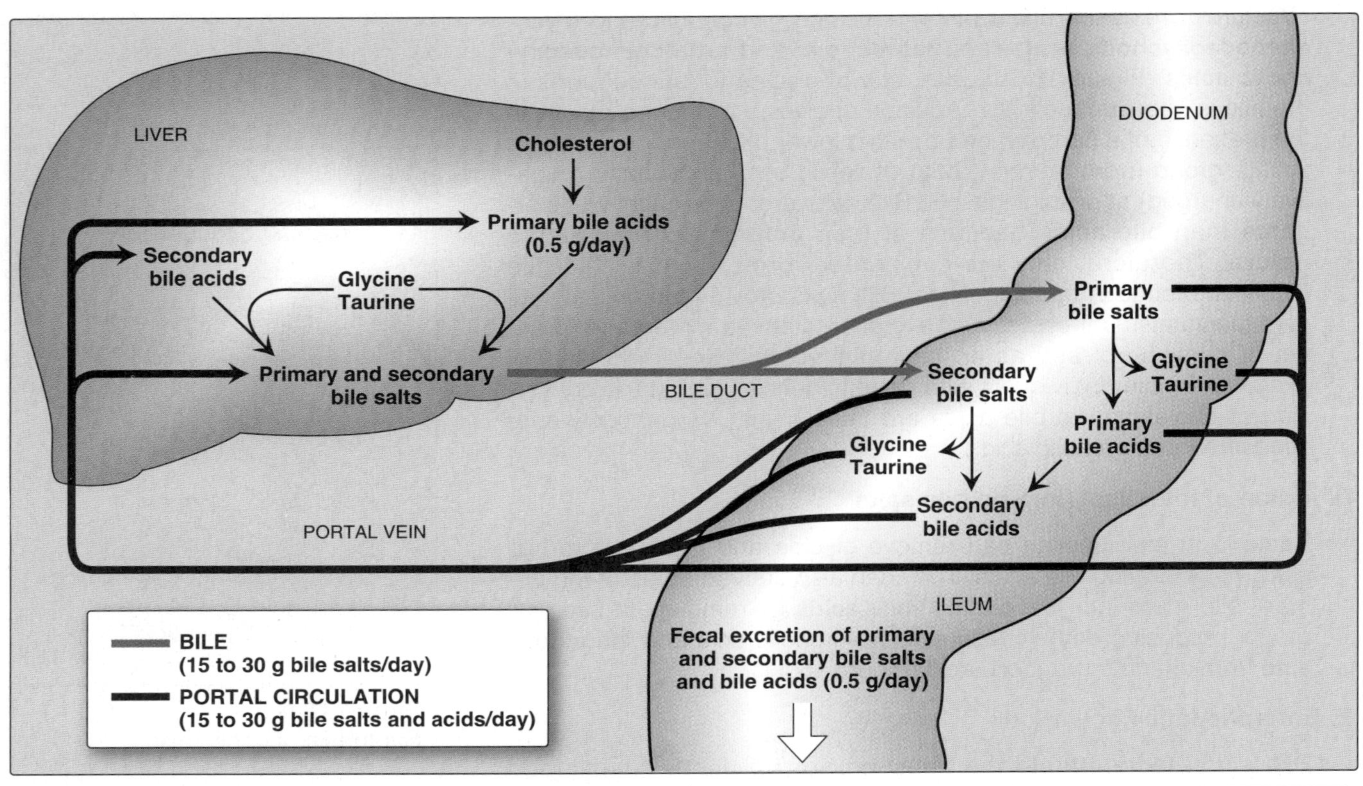

Figure 18.11
Enterohepatic circulation of bile salts and bile acids.

bile salts. If this dual process is disrupted and more cholesterol enters the bile than can be solubilized by the bile salts and lecithin present, the cholesterol may precipitate in the gallbladder, initiating the occurrence of **cholesterol gallstone disease—cholelithiasis** (Figure 18.12). This disorder is typically caused by a decrease of bile acids in the bile, which may result from: 1) gross malabsorption of bile acids from the intestine, as seen in patients with severe ileal disease; 2) obstruction of the biliary tract, interrupting the enterohepatic circulation; 3) severe hepatic dysfunction, leading to decreased synthesis of bile salts, or other abnormalities in bile production; or 4) excessive feedback suppression of bile acid synthesis as a result of an accelerated rate of recycling of bile acids. Cholelithiasis also may result from increased biliary cholesterol excretion, as seen with the use of **fibrates**. [Note: Fibrates, such as gemfibrozil,[3] are derivatives of fibric acid, and are used to reduce triacylglycerol levels in blood.] **Laparoscopic cholecystectomy** (surgical removal of the gallbladder through a small incision) is currently the treatment of choice. However, for patients who are unable to undergo surgery, administration of chenodeoxycholic acid to supplement the body's supply of bile acids results in a gradual (months to years) dissolution of gallstones.

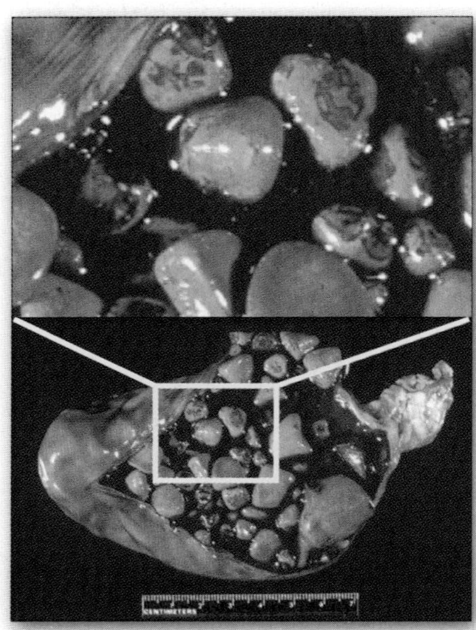

Figure 18.12
Gallbaldder with gallstones.

[3]See Chapter 21 in ***Lippincott's Illustrated Reviews: Pharmacology*** (2nd and 3rd Eds.) for a more detailed discussion of drugs used to treat hyperlipidemia.

VI. PLASMA LIPOPROTEINS

The plasma lipoproteins are spherical macromolecular complexes of lipids and specific proteins (apolipoproteins or apoproteins). The lipoprotein particles include **chylomicrons**, **very-low-density lipoproteins** (**VLDL**), **low-density lipoproteins** (**LDL**), and **high-density lipoproteins** (**HDL**). They differ in lipid and protein composition, size, and density (Figure 18.13). Lipoproteins function both to keep their component lipids soluble as they transport them in the plasma, and also to provide an efficient mechanism for transporting their lipid contents to (and from) the tissues. In humans, the transport system is less perfect than in other animals and, as a result, humans experience a gradual deposition of lipid—especially cholesterol—in tissues. This is a potentially life-threatening occurrence when the lipid deposition contributes to plaque formation, causing the narrowing of blood vessels (**atherosclerosis**).

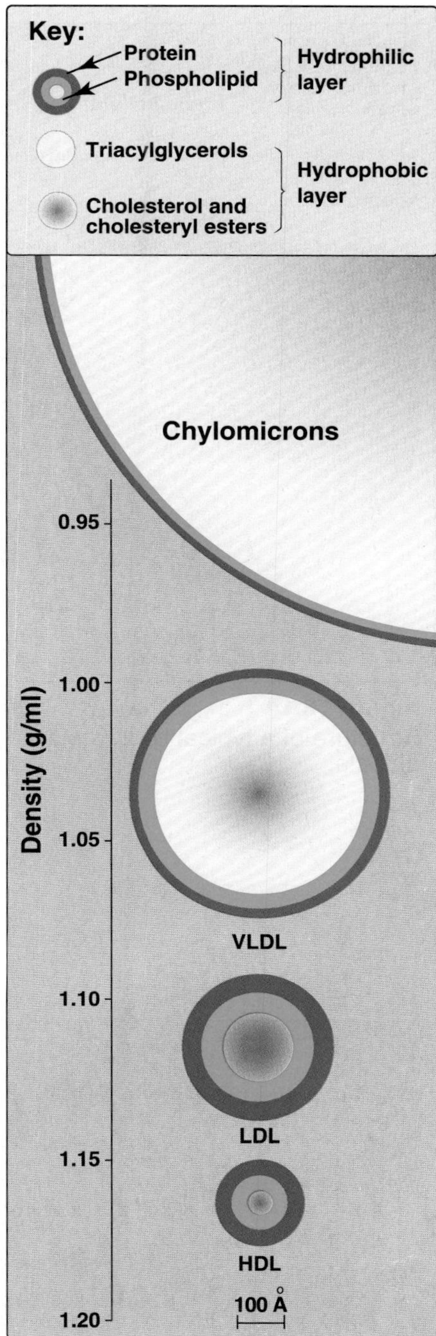

Figure 18.13
Approximate size and density of serum lipoproteins. Each family of lipoproteins exhibits a range of sizes and densities; this figure shows typical values. [Note: The width of the rings approximates the amount of each component.]

A. Composition of plasma lipoproteins

Lipoproteins are composed of a neutral lipid core (containing triacylglycerol, cholesteryl esters) surrounded by a shell of amphipathic apolipoproteins, phospholipid, and nonesterified cholesterol (Figure 18.14). These amphipathic compounds are oriented so that their polar portions are exposed on the surface of the lipoprotein, thus making the particle soluble in aqueous solution. The triacylglycerol and cholesterol carried by the lipoproteins are obtained either from the diet (exogenous source) or de novo synthesis (endogenous source). [Note: Lipoprotein particles constantly interchange lipids and apolipoproteins with each other; therefore, the actual apolipoprotein and lipid content of each class of particles can be somewhat variable.]

1. **Size and density of lipoprotein particles:** Chylomicrons are the lipoprotein particles lowest in density and largest in size, and contain the highest percentage of lipid and the smallest percentage of protein. VLDLs and LDLs are successively denser, having higher ratios of protein to lipid. HDL particles are the densest. Plasma lipoproteins can be separated on the basis of their electrophoretic mobility, as shown in Figure 18.15, or on the basis of their density by ultracentrifugation.

2. **Apolipoproteins:** The apolipoproteins associated with lipoprotein particles have a number of diverse functions, such as providing recognition sites for cell-surface receptors, and serving as activators or coenzymes for enzymes involved in lipoprotein metabolism. Some of the apolipoproteins are required as essential structural components of the particles and cannot be removed (in fact, the particles cannot be produced without them), whereas others are transfered freely between lipoproteins. Apolipoproteins are divided by structure and function into five major classes, A through E, with most classes having subclasses, for example, apo A-I and apo C-II. [Note: Functions of all of the apolipoproteins are not yet known.]

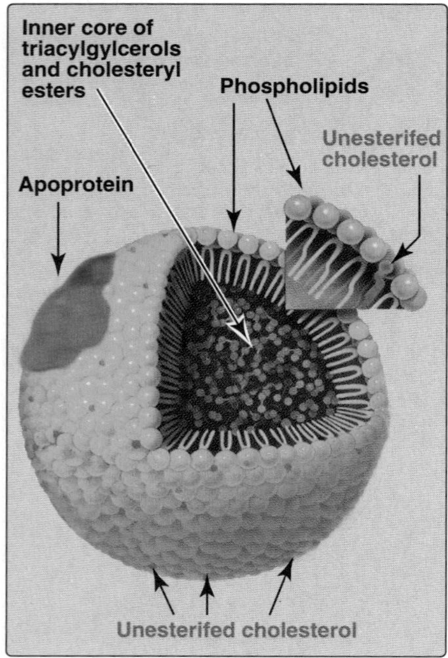

Figure 18.14
Structure of a typical lipoprotein particle.

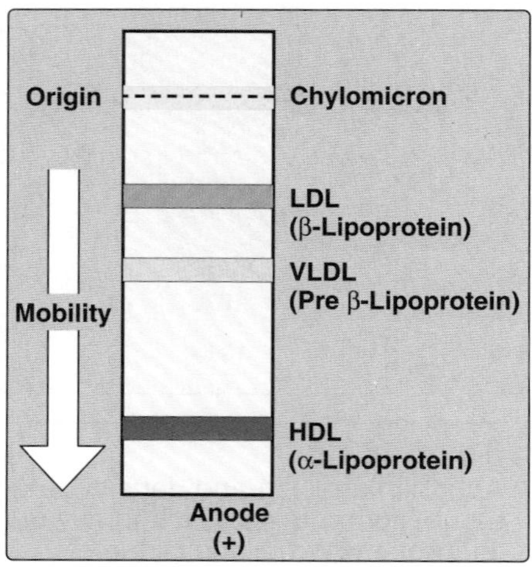

Figure 18.15
Electrophoretic mobility of plasma lipoproteins.

B. Metabolism of chylomicrons

Chylomicrons are assembled in **intestinal mucosal cells** and carry dietary triacylglycerol, cholesterol, fat-soluble vitamins, and cholesteryl esters (plus additional lipids made in these cells) to the peripheral tissues (Figure 18.16).

1. **Synthesis of apolipoproteins:** Apolipoprotein B-48 (apo B-48) synthesis begins on the rough endoplasmic reticulum (RER); it is glycosylated as it moves through the ER and Golgi. [Note: Apo B-48 is unique to chylomicrons. It is so named because it constitutes the N-terminal, 48 percent of the protein coded for by the apo B gene. Apo B-100, which is synthesized by the liver and found in VLDL and LDL, represents the entire protein coded for by the apo B gene. Post-transcriptional editing of a cytosine to a uracil in intestinal apo B-100 mRNA creates a nonsense codon (see p. 431), allowing translation of only 48 percent of the mRNA.].

2. **Assembly of chylomicrons:** The enzymes involved in triacylglycerol, cholesterol, and phospholipid synthesis are located in the smooth ER. Assembly of the apolipoproteins and lipid into chylomicrons requires microsomal **triacylglycerol transfer protein** (see p. 229), which loads apo B-48 with lipid. This occurs during transition from the ER to the Golgi, where the particles are packaged in secretory vesicles. These fuse with the plasma membrane releasing the lipoproteins, which then enter the lymphatic system and, ultimately, the blood.

3. **Modification of nascent chylomicron particles:** The particle released by the intestinal mucosal cell is called a "nascent" chylomicron because it is functionally incomplete. When it reaches the plasma, the particle is rapidly modified, receiving **apo E** (which is recognized by hepatic receptors) and C apolipoproteins. The latter include **apo C-II**, which is necessary for the activation of *lipoprotein lipase*, the enzyme that degrades the triacylglycerol contained in the chylomicron (see below). The source of these apolipoproteins is circulating **HDL** (see Figure 18.16).

4. **Degradation of triacylglycerol by lipoprotein lipase:** *Lipoprotein lipase* is an extracellular enzyme that is anchored by **heparan sulfate** to the capillary walls of most tissues, but predominantly those of **adipose tissue** and **cardiac** and **skeletal muscle**. Adult liver does not have this enzyme. [Note: A *hepatic lipase* is found on the surface of endothelial cells of the liver. However, it does not significantly attack chylomicrons or VLDL triacylglycerol, but rather assists with HDL metabolism (see p. 234).] *Lipoprotein lipase*, activated by **apo C-II** on circulating lipoprotein particles, hydrolyzes the triacylglycerol contained in these particles to yield fatty acids and glycerol. The **fatty acids** are stored (by the **adipose**) or used for energy (by the **muscle**). If they are not immediately taken up by a cell, the long-chain fatty acids are transported by **serum albumin** until their uptake does occur. **Glycerol** is used by **liver**, for example, in lipid synthesis, glycolysis, or gluconeogenesis. [Note: Patients with a deficiency of *lipoprotein lipase* or apo C-II

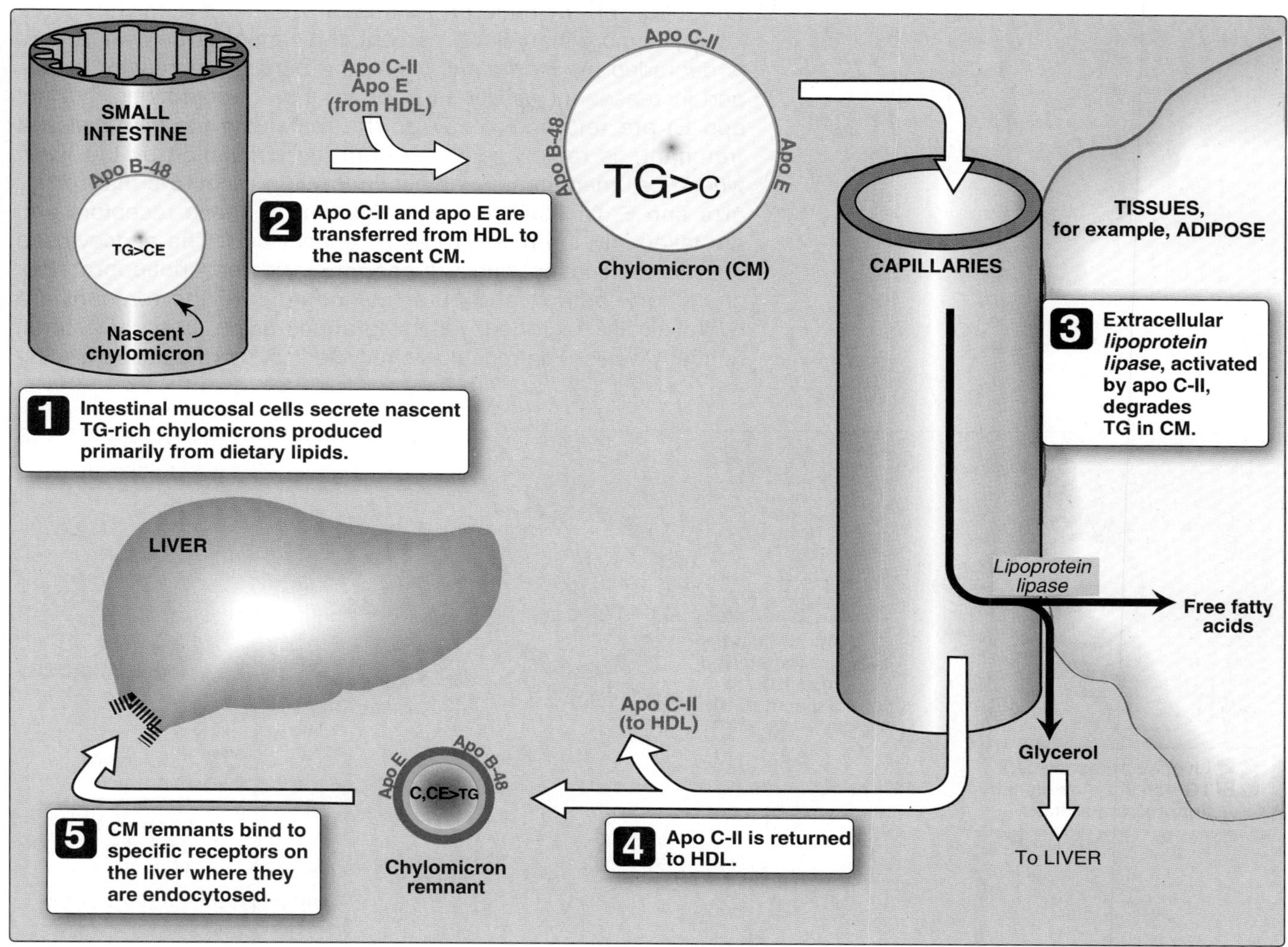

Figure 18.16
Metabolism of chylomicrons. CM = chylomicron; TG = triacylglycerol; C = cholesterol; CE = cholesteryl esters. Apo B-48, apo C-II, and apo E are apolipoproteins found as specific components of plasma lipoproteins. The lipoproteins are not drawn to scale (see Figure 18.13 for details of the size and density of lipoproteins).

(**type 1 hyperlipoproteinemia**, or **familial lipoprotein lipase deficiency**) show a dramatic accumulation of chylomicrons in the plasma (**hypertriacylglycerolemia**).]

5. **Regulation of lipoprotein lipase activity:** *Lipoprotein lipase* synthesis and transfer to the luminal surface of the capillary is stimulated by **insulin** (signifying a fed state, see p. 319). Further, isomers of *lipoprotein lipase* have different K_ms for triacylglycerol (reminiscent of the *hexokinase/glucokinase* story, see p. 96). For example, the adipose enzyme has a large K_m (see p. 59), allowing the removal of fatty acids from circulating lipoprotein particles and their storage as triacylglycerols when plasma lipoprotein concentrations are elevated. Conversely, heart muscle *lipoprotein lipase* has a small K_m, allowing the heart continuing access to the circulating fuel, even when plasma lipoprotein concentrations are low.

6. **Formation of chylomicron remnants:** As the chylomicron circulates and more than ninety percent of the triacylglycerol in its core is degraded by *lipoprotein lipase*, the particle decreases in size and increases in density. In addition, the C apoproteins (but not apo E) are returned to HDLs. The remaining particle, called a "remnant," is rapidly removed from the circulation by the **liver**, whose cell membranes contain **lipoprotein receptors** that recognize **apo E**. Chylomicron remnants bind to these receptors and are taken into the hepatocytes by endocytosis. The endocytosed vesicle then fuses with a **lysosome**, and the apolipoproteins, cholesteryl esters, and other components of the remnant are hydrolytically degraded, releasing amino acids, free cholesterol, and fatty acids. The receptor is recycled. (A more detailed discus-

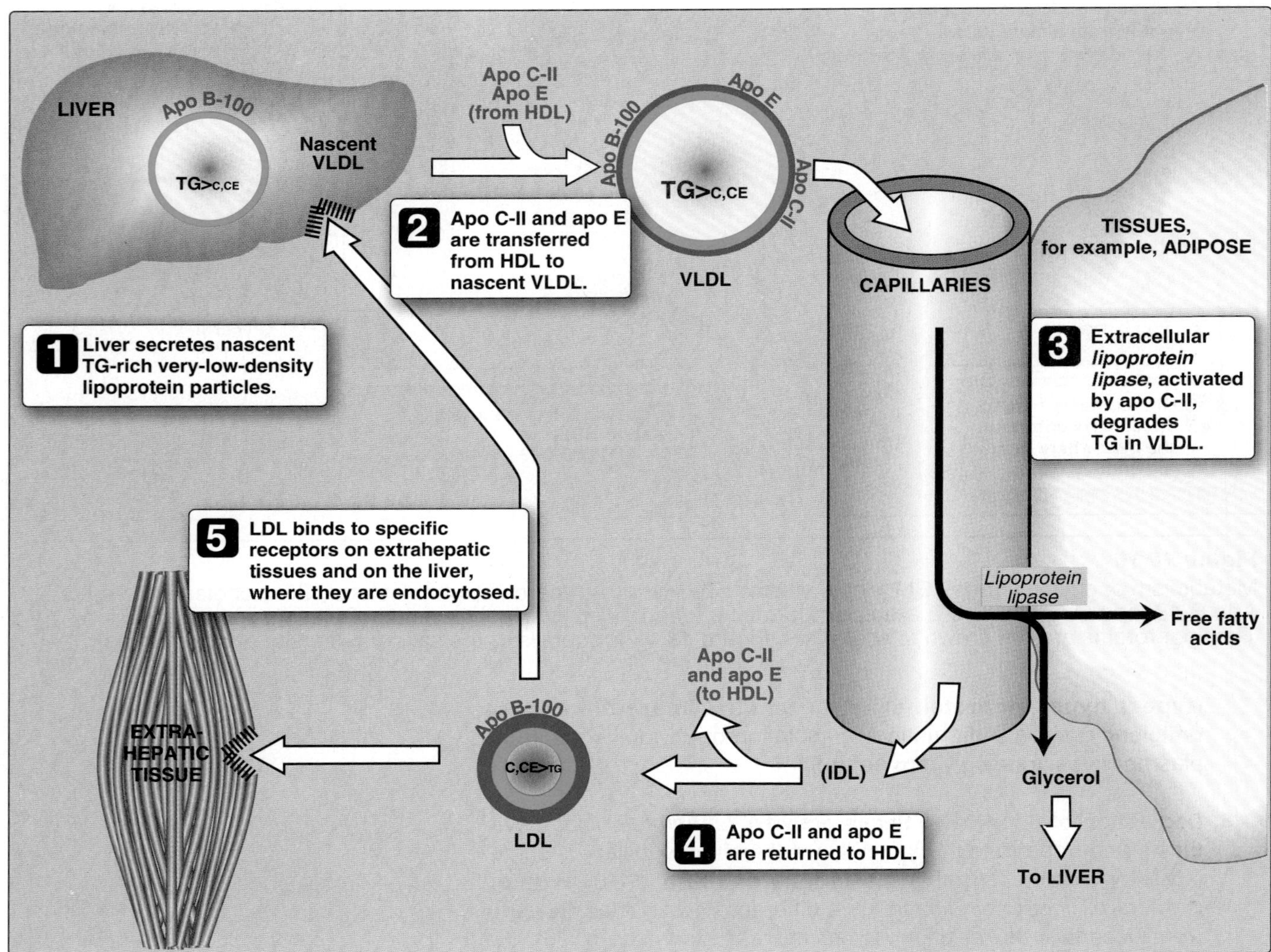

Figure 18.17
Metabolism of VLDL and LDL. TG = triacylglycerol; VLDL = very-low-density lipoprotein; LDL = low-density-lipoprotein; IDL = intermediate-density lipoprotein; C = cholesterol; CE = cholesterol esters. Apo B-100, apo C-II, and apo E are apolipoproteins found as specific components of plasma lipoproteins. Lipoproteins are not drawn to scale (see Figure 18.13 for details of the size and density of lipoproteins).

sion of the mechanism of **receptor-mediated endocytosis** is illustrated for LDL in Figure 18.20.)

C. Metabolism of very low density lipoproteins

VLDLs are produced in the **liver** (Figure 18.17). They are composed predominantly of **triacylglycerol**, and their function is to carry this lipid from the liver to the peripheral tissues. There, the triacylglycerol is degraded by *lipoprotein lipase*, as discussed for chylomicrons (see p. 226). [Note: "**Fatty liver**" **(hepatic steatosis)** occurs in conditions in which there is an imbalance between hepatic triacylglycerol synthesis and the secretion of VLDL. Such conditions include obesity, uncontrolled diabetes mellitus, and chronic ethanol ingestion.]

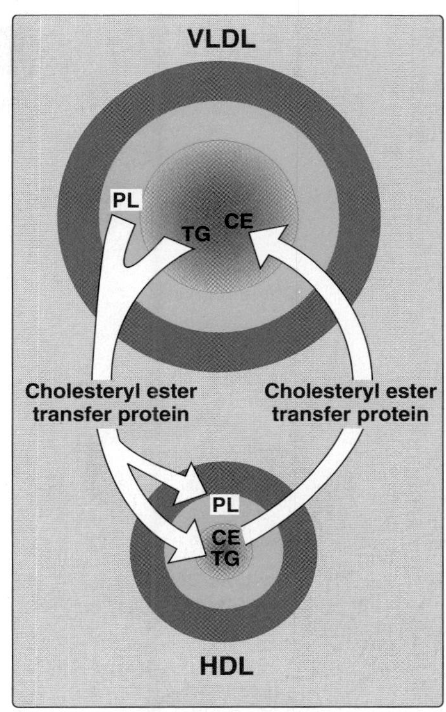

Figure 18.18
Transfer of cholesteryl esters (CE) from HDL to VLDL in exchange for triacylglycerol (TG) or phospholipids (PL).

1. **Release of VLDLs:** VLDLs are secreted directly into the blood by the liver as nascent VLDL particles containing **apolipoprotein B-100**. They must obtain apo C-II and apo E from circulating HDL (see Figure 18.17). As with chylomicrons, apo C-II is required for activation of *lipoprotein lipase*. [Note: **Abetalipoproteinemia** is a rare hypolipoproteinemia caused by a defect in **triacylglycerol transfer protein**, leading to an inability to load apo B with lipid. As a consequence, no chylomicrons or VLDLs are formed, and triacylglycerols accumulate in the liver and intestine.]

2. **Modification of circulating VLDL:** As VLDLs pass through the circulation, triacylglycerol is degraded by *lipoprotein lipase*, causing the VLDL to decrease in size and become denser. Surface components, including the C and E apoproteins, are returned to HDL, but the particles retain apo B-100. Finally, triacylglycerols are transferred from VLDL to HDL in an exchange reaction that concomitantly transfers cholesteryl esters from HDL to VLDL. This exchange is accomplished by **cholesteryl ester transfer protein** (Figure 18.18).

3. **Production of LDL from VLDL in the plasma:** With these modifications, the VLDL is converted in the plasma to LDL. An intermediate-sized particle, the **intermediate-density lipoprotein** (**IDL**) or **VLDL remnant**, is observed during this transition. IDLs can also be taken up by cells through **receptor-mediated endocytosis** that uses **apo E** as the ligand. [Note: Apolipoprotein E is normally present in three isoforms, E2, E3, and E4. Apo E2 binds poorly to receptors, and patients who are homozygotic for apo E2 are deficient in the clearance of chylomicron remants and IDLs. The individuals have **familial type III hyperlipoproteinemia** (**familial dysbetalipoproteinemia**, or **broad beta disease**), with hypercholesterolemia and premature atherosclerosis. Not yet understood is the fact that the E4 isoform confers increased susceptibility to late-onset Alzheimer disease.]

D. Metabolism of low-density lipoproteins

LDL particles contain much less triacylglycerol than their VLDL predecessors, and have a high concentration of cholesterol and cholesteryl esters (Figure 18.19).

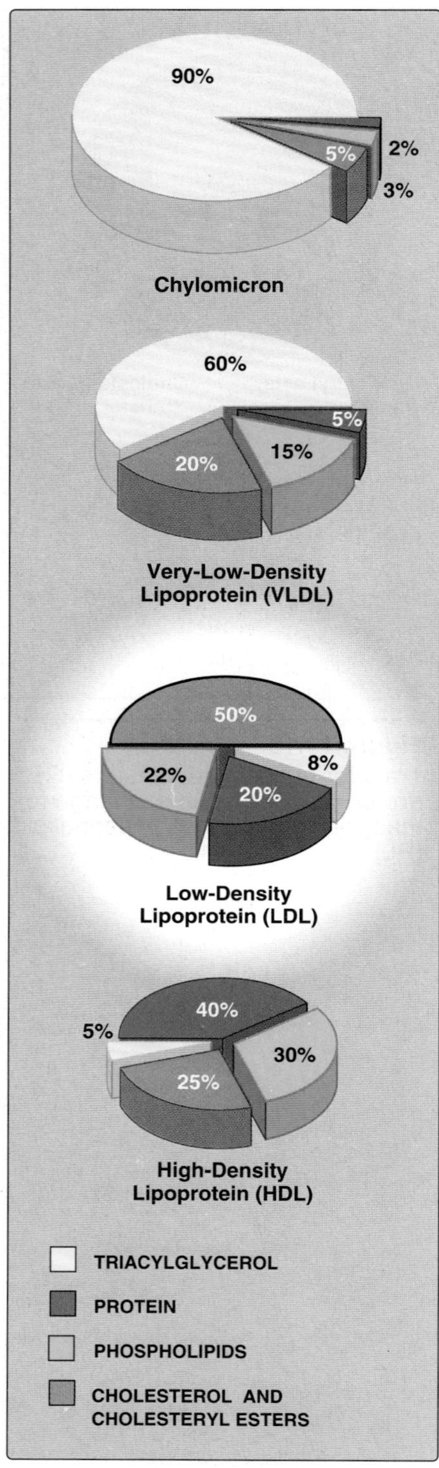

Figure 18.19
Composition of the plasma lipoproteins. Note high concentration of cholesterol and cholesteryl esters in LDL.

1. **Receptor-mediated endocytosis:** The primary function of LDL particles is to provide cholesterol to the peripheral tissues (or return it to the liver). They do so by binding to cell-surface membrane LDL receptors that recognize apolipoprotein B-100 (but not apo B-48). Because these LDL receptors can also bind apo E, they are known as **apo B-100/apo E receptors**. A summary of the uptake and degradation of LDL particles is presented in Figure 18.20. [Note: The numbers in brackets below refer to corresponding numbers on that figure.] A similar mechanism of recepter-mediated endocytosis is used for the cellular uptake and degradation of chylomicron remnants and IDLs by the liver.

[1] LDL receptors are **negatively charged glycoproteins** that are clustered in pits on cell membranes. The intracellular side of the pit is coated with the protein **clathrin,** which stabilizes the shape of the pit.

[2] After binding, the LDL-receptor complex is internalized by **endocytosis**. [Note: A deficiency of functional LDL receptors causes a significant elevation in plasma LDL and, therefore, of plasma cholesterol. Patients with such deficiencies have **type II hyperlipidemia** (**familial hypercholesterolemia**) and premature atherosclerosis. The thyroid hormone, T_3, has a positive effect on the binding of LDL to its receptor. Consequently, **hypothyroidism** is a common cause of hypercholesterolemia.]

[3] The vesicle containing the LDL rapidly loses its clathrin coat and fuses with other similar vesicles, forming larger vesicles called **endosomes**.

[4] The pH of the endosome falls (due to the proton-pumping activity of endosomal *ATPase*), which allows separation of the LDL from its receptor. The receptors then migrate to one side of the endosome, whereas the LDLs stay free within the lumen of the vesicle. [Note: This structure is called CURL—the **Compartment for Uncoupling of Receptor and Ligand**.]

[5] The receptors can be recycled, whereas the lipoprotein remnants in the vesicle are transferred to lysosomes and degraded by lysosomal (hydrolytic) enzymes, releasing free cholesterol, amino acids, fatty acids, and phospholipids. These compounds can be reutilized by the cell. [Note: Rare autosomal recessive deficiencies in the ability to hydrolyze lysosomal cholesteryl esters (**Wolman disease**), or to transport unesterified cholesterol out of the lysosome (**Niemann-Pick disease, type C**) have been identified.]

2. **Effect of endocytosed cholesterol on cellular cholesterol homeostasis:** The chylomicron remnant–, IDL–, and LDL–derived cholesterol affects cellular cholesterol content in several ways (see Figure 18.20). First, *HMG CoA reductase* is inhibited by high cholesterol, as a result of which, de novo cholesterol synthesis decreases. Second, synthesis of new LDL receptor protein is reduced by decreasing the expression of the LDL receptor gene, thus limiting further entry of LDL cholestrol into cells. [Note:

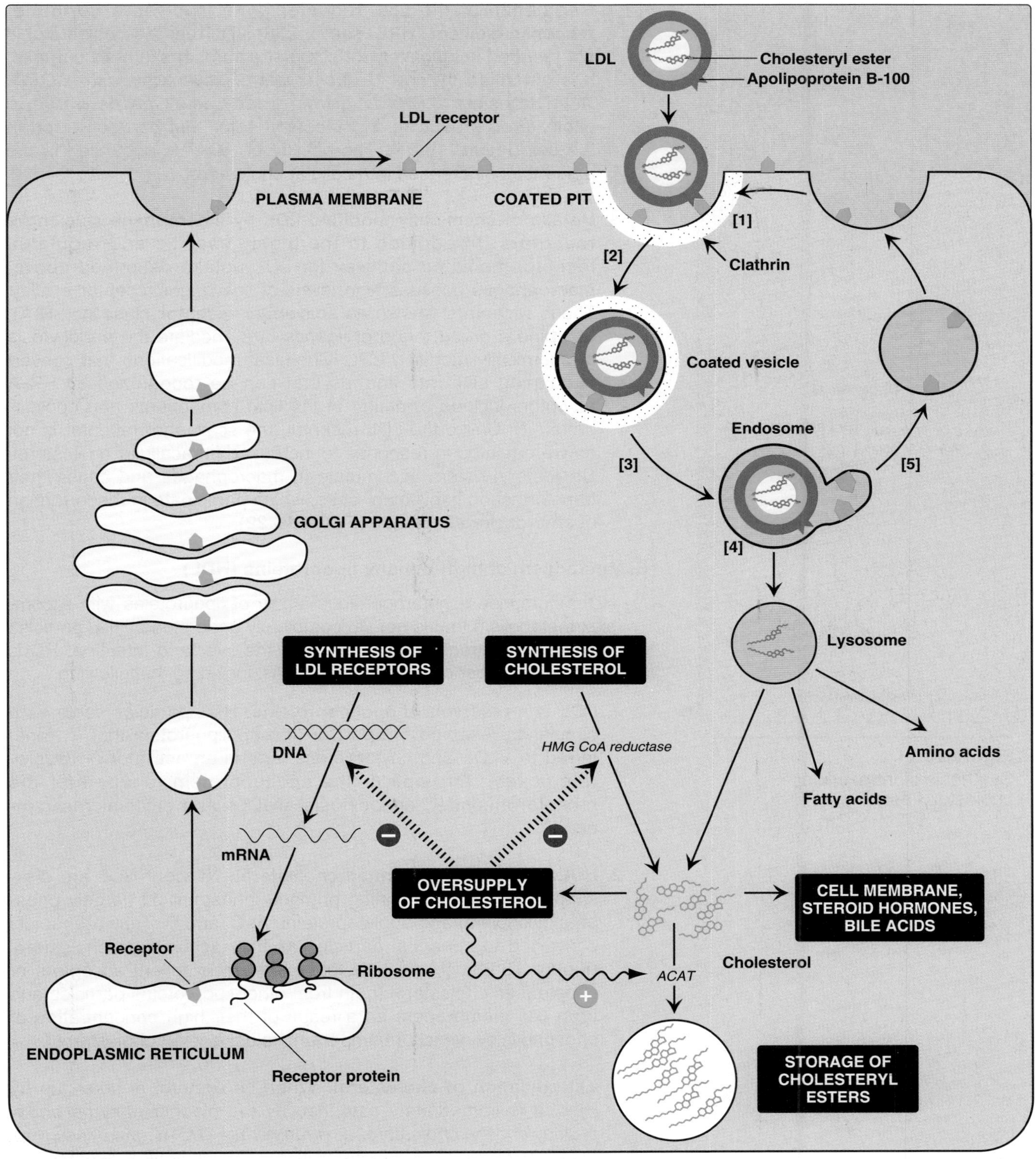

Figure 18.20
Cellular uptake and degradation of LDL. *ACAT = acyl CoA:cholesterol acyltransferase.*

Regulation of the LDL receptor gene involves a **hormone-response element** (**HRE**, see p. 238).] Third, if the cholesterol is not required immediately for some structural or synthetic purpose, it is esterified by *acyl CoA:cholesterol acyltransferase* (*ACAT*). *ACAT* transfers a fatty acid from a fatty acyl CoA derivative to cholesterol, producing a cholesteryl ester that can be stored in the cell (Figure 18.21). The activity of *ACAT* is enhanced in the presence of increased intracellular cholesterol.

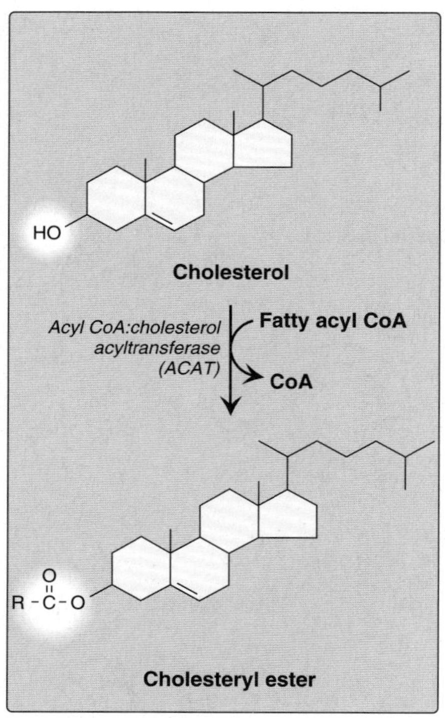

Figure 18.21
Synthesis of intracellular cholesteryl ester.

3. **Uptake of chemically modified LDL by macrophage scavenger receptors:** In addition to the highly specific and regulated receptor-mediated pathway for LDL uptake described above, **macrophages** possess high levels of scavenger receptor activity. These receptors, known as **scavenger receptor class A** (**SR-A**), can bind a broad range of ligands, and mediate the endocytosis of chemically modified LDL. Chemical modifications that convert circulating LDL into ligands that can be recognized by SR-A receptors include oxidation of the lipid components and apolipoprotein B. Unlike the LDL receptor, the scavenger receptor is not down-regulated in response to increased intracellular cholesterol. Cholesteryl esters accumulate in macrophages and cause their transformation into "**foam**" **cells**, which participate in the formation of **atherosclerotic plaque** (Figure 18.22).

E. Metabolism of high-density lipoproteins (HDL)

HDLs comprise a heterogeneous family of lipoproteins with a complex metabolism that is not yet completely understood. HDL particles are secreted directly into blood from the liver and intestine. HDLs perform a number of important functions, including the following:

1. **HDL is a reservoir of apolipoproteins:** HDL particles serve as a circulating reservoir of **apo C-II** (the apolipoprotein that is transferred to VLDL and chylomicrons, and is an activator of *lipoprotein lipase*), and **apo E** (the apolipoprotein required for the receptor-mediated endocytosis of IDLs and chylomicron remnants).

2. **HDL uptake of unesterified cholesterol:** Nascent HDL are disk-shaped particles containing primarily phospholipid (largely phosphatidylcholine) and apolipoproteins A, C, and E. They are rapidly converted to spherical particles as they accumulate cholesterol (Figure 18.23). [Note: HDL particles are excellent acceptors of unesterified cholesterol (both from other lipoproteins particles and from cell membranes) as a result of their high concentration of phospholipids, which are important solubilizers of cholesterol.]

3. **Esterification of cholesterol:** When cholesterol is taken up by HDL, it is immediately esterified by the plasma enzyme *phosphatidylcholine:cholesterol acyltransferase* (*PCAT*, also known as *LCAT*, in which "L" stands for lecithin). This enzyme is synthesized by the **liver**. *PCAT* binds to nascent HDLs, and is activated by apo A-I. *PCAT* transfers the fatty acid from carbon 2 of phosphatidyl-

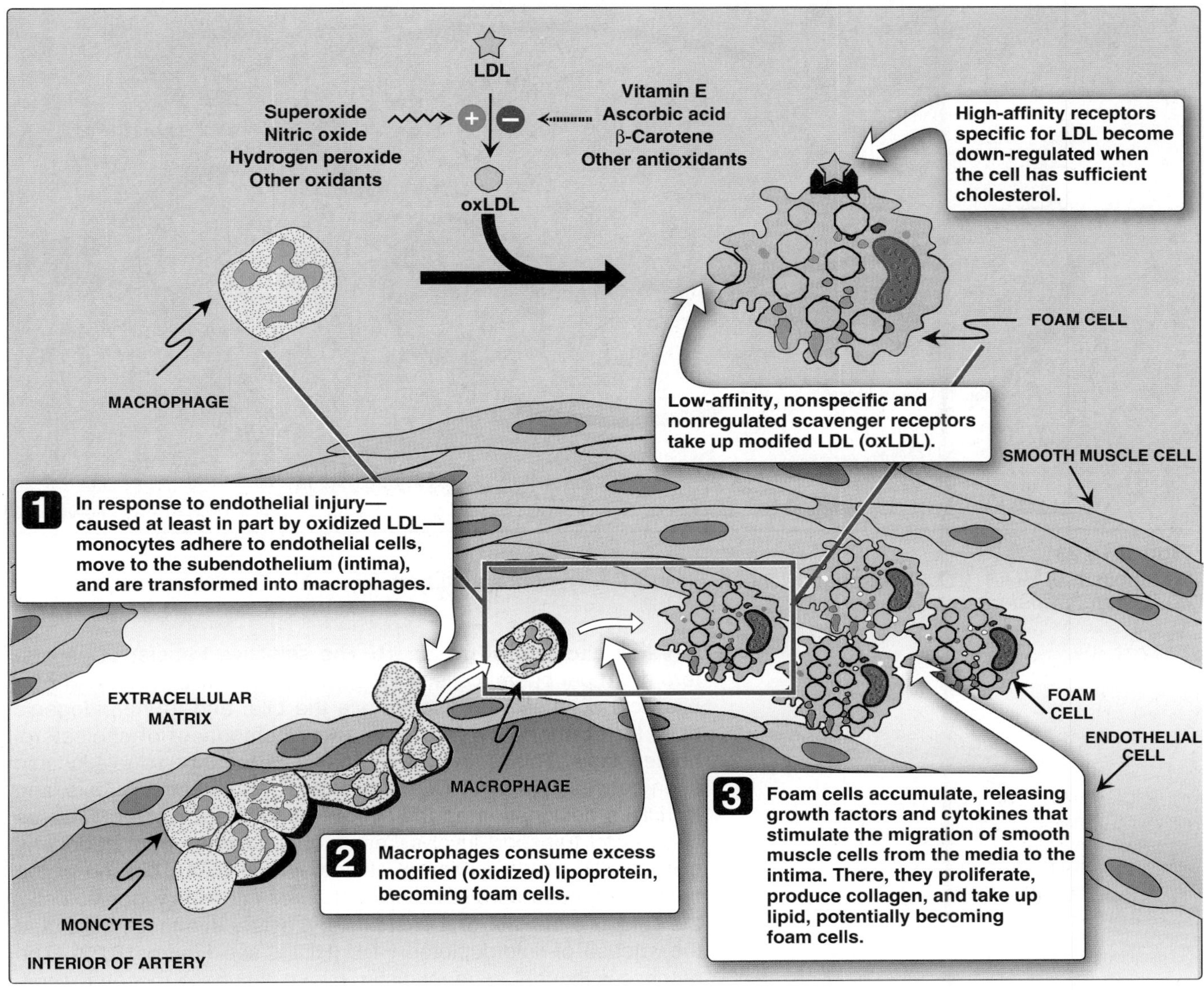

Figure 18.22
Role of oxidized lipoproteins in plaque formation in arterial wall.

choline to cholesterol. This produces a hydrophobic **cholesteryl ester**, which is sequestered in the core of the HDL, and **lysophosphatidylcholine**, which binds to albumin. [Note: Virtually complete (**familial LCAT deficiency**) or partial (**fish eye disease**) absence of *PCAT* results in a marked decrease in HDLs, primarily as a result of the hypercatabolism of lipid-poor HDLs.] As the nascent HDL accumulates cholesteryl esters, it first becomes classified as **HDL_3** and, eventually, becomes a round, micellar-like particle, **HDL_2**. [Note: The **cholesteryl ester transfer protein** (see p. 229) moves some of the cholesteryl esters to VLDLs in exchange for triacylglycerol.]

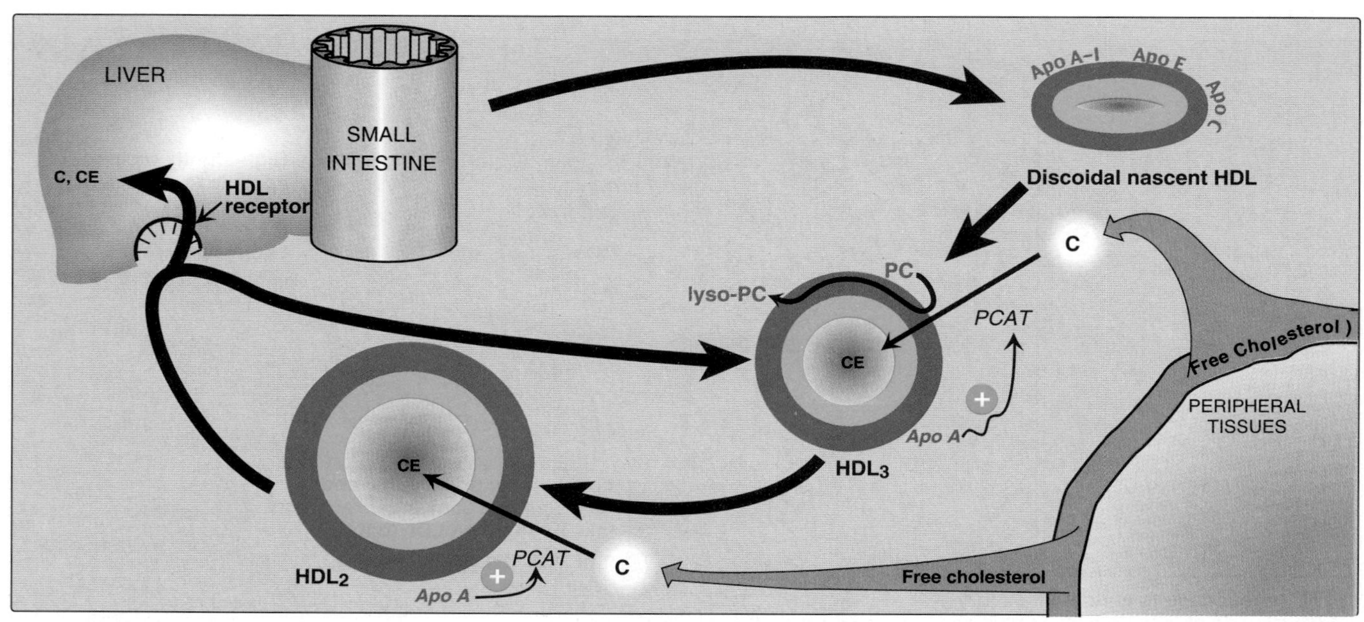

Figure 18.23
Metabolism of HDL. PC = phosphatidylcholine; lyso-PC = lysophosphatidylcholine. *PCAT = Phosphatidylcholine cholesterol transferase.*

4. **Reverse cholesterol transport:** The selective transfer of cholesterol from peripheral cells to HDLs, and from HDLs to the liver for bile acid synthesis or disposal via the bile, and to steroidogenic cells for hormone synthesis, is a key component of cholesterol homeostasis. This is, in part, the basis for the inverse relationship seen between plasma HDL concentration and atherosclerosis, and for HDL's designation as the "**good**" **cholesterol carrier**. Reverse cholesterol transport involves efflux of cholesterol from peripheral cells to HDL, esterification of cholesterol by PCAT, binding of the cholesteryl ester-rich HDL (HDL_2) to liver and steroidogenic cells, the selective transfer of the cholesteryl esters into these cells, and the release of lipid-depleted HDL (HDL_3, see Figure 18.23). The process is thought to be mediated by a cell-surface receptor (**scavenger receptor class B, SR-B1**) that binds HDL. [Note: *Hepatic lipase*, with its ability to degrade both triacylglycerols and phospholipids, participates in the formation of HDL_3.]

F. Role of lipoprotein (a) in heart disease

Lipoprotein (a), or lp(a), is a particle that, when present in large quantities in the plasma, is associated with an increased risk of coronary heart disease. Lipoprotein (a) is nearly identical in structure to an LDL particle. Its distinguishing feature is the presence of an additional apolipoprotein molecule, **apo(a)**, that is covalently linked at a single site to apo B-100. Circulating levels of lp(a) are determined primarily by genetics. However, factors such as diet may play some role, as trans fatty acids have been shown to increase lp(a), and estrogen decreases both LDL and lp(a). [Note: Apo(a) is highly homologous to **plasminogen**—the precursor of a blood protease whose target is fibrin, the main protein component of blood clots. It is hypothesized that elevated lp(a) slows the breakdown of blood clots that trigger heart attacks because it competes with plasminogen for the binding of plasminogen activators.]

VII. STEROID HORMONES

Cholesterol is the precursor of all classes of steroid hormones: **glucocorticoids** (for example, cortisol), **mineralocorticoids** (for example, aldosterone), and **sex hormones**—androgens, estrogens, and progestins (Figure 18.24). [Note: Glucocorticoids and mineralocorticoids are collectively called **corticosteroids**.] Synthesis and secretion occur in the adrenal cortex (cortisol, aldosterone, and androgens), ovaries and placenta (estrogens, progestins), and testes (testosterone). Steroid hormones are transported by the blood from their sites of synthesis to their target organs. Because of their hydrophobicity, they must be complexed with a plasma protein. Plasma albumin can act as a nonspecific carrier, and does carry aldosterone. However, specific steroid-carrier plasma proteins bind the steroid hormones more tightly than does albumin, for example, **corticosteroid-binding globulin** (**transcortin**) is responsible for transporting cortisol, and **sex hormone-binding protein** transports sex steroids. A number of genetic diseases are caused by deficiencies in specific steps in the biosynthesis of steroid hormones. Some representative diseases are described in Figure 18.25.

A. Synthesis of steroid hormones

Synthesis involves shortening the hydrocarbon chain of cholesterol, and hydroxylation of the steroid nucleus. The initial and rate-limiting reaction converts cholesterol to the 21-carbon **pregnenolone**. It is catalyzed by the *cholesterol side chain cleavage enzyme complex* (*desmolase*)—a cytochrome P450 mixed-function oxidase of the inner mitochondrial membrane. NADPH and molecular oxygen are required for the reaction. The cholesterol substrate can be newly synthesized, taken up from lipoproteins, or released from cholesteryl esters stored in the cytosol of steroidogenic tissues. [Note: Steroid hormone synthesis consumes little cholesterol as compared with that required for bile acid synthesis.] Pregnenolone is the parent compound for all steroid hormones (see Figure 18.25). Pregnenolone is oxidized and then isomerized to progesterone, a progestin, which is further modified to the other steroid hormones by hydroxylation reactions that occur in the endoplasmic reticulum and mitochondria. Like *desmolase*, the enzymes are mixed-function oxidases. A defect in the activity or amount of an enzyme in this pathway can lead to a deficiency in the synthesis of hormones beyond the affected step, and to an excess in the hormones or metabolites before that step. Because all members of the pathway have potent biologic activity, serious metabolic imbalances occur if enzyme deficiencies are present (see Figure 18.25).

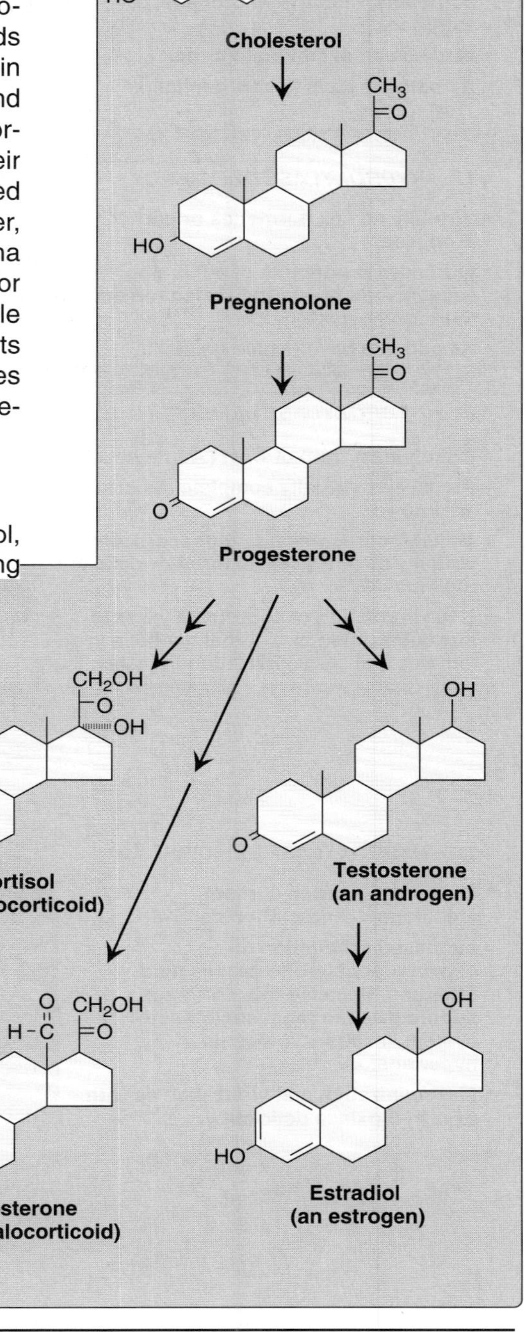

Figure 18.24
Relationships between key steroid hormones.

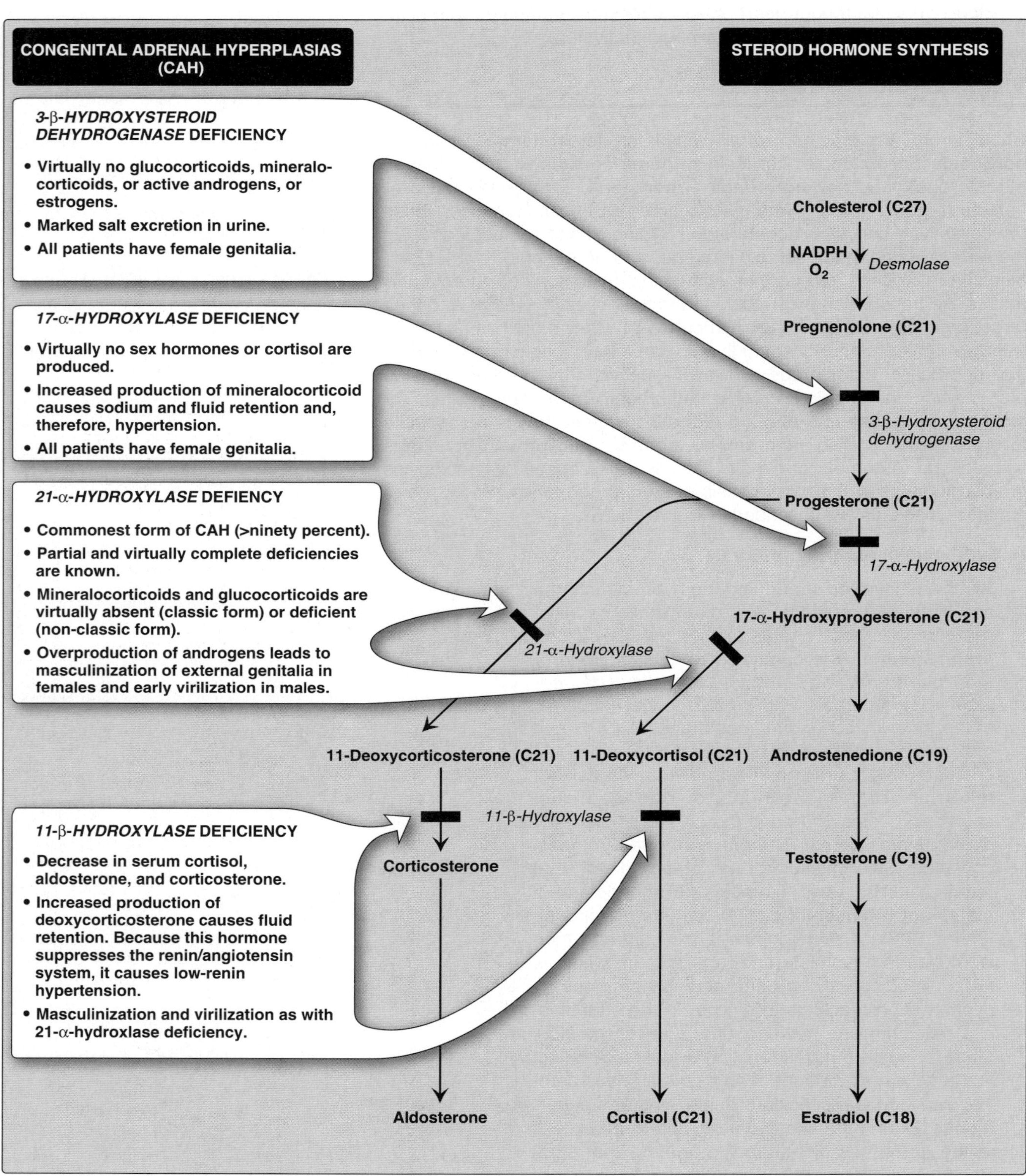

Figure 18.25
Steroid hormone synthesis.

B. Secretion of adrenal cortical steroid hormones

Steroid hormones are secreted on demand from their tissues of origin in response to hormonal signals. The corticosteroids and androgens are made in different regions of the adrenal cortex, and are secreted into blood in response to different signals.

1. **Cortisol:** Its secretion from the **middle layer** of the **adrenal cortex** is controlled by the hypothalamus, to which the pituitary gland is attached (Figure 18.26). In response to severe stress (for example, infection), **corticotropin-releasing hormone** (CRH), produced by the hypothalamus, travels through a network of capillaries to the anterior lobe of the pituitary, where it induces the production and secretion of **adrenocorticotropic hormone** (ACTH, or corticotropin). The polypeptide ACTH, often called the "**stress hormone**," stimulates the adrenal cortex to synthesize and secrete the glucocorticoid cortisol. Cortisol allows the body to respond to stress through its effects on intermediary metabolism and the inflammatory response. As cortisol levels rise, the release of CRH and ACTH is inhibited. [Note: ACTH binds to a plasma membrane receptor. Its intracellular effects are mediated through a second messenger, cAMP (see p. 92).]

2. **Aldosterone:** This hormone's secretion from the **outer layer** of the **adrenal cortex** is induced by a decrease in the plasma Na^+/K^+ ratio, and by the hormone, **angiotensin II**. Angiotensin II is produced from angiotensin I by *angiotensin-converting enzyme* (*ACE*), an enzyme found predominantly in the lungs, but which is also distributed widely in the body. [Note: Angiotensin I, an octapeptide, is produced in the blood by cleavage of an inactive precursor, angiotensinogen, secreted by the liver. Cleavage is accomplished by the enzyme *renin*, made and secreted by the kidney.] Angiotensin II binds to cell-surface receptors. However, in contrast to ACTH, its effects are mediated through the PIP_2 pathway (see p. 203) and not by cAMP. Aldosterone's primary effect is on the kidney tubules, where it stimulates sodium uptake and potassium excretion (Figure 18.27). [Note: An effect of aldosterone is an increase in blood pressure. Competitive inhibitors of ACE are used to treat *renin*-dependent hypertension.[4]]

3. **Androgens:** Both the inner and middle layers of the **adrenal cortex** produce androgens, primarily **dehydroepiandrosterone** and **androstenedione**. Although adrenal androgens themselves are weak, they are converted in peripheral tissues to **testosterone**—a strong androgen—and to **estradiol**.

C. Secretion of steroid hormones from gonads

The **testes** and **ovaries** synthesize hormones necessary for physical development and reproduction. A single hypothalamic-releasing factor, **gonadotropin-releasing hormone**, stimulates the anterior pituitary to release the glycoproteins, **luteinizing hormone** (LH) and

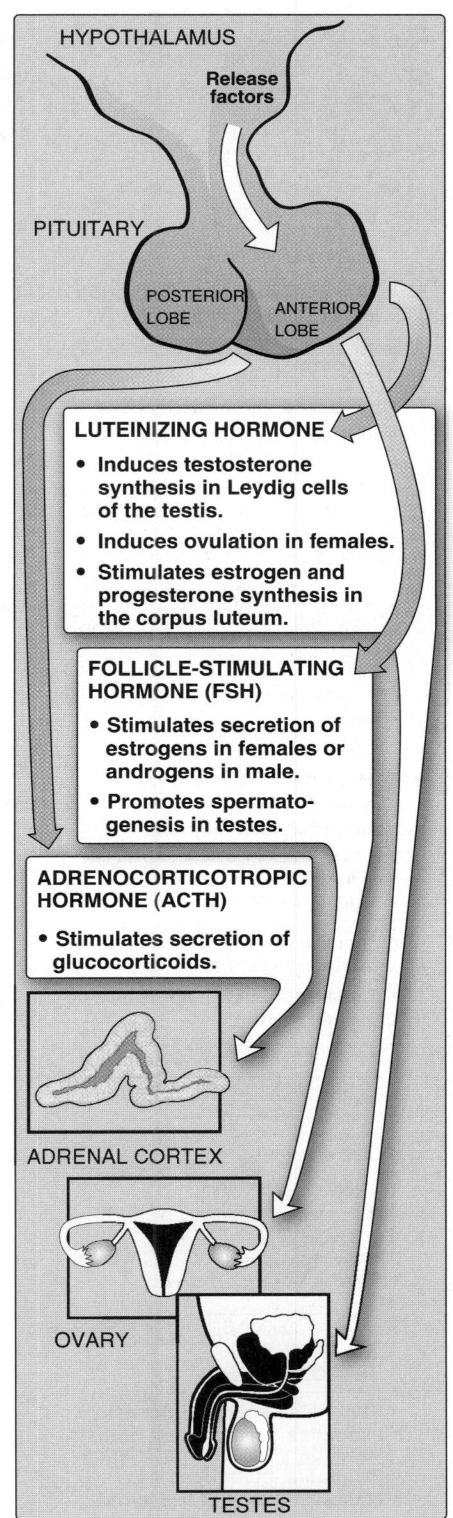

Figure 18.26
Pituitary hormone stimulation of steroid hormone synthesis and secretion.

[4]See Ch. 19 in ***Lippincott's Illustrated Reviews: Pharmacology*** (2nd and 3rd Eds.) for a discussion of hypertension.

follicle-stimulating hormone (FSH). Like ACTH, LH and FSH bind to surface receptors and cause an increase in cAMP. LH stimulates the testes to produce testosterone and the ovaries to produce estrogens and progesterone (see Figure 18.27). FSH regulates the growth of ovarian follicles and stimulates testicular spermatogenesis. [Note: For maximum effect on the male or female gonad, FSH also requires the presence of LH.]

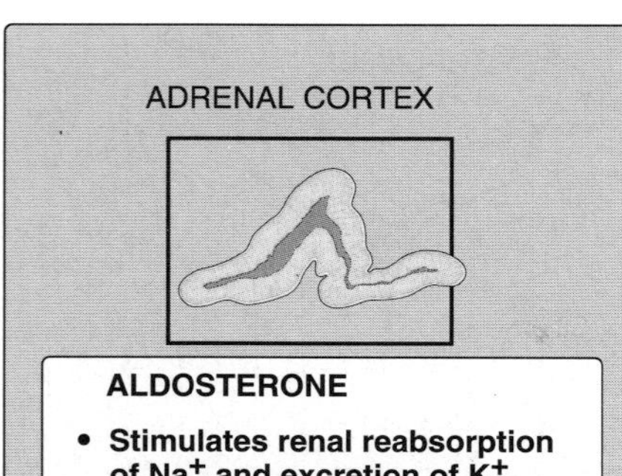

ALDOSTERONE

- **Stimulates renal reabsorption of Na^+ and excretion of K^+.**

CORTISOL

- **Increased gluconeogenesis.**
- **Anti-inflammatory action.**
- **Protein breakdown in muscle.**

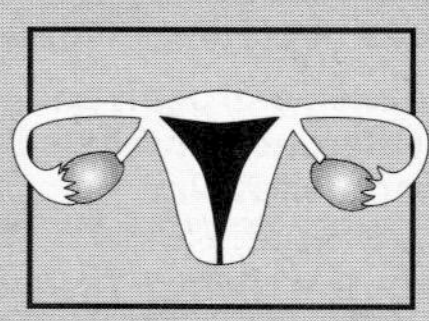

ESTROGENS

- **Control menstrual cycle.**
- **Promote development of female secondary sex characteristics.**

PROGESTERONE

- **Secretory phase of uterus and mammary glands.**
- **Implantation and maturation of fertilized ovum.**

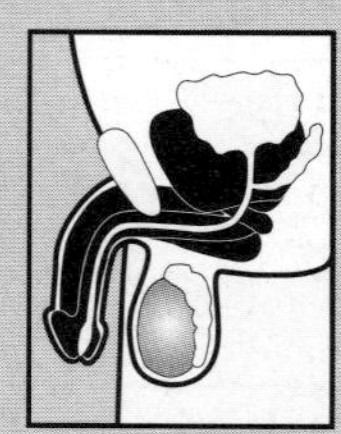

TESTOSTERONE

- **Stimulates spermatogenesis.**
- **Promotes development of male secondary sex characteristics.**
- **Promotes anabolism.**
- **Masculinization of the fetus**

Figure 18.27
Actions of steroid hormones.

D. Mechanism of steroid hormone action

Each steroid hormone diffuses across the plasma membrane of its target cell and binds to a specific cytosolic or nuclear receptor. These receptor-ligand complexes accumulate in the nucleus, dimerize, and bind to specific regulatory DNA sequences (**hormone-response elements**, **HRE**) in association with co-activator proteins, thereby causing promoter activation and increased transcription of targeted genes (Figure 18.28). An HRE is found in an enhancer element (see p. 422) located near genes that respond to a specific steroid hormone, thus ensuring coordinated regulation of these genes. Hormone-receptor complexes can also inhibit transcription in association with co-repressors. [Note: The binding of a hormone to its receptor causes a conformational change in the receptor that uncovers its DNA-binding domain, allowing the complex to interact through a zinc-finger motif with the appropriate sequence on the DNA. It is recognized that the receptors for the diverse group of steroid hormones, plus those for thyroid hormone, retinoic acid (see p. 380), and 1,25-dihydroxycholecalciferol (Vitamin D, see p. 384), are members of a "**superfamily**" of structurally-related gene regulators that function in a similar way.]

E. Further metabolism of steroid hormones

Steroid hormones are generally converted into inactive metabolic excretion products in the liver. Reactions include reduction of unsaturated bonds and the introduction of additional hydroxyl groups. The resulting structures are made more soluble by conjugation with glucuronic acid or sulfate (from PAPS, see p. 160). Approximately twenty to thirty percent of these metabolites are secreted into the bile and then excreted in the feces, whereas the remainder are released into the blood and filtered from the plasma in the kidney, passing into the urine. These conjugated metabolites are fairly water-soluble and do not need protein carriers.

VIII. CHAPTER SUMMARY

Cholesterol is a very **hydrophobic** compound, with a single hydroxyl group—located at carbon 3 of the A ring—to which a fatty acid can be attached, producing a **cholesteryl ester**. Cholesterol is synthesized by virtually all human tissues, although primarily by **liver**, **intestine**, **adrenal cortex**, and **reproductive tissues**. All the carbon atoms in cholesterol are provided by **acetate**, and **NADPH** provides the reducing equivalents. The pathway is driven by hydrolysis of the high-energy thioester bond of acetyl CoA and the terminal phosphate bond of ATP. Cholesterol is synthesized in the **cytoplasm**.The **rate-limiting step** in cholesterol synthesis is cytoplasmic **HMG CoA reductase**, which produces **mevalonic acid** from hydroxymethylglutaryl CoA (HMG CoA). The enzyme is regulated by a number of mechanisms: 1) **Expression of the HMG CoA reductase gene** is activated when cholesterol levels are low, resulting in increased enzyme and, therefore, more cholesterol synthesis. 2) HMG CoA reductase activity is controlled covalently through the actions of a **glucagon-activated protein kinase** (which **inactivates** HMG CoA reductase) and an **insulin-activated protein phosphatase** (which **activates** HMG CoA reductase). 3) Drugs such as **lovastatin** and **mevastatin** are **competitive inhibitors** of HMG CoA reductase. They are used to decrease plasma cholesterol in patients with **hypercholesterolemia**. The ring structure of cholesterol can not be degraded in humans.

Cholesterol can be eliminated from the body either by **conversion to bile salts** or by **secretion into the bile**. Intestinal bacteria can reduce cholesterol to **coprostanol** and **cholestanol**, which together with cholesterol make up the bulk of **neutral fecal sterols**. **Bile salts** and **phosphatidylcholine** are quantitatively the most important organic components of bile. Bile salts are **conjugated bile acids** produced by the **liver**. The **primary bile acids**, **cholic** or **chenodeoxycholic acids**, are **amphipathic**, and can serve as **emulsifying agents**. The rate-limiting step in bile acid synthesis is catalyzed by **cholesterol-7-α-hydroxylase**, which is **activated** by **cholesterol** and **inhibited** by **bile acids**. Before the bile acids leave the liver, they are conjugated to a molecule of either **glycine** or **taurine**, producing the **primary bile salts**: **glycochholic** or **taurocholic acid**, and **glycochenodeoxycholic** or **taurochenodeoxycholic acid**. Bile salts are more amphipathic than bile acids and, therefore, are more effective emulsifiers. In the intestine, bacteria can remove the glycine and taurine, and can remove a hydroxyl group from the steroid nucleus, producing the **secondary bile acids—deoxycholic** and **lithocholic acids**. Bile is secreted into the intestine, and more than 95 percent of the bile acids and salts are efficiently reabsorbed. They are actively transported from the intestinal mucosal cells into the portal blood, where they are carried by albumin back to the liver (**enterohepatic circulation**). In the liver, the primary and secondary bile acids are reconverted to bile salts, and secreted into the bile. If more cholesterol enters the bile than can be solubilized by the available bile salts and phosphatidylcholine, **cholesterol gallstone disease** (**cholelithiasis**) can occur.

The plasma lipoproteins include **chylomicrons**, **very-low-density lipoproteins (VLDL)**, **low-density lipoproteins (LDL)**, and **high-density lipoproteins (HDL)**. They function to keep lipids (primarily **triacylglycerol** and **cholesteryl esters**) soluble as they transport them between tissues. Lipoproteins are composed of a **neutral lipid core** (containing **triacylglycerol**, **cholesteryl esters**, or both) surrounded by a shell of amphipathic **apolipoproteins**, **phospholipid**, and **nonesterified cholesterol**. **Chylomicrons** are assembled in **intestinal mucosal cells** from **dietary lipids** (primarily, **triacylglycerol**) plus additional lipids synthesized in these cells. Each nascent chylomicron particle has one molecule of **apolipoprotein B-48** (**apo B-48**). They are released from the cells into the lymphatic system and travel to the blood, where they receive **apo C-II** and **apo E** from **HDLs**, thus making the chylomicrons functional. Apo C-II activates **lipoprotein lipase**, which degrades the

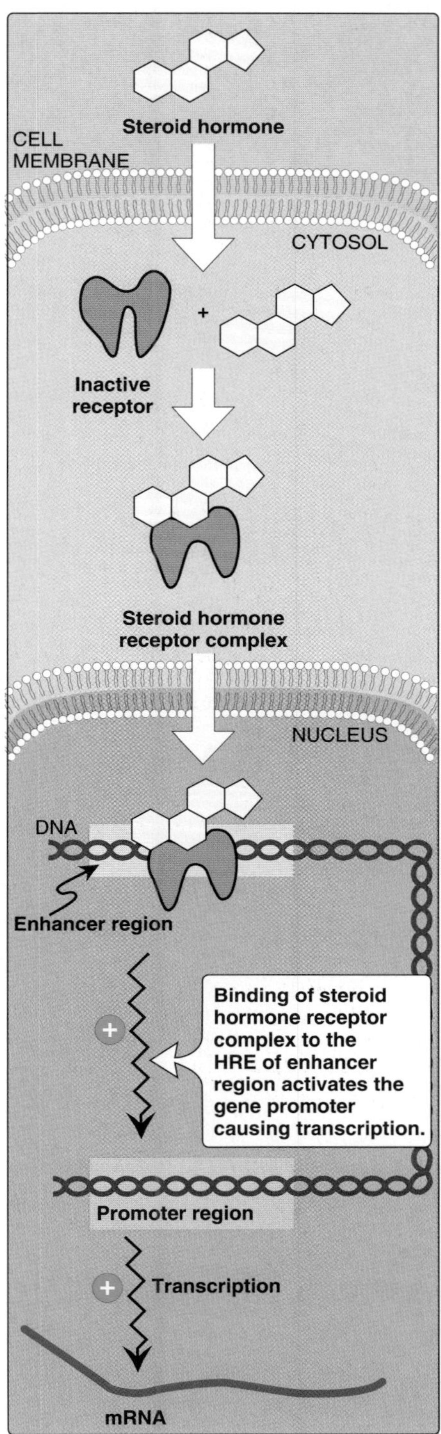

Figure 18.28
Activation of transcription by interaction of steroid hormone-receptor complex with hormone response element (HRE).

chylomicron's **triacylglycerol** to fatty acids and glycerol. The **fatty acids** that are released are **stored** (in the **adipose**) or used for **energy** (by the **muscle**). The **glycerol** is metabolized by the **liver**. Patients with a **deficiency** of **lipoprotein lipase** or **apo C-II** show a dramatic accumulation of chylomicrons in the plasma (**type I hyperlipoproteinemia**, **familial lipoprotein lipase deficiency**, or **hypertriacylglycerolemia**). After most of the triacylglycerol is removed, apo C-II is returned to the HDL, and the **chylomicron remnant**—carrying most of the **dietary cholesterol**—binds to a **receptor** on the **liver** that recognizes **apo E**. The particle is **endocytosed** and its contents degraded by **lysosomal enzymes**. **Nascent VLDLs** are produced in the **liver**, and are composed predominantly of **triacylglycerol**. They contain a single molecule of **apo B-100**. Like nascent chylomicrons, HDLs receive **apo C-II** and **apo E** from **HDLs** in the plasma. The **function** of VLDLs is to **carry triacylglycerol** from the liver to the **peripheral tissues** where **lipoprotein lipase** degrades the lipid. As triacylglycerol is removed from the VLDL, the particle receives **cholesteryl esters** from **HDL**. This process is accomplished by **cholesteryl ester transfer protein**. Eventually, VLDL in the **plasma** is **converted to LDL**—a much smaller, denser particle. Apo C-II and apo E are returned to HDLs, but the LDL retains **apo B-100**, which is recognized by **receptors** on **peripheral tissues** and the **liver**. LDLs undergo **receptor-mediated endocytosis**, and their contents are degraded in the **lysosomes**. A **deficiency of functional LDL receptors** causes **type II hyperlipidemia** (**familial hypercholesterolemia**). The endocytosed cholesterol **inhibits HMG CoA reductase** and **decreases synthesis of LDL receptors**. Some of it can also be esterified by **acyl CoA:cholesterol acyltransferase** and stored. **HDLs** are synthesized by the **liver** and **intestine**. They have a number of functions, including: 1) serving as a **circulating reservoir of apo C-II** and **apo E** for chylomicrons and VLDL; 2) removing **unesterified cholesterol** from cell surfaces and other lipoproteins and **esterifying it** using **phosphatidylcholine:cholesterol acyl transferase**, a liver-synthesized plasma enzyme that is activated by **apo A-1**; and 3) delivering these cholesteryl esters to the liver ("**reverse cholesterol transport**").

Cholesterol is the precursor of all classes of steroid hormones (**glucocorticoids**, **mineralocorticoids**, and **sex hormones—androgens**, **estrogens**, and **progestins**). **Synthesis**, using primarily **mixed-function oxidases**, occurs in the **adrenal cortex** (**cortisol**, **aldosterone**, and **androgens**), **ovaries** and **placenta** (**estrogens** and **progestins**), and **testes** (**testosterone**). Each steroid hormone diffuses across the plasma membrane of its target cell and binds to a specific **cytosolic** or **nuclear receptor**. These **receptor-ligand complexes** accumulate in the nucleus, dimerize, and **bind** to specific regulatory DNA sequences (**hormone-response elements**) in association with co-activator proteins, thereby causing **promoter activation** and **increased transcription** of targeted genes.

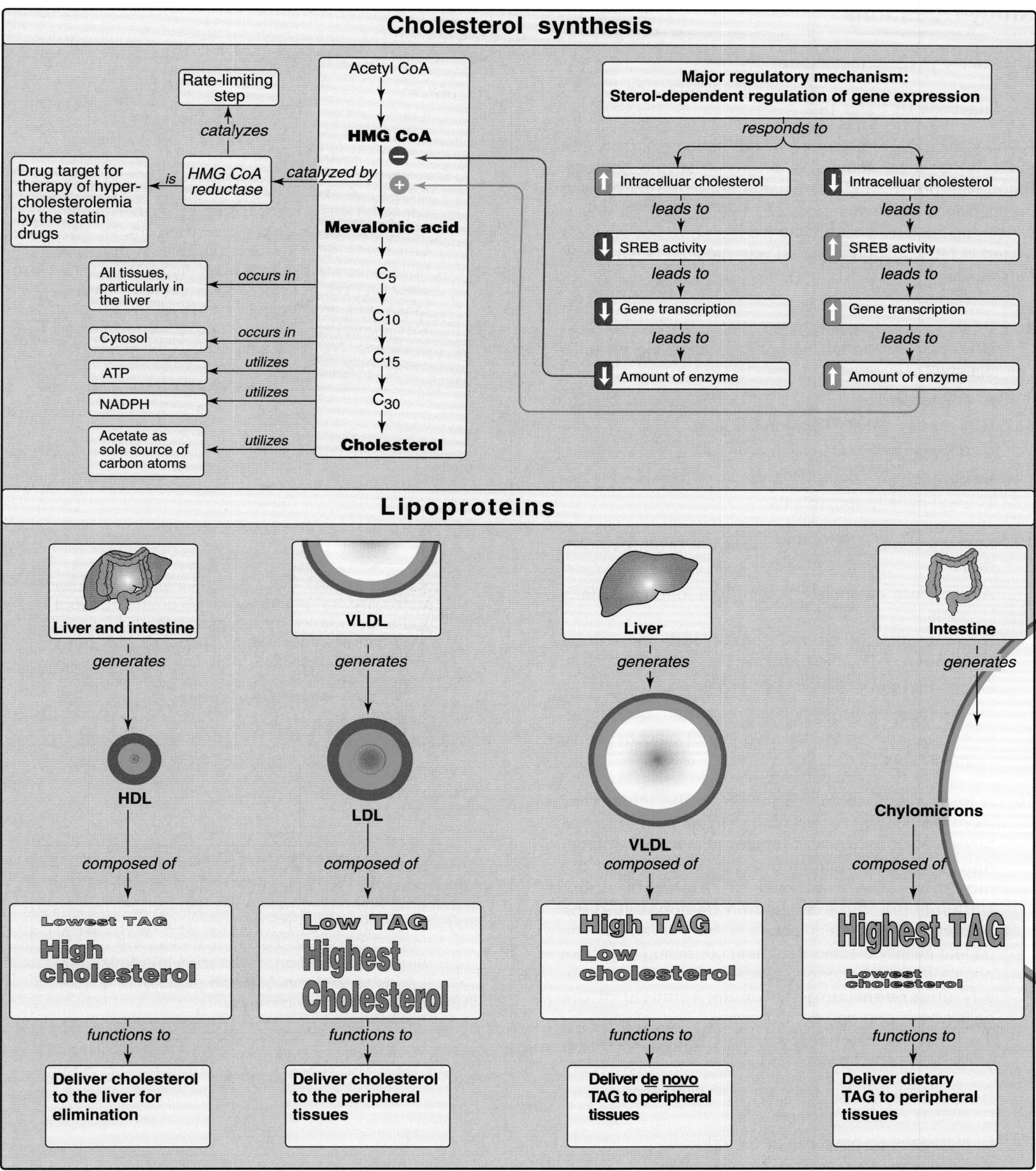

Figure 18.29
Concept map for cholesterol.

Study Questions

Choose the ONE correct answer

For Questions 18.1 and 18.2:

A young girl with a history of severe abdominal pain was taken to her local hospital at 5 a.m. in severe distress. Blood was drawn, and the plasma appeared milky, with the triacylglycerol level in excess of 2000 mg/dl (normal = 4–150 mg/dl). The patient was placed on a diet severely limited in fat, but supplemented with medium-chain length fatty acids.

18.1 Which of the following lipoprotein particles are most likely responsible for the appearance of the patient's plasma?

A. Chylomicrons
B. Very-low-density lipoproteins
C. Intermediate-density lipoproteins
D. Low density-lipoproteins
E. High density-lipoproteins

Correct answer = A. The milky appearance of her blood was a result of triacylglycerol-rich chylomicrons. Because 5 a.m. is presumably several hours after her evening meal, she must have difficulty clearing these lipoprotein particles. IDL, LDL, or HDL contain primarily cholesteryl esters and, if one or more of these particles was elevated, it would cause hypercholesterolemia. VLDLs do not cause the described "milky appearance" in plasma.

18.2 Medium-chain length fatty acids are given because they:

A. are more calorically dense than long-chain fatty acids.
B. enter directly into the portal blood, and can be metabolized by the liver.
C. are activators of lipoprotein lipase.
D. are more efficiently packed into serum lipoproteins.
E. can be converted into a variety of gluconeogenic precursors.

Correct answer = B. Medium-chain length fatty acids are not packaged in chylomicrons, but rather are carried by albumin to the liver where they can be metabolized. They have the same caloric density as long-chain fatty acids, and are generally much more ketogenic than glycogenic. Lipoprotein lipase does not play a role in their metabolism.

18.3 A 35-year-old woman was seen in the emergency room because of recurrent abdominal pain. The history revealed a two-year pattern of pain in the upper right quadrant, beginning several hours after the ingestion of a meal rich in fried/fatty food. Ultrasonographic examination demonstrated the presence of numerous stones in the gallbladder. The patient initially elected treatment consisting of exogenously supplied chenodeoxycholic acid, but eventually underwent surgery for the removal of the gallbladder, and had a full recovery. The rationale for the initial treatment of this patient with chenodeoxycholic acid is that this compound:

A. interferes with the enterohepatic circulation.
B. inhibits cholesterol synthesis.
C. increases de novo bile acid production.
D. increases cholesterol solubility in bile.
E. stimulates VLDL production by the liver.

Correct answer = D. Chenodeoxycholic acid is a bile acid used in the treatment of gallstones. It is an amphipathic molecule that can act like an emulsifying agent and help solubilize cholesterol. The compound will not effect the enterohepatic circulation, interfere with cholesterol synthesis, increase bile acid production, or stimulate VLDL production.

UNIT IV:
Nitrogen Metabolism

Amino Acids: Disposal of Nitrogen

19

I. OVERVIEW

Unlike fats and carbohydrates, amino acids are not stored by the body, that is, no proteins exist whose sole function it is to maintain a supply of amino acids for future use. Therefore, amino acids must be obtained from the diet, synthesized de novo, or produced from normal protein degradation. Any amino acids in excess of the biosynthetic needs of the cell are rapidly degraded. The first phase of catabolism involves the removal of the α-amino groups (usually by transamination and subsequent oxidative deamination), forming ammonia and the corresponding α-ketoacid—the "carbon skeletons" of amino acids. A portion of the free ammonia is excreted in the urine, but most is used in the synthesis of urea (Figure 19.1), which is quantitatively the most important route for disposing of nitrogen from the body. In the second phase of amino acid catabolism, described in Chapter 20, the carbon skeletons of the α-ketoacids are converted to common intermediates of energy producing, metabolic pathways. These compounds can be metabolized to CO_2 and water, glucose, fatty acids, or ketone bodies by the central pathways of metabolism described in Chapters 8 to 13, and 16.

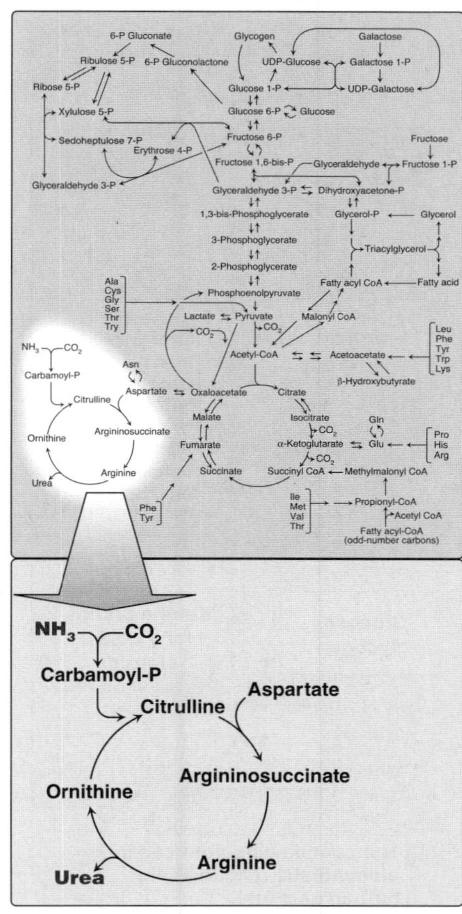

Figure 19.1
Urea cycle shown as part of the essential reactions of energy metabolism. (See Figure 8.2, p. 90, for a more detailed view of the metabolic pathway.)

II. OVERALL NITROGEN METABOLISM

Amino acid catabolism is part of the larger process of whole body nitrogen metabolism. Nitrogen enters the body in a variety of compounds present in food, the most important being amino acids contained in dietary protein. Nitrogen leaves the body as urea, ammonia, and other products derived from amino acid metabolism. The role of body proteins in these transformations involves two important concepts: the **amino acid pool** and **protein turnover**.

Lippincott's Illustrated Reviews: Biochemistry, 3rd Edition
by Pamela C. Champe and Richard A. Harvey.
Lippincott Williams & Wilkins, Baltimore, MD © 2005.

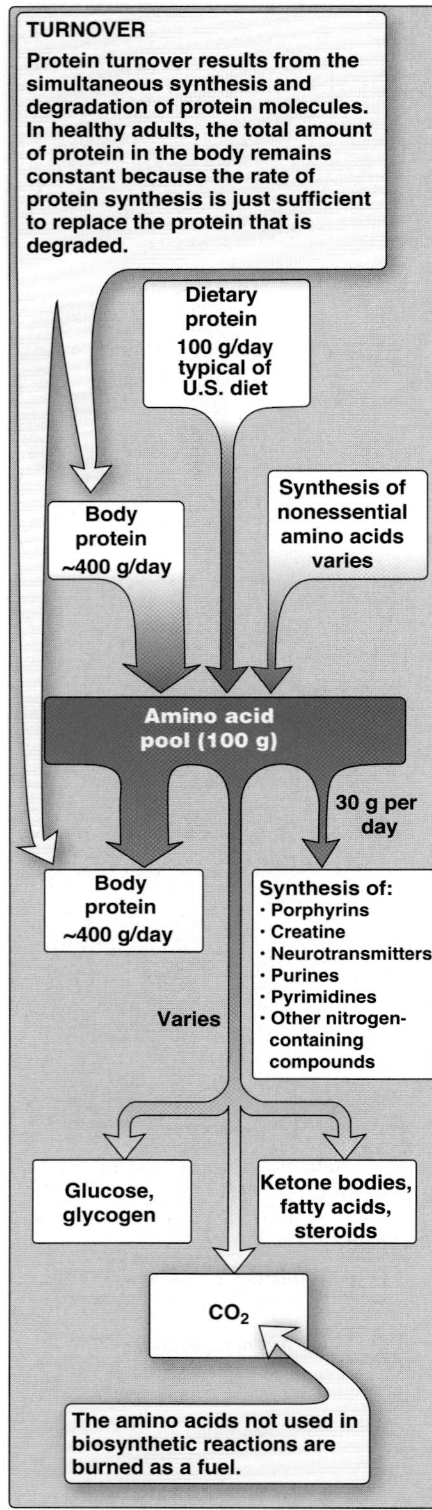

Figure 19.2
Sources and fates of amino acids.

A. Amino acid pool

Amino acids released by hydrolysis of dietary or tissue protein, or synthesized de novo, mix with other free amino acids distributed throughout the body. Collectively, they constitute the amino acid pool (Figure 19.2). The amino acid pool, containing about 100 g of amino acids, is small in comparison with the amount of protein in the body (about 12 kg in a 70 kg man). If the only fate of the amino acid pool were to be used to resynthesize body proteins, adults would not have a significant need for additional dietary protein. However, only about 75 percent of the amino acids obtained through hydrolysis of body protein are recaptured through the biosynthesis of new tissue protein. The remainder are metabolized or serve as precursors for the compounds shown in Figure 19.2, some of which are described in detail in Chapter 21. In well-fed individuals, this metabolic loss of amino acids is compensated for by dietary protein, which contributes to the amino pool.

B. Protein turnover

Most proteins in the body are constantly being synthesized and then degraded, permitting the removal of abnormal or unneeded proteins. For many proteins, regulation of synthesis determines the concentration of protein in the cell, with protein degradation assuming a minor role. For other proteins, the rate of synthesis is constitutive, that is, relatively constant, and cellular levels of the protein are controlled by selective degradation.

1. **Rate of turnover:** In healthy adults, the total amount of protein in the body remains constant, because the rate of protein synthesis is just sufficient to replace the protein that is degraded. This process, called **protein turnover**, leads to the hydrolysis and resynthesis of 300 to 400 g of body protein each day. The rate of protein turnover varies widely for individual proteins. Short-lived proteins (for example, many regulatory proteins and misfolded proteins) are rapidly degraded, having half-lives measured in minutes or hours. Long-lived proteins, with half-lives of days to weeks, constitute the majority of proteins in the cell. Structural proteins, such as collagen, are metabolically stable, and have half-lives measured in months or years.

2. **Protein degradation:** There are two major enzyme systems responsible for degrading damaged or unneeded proteins: the energy-dependent **ubiquitin-proteasome mechanism**, and the non–energy-dependent degradative enzymes of the **lysosomes**. Proteasomes mainly degrade endogenous proteins, that is, proteins that were synthesized within the cell. Lysosomes (see p. 160) primarily degrade extracellular proteins, such as plasma proteins that are taken into the cell by endocytosis, and cell-surface membrane proteins that are used in receptor-mediated endocytosis.

   a. **Ubiquitin-proteasome proteolytic pathway:** Proteins destined for degradation by the ubiquitin-proteasome mechanism are first covalently attached to ubiquitin, a small, globular protein. Ubiquitination of the target substrate occurs through linkage of the α-carboxyl glycine of ubiquitin to a lysine ε-amino group on the protein substrate. The consecutive addition of ubiquitin moieties generates a **polyubiquitin chain**. Proteins tagged with ubiquitin are then recognized by a large, barrel-shaped, proteolytic molecule called a **proteasome**, which functions like a garbage disposal (Figure 19.3). The proteosome cuts the target protein into fragments that are then further degraded to amino acids, which enter the amino acid pool. It is noteworthy that the selective degradation of proteins by the ubiquitin-proteosome complex (unlike simple hydrolysis by proteolytic enzymes) requires ATP, that is, it is energy-dependent.

b. **Chemical signals for protein degradation:** Because proteins have different half-lives, it is clear that protein degradation cannot be random, but rather is influenced by some structural aspect of the protein. For example, some proteins that have been chemically altered by oxidation or tagged with ubiquitin are preferentially degraded. The half-life of a protein is influenced by the nature of the N-terminal residue. For example, proteins that have serine as the N-terminal amino acid are long-lived, with a half-life of more than twenty hours. In contrast, proteins with aspartate as the N-terminal amino acid have a half-life of only three minutes. Further, proteins rich in sequences containing proline, glutamate, serine, and threonine (called **PEST sequences** after the one-letter designations for these amino acids) are rapidly degraded and, therefore, exhibit short intracellular half-lives.

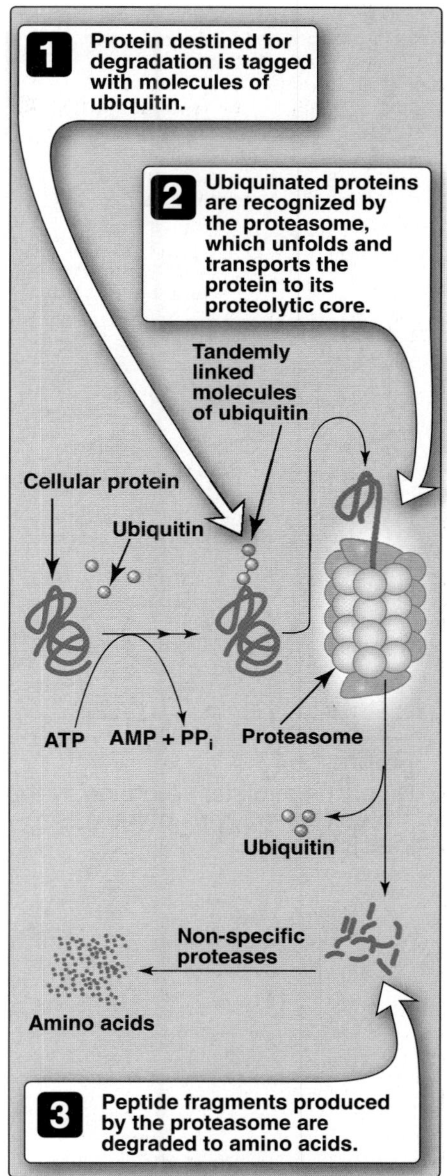

Figure 19.3
The ubiquitin-proteasome degradation pathway of proteins.

III. DIGESTION OF DIETARY PROTEINS

Most of the nitrogen in the diet is consumed in the form of protein, typically amounting from 70 to 100 g/day in the American diet (see Figure 19.2). Proteins are generally too large to be absorbed by the intestine. [Note: An example of an excetion to this rule is that newborns can take up maternal antibodies in breast milk.] They must, therefore, be hydrolyzed to yield their constituent amino acids, which can be absorbed. Proteolytic enzymes responsible for degrading proteins are produced by three different organs: the stomach, the pancreas, and the small intestine (Figure 19.4).

A. Digestion of proteins by gastric secretion

The digestion of proteins begins in the **stomach**, which secretes gastric juice—a unique solution containing hydrochloric acid and the proenzyme, pepsinogen:

1. **Hydrochloric acid:** Stomach acid is too dilute (pH 2 to 3) to hydrolyze proteins. The acid functions instead to kill some bacteria and to denature proteins, thus making them more susceptible to subsequent hydrolysis by proteases.

2. **Pepsin:** This acid-stable endopeptidase is secreted by the serous cells of the stomach as an inactive zymogen (or proenzyme), **pepsinogen**. In general, zymogens contain extra amino acids in

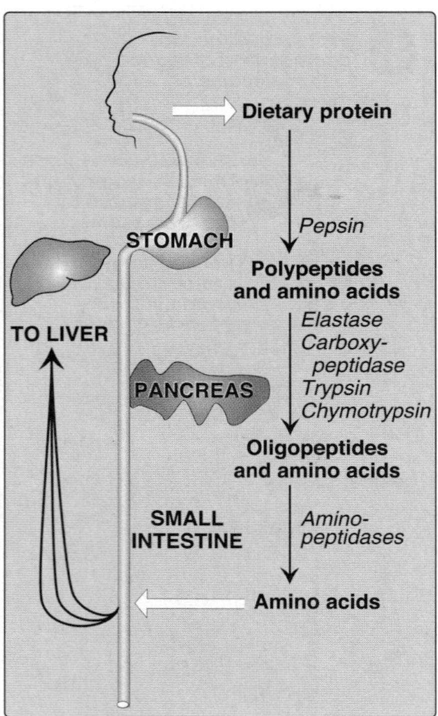

Figure 19.4
Digestion of dietary proteins by the proteolytic enzymes of the gastro-intestinal tract.

their sequences, which prevent them from being catalytically active. [Note: Removal of these amino acids permits the proper folding required for an active enzyme.] Pepsinogen is activated to *pepsin*, either by HCl, or autocatalytically by other *pepsin* molecules that have already been activated. *Pepsin* releases peptides and a few free amino acids from dietary proteins.

B. Digestion of proteins by pancreatic enzymes

On entering the small intestine, large polypeptides produced in the stomach by the action of *pepsin* are further cleaved to oligopeptides and amino acids by a group of pancreatic proteases.

1. **Specificity:** Each of these enzymes has a different specificity for the amino acid R-groups adjacent to the susceptible peptide bond (Figure 19.5). For example, *trypsin* cleaves only when the carbonyl group of the peptide bond is contributed by arginine or lysine. These enzymes, like *pepsin* described above, are synthesized and secreted as inactive zymogens.

2. **Release of zymogens:** The release and activation of the pancreatic zymogens is mediated by the secretion of **cholecystokinin** and **secretin**, two polypeptide hormones of the digestive tract (see p. 174).

3. **Activation of zymogens:** *Enteropeptidase* (formerly called *enterokinase*)—an enzyme synthesized by and present on the luminal surface of intestinal mucosal cells of the brush border membrane—converts the pancreatic zymogen trypsinogen to *trypsin* by removal of a hexapeptide from the NH_2-terminus of trypsinogen. *Trypsin* subsequently converts other trypsinogen molecules to *trypsin* by cleaving a limited number of specific peptide bonds in the zymogen. *Enteropeptidase* thus unleashes a cascade of proteolytic activity, because *trypsin* is the common activator of all the pancreatic zymogens (see Figure 19.5).

4. **Abnormalities in protein digestion:** In individuals with a deficiency in pancreatic secretion (for example, due to **chronic pancreatitis**, **cystic fibrosis**, or **surgical removal** of the pancreas), the digestion and absorption of fat and protein is incomplete. This results in the abnormal appearance of lipids (called **steatorrhea**, see p. 175) and undigested protein in the feces.

C. Digestion of oligopeptides by enzymes of the small intestine

The luminal surface of the intestine contains *aminopeptidase*—an exopeptidase that repeatedly cleaves the N-terminal residue from oligopeptides to produce free amino acids and smaller peptides.

D. Absorption of amino acids and dipeptides

Free amino acids and dipeptides are taken up by the intestinal epithelial cells. There, the dipeptides are hydrolyzed in the cytosol to amino acids before being released into the portal system. Thus, only free amino acids are found in the portal vein after a meal containing protein. These amino acids are either metabolized by the liver or released into the general circulation.

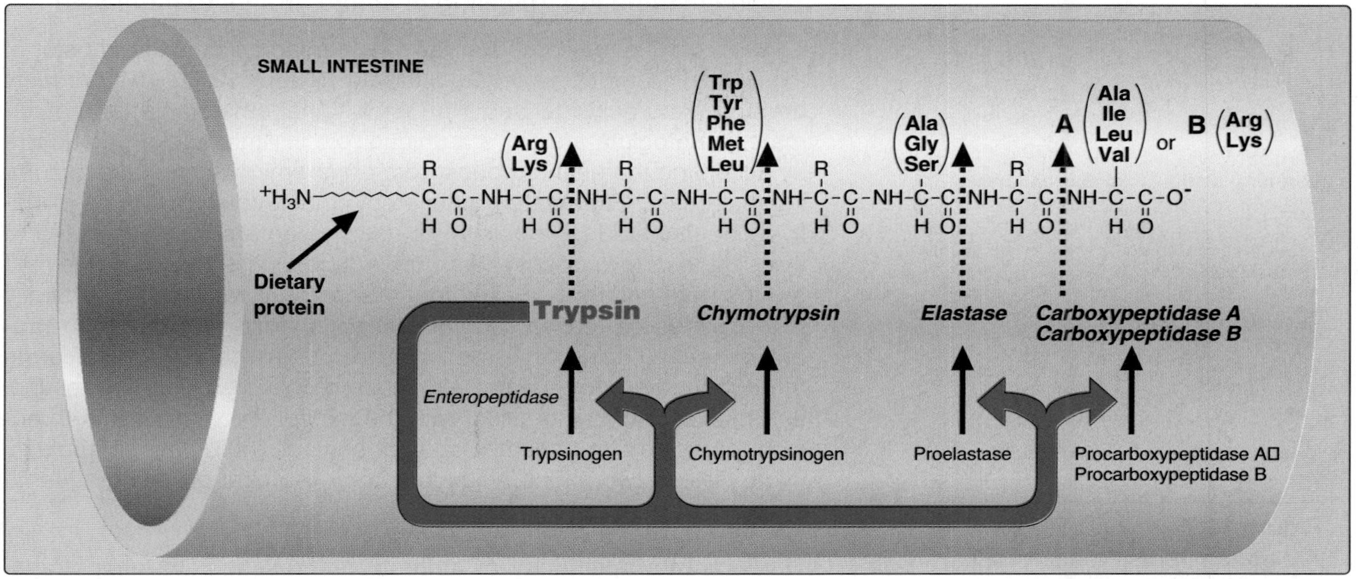

Figure 19.5
Cleavage of dietary protein by proteases from the pancreas. The peptide bonds susceptible to hydrolysis are shown for each of the five major pancreatic proteases. [Note: *Enteropeptidase* is synthesized in the intestine.]

IV. TRANSPORT OF AMINO ACIDS INTO CELLS

The concentration of free amino acids in the extracellular fluids is significantly lower than that within the cells of the body. This concentration gradient is maintained because active transport systems, driven by the hydrolysis of ATP, are required for movement of amino acids from the extracellular space into cells. At least seven different transport systems are known that have overlapping specificities for different amino acids. For example, one transport system is responsible for reabsorption of the amino acids cystine, ornithine, arginine, and lysine in kidney tubules. In the inherited disorder **cystinuria**, this carrier system is defective, resulting in the appearance of all four amino acids in the urine (Figure 19.6). Cystinuria occurs at a frequency of 1 in 7000 individuals, making it one of the most common inherited diseases, and the most common genetic error of amino acid transport. The disease expresses itself clinically by the precipitation of cystine to form kidney stones (calculi), which can block the urinary tract. Oral hydration is an important part of treatment for this disorder.

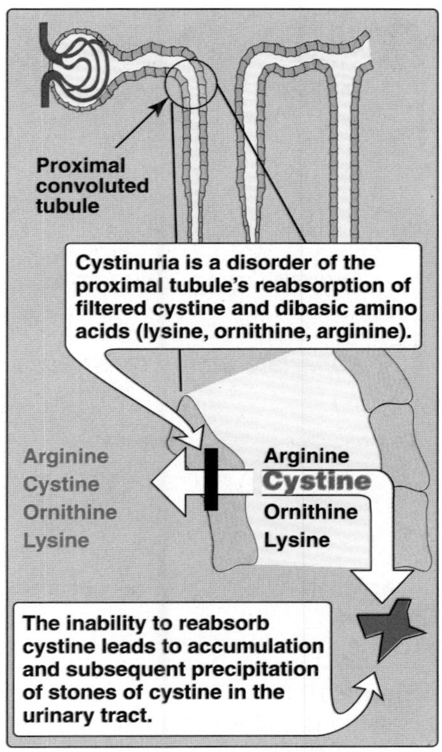

Figure 19.6
Genetic defect seen in cystinuria.

V. REMOVAL OF NITROGEN FROM AMINO ACIDS

The presence of the α-amino group keeps amino acids safely locked away from oxidative breakdown. Removing the α-amino group is essential for producing energy from any amino acid, and is an obligatory step in the catabolism of all amino acids. Once removed, this nitrogen can be incorporated into other compounds or excreted, with the carbon skeletons being metabolized. This section describes transamination and oxidative deamination—reactions that ultimately provide ammonia and aspartate, the two sources of urea nitrogen (see p. 251).

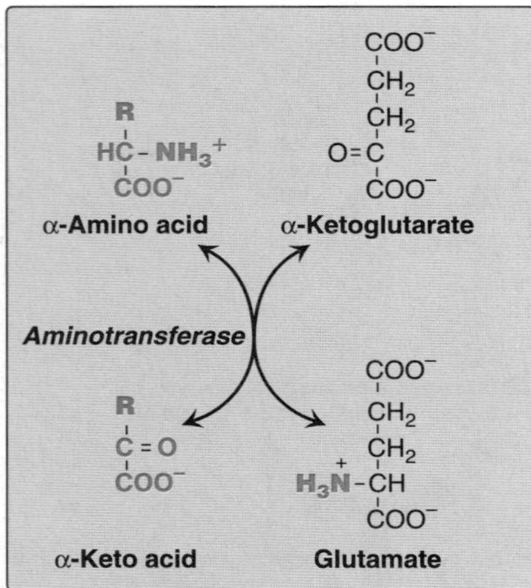

Figure 19.7
Aminotransferase reaction using α-ketoglutarate as the amino-group acceptor.

A. Transamination: the funneling of amino groups to glutamate

The first step in the catabolism of most amino acids is the transfer of their α-amino group to α-ketoglutarate (Figure 19.7). The products are an α-keto acid (derived from the original amino acid) and glutamate. α-Ketoglutarate plays a unique role in amino acid metabolism by accepting the amino groups from other amino acids, thus becoming glutamate. Glutamate produced by transamination can be oxidatively deaminated (see below), or used as an amino group donor in the synthesis of nonessential amino acids. This transfer of amino groups from one carbon skeleton to another is catalyzed by a family of enzymes called **aminotransferases** (formerly called **transaminases**). These enzymes are found in the cytosol of cells throughout the body—especially those of the liver, kidney, intestine, and muscle. All amino acids, with the exception of lysine and threonine, participate in transamination at some point in their catabolism. [Note: These two amino acids lose their α-amino groups by deamination (see p. 264).]

1. **Substrate specificity of aminotransferases:** Each aminotransferase is specific for one or, at most, a few amino group donors. Aminotransferases are named after the specific amino group donor, because the acceptor of the amino group is almost always α-ketoglutarate. The two most important aminotransferase reactions are catalyzed by *alanine aminotransferase* and *aspartate aminotransferase* (Figure 19.8).

   a. **Alanine aminotransferase (ALT)**, formerly called *glutamate-pyruvate transaminase* (*GPT*), is present in many tissues. The enzyme catalyzes the transfer of the amino group of alanine to α-ketoglutarate, resulting in the formation of pyruvate and glutamate. The reaction is readily reversible. However, during amino acid catabolism, this enzyme (like most aminotransferases) functions in the direction of glutamate synthesis. Thus, glutamate, in effect, acts as a "collector" of nitrogen from alanine.

   b. **Aspartate aminotransferase (AST)**, formerly called *glutamate:oxaloacetate transaminase* (*GOT*), is an exception to the rule that aminotransferases funnel amino groups to form glutamate. During amino acid catabolism, *AST* transfers amino groups from glutamate to oxaloacetate, forming aspartate, which is used as a source of nitrogen in the urea cycle (see p. 251).

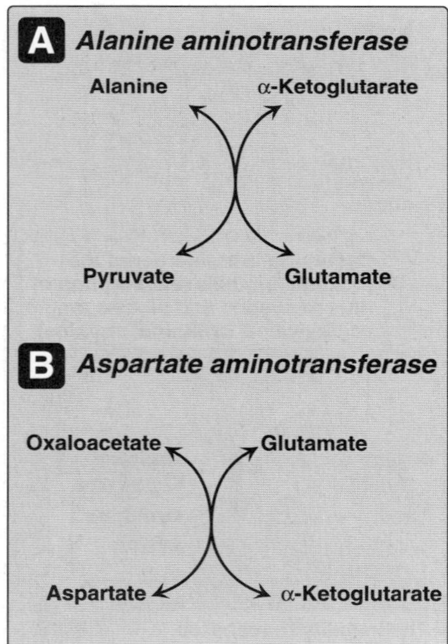

Figure 19.8
Reactions catalyzed during amino acid catabolism.
A. *Alanine aminotransferase.*
B. *Aspartate aminotransferase.*

2. **Mechanism of action of aminotransferases:** All aminotransferases require the coenzyme **pyridoxal phosphate** (a derivative of vitamin B_6, see p. 376), which is covalently linked to the ε-amino group of a specific lysine residue at the active site of the enzyme. Aminotransferases act by transferring the amino group of an amino acid to the pyridoxal part of the coenzyme to generate **pyridoxamine phosphate**. The pyridoxamine form of the coenzyme then reacts with an α-keto acid to form an amino acid, at the same time regenerating the original aldehyde form of the coenzyme. Figure 19.9 shows these two component reactions for the reaction catalyzed by *aspartate aminotransferase.*

3. **Equilibrium of transamination reactions:** For most transamination reactions, the equilibrium constant is near one, allowing the reaction to function in both amino acid degradation through removal of α-amino groups (for example, after consumption of a protein-rich meal), and biosynthesis through addition of amino groups to the carbon skeletons of α-keto acids (for example, when the supply of amino acids from the diet is not adequate to meet the synthetic needs of cells).

4. **Diagnostic value of plasma aminotransferases:** Aminotransferases are normally intracellular enzymes, with the low levels found in the plasma representing the release of cellular contents during normal cell turnover. The presence of elevated plasma levels of aminotransferases indicates damage to cells rich in these enzymes. For example, physical trauma or a disease process can cause cell lysis, resulting in release of intracellular enzymes into the blood. Two aminotransferases—*AST* and *ALT*—are of particular diagnostic value when they are found in the plasma.

   a. **Liver disease:** Plasma *AST* and *ALT* are elevated in nearly all liver diseases, but are particularly high in conditions that cause extensive cell necrosis, such as severe viral hepatitis, toxic injury, and prolonged circulatory collapse. *ALT* is more specific for liver disease than *AST*, but the latter is more sensitive because the liver contains larger amounts of *AST*. Serial enzyme measurements are often useful in determining the course of liver damage. Figure 19.10 shows the early release of *ALT* into the serum, following ingestion of a liver toxin. [Note: Elevated serum bilirubin results from heptocellular damage that decreases the hepatic conjugation and excretion of bilirubin (see p. 282).]

   b. **Nonhepatic disease:** Aminotransferases may be elevated in nonhepatic disease, such as myocardial infarction and muscle disorders. However, these disorders can usually be distinguished clinically from liver disease.

B. Glutamate dehydrogenase: the oxidative deamination of amino acids

In contrast to transamination reactions that transfer amino groups, oxidative deamination by *gutamate dehydrogenase* results in the liberation of the amino group as free ammonia (Figure 19.11). These reactions occur primarily in the liver and kidney. They provide α-ketoacids that can enter the central pathway of energy metabolism, and ammonia, which is a source of nitrogen in urea synthesis.

1. **Glutamate dehydrogenase:** As described above, the amino groups of most amino acids are ultimately funneled to glutamate by means of transamination with α-ketoglutarate. Glutamate is unique in that it is the only amino acid that undergoes rapid oxidative deamination—a reaction catalyzed by *glutamate dehydrogenase* (see Figure 19.10). Therefore, the sequential action of transamination (resulting in the collection of amino groups from other amino acids onto α-ketoglutarate to produce

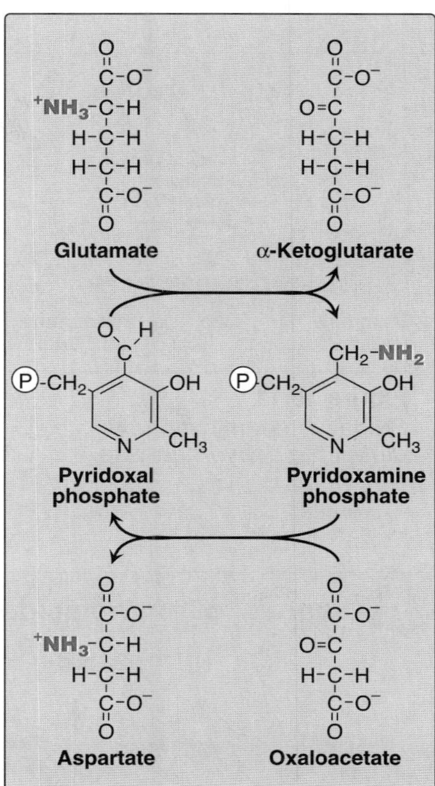

Figure 19.9
Cyclic interconversion of pyridoxal phosphate and pyridoxamine phosphate during the *aspartate aminotransferase* reaction. [Note: Ⓟ = phosphate group.]

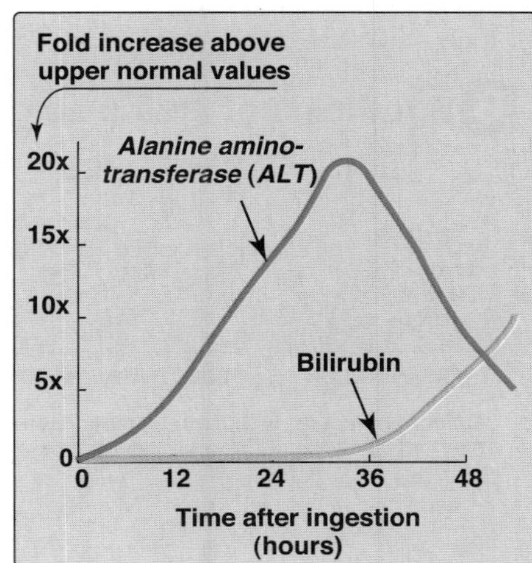

Figure 19.10
Pattern of serum *alanine aminotransferase (ALT)* and bilirubin in the plasma, following poisoning with the toxic mushroom Amanita phalloides.

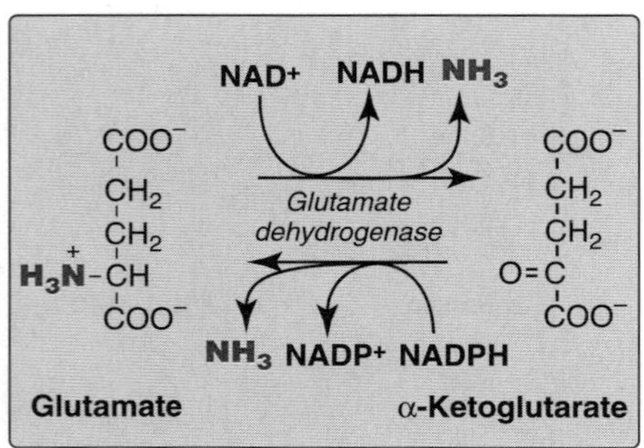

Figure 19.11
Oxidative deamination by *glutamate dehydrogenase*.

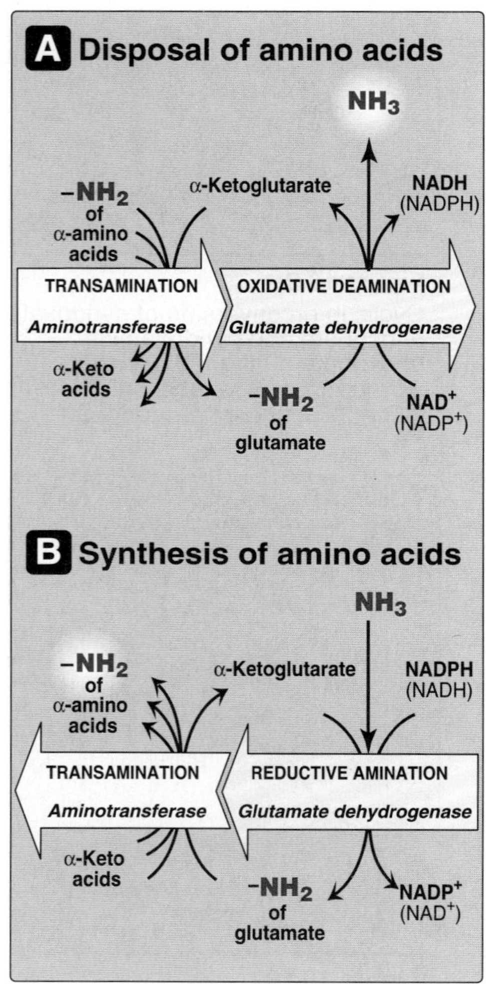

Figure 19.12
Combined actions of *aminotransferase* and *glutamate dehydrogenase* reactions

glutamate) and the subsequent oxidative deamination of that glutamate (regenerating α-ketoglutarate) provide a pathway whereby the amino groups of most amino acids can be released as ammonia.

a. **Coenzymes:** *Glutamate dehydrogenase* is unusual in that it can use either NAD^+ or $NADP^+$ as a coenzyme. NAD^+ is used primarily in oxidative deamination (the simultaneous loss of ammonia coupled with the oxidation of the carbon skeleton, Figure 19.12A), and NADPH is used in reductive amination (the simultaneous gain of ammonia coupled with the reduction of the carbon skeleton, Figure 19.12B).

b. **Direction of reactions:** The direction of the reaction depends on the relative concentrations of glutamate, α-ketoglutarate, and ammonia, and the ratio of oxidized to reduced coenzymes. For example, after ingestion of a meal containing protein, glutamate levels in the liver are elevated, and the reaction proceeds in the direction of amino acid degradation and the formation of ammonia (see Figure 19.11A). [Note: the reaction can also be used to synthesize amino acids from the corresponding α-ketoacids (see Figure 19.11B).]

c. **Allosteric regulators:** ATP and GTP are allosteric inhibitors of *glutamate dehydrogenase*, whereas ADP and GDP are activators of the enzyme. Thus, when energy levels are low in the cell, amino acid degradation by *glutamate dehydrogenase* is high, facilitating energy production from the carbon skeletons derived from amino acids.

2. **D-Amino acid oxidase:** D-Amino acids (see p. 5) are found in plants and in the cell walls of microorganisms, but are not used in the synthesis of mammalian proteins. D-Amino acids are, however, present in the diet, and are efficiently metabolized by the **liver**. *D-Amino acid oxidase* is an FAD-dependent enzyme that catalyzes the oxidative deamination of these amino acid isomers. The resulting α-ketoacids can enter the general pathways of amino acid metabolism, and be reaminated to L-isomers, or catabalized for energy.

C. Transport of ammonia to the liver

Two mechanisms are available in humans for the transport of ammonia from the peripheral tissues to the liver for its ultimate conversion to urea. The first, found in most tissues, uses glutamine synthetase to combine ammonia with glutamate to form glutamine—a nontoxic transport form of ammonia (Figure 19.13). The glutamine is transported in the blood to the liver where is is cleaved by *glutaminase* to produce glutamate and free ammonia (see p. 254). The second transport mechanism, used primarily by muscle, involves transamination of pyruvate (the end-product of aerobic glyclosysis) to form alanine (see Figure 19.8). Alanine is transported by the blood to the liver, where it is converted to pyruvate, again by transamination. In the liver, the pathway of gluconeogenesis can use the pyruvate to synthesize glucose, which can enter the blood and be used by muscle—a pathway called the **glucose-alanine cycle**.

VI. UREA CYCLE

Urea is the major disposal form of amino groups derived from amino acids, and accounts for about ninety percent of the nitrogen-containing components of urine. One nitrogen of the urea molecule is supplied by free NH_3, and the other nitrogen by aspartate. [Note: Glutamate is the immediate precursor of both ammonia (through oxidative deamination by *glutamate dehydrogenase*) and aspartate nitrogen (through transamination of oxaloacetate by *aspartate aminotransferase*).] The carbon and oxygen of urea are derived from CO_2. Urea is produced by the **liver**, and then is transported in the blood to the kidneys for excretion in the urine.

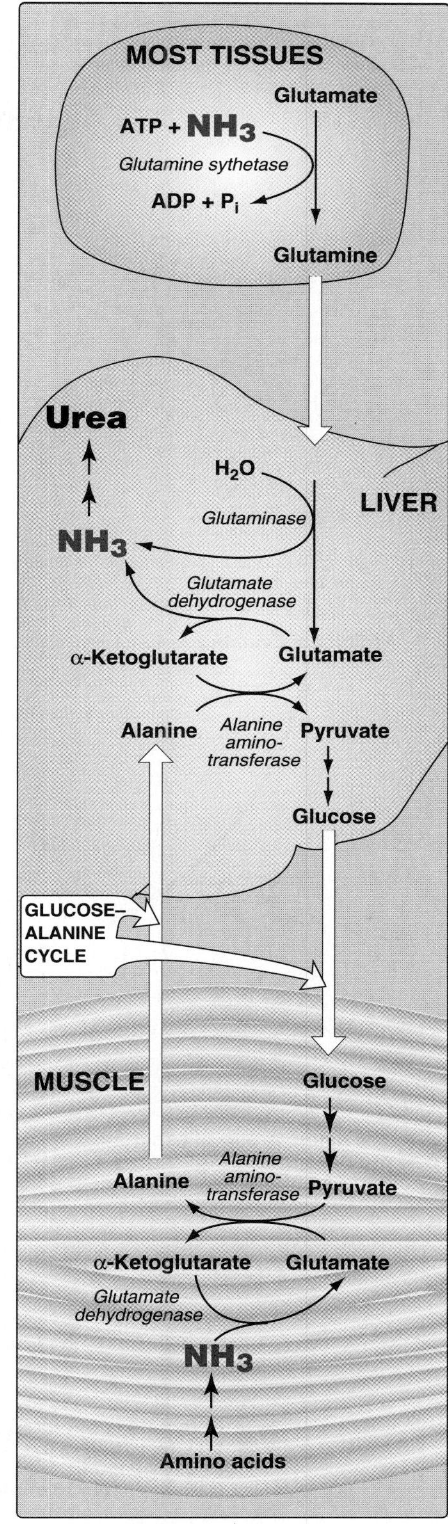

Figure 19.13
Transport of ammonia from peripheral tissues to the liver.

A. Reactions of the cycle

The first two reactions leading to the synthesis of urea occur in the mitochondria, whereas the remaining cycle enzymes are located in the cytosol (Figure 19.14).

1. **Formation of carbamoyl phosphate:** Formation of carbamoyl phosphate by *carbamoyl phosphate synthetase I* is driven by cleavage of two molecules of ATP. Ammonia incorporated into carbamoyl phosphate is provided primarily by the oxidative deamination of glutamate by mitochondrial *glutamate dehydrogenase* (see Figure 19.11). Ultimately, the nitrogen atom derived from this ammonia becomes one of the nitrogens of urea. *Carbamoyl phosphate synthetase I* requires N-acetylglutamate as a positive allosteric activator (see Figure 19.14). [Note: *Carbamoyl phosphate synthetase II* participates in the biosynthesis of pyrimidines (see p. 299). It does not require N-acetylglutamate, and occurs in the cytosol.]

2. **Formation of citrulline:** Ornithine and citrulline are basic amino acids that participate in the urea cycle. [Note: They are not incorporated into cellular proteins, because there are no codons for these amino acids (see p. 429).] Ornithine is regenerated with each turn of the urea cycle, much in the same way that oxaloacetate is regenerated by the reactions of the citric acid cycle (see p. 109). The release of the high-energy phosphate of carbamoyl phosphate as inorganic phosphate drives the reaction in the forward direction. The reaction product, citrulline, is transported to the cytosol.

3. **Synthesis of argininosuccinate:** Citrulline condenses with aspartate to form argininosuccinate. The α-amino group of aspartate provides the second nitrogen that is ultimately incorporated into urea. The formation of argininosuccinate is driven by the cleavage of ATP to AMP and pyrophosphate (PP_i). This is the third and final molecule of ATP consumed in the formation of urea.

4. **Cleavage of argininosuccinate:** Argininosuccinate is cleaved to yield arginine and fumarate. The arginine formed by this reaction serves as the immediate precursor of urea. Fumarate produced in the urea cycle is hydrated to malate, providing a link with several metabolic pathways. For example, the malate can be transported into the mitochondria via the malate shuttle and reenter

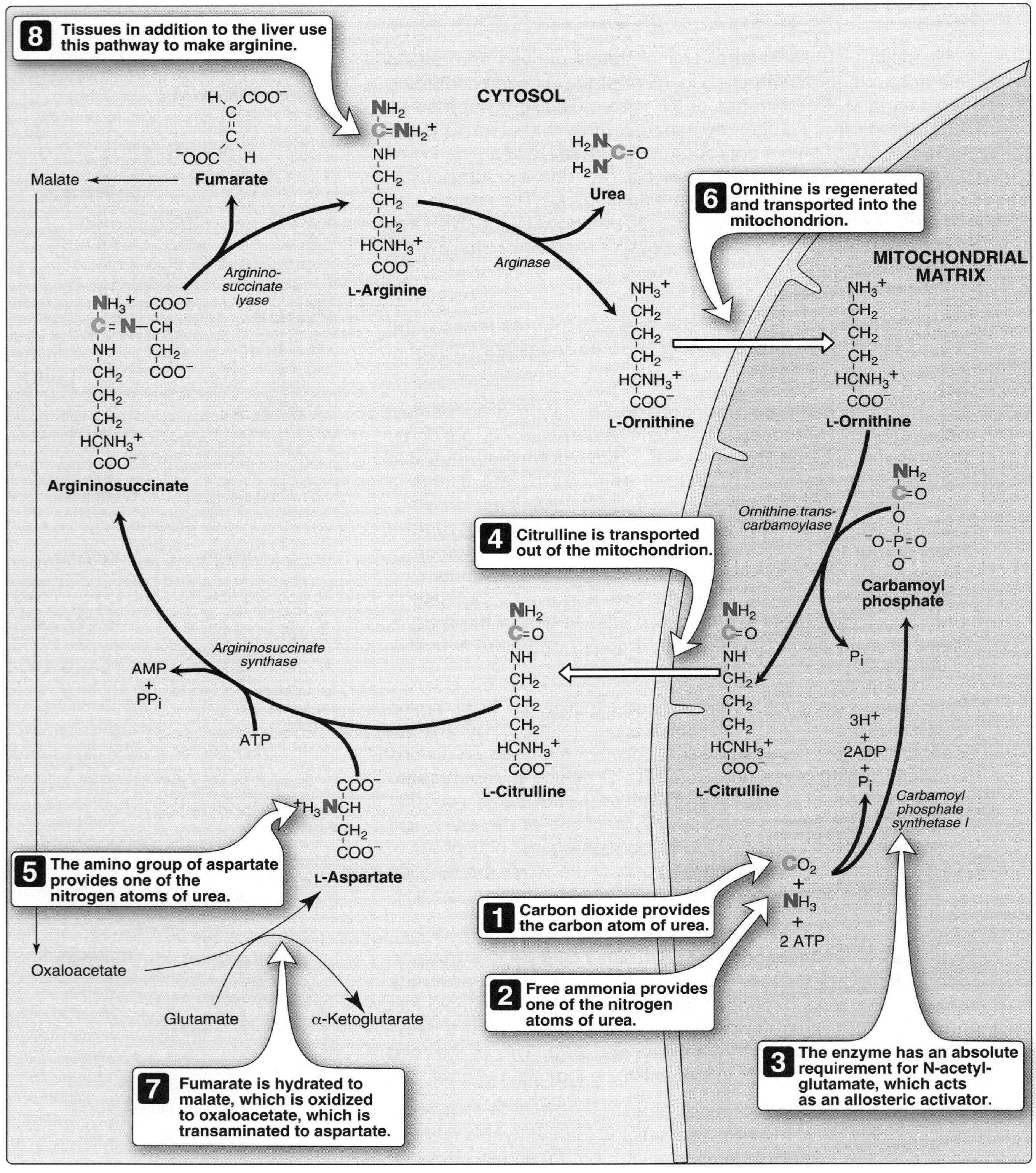

Figure 19.14
Reactions of the urea cycle.

the TCA cycle. Alternatively, cytosolic malate can be oxidized to oxaloacetate, which can be converted to aspartate (see Figure 19.8) or glucose (see p. 185).

5. **Cleavage of arginine to ornithine and urea:** *Arginase* cleaves arginine to ornithine and urea, and occurs almost exclusively in the liver. Thus, whereas other tissues, such as the kidney, can synthesize arginine by these reactions, only the liver can cleave arginine and, thereby, synthesize urea.

6. **Fate of urea:** Urea diffuses from the liver, and is transported in the blood to the kidneys, where it is filtered and excreted in the urine. A portion of the urea diffuses from the blood into the intestine, and is cleaved to CO_2 and NH_3 by bacterial *urease*. This ammonia is partly lost in the feces, and is partly reabsorbed into the blood. In patients with kidney failure, plasma urea levels are elevated, promoting a greater transfer of urea from blood into the gut. The intestinal action of *urease* on this urea becomes a clinically important source of ammonia, contributing to the **hyperammonemia** often seen in these patients. Oral administration of neomycin[1] reduces the number of intestinal bacteria responsible for this NH_3 production.

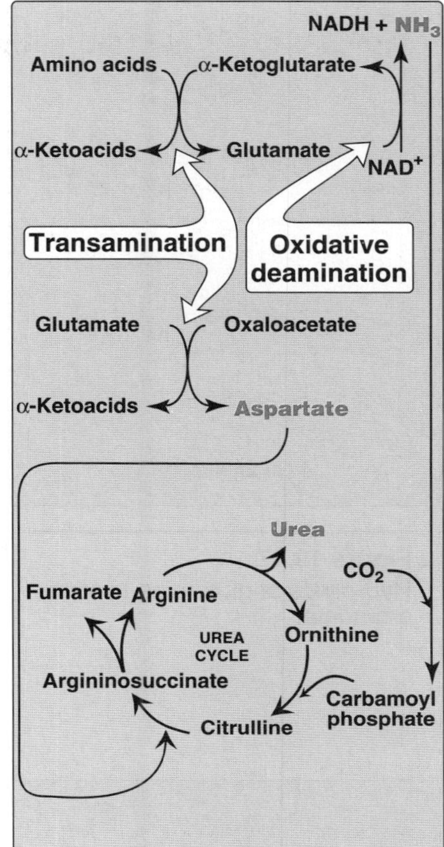

Figure 19.15
Flow of nitrogen from amino acids to urea. Amino groups for urea synthesis are collected in the form of ammonia and aspartate.

B. Overall stoichiometry of the urea cycle

$$\text{Aspartate} + NH_3 + CO_2 + 3\,ATP \longrightarrow$$
$$\text{Urea} + \text{fumarate} + 2\,ADP + AMP + 2\,P_i + PP_i + 3\,H_2O$$

Four high-energy phosphates are consumed in the synthesis of each molecule of urea: two ATP are needed to restore two ADP to two ATP, plus two to restore AMP to ATP. Therefore, the synthesis of urea is irreversible, with a large, negative ΔG (see p. 70). One nitrogen of the urea molecule is supplied by free NH_3, and the other nitrogen by aspartate. Glutamate is the immediate precursor of both ammonia (through oxidative deamination by *glutamate dehydrogenase*) and aspartate nitrogen (through transamination of oxaloacetate by *aspartate aminotransferase*). In effect, both nitrogen atoms of urea arise from glutamate, which, in turn, gathers nitrogen from other amino acids (Figure 19.15).

C. Regulation of the urea cycle

N-Acetylglutamate is an essential activator for *carbamoyl phosphate synthetase I*—the rate-limiting step in the urea cycle (see Figure 19.14). N-Acetylglutamate is synthesized from acetyl CoA and glutamate (Figure 19.16), in a reaction for which arginine is an activator. Therefore, the intrahepatic concentration of N-acetylglutamate increases after ingestion of a protein-rich meal, which provides both the substrate (glutamate) and the regulator of N-acetylglutamate synthesis. This leads to an increased rate of urea synthesis.

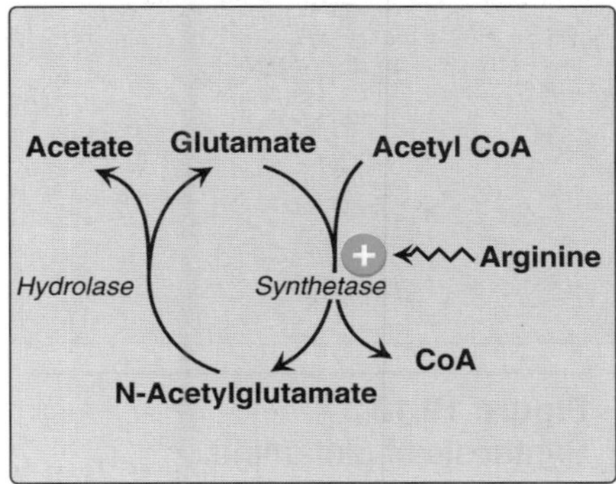

Figure 19.16
Formation and degradation of N-acetylglutamate, an allosteric activator of *carbamoyl phosphate synthetase I*.

[1]See Chapter 33 in ***Lippincott's Illustrated Reviews: Pharmacology*** (3rd Ed.) and Chapter 31 (2nd Ed.) for a discussion of the antibiotic, neomycin.

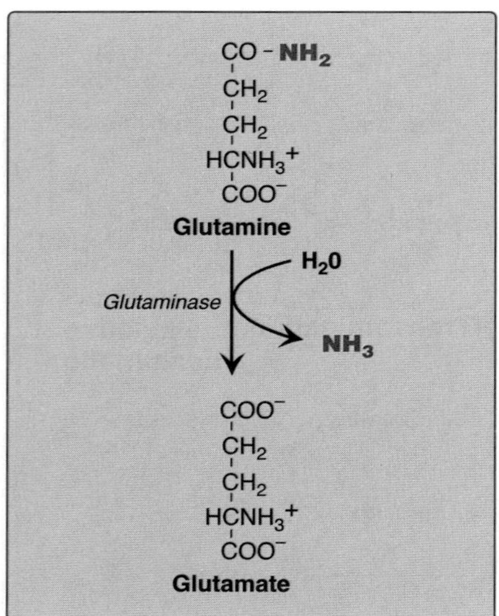

Figure 19.17
Hydrolysis of glutamine to form ammonia.

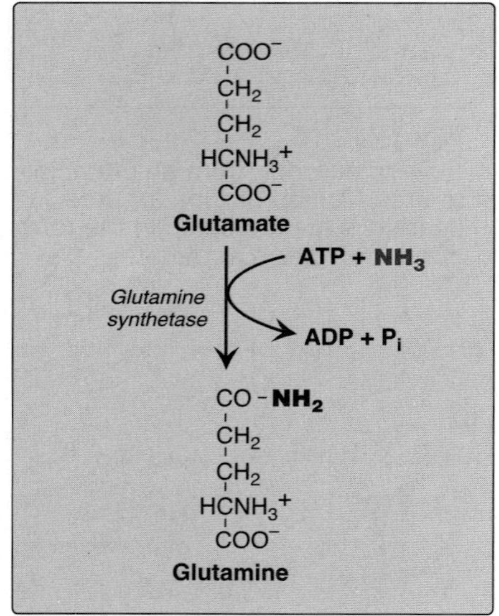

Figure 19.18
Synthesis of glutamine.

VII. METABOLISM OF AMMONIA

Ammonia is produced by all tissues during the metabolism of a variety of compounds, and it is disposed of primarily by formation of urea in the liver. However, the level of ammonia in the blood must be kept very low, because even slightly elevated concentrations (**hyperammonemia**) are toxic to the central nervous system (CNS). There must, therefore, be a metabolic mechanism by which nitrogen is moved from peripheral tissues to the liver for ultimate disposal as urea, while at the same time low levels of circulating ammonia must be maintained.

A. Sources of ammonia

Amino acids are quantitatively the most important source of ammonia, because most Western diets are high in protein and provide excess amino acids, which are deaminated to produce ammonia. However, substantial amounts of ammonia can be obtained from other sources.

1. **From amino acids:** Many tissues, but particularly the liver, form ammonia from amino acids by the *aminotransferase* and *glutamate dehydrogenase* reactions previously described.
2. **From glutamine:** The kidneys form ammonia from glutamine by the action of **renal** *glutaminase* (Figure 19.17). Most of this ammonia is excreted into the urine as NH_4^+, which provides an important mechanism for maintaining the body's acid-base balance. Ammonia is also obtained from the hydrolysis of glutamine by **intestinal** *glutaminase*. The intestinal mucosal cells obtain glutamine either from the blood or from digestion of dietary protein.
3. **From bacterial action in the intestine:** Ammonia is formed from urea by the action of bacterial *urease* in the lumen of the intestine. This ammonia is absorbed from the intestine by way of the portal vein and is almost quantitatively removed by the liver via conversion to urea.
4. **From amines:** Amines obtained from the diet, and monoamines that serve as hormones or neurotransmitters, give rise to ammonia by the action of *amine oxidase* (see p. 284 for the degradation of catecholamines).
5. **From purines and pyrimidines:** In the catabolism of purines and pyrimidines, amino groups attached to the rings are released as ammonia.

B. Transport of ammonia in the circulation

Although ammonia is constantly produced in the tissues, it is present at very low levels in blood. This is due both to the rapid removal of blood ammonia by the liver, and the fact that many tissues, particularly muscle, release amino acid nitrogen in the form of glutamine or alanine, rather than as free ammonia (see Figure 19.13).

1. **Urea:** Formation of urea in the liver is quantitatively the most important disposal route for ammonia. Urea travels in the blood from the liver to the kidneys, where it passes into the glomerular filtrate.

2. **Glutamine:** This amide of glutamic acid provides a nontoxic storage and transport form of ammonia (Figure 19.18). The ATP–requiring formation of glutamine from glutamate and ammonia by *glutamine synthetase* occurs primarily in the muscle and liver, but is also important in the nervous system, where it is the major mechanism for the removal of ammonia in the brain. Glutamine is found in plasma at concentrations higher than other amino acids—a finding consistent with its transport function. Circulating glutamine is removed by the kidneys and deaminated by *glutaminase*. The metabolism of ammonia is summarized in Figure 19.19.

C. Hyperammonemia

The capacity of the hepatic urea cycle exceeds the normal rates of ammonia generation, and the levels of serum ammonia are normally low (5 to 50 µmol/L). However, when the liver function is compromised, due either to genetic defects of the urea cycle, or liver disease, blood levels can rise above 1000 µmol/L. Such hyperammonemia is a medical emergency, because ammonia has a direct neurotoxic effect on the CNS. For example, elevated concentrations of ammonia in the blood cause the symptoms of **ammonia intoxication**, which include tremors, slurring of speech, somnolence, vomiting, cerebral edema, and blurring of vision. At high concentrations, ammonia can cause coma and death. The two major types of hyperammonemia are:

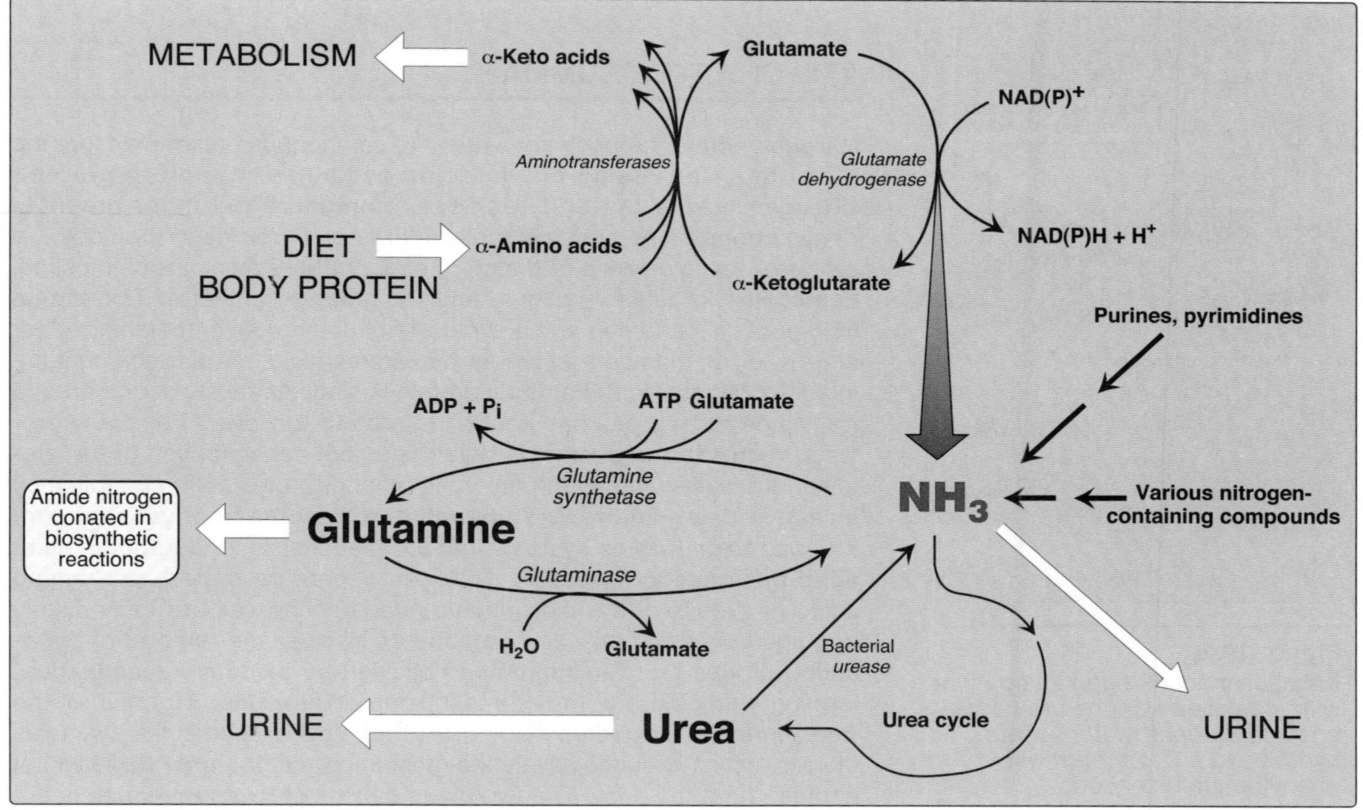

Figure 19.19
Metabolism of ammonia.

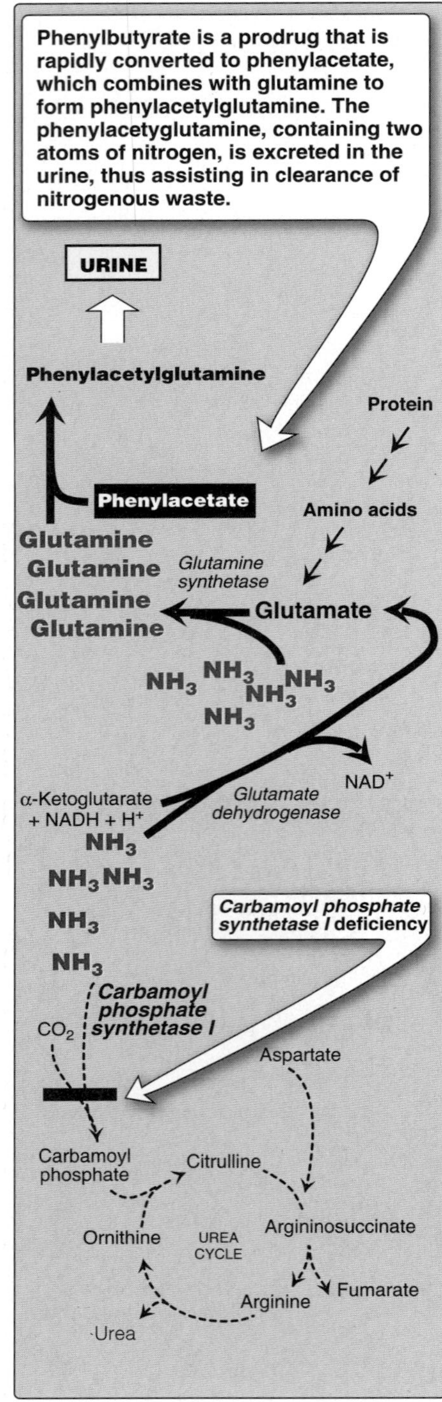

Figure 19.20
Metabolism of nitrogen in a patient with a deficiency in the urea cycle enzyme *carbamoyl phosphate synthetase I.* Treatment with phenylbutyrate converts nitrogenous waste to a form that can be excreted.

1. **Acquired hyperammonemia:** Liver disease is a common cause of hyperammonemia in adults. It may be a result of an acute process, for example, viral hepatitis, ischemia, or hepatotoxins. Cirrhosis of the liver caused by alcoholism, hepatitis, or biliary obstruction may result in formation of collateral circulation around the liver. As a result, portal blood is shunted directly into the systemic circulation and does not have access to the liver. The detoxification of ammonia (that is, its conversion to urea) is, therefore, severely impaired, leading to elevated levels of circulating ammonia.

2. **Hereditary hyperammonemia:** Genetic deficiencies of each of the five enzymes of the **urea cycle** have been described, with an overall prevalence estimated to be 1 in 30,000 live births. *Ornithine transcarbamoylase* deficiency, which is X-linked, is the most common of these disorders, affecting males predominantly, although female carriers have been clinically affected. All of the other urea cycle disorders follow an autosomal recessive inheritence pattern. In each case, the failure to synthesize urea leads to hyperammonemia during the first weeks following birth. All inherited deficiencies of the urea cycle enzymes result in mental retardation. Treatment includes limiting protein in the diet, and administering compounds that bind covalently to amino acids, producing nitrogen-containing molecules that are excreted in the urine. For example, phenylbutyrate given orally is converted to phenylacetate. This condenses with glutamine to form phenylacetylglutamine, which is exceted (Figure 19.20).

VIII. CHAPTER SUMMARY

Nitrogen enters the body in a variety of compounds present in food, the most important being amino acids contained in **dietary protein**. **Nitrogen leaves** the body as **urea**, **ammonia**, and other products derived from amino acid metabolism. Free amino acids in the body are produced by hydrolysis of dietary protein in the stomach and intestine, degradation of tissue proteins, and by de novo synthesis. This **amino acid pool** is consumed in the synthesis of body protein, metabolized for energy, or its members serve as precursors for other nitrogen-containing compounds. Note that body protein is simultaneously degraded and resynthesized—a process known as **protein turnover**. For many proteins, **regulation of synthesis** determines the concentration of the protein in the cell, whereas the amounts of other proteins are controlled by **selective degradation**. The **ubiquitin/proteasome** and **lysosome** are the two major enzyme systems that are responsible for **degrading damaged or unneeded proteins**. Nitrogen cannot be stored, and amino acids in excess of the biosynthetic needs of the cell are immediately degraded. The first phase of **catabolism** involves the removal of the α-amino groups by **transamination**, followed by **oxidative deamination**, forming **ammonia** and the corresponding **α-ketoacids**. A portion of the **free ammonia** is excreted in the **urine**, but most is used in the synthesis of **urea**, which is quantitatively the most important route for disposing of nitrogen from the body. The two major causes of **hyperammonemia** are liver disease and inherited deficiencies of enzymes in the urea cycle.

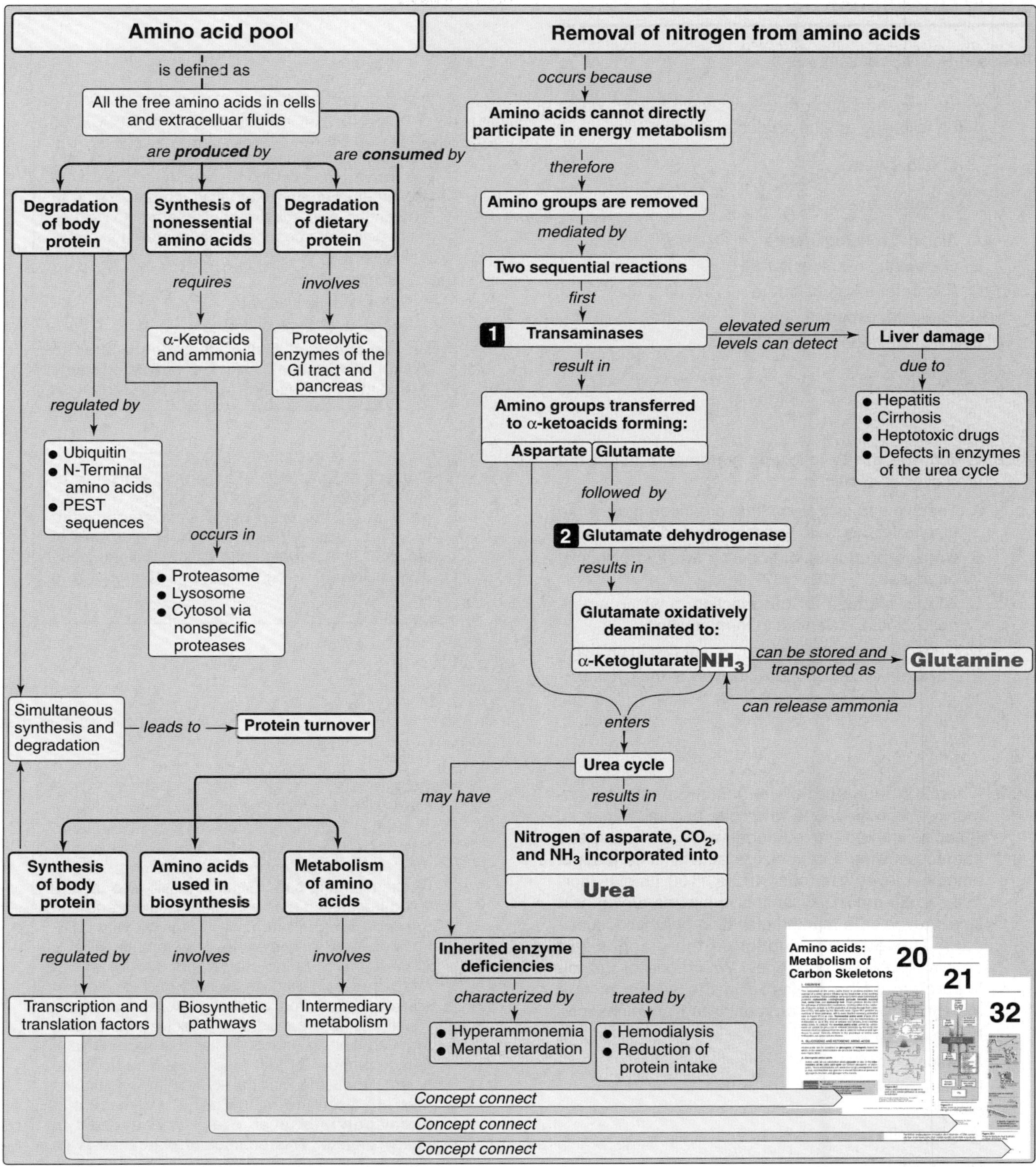

Figure 19.21
Key concept map for nitrogen metabolism.

Study Questions:

Choose the ONE best answer

19.1 In the transamination reaction shown below, which of the following are the products, X and Y?

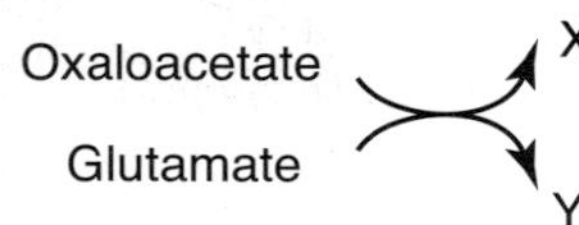

A. Alanine, α-ketoglutarate
B. Glutamate, α-ketoglutarate
C. Asparate, α-ketoglutarate
D. Pyruvate, asparate
E. Pyruvate, alanine

Correct answer = C. Transaminase reactions always have an amino acid and an α-keto acid as substrates. The products of the reaction are also an amino acid (corresponding to the α-keto substrate) and an α-keto acid (corresponding to the amino acid substrate). Three amino acid α-keto acid pairs commonly encountered in metabolism are:

alanine/pyruvate
asparate/oxaloacetate
glutamate/α-ketoglutarate

In this question, glutamate is deaminated to form α-ketoglutarate, and oxaloacetate is aminated to form asparate.

19.2 Which one of the following statements about the urea cycle is correct?

A. The two nitrogen atoms that are incorporated into urea enter the cycle as ammonia and alanine.
B. Urea is produced directly by the hydrolysis of ornithine.
C. ATP is required for the reaction in which argininosuccinate is cleaved to form arginine.
D. Urinary urea is increased by a diet rich in protein.
E. The urea cycle occurs exclusively in the cytosol.

Correct answer = D. The amino nitrogen of dietary protein is excreted as urea. The two nitrogens enter the urea cycle as ammonia and aspartate. Urea is produced by the hydrolysis of arginine. The cleavage of argininosuccinate does not require ATP. The urea cycle occurs partly in the mitochondria.

19.3 A female neonate did well initially until approximately 24 hours of age when she became lethargic. A sepsis workup proved negative. At 56 hours, she started showing focal seizure activity. The plasma ammonia level was found to be 1100 µmol/L (normal 5 to 50 µmol/L. Quantitative plasma amino acid levels revealed a marked elevation of argininosuccinate. These results supported the diagnosis of argininosuccinase deficiency. Which one of the following would be elevated in the serum of this patient, in addition to ammonia and argininosuccinate?

A. Asparagine
B. Glutamine
C. Lysine
D. Urea
E. Uric acid

Correct answer = B. Genetic deficiencies of each of the five enzymes of the urea cycle have been described. In each case, the failure to synthesize urea leads to hyperammonemia during the first weeks following birth. Glutamine will also be elevated because it acts as a non-toxic storage and transport form of ammonia. Thus, elevated glutamine always accompanies hyperammmonemia. Asparagine does not serve this sequesterisng role. Urea would be decreased due to impared activity of the urea cycle. Lysine and uric acid would not be elevated. Treatment of this patient includes limiting protein in the diet and administering compounds that bind covalently to amino acids, producing nitrogen-containing molecules that are excreted in the urine. For example, phenylbutyrate given orally is converted to phenylacetate. This compound condenses with glutamine to form phenylacetylglutamine, which is excreted.

Amino Acid Degradation and Synthesis

20

I. OVERVIEW

The catabolism of the amino acids found in proteins involves the removal of α-amino groups, followed by the breakdown of the resulting carbon skeletons. These pathways converge to form seven intermediate products: **oxaloacetate**, **α-ketoglutarate**, **pyruvate**, **fumarate**, **succinyl CoA**, **acetyl CoA**, and **acetoacetyl CoA**. These products directly enter the pathways of intermediary metabolism, resulting either in the synthesis of glucose or lipid, or in the production of energy through their oxidation to CO_2 and water by the citric acid cycle. Figure 20.1 provides an overview of these pathways, with a more detailed summary presented later in Figure 20.14 (p. 267). **Nonessential amino acids** (Figure 20.2) can be synthesized in sufficient amounts from the intermediates of metabolism or, as in the case of cysteine and tyrosine, from essential amino acids. In contrast, the **essential amino acids** cannot be synthesized (or produced in sufficient amounts) by the body and, therefore, must be obtained from the diet in order for normal protein synthesis to occur. Genetic defects in the pathways of amino acid metabolism can cause serious disease.

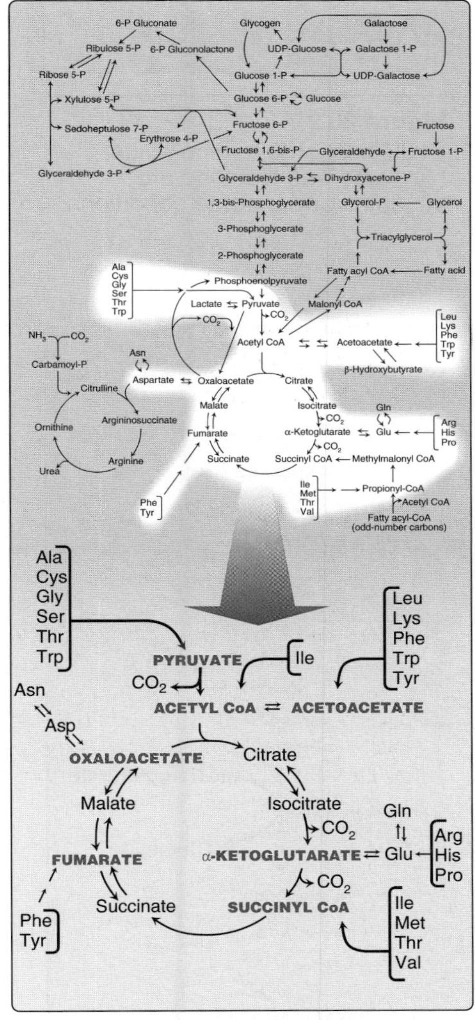

Figure 20.1
Amino acid metabolism shown as a part of the central pathways of energy metabolism.

II. GLUCOGENIC AND KETOGENIC AMINO ACIDS

Amino acids can be classified as **glucogenic** or **ketogenic** based on which of the seven intermediates are produced during their catabolism (see Figure 20.2).

A. Glucogenic amino acids

Amino acids whose catabolism yields **pyruvate** or one of the **intermediates of the citric acid cycle** are termed glucogenic or glycogenic. These intermediates are substrates for gluconeogenesis (see p. 115) and, therefore, can give rise to the net formation of glucose or glycogen in the liver and glycogen in the muscle.

Color-coding used in this chapter:

- BLUE CAPS TEXT = names of seven products of amino acid metabolism
- Red text = names of glucogenic amino acids
- Brown text = names of glucogenic and ketogenic amino acids
- Green text = names of ketogenic amino acids
- Cyan = one-carbon compounds

	Glucogenic	Glucogenic and Ketogenic	Ketogenic
Nonessential	Alanine Arginine* Asparagine Aspartate Cysteine Glutamate Glutamine Glycine Histidine* Proline Serine	Tyrosine	
Essential	Methionine Threonine Valine	Isoleucine Phenyl-alanine Tryptophan	Leucine Lysine

Figure 20.2
Classification of amino acids. *Arginine and histidine are essential under some conditions.

B. Ketogenic amino acids

Amino acids whose catabolism yields either **acetoacetate** or one of its precursor, (**acetyl CoA** or **acetoacetyl CoA**) are termed ketogenic (see Figure 20.2). Acetoacetate is one of the "ketone bodies," which also include 3-hydroxybutyrate and acetone. (See p. 193 for a discussion of ketone body metabolism.) Leucine and lysine are the only exclusively ketogenic amino acids found in proteins. Their carbon skeletons are not substrates for gluconeogenesis and, therefore, cannot give rise to the net formation of glucose or glycogen in the liver, or glycogen in the muscle.

III. CATABOLISM OF THE CARBON SKELETONS OF AMINO ACIDS

The pathways by which amino acids are catabolized are conveniently organized according to which one (or more) of the seven intermediates listed above is produced from a particular amino acid.

A. Amino acids that form oxaloacetate

Asparagine is hydrolyzed by *asparaginase*, liberating ammonia and aspartate (Figure 20.3). [Note: Some rapidly dividing leukemic cells are unable to synthesize sufficient asparagine to support their growth. This makes asparagine an essential amino acid for these cells, which therefore require asparagine from the blood. *Asparaginase*, which hydrolyzes asparagine to aspartate, can be administered systemically to treat leukemic patients.[1] *Asparaginase* lowers the level of asparagine in the plasma and, therefore, deprives cancer cells of a required nutrient.] **Aspartate** loses its amino group by transamination to form oxaloacetate (see Figure 20.3).

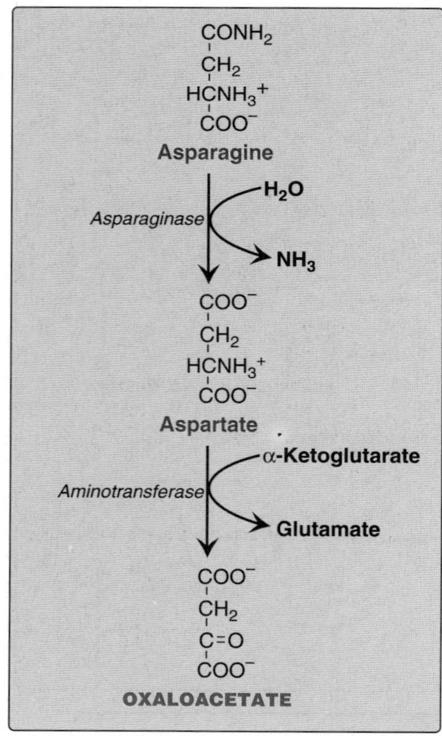

Figure 20.3
Metabolism of asparagine and aspartate.

B. Amino acids that form α-ketoglutarate

1. **Glutamine** is converted to glutamate and ammonia by the enzyme *glutaminase* (see p. 254). **Glutamate** is converted to α-ketoglutarate by transamination, or through oxidative deamination by *glutamate dehydrogenase* (see p. 249).

2. **Proline** is oxidized to glutamate. Glutamate is transaminated or oxidatively deaminated to form α-ketoglutarate.

3. **Arginine** is cleaved by *arginase* to produce ornithine. [Note: This reaction occurs primarily in the liver as part of the **urea cycle** (see p. 253).] Ornithine is subsequently converted to α-ketoglutarate.

4. **Histidine** is oxidatively deaminated by *histidase* to urocanic acid, which subsequently forms **N-formiminoglutamate** (**FIGlu**, Figure 20.4). FIGlu donates its formimino group to tetrahydrofolate, leaving glutamate, which is degraded as described above. [Note: Individuals deficient in folic acid excrete increased amounts of FIGlu in the urine, particularly after ingestion of a large dose of histidine. The **FIGlu excretion test** has been used in diagnosing a deficiency of folic acid.] (See p. 264 for a discussion of folic acid and one-carbon metabolism.)

[1]See Chapter 40 in ***Lippincott's Illustrated Reviews: Pharmacology*** (3rd Ed.) and Chapter 38 (2nd Ed.) for a discussion of the use of asparaginase as an antileukemic drug.

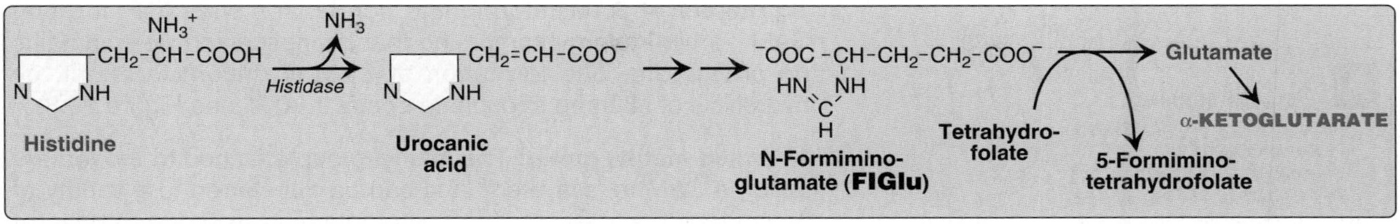

Figure 20.4
Degradation of histidine.

C. Amino acids that form pyruvate

1. **Alanine** loses its amino group by transamination to form pyruvate (Figure 20.5).

2. **Serine** can be converted to glycine and N^5,N^{10}-methylenetetrahydrofolate (Figure 20.6A). Serine can also be converted to pyruvate by *serine dehydratase* (Figure 20.6B). [Note: The role of tetrahydrofolate in the transfer of one-carbon units is presented on p. 265.]

3. **Glycine** can either be converted to serine by addition of a methylene group from N^5,N^{10}-methylenetetrahydrofolic acid (see Figure 20.6A), or oxidized to CO_2 and NH_4^+.

4. **Cystine** is reduced to cysteine, using NADH + H^+ as a reductant. **Cysteine** undergoes desulfuration to yield pyruvate.

5. **Threonine** is converted to pyruvate or to α-ketobutyrate, which forms succinyl CoA.

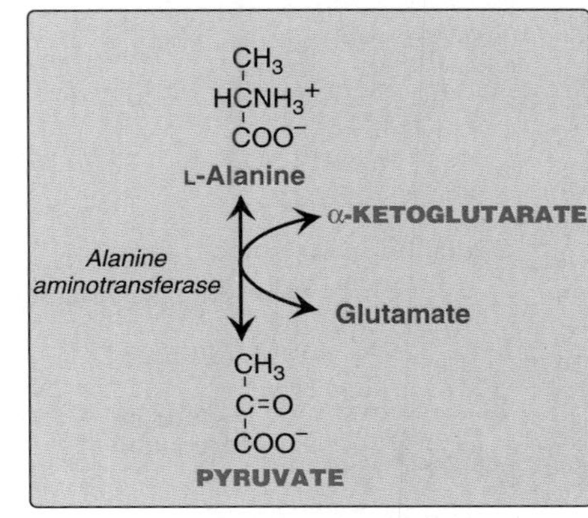

Figure 20.5
Transamination of alanine to form pyruvate.

D. Amino acids that form fumarate

1. **Phenylalanine and tyrosine:** Hydroxylation of phenylalanine leads to the formation of tyrosine (Figure 20.7). This reaction, catalyzed by *phenylalanine hydroxylase*, is the first reaction in the catabolism of phenylalanine. Thus, the metabolism of phenylalanine and tyrosine merge, leading ultimately to the formation of fumarate and acetoacetate. Phenylalanine and tyrosine are, therefore, both glucogenic and ketogenic.

2. **Inherited deficiencies** in the enzymes of phenylalanine and tyrosine metabolism lead to the diseases **phenylketonuria** (see p. 268), and **alkaptonuria** (see p. 272), and the condition of **albinism** (see p. 271).

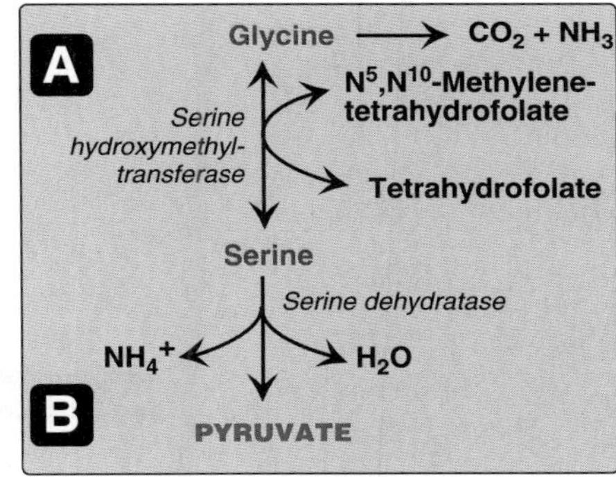

Figure 20.6
A. Interconversion of serine and glycine, and oxidation of glycine. B. Dehydration of serine to form pyruvate.

E. Amino acids that form succinyl CoA: Methionine

Methionine is one of four amino acids that form succinyl CoA. This sulfur-containing amino acid deserves special attention because it is converted to **S-adenosylmethionine (SAM)**, the major methyl-group donor in one-carbon metabolism (Figure 20.8). Methionine is also the source of **homocysteine**—a metabolite associated with **atherosclerotic vascular disease**.

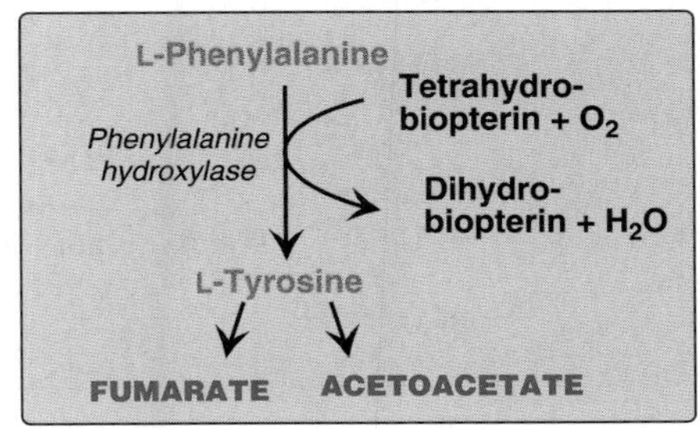

Figure 20.7
Degradation of phenylalanine.

1. **Synthesis of SAM:** Methionine condenses with ATP, forming SAM—a high-energy compound that is unusual in that it contains no phosphate. The formation of SAM is driven, in effect, by hydrolysis of all three phosphate bonds in ATP (see Figure 20.8).

2. **Activated methyl group:** The methyl group attached to the tertiary sulfur in SAM is "activated," and can be transferred to a variety of acceptor molecules, such as ethanolamine in the synthesis of choline (see p. 202). The methyl group is usually transferred to oxygen or nitrogen atoms, but sometimes to carbon atoms. The reaction product, S-adenosylhomocysteine, is a simple thioether, analogous to methionine. The resulting loss of free energy accompanying the reaction makes methyl transfer essentially irreversible.

3. **Hydrolysis of SAM:** After donation of the methyl group, S-adenosylhomocysteine is hydrolyzed to homocysteine and adenosine. **Homocysteine** has two fates. If there is a deficiency of methionine, homocysteine may be remethylated to methionine (see Figure 20.8). If methionine stores are adequate, homocysteine may enter the transsulfuration pathway, where it is converted to cysteine.

   a. **Resynthesis of methionine:** Homocysteine accepts a methyl group from N^5-methyltetrahydrofolate (N^5-methyl-THF) in a reaction requiring **methylcobalamin**, a coenzyme derived from **vitamin B_{12}** (see p. 373;). The methyl group is transferred from the B_{12} derivative to homocysteine, and cobalamin is recharged from N^5-methyl-THF.

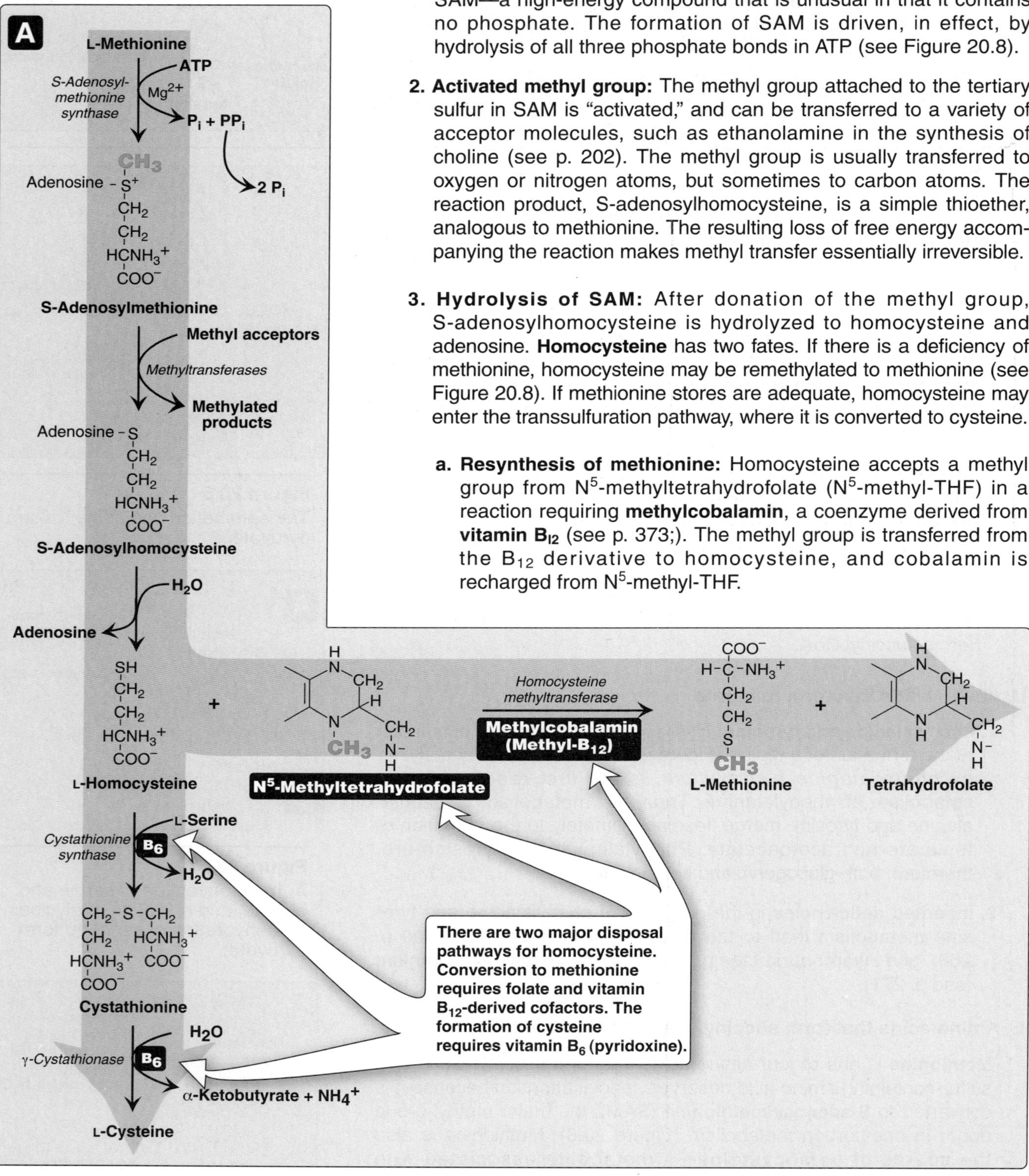

Figure 20.8
Degradation and resynthesis of methionine.

b. Synthesis of cysteine: Homocysteine combines with serine, forming cystathionine, which is hydrolyzed to α-ketobutyrate and cysteine (see Figure 20.8). This sequence has the net effect of converting serine to cysteine, and homocysteine to α-ketobutyrate, which is oxidatively decarboxylated to form propionyl CoA. Propionyl CoA is converted to succinyl CoA (see p. 191). Because homocysteine is synthesized from the essential amino acid methionine, cysteine is not an essential amino acid as long as sufficient methionine is available.

4. **Role of homocysteine in vascular disease:** Elevated plasma homocysteine levels are an independent cardiovascular risk factor that correlates with the severity of coronary artery disease. Dietary supplementation with folate, vitamin B_{12}, and vitamin B_6—the three vitamins involved in the metabolism of homocysteine—leads to a reduction in circulating levels of homocysteine. It is currently unknown if homocysteine-lowering therapy decreases heart disease in the general population. However, the benefits of such therapy can be shown in patients at high risk for vasular disease. For example, vitamin therapy significantly decreases the adverse events, such as reinfarction, in patients undergoing coronary angioplasty, and suggests there is a beneficial effect to reducing homocysteine levels (Figure 20.9). Note also that patients with **homocystinuria** (characterized by high serum levels of homocysteine caused by *cystathionine synthase* deficiency), experience premature vascular disease, and usually die of myocardial infarction, stroke, or pulmonary embolus. Thus, there is an association (but not a proven cause and effect relationship) of elevated homocysteine with cardiovascular disease.

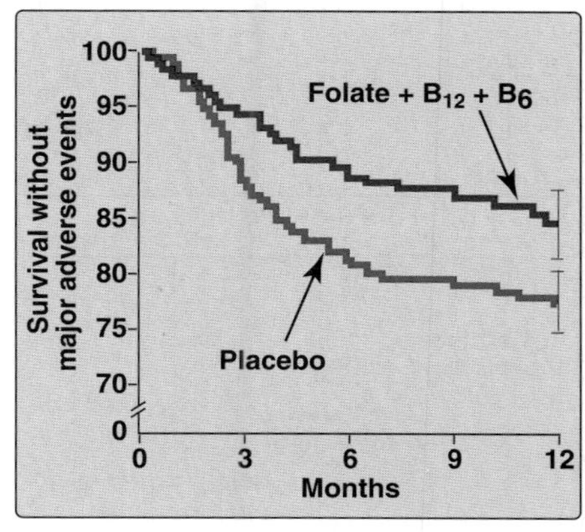

Figure 20.9
Effect of homocysteine-lowering therapy with folic acid, vitamin B_{12}, and vitamin B_6 on clinical outcome after coronary angioplasty. [Note: Balloon angioplasty is a noninvasive procedure in which a balloon-tipped catheter is introduced into a diseased blood vessel. As the balloon is inflated, the vessel opens further, allowing for placement of a stent and improved flow of blood.]

F. Other amino acids that form succinyl CoA

Degradation of valine, isoleucine, and threonine also results in the production of succinyl CoA—a TCA cycle intermediate and glucogenic compound.

1. **Valine** and **isoleucine** are branched-chain amino acids that yield succinyl CoA (Figure 20.10).

2. **Threonine** is dehydrated to α-ketobutyrate, which is converted to propionyl CoA, the precursor of succinyl CoA (see p. 191). [Note: Threonine can also be converted to pyruvate.]

G. Amino acids that form acetyl CoA or acetoacetyl CoA

Leucine, isoleucine, lysine, and tryptophan form acetyl CoA or acetoacetyl CoA directly, without pyruvate serving as an intermediate (through the *pyruvate dehydrogenase* reaction, see p. 107). As mentioned previously, **phenylalanine** and **tyrosine** also give rise to acetoacetate during their catabolism (see Figure 20.7). Therefore, there are a total of **six ketogenic amino acids**.

1. **Leucine** is exclusively **ketogenic** in its catabolism, forming acetyl CoA and acetoacetate (see Figure 20.10). The initial steps in the catabolism of leucine are similar to those of the other branched-chain amino acids, isoleucine and valine (see below).

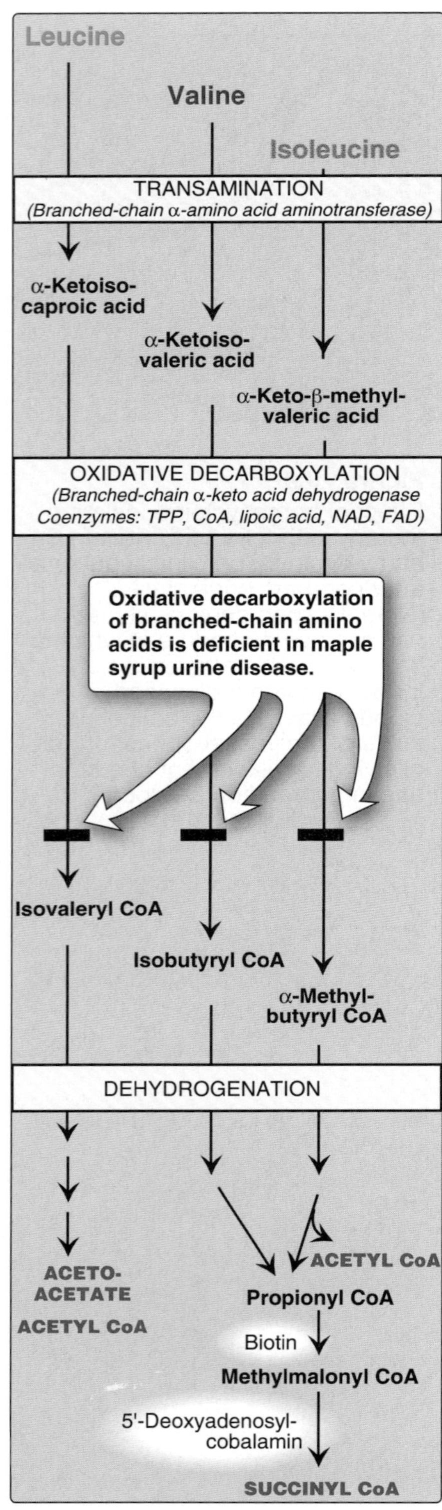

Figure 20.10
Degradation of leucine, valine, and isoleucine. TPP = thiamine pyrophosphate.

2. **Isoleucine** is both **ketogenic** and **glucogenic**, because its metabolism yields acetyl CoA and propionyl CoA. The first three steps in the metabolism of isoleucine are virtually identical to the initial steps in the degradation of the other branched-chain amino acids, valine and leucine (see Figure 20.10).

3. **Lysine**, an exclusively **ketogenic** amino acid, is unusual in that neither of its amino groups undergoes transamination as the first step in catabolism. Lysine is ultimately converted to acetoacetyl CoA.

4. **Tryptophan** is both **glucogenic** and **ketogenic** because its metabolism yields alanine and acetoacetyl CoA.

H. Catabolism of the branched-chain amino acids

The branched-chain amino acids, **isoleucine**, **leucine**, and **valine**, are essential amino acids. In contrast to other amino acids, they are metabolized primarily by the peripheral tissues (particularly muscle), rather than by the liver. Because these three amino acids have a similar route of catabolism, it is convenient to describe them as a group (see Figure 20.10).

1. **Transamination:** Removal of the amino groups of all three amino acids is catalyzed by a single enzyme, *branched-chain α-amino acid aminotransferase.*

2. **Oxidative decarboxylation:** Removal of the carboxyl group of the α-keto acids derived from leucine, valine, and isoleucine is also catalyzed by a single enzyme complex, *branched-chain α-keto acid dehydrogenase complex.* This complex uses thiamine pyrophosphate, lipoic acid, FAD, NAD^+, and coenzyme A as its coenzymes. [Note: This reaction is similar to the conversion of pyruvate to acetyl CoA by *pyruvate dehydrogenase*, (see p. 108) and the oxidation of α-ketoglutarate to succinyl CoA by *α-ketoglutarate dehydrogenase* (see p. 110).] An inherited deficiency of *branched-chain α-keto acid dehydrogenase* results in accumulation of the branched-chain keto acid substrates in the urine. Their sweet odor prompted the name **maple syrup urine disease** (see p. 270).

3. **Dehydrogenation:** Oxidation of the products formed in the above reaction yields **α-β-unsaturated acyl CoA derivatives**. This reaction is analagous to the dehydrogenation described in the β-oxidation scheme of fatty acid degradation (see p. 190).

4. **End products:** The catabolism of isoleucine ultimately yields acetyl CoA and succinyl CoA, rendering it both ketogenic and glucogenic. Valine yields succinyl CoA and is glucogenic. Leucine is ketogenic, being metabolized to acetoacetate and acetyl CoA.

IV. ROLE OF FOLIC ACID IN AMINO ACID METABOLISM

Some synthetic pathways require the addition of single carbon groups. These "one-carbon units" can exist in a variety of oxidation states. These include **methane**, **methanol**, **formaldehyde**, **formic acid**, and **carbonic acid**. It is possible to incorporate carbon units at each of these oxidation states, except methane, into other organic compounds. These single carbon units can be transferred from carrier compounds such as **tetrahydrofolic acid** and **S-adenosylmethionine** to specific structures that are being synthesized or modified. The "**one-carbon pool**" refers to single carbon units attached to this group of carriers. [Note: Carbonic acid—the hydrated form of CO_2—is carried by the vitamin biotin, which participates in carboxylation reactions, but is not considered a member of the one-carbon pool.]

A. Folic acid: a carrier of one-carbon units

The active form of folic acid, **tetrahydrofolic acid** (**THF**), is produced from folate by *dihydrofolate reductase* in a two-step reaction requiring two moles of NADPH. The carbon unit carried by THF is bound to nitrogen N^5 or N^{10}, or to both N^5 and N^{10}. THF allows one-carbon compounds to be recognized and manipulated by biosynthetic enzymes. Figure 20.11 shows the structures of the various members of the THF family, and indicates the sources of the one-carbon units and the synthetic reactions in which the specific members participate.

V. BIOSYNTHESIS OF NONESSENTIAL AMINO ACIDS

Nonessential amino acids are synthesized from intermediates of metabolism or, as in the case of tyrosine and cysteine, from essential amino acids. Two amino acids—histidine and arginine—are generally classified as nonessential. However, their normal concentrations are limitin, and, during periods of tissue growth (for example, in children or in individuals recovering from wasting diseases), histidine and arginine need to be supplemented in the diet. The synthetic reactions for the nonessential amino acids are described below, and are summarized in Figure 20.14. [Note: Some amino acids found in proteins, such as hydroxyproline and hydroxylysine (see p. 45). are modified after their incorporation into the protein (posttranslational modification, see p. 440).]

A. Synthesis from α-keto acids

Alanine, **aspartate**, and **glutamate** are synthesized by transfer of an amino group to the α-keto acids pyruvate, oxaloacetate, and α-ketoglutarate, respectively. These transamination reactions (Figure 20.12, and see p. 248) are the most direct of the biosynthetic pathways. Glutamate is unusual in that it can also be synthesized by the reverse of oxidative deamination, catalyzed by *glutamate dehydrogenase* (see p. 249).

B. Synthesis by amidation

1. **Glutamine:** This amino acid, which contains an amide linkage with ammonia at the γ-carboxyl, is formed from glutamate by *glutamine synthetase* (see Figure 19.18, p. 254) The reaction is driven by the hydrolysis of ATP. In addition to producing glutamine for protein synthesis, the reaction also serves as a major mechanism for the detoxification of ammonia in brain and liver (see p. 254 for a discussion of ammonia metabolism).

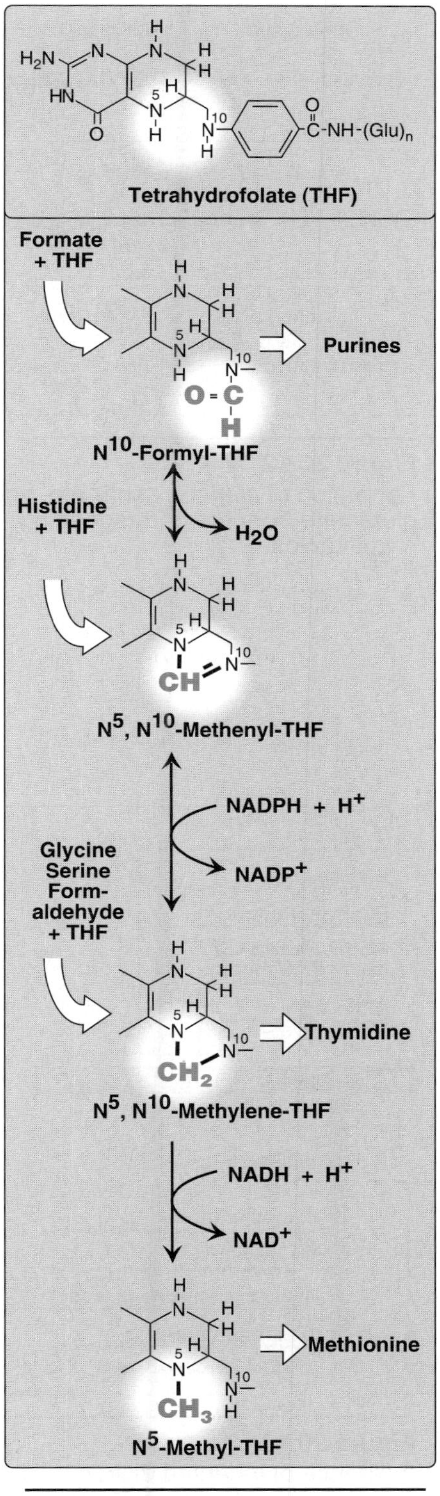

Figure 20.11
Summary of the interconversions and uses of the carrier, tetrahydrofolate.

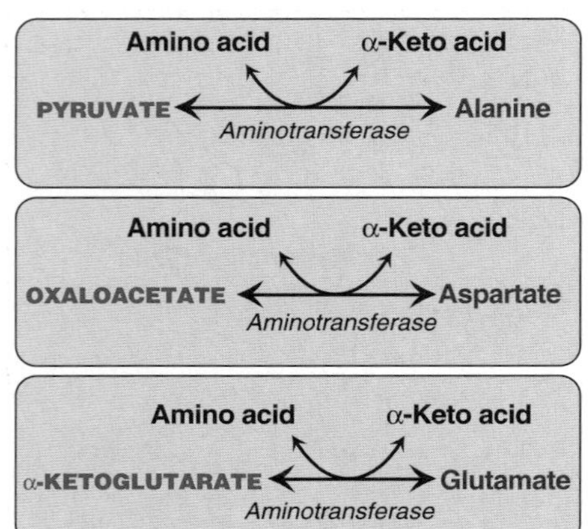

Figure 20.12
Formation of alanine, aspartate, and glutamate from the corresponding α-keto acids.

2. **Asparagine:** This amino acid, which contains an amide linkage with ammonia at the β-carboxyl, is formed from aspartate by *asparagine synthetase,* using glutamine as the amide donor. The reaction requires AT, and, like the synthesis of glutamine, has an equilibrium far in the direction of asparagine synthesis.

C. Proline

Glutamate is converted to proline by cyclization and reduction reactions.

D. Serine, glycine, and cysteine

1. **Serine** arises from 3-phosphoglycerate, an intermediate in glycolysis (see Figure 8.18, p. 99), which is first oxidized to 3-phosphopyruvate, and then transaminated to 3-phosphoserine. Serine is formed by hydrolysis of the phosphate ester. Serine can also be formed from glycine through transfer of a hydroxymethyl group (see Figure 20.6A).

2. **Glycine** is synthesized from serine by removal of a hydroxymethyl group, also by *serine hydroxymethyl transferase* (see Figure 20.6A).

3. **Cysteine** is synthesized by two consecutive reactions in which homocysteine combines with serine, forming cystathionine, which, in turn, is hydrolyzed to α-ketobutyrate and cysteine (see Figure 20.8). Homocysteine is derived from methionine as described on p. 262. Because methionine is an essential amino acid, cysteine synthesis can be sustained only if the dietary intake of methionine is adequate.

E. Tyrosine

Tyrosine is formed from phenylalanine by *phenylalanine hydroxylase.* The reaction requires molecular oxygen and the coenzyme **tetrahydrobiopterin**, which can be synthesized by the body. One atom of molecular oxygen becomes the hydroxyl group of tyrosine, and the other atom is reduced to water. During the reaction, tetrahydrobiopterin is oxidized to dihydrobiopterin. Tetrahydrobiopterin is regenerated from dihydrobiopterin in a separate reaction requiring NADPH. Tyrosine, like cysteine, is formed from an essential amino acid and, is therefore, nonessential only in the presence of adequate dietary phenylalanine.

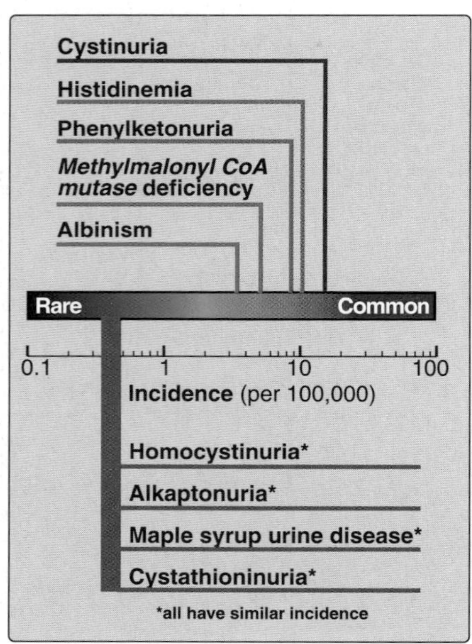

Figure 20.13
Incidence of inherited diseases of amino acid metabolism. [Note: Cystinuria is the most common genetic error of amino acid transport.]

VI. METABOLIC DEFECTS IN AMINO ACID METABOLISM

Inborn errors of metabolism are commonly caused by mutant genes that generally result in abnormal proteins, most often enzymes. The inherited defects may be expressed as a total loss of enzyme activity or, more frequently, as a partial deficiency in catalytic activity. Without treatment, the inherited defects of amino acid metabolism almost invariably result in mental retardation or other developmental abnormalities as a result of harmful accumulation of metabolites. Although more than fifty of these disorders have been described, many are rare, occurring less than 1 per 250,000 in most populations (Figure 20.13). Collectively, however, they constitute a very significant portion of pediatric genetic diseases. Figure 20.14 summarizes some of the more commonly encountered diseases

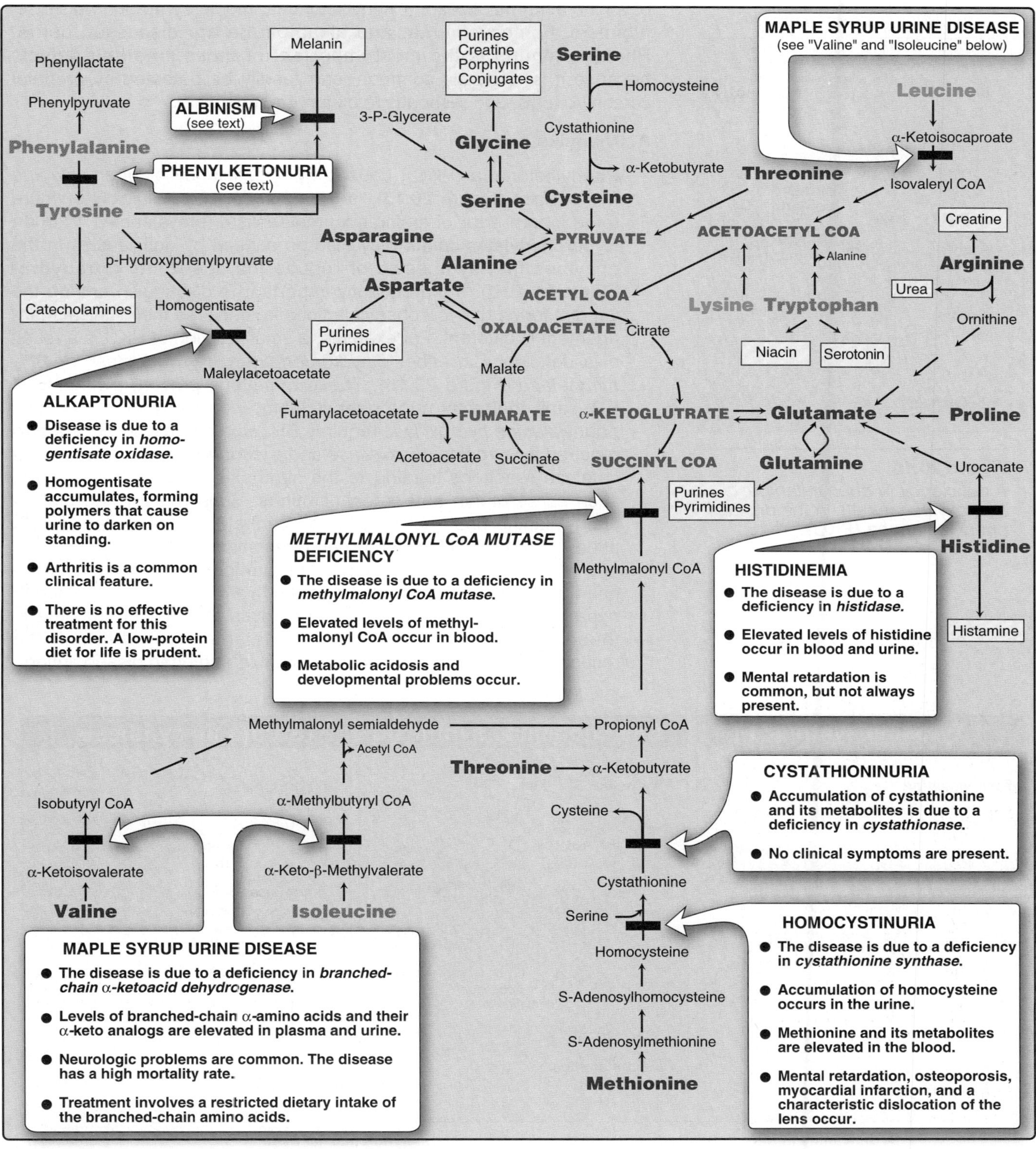

Figure 20.14
Summary of the metabolism of amino acids in humans. Genetically determined enzyme deficiencies are summarized in white boxes. Nitrogen-containing compounds derived from amino acids are shown in small, yellow boxes. Classification of amino acids is color coded: **Red** = glucogenic; **brown** = glucogenic and ketogenic; **green** = ketogenic. Compounds in **BLUE ALL CAPS** are the seven metabolites to which all amino acid metabolism converges.

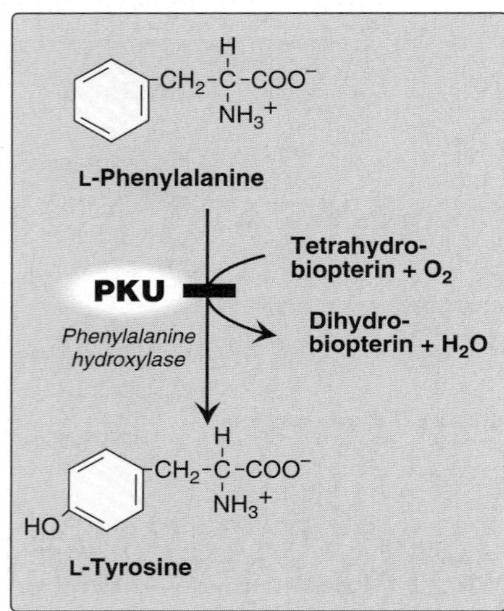

Figure 20.15
A deficiency in *phenylalanine hydroxylase* results in the disease phenylketonuria (PKU).

of amino acid metabolism. Phenylketonuria, maple syrup urine disease, albinism, homocystinuria, and alkaptonuria are discussed below. Phenylketonuria is the most important of these inherited defects because it is relatively common, can readily be detected by prenatal screening tests, and responds to dietary treatment.

A. Phenylketonuria

Phenylketonuria (PKU), caused by a deficiency of *phenylalanine hydroxylase* (Figure 20.15), is the most common clinically encountered inborn error of amino acid metabolism (prevalence 1:11,000). **Hyperphenylalaninemia** may also be caused by deficiencies in the enzymes that synthesize or reduce the coenzyme **tetrahydrobiopterin** (BH_4). It is frequently important to distinguish among the various forms of hyperphenylalaninemia, because their clinical management is different. For example, a small fraction of PKU is a result of a deficiency in either *dihydropteridine (BH_2) reductase* or *BH_2 synthetase* (Figure 20.16). These mutations prevent synthesis of BH_4, and indirectly raise phenylalanine concentrations, because *phenylalanine hydroxylase* requires BH_4 as a coenzyme. BH_4 is also required for *tyrosine hydroxylase* and *tryptophan hydroxylase*, which catalyze reactions leading to the synthesis of neurotransmitters, such as serotonin and catecholamines. Simply restricting dietary phenylalanine does not reverse the central nervous system (CNS) effects due to deficiencies in neurotransmitters. Replacement therapy with BH_4 or 3,4-dihydroxyphenylalanine and 5-hydroxytryptophan (products of the affected *tryosine hydroxylase*- and *tryptophan hydroxylase*-catalyzed reactions) improves the clinical outcome in these variant forms of hyperphenylalaninemia, although the response of these patients is unpredictable and often disappointing.

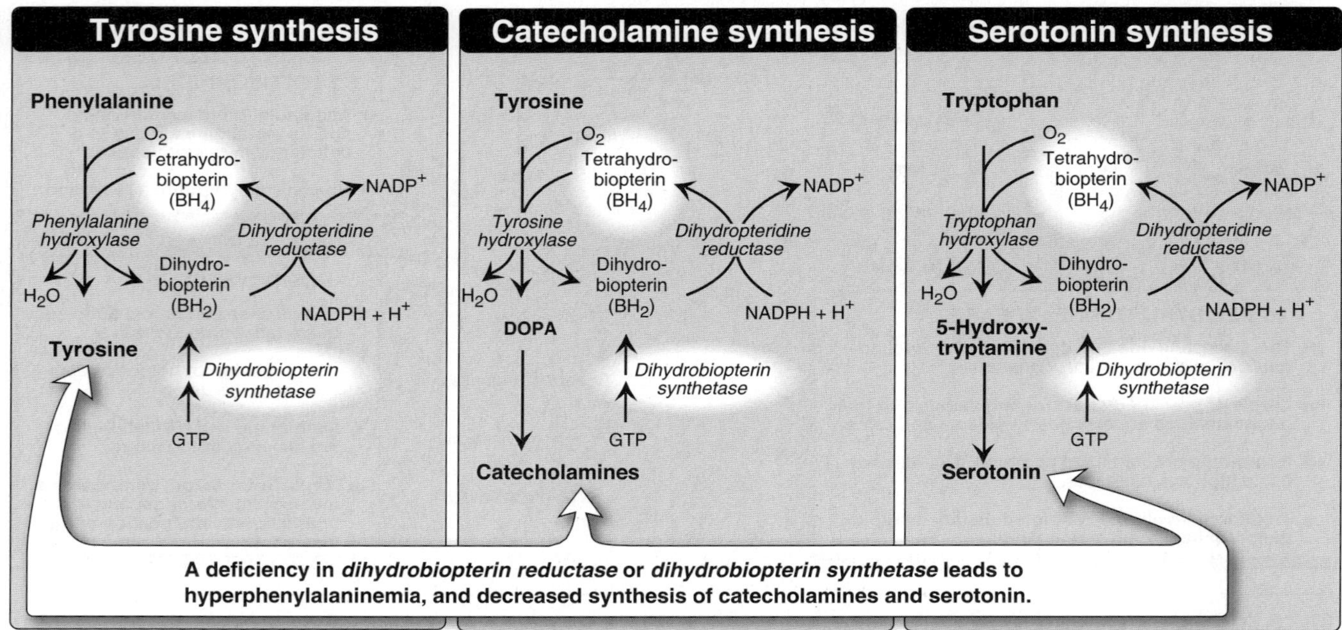

Figure 20.16
Biosynthetic reactions involving amino acids and tetrahydrobiopterin.

1. **Characteristics of PKU:**

   a. **Elevated phenylalanine:** Phenylalanine is present in elevated concentrations in tissues, plasma, and urine. Phenyllactate, phenylacetate, and phenylpyruvate, which are not normally produced in significant amounts in the presence of functional *phenylalanine hydroxylase*, are also elevated in PKU (Figure 20.17). These metabolites give urine a characteristic musty ("mousey") odor. [Note: The disease acquired its name before the phenylketone present in the urine was identified to be phenylpyruvate.]

   b. **CNS symptoms:** Mental retardation, failure to walk or talk, seizures, hyperactivity, tremor, microcephaly, and failure to grow are characteristic findings in PKU. The patient with untreated PKU typically shows symptoms of mental retardation by the age of one year. Virtually all untreated patients show an IQ below fifty (Figure 20.18).

   c. **Hypopigmentation:** Patients with phenylketonuria often show a deficiency of pigmentation (fair hair, light skin color, and blue eyes). The hydroxylation of tyrosine by *tyrosinase*, which is the first step in the formation of the pigment **melanin,** is competitively inhibited by the high levels of phenylalanine present in PKU.

2. **Neonatal diagnosis of PKU:** Early diagnosis of phenylketonuria is important because the disease is treatable by dietary means. Because of the lack of neonatal symptoms, laboratory testing for elevated blood levels of phenylalanine is mandatory for detection. However, the infant with PKU frequently has normal blood levels of phenylalanine at birth because the mother clears increased blood phenylalanine in her affected fetus through the placenta. Thus, tests performed at birth may show false negative results. Normal levels of phenylalanine may persist until the newborn is exposed to at least 24 hours of protein feeding. Blood levels of phenylalanine should be determined on a second blood sample obtained after the infant has ingested protein. Normally, feeding breast milk or formula for 48 hours is sufficient to raise the baby's blood phenylalanine to levels that can be used for diagnosis.

3. **Antenatal diagnosis of PKU:** Classic PKU is a family of diseases caused by any of forty or more different mutations in the gene that codes for *phenylalanine hydroxylase* (*PAH*). The frequency of any given mutation varies among populations, and the disease is often doubly heterozygous, that is, the *PAH* gene has a different mutation in each allele. Despite this complexity, the majority of PKU cases in most populations are caused by a small number of mutations (six to ten). A fetus can be tested in vitro to determine if it carries a PKU mutation (see p. 458).

4. **Treatment of PKU:** Most natural protein contains phenylalanine, and it is impossible to satisfy the body's protein requirement when ingesting a normal diet without exceeding the phenylalanine limit. Therefore, in PKU, blood phenylalanine is maintained in the normal range by feeding synthetic amino acid preparations low in phenylalanine, supplemented with some natural foods (such as

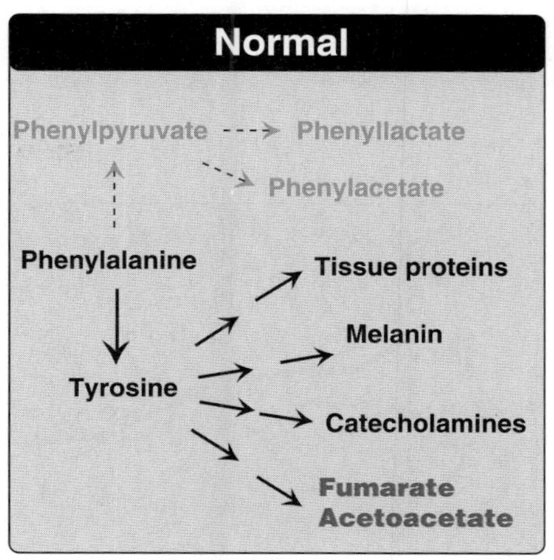

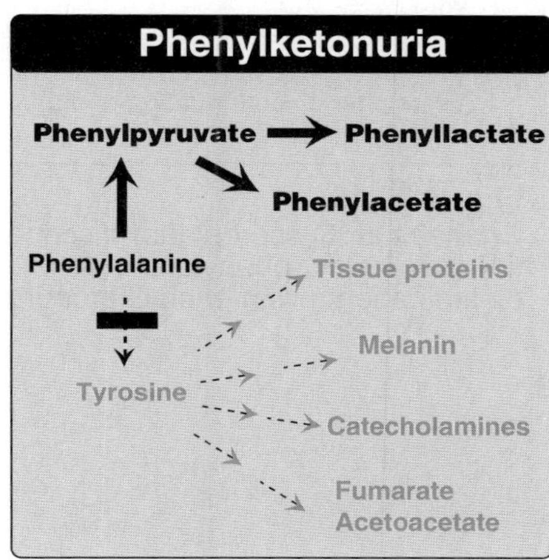

Figure 20.17
Pathways of phenylalanine metabolism in normal individuals and in patients with phenylketonuria.

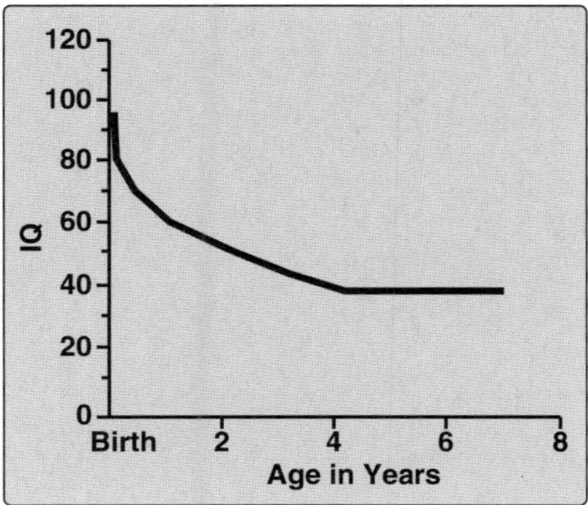

Figure 20.18
Typical intellectual ability in untreated PKU patients of different ages.

fruits, vegetables, and certain cereals) selected for their low phenylalanine content. The amount is adjusted according to the tolerance of the individual as measured by blood phenylalanine levels. The earlier treatment is started, the more completely neurologic damage can be prevented. [Note: Treatment must begin during the first seven to ten days of life to prevent mental retardation.] Because phenylalanine is an essential amino acid, overzealous treatment that results in blood phenylalanine levels below normal should be avoided because this can lead to poor growth and neurologic symptoms. [Note: Even with dietary treatment, patients with PKU show a mildly depressed IQ and an increased incidence of behavioral problems (depressive mood, anxiety, physical complaints, or social isolation).] In patients with PKU, tyrosine cannot be synthesized from phenylalanine and, therefore, it becomes an essential amino acid that must be supplied in the diet. Discontinuance of the phenyalanine-restricted diet before eight years of age is associated with poor performance on IQ tests. Adult PKU patients show deterioration of IQ scores after discontinuation of the diet (Figure 20.19). Life-long restriction of dietary phenylalanine is, therefore, recommended.

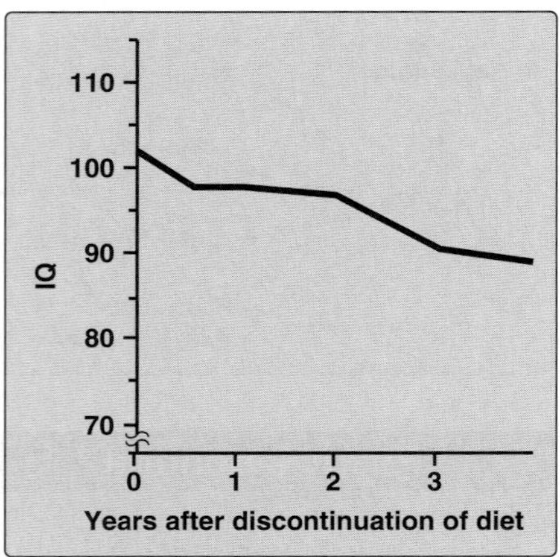

Figure 20.19
Changes in IQ scores after discontinuation of low-phenylalanine diet in patients with phenylketonuria.

5. **Maternal PKU:** When women with PKU who are not on a low phenylalanine diet become pregnant, the offspring are affected with "maternal PKU syndrome." High blood phenylalanine levels in the mother cause microcephaly, mental retardation, and congenital heart abnormalities in the fetus. Some of these developmental responses to high phenylalanine occur during the first months of pregnancy. Thus, dietary control of blood phenylalanine must begin prior to conception, and must be maintained throughout the pregnancy. Children borne to mothers with PKU in metabolic control often show some résidual developmental or behavioral effects, such as hyperactivity.

B. Maple Syrup Urine Disease

Maple syrup urine disease (MSUD) is a recessive disorder in which there is a partial or complete deficiency in *branched-chain α-ketoacid dehydrogenase*, an enzyme that decarboxylates leucine, isoleucine, and valine (see Figure 20.10). These amino acids and their corresponding α-keto acids accumulate in the blood, causing a toxic effect that interferes with brain functions. The disease is characterized by feeding problems, vomiting, dehydration, severe metabolic acidosis, and a characteristic maple syrup odor to the urine. If untreated, the disease leads to mental retardation, physical disabilities, and death.

1. **Classification:** The term maple syrup urine disease includes a classic type and several variant forms of the disorder.

   a. **Classic type:** This is the most common type of MSUD. Leukocytes or cultured skin fibroblasts from these patients show little or no *branched-chain α-ketoacid dehydrogenase* activity. Infants with classic MSUD show symptoms within the first several days of life.

b. Intermediate and intermittent forms: These patients have a higher level of enzyme activity (approximately three to fifteen percent of normal). The symptoms are milder and show an onset from infancy to adulthood.

c. Thiamin-responsive form: Large doses of thiamin can help patients with this rare variant of MSUD achieve increased *branched-chain α-ketoacid dehydrogenase* activity.

2. Treatment: The disease is treated with a synthetic formula that contains limited amounts of leucine, isoleucine, and valine—sufficient to provide the branched-chain amino acids necessary for normal growth and development without producing toxic levels. Infants suspected of having any form of MSUD should be tested within 24 hours of birth. Early diagnosis and treatment is essential if the child with MSUD is to develop normally.

C. Albinism

Albinism refers to a group of conditions in which a defect in tyrosine metabolism results in a deficiency in the production of melanin. These defects result in the partial or full absence of pigment from the skin, hair, and eyes. Albinism appears in different forms, and it may be inherited by one of several modes: autosomal recessive, autosomal dominant, or X-linked. Complete albinism (also called *tyrosinase*-negative **oculocutaneous albinism**) results from a deficiency of *tyrosinase* activity, causing a total absence of pigment from the hair, eyes, and skin (Figure 20.20). It is the most severe form of the condition. Affected people may appear to have white hair, skin, and iris color, and they may have vision defects. They also have photophobia (sunlight is painful to their eyes), they sunburn easily, and do not tan.

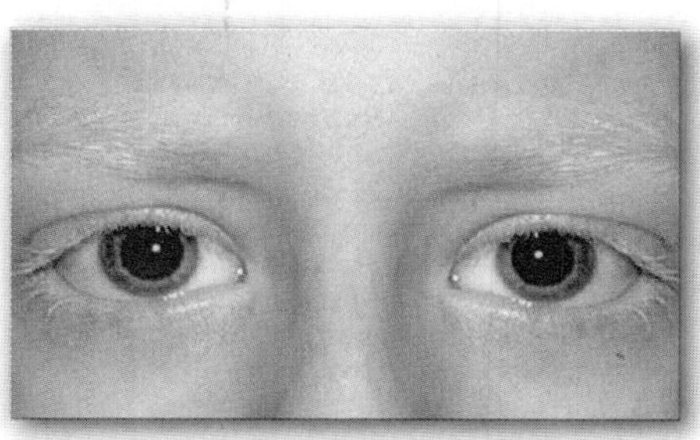

Figure 20.20
Patient with oculocutaneous albinism, showing blond hair and white eyebrows and lashes.

D. Homocystinuria

The homocystinurias are a group of disorders involving defects in the metabolism of homocysteine. The diseases are inherited as autosomal recessive illnesses, characterized by high plasma and urinary levels of homocysteine and methionine and low levels of cysteine. The most common cause of homocystinuria is a defect in the enzyme *cystathionine β-synthase*, which converts homocysteine to cystathionine (Figure 20.21). Individuals who are homozygous for *cystathionine β-synthase* deficiency exhibit ectopia lentis (displacement of the lens of the eye), skeletal abnormalities, premature arterial disease, osteoporosis, and mental retardation. Patients can be responsive or non-responsive to oral administration of pyridoxine (vitamin B_6)—a cofactor of *cystathionine β-synthase.* B_6-responsive patients usually have a milder and later onset of clinical symptoms compared with B_6-non-responsive patients. Treatment includes restriction of methionine intake and supplementation with vitamins B_6, B_{12}, and folate.

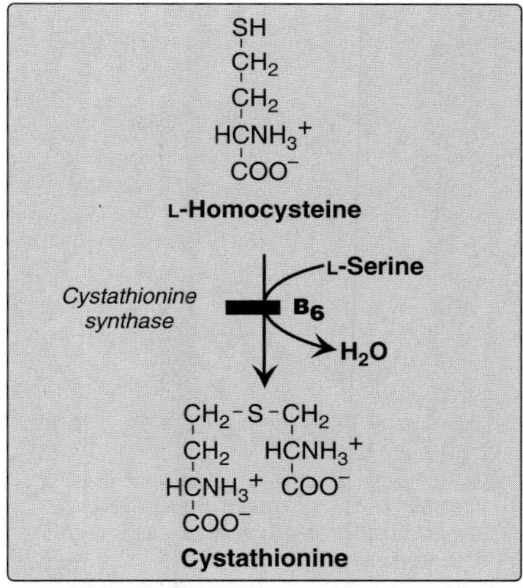

Figure 20.21
Enzyme deficiency in homocystinuria.

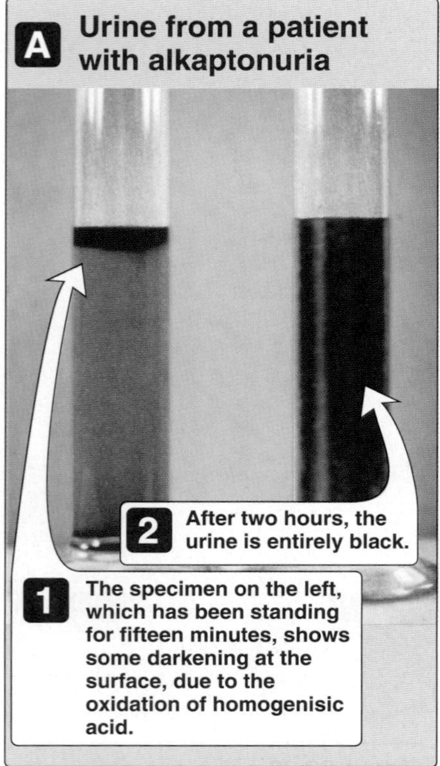

Figure 20.22
A patient with alkaptonuria.
A. Urine. B. Vertebrae.

E. Alkaptonuria

Alkaptonuria is a rare metabolic disease involving a deficiency in *homogentisic acid oxidase*, resulting in the accumulation of homogentisic acid. [Note: This reaction occurs in the degradative pathway of tyrosine.] The illness has three characteristic symptoms: **homogentisic aciduria** (the patient's urine contains elevated levels of homogentisic acid, which is oxidized to a dark pigment on standing, Figure 20.22A), **large joint arthritis**, and black ochronotic **pigmentation of cartilage** and collagenous tissue (Figure 20.22B). Patients with alkaptonuria are usually asymptomatic until about age forty. Dark staining of the diapers sometimes can indicate the disease in infants, but usually no symptoms are present until later in life. Diets low in protein—especially in phenylalanine and tyrosine—help reduce the levels of homogentisic acid, and decrease the amount of pigment deposited in body tissues. Although alkaptonuria is not life-threatening, the associated arthritis may be severely crippling.

VII. CHAPTER SUMMARY

Amino acids whose catabolism yields **pyruvate** or one of the **intermediates of the TCA cycle** are termed **glucogenic**. They can give rise to the net formation of **glucose** or **glycogen** in the **liver**, and **glycogen** in the **muscle**. The glucogenic amino acids are glutamine, glutamate, proline, arginine, histidine, alanine, serine, glycine, cysteine, threonine, phenylalanine, tyrosine, methionine, valine, isoleucine, threonine, aspartate, and asparagine. Amino acids whose catabolism yields either **acetoacetate** or one of its precursors, **acetyl CoA** or **acetoacetyl CoA**, are termed **ketogenic**. Tyrosine, phenylalanine, tryptophan, and isoleucine are both ketogenic and glucogenic. Leucine and lysine are solely ketogenic. **Non-essential** amino acids can be synthesized from metabolic intermediates, or from the carbon skeletons of essential amino acids. Non-essential amino acids include **alanine**, **aspartate**, **glutamate**, **glutamine**, **asparagine**, **proline**, **cysteine**, **serine**, **glycine**, and **tyrosine**. **Essential amino acids** need to be obtained from the **diet**. **Phenylketonuria (PKU)** is caused by a **deficiency** of ***phenylalanine hydroxylase***—the enzyme that converts phenylalanine to tyrosine. **Hyperphenylalaninemia** may also be caused by deficiencies in the enzymes that synthesize or reduce the *hydroxylase's* coenzyme, **tetrahydrobiopterin**. Untreated patients with PKU suffer from **mental retardation**, failure to walk or talk, seizures, hyperactivity, tremor, microcephaly, and failure to grow. Treatment involves controlling dietary phenylalanine. Note that tyrosine becomes an essential dietary component for people with PKU. **Maple syrup urine disease (MSUD)** is a recessive disorder in which there is a partial or complete **deficiency** in ***branched-chain α-ketoacid dehydrogenase***—an enzyme that decarboxylates **leucine**, **isoleucine**, and **valine**. Symptoms include feeding problems, vomiting, dehydration, severe metabolic acidosis, and a characteristic smell of the urine. If untreated, the disease leads to **mental retardation**, physical disabilities, and death. Treatment of MSUD involves a synthetic formula that contains limited amounts of leucine, isoleucine, and valine. Other important genetic diseases associated with amino acid metabolism include **albinism**, **homocystinuria**, **methylmalonyl CoA mutase deficiency**, **alkaptonuria**, **histidinemia**, and **cystathioninuria**.

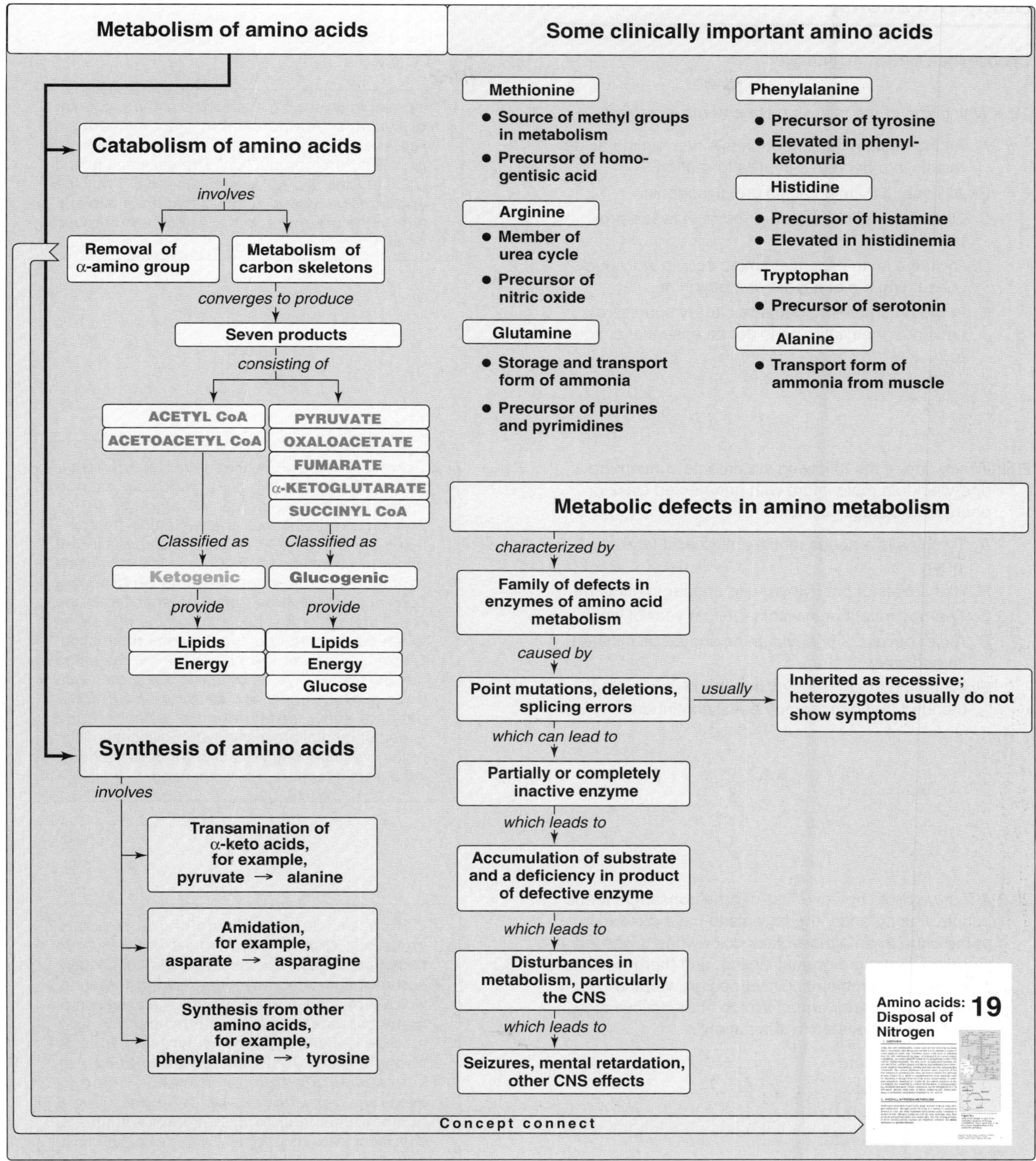

Figure 20.23
Key concept map for amino acid metabolism.

Study Questions:

Choose the ONE correct answer

20.1 Which one of the following statements is correct?

A. An increase in gluconeogenesis from amino acids results in a decrease in urea formation.
B. All essential amino acids are glycogenic.
C. Ornithine and citrulline are found in tissue proteins.
D. Cysteine is an essential amino acid in individuals consuming a diet devoid of methionine.
E. In the presence of adequate dietary sources of tyrosine, phenylalanine is not an essential amino acid.

Correct answer = D. Methionine is the precursor of cysteine. An increase in gluconeogenesis releases increased ammonia and results in increased urea production. The essential amino acids leucine and lysine are ketogenic. Ornithine and citrulline are amino acids that are intermediates in the urea cycle, but are not found in tissue proteins.

20.2 Which one of the following statements concerning a one-week-old male infant with undetected classic phenylketonuria is correct?

A. Tyrosine is a nonessential amino acid for the infant.
B. High levels of phenylpyruvate appear in his urine.
C. Therapy must begin within the first year of life.
D. A diet devoid of phenylalanine should be initiated immediately.
E. When the infant reaches adulthood, it is recommended that diet therapy be discontinued.

Correct answer = B. Phenyllactate, phenylacetate, and phenylpyruvate, which are not normally produced in significant amounts in the presence of functional phenylalanine hydroxylase, are elevated in PKU, and appear in the urine. In patients with PKU, tyrosine cannot be synthesized from phenylalanine and, hence, becomes essential and must be supplied in the diet. Treatment must begin during the first seven to ten days of life to prevent mental retardation. Discontinuance of the phenylalanine-restricted diet before eight years of age is associated with poor performance on IQ tests. Adult PKU patients show deterioration of attention and speed of mental processing after discontinuation of the diet. Life-long restriction of dietary phenylalanine is, therefore, recommended.

20.3 A four-year-old boy of a first-degree consanguineous couple was noted by the parents to have darkening of the urine to an almost black color when it was left standing. He had a normal sibling, and there were no other medical problems. Childhood growth and development were normal. Which of the following is most likely to elevated in this patient?

A. Methylmalonate
B. Homogentisate
C. Phenylpyruvate
D. α-Ketoisovalerate
E. Homocystine

Correct answer = B. Alkaptonuria is a rare metabolic disease involving a deficiency in homogentisic acid oxidase, and the subsequent accumulation of homogentisic acid in the urine, which turns dark upon standing. The elevation of methylmalonate (due to methylmalonyl CoA mutase deficiency), phenylpyruvate (due to phenylalanine hydroxlyase deficiency), α-ketoisovalerate (due to branched-chain α-ketoacid dehydrogenase deficiency), and homocystine (due to cystathionine synthase deficiency) are inconsistent with a healthy child with darkening of the urine.

Conversion of Amino Acids to Specialized Products

21

I. OVERVIEW

In addition to serving as building blocks for proteins, amino acids are precursors of many nitrogen-containing compounds that have important physiologic functions (Figure 21.1). These molecules include porphyrins, neurotransmitters, hormones, purines, and pyrimidines.

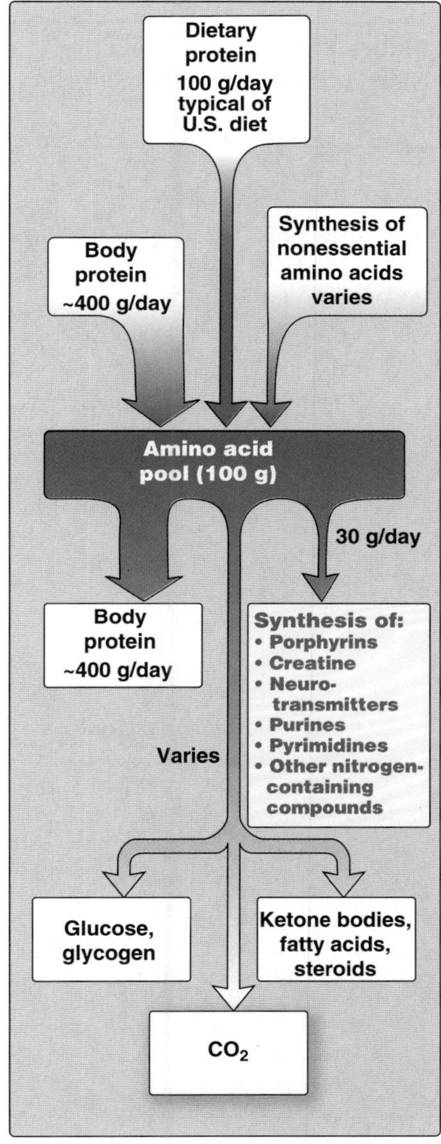

Figure 21.1
Amino acids as precursors of nitrogen-containing compounds.

II. PORPHYRIN METABOLISM

Porphyrins are cyclic compounds that readily bind metal ions—usually Fe^{2+} or Fe^{3+}. The most prevalent **metalloporphyrin** in humans is heme, which consists of one ferrous (Fe^{2+}) iron atom coordinated in the center of the tetrapyrrole ring of protoporphyrin IX (see p. 277). Heme is the prosthetic group for hemoglobin, myoglobin, the cytochromes, *catalase,* and *tryptophan pyrrolase*. These hemeproteins are rapidly synthesized and degraded. For example, 6 to 7g of hemoglobin are synthesized each day to replace heme lost through the normal turnover of erythrocytes. Coordinated with the turnover of hemeproteins is the simultaneous synthesis and degradation of the associated porphyrins, and recycling of the bound iron ions.

A. Structure of porphyrins

Porphyrins are cyclic molecules formed by the linkage of four pyrrole rings through methenyl bridges (Figure 21.2). Three structural features of these molecules are relevant to understanding their medical significance.

1. **Side chains:** Different porphyrins vary in the nature of the side chains that are attached to each of the four pyrrole rings. For example, uroporphyrin contains acetate ($–CH_2–COO^–$) and propionate ($–CH_2–CH_2–COO^–$) side chains, whereas coproporphyrin, is substituted with methyl ($–CH_3$) and propionate groups.

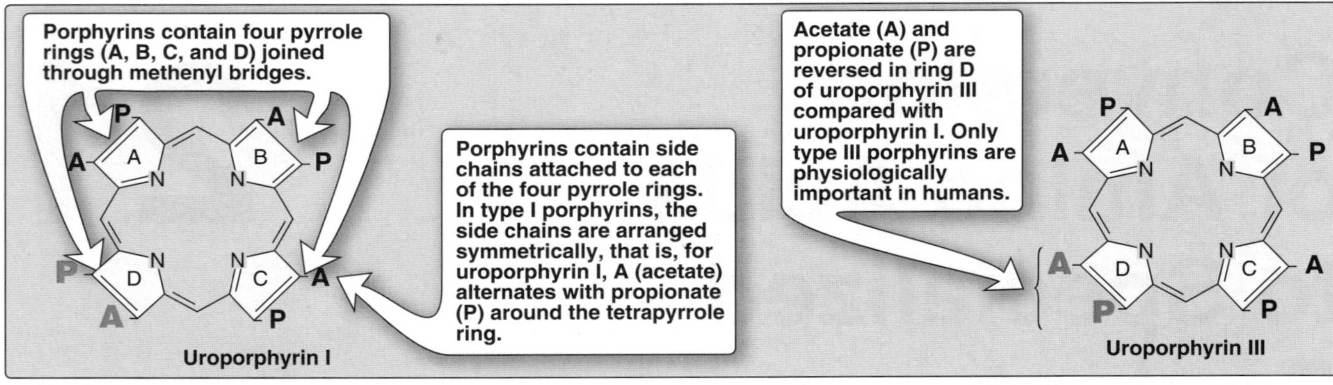

Figure 21.2
Structures of uroporphyrin I and uroporphyrin III. [Note: A = acetate and P = propionate.]

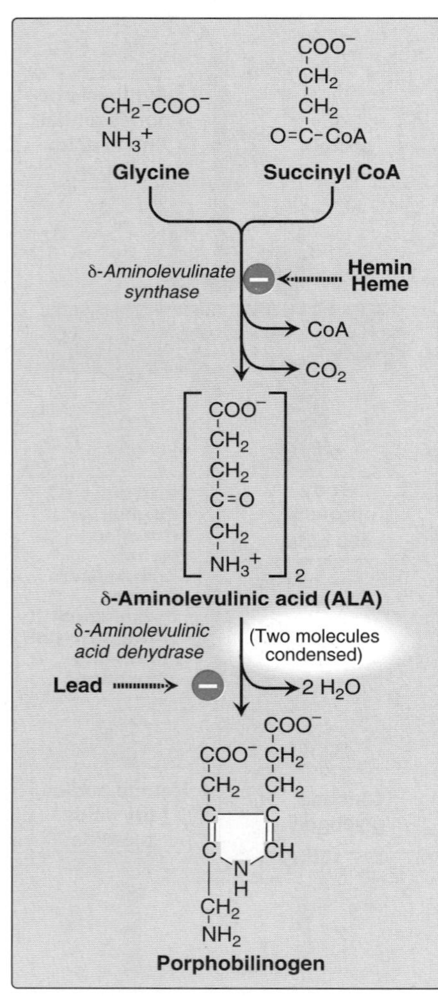

Figure 21.3
Pathway of porphyrin synthesis: formation of porphobilinogen. (Continued in Figure 21.4.)

2. **Distribution of side chains:** The side chains of porphyrins can be ordered around the tetrapyrrole nucleus in four different ways, designated by Roman numerals I to IV. Only type III porphyrins, which contain an asymmetric substitution on ring D (see Figure 21.2), are physiologically important in humans. [Note: In **congenital erythropoietic porphyria** (see Summary Figure 21.7, p. 279), type I porphyrins, which contain a symmetric arrangement of substituents (see Figure 21.2), are synthesized in appreciable quantities.]

3. **Porphyrinogens:** Porphyrin precursors exist in the chemically reduced form called porphyrinogens. In contrast to the porphyrins, which are colored, the porphyrinogens, such as uroporphyrinogen, are colorless. As described in the next section, porphyrinogens serve as intermediates between porphobilinogen and protoporphyrin in the biosynthesis of heme.

B. Biosynthesis of heme

The major sites of heme biosynthesis are the **liver**, which synthesizes a number of heme proteins (particularly, cytochrome P450), and the **erythrocyte-producing cells** of the bone marrow, which are active in hemoglobin synthesis. In the liver, the rate of heme synthesis is highly variable, responding to alterations in the cellular heme pool caused by fluctuating demands for heme proteins. In contrast, heme synthesis in erythroid cells is relatively constant, and is matched to the rate of globin synthesis. The initial reaction and the last three steps in the formation of porphyrins occur in mitochondria, whereas the intermediate steps of the biosynthetic pathway occur in the cytosol (see Summary Figure 21.7). [Note: Mature red blood cells lack mitochondria and are unable to synthesize heme.]

1. **Formation of δ-aminolevulinic acid (ALA):** All the carbon and nitrogen atoms of the porphyrin molecule are provided by two simple building blocks: **glycine** (a nonessential amino acid) and **succinyl CoA** (an intermediate in the citric acid cycle). Glycine and succinyl CoA condense to form ALA in a reaction catalyzed by *ALA synthase* (Figure 21.3) This reaction requires **pyridoxal phosphate** as a coenzyme, and is the **rate-controlling step** in hepatic porphyrin biosynthesis.

a. **End product inhibition by hemin:** When porphyrin production exceeds the availability of globin (or other apoproteins), heme accumulates and is converted to hemin by the oxidation of Fe^{2+} to Fe^{3+}. Hemin decreases the activity of hepatic *ALA synthase* by causing decreased synthesis of the enzyme. [Note: In erythroid cells, heme synthesis is under the control of erythropoietin and the availability of intracellular iron.]

b. **Effect of drugs on ALA synthase activity:** Administration of any of a large number of drugs, such as phenobarbital, griseofulvin or hydantoins, results in a significant increase in hepatic *ALA synthase* activity. These drugs are metabolized by the *microsomal cytochrome P450 monooxygenase* system—a hemeprotein oxidase system found in the liver (see p. 278). In response to these drugs, the synthesis of cytochrome P450 increases, leading to an enhanced consumption of heme—a component of cytochrome P450. This, in turn, causes a decrease in the concentration of heme in liver cells. The lower intracellular heme concentration leads to an increase in the synthesis of *ALA synthase* (derepression), and prompts a corresponding increase in ALA synthesis.

2. **Formation of porphobilinogen:** The dehydration of two molecules of ALA to form porphobilinogen by *δ-aminolevulinic acid dehydrase* is extremely sensitive to inhibition by heavy metal ions (see Figure 21.3, and p. 279). This inhibition is, in part, responsible for the elevation in ALA and the anemia seen in **lead poisoning**.

3. **Formation of uroporphyrinogen:** The condensation of four molecules of porphobilinogen results in the formation of uroporphyrinogen III. The reaction requires *hydroxymethybilane synthase* and *uroporphyrinogen III synthase* (which produces the asymmetric uroporphyrinogen III, Figure 21.4).

4. **Formation of heme:** Uroporphyrinogen III is converted to heme by a series of decarboxylations and oxidations summarized in Figure 21.4. The introduction of Fe^{2+} into protoporphyrin IX occurs spontaneously, but the rate is enhanced by the enzyme *ferrochelatase*—an enzyme that is inhibited by lead (see p. 279).

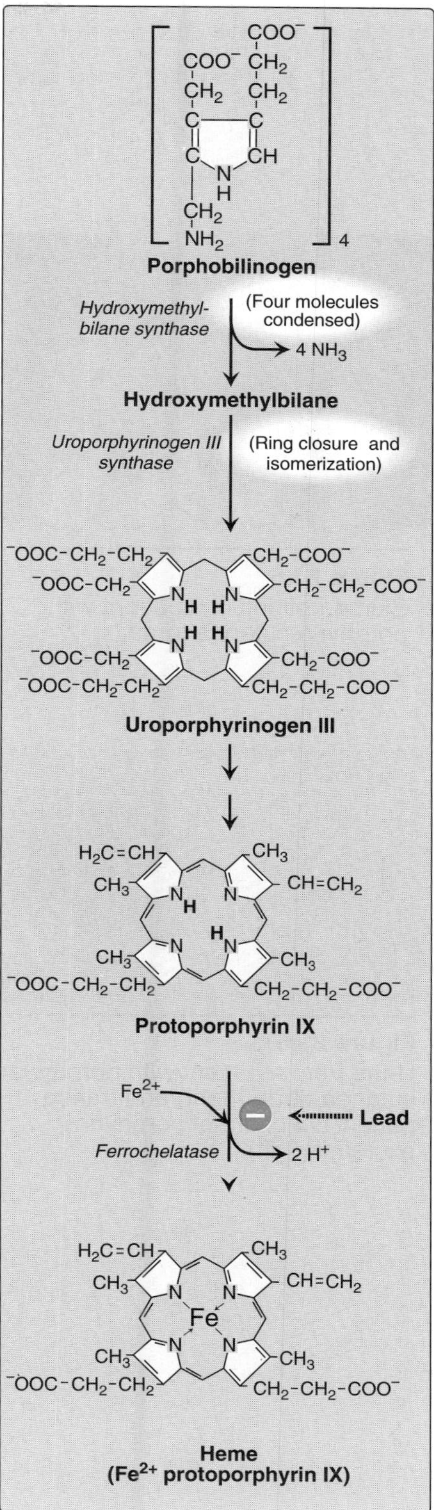

Figure 21.4
Pathway of porphyrin synthesis: formation of heme. (Continued from Figure 21.3.)

C. Porphyrias

Porphyrias are caused by inherited (or occasionally acquired) defects in heme synthesis, resulting in the accumulation and increased excretion of porphyrins or porphyrin precursors (see Summary Figure 21.7). With the exception of congenital erythropoietic porphyria, which is a genetically recessive disease, all porphyrias are inherited as autosomal dominant disorders. The mutations that cause the porphyrias are heterogenous (not all are at the same DNA locus), and nearly every affected family has its own mutation. Each porphyria results in the accumulation of a unique pattern of intermediates caused by the deficiency of an enzyme in the heme synthetic pathway.

1. **Clinical manifestations:** The porphyrias are classified as **erythropoietic** or **hepatic**, depending on whether the enzyme deficiency occurs in the erythropoietic cells of the bone marrow or in the liver. Hepatic porphyrias can be further classified as acute or chronic. Individuals with an enzyme defect leading to the accumulation of

tetrapyrrole intermediates show **photosensitivity**—that is, their skin itches and burns (**pruritis**) when exposed to visible light. [Note: These symptoms are thought to be a result of the porphyrin-mediated formation of superoxide radicals from oxygen. These reactive oxygen species can oxidatively damage membranes, and cause the release of destructive enzymes from lysosomes (see p. 145 for discussion of reactive oxygen intermediates). Destruction of cellular components leads to the photosensitivity.]

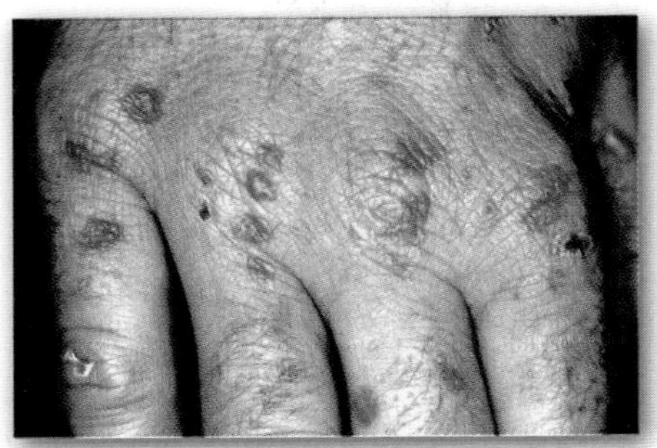

Figure 21.5
Skin eruptions in a patient with porphyria cutanea tarda.

a. **Chronic porphyria: Porphyria cutanea tarda,** the most common porphyria, is a chronic disease of the liver and erythroid tissues. The disease is associated with a deficiency in *uroporphyrinogen decarboxyase,* but clinical expression of the enzyme deficiency is influenced by various factors, such as hepatic iron overload, exposure to sunlight, and the presence of hepatitis B or C, or HIV infections. Clinical onset is typically during the fourth or fifth decade of life. Porphyrin accumulation leads to cutaneous symptoms (Figure 20.5), and urine that is red to brown in natural light (Figure 20.6), and pink to red in fluorescent light.

b. **Acute hepatic porphyrias:** Acute hepatic porphyrias (**acute intermittent porphyria**, **hereditary coproporphyria**, and **variegate porphyria**) are characterized by acute attacks of gastrointestinal, neurologic/psychiatric, and cardiovascular symptoms. Porphyrias leading to accumulation of ALA and porphobilinogen, such as acute intermittent porphyria, cause abdominal pain and neuropsychiatric disturbances. Symptoms of the acute hepatic porphyrias are often precipitated by administration of drugs such as barbiturates and ethanol, which induce the synthesis of the heme-containing cytochrome P450 microsomal drug oxidation system. This further decreases the amount of available heme, which, in turn, promotes the increased synthesis of *ALA synthase*.

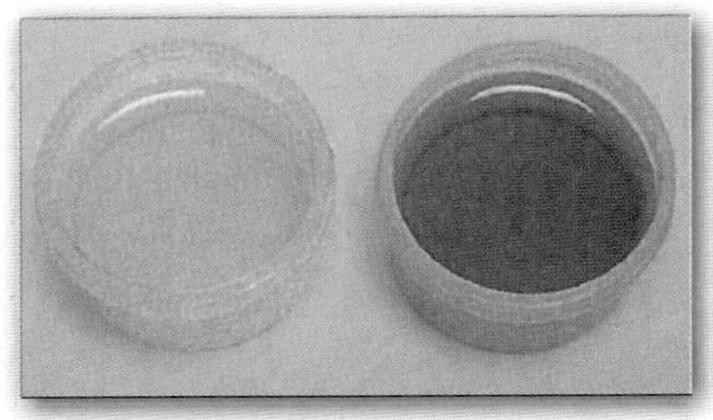

Figure 21.6
Urine from a patient with porphyria cutanea tarda (right) and from a patient with normal porphyrin excretion (left).

c. **Erythropoietic porphyrias:** The erythropoietic porphyrias (**congenital erythropoietic porphyria** and **erythropoietic protoporphyria**) are characterized by skin rashes and blisters that appear in early childhood. The diseases are complicated by cholestatic liver cirrhosis and progressive hepatic failure.

2. **Increased ALA synthase activity:** One common feature of the porphyrias is a decreased synthesis of heme. In the liver, heme normally functions as a repressor of *ALA synthase*. Therefore, the absence of this end product results in an increase in the synthesis of *ALA synthase* (derepression). This causes an increased synthesis of intermediates that occur prior to the genetic block. The accumulation of these toxic intermediates is the major pathophysiology of the porphyrias.

3. **Treatment:** During acute porphyria attacks, patients require medical support, particularly treatment for pain and vomiting. The severity of symptoms of the porphyrias can be diminished by intravenous injection of **hemin**, which decreases the synthesis of *ALA synthase*. Avoidance of sunlight and ingestion of β-carotene (a free-radical scavenger) are also helpful.

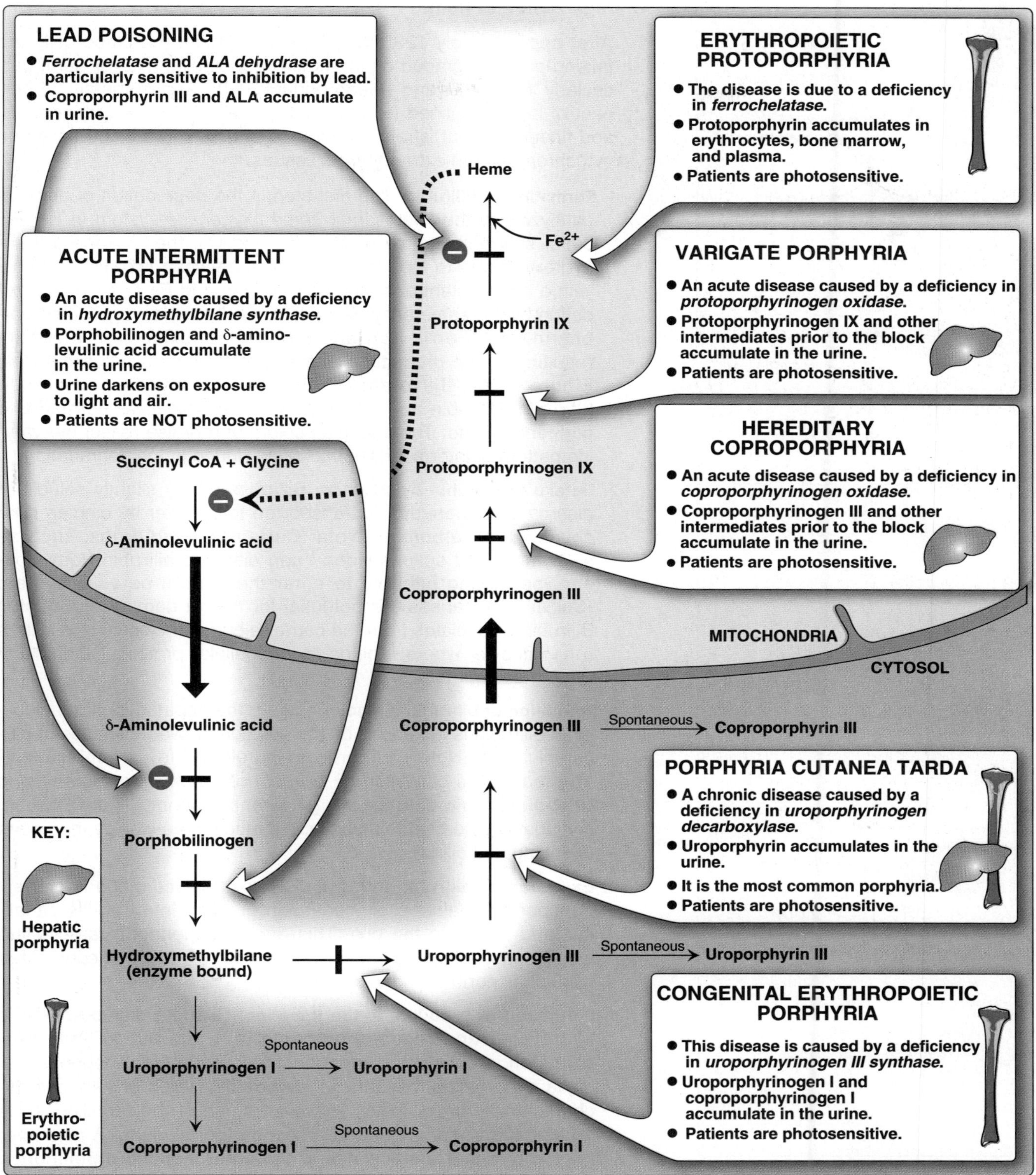

Figure 21.7
Summary of heme synthesis.

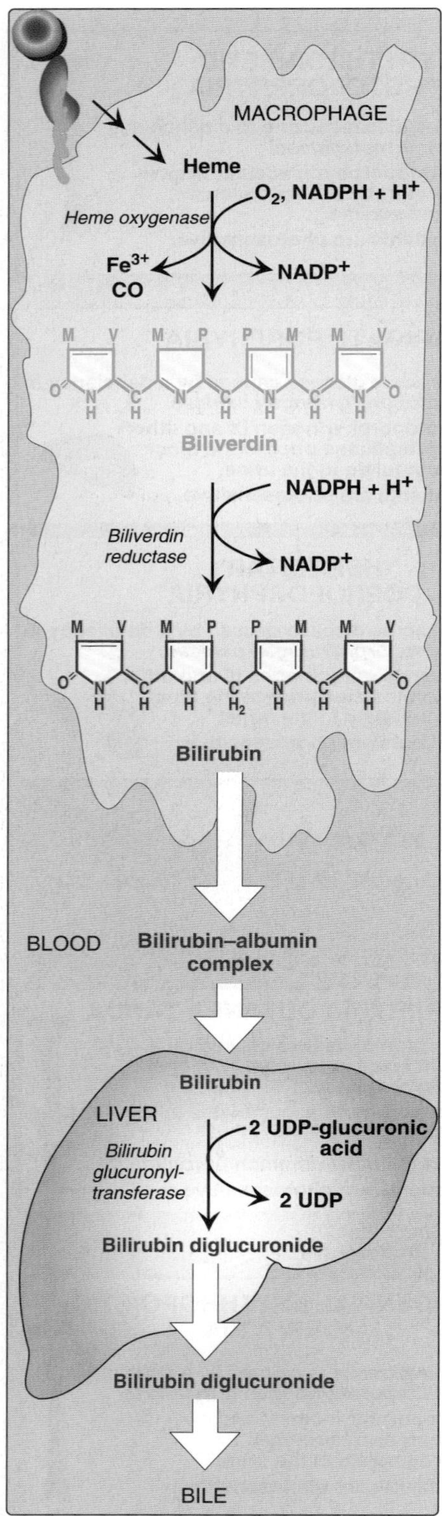

Figure 21.8
Formation of bilirubin from heme.

D. Degradation of heme

After approximately 120 days in the circulation, red blood cells are taken up and degraded by the reticuloendothelial (RE) system, particularly in the liver and spleen (Figure 21.8). Approximately 85 percent of heme destined for degradation comes from red blood cells, and fifteen percent isfrom turnover of immature red blood cells and cytochromes from extraerythroid tissues.

1. **Formation of bilirubin:** The first step in the degradation of heme is catalyzed by the microsomal *heme oxygenase* system of the RE cells. In the presence of NADPH and O_2, the enzyme adds a hydroxyl group to the methenyl bridge between two pyrrole rings, with a concomitant oxidation of ferrous iron to Fe^{3+}. A second oxidation by the same enzyme system results in cleavage of the porphyrin ring. Ferric iron and carbon monoxide are released, resulting in the production of the green pigment **biliverdin** (see Figure 21.8). Biliverdin is reduced, forming the red-orange **bilirubin**. Bilirubin and its derivatives are collectively termed **bile pigments**. [Note: The changing colors of a bruise reflect the varying pattern of intermediates that occur during heme degradation.]

2. **Uptake of bilirubin by the liver:** Bilirubin is only slightly soluble in plasma and, therefore, is transported to the liver by binding noncovalently to **albumin**. [Note: Certain anionic drugs, such as salicylates and sulfonamides,[1] can displace bilirubin from albumin, permitting bilirubin to enter the central nervous system (CNS). This causes the potential for neural damage in infants.] Bilirubin dissociates from the carrier albumin molecule and enters a hepatocyte, where it binds to intracellular proteins, particularly the protein **ligandin**.

3. **Formation of bilirubin diglucuronide:** In the hepatocyte, the solubility of bilirubin is increased by the addition of two molecules of glucuronic acid. [Note: This process is referred to as **conjugation**.] The reaction is catalyzed by *bilirubin glucuronyltransferase* using **UDP-glucuronic acid** as the glucuronate donor. [Note: Bilirubin conjugates also bind to albumin, but much more weakly than does unconjugated bilirubin.]

4. **Excretion of bilirubin into bile:** Bilirubin diglucuronide is actively transported against a concentration gradient into the bile canaliculi and then into the bile. This energy-dependent, rate-limiting step is susceptible to impairment in liver disease. Unconjugated bilirubin is normally not excreted.

5. **Formation of urobilins in the intestine:** Bilirubin diglucuronide is hydrolyzed and reduced by bacteria in the gut to yield **urobilinogen**, a colorless compound. Most of the urobilinogen is oxidized by intestinal bacteria to **stercobilin**, which gives feces the characteristic brown color. However, some of the urobilinogen is reabsorbed from the gut and enters the portal blood. A portion of this urobilinogen participates in the **enterohepatic urobilinogen cycle** in which it is taken up by the liver, and then re-excreted into the bile. The remainder of the urobilinogen is transported by the blood to the kidney, where it is converted to yellow **urobilin** and excreted, giving urine its characteristic color. The metabolism of bilirubin is summarized in Figure 21.9.

[1]See Chapter 34 in ***Lippincott's Illustrated Reviews: Pharmacology*** (3rd Ed.) and Chapter 29 (2nd Ed.) for a discussion of kincterus due to displacement of bilirubin by sulfonamides.

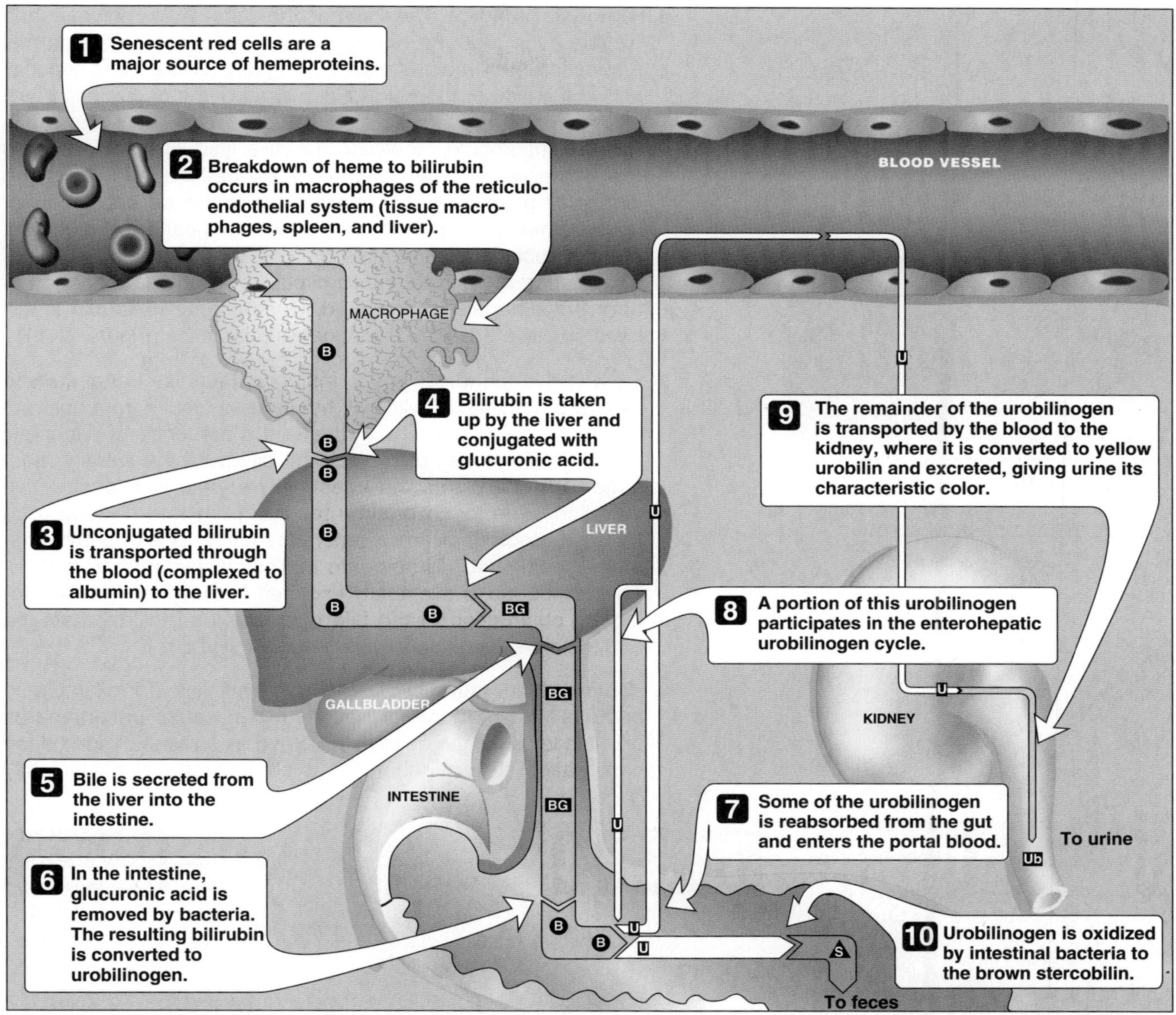

Figure 21.9
Catabolism of heme B = bilirubin; BG = bilirubin diglucuronide; U = urobilinogen; Ub = urobilin; S = stercobilin.

E. Jaundice

Jaundice (also called **icterus**) refers to the yellow color of skin, nail beds, and sclerae (whites of the eyes) caused by deposition of bilirubin, secondary to increased bilirubin levels in the blood (**hyperbilirubinemia**, Figure 21.10). Although not a disease, jaundice is usually a symptom of an underlying disorder.

1. **Types of jaundice:** Jaundice can be classified into three major forms described below. However, in clinical practice, jaundice is often more complex than indicated in this simple classification. For example, the accumulation of bilirubin may be a result of defects at more than one step in its metabolism.

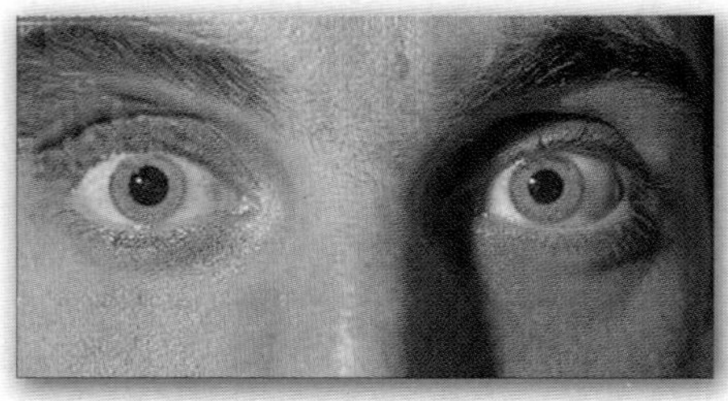

Figure 21.10
Jaundiced patient, with the sclerae of his eyes appearing yellow.

a. **Hemolytic jaundice:** The liver has the capacity to conjugate and excrete over 3000 mg of bilirubin per day, whereas the normal production of bilirubin is only 300 mg/day. This excess capacity allows the liver to respond to increased heme degradation with a corresponding increase in conjugation and secretion of bilirubin diglucuronide. However, massive lysis of red blood cells (for example, in patients with sickle cell anemia, pyruvate kinase or glucose 6-phosphate dehydrogenase deficiency, or malaria) may produce bilirubin faster than it can be conjugated. More bilirubin is excreted into the bile, the amount of urobilinogen entering the enterohepatic circulation is increased, and urinary urobilinogen is increased. Unconjugated bilirubin levels become elevated in the blood, causing jaundice (Figure 21.11).

b. **Obstructive jaundice:** In this instance, jaundice is not caused by overproduction of bilirubin, but instead results from obstruction of the bile duct. For example, the presence of a hepatic tumor or bile stones may block the bile ducts, preventing passage of bilirubin into the intestine. Patients with obstructive jaundice experience gastrointestinal pain and nausea, and produce stools that are a pale, clay color. The liver "regurgitates" conjugated bilirubin into the blood (hyperbilirubinemia). The compound is eventually excreted in the urine. [Note: Prolonged obstruction of the bile duct can lead to liver damage and a subsequent rise in unconjugated bilirubin.]

c. **Hepatocellular jaundice:** Damage to liver cells (for example, in patients with cirrhosis or hepatitis) can cause unconjugated bilirubin levels to increase in the blood as a result of decreased conjugation. The bilirubin that is conjugated is not efficiently

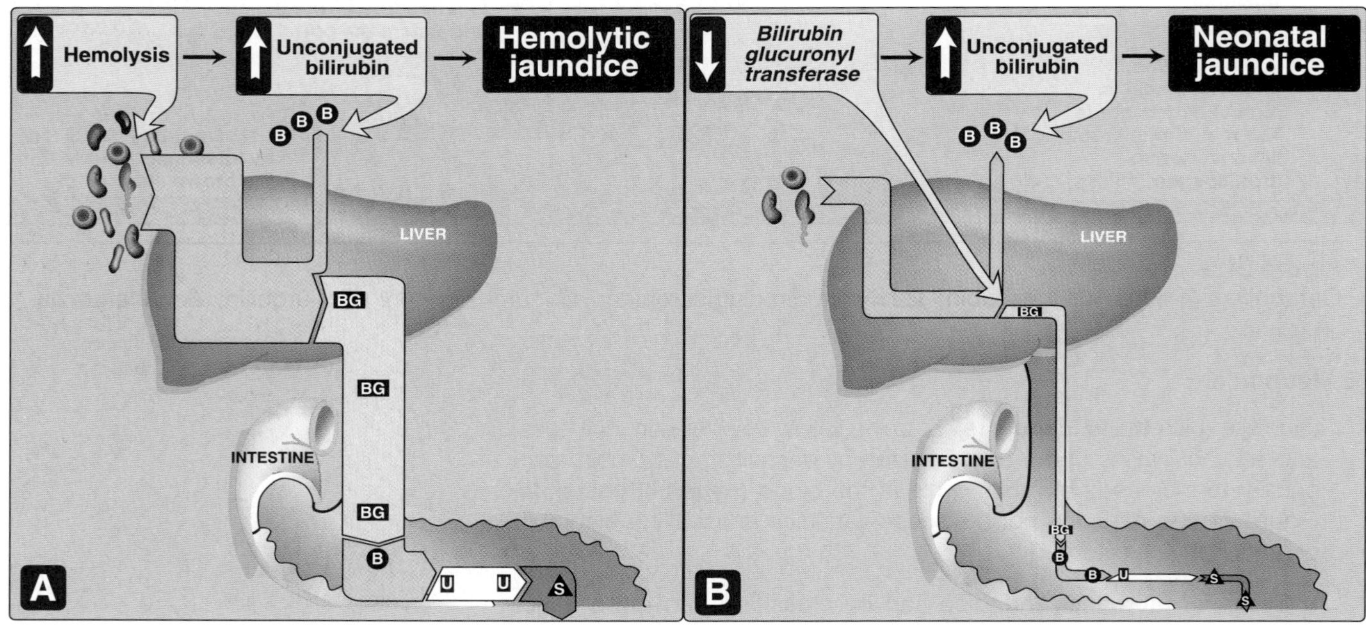

Figure 21.11
Alterations in the metabolism of heme. A. Hemolytic jaundice. B. Neonatal jaundice. [Note: The enterohepatic circulation of urobilinogen is omitted for simplicity.] BG = bilirubin glucuronide; B = bilirubin; U = urobilinogen; S = stercobilin.

secreted into the bile, but instead diffuses ("leaks") into the blood. Urobilinogen is increased in the urine because hepatic damage decreases the enterohepatic circulation of this compound, allowing more to enter the blood, from which it is filtered into the urine. The urine thus becomes dark in color, whereas stools are a pale, clay color. Plasma levels of AST (SGOT) and ALT (SGPT, see p. 249) are elevated, and the patient experiences nausea and anorexia.

2. **Jaundice in newborns:** Newborn infants, particularly premature babies, often accumulate bilirubin, because the activity of hepatic *bilirubin glucuronyl transferase* is low at birth—it reaches adult levels in about four weeks (Figures 21.11B and 21.12). Elevated bilirubin, in excess of the binding capacity of albumin, can diffuse into the basal ganglia and cause toxic encephalopathy (**kernicterus**). Thus, newborns with significantly elevated bilirubin levels are treated with blue fluorescent light (Figure 21.13), which converts bilirubin to more polar and, hence, water-soluble isomers. These photoisomers can be excreted into the bile without conjugation to glucuronic acid. [Note: **Crigler-Najjar syndrome** is caused by a genetic deficiency of hepatic *bilirubin glucuronyl transferase*.]

3. **Determination of bilirubin concentration:** Bilirubin is most commonly determined by the **van den Bergh reaction**, in which diazotized sulfanilic acid reacts with bilirubin to form red azodipyrroles that are measured colorimetrically. In aqueous solution, the water-soluble, **conjugated bilirubin** reacts rapidly with the reagent (within one minute), and is said to be "**direct-reacting**." The unconjugated bilirubin, which is much less soluble in aqueous solution, reacts more slowly. However, when the reaction is carried out in methanol, both conjugated and unconjugated bilirubin are soluble and react with the reagent, providing the **total bilirubin** value. The "**indirect-reacting**" bilirubin, which corresponds to the unconjugated bilirubin, is obtained by subtracting the direct-reacting bilirubin from the total bilirubin. [Note: In normal plasma, only about four percent of the total bilirubin is conjugated.]

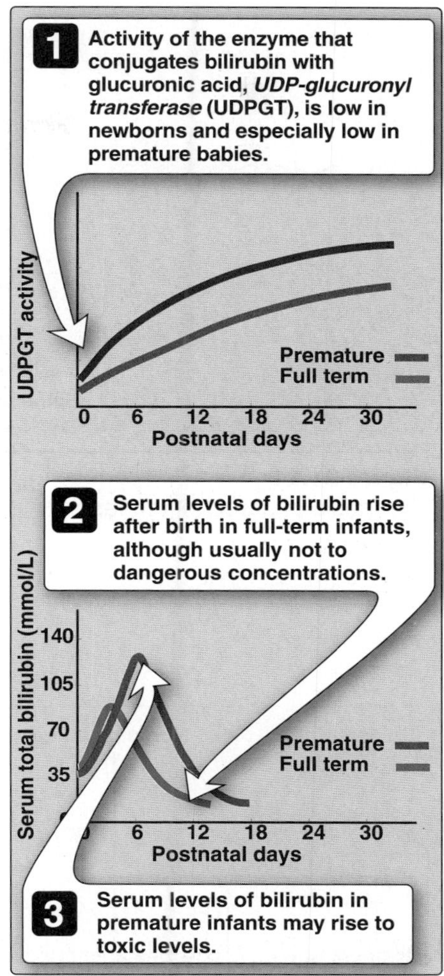

Figure 21.12
Neonatal jaundice.

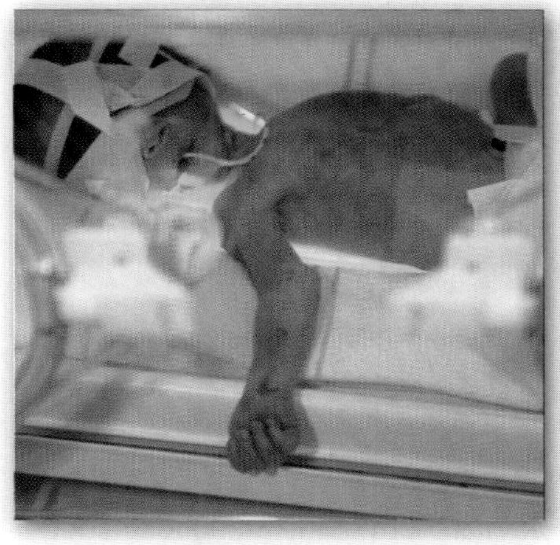

Figure 21.13
Phototherapy in neonatal jaundice.

III. OTHER NITROGEN-CONTAINING COMPOUNDS

A. Catecholamines

Dopamine, **norepinephrine**, and **epinephrine** (**adrenalin**) are biologically active amines that are collectively termed catecholamines. Dopamine and norepinephrine function as neurotransmitters in the brain and the autonomic nervous system. Norepinephrine and epinephrine are also synthesized in the adrenal medulla.

1. **Functions:** Outside the nervous system, norepinephrine and its methylated derivative, epinephrine act as regulators of carbohydrate and lipid metabolism. Norepinephrine and epinephrine are released from storage vesicles in the adrenal medulla in response to fright, exercise, cold, and low levels of blood glucose. They increase the degradation of glycogen and triacylglycerol, as well as increase blood pressure and the output of the heart. These effects are part of a coordinated response to prepare the individual for emergencies, and are often called the "**fight-or-flight**" **reactions**.

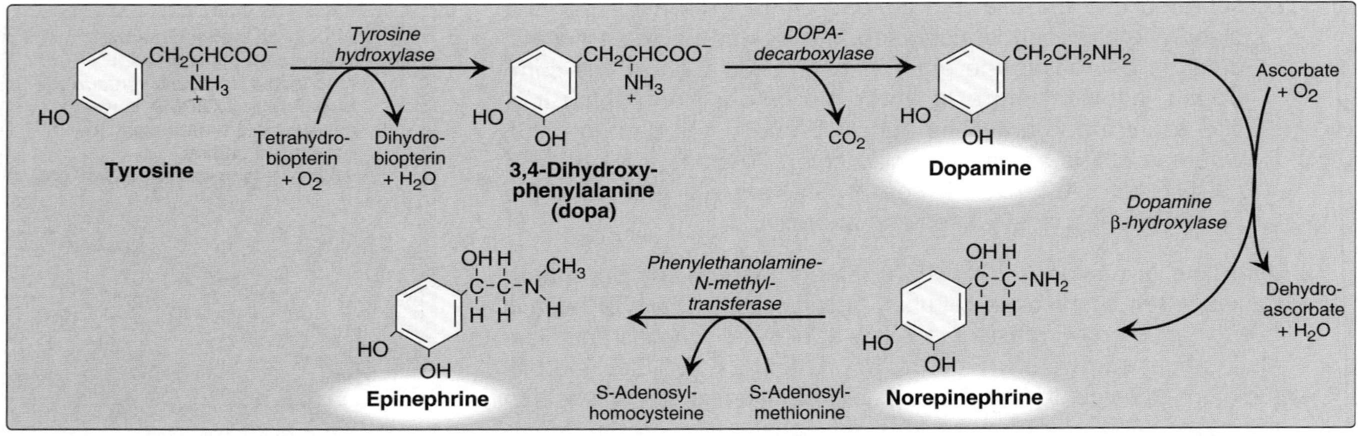

Figure 21.14
Synthesis of catecholamines.

2. **Synthesis of catecholamines:** The catecholamines are synthesized from **tyrosine**,as shown in Figure 21.14. Tyrosine is first hydroxylated by *tyrosine hydroxylase* to form 3,4-dihydroxyphenylalanine (dopa) in a reaction analogous to that described for the hydroxylation of phenylalanine (see p. 266). The enzyme is abundant in the **central nervous system**, the **sympathetic ganglia**, and the **adrenal medulla**, and is the rate-limiting step of the pathway. Dopa is decarboxylated in a reaction requiring pyridoxal phosphate (see p. 376) to form dopamine, which is hydroxylated by the copper-containing *dopamine β-hydroxylase* to yield norepinephrine. Epinephrine is formed from norepinephrine by an N-methylation reaction using **S-adenosylmethionine** as the methyl donor.

3. **Degradation of catecholamines:** The catecholamines are inactivated by oxidative deamination catalyzed by *monoamine oxidase* (*MAO*), and by O-methylation carried out by *catechol-O-methyltransferase* (*COMT*, Figure 21.15). The two reactions can occur in either order. The aldehyde products of the *MAO* reaction are oxidized to the corresponding acids. The metabolic products of these reactions are excreted in the urine as vanillylmandelic acid, metanephrine, and normetanephrine.

4. **Monoamine oxidase inhibitors:** *MAO* is found in neural and other tissues, such as the gut and liver. In the neuron, this enzyme functions as a "safety valve" to oxidatively deaminate and inactivate any excess neurotransmitter molecules (norepinephrine, dopamine, or serotonin) that may leak out of synaptic vesicles when the neuron is at rest. The *MAO* inhibitors[2] may irreversibly or reversibly inactivate the enzyme, permitting neurotransmitter molecules to escape degradation and, therefore, to both accumulate within the presynaptic neuron and to leak into the synaptic space. This causes activation of norepinephrine and serotonin receptors, and may be responsible for the antidepressant action of these drugs.

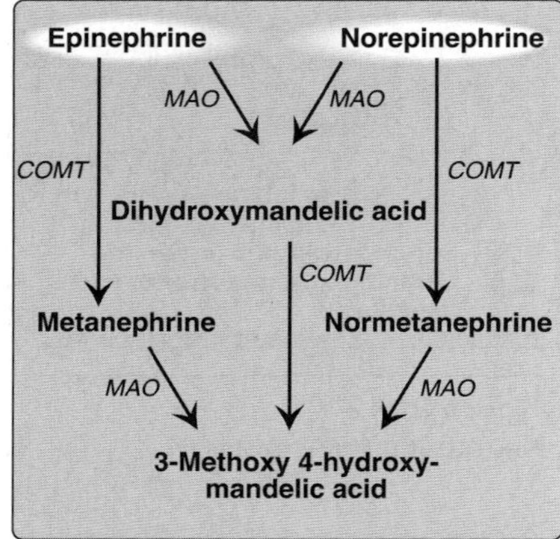

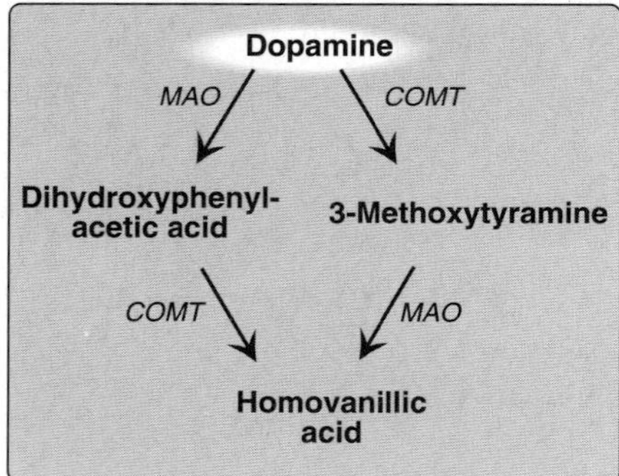

Figure 21.15
Metabolism of the catecholamines by *catechol-O-methyltranferase* (*COMT*) and *monoamine oxidase* (*MAO*).

[2]See Chapter 12 in ***Lippincott's Illustrated Reviews: Pharmacology*** (2nd and 3rd Eds.) for a discussion of the actions of MAO and COMT and the use of MAO inhibitors in the treatment of depression.

B. Creatine

Creatine phosphate (also **called phosphocreatine**), the phosphorylated derivative of creatine found in muscle, is a high-energy compound that can reversibly donate a phosphate group to ADP to form ATP (Figure 21.16). Creatine phosphate provides a small but rapidly mobilized reserve of high-energy phosphates that can be used to maintain the intracellular level of ATP during the first few minutes of intense muscular contraction. [Note: The amount of creatine phosphate in the body is proportional to the muscle mass.]

1. **Synthesis:** Creatine is synthesized from **glycine** and the guanidino group of **arginine**, plus a methyl group from **S-adenosylmethionine** (see Figure 21.16). Creatine is reversibly phosphorylated to creatine phosphate by *creatine kinase*, using ATP as the phosphate donor. [Note: The presence of *creatine kinase* in the plasma is indicative of tissue damage, and is used in the diagnosis of myocardial infarction (see p. 65).]

2. **Degradation:** Creatine and creatine phosphate spontaneously cyclize at a slow, but constant, rate to form **creatinine**, which is excreted in the urine. The amount of creatinine excreted is proportional to the total creatine phosphate content of the body, and thus can be used to estimate muscle mass. When muscle mass decreases for any reason (for example, from paralysis or muscular dystrophy), the creatinine content of the urine falls. In addition, any rise in blood creatinine is a sensitive indicator of kidney malfunction, because creatinine is normally rapidly removed from the blood and excreted. A typical adult male excretes about 15 mmol of creatinine per day. The constancy of this excretion is sometimes used to test the reliability of collected 24-hour urine samples—too little creatinine in the submitted sample may indicate an incomplete sample.

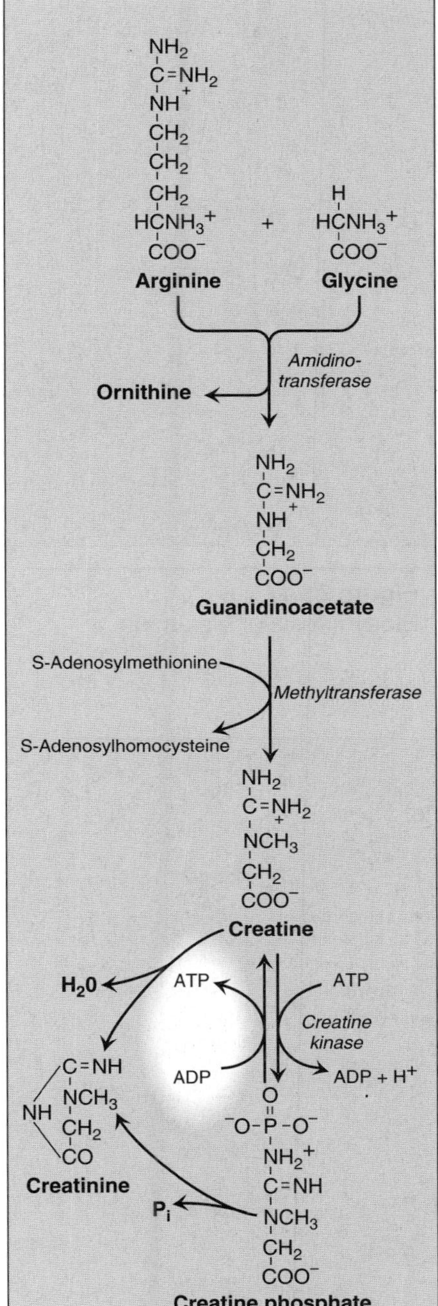

Figure 21.16
Synthesis of creatine.

C. Histamine

Histamine is a chemical messenger that mediates a wide range of cellular responses, including allergic and inflammatory reactions, gastric acid secretion, and possibly neurotransmission in parts of the brain. A powerful vasodilator, histamine is formed by decarboxylation of histidine in a reaction requiring **pyridoxal phosphate** (Figure 21.17). It is secreted by mast cells as a result of allergic reactions or trauma. Histamine has no clinical applications, but agents that interfere with the action of histamine have important therapeutic applications.[3]

D. Serotonin

Serotonin, also called **5-hydroxytryptamine**, is synthesized and stored at several sites in the body (Figure 21.18). By far the largest amount of serotonin is found in cells of the intestinal mucosa. Smaller amounts occur in platelets and in the central nervous system. Serotonin is synthesized from **tryptophan**, which is hydroxylated in a reaction analogous to that catalyzed by *phenylalanine hydroxylase*. The product, 5-hydroxytryptophan, is decarboxylated to serotonin. Serotonin has multiple physiologic roles, including pain perception, affective disorders, and regulation of sleep, temperature, and blood pressure.

[3]See Chapter 43 in ***Lippincott's Illustrated Reviews: Pharmacology*** (3rd Ed.) and Chapter 40 (2nd Ed.) for a discussion of the therapeutic uses of antihistamines.

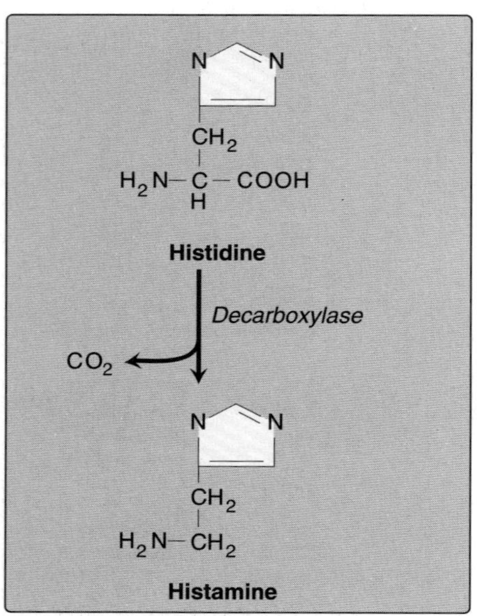

Figure 21.17
Biosynthesis of histamine.

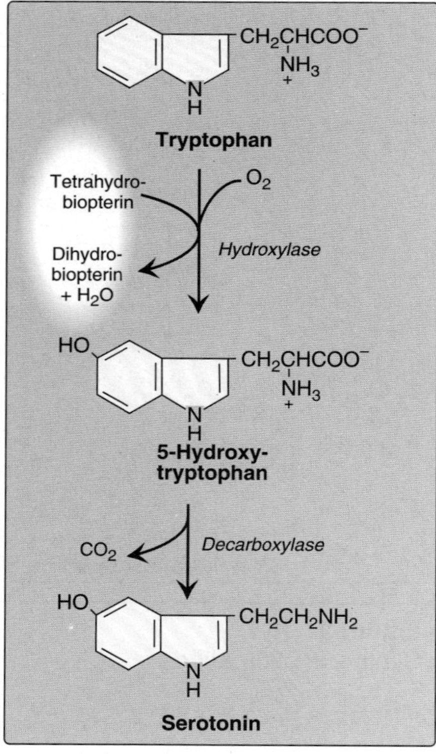

Figure 21.18
Synthesis of serotonin.

E. Melanin

Melanin is a pigment that occurs in several tissues in the body, particularly in the eye, hair, and skin. It is synthesized in the epidermis by pigment-forming cells called **melanocytes**. Its function is to protect underlying cells from the harmful effects of sunlight. The first step in melanin formation from tyrosine is a hydroxylation to form dopa, catalyzed by the copper-containing enzyme *tyrosine hydroxylase* (also called *tyrosinase*, see Figure 21.14). Subsequent reactions leading to the formation of brown and black pigments are also thought to be catalyzed by *tyrosine hydroxylase* or to occur spontaneously.

IV. CHAPTER SUMMARY

Amino acids are precursors of many nitrogen-containing compounds including **porhyrins**, which, in combination with ferrous (Fe^{2+}) iron, form **heme**. The major sites of **heme biosynthesis** are the **liver**, which synthesizes a number of heme proteins (particularly cytochrome P-450), and the **erythrocyte-producing cells** of the bone marrow, which are active in hemoglobin synthesis. In the liver, the rate of heme synthesis is highly variable, responding to alterations in the cellular heme pool caused by fluctuating demands for hemeproteins. In contrast, heme synthesis in erythroid cells is relatively constant, and is matched to the rate of globin synthesis. **Porphyrin synthesis** start with **glycine** and **succinyl CoA**. The **committed step** in heme synthesis is the formation of δ-aminolevulinic acid (ALA). This reaction is catalyzed by **ALA synthase**, and **inhibited** by **hemin** (the oxidized form of heme that accumulates in the cell when it is being underutilized). **Porphyrias** are caused by inherited defects in heme synthesis, resulting in the accumulation and increased excretion of porphyrins or porphyrin precursors. With the exception of **congenital erythropoietic porphyria**, which is a genetically **recessive disease**, all **other porphyrias** are inherited as **autosomal dominant** disorders. **Degradation** of hemeproteins occurs in the **reticuloendothelial system**, particularly in the **liver** and **spleen**. The first step in the degradation of heme is the production of the green pigment **biliverdin**, which is subsequently reduced to **bilirubin**. Bilirubin is transported to the liver, where its solubility is increased by the addition of two molecules of **glucuronic acid**. Bilirubin diglucuronide is transported into the **bile canaliculi**, where it is first hydrolyzed and reduced by bacteria in the gut to yield **urobilinogen**, then oxidized by intestinal bacteria to **stercobilin**. **Jaundice** refers to the yellow color of the skin, nail beds, and sclerae that is caused by deposition of bilirubin, secondary to increased bilirubin levels in the blood. Three commonly encountered type of jaundice are **hepatic jaundice**, **obstructive jaundice**, and **hepatoceullar jaundice**. Other important N-containing compounds derived from amino acids include the **catecholamines** (dopamine, norepinephrine, and epinephrine), **creatine**, **histamine**, **serotonin**, and **melanin**.

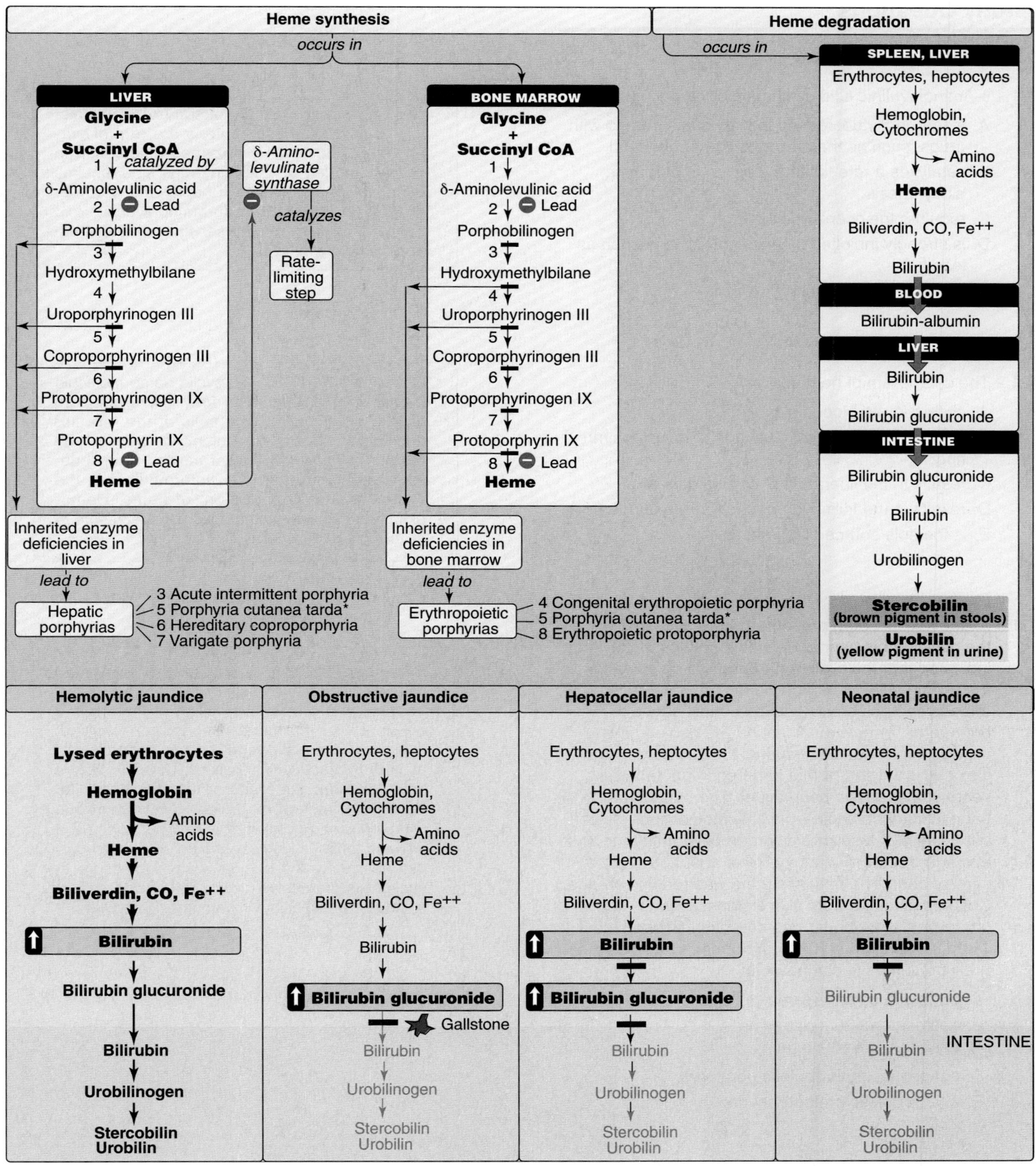

Figure 21.19
Key concept map for heme metabolism. *Note: Porphyria cutanea tarda affects both the liver and erythropoietic cells. ▬ = Block in the pathway.

Study Questions

Choose the ONE best answer

21.1 δ-Aminolevulinic acid synthase activity:

A. is frequently decreased in individuals treated with drugs, such as the barbiturate phenobarbital.
B. catalyzes a rate-limiting reaction in porphyrin biosynthesis.
C. requires the coenzyme biotin.
D. is strongly inhibited by heavy metal ions such as lead.
E. occurs in the cytosol.

Correct answer = B. The activity of δ-aminolevulinic acid synthase controls the rate of porphyrin synthesis. The enzyme is increased in patients treated with certain drugs, and requires pyridoxal phosphate as a coenzyme. Another enzyme in the pathway (δ-aminolevulinic acid dehydrase) is extremely sensitive to the presence of heavy metals.

21.2 The catabolism of hemoglobin:

A. occurs in red blood cells.
B. involves the oxidative cleavage of the porphyrin ring.
C. results in the liberation of carbon dioxide.
D. results in the formation of protoporphyrinogen.
E. is the sole source of bilirubin.

Correct answer = B. The cyclic heme molecule is oxidatively cleaved to form biliverdin. The catabolism occurs in the cells of the reticuloendothelial system, particularly the spleen, and results in the liberation of carbon monoxide. Protoporphyrinogen is an intermediate in the synthesis, not degradation, of heme. Hemoglobin and tissue cytochromes are precursors of bilirubin.

21.3 A fifty-year-old man presented with painful blisters on the backs of his hands. He was a golf instructor, and indicated that the blisters had erupted shortly after the golfing season began. He did not have recent exposure to poison ivy or sumac, new soaps or detergents, or new medications. He denied having previous episodes of blistering. He had partial complex seizure disorder that had begun about three years earlier after a head injury. The patient had been taking phenytoin—his only medication—since the onset of the seizure disorder. He admitted to an average weekly ethanol intake of about 18 12-oz cans of beer. The patient's urine was reddish orange. Cultures obtained from skin lesions failed to grow organisms. A 24-hour urine collection showed elevated uroporphyrin (1000 μg; normal, <27). The most likely presumptive diagnosis is:

A. porphyria cutanea tarda.
B. acute intermittent porphyria.
C. hereditary coproporphyria.
D. congenital erythropoietic porphyria.
E. erythropoietic protoporphyria.

Correct answer = A. The disease is associated with a deficiency in uroporphyrinogen decarboxyase, but clinical expression of the enzyme deficiency is influenced by hepatic iron overload, exposure to sunlight, and the presence of hepatitis B or C and HIV infections. Clinical onset is typically during the fourth or fifth decade of life. Porphyrin accumulation leads to cutaneous symptoms and urine that is red to brown. The disease tends to develop, recur, or worsen during the spring and summer, when exposure to sunlight is greatest. The laboratory and clinical findings are inconsistent with other porphyrias

Nucleotide Metabolism

22

I. OVERVIEW

Ribonucleoside and deoxyribonucleoside phosphates (nucleotides) are essential for all cells. Without them, neither DNA nor RNA can be produced and, therefore, proteins cannot be synthesized or cells proliferate. Nucleotides also serve as carriers of activated intermediates in the synthesis of some carbohydrates, lipids, and proteins, and are structural components of several essential coenzymes, for example, coenzyme A, FAD, NAD^+, and $NADP^+$. Nucleotides, such as cyclic AMP (cAMP) and cyclic GMP (cGMP), serve as second messengers in signal transduction pathways. In addition, nucleotides play an important role as "energy currency" in the cell. Finally, nucleotides are important regulatory compounds for many of the pathways of intermediary metabolism, inhibiting or activating key enzymes. The purine and pyrimidine bases found in nucleotides can be synthesized de novo, or can be obtained through salvage pathways that allow the reuse of the preformed bases resulting from normal cell turnover or from the diet.

II. NUCLEOTIDE STRUCTURE

Nucleotides are composed of a nitrogenous base, a pentose monosaccharide, and one, two, or three phosphate groups. The nitrogen-containing bases belong to two families of compounds: the purines and the pyrimidines.

A. Purine and pyrimidine structures

Both DNA and RNA contain the same purine bases: **adenine** (A) and **guanine** (G). Both DNA and RNA contain the pyrimidine **cytosine** (C), but they differ in their second pyrimidine base: DNA contains **thymine** (T), whereas RNA contains **uracil** (U). T and U differ by only one methyl group, which is present on T but absent on U (Figure 22.1). [Note: **Unusual bases** are occasionally found in some species

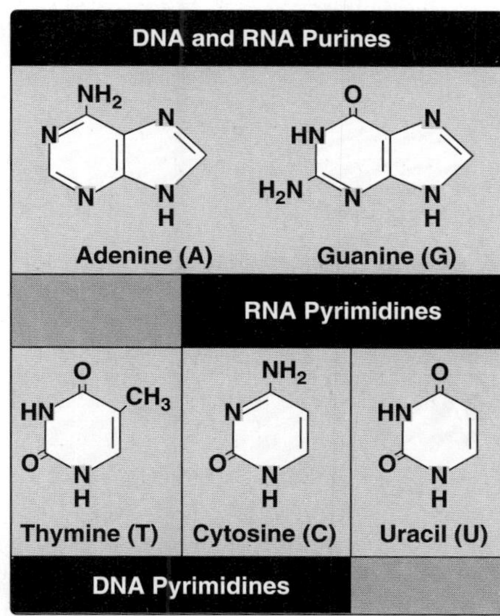

Figure 22.1
Purines and pyrimidines commonly found in DNA and RNA.

Lippincott's Illustrated Reviews: Biochemistry, Third Edition,
by Pamela C. Champe and Richard A. Harvey.
Lippincott Williams & Wilkins, Baltimore, MD © 2005.

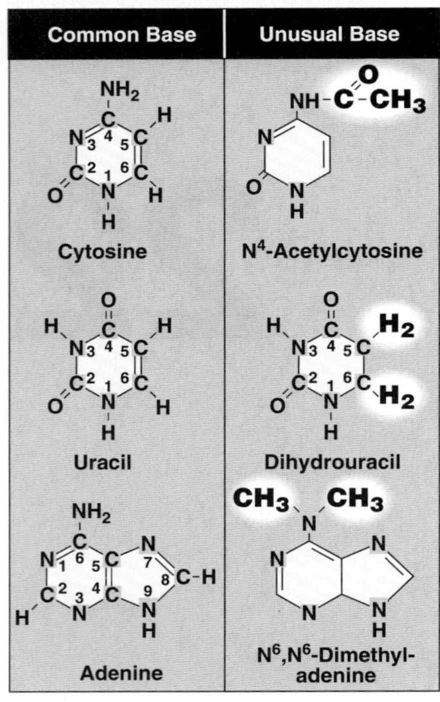

Figure 22.2
Examples of unusual bases.

of DNA and RNA, for example, in some viral DNA, and in transfer RNA. Base modifications include methylation, hydroxymethylation, glycosylation, acetylation, or reduction. Some examples of unusual bases are shown in Figure 22.2.] The presence of an unusual base in a nucleotide sequence may aid in its recognition by specific enzymes, or protect it from being degraded by nucleases.

B. Nucleosides

The addition of a pentose sugar to a base produces a nucleoside. If the sugar is ribose, a ribonucleoside is produced; if the sugar is 2-deoxyribose, a deoxyribonucleoside is produced (Figure 22.3A). The **ribonucleosides** of A, G, C, and U are named **adenosine**, **guanosine**, **cytidine**, and **uridine**, respectively. The **deoxyribonucleosides** of A, G, C, and T have the added prefix, "deoxy-", for example, **deoxyadenosine**. [Note: The compound deoxythymidine is often simply called thymidine, with the "deoxy" prefix being understood.] The carbon and nitrogen atoms in the rings of the base and the sugar are numbered separately (Figure 22.3B). Note that the atoms in the rings of the bases are numbered 1 to 6 in pyrimidines, and 1 to 9 in purines, whereas the carbons in the pentose are numbered 1' to 5'. Thus, when the 5'-carbon of a nucleoside (or nucleotide) is referred to, a carbon atom in the pentose, rather than an atom in the base, is being specified.

C. Nucleotides

Nucleotides are monophosphate, diphosphate, or triphosphate esters of nucleosides. The first phosphate group is attached by an ester linkage to the 5'-OH of the pentose. Such a compound is called a **nucleoside 5'-phosphate** or a **5'-nucleotide**. The type of pentose is denoted by the prefix in the names "5'-ribonucleotide" and "5'-deoxyribonucleotide." If one phosphate group is attached to the 5'-carbon of the pentose, the structure is a **nucleoside monophosphate** (NMP), like AMP or CMP. If a second or third phosphate is added to the nucleoside, a nucleoside diphosphate (for example, ADP) or triphosphate (for example, ATP) results (Figure 22.4). The second and third phosphates are each connected to the nucleotide by a "**high-energy**" **bond**. [Note: The phosphate groups are responsible for the negative charges associated with nucleotides, and cause DNA and RNA to be referred to as "nucleic acids."]

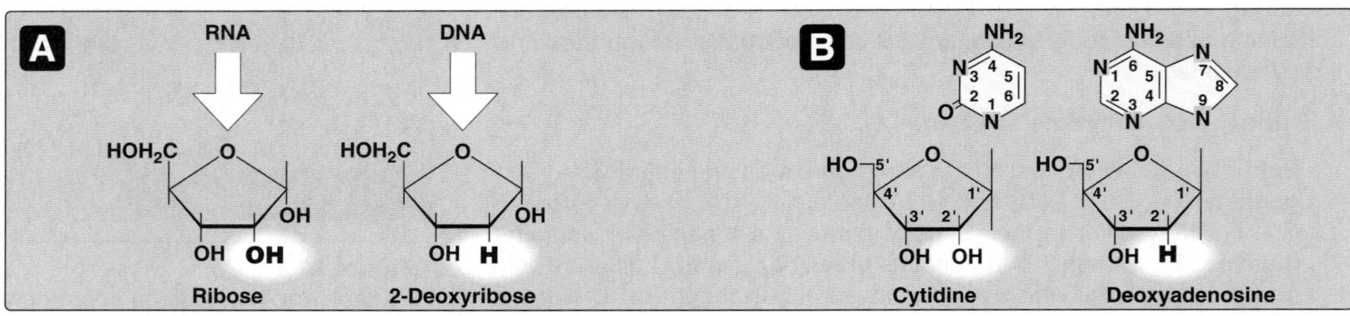

Figure 22.3
A. Pentoses found in nucleic acids. B. Examples of the numbering systems for purine- and pyrimidine-containing nucleosides.

III. SYNTHESIS OF PURINE NUCLEOTIDES

The atoms of the purine ring are contributed by a number of compounds, including amino acids (**aspartic acid**, **glycine**, and **glutamine**), **CO_2**, and **N^{10}–formyltetrahydrofolate** (Figure 22.5). The purine ring is constructed by a series of reactions that add the donated carbons and nitrogens to a preformed ribose 5-phosphate. (See p. 145 for a discussion of ribose 5-phosphate synthesis by the HMP pathway.)

A. Synthesis of 5-phosphoribosyl-1-pyrophosphate (PRPP)

PRPP is an "activated pentose" that participates in the synthesis of purines and pyrimidines, and in the salvage of purine bases (see p. 294). Synthesis of PRPP from ATP and ribose 5-phosphate is catalyzed by *PRPP synthetase* (*ribose phosphate pyrophosphokinase,* Figure 22.6). This enzyme is activated by inorganic phosphate (P_i) and inhibited by purine nucleotides (end-product inhibition). [Note: The sugar moiety of PRPP is ribose, and therefore ribonucleotides are the end products of de novo purine synthesis. When deoxyribonucleotides are required for DNA synthesis, the ribose sugar moiety is reduced (see p. 295).]

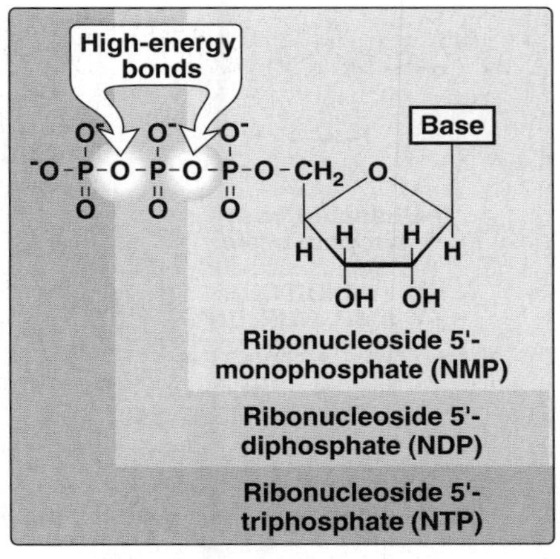

Figure 22.4
Ribonucleoside monophosphate, diphosphate, and triphosphate.

B. Synthesis of 5'-phosphoribosylamine

Synthesis of 5'-phosphoribosylamine from PRPP and glutamine is shown in Figure 22.7. The amide group of glutamine replaces the pyrophosphate group attached to carbon 1 of PRPP. The enzyme, *glutamine:phosphoribosyl pyrophosphate amidotransferase*, is inhibited by the purine 5'-nucleotides AMP, GMP, and IMP—the end products of the pathway. This is the **committed step** in purine nucleotide biosynthesis. The rate of the reaction is also controlled by the intracellular concentration of the substrates glutamine and PRPP. [Note: The intracellular concentration of PRPP is normally far below the K_m for the *amidotransferase*. Therefore, any small change in the PRPP concentration causes a proportional change in the rate of the reaction (see p. 59).]

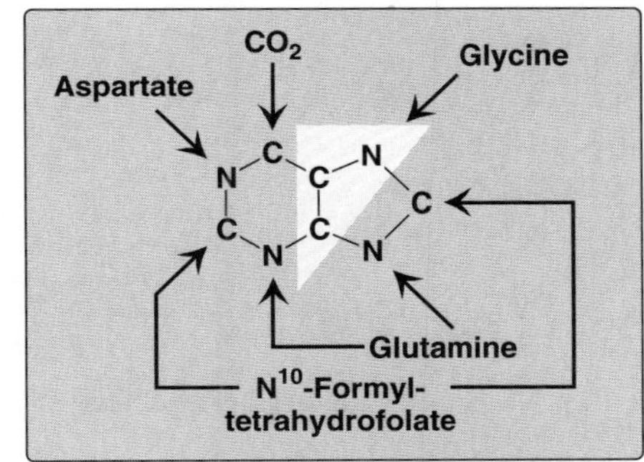

Figure 22.5
Sources of the individual atoms in the purine ring.

C. Synthesis of inosine monophosphate, the "parent" purine nucleotide

The next nine steps in purine nucleotide biosynthesis leading to the synthesis of IMP (whose base is hypoxanthine) are illustrated in Figure 22.7. This pathway requires four ATP molecules as an energy source. Two steps in the pathway require N^{10}-formyltetrahydrofolate.

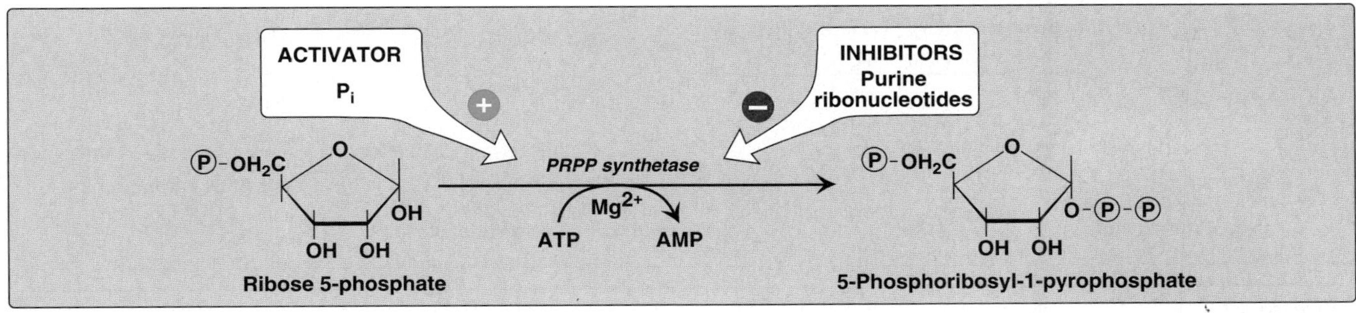

Figure 22.6
Synthesis of 5-phosphoribosyl-1-pyrophosphate (PRPP), showing the activator and inhibitors of the reaction.

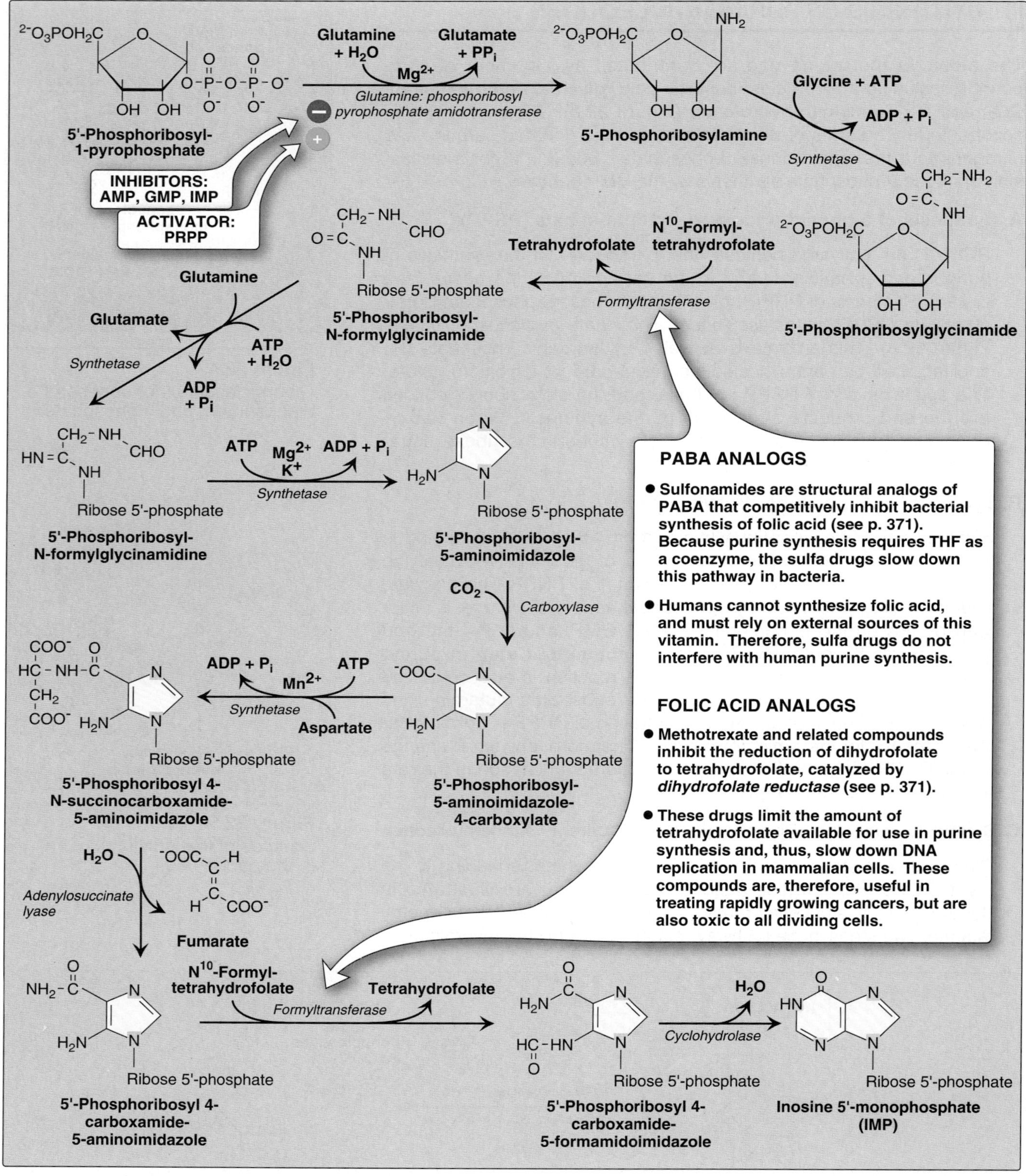

Figure 22.7
Synthesis of purine nucleotides, showing the inhibitory effect of some structural analogs.

D. Synthetic inhibitors of purine synthesis

Some synthetic inhibitors of purine synthesis (for example, the **sulfonamides**[1]), are designed to inhibit the growth of rapidly dividing microorganisms without interfering with human cell functions (see Figure 22.7). Other purine synthesis inhibitors, such as structural analogs of folic acid (for example, **methotrexate**[2]), are used pharmacologically to control the spread of cancer by interfering with the synthesis of nucleotides and, therefore, of DNA and RNA (see Figure 22.7). [Note: Inhibitors of human purine synthesis are extremely toxic to tissues, especially to developing structures such as in a fetus, or to cell types that normally replicate rapidly, including those of bone marrow, skin, gastrointestinal (GI) tract, immune system, or hair follicles. As a result, individuals taking such anti-cancer drugs can experience adverse effects, including anemia, scaly skin, GI tract disturbance, immunodeficiencies, and baldness.] **Trimethoprim**,[3] another folate analog, has potent antibacterial activity because of its selective inhibition of bacterial *dihydrofolate reductase*.

E. Conversion of IMP to AMP and GMP

The conversion of IMP to either AMP or GMP uses a two-step, energy-requiring pathway (Figure 22.8). Note that the synthesis of AMP requires GTP as an energy source, whereas the synthesis of GMP requires ATP. Also, the first reaction in each pathway is inhib-

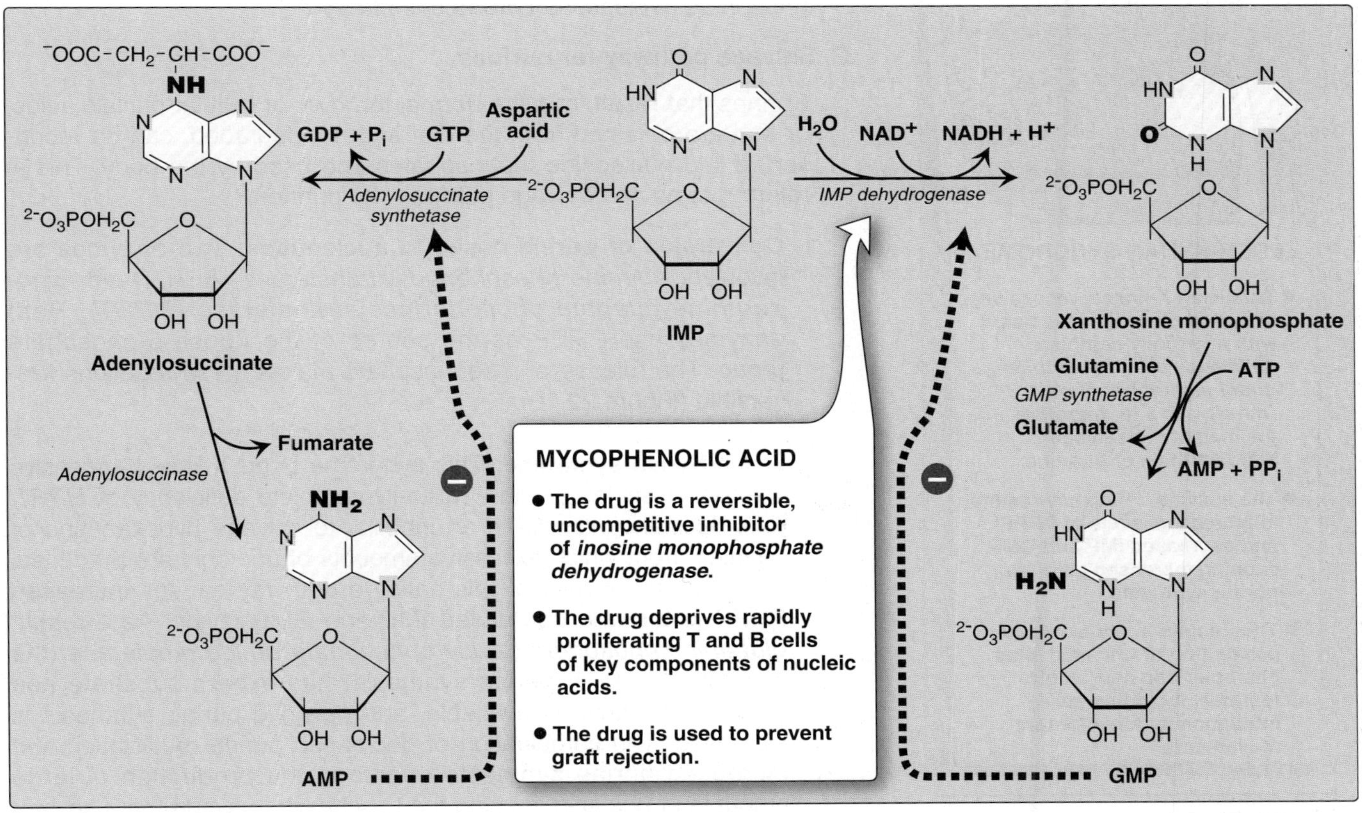

MYCOPHENOLIC ACID

- **The drug is a reversible, uncompetitive inhibitor of *inosine monophosphate dehydrogenase*.**
- **The drug deprives rapidly proliferating T and B cells of key components of nucleic acids.**
- **The drug is used to prevent graft rejection.**

Figure 22.8
Conversion of IMP to AMP and GMP showing feedback inhibition.

[1-3]See Chapters 34 and 40 in ***Lippincott's Illustrated Reviews: Pharmacology*** (3rd Ed.) and Chapters 29 and 38 (2nd Ed.) for a discussion of sulfonamides, trimethoprim, and methotrexate.

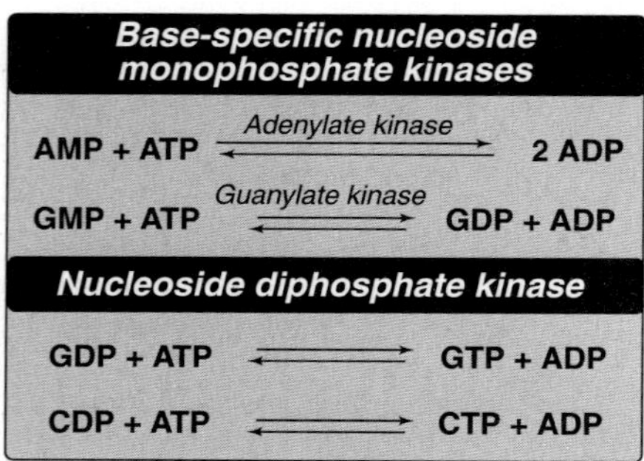

Figure 22.9
Conversion of nucloside monophosphates to nucleoside diphosphates and triphosphates.

ited by the end product of that pathway. This provides a mechanism for diverting IMP to the synthesis of the species of purine present in lesser amounts. If both AMP and GMP are present in adequate amounts, the de novo pathway of purine synthesis is turned off at the *amidotransferase* step. [Note: **Mycophenolic acid** (MPA) is a potent, reversible, uncompetitive inhibitor of *inosine monophosphate dehydrogenase* that is being used successfully in preventing graft rejection. It blocks the de novo formation of guanosine monophosphate (GMP, see Figure 22.8), thus depriving rapidly proliferating cells, including T and B cells, of a key component of nucleic acids.]

F. Conversion of nucleoside monophosphates to nucleoside diphosphates and triphosphates

Nucleoside diphosphates (NDP) are synthesized from the corresponding nucleoside monophosphates (NMP) by **base-specific nucleoside monophosphate kinases** (Figure 22.9). [Note: These kinases do not discriminate between ribose or deoxyribose in the substrate.] ATP is generally the source of the transferred phosphate, because it is present in higher concentrations than the other nucleoside triphosphates. *Adenylate kinase* is particularly active in liver and muscle, where the turnover of energy from ATP is high. Its function is to maintain an equilibrium among AMP, ADP, and ATP. Nucleoside diphosphates and triphosphates are interconverted by *nucleoside diphosphate kinase*—an enzyme that, unlike the monophosphate kinases, has **broad specificity**.

G. Salvage pathway for purines

Purines that result from the normal turnover of cellular nucleic acids, or that are obtained from the diet and not degraded, can be reconverted into nucleoside triphosphates and used by the body. This is referred to as the "salvage pathway" for purines.

1. **Conversion of purine bases to nucleotides:** Two enzymes are involved: *adenine phosphoribosyltransferase* (*APRT*) and *hypoxanthine-guanine phosphoribosyltransferase* (*HPRT*). Both enzymes use PRPP as the source of the ribose 5-phosphate group. The release of pyrophosphate makes these reactions irreversible (Figure 22.10).

2. **Lesch-Nyhan syndrome:** This syndrome is an X-linked, recessive disorder associated with a virtually complete deficiency of *HPRT*. This deficiency results in an inability to salvage hypoxanthine or guanine, from which excessive amounts of uric acid are produced. In addition, the lack of this salvage pathway causes increased PRPP levels and decreased IMP and GMP levels. As a result, *glutamine:phosphoribosylpyrophosphate amidotransferase* (the committed step in purine synthesis) has excess substrate and decreased inhibitors available, and de novo purine synthesis is increased. The combination of decreased purine reutilization and increased purine synthesis results in the production of large amounts of uric acid, making the Lesch-Nyhan syndrome a severe, heritable form of **gout**. Patients with Lesch-Nyhan syndrome tend to produce urate kidney stones. In addition, characteristic neurologic features of the disorder include self-mutilation (Figure 22.11) and involuntary movements.

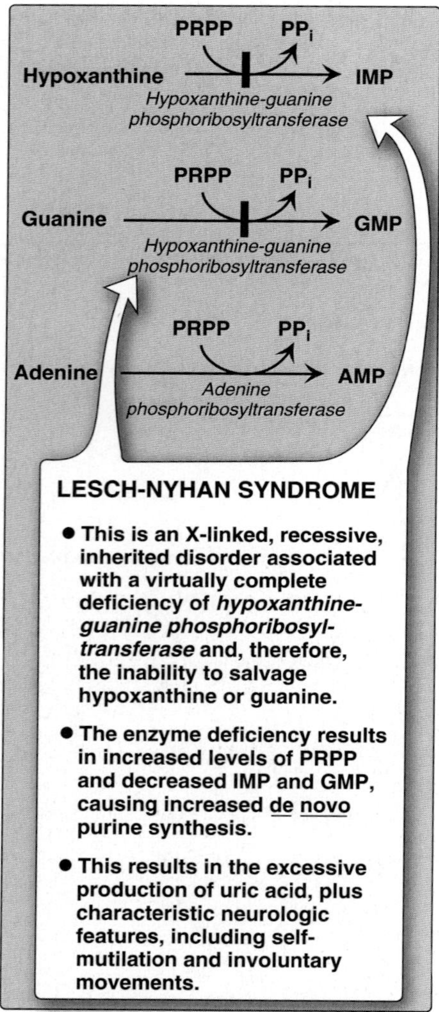

Figure 22.10
Salvage pathways of purine nucleotide synthesis.

IV. SYNTHESIS OF DEOXYRIBONUCLEOTIDES

The nucleotides described thus far in this chapter all contain ribose (**ribonucleotides**). The nucleotides required for DNA synthesis, however, are **2'-deoxyribonucleotides**, which are produced from ribonucleoside diphosphates by the enzyme *ribonucleotide reductase*.

A. Ribonucleotide reductase

Ribonucleotide reductase (*ribonucleoside diphosphate reductase*) is a multisubunit enzyme (two identical B1 subunits and two identical B2 subunits) that is specific for the reduction of nucleoside diphosphates (ADP, GDP, CDP, and UDP) to their deoxy-forms (dADP, dGDP, dCDP, and dUDP). The immediate donors of the hydrogen atoms needed for the reduction of the 2'-hydroxyl group are two sulfhydryl groups on the enzyme itself, which, during the reaction, form a disulfide bond (Figure 22.12).

1. **Regeneration of reduced enzyme:** In order for *ribonucleotide reductase* to continue to produce deoxyribonucleotides, the disulfide bond created during the production of the 2'-deoxy carbon must be reduced. The source of the reducing equivalents is **thioredoxin**—a peptide coenzyme of *ribonucleotide reductase*. Thioredoxin contains two cysteine residues separated by two amino acids in the peptide chain. The two sulfhydryl groups of thioredoxin donate their hydrogen atoms to *ribonucleotide reductase*, in the process forming a disulfide bond (see p. 19).

2. **Regeneration of reduced thioredoxin:** Thioredoxin must be converted back to its reduced form in order to continue to perform its function. The necessary reducing equivalents are provided by NADPH + H^+, and the reaction is catalyzed by *thioredoxin reductase* (see Figure 22.12).

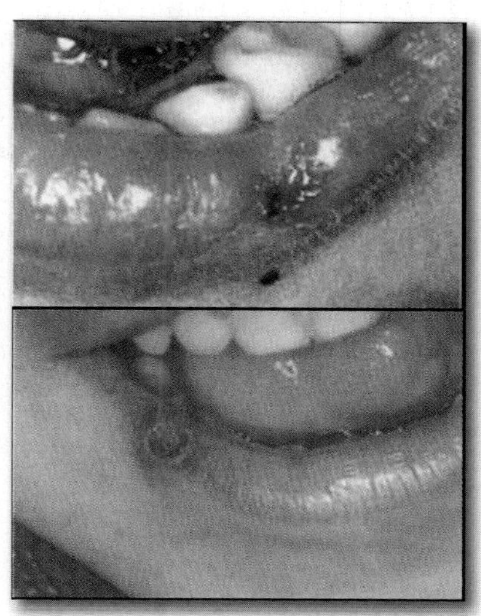

Figure 22.11
Lesions on the lips of Lesch-Nyhan patients caused by self-mutilation.

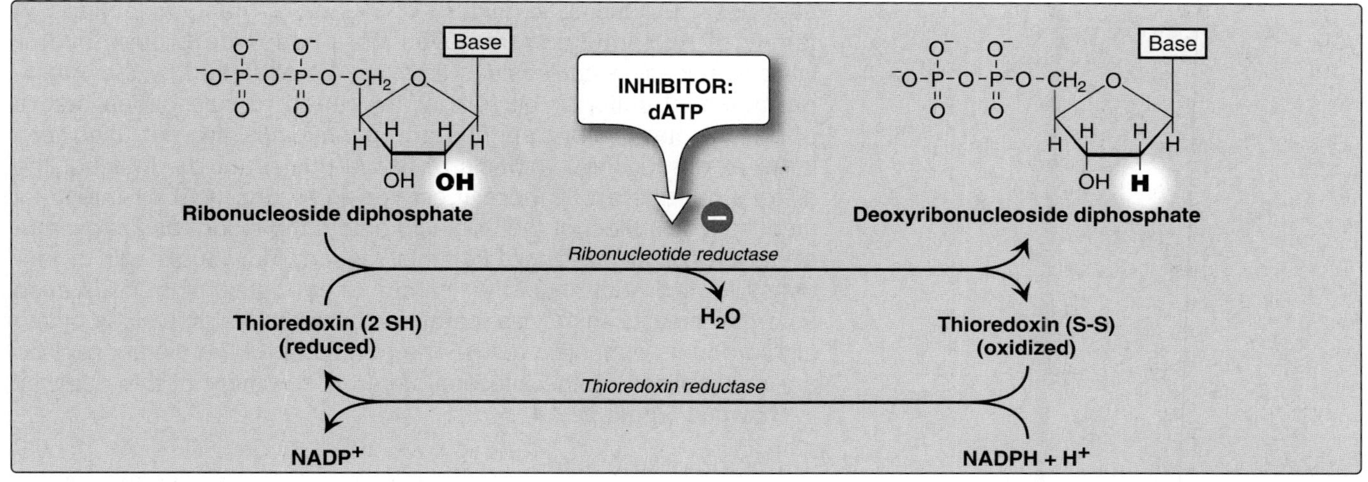

Figure 22.12
Conversion of ribonucleotides to deoxyribonucleotides.

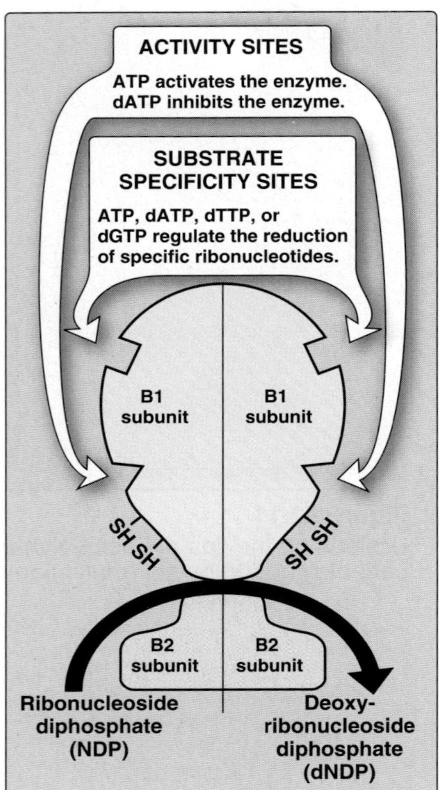

Figure 22.13
Regulation of *ribonucleotide reductase*.

B. Regulation of deoxyribonucleotide synthesis

Ribonucleotide reductase is responsible for maintaining a balanced supply of the deoxyribonucleotides required for DNA synthesis. To achieve this, the regulation of the enzyme is complex. In addition to the single active site, there are two sites on the enzyme involved in regulating its activity (Figure 22.13).

1. **Activity site:** The binding of dATP to an allosteric site (known as the activity site) on the enzyme inhibits the overall catalytic activity of the enzyme and, therefore, prevents the reduction of any of the four nucleoside diphosphates. This effectively prevents DNA synthesis, and explains the toxicity of increased levels of dATP seen in conditions such as *adenosine deaminase* deficiency (see p. 299).

2. **Substrate specificity site:** The binding of nucleoside triphosphates to an additional allosteric site (known as the substrate specificity site) on the enzyme regulates substrate specificity, causing an increase in the conversion of different species of ribonucleotides to deoxyribonucleotides as they are required for DNA synthesis.

V. DEGRADATION OF PURINE NUCLEOTIDES

Degradation of dietary nucleic acids occurs in the small intestine, where a family of pancreatic enzymes hydrolyzes the nucleotides to nucleosides and free bases. Inside cells, purine nucleotides are sequentially degraded by specific enzymes, with **uric acid** as the end product of this pathway. [Note: Mammals other than primates oxidize uric acid further to allantoin, which, in some animals other than mammals, may be further degraded to urea or ammonia.]

A. Degradation of dietary nucleic acids in the small intestine

Ribonucleases and deoxyribonucleases, secreted by the pancreas, hydrolyze RNA and DNA primarily to oligonucleotides. Oligonucleotides are further hydrolyzed by pancreatic phosphodiesterases, producing a mixture of 3'- and 5'-mononucleotides. A family of nucleotidases removes the phosphate groups hydrolytically, releasing nucleosides that may be absorbed by the intestinal mucosal cells, or be further degraded to free bases before uptake. [Note: Dietary purines and pyrimidines are not used to a large extent for the synthesis of tissue nucleic acids. Instead, the dietary purines are generally converted to uric acid by intestinal mucosal cells. Most of the uric acid enters the blood, and is eventually excreted in the urine. For this reason, individuals with a tendency toward gout should be careful about consuming foods such as organ meats, anchovies, sardines, or dried beans, which contain high amounts of nucleic acids. The remainder of the dietary purines are metabolized by the intestinal flora.] A summary of this pathway is shown in Figure 22.14.

B. Formation of uric acid

A summary of the steps in the production of uric acid and genetic diseases associated with deficiencies of specific degradative enzymes are shown in Figure 22.15. [Note: The bracketed numbers refer to specific reactions in the figure.]

[1] An amino group is removed from AMP to produce IMP, or from adenosine to produce inosine (hypoxanthine-ribose) by *AMP* or *adenosine deaminase*.

[2] IMP and GMP are converted into their nucleoside forms—inosine and guanosine—by the action of *5'-nucleotidase*.

[3] *Purine nucleoside phosphorylase* converts inosine and guanosine into their respective purine bases, hypoxanthine and guanine.

[4] Guanine is deaminated to form xanthine.

[5] Hypoxanthine is oxidized by *xanthine oxidase* to xanthine, which is further oxidized by *xanthine oxidase* to uric acid, the final product of human purine degradation. Uric acid is excreted in the urine.

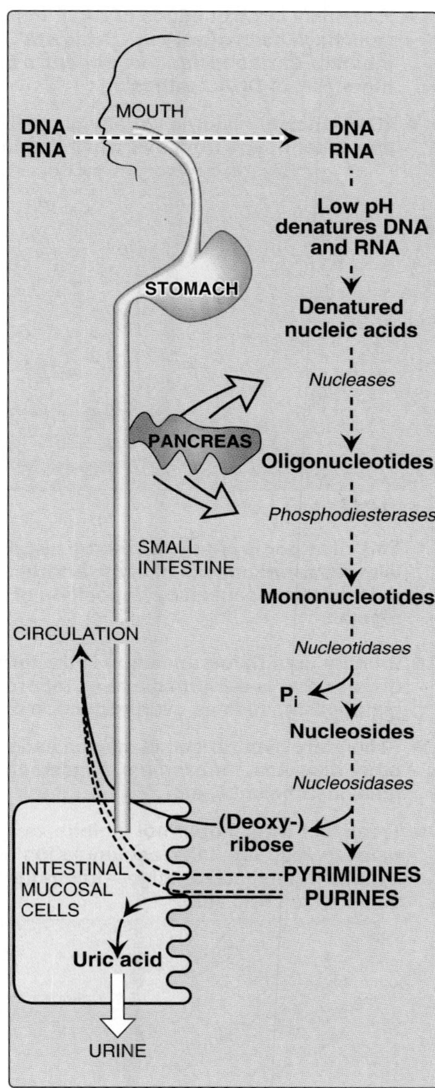

Figure 22.14
Digestion of dietary nucleic acids.

C. Diseases associated with purine degradation

1. Gout: Gout is a disorder characterized by high levels of uric acid in the blood, as a result of either **over-production** or **underexcretion** of uric acid. Hyperuricemia results in the deposition of crystals of **sodium urate**—the end product of purine metabolism—in tissues, especially the kidney and joints, causing first **acute** and progressing to **chronic gouty arthritis**. [Note: **Hyperuricemia** does not always lead to gout, but gout is usually preceded by hyperuricemia.] Figure 22.16 shows a patient with an index finger showing **tophaceous gout**, in which tophi (nodular masses of monosodium urate crystals) are deposited in the soft tissues of the body. The deposition of needle-shaped monosodium urate crystals initiates an **inflammatory process** involving the infiltration of granulocytes that phagocytize the urate crystals. This process generates oxygen metabolites (see p. 145) that damage tissue, resulting in the release of lysosomal enzymes that evoke an inflammatory response. In addition, lactate production in the synovial tissues increases, resulting in a decrease in pH that fosters further deposition of urate crystals. A definitive diagnosis requires aspiration and examination of synovial fluid using polarized light microscopy to confirm the presence of monosodium urate crystals (Figure 22.17).

a. Primary gout: In most patients, gout is caused by the underexcretion of uric acid due to defective renal secretion. However, **overproduction** of uric acid may occur because of an inherited abnormality in the enzymes of purine metabolism. This is defined as "primary gout." For example, several **X-linked mutations** have been identified in the *PRPP synthetase* gene that result in the enzyme having either an increased V_{max} (see p. 58) for the production of PRPP, a lower K_m(see p. 59) for ribose 5-phosphate, or a decreased sensitivity to its purine nucleotide inhibitors. In each case, purine production is increased, resulting in elevated levels of plasma uric acid. **Lesch-Nyhan syndrome** (see p. 294) also causes hyperuricemia, as a result of the decreased salvage of hypoxanthine and guanine bases.

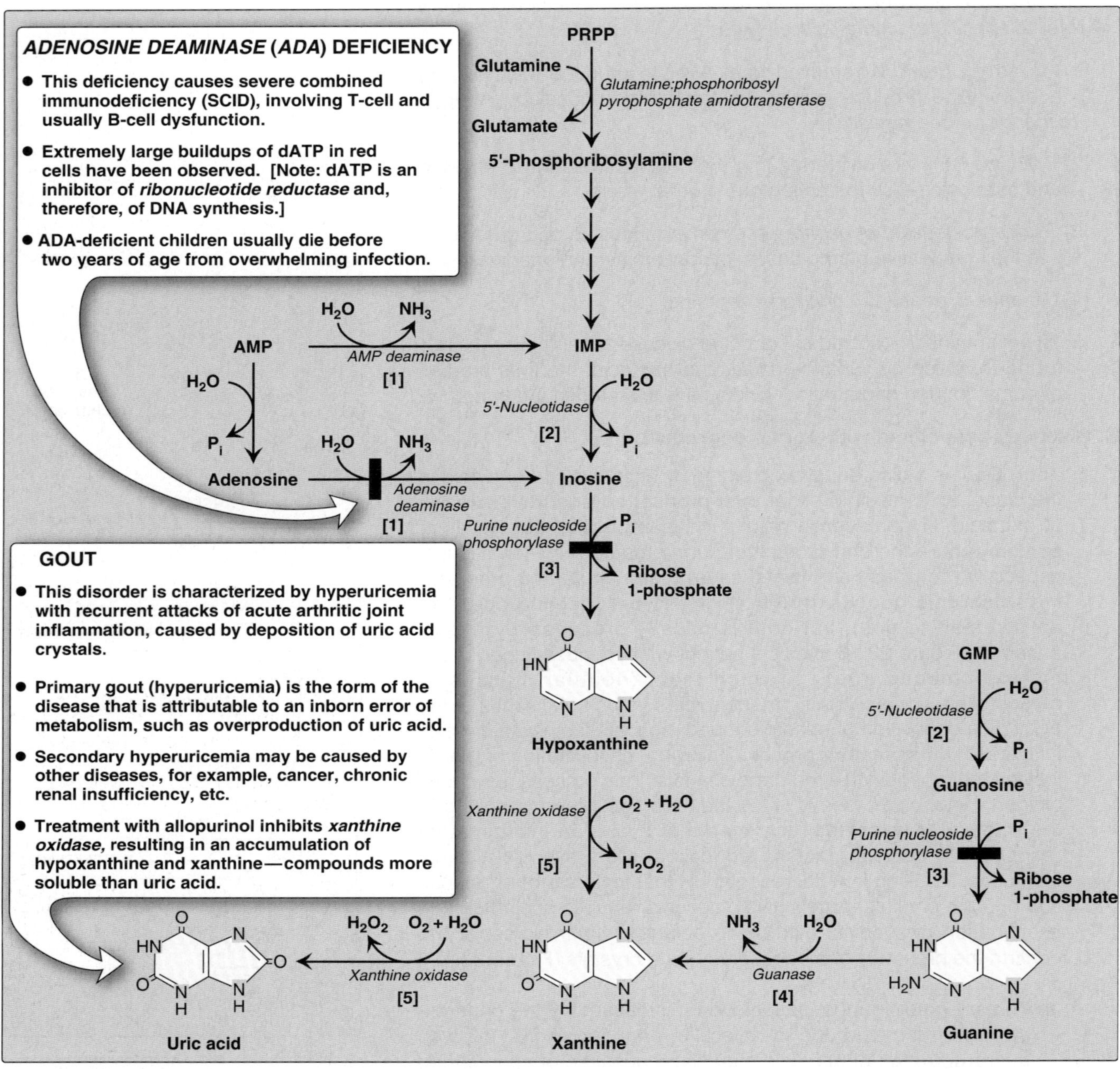

Figure 22.15
The degradation of purine nucleotides to uric acid, illustrating some of the genetic diseases associated with this pathway. [Note: The numbers in brackets refer to the corresponding numbered citations in the text.]

b. **Secondary hyperuricemia:** This form of gout is caused by a variety of disorders and lifestyles, for example, in patients with chronic renal insufficiency, those undergoing chemotherapy, those with myeloproliferative disorders, and those who consume excessive amounts of alcohol or purine-rich foods. Gout can also be an adverse effect of seemingly unrelated metabolic diseases, such as von Gierke disease (see Figure 11.8, p. 128) or fructose intolerance (see p. 136).

c. **Treatment for gout:** Acute attacks are treated with **colchicine** to decrease movement of granulocytes into the affected area, and with anti-inflammatory drugs, such as aspirin, to provide pain relief.[4] Most therapeutic strategies for gout involve lowering the uric acid level below the saturation point, thus preventing the deposition of urate crystals. Uricosuric agents, such as **probenecid** or **sulfinpyrazone**,[5] are used in most patients with gout, because most are "underexcretors" of uric acid. **Allopurinol**—an inhibitor of uric acid synthesis—is more toxic, and is reserved for those patients whose hyperuricemia is a result of overproduction of urate. Allopurinol is converted in the body to oxypurinol, which inhibits *xanthine oxidase*, resulting in an accumulation of hypoxanthine and xanthine (see Figure 22.15)—compounds more soluble than uric acid and, therefore, less likely to initiate an inflammatory response.

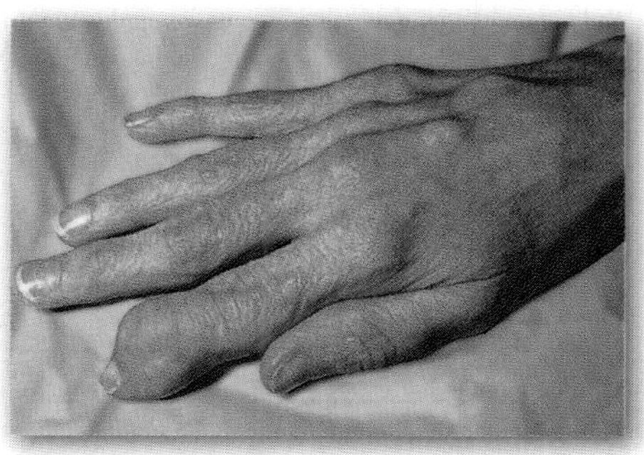

Figure 22.16
Tophaceous gout.

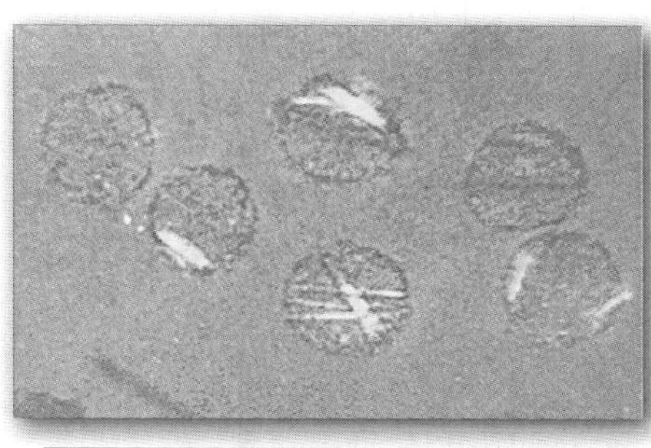

Figure 22.17
Gout can be diagnosed by the presence of negatively birefringent monosodium urate crystals in aspirated synovial fluid examined by polarized-light microscopy. Here, crystals are within polymorphonuclear leukocytes.

2. **Adenosine deaminase deficiency:** *Adenosine deaminase* (*ADA*) is expressed in the cytosol of all cells, but, in humans, lymphocytes have the highest activity of this enzyme. A deficiency of *ADA* results in an accumulation of adenosine, which is converted to its ribonucleotide or deoxyribonucleotide forms by cellular kinases. As dATP levels rise, *ribonucleotide reductase* is inhibited, thus preventing the production of all deoxyribose-containing nucleotides (see p. 295). Consequently, cells cannot make DNA and divide. In its most severe form, this autosomal recessive disorder causes **severe combined immunodeficiency disease** (**SCID**), involving a lack of both T cells and B cells. Children with this disorder must live in a sterile environment (Figure 22.18), and usually die by the age of two. [Note: It is estimated that in the United States, *ADA* deficiency accounts for approximately fourteen percent of all cases of SCID.] Treatment requires either bone marrow replacement or enzyme replacement therapy. [Note: *ADA* deficiency was the first genetic disease successfully treated by gene therapy.]

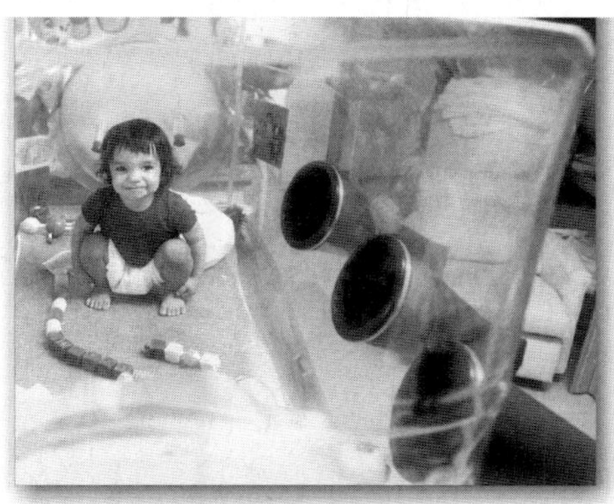

Figure 22.18
Young child born with immune deficiency syndrome plays in the enclosed, germ-free, plastic environment in which he must live to survive.

VI. PYRIMIDINE SYNTHESIS AND DEGRADATION

Unlike the synthesis of the purine ring, in which the ring is constructed on a preexisting ribose 5-phosphate, the pyrimidine ring is synthesized before being attached to ribose 5-phosphate, which is donated by PRPP. The sources of the atoms in the pyrimidine ring are **glutamine**, **CO_2**, and **aspartic acid** (Figure 22.19). [Note: Glutamine and aspartic acid are thus required for both purine and pyrimidine synthesis.]

A. Synthesis of carbamoyl phosphate

The regulated step of this pathway in mammalian cells is the synthesis of carbamoyl phosphate from **glutamine** and **CO_2**, catalyzed by *carbamoyl phosphate synthetase II* (*CPS II*). *CPS II* is inhibited

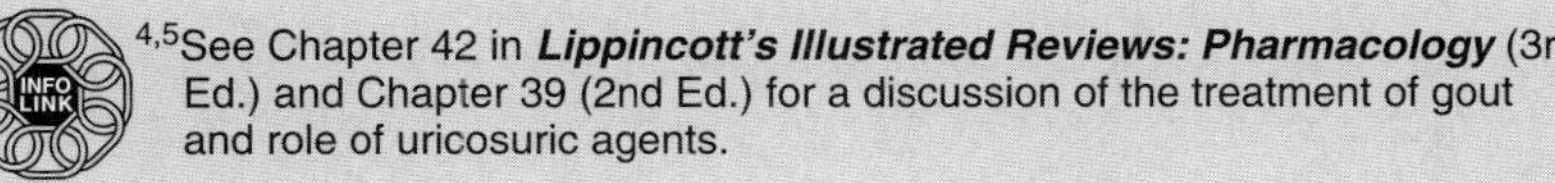

[4,5]See Chapter 42 in ***Lippincott's Illustrated Reviews: Pharmacology*** (3rd Ed.) and Chapter 39 (2nd Ed.) for a discussion of the treatment of gout and role of uricosuric agents.

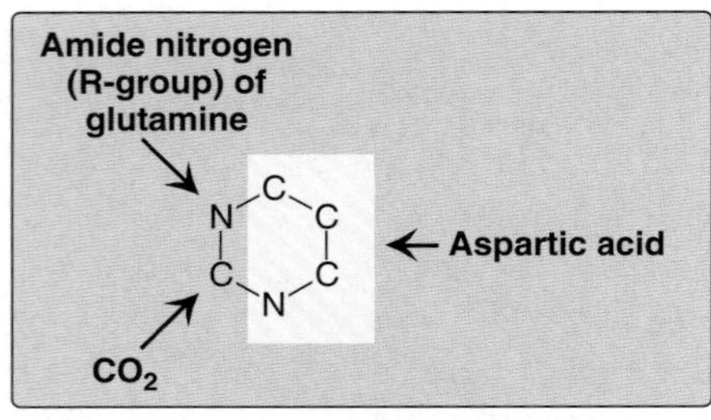

Figure 22.19
Sources of the individual atoms in the pyrimidine ring.

	CPS I	CPS II
Cellular location	Mitochondria	Cytosol
Pathway involved	Urea cycle	Pyrimidine synthesis
Source of nitrogen	Ammonia	γ-Amide group of glutamine
Regulators	Activator: N–acetyl-glutamate	Inhibitor: UTP Activator: ATP

Figure 22.20
Summary of the differences between *carbamoyl phosphate synthetase* (*CPS*) *I* and *II*.

by UTP (the end-product of this pathway, which can be converted into the other pyrimidine nucleotides), and is activated by ATP and PRPP. [Note: Carbamoyl phosphate, synthesized by *CPS I*, is also a precursor of urea (see p. 251). Unlike other carboxylating enzymes, neither *CPS* requires biotin as a coenzyme. A comparison of the two enzymes is presented in Figure 22.20.]

B. Synthesis of orotic acid

The second step in pyrimidine synthesis is the formation of carbamoylaspartate, catalyzed by *aspartate transcarbamoylase*. The pyrimidine ring is then closed hydrolytically by *dihydroorotase*. The resulting dihydroorotate is oxidized to produce orotic acid (orotate, Figure 22.21). The enzyme that produces orotate, *dihydroorotate dehydrogenase*, is located inside the mitochondria. All other reactions in pyrimidine biosynthesis are cytosolic. [Note: The first three enzymes in this pathway (*CPS II*, *aspartate transcarbamoylase*, and *dihydroorotase*) are all domains of the same polypeptide chain. (See p. 19 for a discussion of domains.) This is an example of a multifunctional or multicatalytic polypeptide that facilitates the ordered synthesis of an important compound.]

C. Formation of a pyrimidine nucleotide

The completed pyrimidine ring is converted to the nucleotide **orotidine 5'-monophosphate** (**OMP**) in the second stage of pyrimidine nucleotide synthesis (see Figure 22.21). PRPP is again the ribose 5-phosphate donor. The enzyme *orotate phosphoribosyltransferase* produces OMP and releases pyrophosphate, thereby making the reaction biologically irreversible. [Note: Both purine and pyrimidine synthesis thus require glutamine and PRPP as essential precursors.] OMP, the parent pyrimidine mononucleotide, is converted to **uridine monophosphate** (**UMP**) by *orotidylate decarboxylase*, which removes the acidic carboxyl group. *Orotate phosphoribosyltransferase* and *orotidylate decarboxylase* are also domains of a single polypeptide chain called *UMP synthase*. **Orotic aciduria**—a rare genetic defect—is caused by a deficiency of this bifunctional enzyme, resulting in orotic acid in the urine (see Figure 22.21).

D. Synthesis of uridine triphosphate and cytidine triphosphate

Cytidine triphosphate (CTP) is produced by amination of UTP by *CTP synthetase* (Figure 22.22). [Note: The nitrogen is provided by glutamine—another example of a reaction in nucleotide biosynthesis in which this amino acid is required.]

E. Syntheisis of thymidine monophosphate from dUMP

dUMP is converted to dTMP by *thymidylate synthase*, which uses **N^5,N^{10}–methylene tetrahydrofolate** as the source of the methyl group (see p. 265 for a discussion of this coenzyme). This is an unusual reaction in that tetrahydrofolate (THF) contributes not only a carbon unit but also two hydrogen atoms from the pteridine ring, resulting in the oxidation of THF to dihydrofolate (DHF) (Figure 22.23). Inhibitors of *thymidylate synthase* include thymine analogs such as **5-fluorouracil**, which serve as successful antitumor

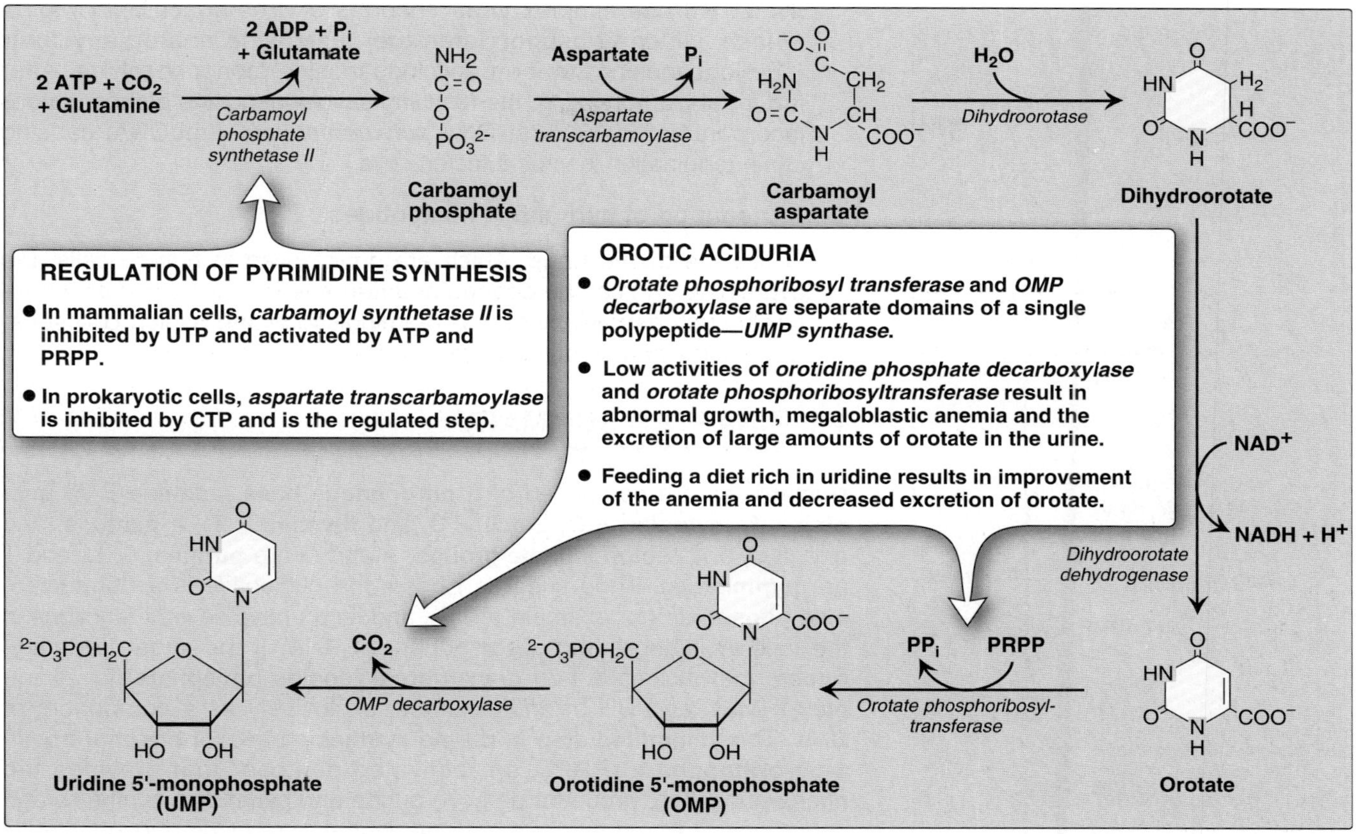

REGULATION OF PYRIMIDINE SYNTHESIS

- In mammalian cells, *carbamoyl synthetase II* is inhibited by UTP and activated by ATP and PRPP.
- In prokaryotic cells, *aspartate transcarbamoylase* is inhibited by CTP and is the regulated step.

OROTIC ACIDURIA

- *Orotate phosphoribosyl transferase* and *OMP decarboxylase* are separate domains of a single polypeptide—*UMP synthase.*
- Low activities of *orotidine phosphate decarboxylase* and *orotate phosphoribosyltransferase* result in abnormal growth, megaloblastic anemia and the excretion of large amounts of orotate in the urine.
- Feeding a diet rich in uridine results in improvement of the anemia and decreased excretion of orotate.

Figure 22.21
De novo pyrimidine synthesis.

agents.[6] 5-Fluorouracil is metabolically converted to 5-FdUMP, which becomes permanently bound to the inactivated *thymidylate synthase*; for this reason, the drug is called a "suicide" inhibitor. DHF can be reduced to THF by *dihydrofolate reductase* (see Figure 28.3, p. 372), an enzyme that is inhibited in the presence of drugs such as **methotrexate**. By decreasing the supply of THF, these folate analogs not only inhibit purine synthesis (see Figure 22.7), but, by preventing methylation of dUMP to dTMP, they also lower the cellular concentration of this essential component of DNA. DNA synthesis is, therefore, inhibited and cell growth slowed. Because of their ability to slow the replication of DNA by decreasing the availability of nucleotide precursors, drugs such as those described above are used to decrease the growth rate of cancer cells.

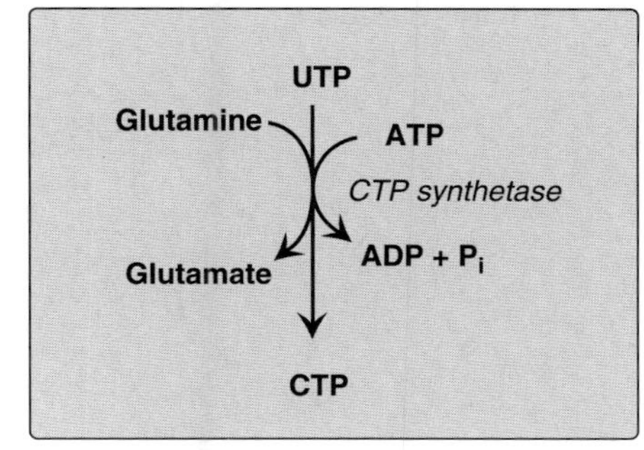

Figure 22.22
Synthesis of CTP from UTP.

F. Salvage of pyrimidines

Few pyrimidine bases are salvaged in human cells. However, the pyrimidine nucleosides **uridine** and **cytidine** can be salvaged by *uridine-cytidine kinase*, **deoxycytidine** can be salvaged by *deoxycytidine kinase*, and **thymidine** can be salvaged by the enzyme *thymidine kinase*. Each of these enzymes catalyzes the phosphorylation of a nucleoside(s) utilizing ATP, and forming UMP, CMP, dCMP, and TMP.

[6]See Chapter 40 ***Lippincott's Illustrated Reviews: Pharmacology*** (3rd Ed.) and Chapter 38 (2nd Ed.) for a discussion of the treatment of 5-fluorouracil as an anticancer drug.

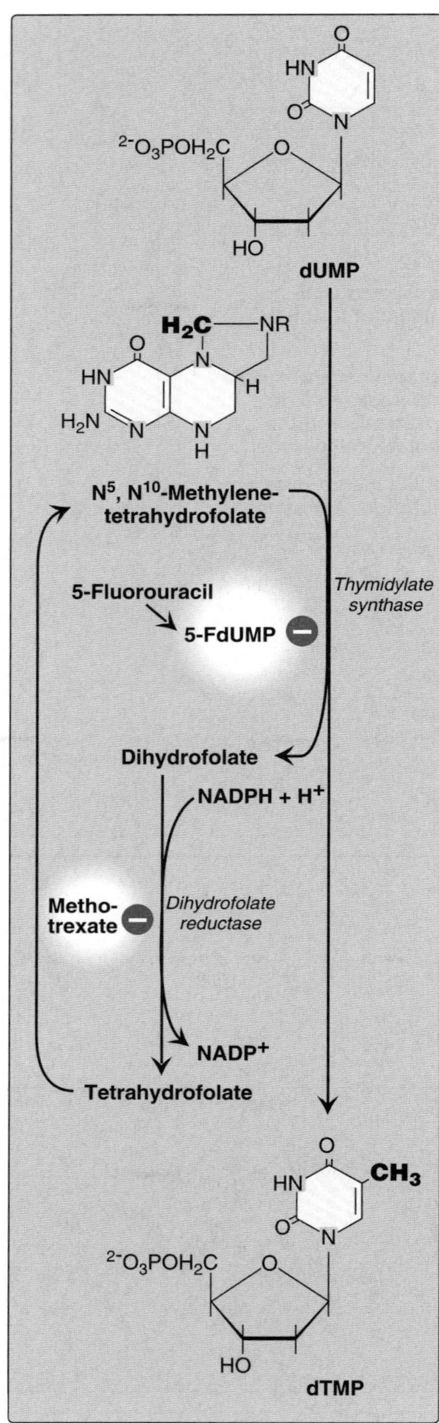

Figure 22.23
Synthesis of dTMP from dUMP, illustrating sites of action of antineoplastic drugs.

[Note: Herpes simplex virus encodes a virus-specific *thymidine kinase*, which phosphorylates the nucleoside analog **acyclovir** (acycloguanosine) to form acycloguanosine monophosphate. After further phosphorylation, the resulting acycloguanosine triphosphate is incorporated by the viral *DNA polymerase* into viral DNA, causing chain termination in virus-infected cells.]

G. Degradation of pyrimidine nucleotides

Unlike the purine rings, which are not cleaved in human cells, the pyrimidine ring can be opened and degraded to highly soluble structures, such as β-alanine, and β-aminoisobutyrate, which can serve as precursors of acetyl CoA and succinyl CoA, respectively.

VII. CHAPTER SUMMARY

Nucleotides are composed of a **nitrogenous base** (**adenine** = **A**, **guanine** = **G**, **cytosine** = **C**, **uracil** = **U**, and **thymine** = **T**), a **pentose**, and one, two, or three **phosphate groups**. A and G are **purines**, C, U, and T are **pyrimidines**. If the sugar is **ribose**, the nucleotide is a **ribonucleoside phosphate** (for example, AMP), and it can have several functions in the cell, including being a component of **RNA**. If the sugar is **deoxyribose**, the nucleotide is a **deoxyribonucleoside phosphate** (for example, dAMP), and will be found almost exclusively as a component of **DNA**. The **committed step** in **purine synthesis** uses **5-phosphoribosyl-1-pyrophosphate** (PRPP, an "activated pentose" that provides the ribose-phosphate group for de novo purine and pyrimidine synthesis and purine salvage) and nitrogen from **glutamine** to produce phosphoribosyl amine. The enzyme is **glutamine:PRPP amidotransferase**, and is **inhibited** by **AMP**, **GMP**, and **IMP** (the end products of the pathway). Purine nucleotides can also be produced from preformed purine bases by using **salvage reactions** catalyzed by **adenine phosphoribosyltransferase** and **hypoxanthine–guanine phosphoribosyltransferase** (**HPRT**). A deficiency of HPRT causes **Lesch-Nyhan syndrome**—a severe, heritable form of gout, accompanied by compulsive self-mutilization. All deoxyribonucleotides are synthesized from ribonucleotides by the enzyme **ribonucleotide reductase**. This enzyme is highly regulated, for example, it is **strongly inhibited by dATP**—a compound that is over-produced in bone marrow cells in individuals having **adenosine deaminase deficiency**. This syndrome causes **severe combined immunodeficiency disease**. The end product of purine degradation is **uric acid**—a compound whose over-production or under-secretion causes **gout**. The first step in **pyrimidine synthesis**—the production of carbamoyl phosphate by **carbamoyl phosphate synthetase II**—is the **regulated** step in this pathway (it is **inhibited by UTP** and **activated by ATP** and **PRPP**). The UTP produced by this pathway can be converted to CTP. dUMP can be converted to **dTMP** using **thymidylate synthase**—an enzyme targeted by anti-cancer drugs such as 5-fluorouracil.

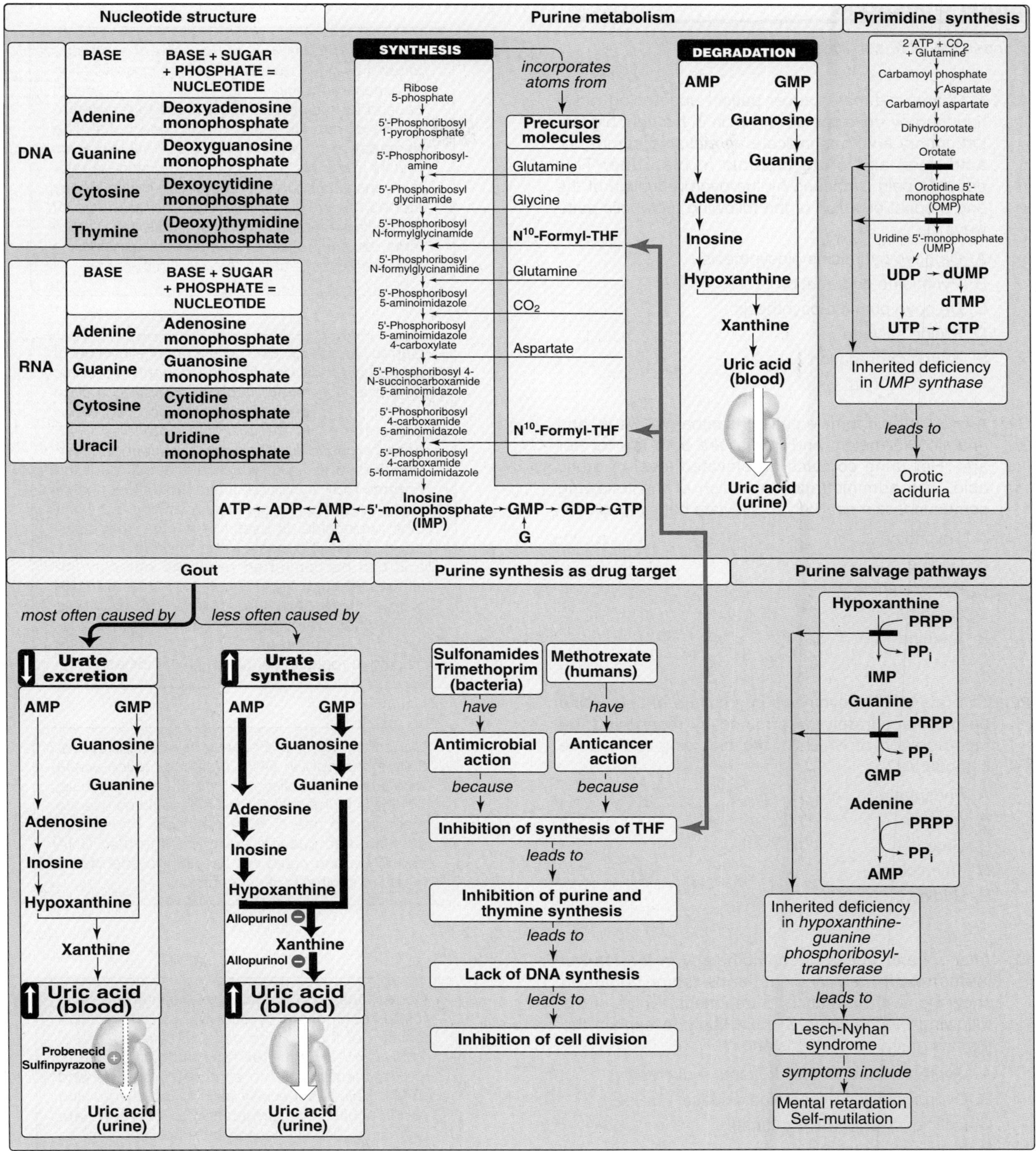

Figure 22.24
Key concept map for nucleotide metabolism.

Study Questions

Choose the ONE correct answer

22.1 A 42-year-old male cancer patient undergoing radiation therapy develops severe pain in his right big toe. Laboratory analyses indicate an elevated serum uric acid level and urate crystals in his urine. This patient's pain is caused by the overproduction of the end product of which of the following metabolic pathways?

A. De novo pyrimidine biosynthesis
B. Pyrimidine degradation
C. De novo purine biosynthesis
D. Purine salvage
E. Purine degradation

Correct answer = E. The patient's pain is caused by gout, resulting from the crystallization of excess uric acid in his joints. The cell death caused by radiation therapy leads to the degradation of nucleic acids from those cells. The degradation of purines from these nucleic acids results in excess production of uric acid—a relatively insoluble compound that can cause kidney stones, as well as gout. The end products of pyrimidine degradation do not cause these problems, because they are all soluble compounds that can be more easily excreted in the urine.

22.2 A one-year-old female patient is lethargic, weak, and anemic. Her height and weight are both low for her age. Her urine contains an elevated level of orotic acid. The administration of which of the following compounds is most likely to alleviate her symptoms?

A. Thymidine
B. Uridine
C. Hypoxanthine
D. Guanine
E. Adenine

Correct answer = B. The elevated excretion of orotic acid indicates that the patient has orotic aciduria, a genetic disorder affecting the de novo pyrimidine biosynthetic pathway. Deficiencies in the enzyme activities OMP decarboxylase and/or orotate phosphoribosyltransferase (both of which are domains of the enzyme UMP synthase) leave the patient unable to synthesize any pyrimidines. Uridine, a pyrimidine nucleoside, is useful in treating this disorder because it bypasses the missing enzymes, and can be converted to all the other pyrimidines. Although thymidine is a pyrimidine nucleoside, it cannot be converted to other pyrimidines. Hypoxanthine, guanine, and adenine are all purine bases that have no value in helping to replace the missing pyrimidines.

22.3 The rate of DNA synthesis in a culture of cells could be most accurately determined by measuring the incorporation of which of the following radioactive compounds?

A. Phosphate
B. Adenine
C. Guanine
D. Thymidine
E. Uridine

Correct answer = D. Because thymidine is essentially found only in DNA, its incorporation would most accurately reflect the rate of DNA synthesis. Uridine is found only in RNA and could be used to measure the rate of RNA synthesis. Phosphate, adenine, and guanine are present in both DNA and RNA, and could not be used to specifically measure synthesis of either one.

22.4 After several weeks of chemotherapy in the form of methotrexate, a cancer patient's tumor begins to show signs of resistance to treatment. Which of the following mechanisms is most likely to explain the tumor's methotrexate resistance?

A. Overproduction of dihydrofolate reductase
B. Overproduction of xanthine oxidase
C. Deficiency of PRPP synthase
D. Deficiency of thymidine kinase
E. Deficiency of thymidylate synthase

Correct answer = A. Methotrexate interferes with folate metabolism by acting as a competitive inhibitor of the enzyme *dihydrofolate reductase*. This starves cells for tetrahydrofolate, and makes them unable to synthesize purines and dTMP. This is especially toxic to rapidly-growing cancer cells. Overproduction of dihydrofolate reductase, usually caused by amplification of its gene, can overcome the inhibition of the enzyme at the methotrexate concentrations used for chemotherapy, and can result in resistance of the tumor to treatment by this drug.

UNIT V:
Integration of Metabolism

Metabolic Effects of Insulin and Glucagon

23

I. OVERVIEW

Four major organs play a dominant role in fuel metabolism: liver, adipose, muscle, and brain. These tissues contain unique sets of enzymes, such that each organ is specialized for the storage, use, and generation of specific fuels. These tissues do not function in isolation, but rather form part of a community in which one tissue may provide substrates to another, or process compounds produced by other organs. Communication between tissues is mediated by the nervous system, by the availability of circulating substrates, and by variation in the levels of plasma hormones (Figure 23.1). The integration of energy metabolism is controlled primarily by the actions of two hormones: insulin and glucagon, with the catecholamines epinephrine and norepinephrine playing a supporting role. Changes in the circulating levels of these hormones allow the body to store energy when food is available in abundance, or to make stored energy available, for example, during "survival crises," such as famine, severe injury, and "fight-or-flight" situations. This chapter describes the structure, secretion, and metabolic effects of the two hormones that most profoundly affect energy metabolism.

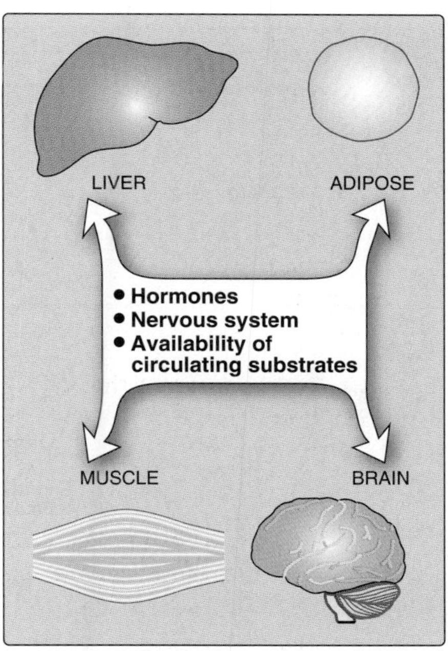

Figure 23.1
Mechanisms of communication between four major tissues.

II. INSULIN

Insulin is a polypeptide hormone produced by the β **cells** of the **islets of Langerhans**—clusters of cells that are embedded in the exocrine portion of the pancreas (Figure 23.2). The islets of Langerhans make up only about one to two percent of the total cells of the pancreas. Insulin

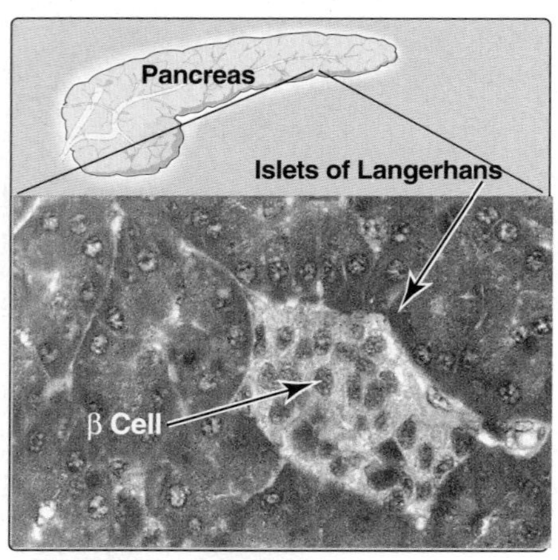

Figure 23.2
Islets of Langerhans.

is the most important hormone coordinating the use of fuels by tissues. Its metabolic effects are **anabolic**, favoring, for example, synthesis of glycogen, triacylglycerols, and protein.

A. Structure of insulin

Insulin is composed of 51 amino acids arranged in two polypeptide chains, designated A and B, which are linked together by two disulfide bridges (Figure 23.3A). The insulin molecule also contains an intramolecular disulfide bridge between amino acid residues of the A chain. Beef insulin differs from human insulin at three amino acid positions, whereas pork insulin varies at only one position.

B. Synthesis of insulin

The processing and transport of intermediates that occur during the synthesis of insulin are shown in Figures 23.3B and 23.4. Note that the biosynthesis involves two inactive precursors, preproinsulin and proinsulin, which are sequentially cleaved to form the active hormone plus the C-peptide (see Figure 23.4). [Note: The C-peptide is essential for proper insulin folding. Also, because of its longer half-life in the plasma, the C-peptide is a good indicator of insulin production and secretion in early diabetes.] Insulin is stored in the cytosol in granules that, given the proper stimulus (see below), are released by exocytosis. (See p. 164 for a discussion of the synthesis of proteins destined for secretion.) Insulin is degraded by the enzyme *insulinase* present in the liver and, to a lesser extent, in the kidneys. Insulin has a plasma half-life of approximately six minutes. This short duration of action permits rapid changes in circulating levels of the hormone.

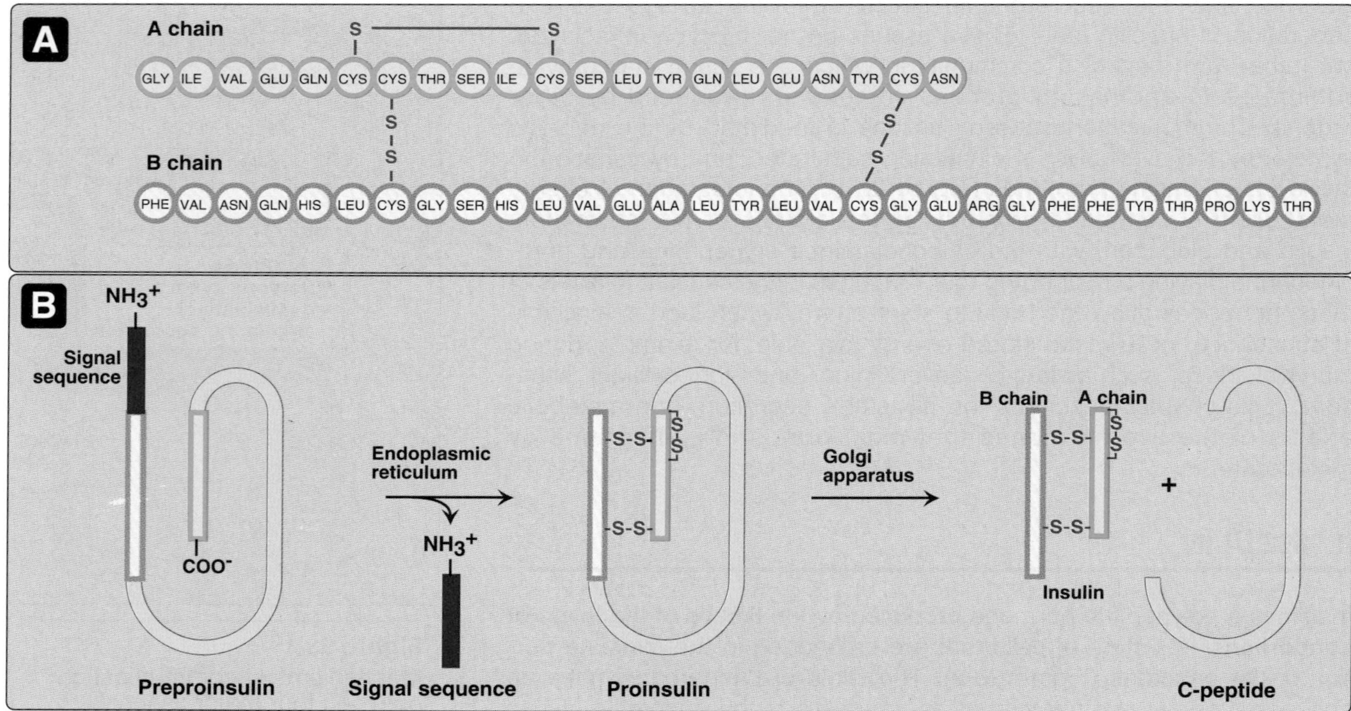

Figure 23.3
A. Structure of insulin. B. Formation of human insulin from preproinsulin.

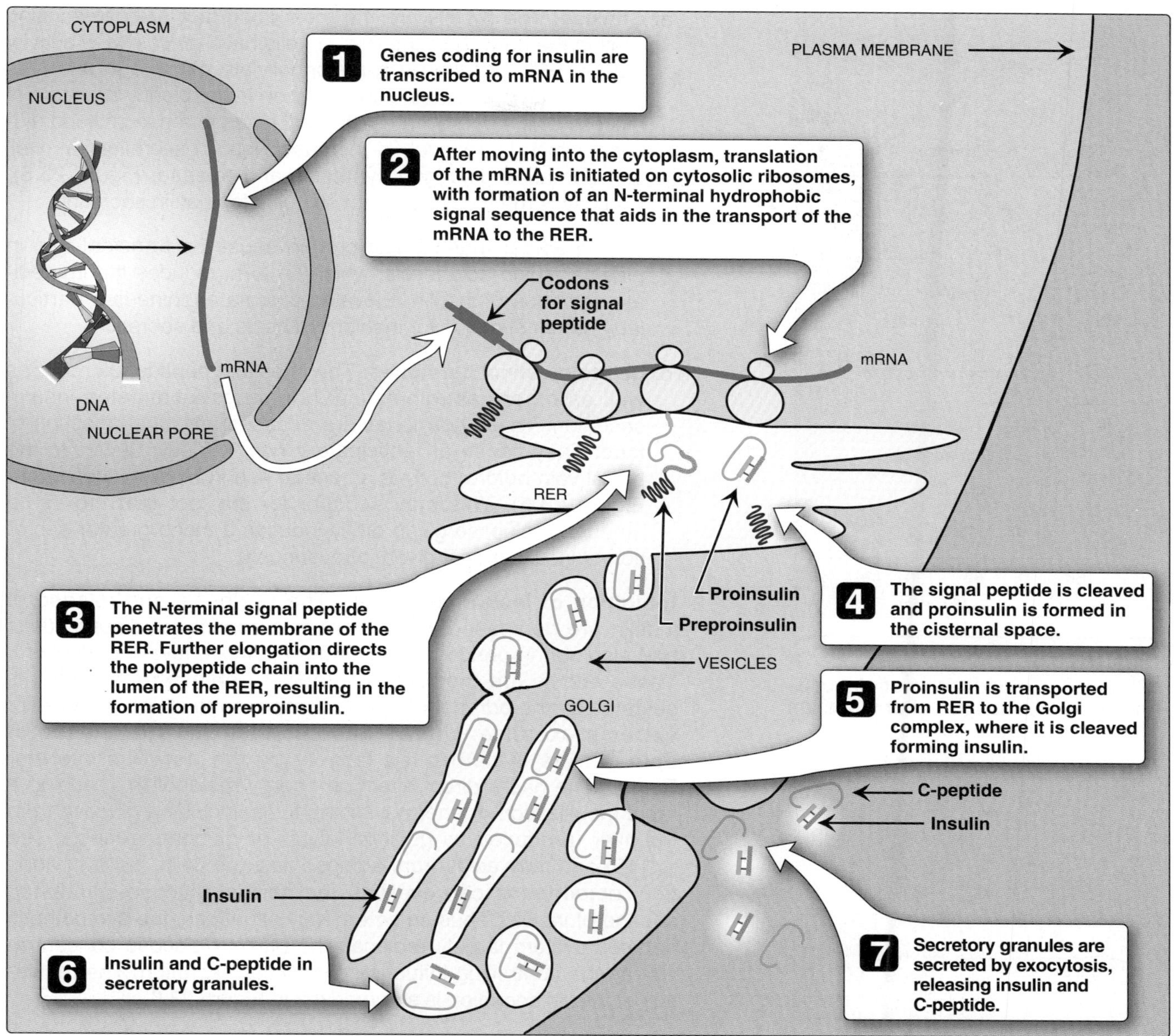

Figure 23.4
Intracellular movements of insulin and its precursors. RER = rough endoplasmic reticulum.

C. Regulation of insulin secretion

1. **Stimulation of insulin secretion:** Insulin secretion by the β cells of the islets of Langerhans of the pancreas is closely coordinated with the release of glucagon by pancreatic α cells. The relative amounts of insulin and glucagon released by the pancreas are regulated so that the rate of hepatic glucose production is kept equal to the use of glucose by peripheral tissues. In view of its coordinating role, it is not surprising that the β cell responds to a variety of stimuli. In particular, insulin synthesis and secretion are increased by:

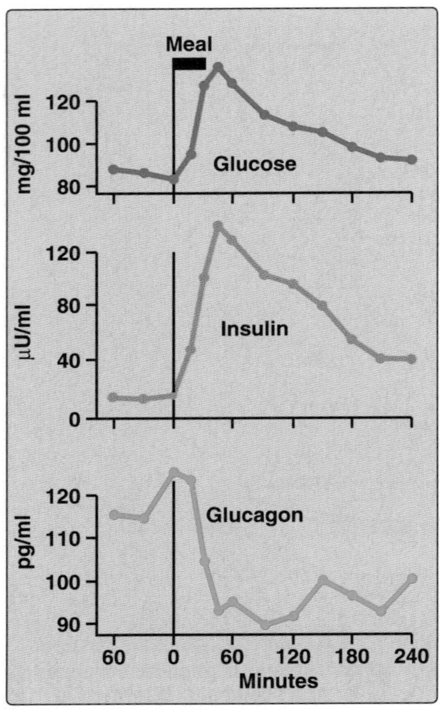

Figure 23.5
Changes in blood levels of glucose, insulin, and glucagon after ingestion of a carbohydrate-rich meal.

a. **Glucose:** The β cells are the most important glucose-sensing cells in the body. Like the liver, β cells have *glucokinase* activity (see p. 96), and thus can phosphorylate glucose in amounts proportional to its actual concentration in the blood. Ingestion of glucose or a carbohydrate-rich meal leads to a rise in blood glucose, which is a signal for increased insulin secretion (as well as decreased glucagon synthesis and release, Figure 23.5). Glucose is the most important stimulus for insulin secretion.

b. **Amino acids:** Ingestion of protein causes a transient rise in plasma amino acid levels, which, in turn, induces the immediate secretion of insulin. Elevated plasma arginine is a particularly potent stimulus for insulin synthesis and secretion.

c. **Gastrointestinal hormones:** The intestinal peptide **secretin**, as well as other gastrointestinal hormones, stimulates insulin secretion. These hormones are released after the ingestion of food. They cause an anticipatory rise in insulin levels in the portal vein before there is an actual rise in blood glucose (see Figure 23.5). This may account for the fact that the same amount of glucose given orally induces a much greater secretion of insulin than if given intravenously.

2. **Inhibition of insulin secretion:** The synthesis and release of insulin are decreased when there is a scarcity of dietary fuels, and also during periods of stress (for example, fever or infection). These effects are mediated primarily by **epinephrine**, which is secreted by the adrenal medulla in response to stress, trauma, or extreme exercise. Under these conditions, the release of epinephrine is controlled largely by the nervous system. Epinephrine has a direct effect on energy metabolism, causing a rapid mobilization of energy-yielding fuels, including glucose from the liver (produced by glycogenolysis or gluconeogenesis, see p. 119) and fatty acids from adipose tissue (see p. 187). In addition, epinephrine can override the normal glucose-stimulated release of insulin. Thus, in emergency situations, the sympathetic nervous system largely replaces the plasma glucose concentration as the controlling influence over β cell secretion. The regulation of insulin secretion is summarized in Figure 23.6.

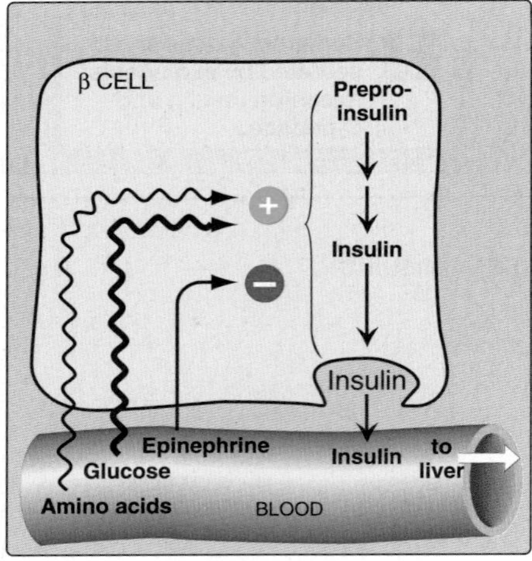

Figure 23.6
Regulation of insulin release from pancreatic β cells.

D. Metabolic effects of insulin

1. **Effects on carbohydrate metabolism:** The effects of insulin on glucose metabolism are most prominent in three tissues: liver, muscle, and adipose. In the **liver**, insulin decreases the production of glucose by inhibiting gluconeogenesis and the breakdown of glycogen. In the **muscle and liver**, insulin increases glycogen synthesis. In the **muscle and adipose tissue**, insulin increases glucose uptake by increasing the number of glucose transporters in the cell membrane (see p. 310). The intravenous administration of insulin thus causes an immediate decrease in the concentration of blood glucose.

2. **Effects on lipid metabolism:** Adipose tissue responds within minutes to administration of insulin, which causes a significant reduction in the release of fatty acids:

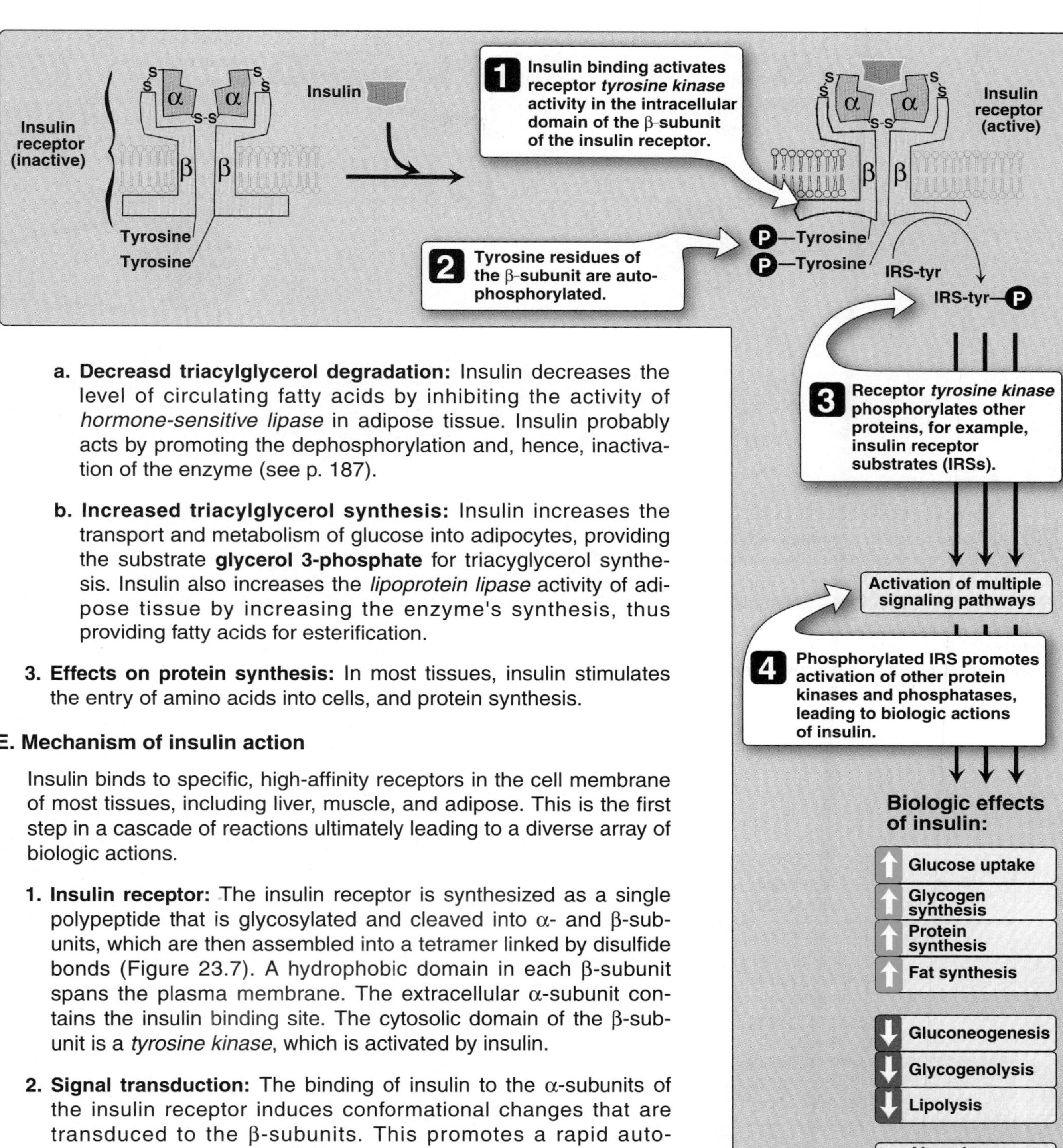

a. **Decreasd triacylglycerol degradation:** Insulin decreases the level of circulating fatty acids by inhibiting the activity of *hormone-sensitive lipase* in adipose tissue. Insulin probably acts by promoting the dephosphorylation and, hence, inactivation of the enzyme (see p. 187).

b. **Increased triacylglycerol synthesis:** Insulin increases the transport and metabolism of glucose into adipocytes, providing the substrate **glycerol 3-phosphate** for triacyglycerol synthesis. Insulin also increases the *lipoprotein lipase* activity of adipose tissue by increasing the enzyme's synthesis, thus providing fatty acids for esterification.

3. **Effects on protein synthesis:** In most tissues, insulin stimulates the entry of amino acids into cells, and protein synthesis.

E. Mechanism of insulin action

Insulin binds to specific, high-affinity receptors in the cell membrane of most tissues, including liver, muscle, and adipose. This is the first step in a cascade of reactions ultimately leading to a diverse array of biologic actions.

1. **Insulin receptor:** The insulin receptor is synthesized as a single polypeptide that is glycosylated and cleaved into α- and β-subunits, which are then assembled into a tetramer linked by disulfide bonds (Figure 23.7). A hydrophobic domain in each β-subunit spans the plasma membrane. The extracellular α-subunit contains the insulin binding site. The cytosolic domain of the β-subunit is a *tyrosine kinase*, which is activated by insulin.

2. **Signal transduction:** The binding of insulin to the α-subunits of the insulin receptor induces conformational changes that are transduced to the β-subunits. This promotes a rapid autophosphorylation of a specific tyrosine residue on each β-subunit (see Figure 23.7). Autophosphorylation initiates a cascade of cell-signaling responses, including phosphorylation of a family of proteins called **insulin receptor substrate** (**IRS**) **proteins**. At least four IRSs have been identified that show similar structures but different tissue distributions. The actions of insulin are terminated by dephosphorylation of the receptor.

Figure 23.7
Insulin receptor. IRS = Insulin receptor substrate.

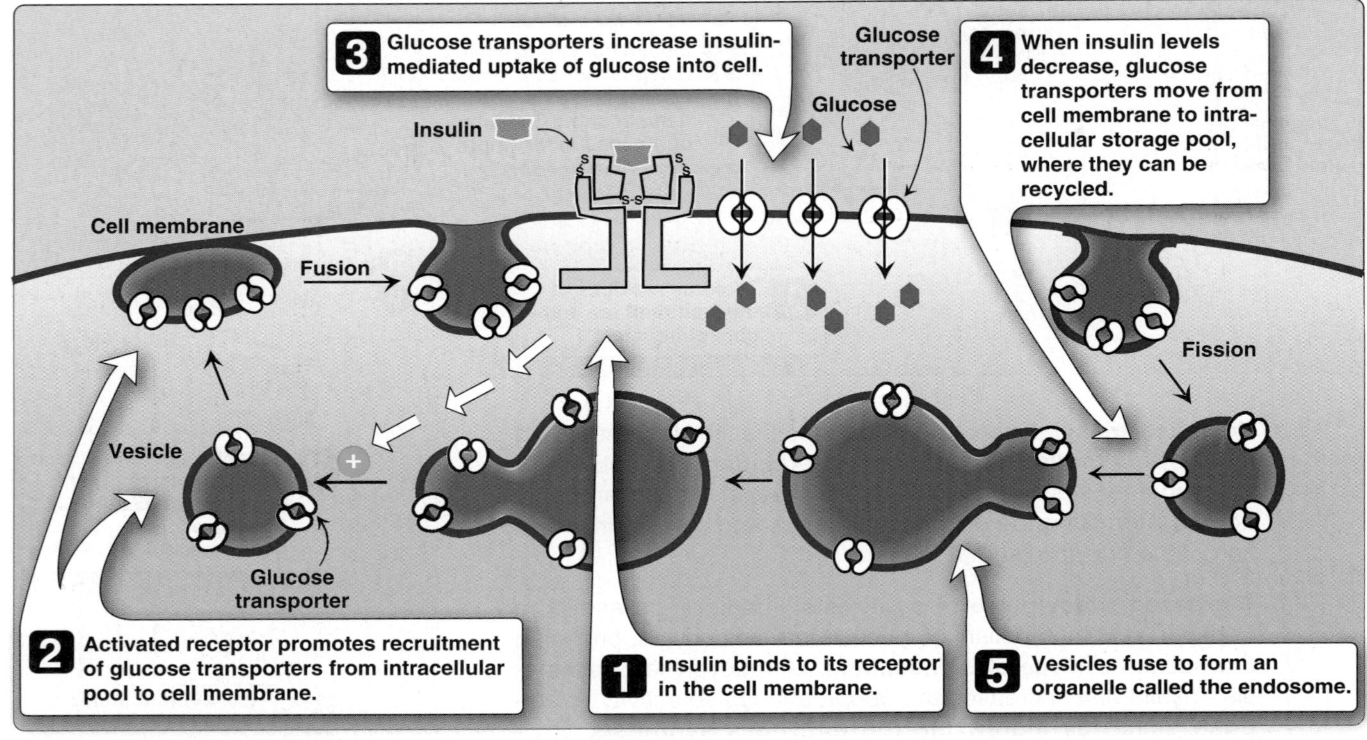

Figure 23.8
Insulin causes cells to recruit transporters from intracellular stores.

	Active transport	Facilitated transport
Insulin-sensitive		Most tissues (for example, skeletal muscle and adipose tissue)
Insulin-insensitive	Epithelia of intestine Renal tubules Choroid plexus	Erythrocytes Leukocytes Lens of eye Cornea Liver Brain

Figure 23.9
Characteristics of glucose transport in various tissues.

3. **Membrane effects of insulin:** Glucose transport in some tissues, such as skeletal muscle and adipocytes, increases in the presence of insulin (Figure 23.8). Insulin promotes the recruitment of **insulin-sensitive glucose transporters** (**GLUT-4**, see p. 95) from a pool located in intracellular vesicles. [Note: Some tissues have insulin-independent systems for glucose transport (Figure 23.9). For example, hepatocytes, erythrocytes, and cells of the nervous system, intestinal mucosa, renal tubules, and cornea do not require insulin for glucose uptake.]

4. **Receptor regulation:** Binding of insulin is followed by internalization of the hormone-receptor complex. Once inside the cell, insulin is degraded in the lysosomes. The receptors may be degraded but most are recycled to the cell surface. Elevated levels of insulin promote the degradation of receptors, thus decreasing the number of surface receptors. This is one type of "down-regulation."

5. **Time course of insulin actions:** The binding of insulin provokes a wide range of actions. The most immediate response is an increase in glucose transport into adipocytes and skeletal muscle cells, which occurs within seconds of insulin binding to its membrane receptor. Insulin-induced changes in enzymic activity in many cell types occur over minutes to hours, and reflect changes in the phosphorylation states of existing proteins. Insulin also initiates an increase in the amount of many enzymes, such as

glucokinase, *phosphofructokinase*, and *pyruvate kinase,* which requires hours to days. These changes reflect an increase in gene transcription, mRNA, and enzyme synthesis.

III. GLUCAGON

Glucagon is a polypeptide hormone secreted by the α **cells** of the pancreatic **islets of Langerhans**. Glucagon, along with epinephrine, cortisol, and growth hormone (the "**counterregulatory hormones**"), opposes many of the actions of insulin. Most importantly, glucagon acts to maintain blood glucose levels by activation of hepatic glycogenolysis and gluconeogenesis. Glucagon is composed of 29 amino acids arranged in a single polypeptide chain. [Note: Unlike insulin, the amino acid sequence of glucagon is the same in all mammalian species examined to date.] Glucagon is synthesized as a large precursor molecule that is converted to glucagon through a series of selective proteolytic cleavages, similar to those described for insulin biosynthesis (see Figure 23.3).

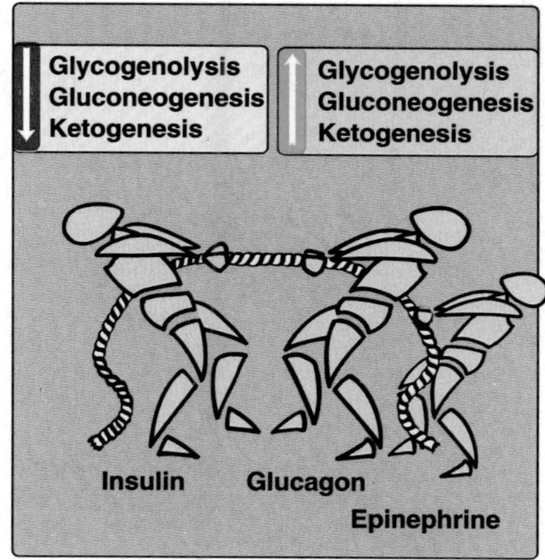

Figure 23.10
Opposing actions of insulin and glucagon plus epinephrine.

A. Stimulation of glucagon secretion

The α cell is responsive to a variety of stimuli that signal actual or potential hypoglycemia (Figure 23.10). Specifically, glucagon secretion is increased by:

1. **Low blood glucose:** A decrease in plasma glucose concentration is the primary stimulus for glucagon release. During an overnight or prolonged fast, elevated glucagon levels prevent hypoglycemia (see below for a discussion of hypoglycemia).

2. **Amino acids:** Amino acids derived from a meal containing protein stimulate the release of both glucagon and insulin. The glucagon effectively prevents hypoglycemia that would otherwise occur as a result of increased insulin secretion that occurs after a protein meal.

3. **Epinephrine:** Elevated levels of circulating epinephrine produced by the adrenal medulla, or norepinephrine produced by sympathetic innervation of the pancreas, or both, stimulate the release of glucagon. Thus, during periods of stress, trauma, or severe exercise, the elevated epinephrine levels can override the effect on the α cell of circulating substrates. In these situations—regardless of the concentration of blood glucose—glucagon levels are elevated in anticipation of increased glucose use. In contrast, insulin levels are depressed.

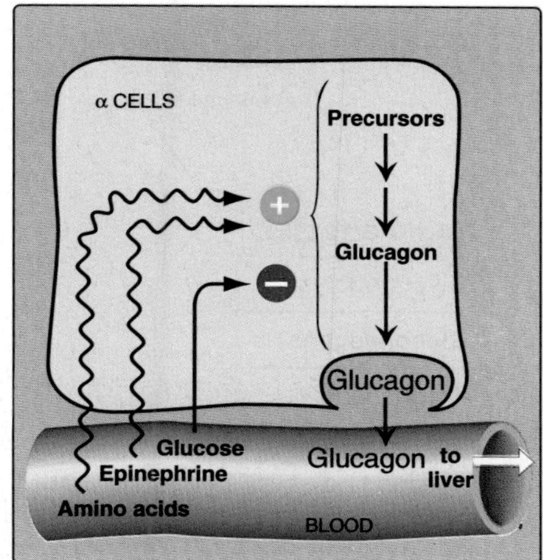

Figure 23.11
Regulation of glucagon release from pancreatic α cells.

B. Inhibition of glucagon secretion

Glucagon secretion is significantly decreased by elevated blood glucose and by insulin. Both substances are increased following ingestion of glucose or a carbohydrate-rich meal (see Figure 23.5). The regulation of glucagon secretion is summarized in Figure 23.11.

C. Metabolic effects of glucagon

1. **Effects on carbohydrate metabolism:** The intravenous administration of glucagon leads to an immediate rise in blood glucose. This results

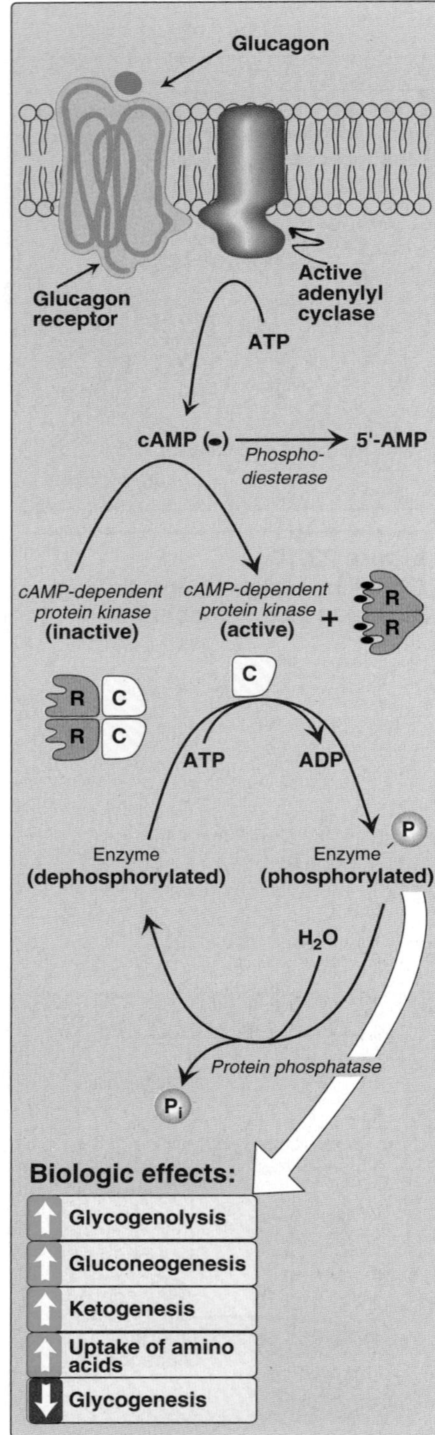

Figure 23.12
Mechanism of action of glucagon. [Note: For clarity, G-protein activation of *adenylyl cyclase* has been omitted.] R = regulatory subunit; C = catalytic subunit.

from an increase in the breakdown of liver (not muscle) glycogen and an increase in gluconeogenesis.

2. **Effects on lipid metabolism:** Glucagon favors hepatic oxidation of fatty acids and the subsequent formation of ketone bodies from acetyl CoA. The lipolytic effect of glucagon in adipose tissue is minimal in humans.

3. **Effects on protein metabolism:** Glucagon increases uptake of amino acids by the liver, resulting in increased availability of carbon skeletons for gluconeogenesis. As a consequence, plasma levels of amino acids are decreased.

D. Mechanism of action of glucagon

Glucagon binds to high-affinity receptors on the cell membrane of the hepatocyte. The receptors for glucagon are distinct from those that bind insulin or epinephrine. Glucagon binding results in activation of *adenylyl cyclase* in the plasma membrane (Figure 23.12, and see p. 92). This causes a rise in cAMP (the "second messenger"), which, in turn, activates *cAMP-dependent protein kinase* and increases the phosphorylation of specific enzymes or other proteins. This cascade of increasing enzymic activities results in the phosphorylation-mediated activation or inhibition of key regulatory enzymes involved in carbohydrate and lipid metabolism. An example of such a cascade is presented for the case of glycogen degradation on p. 129 and Figure 11.11, p. 131.

IV. HYPOGLYCEMIA

Hypoglycemia is characterized by: 1) central nervous system symptoms, including confusion, aberrant behavior, or coma; 2) a simultaneous blood glucose level equal to or less than 40 mg/dl; and 3) symptoms being resolved within minutes following the administration of glucose. Hypoglycemia is a medical emergency because the central nervous system (CNS) has an absolute requirement for a continuous supply of blood-borne glucose to serve as fuel for energy metabolism. Transient hypoglycemia can cause cerebral dysfunction, whereas severe, prolonged hypoglycemia causes brain death. It is, therefore, not surprising that the body has multiple overlapping mechanisms to prevent or correct hypoglycemia. The most important hormone changes in combating hypoglycemia are elevated glucagon and epinephrine, combined with the diminished release of insulin.

A. Symptoms of hypoglycemia

The symptoms of hypoglycemia can be divided into two categories. **Adrenergic symptoms**—anxiety, palpitation, tremor, and sweating—are mediated by epinephrine release regulated by the hypothalamus in response to hypoglycemia. Usually adrenergic symptoms (that is, symptoms mediated by elevated epinephrine) occur when blood glucose levels fall abruptly. The second category of hypoglycemic symptoms is neuroglycopenic. **Neuroglycopenia**—the impaired delivery of glucose to the brain—results in impairment of brain function, causing headache, confusion, slurred speech, seizures, coma, and death. Neuroglycopenic symptoms often result from a gradual decline in blood glucose, often to levels below 40 mg/dl. The slow decline in glucose deprives the CNS of fuel, but fails to trigger an adequate epinephrine response.

B. Glucoregulatory systems

Humans have two overlapping glucose-regulating systems that are activated by hypoglycemia: 1) the islets of Langerhans, which release glucagon; and 2) receptors in the hypothalamus, which respond to abnormally low concentrations of blood glucose. The hypothalamic glucoreceptors can trigger both the secretion of epinephrine (mediated by the autonomic nervous system) and release of ACTH and growth hormone (GH) by the anterior pituitary (Figure 23.13). Glucagon, epinephrine, cortisol, and growth hormones are sometimes called the "counterregulatory" hormones because each opposes the action of insulin on glucose use.

1. **Glucagon and epinephrine:** Hypoglycemia is combatted by decreased release of insulin and increased secretion of glucagon, epinephrine, cortisol, and growth hormone (see Figure 23.13). Glucagon and epinephrine are most important in the acute, short-

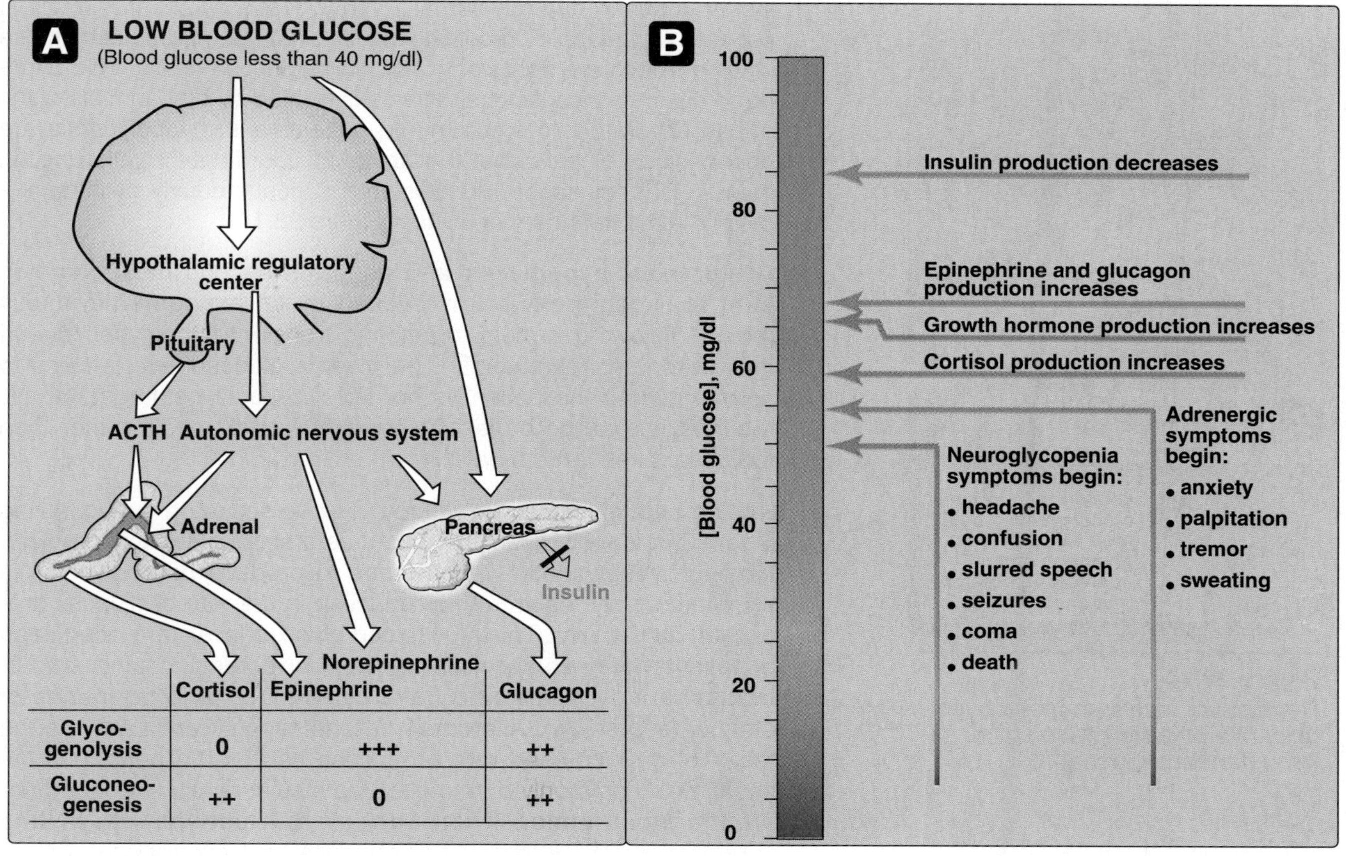

	Cortisol	Epinephrine	Glucagon
Glyco-genolysis	0	+++	++
Gluconeo-genesis	++	0	++

Figure 23.13
A. Actions of some of the glucoregulatory hormones in response to low blood glucose. B. Glycemic thresholds for the various resposes to hypoglycemia. + = Weak stimulation; ++ = moderate stimulation; +++ = strong stimulation; 0 = no effect.

term regulation of blood glucose levels. Glucagon stimulates hepatic glycogenolysis and gluconeogenesis. Epinephrine promotes glycogenolysis and lipolysis, inhibits insulin secretion, and inhibits the insulin-mediated uptake of glucose by peripheral tissues. Epinephrine is not normally essential in combating hypoglycemia, but it can assume a critical role when glucagon secretion is deficient, for example, in the late stages of type 1 (insulin-dependent) diabetes mellitus (see p. 339). The prevention or correction of hypoglycemia fails when the secretion of both glucagon and epinephrine are deficient.

2. **Cortisol and growth hormone:** These hormones are less important in the short-term maintenance of blood glucose concentrations. They do, however, play a role in the long-term management of glucose metabolism.

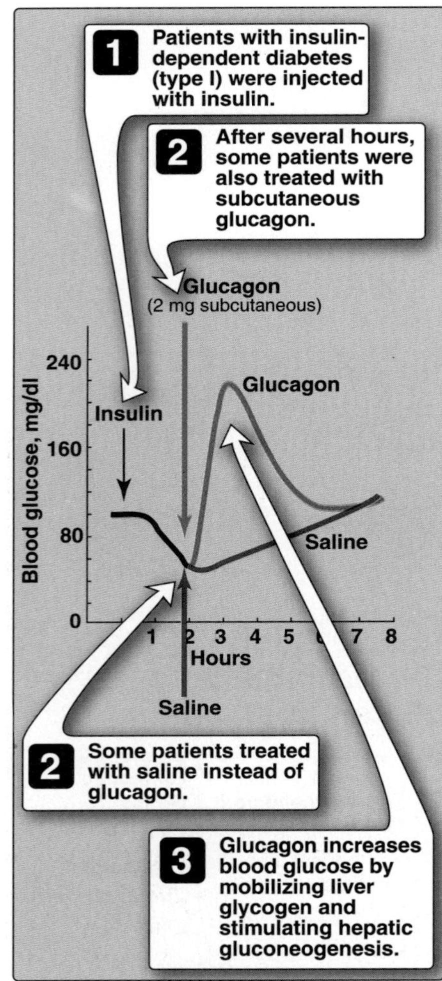

Figure 23.14
Reversal of insulin-induced hypoglycemia by admistration of subcutaneous glucagon.

C. Types of hypoglycemia

Hypoglycemia may be divided into three groups: 1) insulin-induced; 2) postprandial (sometimes called reactive hypoglycemia); and 3) fasting hypoglycemia. [Note: Alcohol intoxication in fasting individuals can also be associated with hypoglycemia.]

1. **Insulin-induced hypoglycemia:** Hypoglycemia occurs frequently in patients with diabetes receiving insulin treatment, particularly those striving to achieve tight control of blood glucose levels. Mild hypoglycemia in fully conscious patients is treated by oral administration of carbohydrate. More commonly, patients with hypoglycemia are unconscious or have lost the ability to coordinate swallowing. In these cases, glucagon, administered subcutaneously or intramuscularly, is the treatment of choice (Figure 23.14).

2. **Postprandial hypoglycemia:** This is the second most common form of hypoglycemia. It is caused by an exaggerated insulin release following a meal, prompting transient hypoglycemia with mild adrenergic symptoms. The plasma glucose level returns to normal even if the patient is not fed. The only treatment usually required is that the patient eat frequent small meals rather than the usual three large meals.

3. **Fasting hypoglycemia:** Low blood glucose occurring during fasting is rare, but is more likely to present as a serious medical problem. Fasting hypoglycemia, which tends to produce neuroglycopenia symptoms, may result from a reduction in the rate of glucose production by the liver. Thus, low blood glucose levels are often seen in patients with hepatocellular damage or adrenal insufficiency, or in fasting individuals who have consumed large quantities of ethanol (see below). Alternately, fasting hypoglycemia may be the result of an increased rate of glucose use by the peripheral tissues, most commonly due to elevated insulin resulting from a pancreatic β cell tumor. If left untreated, a patient with fasting hypoglycemia may lose consciousness and experience convulsions and coma.

4. **Hypoglycemia and alcohol intoxication:** Alcohol is metabolized in the liver by two oxidation reactions (Figure 23.15). Ethanol is first

converted to acetaldehyde by *alcohol dehydrogenase*. Acetaldehyde is subsequently oxidized to acetate by *aldehyde dehydrogenase*. [Note: This enzyme is inhibited by **disulfiram**, a drug that has found some use in patients desiring to stop alcohol ingestion.[1] It causes the accumulation of acetaldehyde in the blood, which results in flushing, tachycardia, hyperventilation, and nausea.] In each reaction, electrons are transferred to NAD^+, resulting in a massive increase in the concentration of cytosolic NADH. The abundance of NADH favors the reduction of pyruvate to lactate, and oxaloacetate to malate. Recall from p. 116 that pyruvate and oxaloacetate are both intermediates in the synthesis of glucose by gluconeogenesis. Thus, the ethanol-mediated increase in NADH causes the intermediates of gluconeogenesis to be diverted into alternate reaction pathways, resulting in the decreased synthesis of glucose. This can precipitate hypoglycemia, particularly in individuals who have depleted their stores of liver glycogen. [Note: Recall from p. 123 that the mobilization of liver glycogen is the body's first defense against hypoglycemia. Thus, individuals who are fasting or malnourished have depleted glycogen stores, and must rely on gluconeogenesis to maintain their blood glucose levels.] Hypoglycemia can produce many of the behaviors associated with alcohol intoxication—agitation, impaired judgement, and combativeness.

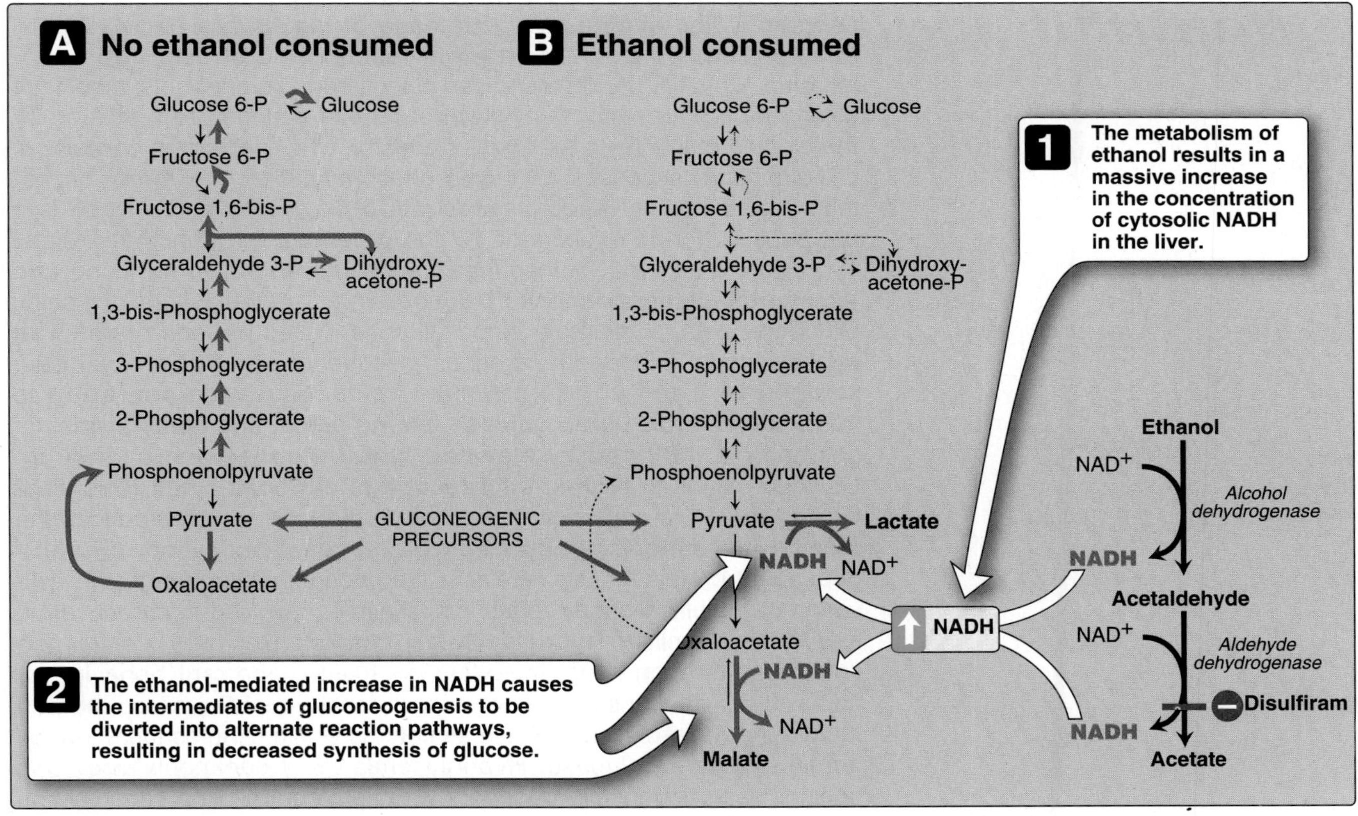

Figure 23.15
A. Normal gluconeogenesis in the absence of ethanol consumption. B. Inhibition of gluconeogenesis resulting from hepatic metabolism of ethanol.

INFO LINK [1]See Chapter 9 in ***Lippincott's Illustrated Reviews: Pharmacology*** (2nd and 3rd Eds.) for a discussion of the use of disulfiram in the treatment of alcoholism.

Thus, alcohol consumption in vulnerable individuals—those who are fasted or have engaged in prolonged, strenuous exercise—can produce hypoglycemia that may contribute to the behavioral effects of alcohol. Alcohol consumption can also increase the risk for hypoglycemia in patients using insulin. Any patient using insulin in an aggressive treatment protocol should be advised about the increased risk of hypoglycemia that does not generally occur during the time when the alcohol is being consumed but, instead, many hours afterward.

V. CHAPTER SUMMARY

The integration of energy metabolism is controlled primarily by **insulin** and the opposing actions of **glucagon** and **epinephrine**. Changes in the circulating levels of these hormones allow the body to store energy when food is available in abundance, or to make stored energy available, for example, during "survival crises," such as famine, severe injury, and "fight-or-flight" situations. **Insulin** is a polypeptide hormone produced by the β **cells** of the **islets of Langerhans** of the pancreas. The biosynthesis involves two inactive precursors, **preproinsulin** and **proinsulin**, which are sequentially cleaved to form the active hormone. A rise in blood glucose is the most important signal for increased insulin secretion. The synthesis and release of insulin are decreased by epinephrine, which is secreted in response to stress, trauma, or extreme exercise. Insulin increases glucose uptake and the synthesis of glycogen, protein, and triacylglycerol. These actions are mediated by the binding of insulin to the insulin receptor, which initiates a cascade of cell-signaling responses, including phosphorylation of a family of proteins called **insulin receptor substrate** (**IRS**) **proteins**. Glucagon is a polypeptide hormone secreted by the α **cells** of the pancreatic islets. Glucagon, along with epinephrine, cortisol, and growth hormone (the "**counterregulatory hormones**"), opposes many of the actions of insulin. Glucagon acts to maintain blood glucose during periods of potential hypoglycemia. Glucagon increases glycogenolysis, gluconeogenesis, ketogenesis, and uptake of amino acids. Glucagon **secretion** is **stimulated** by **low blood glucose**, **amino acids**, and **epinephrine**. Its secretion is **inhibited** by **elevated blood glucose** and by **insulin**. Glucagon binds to **high-affinity receptors** of **hepatocytes**. This binding results in the activation of **adenylate cyclase**, which produces the second messenger, **cyclic AMP**. Subsequent activation of cAMP-dependent protein kinase results in the phosphorylation-mediated activation or inhibition of key regulatory enzymes involved in carbohydrate and lipid metabolism. **Hypoglycemia** is characterized by: 1) central nervous system symptoms, including confusion, aberrant behavior, or coma; 2) a simultaneous blood glucose level equal to or less than 40 mg/dl; and 3) resolution of these symptoms within minutes following the administration of glucose. Hypoglycemia most commonly occurs in patients receiving insulin treatment with tight control. The consumption and subsequent metabolism of **ethanol** inhibits gluconeogenesis, leading to hypoglycemia in individuals with depleted stores of liver glycogen. Alcohol consumption can also increase the risk for hypoglycemia in patients using insulin.

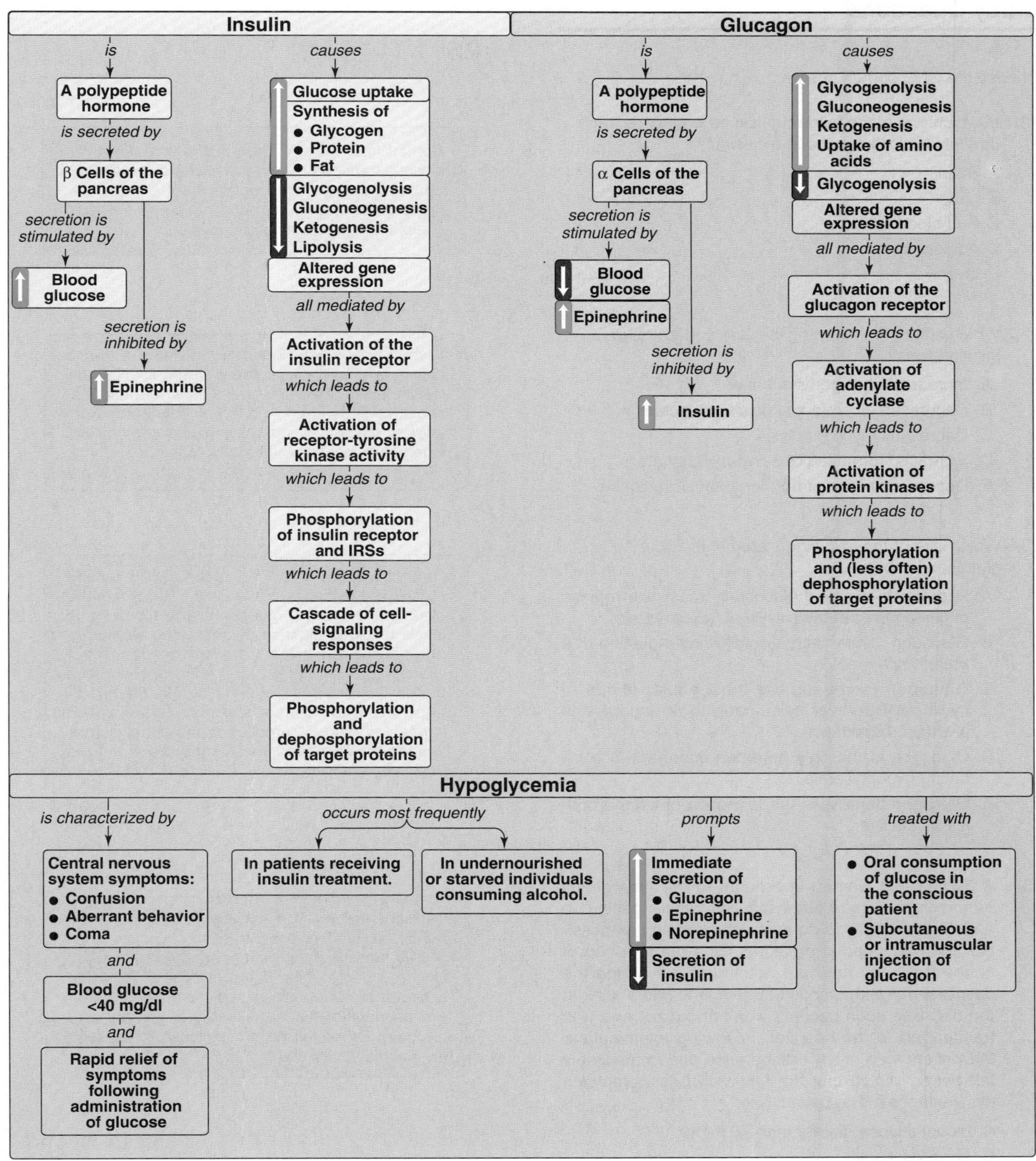

Figure 23.16
Key concept map for integration of energy metabolism.

Study Questions

Choose the ONE correct answer

23.1 In which one of the following tissues is glucose transport into the cell enhanced by insulin?

A. Brain
B. Lens
C. Red blood cells
D. Adipose tissue
E. Liver

Correct answer = D. The major tissues in which glucose transport requires insulin are muscle and adipose tissue. The metabolism of the liver responds to insulin, but hepatic glucose transport is rapid and does not require insulin.

23.2 Which one of the following is characteristic of low insulin levels?

A. Increased glycogen synthesis
B. Decreased gluconeogenesis from lactate
C. Decreased glycogenolysis
D. Increased formation of 3-hydroxybutyrate
E. Decreased action of hormone-sensitive lipase

Correct answer = D. 3-hydroxybutyrate—a ketone body—synthesis is enhanced in the liver by low insulin levels, which favor activation of hormone-sensitive lipase and release of fatty acids from adipose tissue. Glycogen synthesis is decreased, whereas gluconeogenesis is increased.

23.3 Which one of the following statements about glucagon is correct ?

A. High levels of blood glucose increase the release of glucagon from the α cells of the pancreas.
B. Glucagon levels decrease following ingestion of a protein-rich meal.
C. Glucagon increases the intracellular levels of cyclic AMP in liver cells, causing an increase in glycogen breakdown.
D. Glucagon is the only hormone important in combating hypoglycemia.
E. Glucagon depresses the formation of ketone bodies by the liver.

Correct answer = C. The cyclic AMP cascade initiated by glucagon causes the liver to degrade glycogen, releasing glucose to the blood. High levels of blood glucose decrease the release of glucagon from the α cells of the pancreas. Glucagon levels increase following ingestion of a protein-rich meal. In addition to glucagon, epinephrine and cortisol are also important in increasing glucose production in hypoglycemia. Glucagon increases the formation of ketone bodies by the liver.

23.4 A 39-year-old woman is brought to the emergency room complaining of dizziness. She recalls getting up early that morning to do as much shopping as possible and had skipped breakfast. She drank a cup of coffee for lunch and had nothing to eat during the day. She met with friends at 8 p.m. and had a drink at the bar. She soon became weak and dizzy and was transported to the hospital. Following examination, the patient was given orange juice and immediately felt better. Which one the following best completes this sentence? "The patient has:"

A. blood glucose greater than 70 mg/dl.
B. elevated insulin.
C. elevated glucagon.
D. elevated liver glycogen.
E. presence of an insulinoma.

Correct answer = C. The patient's glucagon level will be elevated in response to the hypoglycemia. She is most likely experiencing alcohol-induced fasting hypoglycemia. Blood glucose is expected to be 40 mg/dl or less, insulin secretion depressed because of the low blood glucose, and liver glycogen levels will be low because of the fast. Insulinoma is unlikely.

The Feed/Fast Cycle

24

I. OVERVIEW

The absorptive state is the two- to four-hour period after ingestion of a normal meal. During this interval, transient increases in plasma glucose, amino acids, and triacylglycerols occur, the latter primarily as components of chylomicrons synthesized by the intestinal mucosal cells (see p. 226). Islet tissue of the pancreas responds to the elevated levels of glucose and amino acids with an increased secretion of insulin and a drop in the release of glucagon. The elevated insulin to glucagon ratio and the ready availability of circulating substrates make the two to four hours after ingestion of a meal into an anabolic period characterized by increased synthesis of triacylglycerols, glycogen, and protein. During this absorptive period, virtually all tissues use glucose as a fuel, and the metabolic response of the body is dominated by alterations in the metabolism of liver, adipose tissue, muscle, and brain. In this chapter, an "organ map" is introduced that traces the movement of metabolites between tissues. The goal is to create an expanded and clinically useful vision of whole body metabolism.

II. ENZYMIC CHANGES IN THE FED STATE

The flow of intermediates through metabolic pathways is controlled by four mechanisms: 1) the availability of substrates; 2) allosteric activation and inhibition of enzymes; 3) covalent modification of enzymes; and 4) induction-repression of enzyme synthesis. This scheme may at first seem unnecessarily redundant; however, each mechanism operates on a different timescale (Figure 24.1), and allows the body to adapt to a wide variety of physiologic situations. In the fed state, these regulatory mechanisms ensure that available nutrients are captured as glycogen, triacylglycerol, and protein.

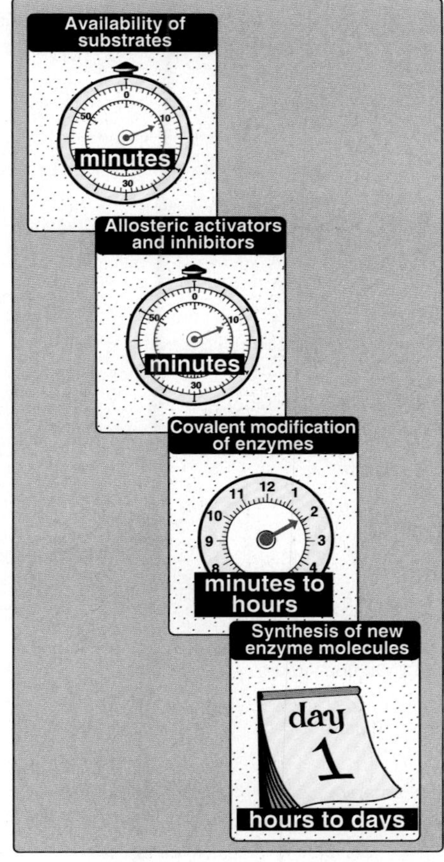

Figure 24.1
Control mechanisms of metabolism and some typical response times. [Note: Response times may vary according to the nature of the stimulus and from tissue to tissue.]

A. Allosteric effects

Allosteric changes usually involve rate-determining reactions. For example, glycolysis in the liver is stimulated following a meal by an increase in fructose 2,6-bisphosphate, an allosteric activator of *phosphofructokinase* (see p. 98). Gluconeogenesis is inhibited by fructose 2,6-bisphosphate, an inhibitor of *fructose 1,6-bisphosphatase* (see p. 118).

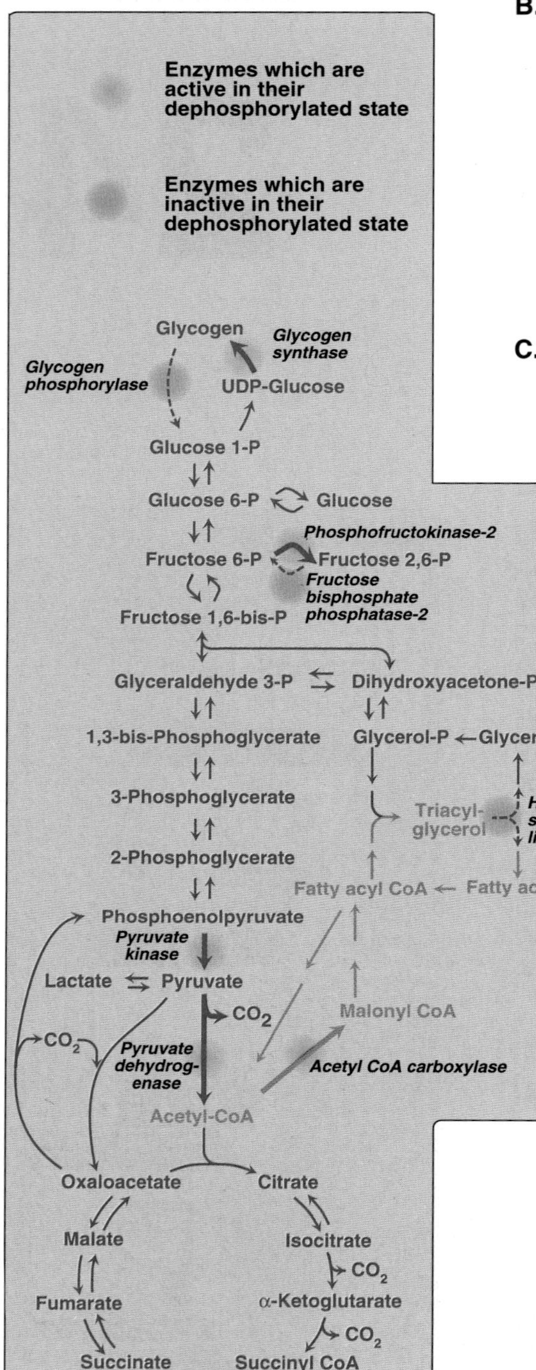

Figure 24.2
Important reactions of intermediary metabolism regulated by enzyme phosphorylation. Key: **Blue text** = intermediates of carbohydrate metabolism; **Brown text** = intermediates of lipid metabolism.

B. Regulation of enzymes by covalent modification

Many enzymes are regulated by covalent modification, most frequently by the addition or removal of phosphate groups from specific serine, threonine, or tyrosine residues of the enzyme. In the fed state, most of the enzymes regulated by covalent modification are in the dephosphorylated form and are active (see Figure 24.2). Three exceptions are *glycogen phosphorylase* (see p. 129), *fructose bisphosphate phosphatase-2* (see p. 98), and *hormone-sensitive lipase* of adipose tissue (see p. 187), which are inactive in their dephosphorylated state.

C. Induction and repression of enzyme synthesis

Increased (induction of) or decreased (repression of) protein synthesis leads to an alteration in the total population of active sites, rather than influencing the efficiency of existing enzyme molecules. Enzymes subject to regulation of synthesis are often those that are needed at only one stage of development or under selected physiologic conditions. For example, in the fed state, elevated insulin levels result in an increase in the synthesis of key enzymes involved in anabolic metabolism.

III. LIVER: NUTRIENT DISTRIBUTION CENTER

The liver is uniquely situated to process and distribute dietary nutrients because the venous drainage of the gut and pancreas passes through the hepatic portal vein before entry into the general circulation. Thus, after a meal, the liver is bathed in blood containing absorbed nutrients and elevated levels of insulin secreted by the pancreas. During the absorptive period, the liver takes up carbohydrates, lipids, and most amino acids. These nutrients are then metabolized, stored, or routed to other tissues. Thus, the liver smooths out potentially broad fluctuations in the availability of nutrients for the peripheral tissues.

A. Carbohydrate metabolism

Liver is normally a glucose-producing rather than a glucose-using tissue. However, after a meal containing carbohydrate, the liver becomes a net consumer of glucose, retaining roughly 60 of every 100 g of glucose presented by the portal system. This increased use of glucose is not a result of stimulated glucose transport into the hepatocyte, because this process is normally rapid and not influenced by insulin. Rather, hepatic metabolism of glucose is increased by the following mechanisms. [Note: The numbers in colored circles in the text refer to Figure 24.3.]

1. **Increased phosphorylation of glucose:** Elevated levels of intracellular glucose in the hepatocyte allow *glucokinase* to phosphorylate glucose to glucose 6-phosphate. This contrasts with the postabsorptive state in which hepatic glucose levels are lower and *glucokinase* is largely dormant because of its low affinity (high K_m) for glucose (see Figure 24.3, ❶).

2. **Increased glycogen synthesis:** The conversion of glucose 6-phosphate to glycogen is favored by inactivation of *glycogen phosphorylase* and activation of *glycogen synthase* (see p. 129 and Figure 24.3, ❷).

3. **Increased activity of the hexose monophosphate pathway (HMP):** The increased availability of glucose 6-phosphate in the well-fed state, combined with the active use of NADPH in hepatic lipogenesis, stimulate the HMP (see Chapter 12, p. 143). This pathway typically accounts for five to ten percent of the glucose metabolized by the liver (see Figure 24.3, ❸).

4. **Increased glycolysis:** In liver, glycolytic metabolism of glucose is significant only during the absorptive period following a carbohydrate-rich meal. The conversion of glucose to acetyl CoA is stimulated by the elevated insulin to glucagon ratio that activates the rate-limiting enzymes of glycolysis, for example, *phosphofructokinase* (see p. 98). Acetyl CoA is used as either a building block for fatty acid synthesis, or it provides energy by oxidation in the TCA cycle (see Figure 24.3, ❹).

5. **Decreased gluconeogenesis:** Although glycolysis is stimulated in the absorptive state, gluconeogenesis is decreased. *Pyruvate carboxylase*, which catalyzes the first step in gluconeogenesis, is largely inactive due to low levels of acetyl CoA—an allosteric effector essential for enzyme activity (see p. 117). The high insulin to glucagon ratio observed in the absorptive period also favors inactivation of other enzymes unique to gluconeogenesis, such as *fructose 1,6-bisphosphatase* (see Figure 8.17, p. 98).

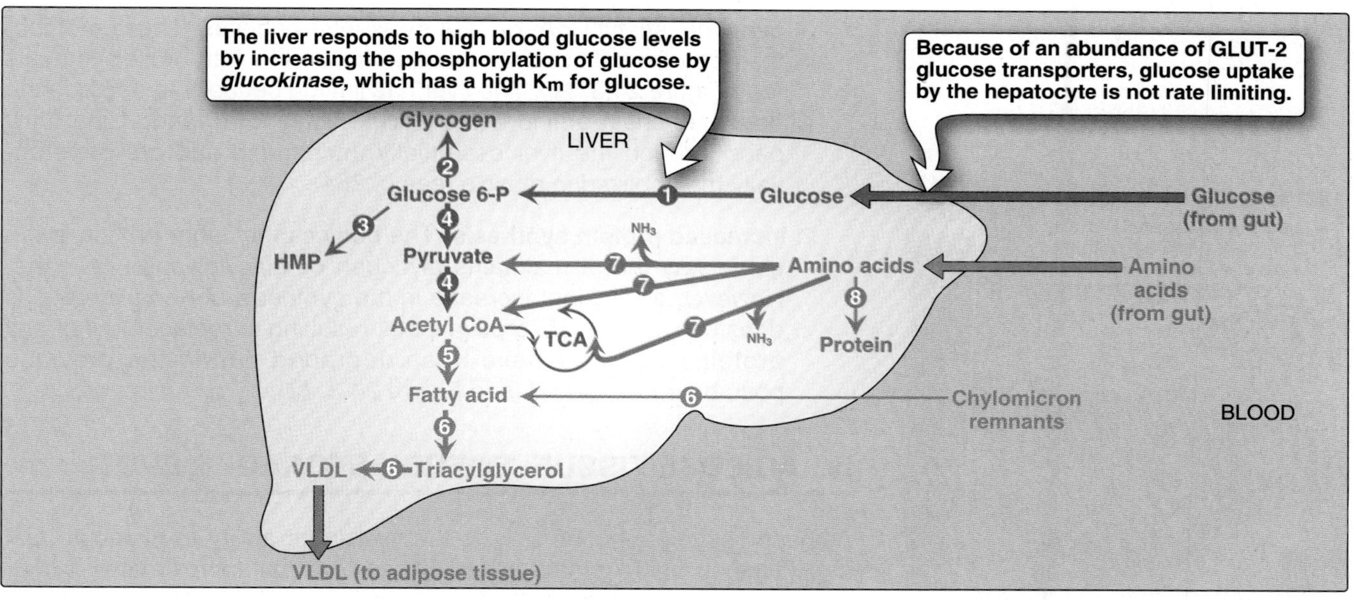

Figure 24.3
Major metabolic pathways in liver in the absorptive state. [Note: The numbers in circles, which appear both on the figure and in the text, indicate important pathways for carbohydrate, fat, or protein metabolism.]
Key: **Blue text** = intermediates of carbohydrate metabolism; **Brown text** = intermediates of lipid metabolism; **Green text** = intermediates of protein metabolism.

B. Fat metabolism

1. **Increased fatty acid synthesis:** Liver is the primary tissue for de novo synthesis of fatty acids (see Figure 24.3, ❺). This pathway occurs in the absorptive period, because the dietary energy intake exceeds energy expenditure by the body. Fatty acid synthesis is favored by the availability of substrates (acetyl CoA and NADPH derived from the metabolism of glucose) and by the activation of *acetyl CoA carboxylase*. This enzyme catalyzes the formation of malonyl CoA from acetyl CoA—a reaction that is rate-limiting in fatty acid synthesis (see p. 181).

2. **Increased triacylglycerol synthesis:** Triacylglycerol synthesis is favored because fatty acyl CoA is available both from de novo synthesis from acetyl CoA and from hydrolysis of the triacylglycerol component of chylomicron remnants removed from the blood by hepatocytes (see p. 176). Glycerol 3-phosphate, the backbone for triacylglycerol synthesis, is provided by the glycolytic metabolism of glucose (see p. 186). The liver packages triacylglycerols into very low density lipoprotein (VLDL) particles that are secreted into the blood for use by extrahepatic tissues, particularly adipose and muscle tissue (see Figure 24.3, ❻).

C. Amino acid metabolism

1. **Increased amino acid degradation:** In the absorptive period, more amino acids are present than the liver can use in the synthesis of proteins and other nitrogen-containing molecules. The surplus amino acids are not stored, but are either released into the blood for all tissues to use in protein synthesis, or they are deaminated, with the resulting carbon skeletons being degraded by the liver to pyruvate, acetyl CoA, or TCA cycle intermediates. These metabolites can be oxidized for energy or used in fatty acid synthesis (see Figure 24.3, ❼). The liver has limited capacity to degrade the branched-chain amino acids leucine, isoleucine, and valine; they pass through the liver essentially unchanged and are preferentially metabolized in muscle (see p. 264).

2. **Increased protein synthesis:** The body cannot store protein in the same way that it maintains glycogen or triacylglycerol reserves. However, a transient increase in the synthesis of hepatic proteins does occur in the absorptive state, resulting in replacement of any proteins that may have been degraded during the previous postabsorptive period (see Figure 24.3, ❽).

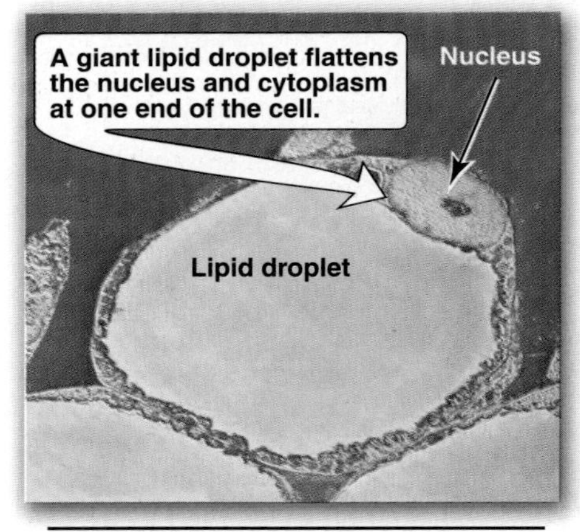

Figure 24.4
Colorized transmission electron micrograph of adipocytes.

IV. ADIPOSE TISSUE: ENERGY STORAGE DEPOT

Adipose tissue is second only to the liver in its ability to distribute fuel molecules. In a 70 kg man, adipose tissue weighs approximately 14 kg, or about half as much as the total muscle mass. In obese individuals adipose tissue can constitute up to seventy percent of body weight. Nearly the entire volume of each adipocyte can be occupied by a droplet of triacylglycerol (Figure 24.4).

A. Carbohydrate metabolism

1. **Increased glucose transport:** Glucose transport into adipocytes is very sensitive to the concentration of insulin in the blood. Circulating insulin levels are elevated in the absorptive state, resulting in an influx of glucose into adipocytes (Figure 24.5, ❶).

2. **Increased glycolysis:** The increased intracellular availability of glucose results in an enhanced rate of glycolysis (see Figure 24.5, ❷). In adipose tissue, glycolysis serves a synthetic function by supplying glycerol phosphate for triacylglycerol synthesis (see p. 186).

3. **Increased activity in the hexose monophosphate pathway (HMP):** Adipose tissue can metabolize glucose by means of the HMP, thereby producing NADPH, which is essential for fat synthesis (see p. 184 and Figure 24.5, ❸). However in humans, de novo synthesis is not a major source of fatty acids in adipose tissue.

B. Fat metabolism

1. **Increased synthesis of fatty acids:** De novo synthesis of fatty acids from acetyl CoA in adipose tissue is nearly undetectable in humans, except when refeeding a previously fasted individual. At other times, fatty acid synthesis in adipose tissue is not a major pathway (see Figure 24.5, ❹). Instead, most of the fatty acids added to the lipid stores of adipocytes are provided by dietary fat (in the form of chylomicrons), with a lesser amount is supplied by VLDL from the liver (see p. 229).

2. **Increased triacylglycerol synthesis:** After consumption of a lipid-containing meal, hydrolysis of the triacylglycerol of chylomicrons (from the intestine) and VLDL (from the liver) provides adipose tissue with fatty acids (see Figure 24.5, ❺). These exogenous fatty acids are released by the action of *lipoprotein lipase,* an extracellular enzyme attached to the capillary walls in many tissues—particularly adipose and muscle. Because adipocytes lack *glycerol kinase,* glycerol 3-phosphate used in triacylglycerol synthesis must come from the metabolism of glucose (see p. 186). Thus, in the well-fed state, elevated levels of glucose and insulin favor storage of triacylglycerol (see Figure 24.5, ❻).

3. **Decreased triacylglycerol degradation:** Elevated insulin favors the dephosphorylated (inactive) state of *hormone-sensitive lipase* (see p. 187). Triacylglycerol degradation is thus inhibited in the well-fed state.

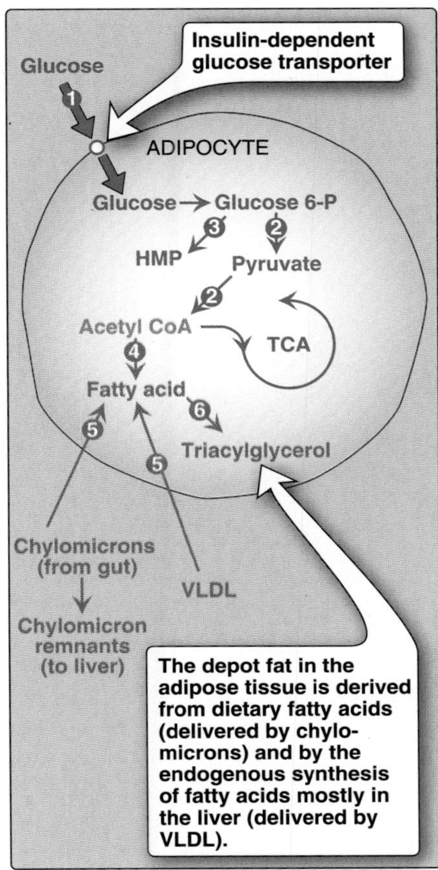

Figure 24.5
Major metabolic pathways in adipose tissue in the absorptive state. [Note: The numbers in the circles, which appear both on the figure and in the corresponding text, indicate important pathways for adipose tissue metabolism.]

V. RESTING SKELETAL MUSCLE

The energy metabolism of skeletal muscle is unique in being able to respond to substantial changes in the demand for ATP that accompanies muscle contraction. At rest, muscle accounts for approximately thirty percent of the oxygen consumption of the body, whereas during vigorous exercise, it is responsible for up to ninety percent of the total oxygen consumption. This graphically illustrates the fact that skeletal muscle, despite its potential for transient periods of anaerobic glycolysis, is an oxidative tissue. [Note: **Heart muscle** differs from skeletal muscle in three important ways: 1) the heart is continuously active, whereas skeletal muscle contracts intermittently on demand; 2) the heart has a completely aerobic metabolism; and 3) the heart contains negligible energy stores, such as glycogen or lipid. Thus, any interruption of the vascular supply, for example, as occurs during a myocardial infarction, results in rapid death of the myocardial cells. Heart muscle uses glucose, free fatty acids, and ketone bodies as fuels.]

A. Carbohydrate metabolism

1. **Increased glucose transport:** The transient increase in plasma glucose and insulin after a carbohydrate-rich meal leads to an increase in glucose transport into the cells (see p. 95, and Figure 24.6, ❶). Glucose is phosphorylated to glucose 6-phosphate by *hexokinase*, and metabolized to provide the energy needs of the cells. This contrasts with the postabsorptive state in which ketone bodies and fatty acids are the major fuels of resting muscle.

2. **Increased glycogen synthesis:** The increased insulin to glucagon ratio and the availability of glucose 6-phosphate favor glycogen synthesis, particularly if glycogen stores have been depleted as a result of exercise (see p. 124, and Figure 24.6, ❷).

B. Fat metabolism

Fatty acids are released from chylomicrons and VLDL by the action of *lipoprotein lipase* (see pp. 226, 229). However, fatty acids are of secondary importance as a fuel for muscle in the well-fed state, in which glucose is the primary source of energy.

C. Amino acid metabolism

1. **Increased protein synthesis:** A spurt in amino acid uptake and protein synthesis occurs in the absorptive period after ingestion of a meal containing protein (see Figure 24.6, ❸ and ❹). This synthesis replaces protein degraded since the previous meal.

2. **Increased uptake of branched-chain amino acids:** Muscle is the principal site for degradation of branched-chain amino acids (see

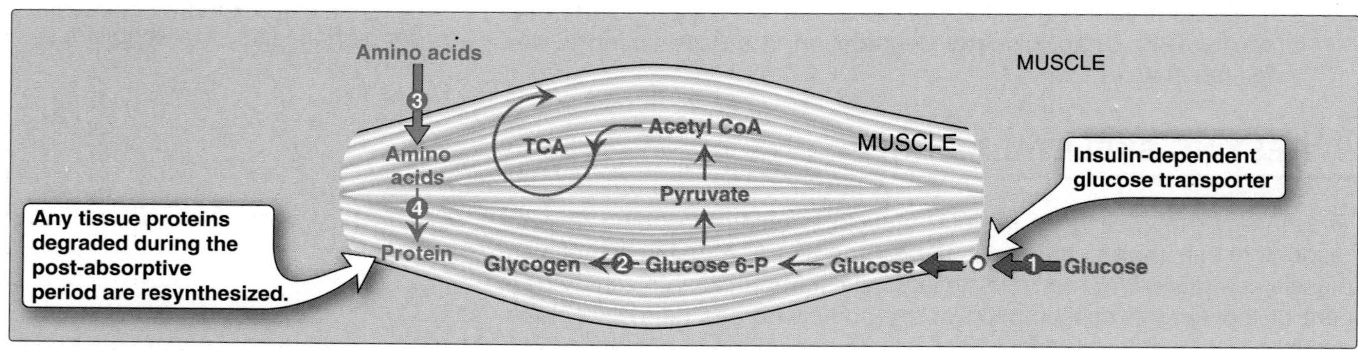

Figure 24.6
Major metabolic pathways in skeletal muscle in the absorptive state. [Note: The numbers in circles, which appear both on the figure and in the text, indicate important pathways for carbohydrate or protein metabolism.]

p. 264). The branched chain amino acids, **leucine**, **isoleucine**, and **valine**, escape metabolism by the liver, and are taken up by muscle, where they are used for protein synthesis and as sources of energy (see Figure 24.6, ❸).

VI. BRAIN

Although contributing only two percent of the adult weight, the brain accounts for twenty percent of the basal oxygen consumption of the body at rest. The brain uses energy at a constant rate. Because the brain is vital to the proper functioning of all organs of the body, special priority is given to its fuel needs. To provide energy, substrates must be able to cross the endothelial cells that line the blood vessels in the brain (sometimes called the "blood–brain barrier"). Normally, glucose serves as the primary fuel, because the concentration of ketone bodies in the fed state is too low to serve as an alternate energy source. If blood glucose levels fall below approximately 30 mg/100 ml (normal blood glucose is 70 to 90 mg/100 ml), cerebral function is impaired. If the hypoglycemia occurs for even a short time, severe and potentially irreversible brain damage may occur. Note, however, that ketone bodies play a significant role as a fuel during a fast (see p. 193).

A. Carbohydrate metabolism

In the well-fed state, the brain uses glucose exclusively as a fuel, completely oxidizing approximately 140 g/day to carbon dioxide and water. The brain contains no significant stores of glycogen and is, therefore, completely dependent on the availability of blood glucose (Figure 24.7, ❶).

B. Fat metabolism

The brain has no significant stores of triacylglycerol, and the oxidation of fatty acids obtained from blood makes little contribution to energy production because fatty acids do not efficiently cross the blood-brain barrier. The intertissue exchanges characteristic of the absorptive period are summarized in Figure 24.8.

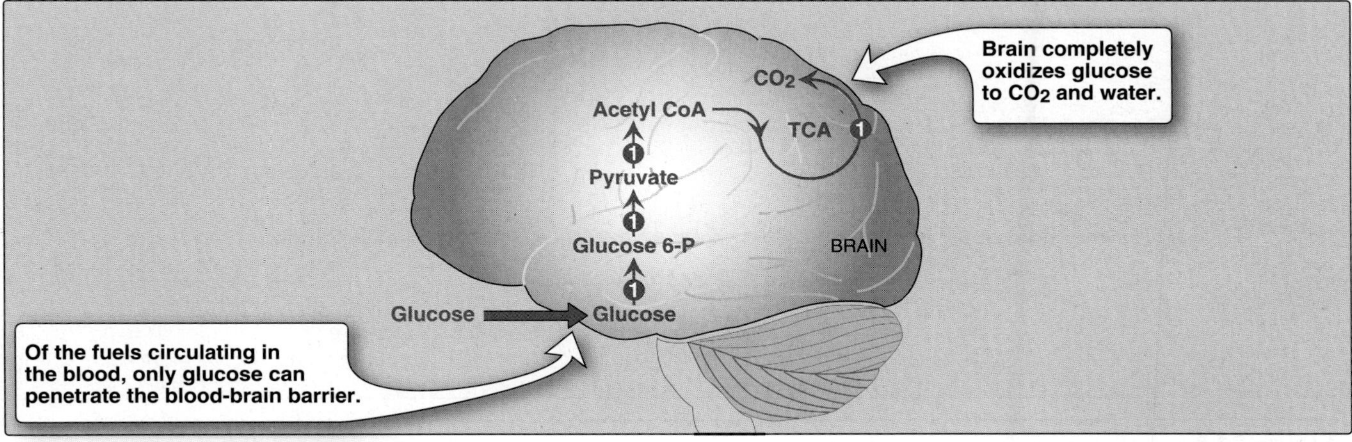

Figure 24 .7
Major metabolic pathways in brain in the absorptive state. [Note: The numbers in circles, which appear both on the figure and in the text, indicate important pathways for carbohydrate metabolism.]

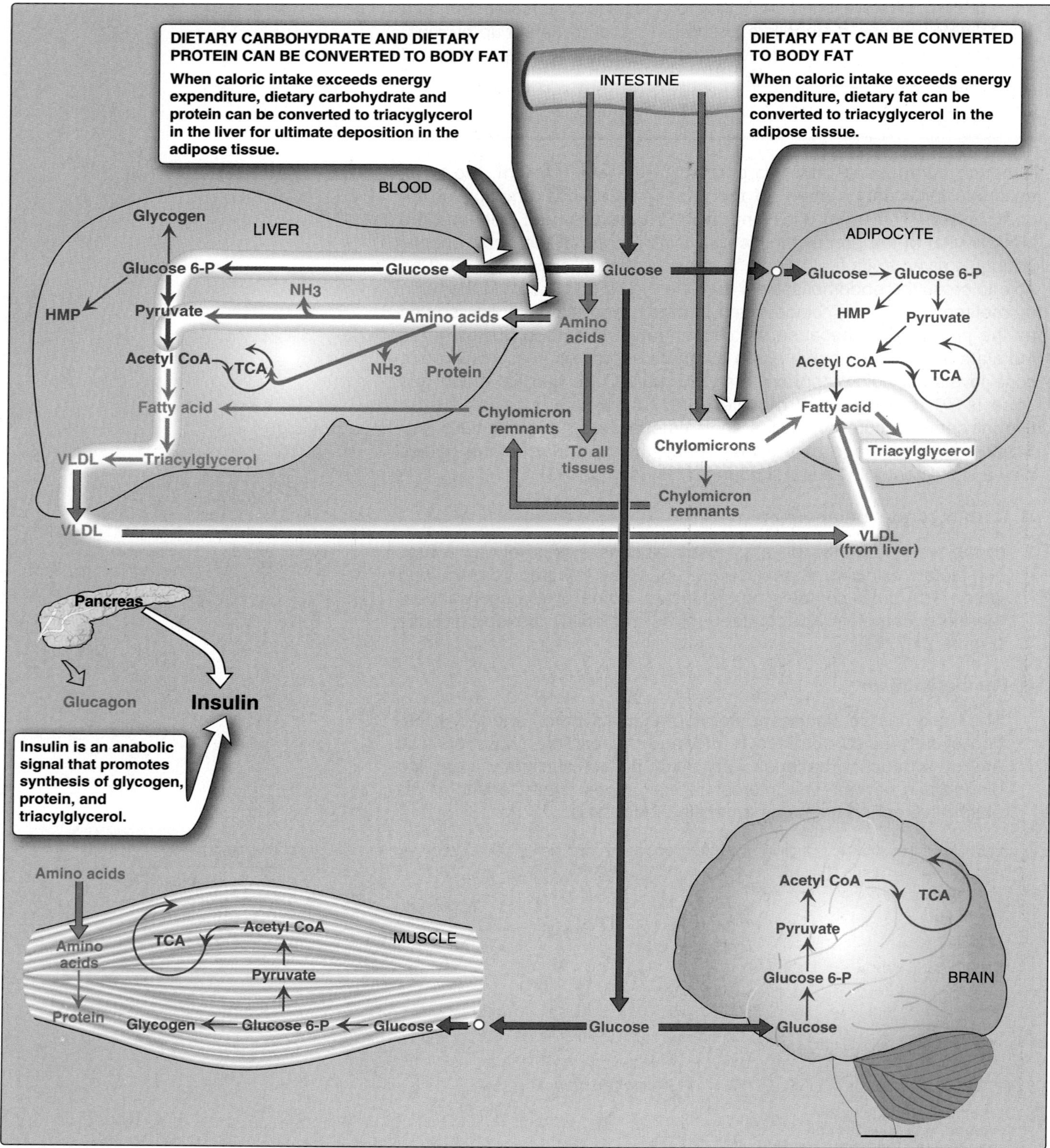

Figure 24.8
Intertissue relationships in the absorptive state. [Note: Small circles on perimeter of tissues indicate insulin-dependent transport systems.]

VII. OVERVIEW OF FASTING

Fasting may result from an inability to obtain food, from the desire to lose weight rapidly, or in clinical situations in which an individual cannot eat because of trauma, surgery, neoplasms, burns, and so forth. In the absence of food, plasma levels of glucose, amino acids, and triacylglycerols fall, triggering a decline in insulin secretion and an increase in glucagon release. The decreased insulin to glucagon ratio, and the decreased availability of circulating substrates, makes the period of nutrient deprivation a **catabolic period** characterized by degradation of triacylglycerol, glycogen, and protein. This sets into motion an exchange of substrates between liver, adipose tissue, muscle, and brain that is guided by two priorities: 1) the need to maintain adequate plasma levels of glucose to sustain energy metabolism of the brain and other glucose-requiring tissues, and 2) the need to mobilize fatty acids from adipose tissue, and the synthesis and release of ketone bodies from the liver, to supply energy to all other tissues.

A. Fuel stores

The metabolic fuels available in a normal 70 kg man at the beginning of a fast are shown in Figure 24.9. Note the enormous caloric stores available in the form of triacylglycerols compared with those contained in glycogen. [Note: Although protein is listed as an energy source, each protein also has a function, for example, as a structural component of the body, an enzyme, and so forth. Therefore, only about one third of the body's protein can be used for energy production without fatally compromising vital functions.]

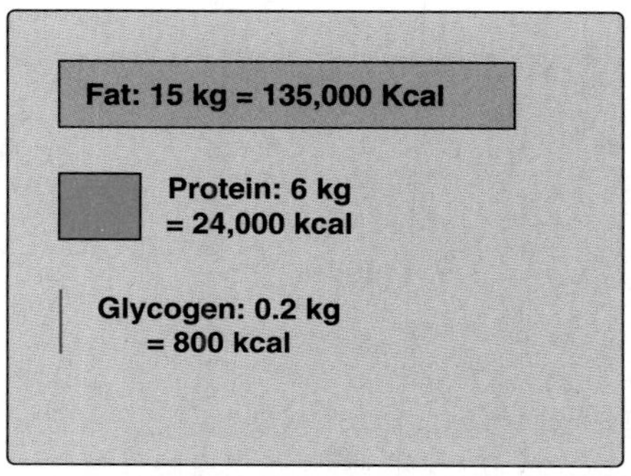

Figure 24.9
Metabolic fuels present in a 70 kg man at the beginning of a fast.

B. Enzymic changes in fasting

In fasting (as well as in the fed state), the flow of intermediates through the pathways of energy metabolism is controlled by four mechanisms: 1) the availability of substrates, 2) allosteric activation and inhibition of enzymes, 3) covalent modification of enzymes, and 4) induction-repression of enzyme synthesis. The metabolic changes observed in fasting are generally opposite to those described for the well-fed state (see Figure 24.8). For example, most of the enzymes regulated by covalent modification are in the phosphorylated state and are inactive, whereas in the fed state, they are dephosphorylated and active. Three exceptions are *glycogen phosphorylase* (see p. 129), *fructose bisphosphate phosphatase 2* (see p. 98), and *hormone-sensitive lipase* of adipose tissue (see p. 187), which are inactive in their dephosphorylated states. In fasting, substrates are not provided by the diet, but are available from breakdown of tissues, for example, degradation of triacylglycerols and release of fatty acids from adipose tissue. Recognition that the changes in fasting are the reciprocal of those in the fed state is helpful in understanding the ebb and flow of metabolism.

VIII. LIVER IN FASTING

The primary role of liver in energy metabolism during fasting is the synthesis and distribution of fuel molecules for use by other organs. Thus, one speaks of "hepatic metabolism" and "extraheptic" or "peripheral" metabolism.

A. Carbohydrate metabolism

The liver first uses glycogen degradation, then gluconeogenesis, to maintain blood glucose levels to sustain energy metabolism of the brain and other glucose-requiring tissues.

1. **Increased glycogen degradation:** Figure 24.10 shows the sources of blood glucose after ingestion of 100 g of glucose. During the brief absorptive period, glucose from the diet is the major source of blood glucose. Several hours after the meal, blood glucose levels have declined sufficiently to cause increased secretion of glucagon and decreased release of insulin. The increased glucagon to insulin ratio causes a rapid mobilization of liver glycogen stores (which contain about 100 g of glycogen in the well-fed state). Note that liver glycogen is nearly exhausted after ten to eighteen hours of fasting; therefore, hepatic glycogenolysis is a transient response to early fasting. Figure 24.11, ❶, shows glycogen degradation as part of the overall metabolic response of the liver during fasting.

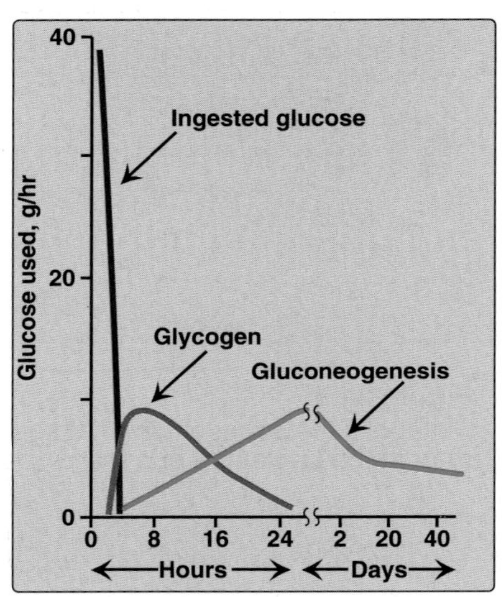

Figure 24.10
Sources of blood glucose after ingestion of 100 g of glucose.

2. **Increased gluconeogenesis:** The synthesis of glucose and its subsequent release into the circulation are vital hepatic functions during fasting (see Figure 24.11, ❷). The carbon skeletons for gluconeogenesis are derived primarily from amino acids, glycerol, and lactate. Gluconeogenesis begins four to six hours after the last meal and becomes fully active as stores of liver glycogen are depleted (see Figure 24.10). Gluconeogenesis plays an essential role in maintaining blood glucose during both overnight and prolonged fasting.

B. Fat metabolism

1. **Increased fatty acid oxidation:** The oxidation of fatty acids derived from adipose tissue is the major source of energy in hepatic tissue in the postabsorptive state (see Figure 24.11, ❸).

2. **Increased synthesis of ketone bodies:** The liver is unique in being able to synthesize and release ketone bodies (primarily **3-hydroxybutyrate**, (formerly called β-hydroxybutyrate) for use as fuels by peripheral tissues (see p. 193). [Note: The liver cannot use ketone bodies as a fuel (see p. 194).] Ketone body synthesis is favored when the concentration of acetyl CoA, produced from fatty acid metabolism, exceeds the oxidative capacity of the TCA cycle. Significant synthesis of ketone bodies starts during the first days of fasting (Figure 24.12). [Note: Unlike fatty acids, ketone bodies are water-soluble, and appear in the blood and urine by the second day of a fast.] The availability of circulating ketone bodies is important in fasting because they can be used for fuel by most tissues, including brain tissue, once their level in the

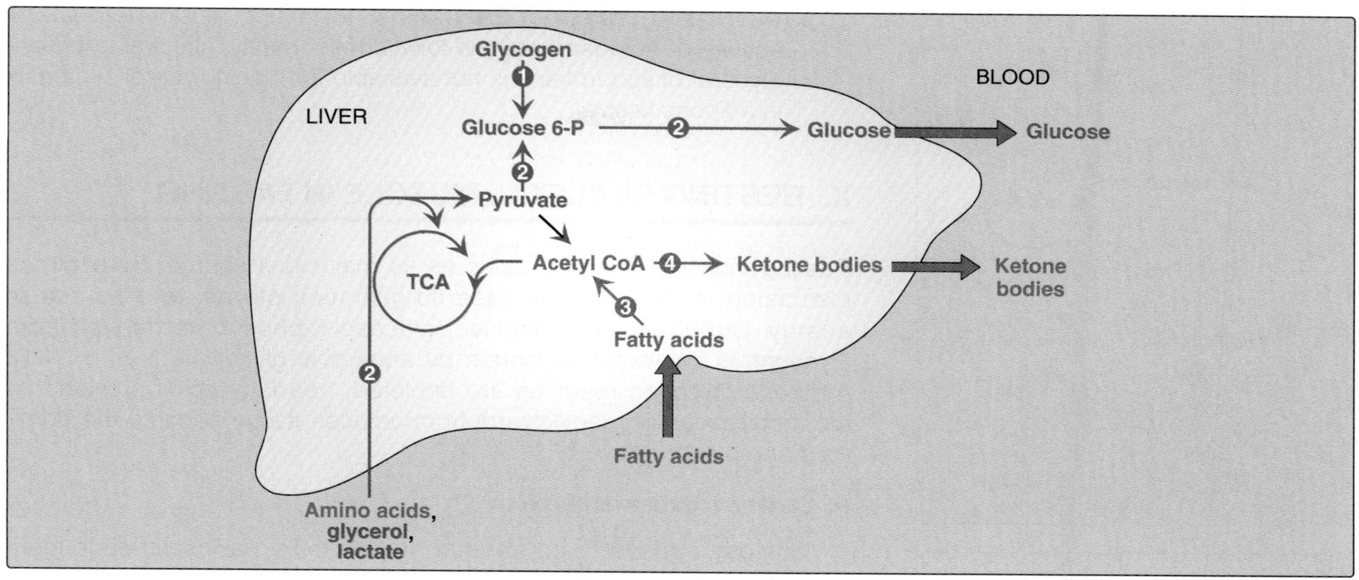

Figure 24.11
Major metabolic pathways in liver during starvation. The numbers in circles, which appear both on the figure and in the corresponding citation in the text, indicate important metabolic pathways for carbohydrate or fat.

blood is sufficiently high. This reduces the need for gluconeogenesis from amino acid carbon skeletons, thus slowing the loss of essential protein. Ketone body synthesis as part of the overall hepatic response to fasting is shown in Figure 24.11, ❹.

IX. ADIPOSE TISSUE IN FASTING

A. Carbohydrate metabolism

Glucose transport into the adipocyte and its subsequent metabolism are depressed owing to low levels of circulating insulin. This leads to a decrease in fatty acid and triacylglycerol synthesis.

B. Fat metabolism

1. **Increased degradation of triacylglycerols:** The activation of *hormone-sensitive lipase* (see p. 187) and subsequent hydrolysis of stored triacylglycerol are enhanced by the elevated catecholamines epinephrine and, particularly, norepinephrine. These compounds, which are released from the sympathetic nerve endings in adipose tissue, are physiologically important activators of *hormone-sensitive lipase* (Figure 24.13, ❶).

2. **Increased release of fatty acids:** Fatty acids obtained from hydrolysis of stored triacylglycerol are released into the blood (see Figure 24.13, ❷). Bound to albumin, they are transported to a variety of tissues for use as fuel. The glycerol produced following triacylglycerol degradation is used as a gluconeogenic precursor by the liver. [Note: Fatty acids are also converted to acetyl CoA, which can enter the TCA cycle, thus producing energy for the adipocyte.]

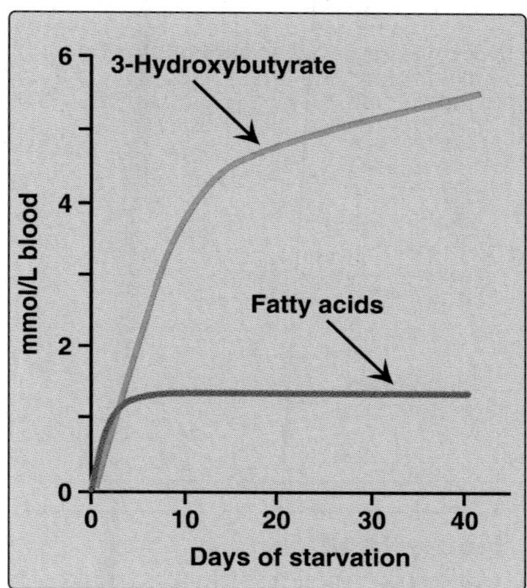

Figure 24.12
Concentrations of fatty acids and 3-hydroxybutyrate in the blood during fasting.

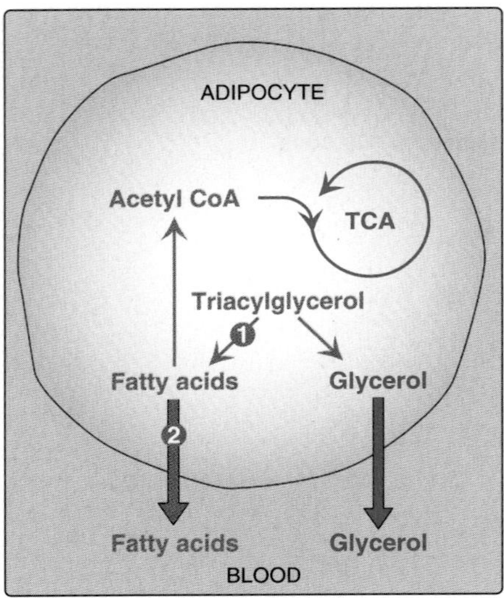

Figure 24.13
Major metabolic pathways in adipose tissue during starvation. The numbers in the circles, which appear both on the figure and in the corresponding citation in the text, indicate important pathways for fat metabolism.

3. **Decreased uptake of fatty acids:** In fasting, *lipoprotein lipase* activity of adipose tissue is low. Consequently, circulating triacylglycerol of lipoproteins is not available for triacylglycerol synthesis in adipose tissue.

X. RESTING SKELETAL MUSCLE IN FASTING

Resting muscle uses fatty acids as its major fuel source. By contrast, exercising muscle initially uses its glycogen stores as a source of energy. During intense exercise, glucose 6-phosphate derived from glycogen is converted to lactate by anaerobic glycolysis (see p. 101). As these glycogen reserves are depleted, free fatty acids provided by the mobilization of triacylglyerol from adipose tissue become the dominant energy source.

A. Carbohydrate metabolism

Glucose transport into skeletal muscle cells via insulin-dependent glucose transport proteins in the plasma membrane (see p. 95) and subsequent glucose metabolism are depressed because of low levels of circulating insulin.

B. Lipid metabolism

During the first 2 weeks of fasting, muscle uses fatty acids from adipose tissue and ketone bodies from the liver as fuels (Figure 24.14, ❶ and ❷). After about three weeks of fasting, muscle decreases its use of ketone bodies and oxidizes fatty acids almost exclusively. This leads to a further increase in the already elevated level of circulating ketone bodies. [Note: The increased use of ketone bodies by the brain as a result of their increased concentration in the blood is correlated with the decreased use of these compounds by the muscle.]

C. Protein metabolism

During the first few days of fasting, there is a rapid breakdown of muscle protein, providing amino acids that are used by the liver for gluconeogenesis (see Figure 24.14, ❸). [Note: Alanine and glutamine are quantitatively the most important gluconeogenic amino acids released from muscle.] After several weeks of fasting, the rate of muscle proteolysis decreases because there is a decline in the need for glucose as a fuel for brain, which has begun using ketone bodies as a source of energy.

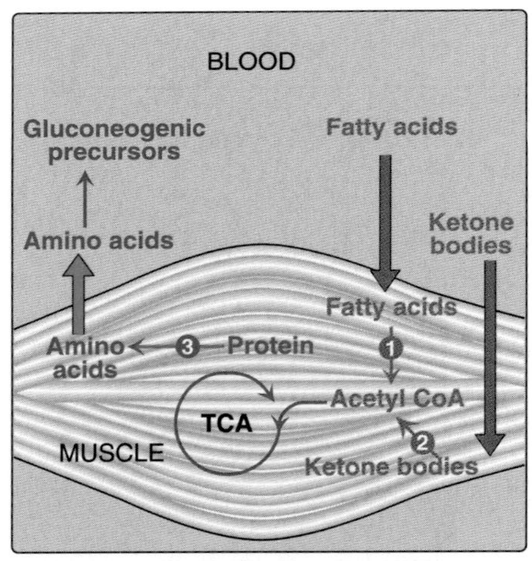

Figure 24.14
Major metabolic pathways in skeletal muscle during starvation. The numbers in the circles, which appear both on the figure and in the corresponding citation in the text, indicate important pathways for fat or protein metabolism.

XI. BRAIN IN FASTING

During the first days of fasting, the brain continues to use glucose exclusively as a fuel (Figure 24.15, ❶). [Note: Blood glucose is maintained by hepatic gluconeogenesis from glucogenic precursors, such as amino acids provided by the rapid breakdown of muscle protein.] In prolonged fasting (greater than two to three weeks), plasma ketone bodies (see Figure 24.12) reach significantly elevated levels, and are used in addition to glucose as a fuel by the brain (see Figure 24.15, ❷). This reduces the need for protein catabolism for gluconeogenesis. The metabolic changes that occur during fasting ensure that all tissues have an adequate supply of fuel molecules. The response of the major tissues involved in energy metabolism during fasting is summarized in Figure 24.16.

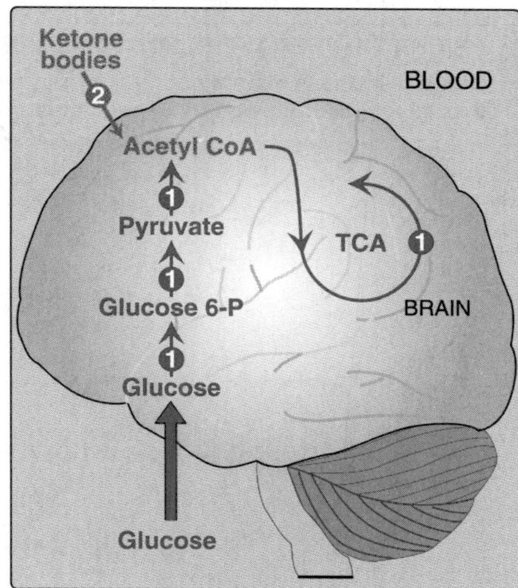

Figure 24.15
Major metabolic pathways in the brain during starvation. The numbers in the circles, which appear both on the figure and in the corresponding citation in the text, indicate important pathways for metabolism of fat or carbohydrates.

XII. CHAPTER SUMMARY

The flow of intermediates through metabolic pathways is controlled by four mechanisms: 1) the availability of substrates; 2) allosteric activation and inhibition of enzymes; 3) covalent modification of enzymes; and 4) induction-repression of enzyme synthesis. In the fed state, these regulatory mechanisms ensure that available nutrients are captured as **glycogen**, **triacylglycerol**, and **protein**. The **absorptive state** is the two- to four-hour period after ingestion of a normal meal. During this interval, transient increases in plasma glucose, amino acids, and triacylglycerols occur, the last primarily as components of chylomicrons synthesized by the intestinal mucosal cells. The **pancreas** responds to the elevated levels of glucose and amino acids with an **increased secretion of insulin** and a **drop in the release of glucagon** by the **islets of Langerhans**. The elevated insulin to glucagon ratio and the ready availability of circulating substrates make the two to four hours after ingestion of a meal into an **anabolic period**. During this absorptive period, virtually all tissues use **glucose** as a fuel. In addition, the **liver** replenishes its **glycogen** stores, replaces any needed **hepatic proteins**, and increases **triacylglycerol** synthesis. The latter are packaged in **very-low-density glycoproteins**, which are exported to the peripheral tissues. The **adipose** increases **triacylglycerol** synthesis and storage, whereas the **muscle** increases **protein** synthesis to replace protein degraded since the previous meal. In the well-fed state, the **brain** uses glucose exclusively as a fuel. In the **absence of food**, plasma levels of glucose, amino acids, and triacylglycerols fall, triggering a **decline** in **insulin secretion** and an **increase in glucagon** and **epinephrine release**. The decreased insulin to glucagon ratio, and the decreased availability of circulating substrates, makes the period of nutrient deprivation a **catabolic period**. This sets into motion an **exchange of substrates** among liver, adipose tissue, muscle, and brain that is guided by two priorities: 1) the need to maintain adequate plasma levels of glucose to sustain energy metabolism of the brain and other glucose-requiring tissues; and 2) the need to mobilize fatty acids from adipose tissue and ketone bodies from liver to supply energy to all other tissues. To accomplish these goals, the **liver** degrades **glycogen** and initiates **gluconeogenesis**, using **increased fatty acid oxidation** both as a source of energy, and to supply the acetyl CoA building blocks for **ketone body synthesis**. The **adipose** degrades stored **triacylglycerols**, thus providing **fatty acids** and **glycerol** to the liver. The **muscle** can also use **fatty acids** as fuel, as well as **ketone bodies** supplied by the liver. Muscle **protein** is **degraded** to supply **amino acids** for the liver to use in gluconeogenesis. The **brain** can use both **glucose** and **ketone bodies** as fuels.

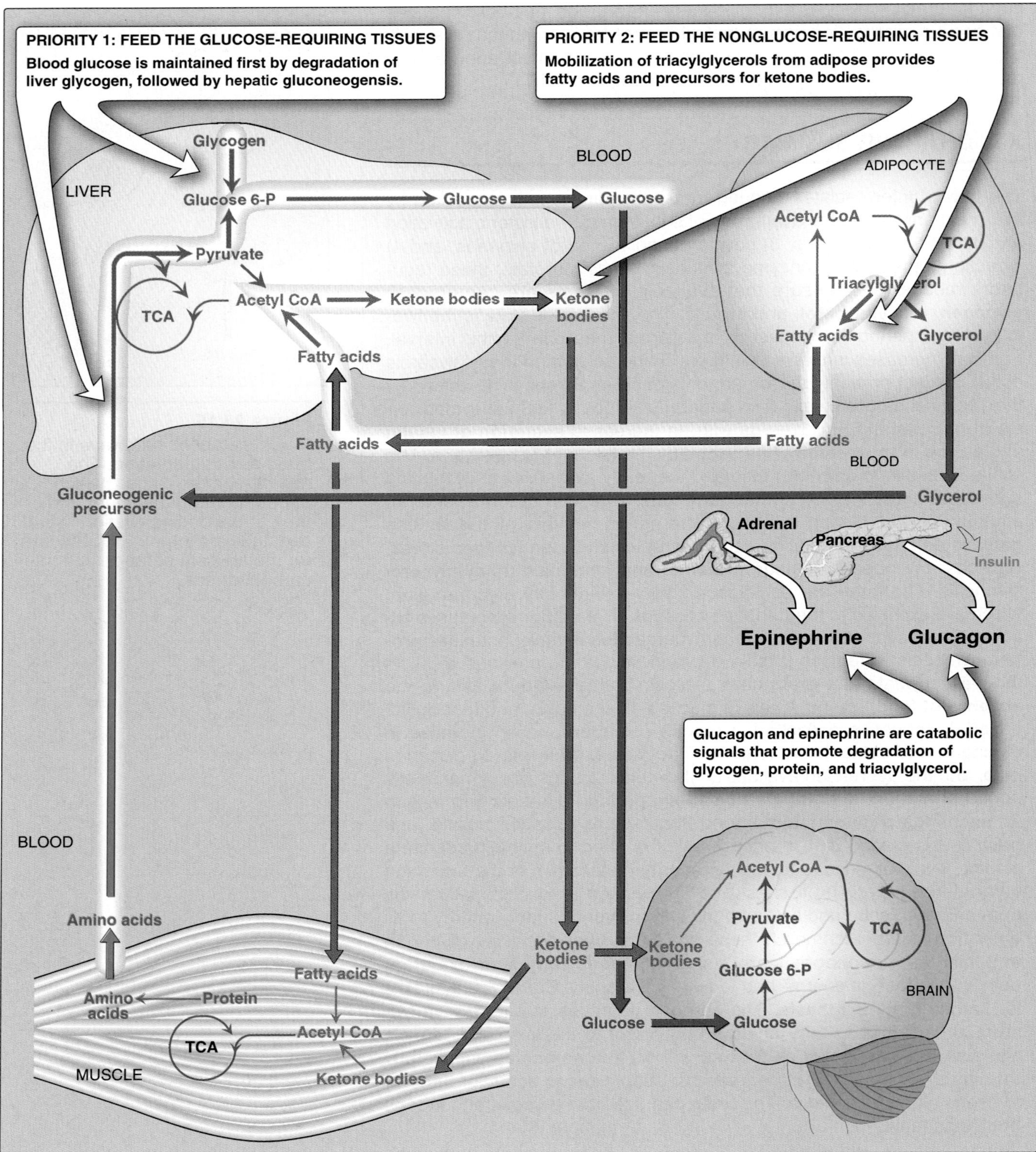

Figure 24.16
Intertissue relationships during starvation.

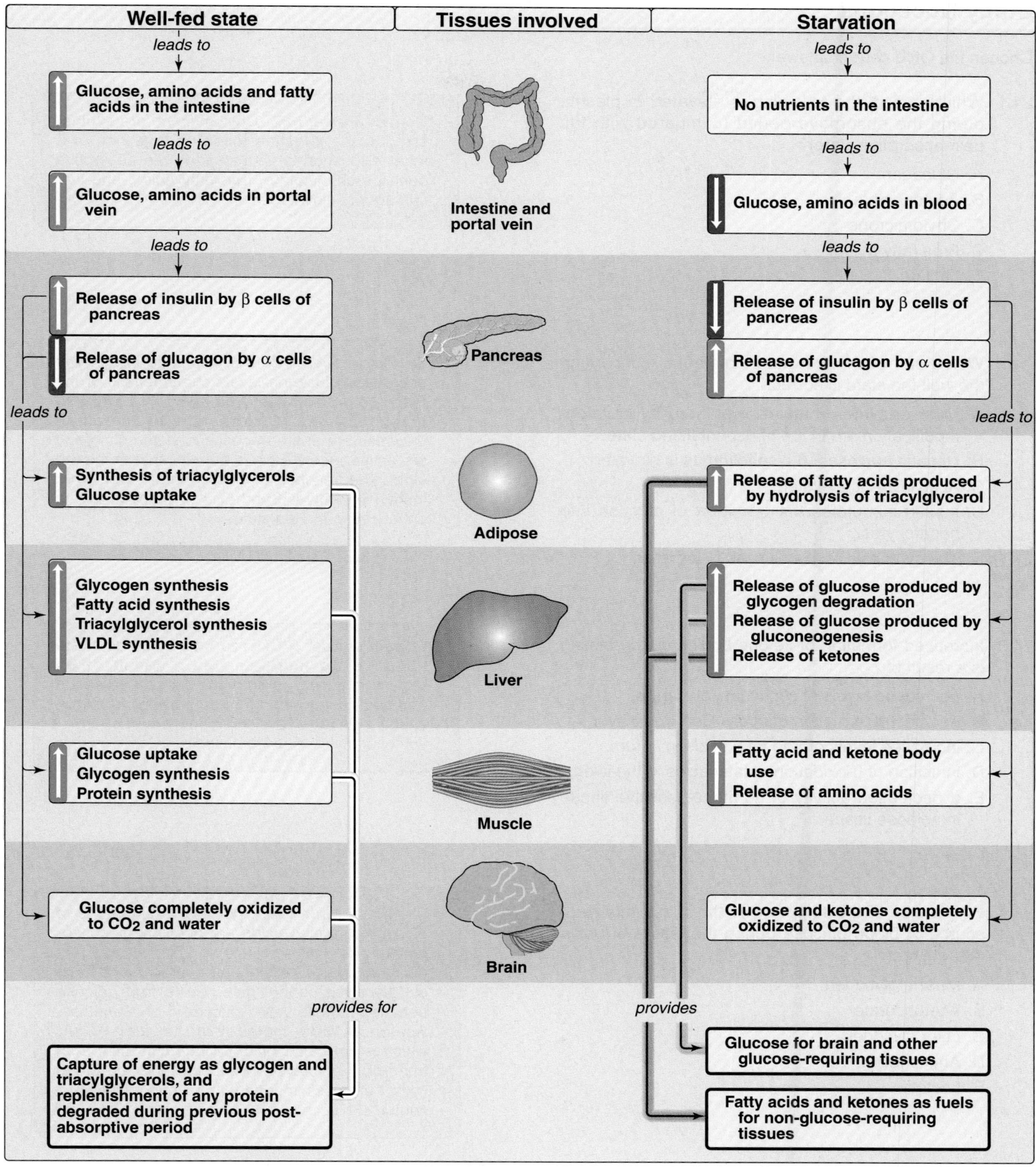

Figure 24.17
Key concept map for feed/fast cycle.

Study Questions

Choose the ONE correct answer

24.1 Which one of the following is elevated in plasma during the absorptive period (compared with the post-absorptive state)?

A. Glucagon
B. Acetoacetate
C. Chylomicrons
D. Free fatty acids
E. Lactate

Correct answer = C. Chylomicrons are synthesized in the intestine following ingestion of a meal. Glucagon is depressed in the absorptive period. Acetoacetate, free fatty acids, and lactate are not elevated.

24.2 Which one of the following statements concerning the well-fed state is correct?

A. Most enzymes that are regulated by covalent modification are in the phosphorylated state.
B. Hepatic fructose 2,6-bisphosphate is elevated.
C. Acetyl CoA is elevated.
D. Insulin stimulates the transport of glucose into hepatocytes.
E. The synthesis of glucokinase is repressed.

Correct answer = B. The increased insulin and decreased glucagon levels characteristic of the fed state promote the synthesis of fructose 2,6-bisphosphate. Most covalently modified enzymes are in the dephosphorylated state and are active. Acetyl CoA is not elevated in the fed state. The transport of glucose in the liver is not insulin sensitive. Synthesis of glucokinase is enhanced in the fed state.

24.3 Increased formation of ketone bodies during fasting is a result of:

A. decreased levels of circulating glucagon.
B. decreased formation of acetyl CoA in the liver.
C. increased levels of free fatty acids in serum.
D. inhibition of β-oxidation of fatty acids in the liver.
E. a decreased activity of hormone-sensitive lipase in adipose tissue.

Correct answer = C. Free fatty acids bound to albumin are increased as a result of an increased activity of hormone-sensitive lipase in adipose tissue. Hepatic ketogenesis is stimulated by elevated levels of glucagon. The formation of acetyl CoA is increased.

24.4 Which one of the following is the most important source of blood glucose during the last hours of a 48-hour fast?

A. Muscle glycogen
B. Acetoacetate
C. Liver glycogen
D. Amino acids
E. Lactate

Correct answer = D. The carbon skeletons of glycogenic amino acids are used by gluconeogenesis to produce glucose. Liver glycogen is depleted by twelve hours after a meal, and muscle glycogen cannot give rise to free glucose because muscle lacks glucose 6-phosphatase. Acetoacetate is metabolized to acetyl CoA, which is not glucogenic. Lactate can arise from anaerobic glycolysis in muscle and red blood cells, but is less important than amino acids as a source of glucose.

Diabetes Mellitus 25

I. OVERVIEW OF DIABETES MELLITUS

Diabetes is not one disease, but rather is a heterogeneous group of syndromes characterized by an elevation of fasting blood glucose caused by a relative or absolute deficiency in insulin. Diabetes mellitus is the leading cause of adult blindness and amputation, and a major cause of renal failure, heart attacks, and strokes. Most cases of diabetes mellitus can be separated into two groups (Figure 25.1), **type 1** (formerly called **insulin-dependent diabetes mellitus**) and **type 2** (formerly called **non-insulin-dependent diabetes mellitus**). Approximately 30,000 newly-diagnosed cases of type 1 and 625,000 cases of type 2 diabetes mellitus are estimated to occur yearly in the United States. The incidence and prevalence of type 2 disease is increasing because of the aging of the United States population, and the increasing prevalence of obesity and sedentary lifestyles. This increase in children with type 2 diabetes is particularly disturbing.

	Type 1 Diabetes	Type 2 Diabetes
AGE OF ONSET	Usually during childhood or puberty; symptoms develop rapidly	Frequently after age 35; symptoms develop gradually
NUTRITIONAL STATUS AT TIME OF DISEASE ONSET	Frequently undernourished	Obesity usually present
PREVALENCE	900,000 = 10% of diagnosed diabetics	10 Million = 90% of diagnosed diabetics
GENETIC PREDISPOSITION	Moderate	Very strong
DEFECT OR DEFICIENCY	β Cells are destroyed, eliminating production of insulin	Insulin resistance combined with inability of β cells to produce appropriate quantities of insulin
FREQUENCY OF KETOSIS	Common	Rare
PLASMA INSULIN	Low to absent	High early in disease; low in disease of long duration
ACUTE COMPLICATIONS	Ketoacidosis	Hyperosmolar coma
TREATMENT WITH ORAL HYPOGLYCEMIC DRUGS	Unresponsive	Responsive
TREATMENT	Insulin is always necessary	Diet, exercise, oral hypoglycemic drugs, +/– insulin

Figure 25.1
Comparison of type 1 and type 2 diabetes.

II. TYPE 1 DIABETES

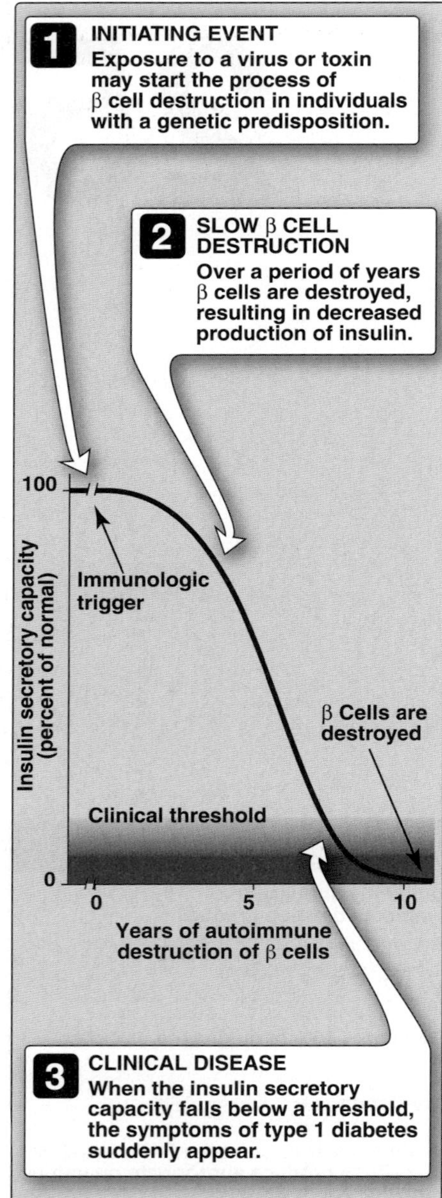

Figure 25.2
Insulin secretory capacity during onset of type 1 diabetes. [Note: Rate of autoimmune destruction of β cells may be faster or slower than shown.]

Persons with type 1 diabetics constitute approximately ten percent of the ten million diabetics in the United States. The disease is characterized by an **absolute deficiency of insulin** caused by an **autoimmune attack** on the **β cells of the pancreas**. In type 1 diabetes, the **islets of Langerhans** become infiltrated with activated T lymphocytes, leading to a condition called **insulitis**. Over a period of years, this autoimmune attack leads to gradual depletion of the β cell population (Figure 25.2). However, symptoms appear abruptly when eighty to ninety percent of the β cells have been destroyed. At this point, the pancreas fails to respond adequately to ingestion of glucose, and insulin therapy is required to restore metabolic control and prevent life-threatening ketoacidosis. This destruction requires both a stimulus from the **environment** (such as a viral infection) and a **genetic determinant** that allows the β cells to be recognized as "non-self." [Note: Among monozygotic (identical) twins, if one sibling develops type 1 diabetes melllitus, the other twin has only a thirty to fifty percent chance of developing the disease. This contrasts with type 2 disease (see 340), in which the genetic influence is stronger, and in virtually all monozygotic twinships, the disease develops in both individuals.]

A. Diagnosis of type 1 diabetes

The onset of type 1 diabetes is typically during childhood or puberty, and symptoms develop rapidly. Patients with type 1 diabetes can usually be recognized by the abrupt appearance of **polyuria** (frequent urination), **polydipsia** (excessive thirst), and **polyphagia** (excessive hunger), often triggered by stress or an illness. These symptoms are usually accompanied by fatigue, weight loss, and weakness.The diagnosis is confirmed by a fasting blood glucose greater than 126 mg/dl, commonly accompanied by ketoacidosis.

1. **Glucose tolerance test:** In the past, the glucose tolerance test was a commonly employed diagnostic test for diabetes mellitus. The patient is given 75 g of glucose orally following an eight hour fast. Blood glucose concentrations are determined at thirty minute intervals for the next three hours. Fasting blood glucose is initially high (greater than 126 mg/dl) in the diabetic, and it rises to concentrations greater than 200 mg/dl following the oral administration of glucose. The rate of glomerular filtration of glucose exceeds that of tubular reabsorption in the kidney, and glucose appears in the urine. In contrast, normal individuals show fasting blood glucose levels of <110 mg/dl, and a rise to less than 140 mg/dl after a glucose load.

2. **False-positive results:** The glucose tolerance test gives many false-positive results because the test itself can be stress-inducing, causing epinephrine release. This hormone decreases the release of insulin from the β cells (see p. 308) and, thus, impairs the response to a glucose load. As a result, the glucose tolerance test is usually used only in situations in which diagnosis is uncertain or as a test for gestational diabetes. For the general population, a fasting blood glucose (FBG) test is the more commonly used diagnostic tool.

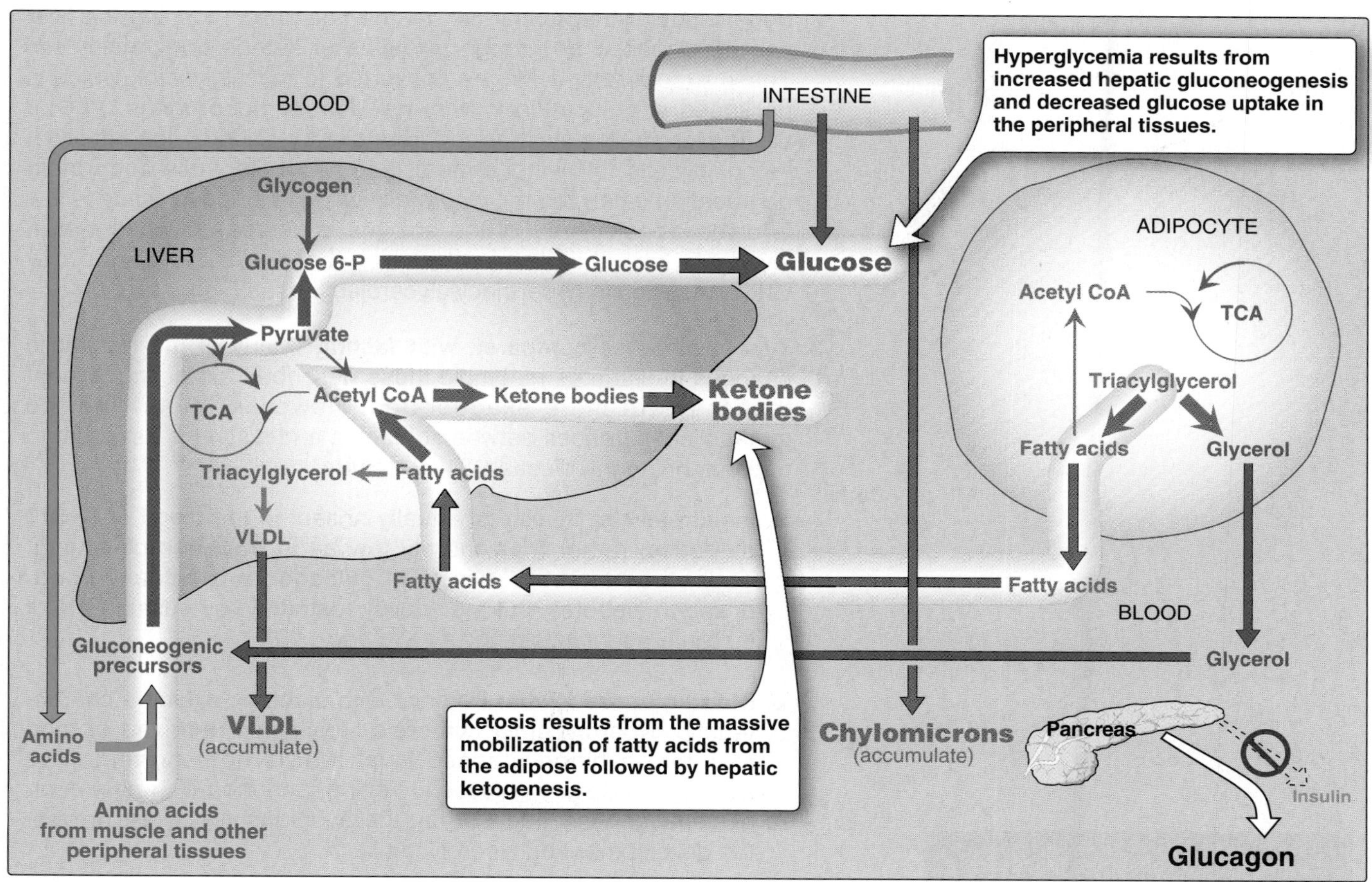

Figure 25.3
Intertissue relationships in type 1 diabetes.

B. Metabolic changes in type 1 diabetes

The metabolic abnormalities of diabetes mellitus result from a deficiency of insulin and a relative excess of glucagon. These aberrant hormonal levels most profoundly affect metabolism in three tissues: liver, muscle, and adipose tissue (Figure 25.3).

1. **Hyperglycemia and ketoacidosi**s: Elevated levels of blood glucose and ketones are the hallmarks of untreated diabetes mellitus. Hyperglycemia is caused by increased hepatic production of glucose, combined with diminished peripheral use due to an inability of muscle and adipose cells to take up glucose (see Figure 25.3). Ketosis results from increased mobilization of fatty acids from adipose tissue, combined with accelerated hepatic synthesis of 3-hydroxybutyrate and acetoacetate. **Diabetic keotacidosis** occurs in twenty-five to forty percent of those newly diagnosed with type 1 diabetes, and may recur if the patient becomes ill (most commonly with an infection) or does not comply with therapy. Ketoacidosis is treated by replacing fluid and electrolytes, followed by administration of low-dose insulin to gradually correct hyperglycemia without precipitating hypoglycemia.

2. **Hypertriacylglycerolemia:** Not all the fatty acids flooding the liver can be disposed of through oxidation or ketone body synthesis. These excess fatty acids are converted to triacylglycerol, which is packaged and secreted in **very-low-density lipoproteins (VLDL)**. **Chylomicrons** are synthesized from dietary lipids by the intestinal mucosal cells following a meal (see 175). Because lipoprotein degradation catalyzed by *lipoprotein lipase* in adipose tissue is low in diabetics (synthesis of the enzyme is decreased when insulin levels are low), the plasma chylomicron and VLDL levels are elevated, resulting in hypertriacylglycerolemia (see Figure 25.3).

3. **Type 1 diabetes compared with fasting:** Many of the metabolic changes in diabetes resemble those described for fasting, except that they are more exaggerated. However, identifying the metabolic differences between diabetes and fasting is essential to understanding the disease. They include:

   a. **Insulin levels:** Insulin is virtually absent in the blood of type 1 diabetics, rather than merely low as in the case of fasting. Thus, the metabolic effects of glucagon are virtually unopposed in diabetes, but are weakly restrained by basal levels of insulin present in fasting.

   b. **Blood glucose levels:** Persons with diabetes exhibit a characteristic hyperglycemia, whereas individuals deprived of food maintain a blood glucose level that is near normal. The absence of dietary glucose in fasting, and the mild restraint on gluconeogenesis imposed by the basal insulin levels, prevent the development of hyperglycemia.

   c. **Ketosis:** The mobilization of fatty acids from adipose tissue and hepatic ketogenesis are greater in diabetes than in fasting. As a result, the ketoacidosis observed in diabetes is much more severe than that observed during fasting.

   d. **Hypertriacylglycerolemia:** In diabetes, the significantly elevated concentration of free fatty acids being released from adipocytes in response to low insulin levels (see p. 187) promotes the hepatic synthesis of triacylglycerols. Dietary fats also contribute to the hypertriacylglycerolemia in diabetes, whereas in fasting, dietary fats are not an issue, and stored triacylglycerols are degraded only as needed by the body. Therefore, hypertriacylglycerolemia does not occur.

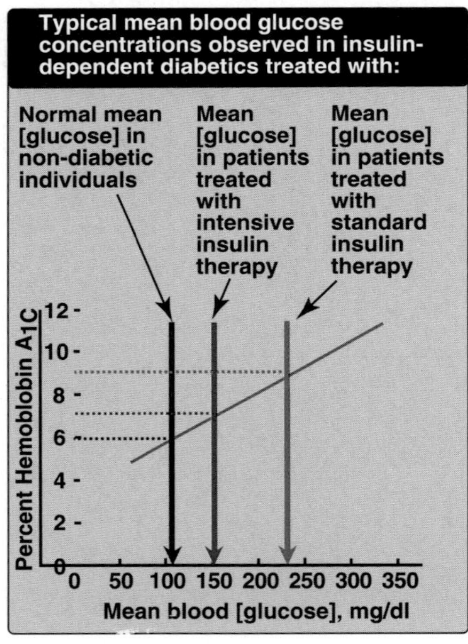

Figure 25.4
Correlation between mean blood glucose and HbA_{1C} in patients with type 1 diabetes.

C. Treatment of type 1 diabetes

Patients with type 1 diabetes have virtually no functional β cells, and can neither respond to variations in circulating fuels nor maintain a basal secretion of insulin. This type of diabetic must rely on **exogenous insulin** injected subcutaneously to control the hyperglycemia and ketoacidosis. Two therapeutic regimens are currently in use—standard and intensive insulin treatment.

1. **Standard treatment versus intensive treatment:** Standard treatment, which has as its therapetic goal the clinical well-being of the patient, typically consists of one or two daily injections of insulin. Mean blood glucose levels obtained are typically in the 225 to 275 mg/dl range, with an HbA_{1C} (see p. 34) of eight to nine percent of the total hemoglobin (blue arrow, Figure 25.4). [Note: The rate of formation of HbA_{1C} is proportional to the average blood glucose concentration over the previous several months. Thus, HbA_{1C} provides a measure of how well treatment has normalized blood glucose in the diabetic over time.] In contrast to standard therapy, intensive treatment seeks to normalize blood glucose through more frequent monitoring, and subsequent injections of insulin—typically three or more times a day. Mean blood glucose levels of 150 mg/dl can be achieved, with HbA_{1C} approximately seven percent of the total hemoglobin (red arrow, see Figure 25.4). [Note: Normal mean blood glucose is approximately 110 mg/dl and HbA_{1C} is six percent or less (black arrow, see Figure 25.4).] Thus, normalization of glucose values is not achieved even in intensively treated patients. Nonetheless, patients on intensive therapy showed a sixty percent reduction in the long-term complications of diabetes—retinopathy, nephropathy, and neuropathy—compared with patients receiving standard care. This confirms that the complications of diabetes are related to an elevation of plasma glucose.

2. **Hypoglycemia in type 1 diabetes:** One of the therapeutic goals of diabetes is to decrease blood glucose levels in an effort to minimize the development of the long-term complications of the disease (see p. 343 for a discussion of the chronic complications of diabetes). However, appropriate dosage is difficult to achieve in all patients, and hypoglycemia caused by excess insulin is the most common complication of insulin therapy, occurring in more than ninety percent of patients. The frequency of hypoglycemic episodes, coma, and seizures is particularly high with intensive treatment regimens designed to achieve tight control of blood glucose (Figure 25.5). Recall that in normal individuals hypoglycemia triggers a compensatory secretion of counterregulatory hormones, most notably glucagon and epinephrine, which promote hepatic production of glucose. However, patients with type 1 diabetes also develop a deficiency of glucagon secretion. This defect occurs early in the disease and is almost universally present four years after diagnosis. These patients thus rely on epinephrine secretion to prevent severe hypoglycemia. However, as the disease progresses, type 1 diabetes patients show diabetic autonomic neuropathy and impaired ability to secrete epinephrine in response to hypoglycemia. The combined deficiency of glucagon and epinephrine secretion creates a condition sometimes called "**hypoglycemia unawareness**." Thus, patients with longstanding diabetes are particularly vulnerable to hypoglycemia. Hypoglycemica can also be caused by strenous exercise. Exercise promotes glucose uptake into muscle and decreases the need for exogenous insulin. Patients should, therefore, check blood glucose levels before or after intensive exercise to prevent or abort hypoglycemia.

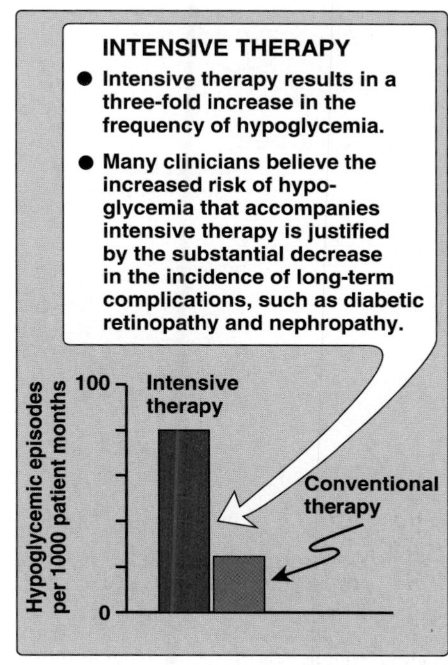

Figure 25.5
Effect of tight glucose control on hypoglycemic episodes in a population of patients on intensive therapy or conventional therapy.

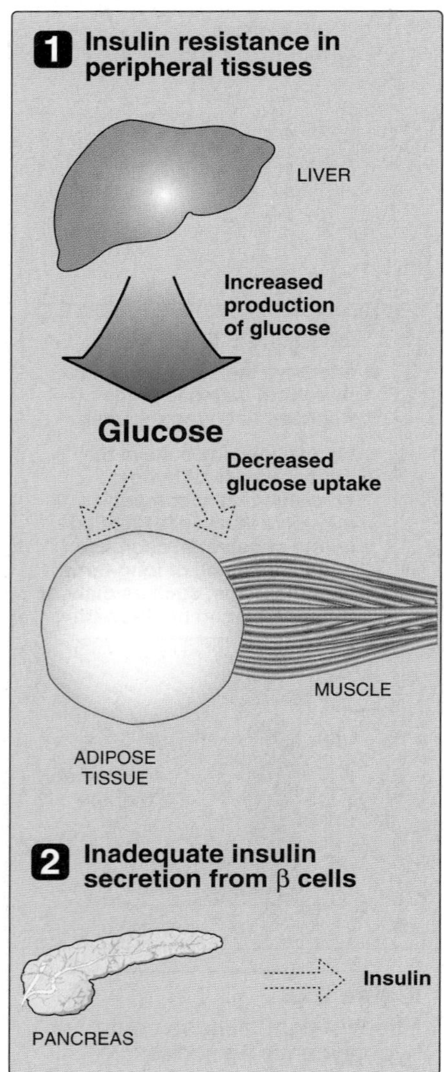

Figure 25.6
Major factors contributing to hyperglycemia observed in type 2 diabetes.

III. TYPE 2 DIABETES

Type 2 diabetes is the most common form of the disease, afflicting approximately ninety percent of the diabetic population in the United States. Typically, type 2 diabetes develops gradually without obvious symptoms. The disease is often detected by routine screening tests. However, many individuals with type 2 diabetes have symptoms of polyuria and polydipsia of several weeks duration. Polyphagia may be present, but is less common. Patients with type 2 diabetes have a combination of **insulin resistance** and **dysfunctional β cells** (Figure 25.6), but do not require insulin to sustain life, although insulin may be required to control hyperglycemia in some patients. The metabolic alterations observed in type 2 diabetes are milder than those described for the insulin-dependent form of the disease, in part, because insulin secretion in type 2 diabetes—although not adequate—does restrain ketogenesis and blunts the development of diabetic ketoacidosis. Diagnosis is based most commonly on the presence of hyperglycemia—that is, a blood glucose concentration of greater than 126 mg/dl. The occurrence of the type 2 disease is almost completely determined by **genetic factors** (see Figure 25.1). For example, in virtually all monozygotic twinships, the disease develops in both individuals. The disease does not involve viruses or autoimmune antibodies.

A. Insulin resistance

Insulin resistance is the decreased ability of target tissues, such as liver, adipose, and muscle, to respond properly to normal circulating concentrations of insulin. For example, insulin resistance is characterized by uncontrolled hepatic glucose production, and decreased glucose uptake by muscle and adipose tissue.

1. **Insulin resistance and obesity:** Obesity is the most common cause of insulin resistance. Most people with obesity and insulin resistance do not become diabetic. In the absence of a defect in β cell function, non-diabetic, obese individuals can compensate for insulin resistance with elevated levels of insulin. For example, Figure 25.7A shows that insulin secretion is two to three times higher in obese subjects than it is in lean individuals. This higher insulin concentration compensates for the diminished effect of the hormone (as a result of insulin resistance), and produces blood glucose levels similar to those observed in lean individuals (Figure 25.7B).

2. **Insulin resistance and type 2 diabetes:** Insulin resistance alone will not lead to type 2 diabetes. Rather, type 2 diabetes develops in insulin-resistant individuals who also show impaired β cell function. Insulin resistance and subsequent development of type 2 diabetes is commonly observed in the elderly, and in individuals who are obese, physically inactive, or in women who are pregnant. These patients are unable to sufficiently compensate for insulin resistance with increased insulin release. Figure 25.8 shows the time course for the develpment of hyperglycemia and the destruction of β cells.

3. **Causes of insulin resistance:** Insulin resistance increases with weight gain and, conversely, diminishes with weight loss. This

A Insulin level in blood
Higher insulin levels are required to control blood glucose in the insulin-resistant, obese individual.
Blood insulin rises from basal levels after each meal.
Insulin (μU/ml)
160
120
80
40
0
Obese
Normal
8
12 noon
4
8
12 midnight
4
B Glucose level in blood
Blood glucose is kept within the same narrow range thoughout the day in both normal weight and obese individuals.
Glucose (mg/dl)
160
140
120
100
80
Obese
Normal
8
12 noon
4
8
12 midnight
4

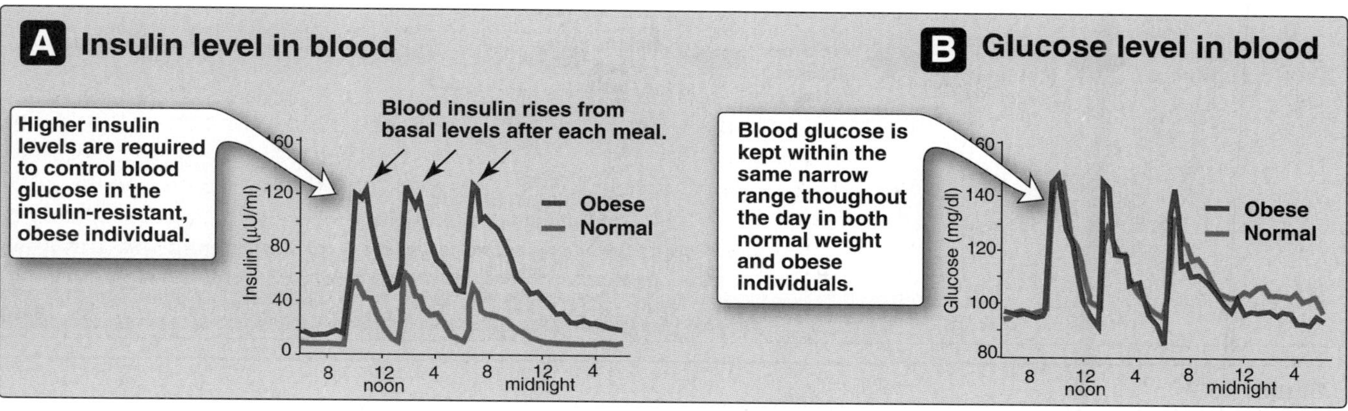

Figure 25.7
Blood insulin and glucose levels in normal weight and obese subjects.

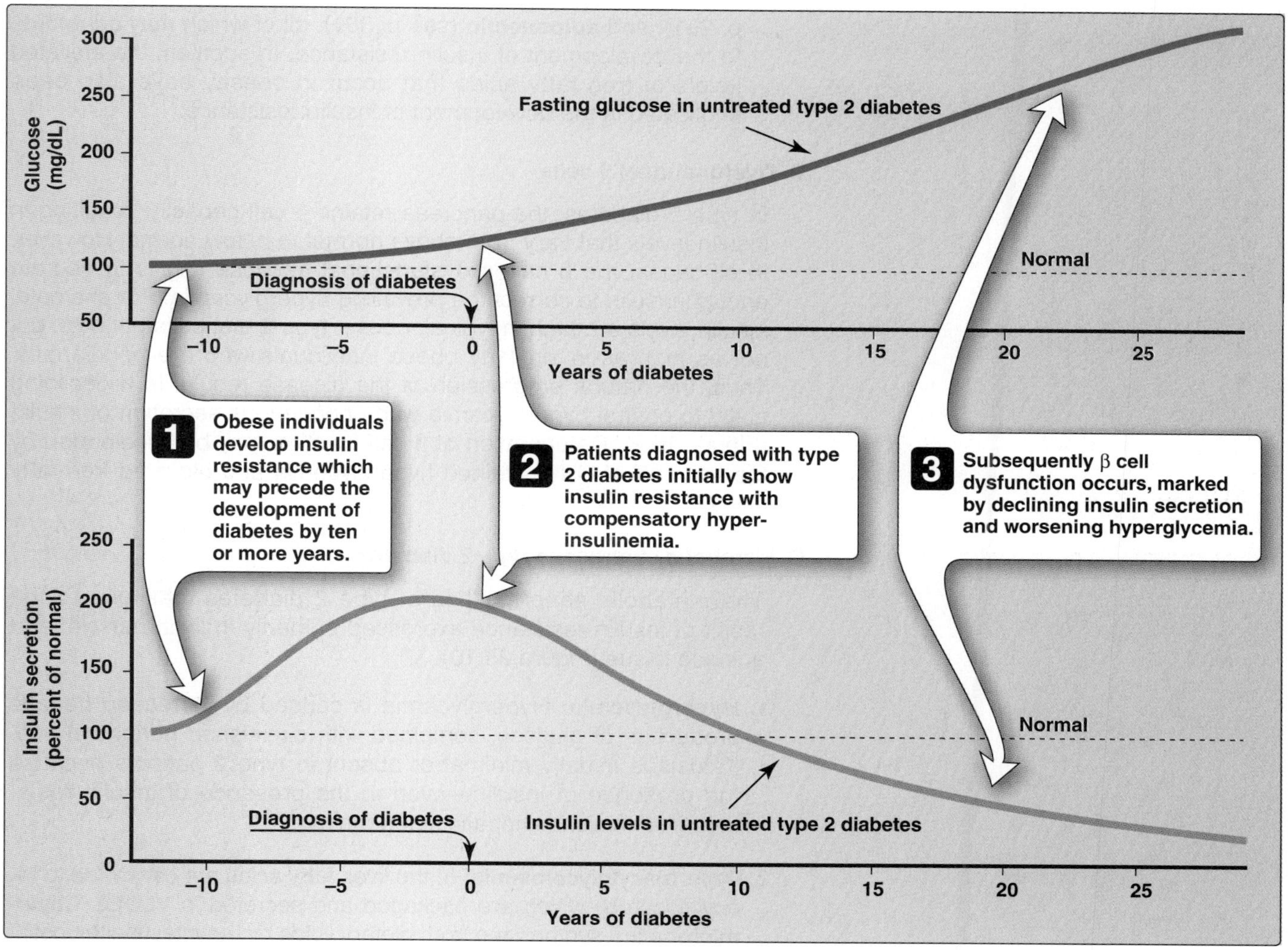

Glucose (mg/dL)
300
250
200
150
100
50
Fasting glucose in untreated type 2 diabetes
Normal
Diagnosis of diabetes
–10
–5
0
5
10
15
20
25
Years of diabetes
1 Obese individuals develop insulin resistance which may precede the development of diabetes by ten or more years.
2 Patients diagnosed with type 2 diabetes initially show insulin resistance with compensatory hyper-insulinemia.
3 Subsequently β cell dysfunction occurs, marked by declining insulin secretion and worsening hyperglycemia.
Insulin secretion (percent of normal)
250
200
150
100
50
0
Normal
Diagnosis of diabetes
Insulin levels in untreated type 2 diabetes
–10
–5
0
5
10
15
20
25
Years of diabetes

Figure 25.8
Progression of blood glucose and insulin levels in patients with type 2 diabetes.

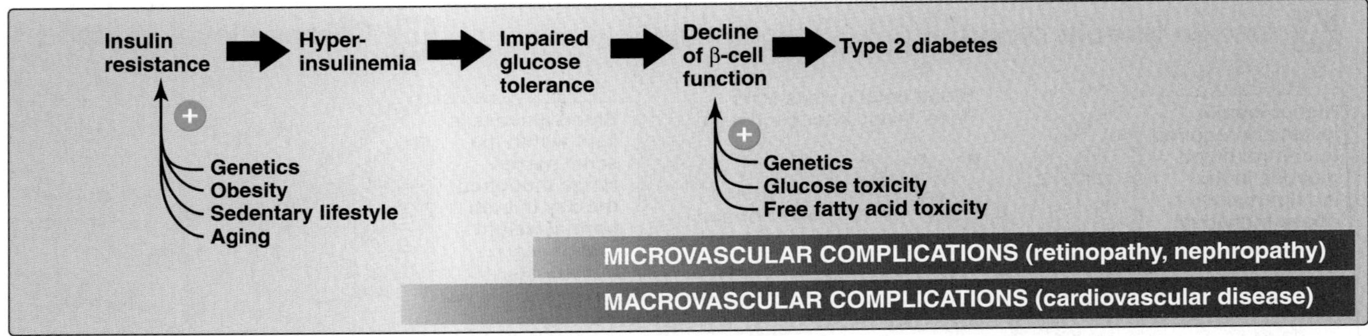

Figure 25.9
Typical progression of type 2 diabetes.

suggests that fat accumulation is important in the development of insulin resistance. Adipose tissue is not simply an energy storage organ, but also a secretory organ. Regulatory substances produced by adipocytes include **leptin** (see p. 350), **resistin** (see p. 351), and **adiponectin** (see p. 351), all of which may contribute to the development of insulin resistance. In addition, the elevated levels of free fatty acids that occur in obesity have also been implicated in the development of insulin resistance.

B. Dysfunctional β cells

In type 2 diabetes, the pancreas retains β cell capacity, resulting in insulin levels that vary from above normal to below normal However, in all cases, the β cell is dysfunctional because it fails to secrete enough insulin to correct the prevailing hyperglycemia. For example, insulin levels are high in typical, obese, type 2 diabetics patients, but not as high as in similarly obese individuals who are nondiabetic. Thus, the natural progression of the disease results in a declining abilty to control hyperglycemia with endogenous secretion of insulin (Figure 25.9). Deterioration of β cell function may be accelerated by the toxic effects of sustained hyperglycemia and elevated free fatty acids.

C. Metabolic changes in type 2 diabetes

The metabolic abnormalities of type 2 diabetes mellitus are the result of insulin resistance expressed primarily in liver, muscle, and adipose tissue (Figure 25.10).

1. **Hyperglycemia:** Hyperglycemia is caused by increased hepatic production of glucose, combined with diminished peripheral use. Ketosis is usually minimal or absent in type 2 patients because the presence of insulin—even in the presence of insulin resistance—diminishes hepatic ketogenesis.

2. **Hypertriacylglycerolemia:** In the liver, fatty acids are converted to triacylglycerols, which are packaged and secreted in **VLDLs**. **Chylomicrons** are synthesized from dietary lipids by the intestinal mucosal cells following a meal (see p. 175). Because lipoprotein degradation catalyzed by *lipoprotein lipase* in adipose tissue (see p. 226) is low in diabetics, the plasma chylomicron and VLDL levels are elevated, resulting in hypertriacylglycerolemia (see Figure 25.10).

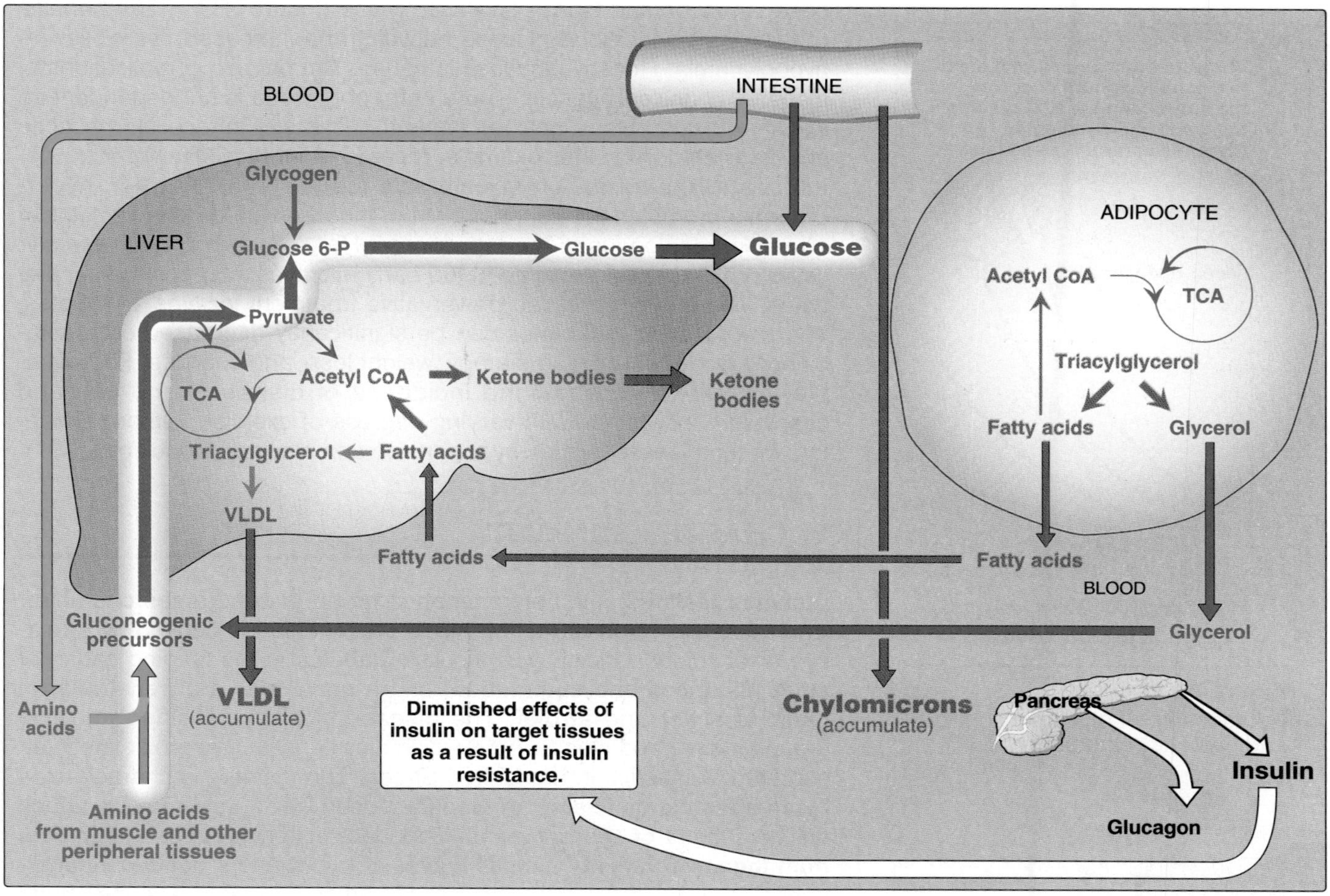

Figure 25.10
Intertissue relationships in type 2 diabetes.

D. Treatment of type 2 diabetes

The goal in treating type 2 diabetes is to maintain blood glucose concentrations within normal limits, and to prevent the development of long-term complications. Weight reduction, exercise, and dietary modifications often correct the hyperglycemia of type 2 diabetes. Hypoglycemic agents[1] or insulin therapy may be required to achieve satisfactory plasma glucose levels.

IV. CHRONIC EFFECTS AND PREVENTION OF DIABETES

As noted previously, available therapies moderate the hyperglycemia of diabetes, but fail to completely normalize metabolism. The long-standing elevation of blood glucose causes the chronic complications of diabetes—premature atherosclerosis, retinopathy, nephropathy, and neuropathy. Intensive treatment with insulin (see p. 339) delays the onset and slows the progression of these long-term complications. For example, the incidence of retinopathy decreases as control of blood glu-

[1]See Chapter 24 in ***Lippincott's Illustrated Reviews: Pharmacology*** (3rd ed.) and Chapter 26 (2nd ed.) for a discussion of the use of hypoglycemic agents in the treatment of diabetes.

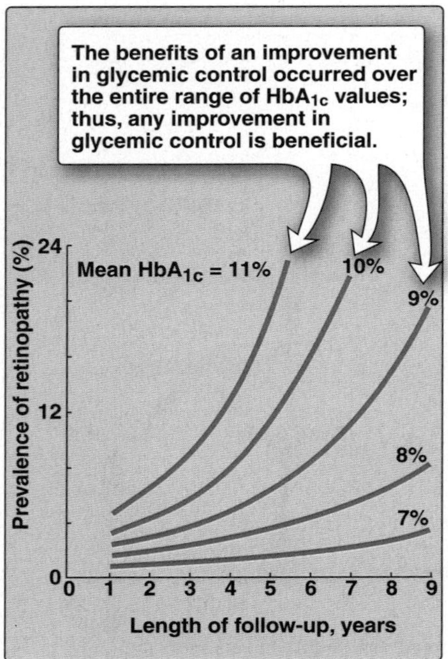

Figure 25.11
Relationship of glycemic control and diabetic retinopathy.

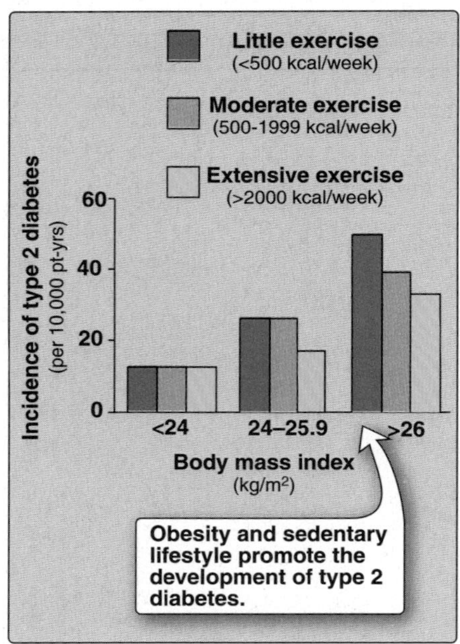

Figure 25.12
Effect of body weight and excercise on the development of type 2 diabetes.

cose improves and HbA_{1C} levels decrease (Figure 25.11).The benefits of tight control of blood glucose outweigh the increased risk of severe hypoglycemia. How hyperglycemia causes the chronic complications of diabetes is unclear. In cells where entry of glucose is not dependent on insulin, elevated blood glucose leads to increased intracellular glucose and its metabolites. For example, increased intracellular sorbitol contributes to the formation of cataracts (see p. 138). Further, hyperglycemia promotes the condensation of glucose with cellular proteins in a reaction analogous to the formation of HbA_{1C} (see p. 34). These glycated proteins mediate some of the early microvascular changes of diabetes. There is currently no preventative treatment for type 1 diabetes. The risk for type 2 diabetes can be significantly decreased by a combined regimen of nutrition therapy, weight loss, and exercise. For example, Figure 25.12 show the incidence of disease in normal and overweight individuals with varying degress of exercise. Other risk factors for the disease include hypertension and elevated blood lipids.

V. CHAPTER SUMMARY

Diabetes mellitus is a heterogeneous group of syndromes characterized by an **elevation of fasting blood glucose** that is caused by a relative or absolute deficiency in insulin. Diabetes is the leading cause of **adult blindness** and **amputation**, and a major cause of **renal failure**, **heart attacks**, and **stroke**. The disease can be classified into two groups, type 1 and type 2. **Type 1** diabetics constitute approximately ten percent of diabetics in the United States. The disease is characterized by an **absolute deficiency of insulin** caused by an **autoimmune attack** on the **β cells of the pancreas**. This destruction requires a **stimulus from the environment** (such as a viral infection) and a **genetic determinant** that allows the β cell to be recognized as "non-self." The **metabolic abnormalities** of type 1 diabetes mellitus include **hyperglycemia**, **ketoacidosis**, and **hypertriglyceridemia**. They result from a deficiency of insulin and a relative excess of glucagon. Type 1 diabetics must rely on **exogenous insulin** injected subcutaneously to control hyperglycemia and ketoacidosis. **Type 2** diabetes has a strong **genetic** component. It results from a combination of **insulin resistance** and **dysfunctional β cells**. Insulin resistance is the decreased ability of target tissues, such as liver, adipose tissue and muscle, to respond properly to normal circulating concentrations of insulin. **Obesity** is the most common cause of insulin resistance. However, mostf people with obesity and insulin resistance do not become diabetic. In the absence of a defect in β cell function, **non-diabetic**, **obese individuals** can compensate for insulin resistant with **elevated levels of insulin**. Insulin resistance alone will not lead to type 2 diabetes. Rather, type 2 diabetes develops in insulin-resistant individuals who also show impaired β cell function. The **metabolic alterations** observed in type 2 diabetes are **milder** than those described for the insulin-dependent form of the disease, in part, because insulin secretion in type 2 diabetes—although not adequate—does restrain ketogenesis and blunts the development of diabetic ketoacidosis. Available treatments for diabetes moderate the hyperglycemia, but fail to completely normalize metabolism. The long-standing elevation of blood glucose causes the **chronic complications of diabetes—premature atherosclerosis**, **retinopathy**, **nephropathy**, and **neuropathy**.

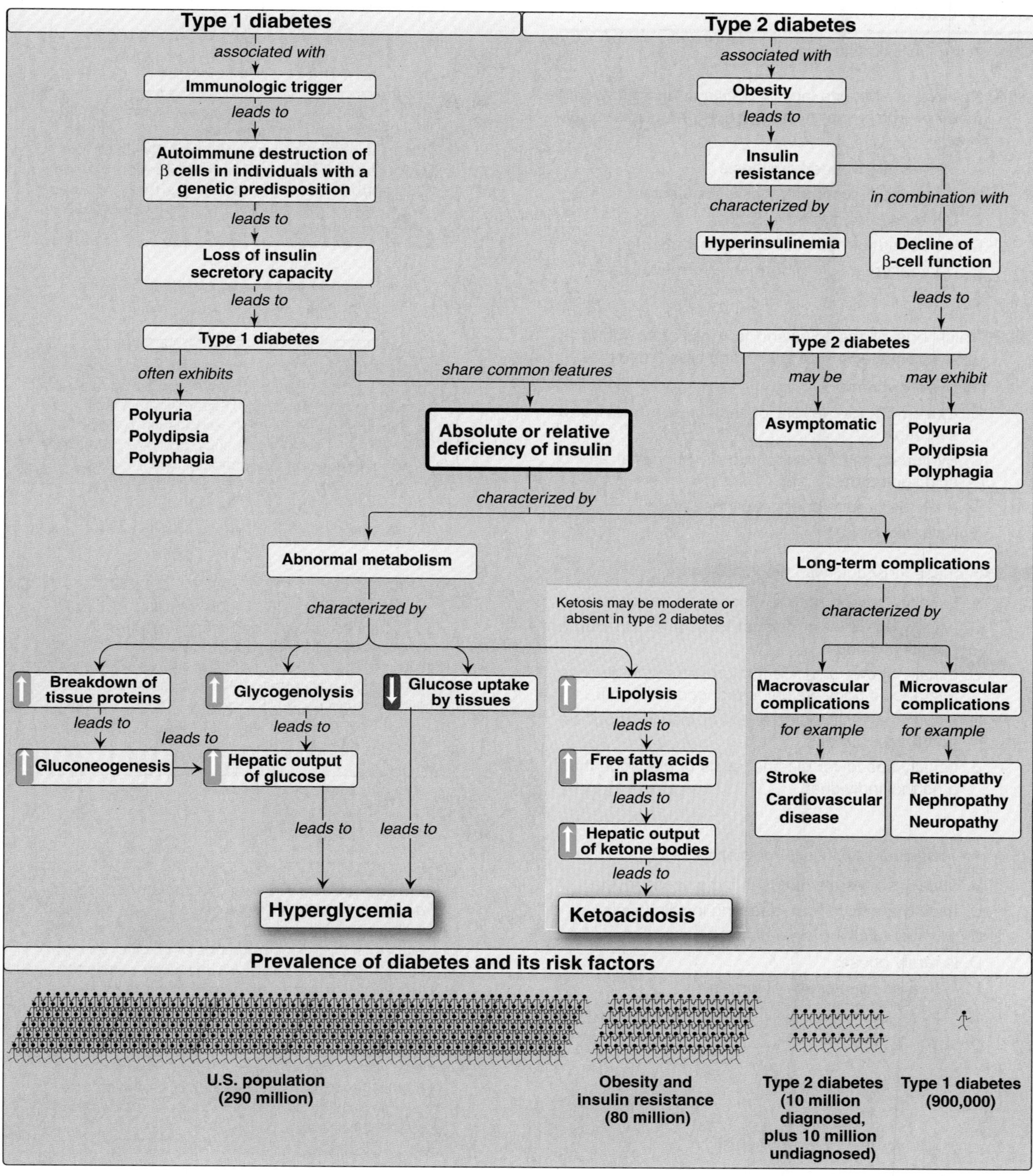

Figure 25.13
Key concept map for diabetes.

Study Questions

Choose the ONE correct answer

25.1 Relative or absolute lack of insulin in humans would result in which one of the following reactions in the liver?

A. Increased glycogen synthesis
B. Decreased gluconeogenesis from lactate
C. Decreased glycogenolysis
D. Increased formation of 3-hydroxybutyrate
E. Decreased action of hormone-sensitive lipase

Correct answer = D. Low insulin levels favor the liver producing ketone bodies, using acetyl CoAs it obtained from excess fatty acids provided by the adipose. Low insulin also causes activation of hormone-sensitive lipase, decreased glycogen synthesis, and increased gluconeogenesis.

25.2 Which one of the following is most often found in untreated patients with type 1 and type 2 diabetes?

A. Hyperglycemia
B. Extremely low levels of insulin synthesis and secretion
C. Synthesis of an insulin with an abnormal amino acid sequence
D. A simple pattern of genetic inheritance
E. Ketoacidosis

Correct answer = A. Elevated blood glucose occurs in type 1 diabetes as a result of a lack of insulin. In type 2 diabetes, hyperglycemia is due to a defect in β cell function and insulin resistance. Both forms of the disease show complex genetics. Ketoacidosis is more common in type 1 disease.

25.3 An obese individual with type 2 diabetes:

A. usually shows a normal glucose tolerance test.
B. usually has a lower plasma level of insulin than a normal individual.
C. usually shows significant improvement in glucose tolerance if body weight is reduced to normal.
D. usually benefits from receiving insulin about six hours after a meal.
E. usually has lower plasma levels of glucagon than a normal individual.

Correct answer = C. Eighty percent of type 2 diabetics are obese, and almost all show some improvement in blood glucose with weight reduction. These patients show an abnormal glucose tolerance test, have elevated insulin levels, and usually do not require insulin (certainly not six hours after a meal). Glucagon levels are typically normal.

25.4 An individual with insulin resistance:

A. usually shows elevated fasting glucose levels.
B. usually shows elevated fasting insulin levels.
C. will eventually become diabetic.
D. is rarely obese.
E. is treated by injection of insulin.

Correct answer = B. Insulin resistance is the decreased ability of target tissues, such as liver, adipose, and muscle, to respond properly to normal circulating concentrations of insulin. Obesity is the most common cause of insulin resistance. Most of the people with obesity and insulin resistance do not become diabetic. In the absence of a defect in β cell function, non-diabetic, obese individuals can compensate for insulin resistant with elevated levels of insulin. The elevated insulin levels normalize fasting blood glucose levels. Insulin resistance without overt diabetes requires no treatment.

Obesity

26

I. OVERVIEW

Obesity is a disorder of body weight regulatory sysems characterized by an accumulation of excess body fat. In primitive societies, in which daily life required a high level of physical activity and food was only available intermittently, a genetic tendency favoring storage of excess calories as fat had a survival value. However, the current abundance of food has encouraged Americans to eat more (Figure 26.1). This, in combination with reduced activity levels found in industrialized societies, has resulted in a tendency for the sustained deposition of fat. For example, this has caused an epidemic of obesity in the United States (Figure 26.2). The prevalence of obesity increases with age and is more common among poorer persons and in individuals with only a high school education or less. Particularly alarming is the explosion of childhood obesity, which has shown a three-fold increase in prevalence over the last two decades. Obesity is not limited to the United States, but rather has increased globally. In fact, by some estimates, there are more obese than undernourished individuals worldwide.

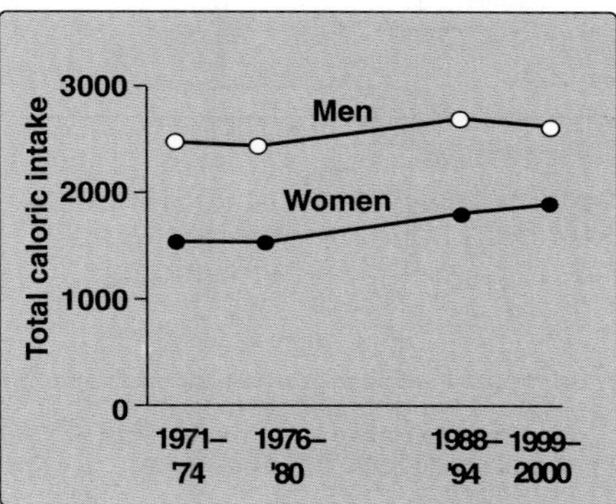

Figure 26.1
Total caloric consumption in adults.

II. ASSESSMENT OF OBESITY

The amount of body fat is difficult to measure directly, and is usually determined from an indirect measure—the **body mass index (BMI)**—which has been shown to correlate with the amount of body fat in most individuals. (Notable exceptions are athletes who have large amounts of lean muscle mass.)

A. The body mass index

BMI gives a measure of relative weight, adjusted for height. This allows comparisons both within and between populations. The BMI is calculated in both men and women as follows:

$$BMI = (\text{weight in kg})/(\text{height in meters})^2$$

$$= (\text{weight in lb})/(\text{height in inches})^2 \times 703$$

The healthy range for the BMI is between 19.5 and 25.0. Individuals with a BMI **between 25 and 29.9** are considered **overweight**, those whose BMI is **equal to or greater than 30** are defined as **obese**.

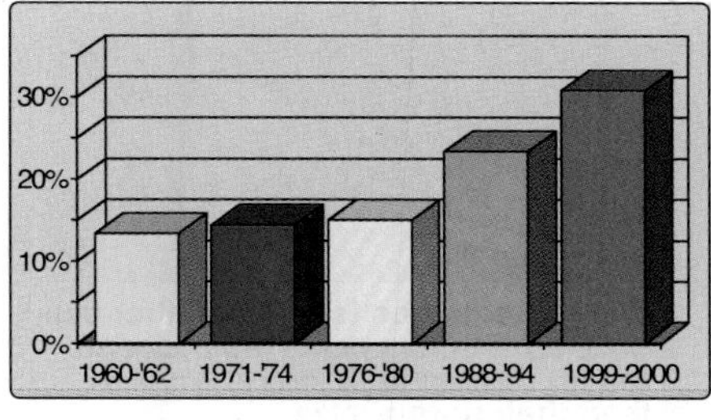

Figure 26.2
Prevalence of obesity in adults, ages 20 to 74 years, in the United States.

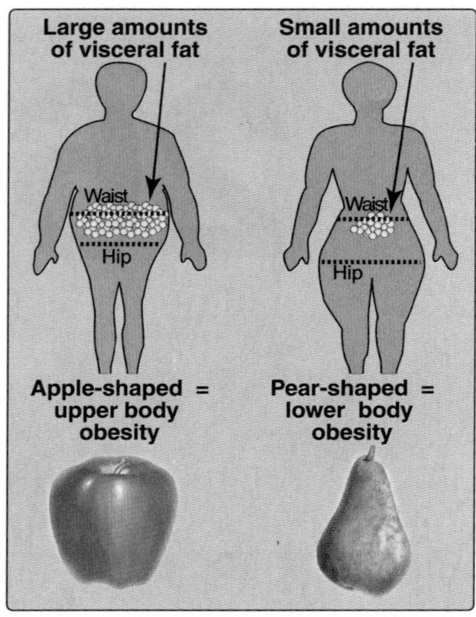

Figure 26.3
Individuals with more upper body fat (left) have greater health risks than pear-shaped individuals (right).

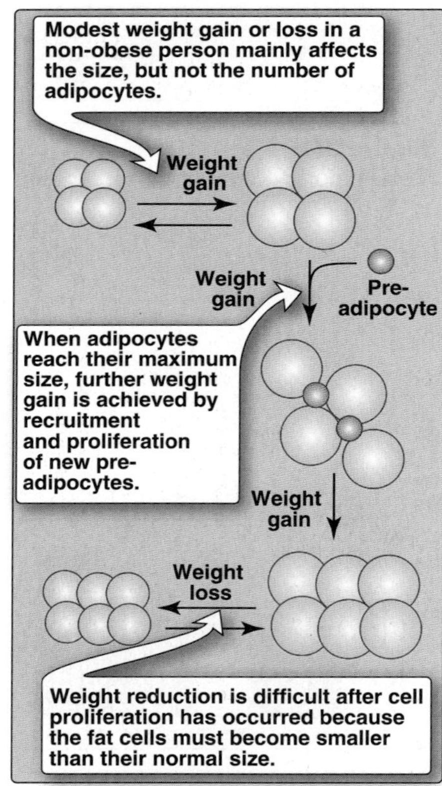

Figure 26.4
Hypertrophic and hyperplastic changes thought to occur in severe obesity.

Nearly two-thirds of American adults are overweight, and more than thirty percent are obese (Figure 26.2).

B. Anatomic differences in fat deposition

The anatomic distribution of body fat has a major influence on associated health risks. Excess fat located in the central abdominal area of the body is called **android**, "**apple-shaped**," or **upper body obesity** (Figure 26.3), and is associated with a greater risk for hypertension, insulin resistance, diabetes, dyslipidemia, and coronary heart disease (see p. 352). It is defined as a waist to hip ratio of more than 0.8 for women and more than 1.0 for men. In contrast, fat distributed in the lower extremities around the hips or gluteal region is call **gynoid**, "**pear-shaped**," or **lower body obesity.** It is defined as a waist to hip ratio of less than 0.8 for women and less than 1.0 for men. The pear shape is relatively benign healthwise, and is commonly found in females. Alternatively, some experts feel that waist circumference is better predictor of risk. Values greater than 40 inches in men and greater than 35 inches in women are at increased risk for coronary artery disease.

C. Biochemical differences in regional fat depots

The regional types of fat described above are biochemically different. Abdominal fat cells are much larger and have a higher rate of fat turnover than do lower body fat cells. The abdominal adipocytes are also hormonally more responsive than are fat cells in the legs and buttocks. Because men tend to accumulate the readily mobilizable abdominal fat, they generally lose weight more readily than do women, who accumulate lower body fat. Further, substances released from abdominal fat are absorbed via the portal vein and, thus, have direct access to the liver. Fatty acids taken up by the liver may lead to insulin resistance (see p. 340) and increased synthesis of triacylglycerols, which are released as VLDL (see p. 351). By contrast, free fatty acids from gluteal fat enter the general circulation, and have no preferential action on hepatic metabolism.

D. Number of fat cells

When triacylglycerols are deposited in adipocytes, the cells initially show a modest increase in size (Figure 26.4). However, the ability of a fat cell to expand is limited, and when its maximal size is reached, it divides. Most obesity is, therefore, thought to involve an increase in both the number and size of adipocytes. Fat cells, once gained, are never lost. Thus, when an obese individual loses weight, the size of the fat cells is reduced, but the number of fat cells is not affected. An obese individual, with increased numbers of adipocytes, will have to reduce the size of those fat cells in order to normalize fat stores. These individuals will be in the doubly abnormal state of having too many, too small fat cells. This, in part, explains why formerly obese patients have a particularly difficult time maintaining their reduced body weight. The observation that fat cells are never lost emphasizes the importance of preventing obesity in the first place.

III. BODY WEIGHT REGULATION

The body weight of most individuals tends to range within ten percent of a set value. This observation prompted the theory that each individual has a biologically predetermined "**set point**" for body weight. The body attempts to add adipose tissue when the body weight falls below the set point, and to lose weight when the body weight is higher than the set point. For example, with weight loss, appetite increases and energy expenditure falls, whereas with overfeeding, appetite falls and energy expenditure increases (Figure 26.5). However, a strict set point model does not explain why some individuals fail to revert to their starting weight after a period of overeating, or the current epidemic of obesity. Body weight, rather than being irrevocably set, seems to drift around a natural "**settling point**," which reflects a balance between factors that influence food intake and energy expenditure. In this context, body weight is stable as long as the behavorial and environmental factors that influence energy balance are constant.

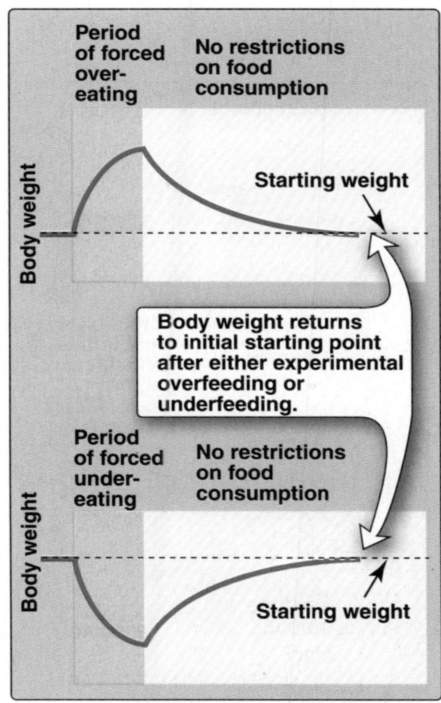

Figure 26.5
Weight changes following episodes of overfeeding or underfeeding followed by feeding with no restrictions.

A. Genetic contributions to obesity

Despite the widely held belief that obesity is a result of uncontrolled, gluttonous eating behavior, it is now evident that genetic mechanisms play a major role in determining body weight, rather than a lack of will power. For example, obesity is often observed clustered in families. If both parents are obese, there is a seventy to eighty percent chance of the children being obese. In contrast, only nine percent of children were fat when both parents were lean. The inheritance of obesity is not simple Mendelian genetics as would be expected if the disease were a result of a defect in a single gene. Rather, obesity behaves as a complex polygenic disease involving interactions between multiple genes and the environment. The importance of genetics as a determinant of obesity is also indicated by the observation that children who are adopted usually show a body weight that correlates with their biologic rather than adoptive parents Further, identical twins have very similar BMIs (Figure 26.6), whether reared together or apart, and their BMIs are more similar than those of nonidentical, dizygotic twins.

Figure 26.6
Identical twins with combined weight of 1300 pounds. Note similarity in body shape.

B. Environmental and behavioral contributions

The epidemic of obesity occurring over the last decade cannot be explained by changes in genetic factors, which are stable on this short time scale. Clearly, environmental factors, such as the ready availability of palatable, energy-dense foods, plays a role in the increased prevalence of obesity. Further, sedentary lifestyles encouraged by TV watching, automobiles, computer usage, and energy-sparing devices in the workplace and at home, decrease physical activity and enhance the tendency to gain weight. The importance of lifestyle in the development of obesity is reinforced by the observation that when Japanese or Chinese populations migrate to the United States, their BMI increases. For example, men in Japan (aged 46 to 49 years) are lean, with an average BMI of 20, whereas Japanese men of the same age living in California are heavier, with an average BMI of 24. Eating behaviors, such as snacking, portion size, variety of foods consumed, an individual's unique food preferences, and the number of people with whom one eats also influence food consumption and the tendency toward obesity.

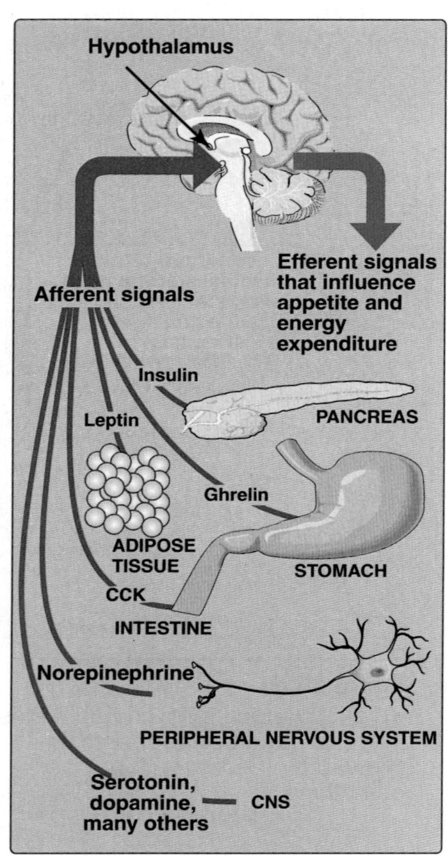

Figure 26.7
Some afferent signals reflecting the nutritional state of the body. CCK = cholecystokinin.

IV. MOLECULES THAT INFLUENCE OBESITY

The cause of obesity can be summarized in a deceptively simple statement of the first law of thermodynamics: obesity results when energy intake exceeds energy expenditure. However, unraveling the mechanism underlying this imbalance involves a complex interaction of biochemical, neurologic, environmental, and psychological factors. For example, appetite is influenced by afferent, or incoming signals—neural signals, circulating hormones, and metabolites—that impinge on the hypothalamus (Figure 26.7). These diverse signals prompt release of hypothalamic peptides, and activate outgoing efferent neural signals. Some important **afferent signaling molecules** regulating appetite and energy consumption include:

A. Hormones of adipose tissue

Although the adipocyte's primary role is to store fat, it also functions as an endocrine cell that releases numerous regulatory molecules, such as leptin, adiponectin, and resistin.

1. **Leptin:** Studies of the molecular genetics of mouse obesity have led to the isolation of at least six genes associated with obesity. The most well-known mouse gene, named Ob (for obesity), leads to severe hereditary obesity in mice. It has been identified and cloned. In one strain of fat mice, the gene was completely absent, indicating that the gene's protein product is required to keep the animals' weight under control. The product of the Ob gene is a hormone called leptin. Leptin is produced proportionally to the adipose mass and, thus, informs the brain of the fat store level (Figure 26.8). It is secreted by fat cells, and acts on the hypothalamus of the brain to regulate the amount of body fat through the control of appetite and energy expenditure. Leptin's secretion is suppressed by depletion of fat stores (starvation) and enhanced by expansion of fat stores (well-fed state). Daily injection of leptin

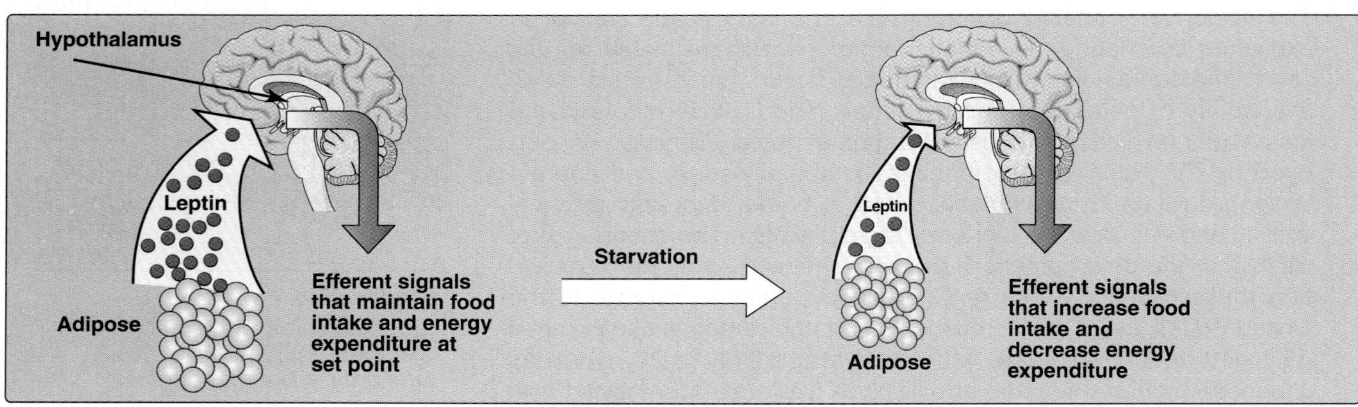

Figure 26.8
Action of leptin in maintaining adequate fat stores.

causes overweight mice to lose weight and maintain weight loss. The protein also causes weight loss in mice that are not obese. In humans, leptin increases the metabolic rate and decreases appetite. However, plasma leptin in obese humans is usually normal for their fat mass, suggesting that resistance to leptin, rather than its deficiency, occurs in human obesity. The receptor for leptin in the hypothalamus has been cloned and is produced by a gene known as db. In rodents, mutation in the db gene produces leptin resistance. However, the mutations thus far described in rodents do not appear to account for most human obesity. Therefore, current research is focused on other possible defects in leptin signal transduction in humans. Other hormones released by adipose tissue, such as **adiponectin** and **resistin**, may mediate insulin resistance observed in obesity.

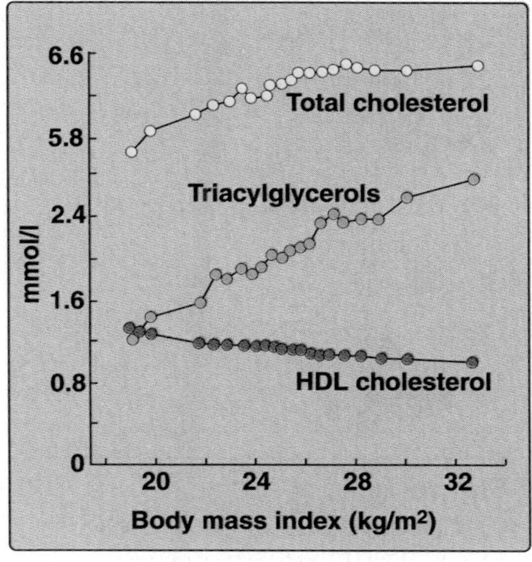

Figure 26.9
Body mass index and changes in blood lipids.

B. Other hormones influencing obesity

Ghrelin, a peptide secreted primarily by the stomach, is the only known appetite-stimulating hormone. Injection of ghrelin increases short-term food intake in rodents, and may decrease energy expenditure and fat catabolism. Peptides, such as **cholecystokinin (CCK),** released from the gut following ingestion of a meal can act as satiety signals to the brain. **Insulin** not only influences metabolism, but also promotes decreased energy intake.

V. METABOLIC CHANGES OBSERVED IN OBESITY

The metabolic abnormalities of obesity reflect molecular signals originating from the increased mass of adipocytes. The predominant effects of obesity include dyslipidemias, glucose intolerance, and insulin resistance, expressed primarily in the liver, muscle, and adipose tissue.

A. Metabolic syndrome

Abdominal obesity is associated with a threatening combination of metabolic abnormalities that includes glucose intolerance, insulin resistance, hyperinsulinemia, dyslipidemia (low HDL and elevated VLDL), and hypertension. This clustering of metabolic abnormalities has been referred to as **syndrome X**, the **insulin resistance syndrome**, or the **metabolic syndrome**. Individuals with this syndrome have a significantly increased risk for developing diabetes mellitus and cardiovascular disorders. For example, men with the syndrome are three to four times more likely to die of cardiovascular disease.

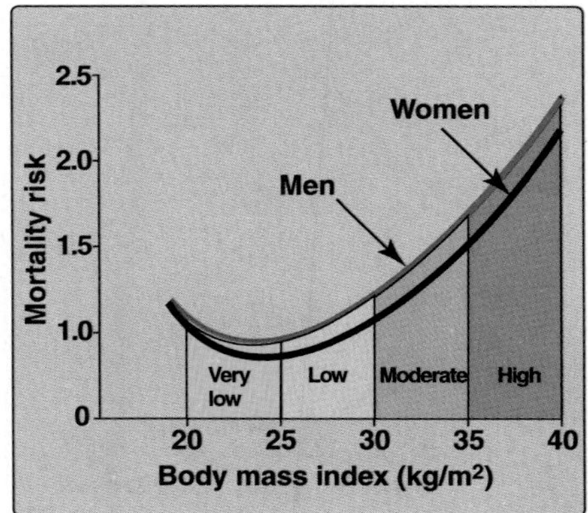

Figure 26.10
Body mass index and the relative risk of death.

B. Dyslipidemia

Insulin resistance in obese individuals leads to increased production of insulin in an effort by the body to maintain blood glucose levels. Insulin resistance in adipose tissue causes increased activity of *hormone-sensitive lipase*, resulting in increased levels of circulating fatty acids. These fatty acids are carried to the liver and converted to triacyglycerol and cholesterol. Excess triacyglycerol and cholesterol are released as VLDL, resulting in elevated serum triacylglycerols (Figure 26.9). Concomitantly, HDL levels are decreased.

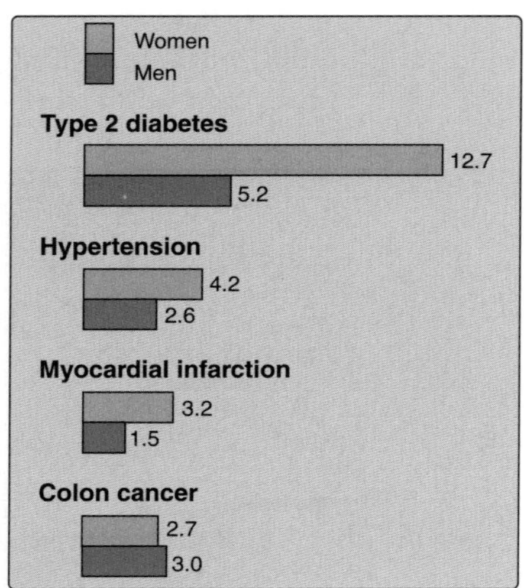

Figure 26.11
The relative risk of developing associated diseases in obese women and men compared with non-obese individuals in which the risk = 1.0.

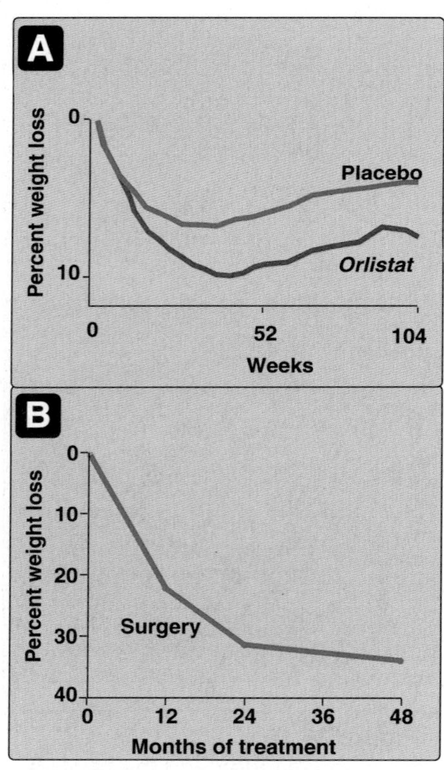

Figure 26.12
Effect of orlistat treatment and surgery on weight loss.

VI. OBESITY AND HEALTH

Obesity is correlated with an increased risk of death (Figure 26.10), and is a risk factor for a number of chronic conditions, including adult onset diabetes, hypercholesterolemia, high plasma triacylglycerols, hypertension, heart disease, some cancers, gallstones, arthritis, and gout (Figure 26.11). The relationship between obesity and associated morbidities is stronger among individuals younger than 55 years. After age 74, there is no longer an association between increased BMI and mortality. Weight loss in obese individuals leads to decreased blood pressure, serum triacylglycerols, and blood glucose levels. HDL levels increase. Mortality decreases, particularly deaths due to cancer. Some obesity experts suggest that moderately overweight and otherwise healthy individuals should not obsess about weight loss, but rather should direct their energies to a healthier lifestyle, particularly including some exercise in their weekly routine. The increased mortality associated with individuals who are only moderately overweight but otherwise healthy may result from a sedentary lifestyle that is associated with obesity. For example, unfit, lean men with BMIs of 25 or less have twice the risk of mortality from all causes than fit overweight men with BMIs of 27.8 or greater.

VII. WEIGHT REDUCTION

The goals of weight management in the obese patient are first, to induce a negative energy balance to reduce body weight, that is, decrease caloric intake and/or increase energy expenditure. The second aim is to maintain a lower body weight over the longer term.

A. Physical activity

An increase in physical activity can create an energy deficit, and is an important component of weight loss treatments. In addition, physical activity increases cardiorespiratory fitness and reduces the risk of cardiovascular disease, independent of weight loss. Persons who combine caloric restriction and exercise with behavioral treatment may expect to lose about five to ten percent of preintervention body weight over a period of four to six months. Exercise is an essential component of maintaining weight reduction.

B. Caloric restriction

Dieting is the most commonly practiced approach to weight control. Since one pound of adipose tissue corresponds to approximately 3500 kcal, one can estimate the effect of caloric restriction on the reduction in adipose tissue. Weight loss on calorie-restricted diets is determined primarily by energy intake and not nutrient composition. Caloric restriction is ineffective over the long term for many individuals. More than ninety percent of people who attempt to lose weight regain the lost weight when dietary intervention is suspended. Nonetheless, it is important to recognize that, although few individuals will reach their ideal weight with treatment, weight losses of ten percent of body weight over a six-month period often reduce blood pressure and lipid levels, and enhance control of type 2 diabetes. The health benefits of relatively small weight losses should, therefore, be emphasized to the the patient.

C. Pharmacologic and surgical treatment

Two weight-loss medications are currently approved by the FDA for use in adults who have a BMI of 30 or higher. The first, **sibutramine**[1] is an appetite suppressant that inhibits the reuptake of both serotonin and norepinephrine. The second, **orlistat**,[2] is a lipase inhibitor that inhibits gastric and pancreatic lipases, thus decreasing the breakdown of dietary fat into smaller molecules (Figure 26.12A). Surgical procedures designed to reduce food consumption are an option for the severely obese patient who has not responded to other treatments (Figure 26.12B). Surgery produces greater and more sustained weight loss than dietary or pharmacologic therapy, but has substantial risks for complications.

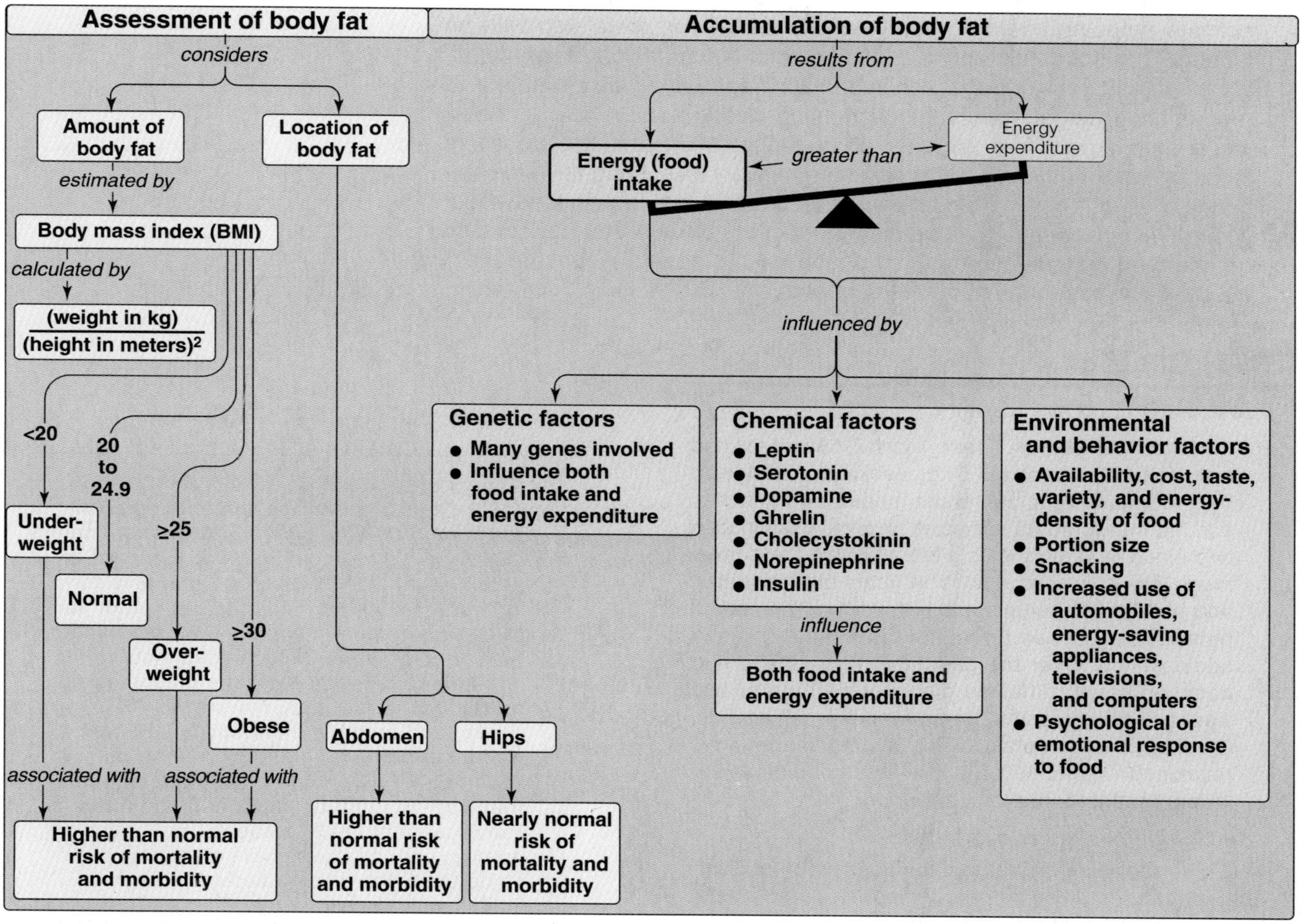

Assessment of body fat

considers

Amount of body fat

estimated by

Body mass index (BMI)

calculated by

$$\frac{\text{(weight in kg)}}{\text{(height in meters)}^2}$$

<20 → Underweight

20 to 24.9 → Normal

≥25 → Overweight

≥30 → Obese

Underweight *associated with* Higher than normal risk of mortality and morbidity

Overweight, Obese *associated with* Higher than normal risk of mortality and morbidity

Location of body fat

Abdomen → Higher than normal risk of mortality and morbidity

Hips → Nearly normal risk of mortality and morbidity

Accumulation of body fat

results from

Energy (food) intake ← *greater than* → Energy expenditure

influenced by

Genetic factors

- Many genes involved
- Influence both food intake and energy expenditure

Chemical factors

- Leptin
- Serotonin
- Dopamine
- Ghrelin
- Cholecystokinin
- Norepinephrine
- Insulin

influence

Both food intake and energy expenditure

Environmental and behavior factors

- Availability, cost, taste, variety, and energy-density of food
- Portion size
- Snacking
- Increased use of automobiles, energy-saving appliances, televisions, and computers
- Psychological or emotional response to food

Figure 26.13
Key concept map for obesity.

INFO LINK [1,2]See Chapter 28 in ***Lippincott's Illustrated Reviews: Pharmacology*** (3rd Ed.) and Chapter 42 (2nd Ed.) for a more detailed discussion of drugs used to treat obesity.

VIII. CHAPTER SUMMARY

Obesity—the accumulation of excess body fat—results when energy intake exceeds energy expenditure. Obesity is increasing in industrialized countries because of a reduction in daily energy expenditure, and an increase in energy intake resulting from the increasing availability of palatable, energy dense food. The **body mass index (BMI)** is a measure of body fat. Nearly two thirds of American adults are **overweight** (BMI >25 kg/m^2) and more than thirty percent are **obese** (BMI >30 kg/m^2). Excess fat located in the central abdominal area of the body is associated with greater risk for hypertension, insulin resistance, diabetes, dyslipidemia and coronary heart disease than is fat located in the hip and thighs. The body attempts to add adipose tissue when the body weight falls below a **set point**, and to lose weight when the body weight is higher than the set point, The **weight** is determined by **genetic** and **environmental factors**. Appetite is influenced by afferent, or incoming signals—neural signals, circulating hormones, and metabolites—that impinge on the **hypothalamus**. These diverse signals prompt release of hypothalamic peptides and activate outgoing efferent neural signals. Obesity is correlated with an increased risk of death, and is a risk factor for a number of chronic conditions. **Weight reduction** is achieved with **negative energy balance** to reduce body weight, that is, by decreasing caloric intake and/or increasing energy expenditure. Virtually all diets that limit particular groups of foods or macronutrients lead to short-term weight loss. Long-term maintenance of weight loss is difficult to achieve. Modest reduction in food intake occurs with pharmacologic treatment. **Surgical procedures** designed to reduce food consumption are an option for the severely obese patient who has not responded to other treatments.

Study Question

Choose the ONE correct answer

26.1 A 40-year-old woman, 5 feet, 1 inch (155 cm) tall and weighing 188 pounds (85.5 kg), seeks your advice on how to lose weight. Her waist measured 41 inches and hip measured 39 inches. A physical examination and blood laboratory data were all within the normal range. Her only child, who is 14 years old, her sister, and both of her parents are overweight. The patient recalls being obese throughout her childhood and adolescence. Over the past fifteen years she had been on seven different diets for periods of two weeks to three months, losing from 5 to 25 pounds. On discontinuation of each diet, she regained weight, returning to 185 to 190 pounds. Which one of the following best describes this patient?

A. She is classified as overweight.

B. She shows an "apple" (android) pattern of fat distribution.

C. She has approximately the same number of fat cells as a normal weight individual, but each adipocyte is larger.

D. She would be expected to show lower than normal levels of circulating leptin.

E. She would be expected to show lower than normal levels of circulating triacylglycerols.

Correct answer = B. Her waist to hip ratio is 41/39 = 1.05. Apple shape is defined as a waist to hip ratio of more than 0.8 for women, and more than 1.0 for men. She has, therefore, an apple pattern of fat distribution, commonly seen in males. Compared with other women of the same body weight who have a gynecoid fat pattern, the presence of increased visceral or intraabdominal adipose tissue places her at greater risk for diabetes, hypertension, dyslipidemia, and coronary heart disease. For this patient BMI = weight (kg)/height (m^2) = 85.5/(1.55)^2 = 35.6 kg/m^2. The result indicates that the patient is classified as obese. Individuals with marked obesity and a history dating to early childhood, have an adipose depot made up of too many adipocytes, each fully loaded with triacyglycerols. Plasma leptin in obese humans is usually normal for their fat mass, suggesting that resistance to leptin, rather than its deficiency, occurs in human obesity. The elevated circulating fatty acids characteristic of obesity are carried to the liver and converted to triacyglycerol and cholesterol. Excess triacyglycerol and cholesterol are released as VLDL, resulting in elevated serum triacylglycerols.

Nutrition

27

I. OVERVIEW

Nutrients are the constituents of food necessary to sustain the normal functions of the body. All energy is provided by three classes of nutrients: fats, carbohydrates, protein, and in some diets, ethanol (Figure 27.1). The intake of these energy-rich molecules is larger than that of the other dietary nutrients. Therefore, they are called the **macronutrients**. This chapter focuses on the kinds and amounts of macronutrients that are needed to maintain optimal health and prevent chronic disease in adults. Those nutrients needed in lesser amounts, such as vitamins and minerals, are called the **micronutrients**, and are considered in Chapter 28.

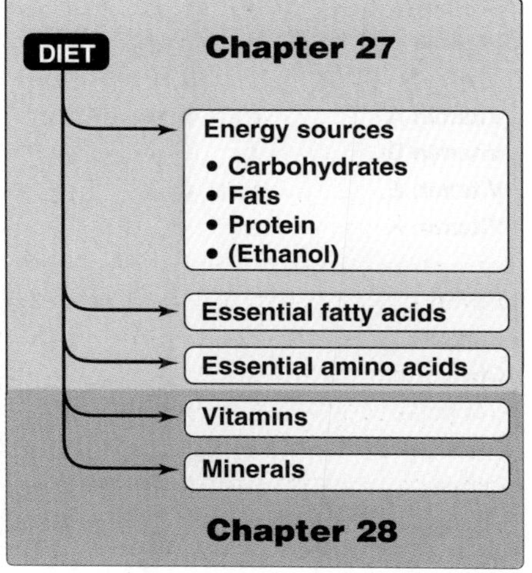

Figure 27.1
Essential nutrients obtained from the diet. [Note: Ethanol is not considered an essential component of the diet but may provide a significant contribution to the daily caloric intake of some individuals.]

II. DIETARY REFERENCE INTAKES

Committees of experts organized by the Food and Nutrition Board of the National Academy of Sciences have compiled Dietary Reference Intakes (DRIs)—estimates of the amounts of nutrients required to prevent deficiencies and maintain optimal health. DRIs replace and expand on Recommended Dietary Allowances (RDAs), which have been published with periodic revisions since 1941. Unlike the RDAs, the DRIs establish upper limits on the consumption of some nutrients, and incorporate the role of nutrients in lifelong health, going beyond deficiency diseases. Both the DRIs and the RDAs refer to long-term average daily nutrient intakes, because it is not necessary to consume the full RDA every day.

A. Definition of the DRIs

The DRIs consist of four dietary reference standards for the intake of nutrients designated for specific age-groups, physiologic states, and sexes (Figure 27.2).

1. **Estimated Average Requirement (EAR):** The EAR is the average daily nutrient intake level estimated to meet the requirement of **one half** the healthy individuals in a particular life stage and gender group. It is useful in estimating the actual requirements in groups and individuals.

2. **The Recommended Dietary Allowance (RDA):** The RDA is the average daily dietary intake level that is sufficient to meet the nutrient requirements of **nearly all** (97 to 98 percent) individuals in a life stage and gender group. The RDA is not the minimal requirement for healthy individuals; rather, it is intentionally set to provide a margin of safety for most individuals. The EAR serves as the foundation for setting the RDA. If the standard deviation (SD) of the

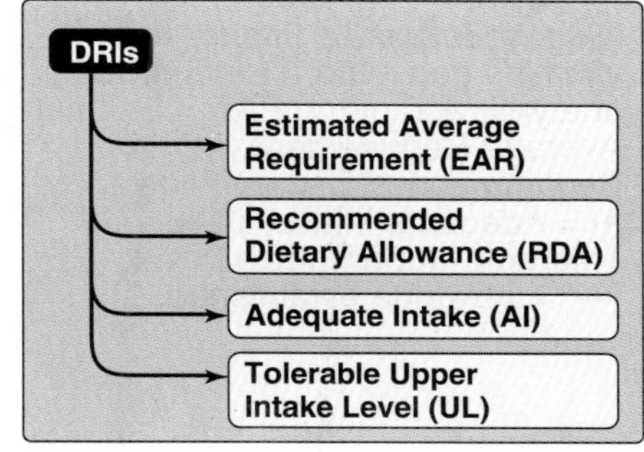

Figure 27.2
Components of the Dietary Reference Intakes (DRIs).

NUTRIENT	EAR, RDA or AI	UL
Thiamine	EAR, RDA	—
Riboflavin	EAR, RDA	—
Niacin	EAR, RDA	UL
Vitamin B_6	EAR, RDA	UL
Folate	EAR, RDA	UL
Vitamin B_{12}	EAR, RDA	—
Pantothenic acid	AI	—
Biotin	AI	—
Choline	AI	UL
Vitamin C	EAR, RDA	UL
Vitamin A	EAR, RDA	UL
Vitamin D	AI	UL
Vitamin E	EAR, RDA	UL
Vitamin K	AI	—
Boron	—	UL
Calcium	AI	UL
Chromium	AI	—
Copper	EAR, RDA	UL
Fluoride	AI	UL
Iodine	EAR, RDA	UL
Iron	EAR, RDA	UL
Magnesium	EAR, RDA	UL
Manganese	AI	UL
Molybdenum	EAR, RDA	UL
Nickel	—	UL
Phosphorus	EAR, RDA	UL
Selenium	EAR, RDA	UL
Vanadium	—	UL
Zinc	EAR, RDA	UL

Figure 27.3
Dietary Reference Intakes for vitamins and minerals in individuals one year and older. EAR = Estimated Average Requirement; RDA = Recommended Dietary Allowance; AI = Adequate Intake; UL = Tolerable Upper Intake Level; "—" = no value established.

EAR is available and the requirement for the nutrient is normally distributed, the RDA is set at two SDs above the EAR, that is, RDA = EAR + 2 SD_{EAR}.

3. **Adequate Intake (AI):** The AI is set instead of an RDA if sufficient scientific evidence is not available to calculate an EAR or RDA. The AI is based on estimates of nutrient intake by a group (or groups) of apparently healthy people that are assumed to be adequate. For example, the AI for young infants, for whom human milk is the recommended sole source of food for the first four to six months, is based on the estimated daily mean nutrient intake supplied by human milk for healthy, full-term infants who are exclusively breast-fed.

4. **Tolerable Upper Intake Level (UL):** UL is the highest average daily nutrient intake level that is likely to pose no risk of adverse health effects to almost all individuals in the general population. As intake increases above the UL, the potential risk of adverse effects may increase. The UL is not intended to be a recommended level of intake. ULs are useful because of the increased availability of fortified foods and the increased use of dietary supplements. The UL applies to chronic daily use. For some nutrients, there may be insufficient data on which to develop a UL.

B. Using the DRIs

Most nutrients have a set of DRIs (Figure 27.3). Usually a nutrient has an EAR and a corresponding RDA. Most are set by age and gender, and may be influenced by special factors, such as pregnancy and lactation in women. When the data are not sufficient to estimate an EAR (or an RDA), then an AI is designated. The AI is judged by experts to meet the needs of all individuals in a group, but is based on less data than in establishing an EAR and RDA. Intakes below the EAR need to be improved because the probability of adequacy is fifty percent or less (Figure 27.4). Intakes between the EAR and RDA probably need to be improved because the probability of adequacy is less than 98 percent, and intakes at or above the RDA can be considered adequate. Intake above the AI can be considered adequate. Intakes between the UL and the RDA can be considered at no risk for adverse effects.

III. ENERGY REQUIREMENT IN HUMANS

The **Estimated Energy Requirement** is the average dietary energy intake predicted to maintain an energy balance (that is, when calories consumed are equal to the energy expended) in a healthy adult of a defined age, gender, and height whose weight and level of physical activity are consistent with good health. Differences in the genetics, metabolism, and behavior of individuals make it difficult to accurately predict a person's caloric requirements. However, some simple approximations can provide useful estimates: for example, sedentary adults require about 30 kcal/kg/day to maintain body weight; moderately active adults require 35 kcal/kg/day; and very active adults require 40 kcal/kg/day. [Note: The daily average requirement for energy that is listed on food labels is 2000 kcal/day.]

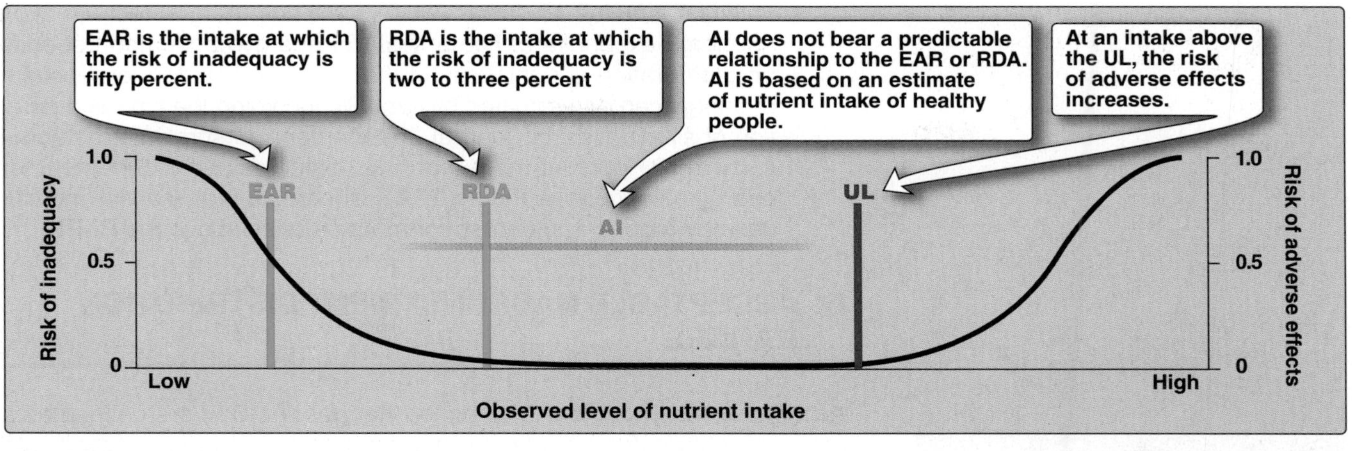

Figure 27.4
Comparison of the components of the Dietary Reference Intakes. EAR = Estimated Average Requirement; RDA = Recommended Dietary Allowance; AI = Adequate Intake; UL = Tolerable Upper Intake Level.

A. Energy content of food

The energy content of food is calculated from the heat released by the total combustion of food in a calorimeter. It is expressed in kilocalories (kcal, or Cal). The standard conversion factors for determining the metabolic caloric value of fat, protein, and carbohydrate are shown in Figure 27.5. Note that the energy content of fat is more than twice that of carbohydrate or protein, whereas the energy content of ethanol is intermediate between fat and carbohydrate. [Note: The joule is a unit of energy widely used in countries other than the United States. For uniformity, many scientists are promoting the use of joules (J), rather than calories (1 cal = 4.128 J). However, kcal still predominates and is used throughout this text.]

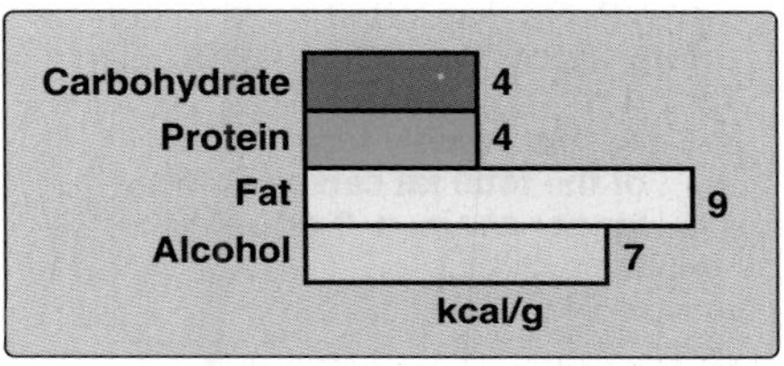

Figure 27.5
Energy available from the major food components.

B. How energy is used in the body

The energy generated by metabolism of the macronutrients is used for three energy-requiring processes that occur in the body: resting metabolic rate, thermic effect of food (formerly termed specific dynamic action), and physical activity.

1. **Resting metabolic rate:** The energy expended by an individual in a resting, postabsorptive state is called the **resting** (formerly, **basal**) **metabolic rate** (**RMR**). It represents the energy required to carry out the normal body functions, such as respiration, blood flow, ion transport, and maintenance of cellular integrity. In an adult, the RMR is about 1800 kcal for men (70 kg) and 1300 kcal for women (50 kg). From fifty to seventy percent of the daily energy expenditure in sedentary individuals is attributable to the RMR (Figure 27.6).

2. **Thermic effect of food:** The production of heat by the body increases as much as thirty percent above the resting level during the digestion and absorption of food. This effect is called the thermic effect of food or **diet-induced thermogenesis**. Over a 24–hour period, the thermic response to food intake may amount to five to ten percent of the total energy expenditure.

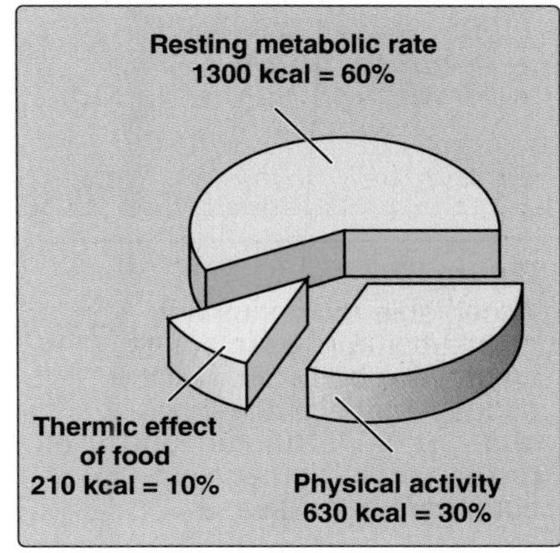

Figure 27.6
Estimated total energy expenditure in a typical 20-year-old woman, 165 cm (5 feet, 4 inches) tall, weighing 50 kg (110 lb), and engaged in light activity.

3. **Physical activity:** Muscular activity provides the greatest variation in energy expenditure. The amount of energy consumed depends on the duration and intensity of the exercise. The daily expenditure of energy can be estimated by carefully recording the type and duration of all activities. In general, a sedentary person requires about thirty to fifty percent more than the resting caloric requirement for energy balance (see Figure 27.6), whereas a highly active individual may require 100 percent or more calories above the RMR.

IV. ACCEPTABLE MACRONUTRIENT DISTRIBUTION RANGES

Acceptable Macronutrient Distribution Ranges (AMDR) are defined as a range of intakes for a particular macronutrient that is associated with reduced risk of chronic disease while providing adequate amounts of essential nutrients. The AMDR for adults is 45 to 65 percent of their total calories from carbohydrates, 20 to 35 percent from fat, and 10 to 35 percent from protein (Figure 27.7). Note that there is a range of acceptable intakes for the macronutrients. This is, in part, due to the fact that fats and carbohydrates (and, to a limited extent, protein) can substitute for one another to meet the body's energy needs. The AMDR thus represents a balance designed to avoid risks associated with excess consumption of any particular macronutrient. For example, very high-fat diets are associated with weight gain and an increased intake of saturated fats that can raise the plasma low-density lipoprotein (LDL) cholesterol concentration (see p. 229) and increase the risk of coronary heart disease (CHD). Conversely, very high-carbohydrate diets are associated with a reduction in plasma high-density lipoprotein (HDL) cholesterol, an increase in plasma triacylglycerol concentration, and an increased risk of CHD. The AMDR for protein ensures an adequate supply of amino acids for tissue growth, maintenance, and repair. The biologic properties of dietary fat, carbohydrate, and protein are described below.

MACRONUTRIENT	RANGE (percent of energy)
Fat	**20–35**
n–6 Polyunsaturated fatty acids	**5–10**
n–3 Polyunsaturated fatty acids	**0.6–1.2***
(Approximately ten percent of the total fat can come from longer-chain, n–3 or n–6 fatty acids.)	
Carbohydrate • **No less than 130 g/day**	**45–65**
(No more than 25 percent of total calories should come from added sugars.)	
Fiber • **Men: 38 g** • **Women: 25 g**	
Protein	**10–35**

Figure 27.7
Acceptable macronutrient distribution ranges in adults. *A growing body of evidence suggest that higher levels of n–3 polyunsaturated fatty acids provide protection against coronary heart disease.

V. DIETARY FATS

The incidence of a number of chronic diseases are significantly influenced by the kinds and amounts of nutrients consumed (Figure 27.8). The role of dietary fats and the risk for CHD have been thoroughly documented, and are the focus of this section.

A. Plasma cholesterol and coronary heart disease

Plasma cholesterol may arise from the diet or from endogenous biosynthesis. In either case, cholesterol is transported between the tissues in combination with protein and phospholipids as lipoproteins.

1. **LDL and HDL:** The level of plasma cholesterol is not precisely regulated, but rather varies in response to the diet. Elevated levels result in an increased risk for cardiovascular disease (Figure 27.9). The risk increases progressively with higher values for serum total cholesterol. A much stronger correlation exists between the levels of blood LDL cholesterol and heart disease. In contrast, high levels of HDL cholesterol have been associated

with a decreased risk for heart disease. Abnormal levels of plasma lipids (dyslipidemias) act in combination with smoking, obesity, sedentary lifestyle, and other risk factors to increase the risk of CHD. Elevated plasma triacylglycerols are also a risk factor for CHD, but the association is weaker than that of LDL cholesterol with CHD.

2. **Beneficial effect of lowering plasma cholesterol:** Clinical trials have demonstrated that dietary or drug treatment of hypercholesterolemia is effective in decreasing LDLs, increasing HDLs, and reducing the risk for cardiovascular events. The diet-induced changes of plasma lipoprotein concentrations are modest, typically ten to twenty percent, whereas treatment with "statin" drugs[1] decreases plasma cholesterol by thirty to forty percent.

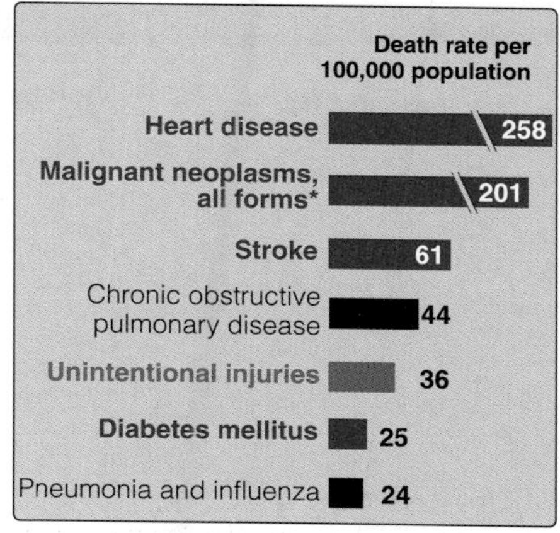

Figure 27.8
Influence of nutrition on some common causes of death in the United States in the year 2000. Red indicates causes of death in which the diet plays a significant role. Blue indicates causes of death in which excessive alcohol consumption plays a part. (*Diet plays a role in only some forms of cancer.)

B. Dietary fats and plasma lipids

Triacylglycerols are quantitatively the most important class of dietary fats. Their biologic properties are determined by the chemical nature of the constituent fatty acids, in particular, the presence or absence of double bonds, the number and location of the double bonds, and the cis–trans configuration of the unsaturated fatty acids.

1. **Saturated fat:** Triacylglycerols containing primarily fatty acids whose side chains do not contain any double bonds are referred to as saturated fats. Consumption of saturated fats is strongly associated with high levels of total plasma cholesterol and LDL cholesterol, and an increased risk of coronary heart disease. The main sources of saturated fatty acids are dairy and meat products and some vegetable oils, such as coconut and palm oils (a major source of fat in Latin American and Asia, although not in the United States, Figure 27.10). Most experts strongly advise limiting intake of saturated fats.

2. **Monounsaturated fats:** Triacylglycerols containing primarily fatty acids with one double bond are referred to as monounsaturated fat. Unsaturated fatty acids are generally derived from vegetables and fish. When substituted for saturated fatty acids in the diet, monounsaturated fats lower both total plasma cholesterol and LDL cholesterol, but increase HDLs. This ability of monounsaturated fats to favorably modify lipoprotein levels may explain, in part, the observation that Mediterranean cultures, with diets rich in olive oil (high in monounsaturated oleic acid), show a low incidence of coronary heart disease.

   a. **The Mediterranean diet:** The Mediterranean diet is an example of a diet rich in monounsaturated fatty acids (from olive oil) and n–3 fatty acids (from fish oils and some nuts), but low in saturated fat. For example, Figure 27.11 shows the composition of the Mediterranean diet in comparison with both a "Western" diet similar to that consumed in the United States and a typical low-fat diet. The Mediterranean diet contains seasonally fresh food, with an abundance of plant material, low amounts of red meat, and olive oil as the principal source of fat. The Mediterranean diet is associated with decreased serum total cholesterol and

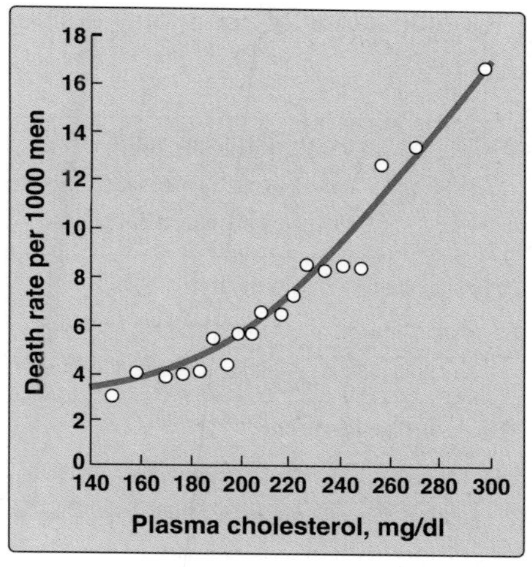

Figure 27.9
Correlation of the death rate from coronary heart disease with the concentration of plasma cholesterol. [Note: The data were obtained from a six year study of men with the death rate adjusted for age.]

[1]See Chapter 21 in ***Lippincott's Illustrated Reviews: Pharmacology*** (2nd and 3rd eds.) for a more detailed discussion of antihyperlipidemic drugs.

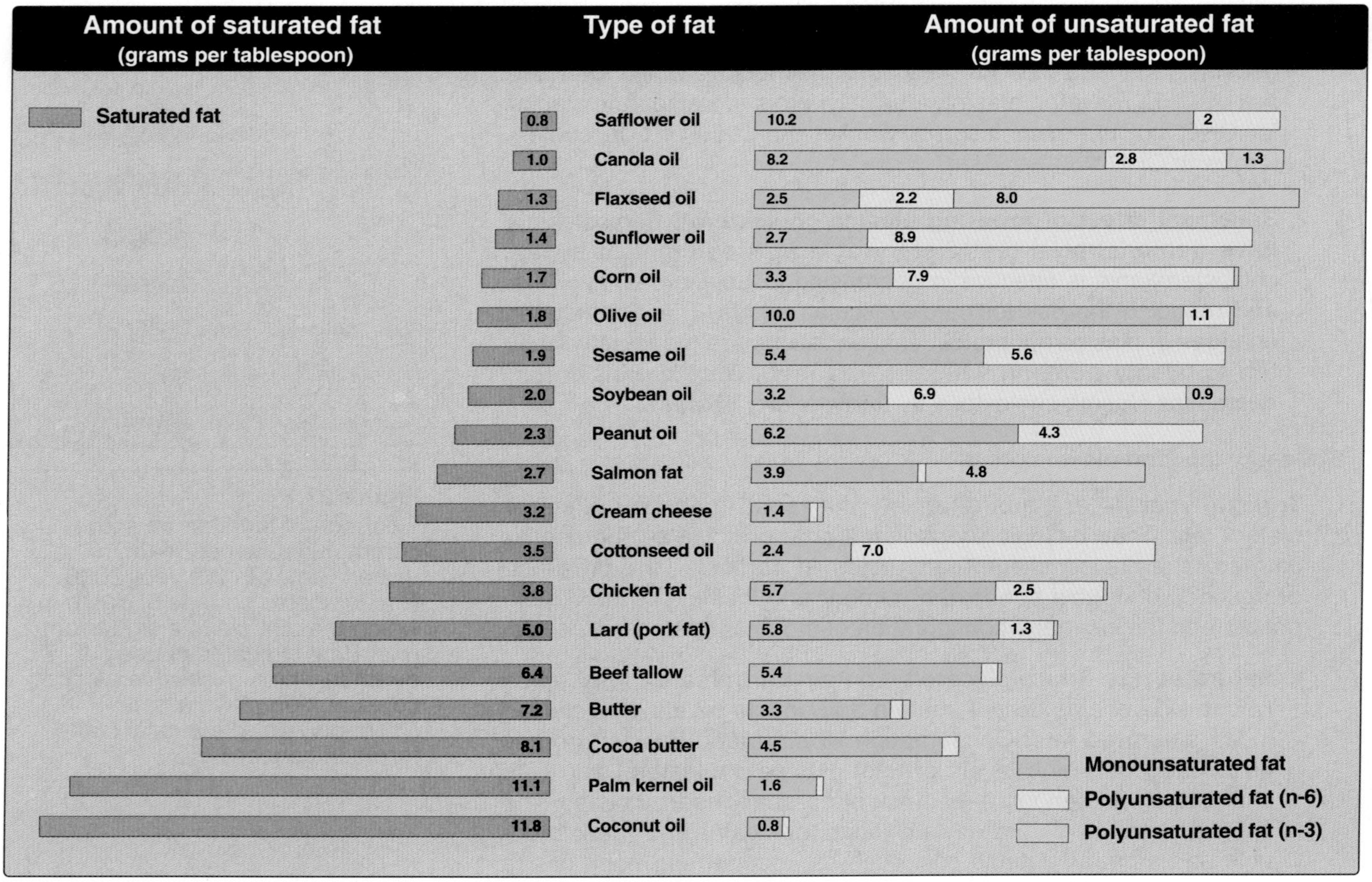

Amount of saturated fat (grams per tablespoon)	Type of fat	Amount of unsaturated fat (grams per tablespoon)
0.8	Safflower oil	10.2, 2
1.0	Canola oil	8.2, 2.8, 1.3
1.3	Flaxseed oil	2.5, 2.2, 8.0
1.4	Sunflower oil	2.7, 8.9
1.7	Corn oil	3.3, 7.9
1.8	Olive oil	10.0, 1.1
1.9	Sesame oil	5.4, 5.6
2.0	Soybean oil	3.2, 6.9, 0.9
2.3	Peanut oil	6.2, 4.3
2.7	Salmon fat	3.9, 4.8
3.2	Cream cheese	1.4
3.5	Cottonseed oil	2.4, 7.0
3.8	Chicken fat	5.7, 2.5
5.0	Lard (pork fat)	5.8, 1.3
6.4	Beef tallow	5.4
7.2	Butter	3.3
8.1	Cocoa butter	4.5
11.1	Palm kernel oil	1.6
11.8	Coconut oil	0.8

Key: Saturated fat; Monounsaturated fat; Polyunsaturated fat (n-6); Polyunsaturated fat (n-3)

Figure 27.10
Compositions of commonly encountered dietary lipids.

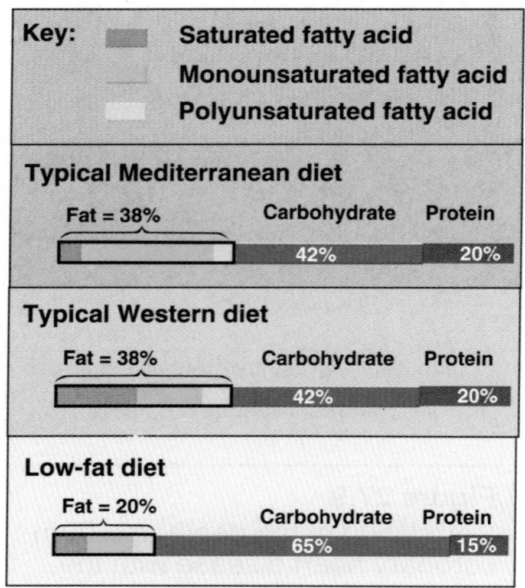

Figure 27.11
Composition of typical Mediterranean, Western, and low-fat diets.

LDL—but little change in HDL—when compared with a typical Western diet higher in saturated fats. Plasma triacylglycerols are unchanged.

3. **Polyunsaturated fatty acids:** Triacylglycerols containing primarily fatty acids with more than one double-bond are referred to as polyunsaturated fats. The effects of polyunsaturated fatty acids on cardiovascular disease is influenced by the location of the double bonds within the molecule.

   a. **n-6 Fatty acids:** These are long-chain, polyunsaturated fatty acids, with the first double bond beginning at the sixth carbon atom (when counting from the methyl end of the fatty acid molecule, Figure 27.12). [Note: They are also called ω-6 (omega-6) fatty acids.] Consumption of fats containing n-6 polyunsaturated fatty acids, principally linoleic acid (18:22; Δ9, 12) obtained from vegetable oils, lowers plasma cholesterol when substituted for saturated fats. Plasma LDLs are lowered, but HDLs, which protect against coronary heart disease, are also lowered. The powerful benefits of lowering LDLs are only partially offset because of the decreased HDLs. Nuts, avocados, olives, soybeans, and various oils, including sesame, cottonseed, and corn oil, are common sources of these fatty acids (see Figure 27.10). Linoleic acid, along with linolenic acid (18:3, Δ9,12,15, an n-3 fatty acid, see below), are **essential fatty acids** required for fluidity of membrane structure and synthesis of eicosanoids (see p. 211). [Note: A deficiency of essential fatty acids is characterized by scaly dermatitis, hair loss, and poor wound healing.] A lower boundary level of five percent of calories meets the AI set for linoleic acid. An upper boundary for linoleic acid is set at ten percent of total calories because of concern that oxidation of these polyunsaturated fatty acids may lead to deleterious products.

b. n-3 Fatty acids: These are long-chain, polyunsaturated fatty acids, with the first double bond beginning at the third carbon atom (when counting from the methyl end of the fatty acid molecule, see Figure 27.12). Dietary n-3 polyunsaturated fats suppress cardiac arrhythmias, reduce serum triacylglycerols, decrease the tendency to thrombosis, and substantially reduce risk of cardiovascular mortality, but they have little effect on LDL or HDL cholesterol levels. The n-3 polyunsaturated fats are found in plants (mainly α-linolenic acid—an essential fatty acid), and in fish oil containing docosahexaenoic acid (DHA) and eicosapentaenoic acid (EPA). The acceptable range for α-linolenic acid is 0.6 to 1.2 percent of total calories, although emerging data suggest that higher values may provide protection against CHD.

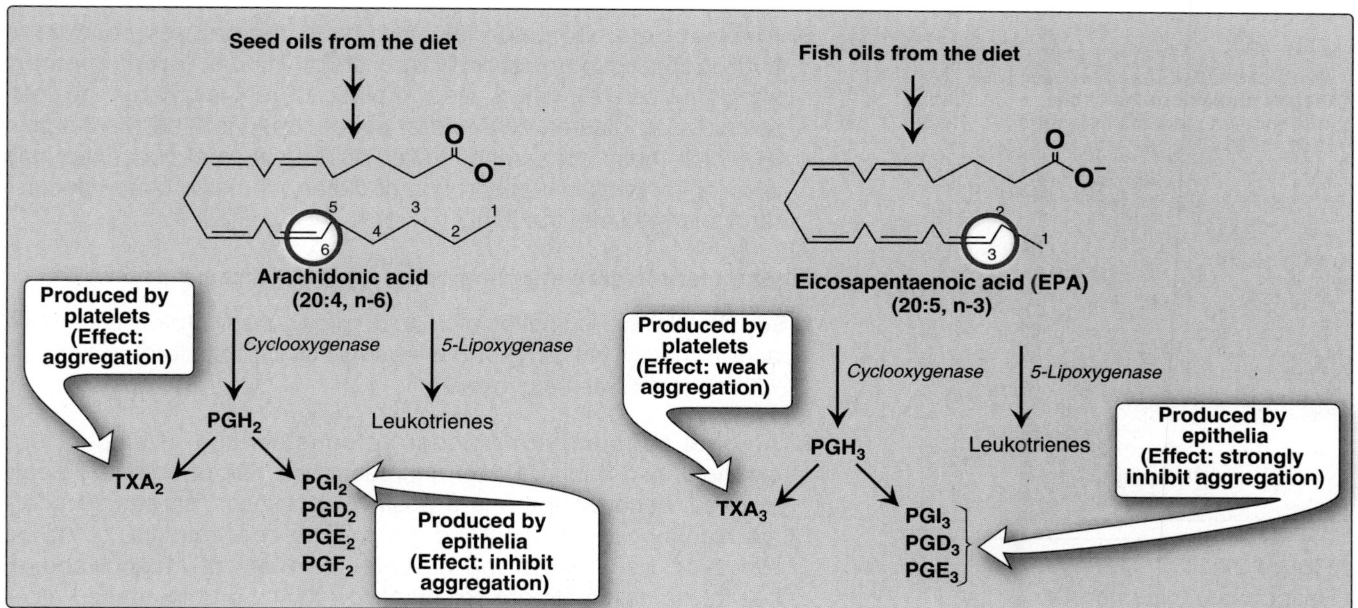

Figure 27.12
Forms of prostaglandins (PG), thromboxanes (TX), and leukotrienes synthesized from n-6 and n-3 polyunsaturated fatty acids. [Note the positions of the first double bonds in the n-6 and the n-3 polyunsaturated fatty acids (red circles).]

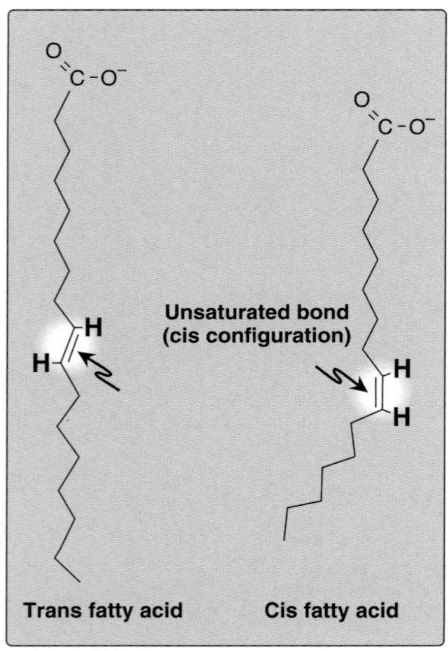

Figure 27.13
Structure of cis and trans fatty acids.

c. **Antithrombotic effects of n-3 fatty acids:** The reduced blood platelet reactivity observed with increased consumption of EPA and DHA n-3 fatty acids results from inhibition of the conversion of arachidonic acid to thromboxane A_2 (TXA_2) by platelets (see p. 211 for a discussion of eicosanoid biosynthesis). Instead, the n-3 fatty acids are converted to TXA_3, which is less thrombogenic than TXA_2 (see Figure 27.12). Thus, the products of fish oils decrease platelet aggregation and, therefore, are antithrombogenic. In addition to these effects, n-3 fatty acids decrease arrhythmias, and affect a variety of membrane functions. The fatty fish can be remembered as SMASH: **s**almon, **m**ackerel, **a**nchovies, **s**ardines, and **h**erring.] [Note: Generally speaking, Western diets contain excess dietary n-6 fatty acids that compete with the formation of eicosanoids derived from n-3 fatty acids.]

4. **Trans fatty acids:** Trans fatty acids (Figure 27.13) are chemically classified as unsaturated fatty acids, but behave more like saturated fatty acids in the body, that is, they elevate serum LDL (but not HDL), and they increase the risk of CHD. Trans fatty acids do not occur naturally in plants and only occur in small amounts in animals. However, trans fatty acids are formed during the hydrogenation of liquid vegetable oils, for example, in the manufacture of margarine.

5. **Dietary cholesterol:** Cholesterol is found only in animal products. The effect of dietary cholesterol on plasma cholesterol (Figure 27.14) is less important than the amount and types of fatty acids consumed.

6. **Plant sterols:** Commercially available margarines containing hydrogenated plant sterols and sterol esters (predominantly sitostanol esters), when used in place of regular margarine, can reduce LDL plasma cholesterol concentrations. The mechanism by which these compounds lower LDL cholesterol concentrations is to inhibit intestinal absorption of dietary cholesterol and cholesterol secreted into the bile.

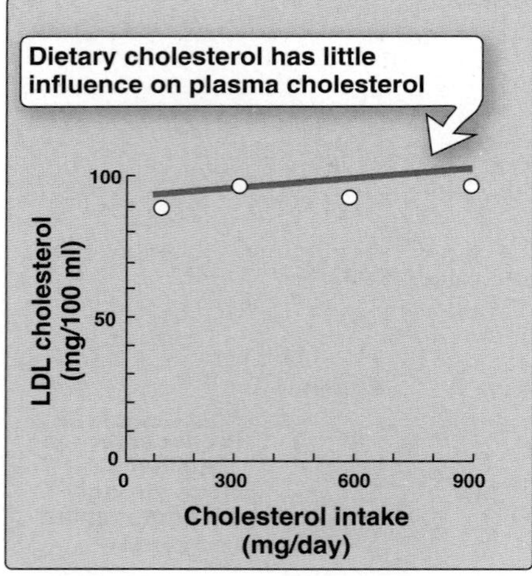

Figure 27.14
Response of plasma LDL concentrations to an increase in dietary cholesterol intake.

C. Other dietary factors affecting coronary heart disease

1. **Soy protein:** Consumption of 25 to 50 g/day of soy protein causes an approximately ten percent decrease in LDL cholesterol in patients with elevated plasma cholesterol.

2. **Alcohol consumption:** Moderate consumption of alcohol (for example, two drinks a day) decreases the risk of coronary heart disease, because there is a positive correlation between moderate alcohol consumption and the plasma concentration of HDLs. However, because of the potential dangers of alcohol abuse, health professionals are reluctant to recommend increased alcohol consumption to their patients. Red wine may provide cardioprotective benefits in addition to those resulting from its alcohol content, for example, red wine contains phenolic compounds that inhibit lipoprotein oxidation (see p. 233). [Note: These antioxidants are also present in raisins and grape juice.]

3. **Vitamins B_6, B_{12}, and folate:** An elevated plasma homocysteine level is associated with increased cardiovascular risk (see p. 263). Homocysteine, which is thought to be toxic to the vascular endothelium, is converted into harmless amino acids by the action of enzymes that require the B vitamins—folate, B_6 (pyridoxine), and B_{12} (cobalamin). Ingesting foods rich in these vitamins can lower homocysteine levels and possibly decrease the risk of cardiovascular disease. Folate and B_6 are found in leafy green vegetables, whole grains, some fruits, and fortified breakfast cereals. B_{12} comes from animal food, for example, meat, fish, and eggs.

VI. DIETARY CARBOHYDRATES

The primary role of dietary carbohydrate is to provide energy. Although caloric intake in the United States has shown a modest increase since 1971 (Figure 27.15), the incidence of obesity has dramatically increased (see Figure 26.2, p. 347). During this same period, carbohydrate consumption has significantly increased, leading some uncritical observers to link obesity with carbohydrate consumption. However, obesity has been more directly related to increasingly inactive lifestyles, and to calorie-dense foods served in expanded portion size. Carbohydrates are not inherently fattening.

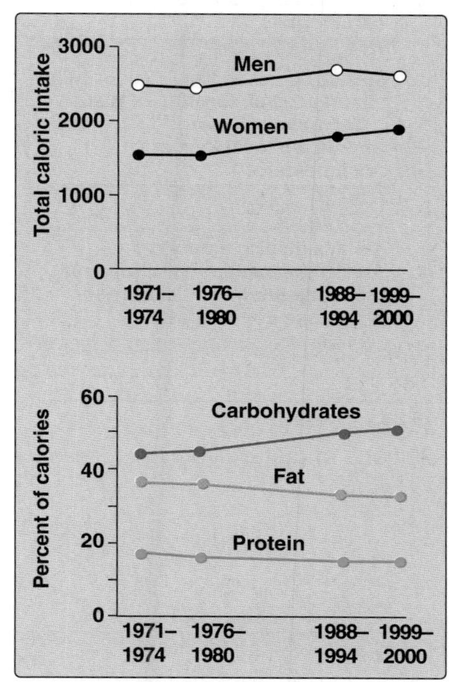

Figure 27.15
Total caloric consumption and distribution of calories between the macronutrients in adults.

A. Classification of carbohydrates

Carbohydrates in the diet are classified as either monosaccharides and disaccharides (simple sugars), polysaccharides (complex sugars), or fiber.

1. **Monosaccharides: Glucose** and **fructose** are the principal monosaccharides found in food. Glucose is abundant in fruits, sweet corn, corn syrup, and honey. Free fructose is found together with free glucose and sucrose in honey and fruits.

2. **Disaccharides:** The most abundant disaccharides are **sucrose** (glucose + fructose), **lactose** (glucose + galactose), and **maltose** (glucose + glucose). Sucrose is ordinary "table sugar," and is abundant in molasses and maple syrup. Lactose is the principal sugar found in milk. Maltose is a product of enzymic digestion of polysaccharides. It is also found in significant quantities in beer and malt liquors. The term "**sugar**" refers to monosaccharides and disaccharides. "**Added sugars**" are those sugars and syrups added to foods during processing or preparation.

3. **Polysaccharides:** Complex carbohydrates are polysaccharides (most often polymers of glucose), which do not have a sweet taste. **Starch** is an example of a complex carbohydrate that is found in abundance in plants. Common sources include wheat and other grains, potatoes, dried peas and beans, and vegetables.

4. **Fiber: Dietary fiber** is defined as the nondigestible carbohydrates and lignin (a complex polymer of phenylpropanoid subunits) present in plants. Several different terms are used to described this complex group of compounds. For example, **functional fiber** is the

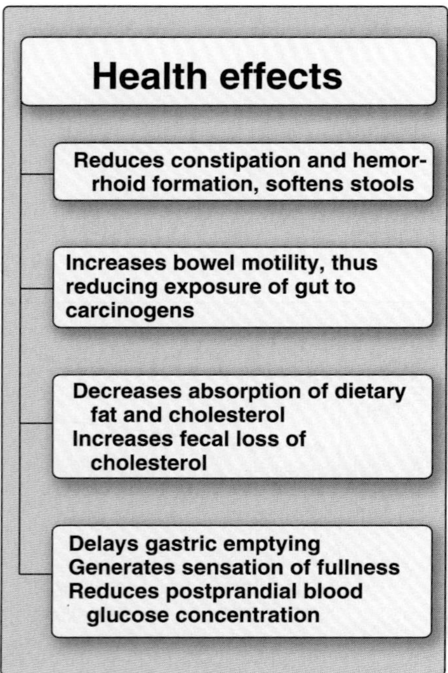

Figure 27.16
Actions of dietary fiber.

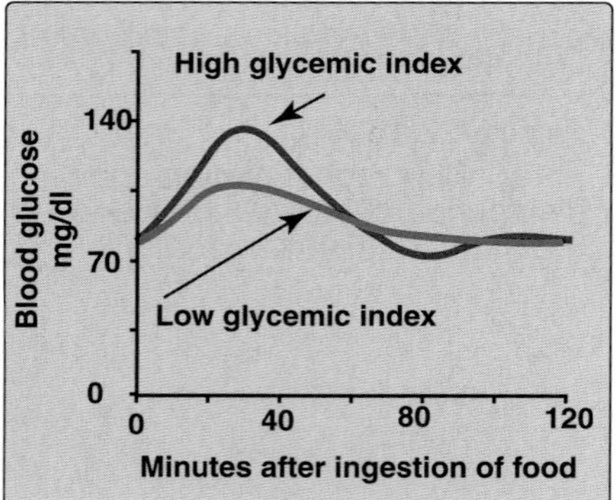

Figure 27.17
Blood glucose concentrations following ingestion of food with low or high glycemic index.

isolated, extracted, or synthetic fiber that has proven health benefits. **Total fiber** is the sum of dietary fiber and functional fiber. **Soluble fiber** refers to fibers that form a viscous gel when mixed with a liquid. **Insoluble fiber** passes through the digestive track largely intact. Dietary fiber provides little energy but has several beneficial effects. First, it adds bulk to the diet (Figure 27.16). Fiber can absorb ten to fifteen times its own weight in water, drawing fluid into the lumen of the intestine and increasing bowel motility. Soluble fiber delays gastric emptying and can result in a sensation of fullness. This delayed emptying also results in reduced peaks of blood glucose following a meal. Second, consumption of soluble fiber has now been shown to lower LDL cholesterol levels by increasing fecal bile acid excretion and interfering with bile acid reabsorption. For example, diets rich in the soluble fiber oat bran (25 to 50 g/day) are associated with a modest, but significant, reduction in risk for cardiovascular disease by lowering total and LDL cholesterol levels. Also, fiber-rich diets decrease the risk for constipation, hemorrhoids, diverticulosis, and colon cancer. The recommended daily fiber intake (AI) is 25 g/day for women and 38 g/day for men. However, most American diets are far lower in fiber—approximately 11 g/day.

B. Dietary carbohydrate and blood glucose

Some carbohydrate-containing foods produce a rapid rise followed by a steep fall in blood glucose concentration, whereas others result in a gradual rise followed by a slow decline. The **glycemic index** has been proposed to quantitate these differences in the time course of postprandial glucose concentrations (Figure 27.17). Glycemic index is defined as the area under the blood glucose curves seen after ingestion of a meal with carbohydrate-rich food, compared with the area under the blood glucose curve observed after a meal consisting of the same amount of carbohydrate in the form of glucose or white bread. The clinical importance of glycemic index is controversial. Food with a low glycemic index tends to create a sense of satiety over a longer period of time, and may be helpful in limiting caloric intake. However, many experts feel that high nutrient and fiber content, such as occurs in whole grains, fruits, and vegetables, is a better guide for selecting dietary carbohydrates.

C. Requirements for carbohydrate

Carbohydrates are not essential nutrients, because the carbon skeletons of amino acids can be converted into glucose (see p. 259). However, the absence of dietary carbohydrate leads to ketone body production (see p. 260), and degradation of body protein whose constituent amino acids provide carbon skeletons for gluconeogenesis (see p. 116). The RDA for carbohydrate is set at 130 g/day for adults and children, based on the amount of glucose used by carbohydrate-dependent tissues, such as the brain and erythrocytes. However, this level of intake is usually exceeded to meet energy needs. Adults should consume 45 to 65 percent of their total calories from carbohydrates. It is recommended that added sugar represent no more than 25 percent of total energy because of concerns that sugar may displace nutrient-rich foods from the diet, potentially leading to deficiencies of certain micronutrients

D. Simple sugars and disease

There is no direct evidence that the consumption of simple sugars is harmful. Contrary to folklore, diets high in sucrose do not lead to diabetes or hypoglycemia. Also contrary to popular belief, carbohydrates are not inherently fattening. They yield 4 kcal/g (the same as protein and less than half that of fat, see Figure 27.5), and result in fat synthesis only when consumed in excess of the body's energy needs. However, there is an association between sucrose consumption and dental caries, particularly in the absence of fluoride treatment.

VII. DIETARY PROTEIN

Humans have no dietary requirement for protein, per se, but, the protein in food does provide essential amino acids (see Figure 20.2, p. 260). Ten of the twenty amino acids needed for the synthesis of body proteins are essential—that is, they cannot be synthesized in humans at an adequate rate. Of these ten, eight are essential at all times, whereas two (arginine and histidine) are required only during periods of rapid tissue growth characteristic of childhood or recovery from illness.

A. Quality of proteins

The quality of a dietary protein is a measure of its ability to provide the essential amino acids required for tissue maintenance. Most government agencies have adopted the Protein Digestibility-Corrected Amino Acid Scoring (PDCAAS) as the standard by which to evaluate protein quality. PDCAAS is based on the profile of essential amino acids and the digestibility of the protein. The highest possible score under these guidelines is 1.00. This amino acid score provides a method to balance intakes of poorer quality proteins by vegetarians and others who consume limited quantities of high-quality dietary proteins.

1. **Proteins from animal sources:** Proteins from animal sources (meat, poultry, milk, fish) have a high quality because they contain all the essential amino acids in proportions similar to those required for synthesis of human tissue proteins (Figure 27.18). [Note: Gelatin prepared from animal collagen is an exception; it has a low biologic value as a result of deficiencies in several essential amino acids.]

2. **Proteins from plant sources:** Proteins from wheat, corn, rice, and beans have a lower quality than do animal proteins. However, proteins from different plant sources may be combined in such a way that the result is equivalent in nutritional value to animal protein. For example, wheat (lysine-deficient but methionine-rich) may be combined with kidney beans (methionine-poor but lysine-rich) to produce a complete protein of improved biologic value. Thus, eating foods with different limiting amino acids at the same meal (or at least during the same day) can result in a dietary combination with a higher biologic value than either of the component proteins (Figure 27.19). [Note: Animal proteins can also complement the biologic value of plant proteins.]

Source	PDCAAS value
Animal proteins	
Egg	1.00
Milk protein	1.00
Beef/poultry/fish	0.82–0.92
Gelatin	0.08
Plant proteins	
Soybean protein	1.00
Kidney beans	0.68
Whole wheat bread	0.40

Figure 27.18
Relative quality of some common dietary proteins.

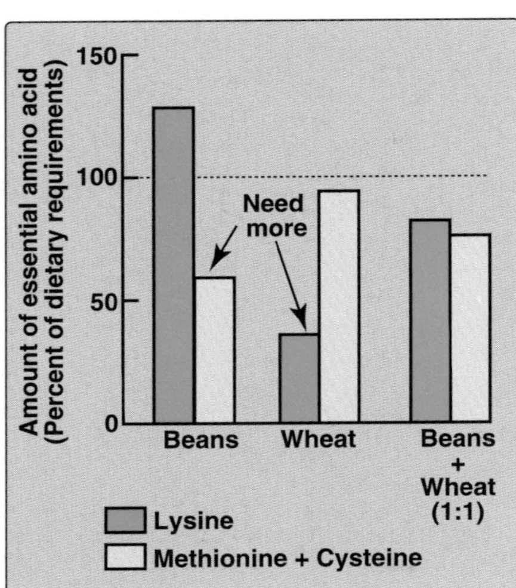

Figure 27.19
Combining two incomplete proteins that have complementary amino acid deficiencies results in a mixture with a higher biologic value.

B. Nitrogen balance

Nitrogen balance occurs when the amount of nitrogen consumed equals that of the nitrogen excreted in the urine, sweat, and feces. Most healthy adults are normally in nitrogen balance.

1. **Positive nitrogen balance:** This occurs when nitrogen intake exceeds nitrogen excretion. It is observed in situations in which tissue growth occurs, for example, in children, pregnancy, or during recovery from an emaciating illness.

2. **Negative nitrogen balance:** This occurs when nitrogen loss is greater than nitrogen intake. It is associated with inadequate dietary protein, lack of an essential amino acid, or during physiologic stresses such as trauma, burns, illness, or surgery.

C. Requirement for protein in humans

The amount of dietary protein required in the diet varies with its biologic value. The greater the proportion of animal protein included in the diet, the less protein is required. Recommended Dietary Allowance (RDA) for protein is computed for proteins of mixed biologic value at 0.8 g/kg of body weight for adults, or about 56 g of protein for a 70 kg individual. People who exercise strenuously on a regular basis may benefit from extra protein to maintain muscle mass; a daily intake of about 1 g/kg has been recommended for athletes. Women who are pregnant or lactating require up to 30 g/day in addition to their basal requirements. To support growth, children should consume 2 g/kg/day.

1. **Consumption of excess protein:** There is no physiologic advantage to the consumption of more protein than the RDA. Protein consumed in excess of the body's needs is deaminated, and the resulting carbon skeletons metabolized to provide energy or acetyl CoA for fatty acid synthesis. When excess protein is eliminated from the body as urinary nitrogen, it is often accompanied by increased urinary calcium, increasing the risk of nephrolithiasis and osteoporosis.

2. **The protein-sparing effect of carbohydrate:** The dietary protein requirement is influenced by the carbohydrate content of the diet. When the intake of carbohydrates is low, amino acids are deaminated to provide carbon skeletons for the synthesis of glucose that is needed as a fuel by the central nervous system. If carbohydrate intake is less than 130 g/day, substantial amounts of protein are metabolized to provide precursors for gluconeogenesis. Therefore, carbohydrate is considered to be "protein-sparing," because it allows amino acids to be used for repair and maintenance of tissue protein rather than for gluconeogenesis.

D. Protein-calorie malnutrition

In developed countries, protein-calorie malnutrition is seen most frequently in hospital patients with chronic illness, or in individuals who suffer from major trauma, severe infection, or the effects of major surgery. Such highly catabolic patients frequently require intravenous administration of nutrients (see p. 308 for metabolic changes elicited by trauma). In developing countries, an inadequate intake of protein and/or energy may be observed. Affected individuals show a variety of symptoms, including a depressed immune system with a reduced ability to resist infection. Death from secondary infection is common. Two extreme forms of malnutrition are kwashiorkor and marasmus.

1. **Kwashiorkor:** Kwashiorkor occurs when protein deprivation is relatively greater than the reduction in total calories. Unlike marasmus, significant protein deprivation is associated with severe loss of visceral protein. Kwashiorkor is frequently seen in children after weaning at about one year of age, when their diet consists predominantly of carbohydrates. Typical symptoms include stunted growth, edema, skin lesions, depigmented hair, anorexia, enlarged fatty liver, and decreased plasma albumin concentration. Edema results from the lack of adequate plasma proteins to maintain the distribution of water between blood and tissues. A child with kwashiorkor frequently shows a deceptively plump belly as a result of edema (Figure 27.20).

2. **Marasmus:** Marasmus occurs when calorie deprivation is relatively greater than the reduction in protein. Marasmus usually occurs in children younger than one year of age when the mother's breast milk is supplemented with thin watery gruels of native cereals, which are usually deficient in protein and calories. Typical symptoms include arrested growth, extreme muscle wasting (emaciation), weakness, and anemia. Victims of marasmus do not show the edema or changes in plasma proteins observed in kwashiorkor.

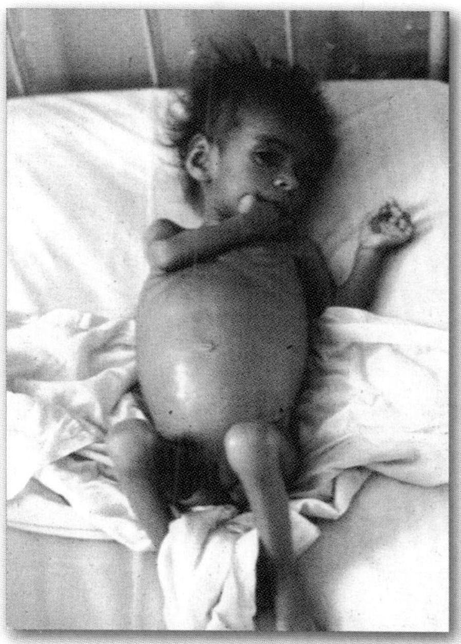

Figure 27.20
Child with kwashiorkor.

VIII. DIET AND CANCER

Diet influences the risk for certain forms of cancer, especially cancer of the esophagus, stomach, large bowel, breast, lung, and prostate. As with most chronic diseases that are influenced by nutritional factors, the incidence of cancer is also influenced by genetic and environmental factors. High intakes of saturated fats are associated with increased risk of certain cancers, especially cancer of the colon, prostate, and breast. For example, Figure 27.21 shows the correlation between the relative risk for colon cancer and consumption of animal fat in women. The data show that those women whose diets were rich in animal fat have significantly increased risk for colon cancer. However, whereas these studies show association between fat and cancer, they do not establish fat as the cause of cancer. In general, populations consuming diets rich in fruits and vegetables have lower incidence of many kinds of cancer. However, studies investigating the prophylactic effects of compounds isolated from fruits and vegetables, such as vitamins C, E or β-carotene, have been disappointing. High fiber diets are associated with a lower risk of colon cancer and diverticulitis.

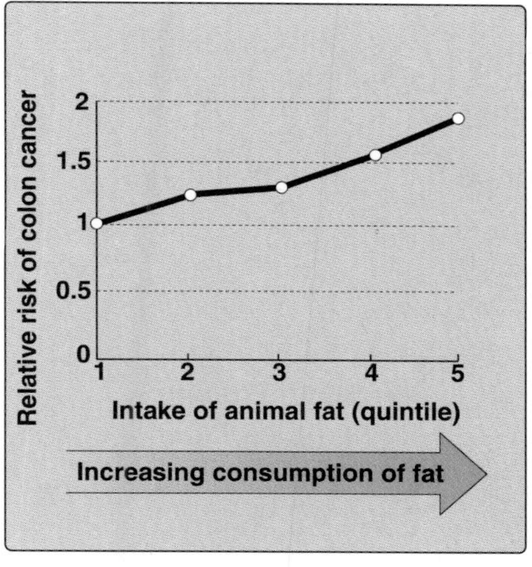

Figure 27.21
Relative risk of colon cancer according to the intake of animal fat. First quintile (1 on the graph) indicates the relative risk of twenty percent of the population (88,751 women) who consumed the least animal fat. Individuals in the fifth quintile (5 on the graph) consumed the most animal fat.

IX. CHAPTER SUMMARY

Estimated Average Requirement (EAR) is the average daily nutrient intake level estimated to meet the requirement of one half the healthy individuals in a particular life stage and gender group. The **Recommended Dietary Allowance** (**RDA**) is the average daily dietary intake level that is sufficient to meet the nutrient requirements of nearly all (97 to 98 percent) individuals. **Adequate Intake** (**AI**) is set instead of an RDA if sufficient scientific evidence is not available to calculate the RDA. The **Tolerable Upper Intake Level** (**UL**) is the highest average daily nutrient intake level that is likely to pose no risk of adverse health effects to almost all individuals in the general population. The energy generated by the metabolism of the **macronutrients** is used for three energy-requiring processes that occur in the body: **resting metabolic rate**, **thermic effect of food**, and **physical activity**. **Acceptable Macronutrient Distribution Ranges** (**AMDR**) are defined as the ranges of intake for a particular macronutrient that is associated with reduced risk of chronic disease while providing adequate amounts of essential nutrients. Adults should consume **45 to 65 percent** of their **total calories** from **carbohydrates**, **20 to 35 percent** from **fat**, and **10 to 35 percent** from **protein**. Elevated levels of total cholesterol or LDL cholesterol result in increased risk for **cardiovascular disease**. In contrast, high levels of HDL cholesterol have been associated with a decreased risk for heart disease. Dietary or drug treatment of **hypercholesterolemia** are effective in decreasing LDLs, increasing HDLs, and reducing the risk for cardiovascular events. Consumption of **saturated fats** is strongly associated with high levels of total plasma cholesterol and LDL cholesterol. When substituted for saturated fatty acids in the diet, **monounsaturated fats** lower both total plasma cholesterol and LDL cholesterol, but increase HDLs. Consumption of fats containing **n-6 polyunsaturated fatty acids** lowers plasma LDLs, but HDLs, which protect against coronary heart disease, are also lowered. Dietary **n-3 polyunsaturated fats** suppress cardiac arrhythmias and reduce serum triacylglycerols, decrease the tendency to thrombosis, and substantially reduce the risk of cardiovascular mortality. **Carbohydrates** provide **energy** and **fiber** to the diet. When they are consumed as part of a diet in which caloric intake is equal to energy expenditure they do not promote obesity. Dietary **protein** provides **essential amino acids**. The **quality of a protein** is a measure of its ability to provide the essential amino acids required for tissue maintenance. Proteins from animal sources, in general, have a higher quality protein than that derived from plants. However, proteins from different plant sources may be combined in such a way that the result is equivalent in nutritional value to animal protein. **Positive nitrogen balance** occurs when nitrogen intake exceeds nitrogen excretion. It is observed in situations in which tissue growth occurs, for example, in children, pregnancy, or during recovery from an emaciating illness. **Negative nitrogen balance** occurs when nitrogen losses are greater than nitrogen intake. It is associated with inadequate dietary protein, lack of an essential amino acid, or during physiologic stresses such as trauma, burns, illness, or surgery. **Kwashiorkor** is caused by inadequate intake of protein. **Marasmus** results from chronic deficiency of calories. Populations consuming diets rich in fruits and vegetables have a lower incidence of many kinds of cancer.

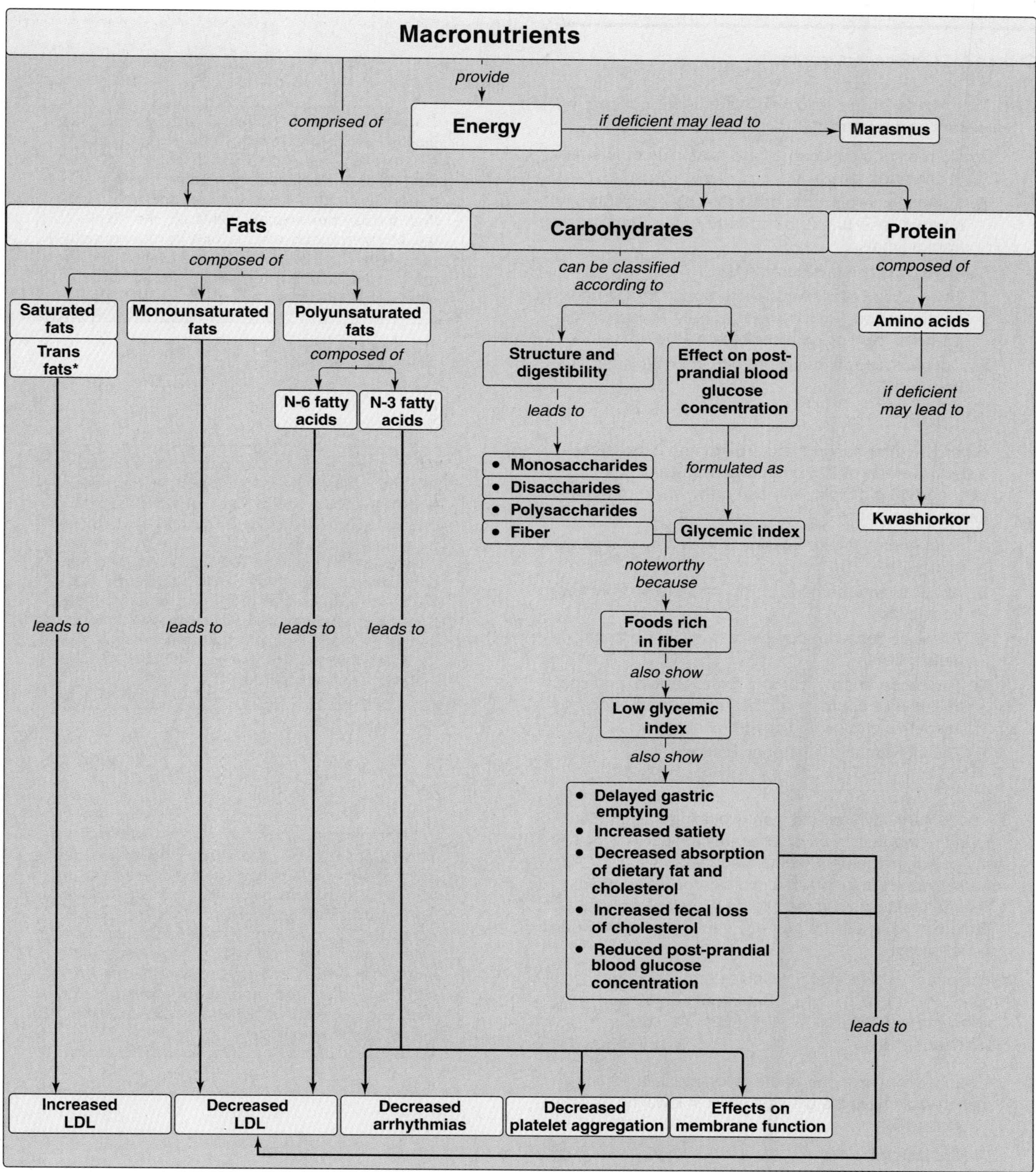

Figure 27.22
Key concept map for the macronutrients. *Note: Trans fatty acids are chemically classified as monounsaturated.

Study Questions

Choose the ONE correct answer

27.1 Which one of the following statements concerning dietary lipid is correct?

A. Corn oil and soybean oil are examples of fats rich in saturated fatty acids.
B. Triacylglycerols obtained from plants generally contain less unsaturated fatty acids than those from animals.
C. Olive oil is rich in saturated fats.
D. Fatty acids containing double bonds in the trans-configuration, unlike the naturally occurring cis isomers, raise plasma cholesterol levels.
E. Coconut and palm oils are rich in polyunsaturated fats.

Correct answer = D. Trans fatty acids raise plasma cholesterol levels. Corn oil and soybean oil are examples of fats rich in polyunsaturated fatty acids.Triacylglycerols obtained from plants generally contain more unsaturated fatty acids than those from animals. Olive oil, the staple of the Mediterranean diet, is rich in monounsaturated fats. Trans fatty acids raise plasma cholesterol levels. Coconut and palm oils are unusual plant oils in that they are rich in saturated fats.

27.2 Given the information that a 70-kg man is consuming a daily average of 275 g of carbohydrate, 75 g of protein, and 65 g of lipid, one can draw which of the following conclusions?

A. Total energy intake per day is approximately 3000 kcal.
B. About twenty percent of the calories are derived from lipids.
C. The diet does not contain a sufficient amount of dietary fiber.
D. The proportions of carbohydrate, protein, and lipid in the diet conform to the recommendations of academic groups and government agencies.
E. The individual is in nitrogen balance.

Correct answer = D. The total energy intake is (275 g carbohydrate × 4 kcal/g) + (75 g protein × 4 kcal/g) + (65 g lipid × 9 kcal/g) = 1100 + 300 + 585 = 1985 total kcal/day. The percent calories from carbohydrate is 1100/1985 = 55; percent calories from protein is 300/1985 = 15; and percent calories derived from lipid is 585/1985 = 30. These are very close to current recommendations. The amount of fiber or nitrogen balance cannot be deduced from the data presented. If the protein were of low biologic value, a negative nitrogen balance is possible.

27.3 A sedentary fifty-year-old man, weighing 80 kg (176 pounds), requests a physical examination. He denies any health problems. Routine blood analysis is unremarkable except for plasma cholesterol of 280 mg/dl. The man refuses drug therapy for his hypercholesterolemia. Analysis of a one-day dietary recall showed the following:

Kilocalories	3475 kcal	Cholesterol	822 mg
Protein	102 g	Saturated Fat	69 g
Carbohydrate	383 g	Total Fat	165 g
Fiber-Crude	6 g		

Changes in which one of the following dietary components would have the greatest effect in lowering plasma cholesterol?

A. Cholesterol
B. Saturated fat
C. Polyunsaturated fat
D. Monounsaturated fat
E. Carbohydrate

Correct answer = B. The intake of saturated fat most strongly influences plasma cholesterol in this diet. The patient is consuming a high-calorie, high-fat diet with forty percent of the fat as saturated fat. The most important dietary recommendations are: lower total caloric intake, substitute monounsaturated and polyunsaturated fats for saturated fats, and increase dietary fiber. A decrease in dietary cholesterol would be helpful, but not a primary objective.

Vitamins

28

I. OVERVIEW

Vitamins are chemically unrelated organic compounds that cannot be synthesized by humans and, therefore, must must be supplied by the diet. Nine vitamins (folic acid, cobalamin, ascorbic acid, pyridoxine, thiamine, niacin, riboflavin, biotin, and pantothenic acid) are classified as **water-soluble**, whereas four vitamins (vitamins A, D, K, and E) are termed **fat-soluble** (Figure 28.1). Vitamins are required to perform specific cellular functions, for example, many of the water-soluble vitamins are precursors of coenzymes for the enzymes of intermediary metabolism. In contrast to the water-soluble vitamins, only one fat soluble vitamin (vitamin K) has a coenzyme function. These vitamins are released, absorbed, and transported with the fat of the diet. They are not readily excreted in the urine, and significant quantities are stored in the liver and adipose tissue. In fact, consumption of vitamins A and D in excess of the recommended dietary allowances can lead to accumulation of toxic quantities of these compounds.

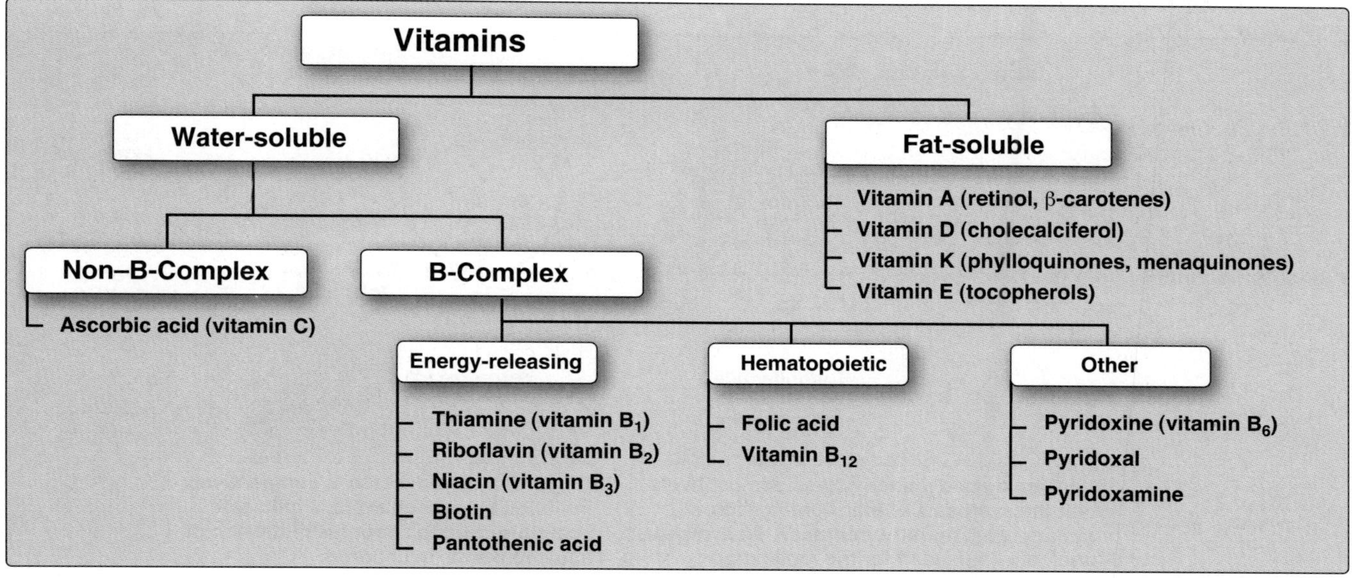

Figure 28.1
Classification of the vitamins.

Lippincott's Illustrated Reviews: Biochemistry, 3rd Edition
by Pamela C. Champe and Richard A. Harvey.
Lippincott Williams & Wilkins, Baltimore, MD © 2005.

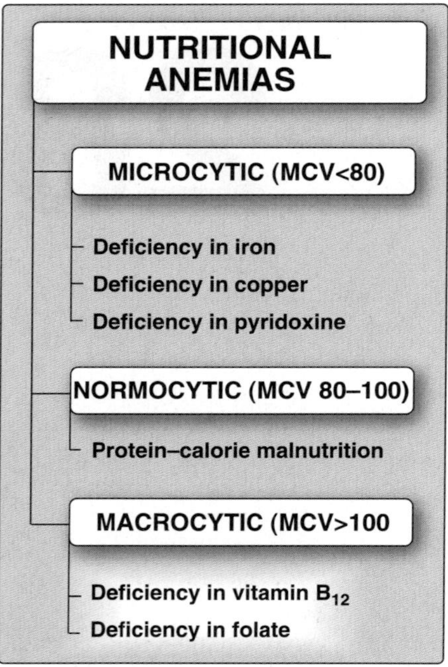

Figure 28.2
Classification of nutritional anemias by cell size. MCV = Mean corpuscular volume. The normal MCV level for people older than age 18 is between 80 and 100 μm.[3]

II. FOLIC ACID

Folic acid (or folate), which plays a key role in one-carbon metabolism, is essential for the biosynthesis of several compounds. Folic acid deficiency is probably the most common vitamin deficiency in the United States, particularly among pregnant women and alcoholics.

A. Function of folic acid

Tetrahydrofolate receives one-carbon fragments from donors such as serine, glycine, and histidine and transfers them to intermediates in the synthesis of amino acids, purines, and thymine—a pyrimidine found in DNA .

B. Nutritional anemias

Anemia is a condition in which the blood has a lower than normal concentration of hemoglobin, which results in a reduced ability to transport oxygen. Nutritional anemias—those caused by inadequate intake of one or more essential nutrients—can be classified according to the size of the red blood cells or mean corpuscular volume observed in the individual (Figure 28.2). **Microcytic anemia**, caused by lack of iron, is the most common form of nutritional anemia. The second major category of nutritional anemia results from a deficiency in folic acid or vitamin B_{12}. [Note: These **macrocytic anemias** are commonly called **megaloblastic** because a deficiency of folic acid or vitamin B_{12} causes accumulation of large, immature red cell precursors, known as megaloblasts, in the the bone marrow.

1. **Folate and anemia:** Inadequate serum levels of folate can be caused by increased demand (for example, pregnancy and lactation), poor absorption caused by pathology of the small intestine,

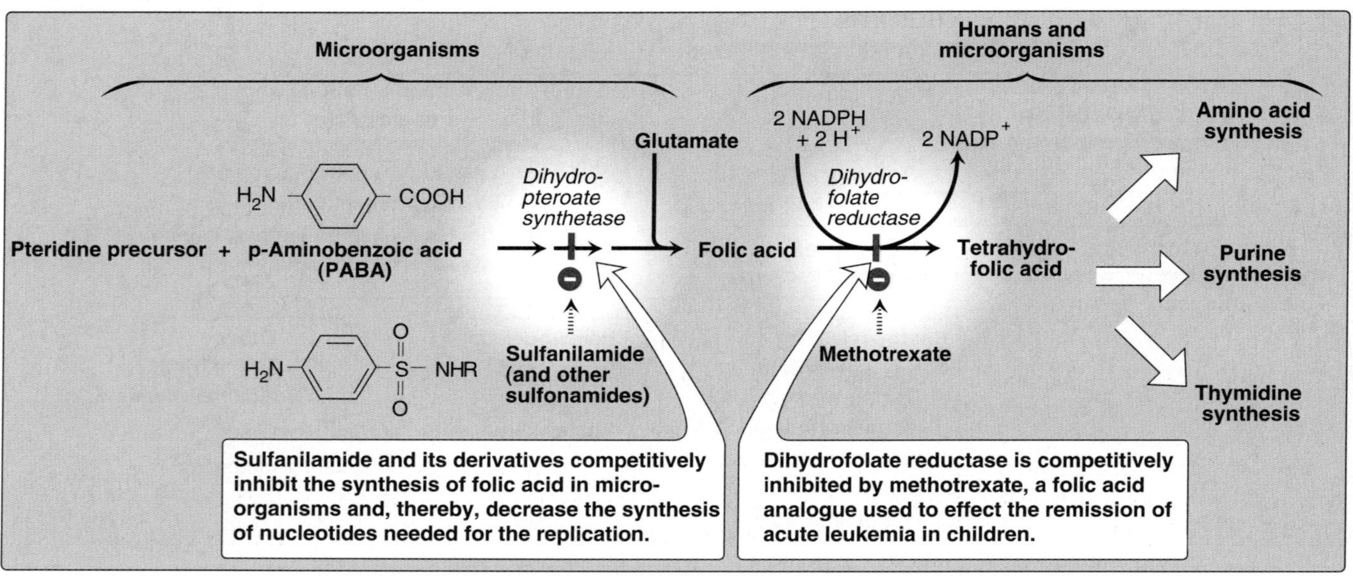

Figure 28.3
Inhibition of tetrahydrofolate synthesis by sulfonamides and trimethoprim.

alcoholism, or treatment with drugs that are *dihydrofolate reductase* inhibitors for example, methotrexate (Figure 28.3). A folate-free diet can cause a deficiency within a few weeks. A primary result of folic acid deficiency is **megaloblastic anemia** (Figure 28.4), caused by diminished synthesis of purines and thymidine, which leads to an inability of cells to make DNA and, therefore, they cannot divide. [Note: It is important to evaluate the cause of the megaloblastic anemia prior to instituting therapy, because vitamin B_{12} deficiency indirectly causes symptoms of this disorder (see p. 374).]

2. **Folate and neural tube defects in the fetus:** Spina bifida and anencephaly, the most common neural tube defects, affect approximately 4000 pregnancies in the United State annually. Folic acid supplementation before conception and during the first trimester has been shown to virtually eliminate the defects. Therefore, all women of childbearing age should consume 0.4 mg/day of folic acid to reduce the risk of having a pregnancy affected by neural tube defects. Adequate folate nutrition must occur at the time of conception because critical folate-dependent development occurs in the first weeks of fetal life—at a time when many women are not yet aware of their pregnancy. The U.S. Food and Drug Administration has authorized the addition of folic acid to enriched grain products, resulting in a dietary supplementation of about 0.1 mg/day. It is estimated that this supplementation will allow approximately fifty percent of all reproductive-aged women to receive 0.4 mg of folate from all sources. However, folic acid intake should not exceed approximately 1 mg/day to avoid complicating the diagnosis of vitamin B_{12} deficiency.

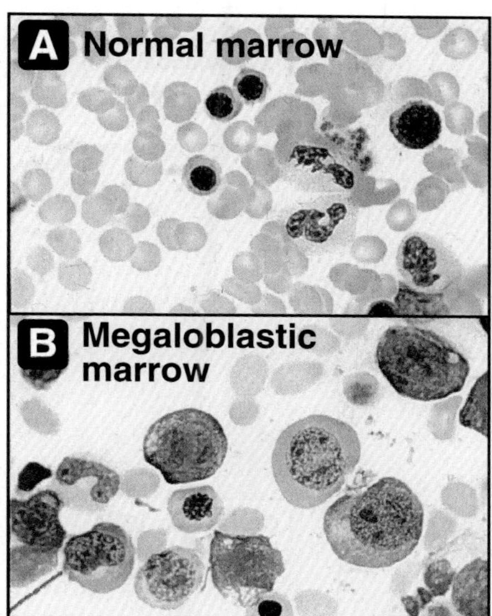

Figure 28.4
Bone marrow histology in normal and folate-deficient individuals.

III. COBALAMIN (VITAMIN B_{12})

Vitamin B_{12} is required in humans for two essential enzymatic reactions: the synthesis of methionine and the isomerization of methylmalonyl CoA that is produced during the degradation of some amino acids, and fatty acids with odd numbers of carbon atoms (Figure 28.5). When the vitamin is deficient, abnormal fatty acids accumulate and become incorporated into cell membranes, including those of the nervous system. This may account for some of the neurologic manifestations of vitamin B_{12} deficiency.

A. Structure of cobalamin and its coenzyme forms

Cobalamin contains a corrin ring system that differs from the porphyrins in that two of the pyrrole rings are linked directly rather than through a methene bridge. Cobalt is held in the center of the corrin ring by four coordination bonds from the nitrogens of the pyrrole groups. The remaining coordination bonds of the cobalt are with the nitrogen of 5,6-dimethylbenzimidazole and with cyanide in commercial preparations of the vitamin in the form of cyanocobalamin (Figure 28.6). The coenzyme forms of cobalamin are **5'-deoxyadenosylcobalamin**, in which cyanide is replaced with 5'-deoxyadenosine (forming an unusual carbon–cobalt bond), and **methylcobalamin**, in which cyanide is replaced by a methyl group (see Figure 28.6).

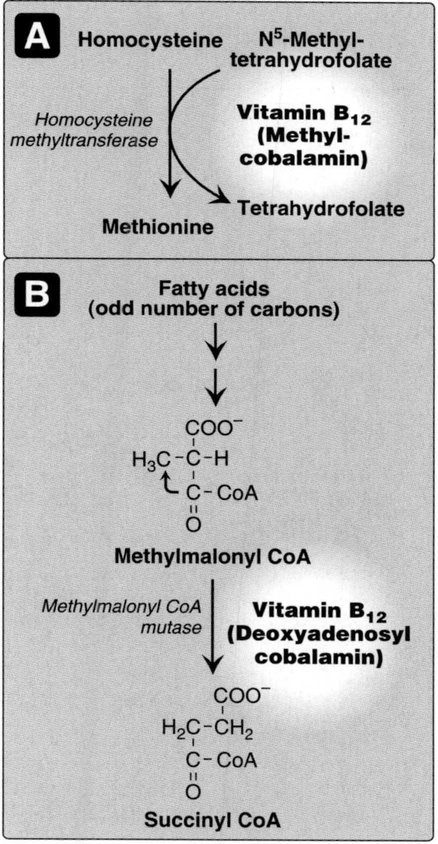

Figure 28.5
Reactions requiring cofactor forms of vitamin B_{12}.

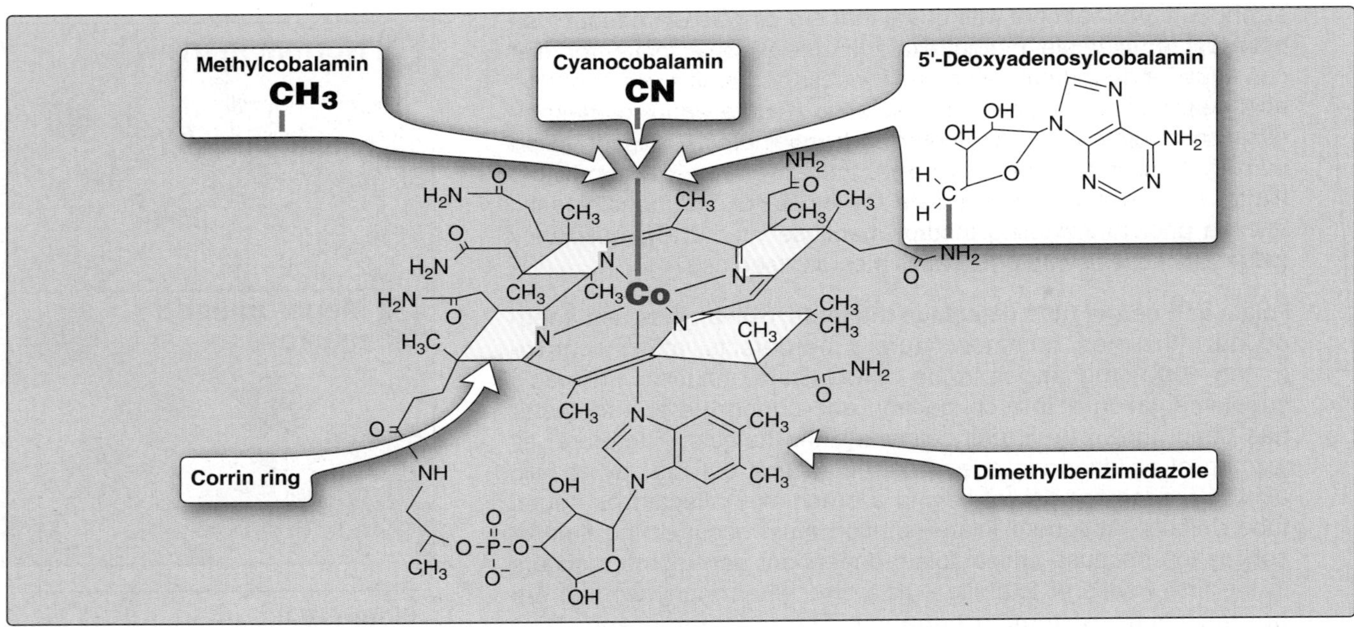

Figure 28.6
Structure of vitamin B_{12} (cyanocobalamin) and its coenzyme forms (methylcobalamin and 5'-deoxyadenosylcobalamin).

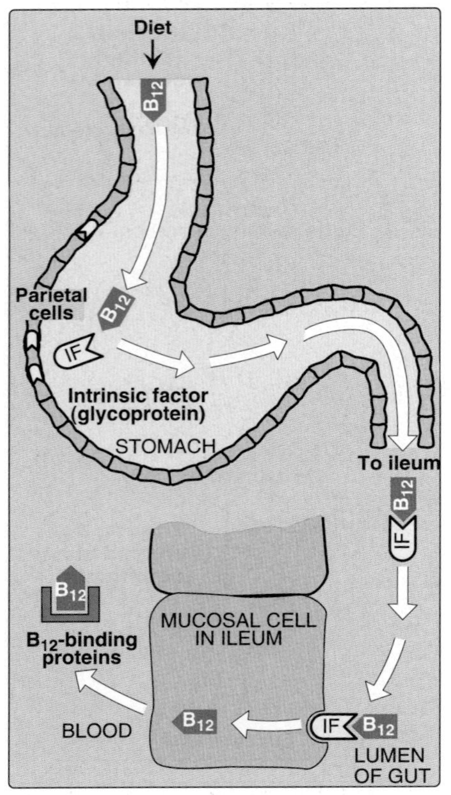

Figure 28.7
Absorption of vitamin B_{12}.

B. Distribution of cobalamin

Vitamin B_{12} is synthesized only by microorganisms; it is not present in plants. Animals obtain the vitamin preformed from their natural bacterial flora or by eating foods derived from other animals. Cobalamin is present in appreciable amounts in liver, whole milk, eggs, oysters, fresh shrimp, pork, and chicken.

C. Folate trap hypothesis

The effects of cobalamin deficiency are most pronounced in rapidly dividing cells, such as the erythropoietic tissue of bone marrow and the mucosal cells of the intestine. Such tissues need both the N^5-N^{10}-methylene and N^{10}-formyl forms of tetrahydrofolate for the synthesis of nucleotides required for DNA replication (see pp. 291, 301). However, in vitamin B_{12} deficiency, the N^5-methyl form of tetrahydrofolate is not efficiently used. Because the methylated form cannot be converted directly to other forms of tetrahydrofolate, the N^5-methyl form accumulates, whereas the levels of the other forms decrease. Thus, cobalamin deficiency is hypothesized to lead to a deficiency of the tetrahydrofolate forms needed in purine and thymine synthesis, resulting in the symptoms of megaloblastic anemia.

D. Clinical indications for vitamin B_{12}

In contrast to other water-soluble vitamins, significant amounts (4 to 5 mg) of vitamin B_{12} are stored in the body. As a result, it may take several years for the clinical symptoms of B_{12} deficiency to develop in individuals who have had a partial or total gastrectomy (who, therefore, become intrinsic factor-deficient) and can no longer absorb the vitamin.

1. **Pernicious anemia:** Vitamin B_{12} deficiency is rarely a result of an absence of the vitamin in the diet. It is much more common to find deficiencies in patients who fail to absorb the vitamin from the intestine, resulting in pernicious anemia. The disease is most commonly a result of an autoimmune destruction of the gastric parietal cells that are responsible for the synthesis of a glycoprotein called **intrinsic factor**. Normally, vitamin B_{12} obtained from the diet binds to intrinsic factor in the intestine (Figure 28.7). The cobalamin—intrinsic factor complex travels through the gut and eventually binds to specific receptors on the surface of mucosal cells of the ileum. The bound cobalamin is transported into the mucosal cell and, subsequently, into the general circulation, where it is carried by B_{12}-binding proteins. Lack of intrinsic factor prevents the absorption of vitamin B_{12}, resulting in pernicious anemia. Patients with cobalamin deficiency are usually anemic, but later in the development of the disease they show neuropsychiatric symptoms. However, central nervous system (CNS) symptoms may occur in the absence of anemia. The CNS effects are irreversible and occur by mechanisms that appear to be different from those described for megaloblastic anemia. The disease is treated by giving high-dose B_{12} orally, or intramuscular injection of cyanocobalamin. Therapy must be continued throughout the lives of patients with pernicious anemia. [Note: Folic acid administration alone reverses the hematologic abnormality and, thus, masks the B_{12} deficiency, which can then proceed to severe neurologic dysfunction and pathology; therefore, megaloblastic anemia should not be treated with folic acid alone, but rather with a combination of folate and vitamin B_{12}.]

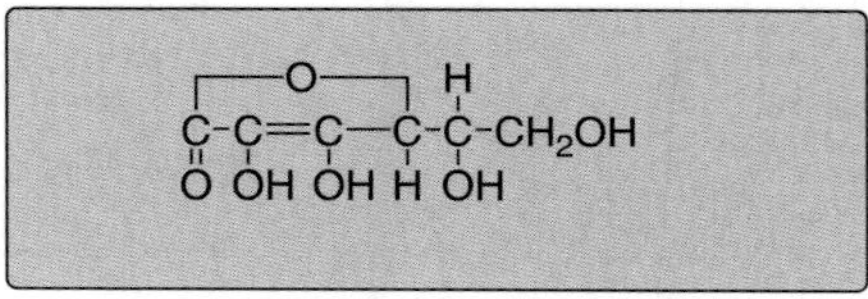

Figure 28.8
Structure of ascorbic acid.

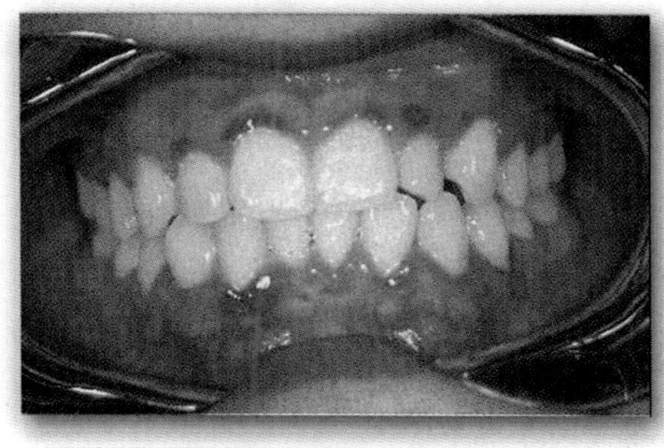

Figure 28.9
Hemorrhage and swollen gums of a patient with scurvy.

IV. ASCORBIC ACID (VITAMIN C)

The active form of vitamin C is ascorbate acid (Figure 28.8). The main function of ascorbate is as a reducing agent in several different reactions. Vitamin C has a well-documented role as a coenzyme in hydroxylation reactions, for example, hydroxylation of prolyl- and lysyl-residues of collagen (see p. 47). Vitamin C is, therefore, required for the maintenance of normal connective tissue, as well as for wound healing. Vitamin C also facilitates the absorption of dietary iron from the intestine.

A. Deficiency of ascorbic acid

A deficiency of ascorbic acid results in scurvy, a disease characterized by sore, spongy gums, loose teeth, fragile blood vessels, swollen joints, and anemia (Figure 28.9). Many of the deficiency symptoms can be explained by a deficiency in the hydroxylation of collagen, resulting in defective connective tissue.

B. Prevention of chronic disease:

Vitamin C is one of a group of nutrients that includes vitamin E (see p. 389) and β-carotene (see p. 380), which are known as **antioxidants**. Consumption of diets rich in these compounds is associated with a decreased incidence of some chronic diseases, such as coronary heart disease and certain cancers. However, clinical trials involving supplementation with the isolated antioxidants have failed to determine any convincing beneficial effects.

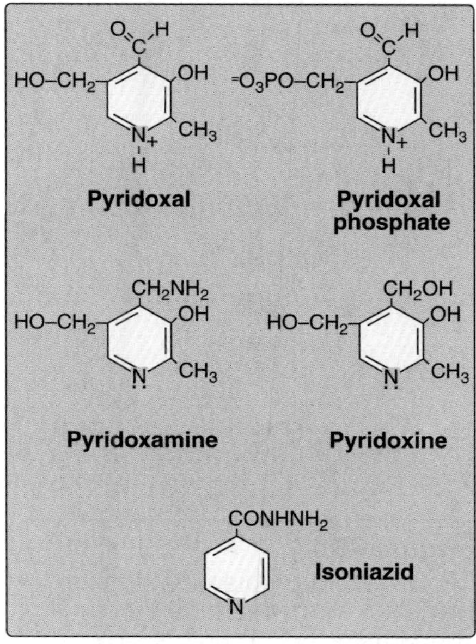

Figure 28.10
Structures of vitamin B_6 and and the anti-tuberculosis drug isoniazid.

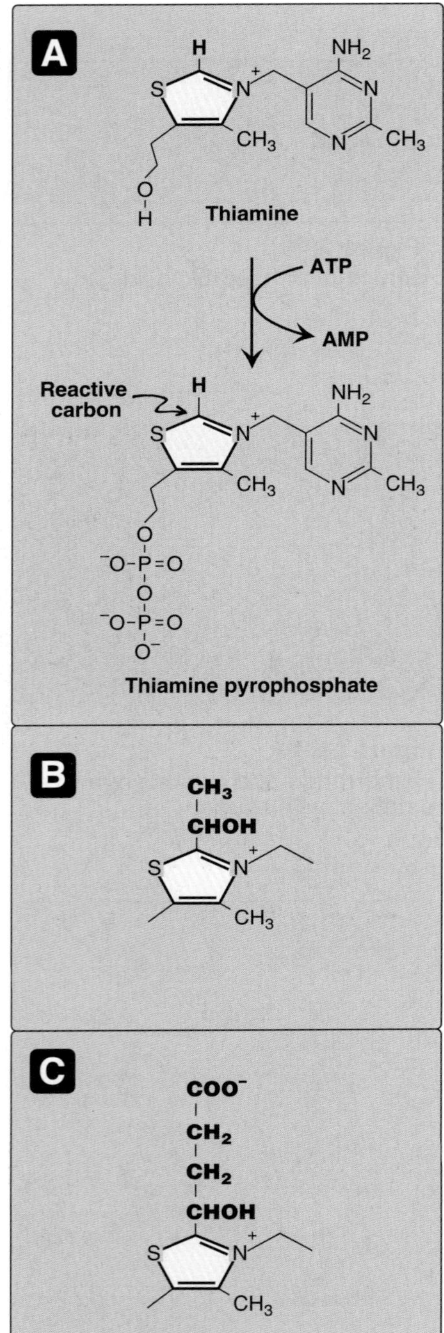

Figure 28.11
A. Structure of thiamine and its cofactor form, thiamine pyrophosphate. B. Structure of intermediate formed in the reaction catalyzed by *pyruvate dehydrogenase*. C. Structure of intermediate formed in the reaction catalyzed by *α-ketoglutarate dehydrogenase*.

V. PYRIDOXINE (VITAMIN B_6)

Vitamin B_6 is a collective term for **pyridoxine**, **pyridoxal**, and **pyridoxamine**, all derivatives of pyridine. They differ only in the nature of the functional group attached to the ring (Figure 28.10). Pyridoxine occurs primarily in plants, whereas pyridoxal and pyridoxamine are found in foods obtained from animals. All three compounds can serve as precursors of the biologically active coenzyme, **pyridoxal phosphate**. Pyridoxal phosphate functions as a coenzyme for a large number of enzymes, particularly those that catalyze reactions involving amino acids.

Reaction type	Example
Transamination	Oxaloacetate + glutamate ⇄ aspartate + α-ketoglutarate
Deamination	Serine → pyruvate + NH_3
Decarboxylation	Histidine → histamine + CO_2
Condensation	Glycine + succinyl CoA → δ-aminolevulinic acid

A. Clinical indications for pyridoxine:

Isoniazid (isonicotinic acid hydrazide), a drug frequently used to treat tuberculosis, can induce a B_6 deficiency by forming an inactive derivative with pyridoxal phosphate. Dietary supplementation with B_6 is, thus, an adjunct to isoniazide treatment. Otherwise, dietary deficiencies in pyridoxine are rare but have been observed in newborn infants fed formulas low in vitamin B_6, in women taking oral contraceptives, and in alcoholics.

B. Toxicity of pyridoxine

Neurologic symptoms have been observed at intakes of greater than 2 g/day. Substantial improvement, but not complete recovery, occurs when the vitamin is discontinued.

VI. THIAMINE (VITAMIN B_1)

Thiamine pyrophosphate (TPP) is the biologically active form of the vitamin, formed by the transfer of a pyrophosphate group from ATP to thiamine (Figure 28.11). Thiamine pyrophosphate serves as a coenzyme in the formation or degradation of α-ketols by transketolase (Figure 28.12A), and in the oxidative decarboxylation of α-keto acids (Figure 28.12B).

A. Clinical indications for thiamine

The oxidative decarboxylation of pyruvate and α-ketoglutarate, which plays a key role in energy metabolism of most cells, is particularly important in tissues of the nervous system. In **thiamine deficiency**, the activity of these two dehydrogenase reactions is decreased, resulting in a decreased production of ATP and, thus, impaired cellular function. [Note: Thiamine deficiency is diagnosed by an increase in erythrocyte *transketolase* activity observed on addition of thiamine pyrophosphate.]

1. **Beriberi:** This is a severe thiamine-deficiency syndrome found in areas where polished rice is the major component of the diet. Signs of infantile beriberi include tachycardia, vomiting, convulsions, and, if not treated, death. The deficiency syndrome can have a rapid onset in nursing infants whose mothers are deficient in thiamine. Adult beriberi is characterized by dry skin, irritability, disorderly thinking, and progressive paralysis.

2. **Wernicke-Korsakoff syndrome:** In the United States, thiamine deficiency, which is seen primarily in association with chronic alcoholism, is due to dietary insufficiency or impaired intestinal absorption of the vitamin. Some alcoholics develop Wernicke-Korsakoff syndrome—a deficiency state characterized by apathy, loss of memory, and a rhythmical to-and-fro motion of the eyeballs.

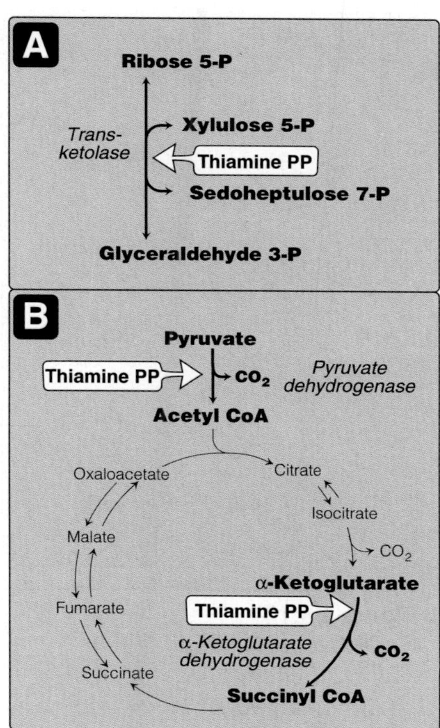

Figure 28.12
Reactions that use thiamine pyrophosphate (thiamine-PP) as coenzyme. A. *Transketolase*. B. *Pyruvate dehydrogenase* and *α-ketoglutarate dehydrogenase*.

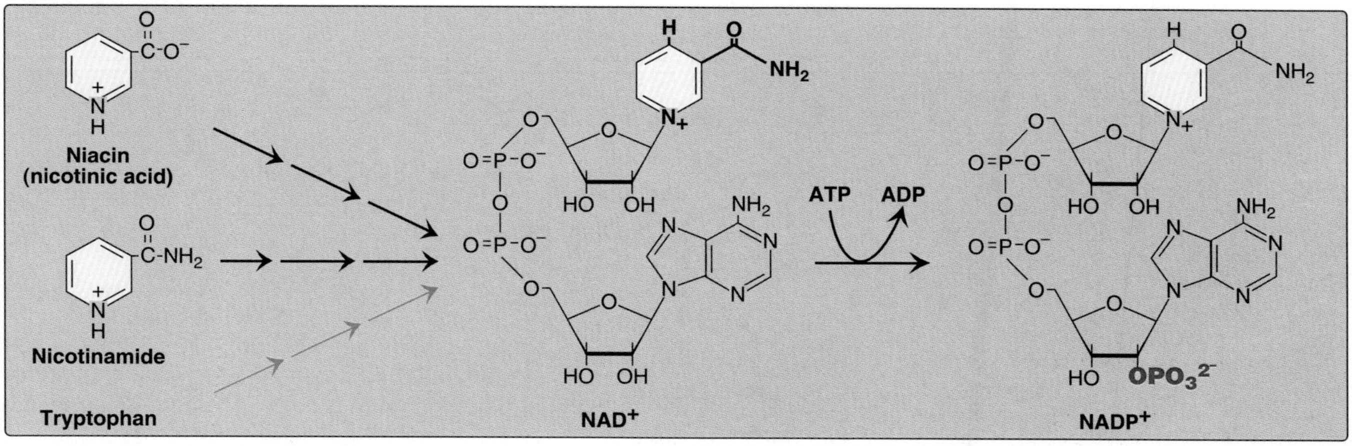

VII. NIACIN

Niacin, or **nicotinic acid**, is a substituted pyridine derivative. The biologically active coenzyme forms are **nicotinamide adenine dinucleotide** (**NAD⁺**) and its phosphorylated derivative, **nicotinamide adenine dinucleotide phosphate** (**NADP⁺**; Figure 28.13). Nicotinamide, a derivative of nicotinic acid that contains an amide instead of a carboxyl group, also occurs in the diet. Nicotinamide is readily deaminated in the body and, therefore, is nutritionally equivalent to nicotinic acid. NAD⁺ and NADP⁺ serve as coenzymes in oxidation-reduction reactions in which the coenzyme undergoes reduction of the pyridine ring by accepting a hydride ion (hydrogen atom plus one electron; Figure 28.14). The reduced forms of NAD⁺ and NADP⁺ are NADH and NADPH, respectively.

Figure 28.13
Structure and biosynthesis of NAD⁺ and NADP⁺.

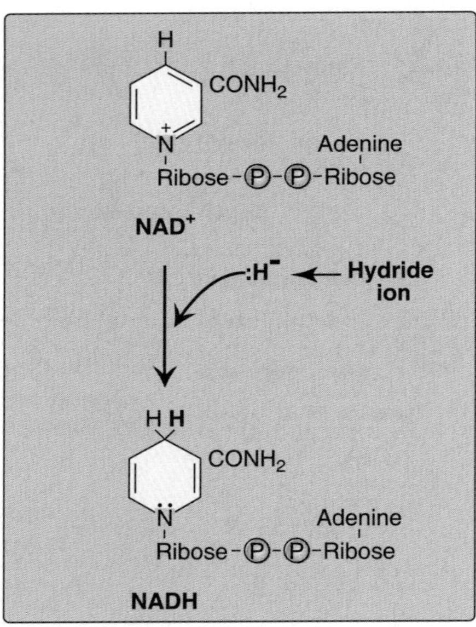

Figure 28.14
Reduction of NAD^+ to NADH.

A. Distribution of niacin

Niacin is found in unrefined and enriched grains and cereal, milk, and lean meats, especially liver. Limited quantities of niacin can also be obtained from the metabolism of tryptophan. [Note: The pathway is inefficient in that only about 1 mg of nicotinic acid is formed from 60 mg of tryptophan. Further, tryptophan is metabolized to niacin only when there is a relative abundance of the amino acid—that is, after the needs for protein synthesis and energy production have been met.]

B. Clinical indications for niacin

1. **Deficiency of niacin:** A deficiency of niacin causes **pellagra**, a disease involving the skin, gastrointestinal (GI) tract, and CNS. The symptoms of pellagra progress through the three **D**s: **d**ermatitis, **d**iarrhea, **d**ementia, and, if untreated, death.

2. **Treatment of hyperlipidemia:** Niacin (at doses of 1.5 g/day or 100 times the RDA) strongly inhibits lipolysis in adipose tissue—the primary producer of circulating free fatty acids. The liver normally uses these circulating fatty acids as a major precursor for triacylglycerol synthesis. Thus, niacin causes a decrease in liver triacylglycerol synthesis, which is required for very-low-density lipoprotein (VLDL, see p. 229) production. Low-density lipoprotein (LDL, the cholesterol-rich lipoprotein) is derived from VLDL in the plasma. Thus, both plasma triacylglycerol (in VLDL) and cholesterol (in VLDL and LDL) are lowered. Therefore, niacin is particularly useful in the treatment of **type IIb hyperlipoproteinemia**, in which both VLDL and LDL are elevated.

VIII. RIBOFLAVIN (VITAMIN B_2)

The two biologically active forms are **flavin mononucleotide** (**FMN**) and **flavin adenine dinucleotide** (**FAD**), formed by the transfer of an AMP moiety from ATP to FMN (Figure 28.15). FMN and FAD are each capable of reversibly accepting two hydrogen atoms, forming $FMNH_2$ or $FADH_2$. FMN and FAD are bound tightly—sometimes covalently—to flavoenzymes that catalyze the oxidation or reduction of a substrate.

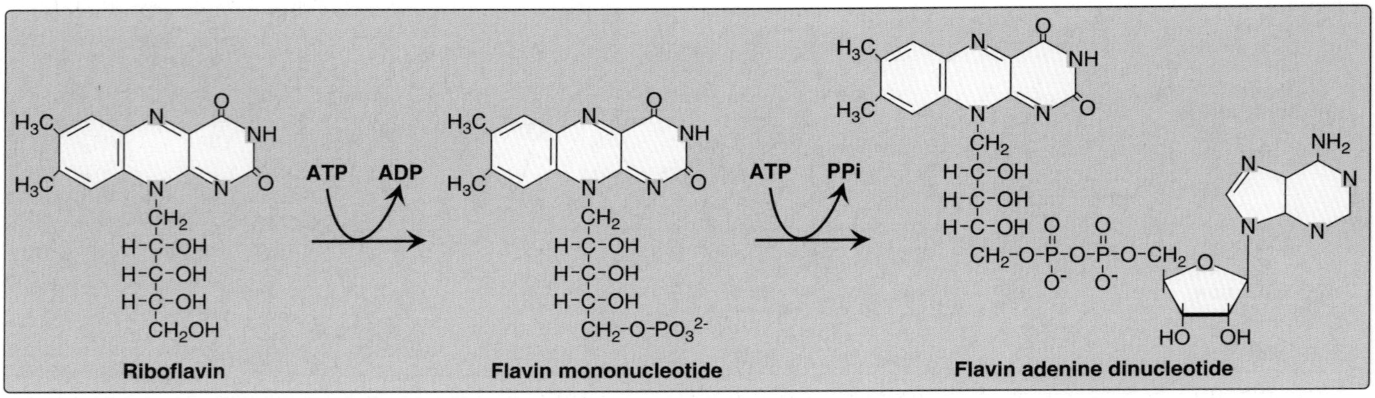

Riboflavin → (ATP → ADP) → Flavin mononucleotide → (ATP → PPi) → Flavin adenine dinucleotide

Figure 28.15
Structure and biosynthesis of flavin mononucleotide (FMN) and flavin adenine dinucleotide (FAD).

Riboflavin deficiency is not associated with a major human disease, although it frequently accompanies other vitamin deficiencies. Deficiency symptoms include dermatitis, cheilosis (fissuring at the corners of the mouth), and glossitis (the tongue appearing smooth and purplish).

IX. BIOTIN

Biotin is a coenzyme in carboxylation reactions, in which it serves as a carrier of activated carbon dioxide (see Figure 10.3, p. 117 for the mechanism of biotin-dependent carboxylations). Biotin is covalently bound to the ε-amino groups of lysine residues of biotin-dependent enzymes (Figure 28.16). Biotin deficiency does not occur naturally because the vitamin is widely distributed in food. Also, a large percentage of the biotin requirement in humans is supplied by intestinal bacteria. However, the addition of raw egg-white to the diet as a source of protein induces symptoms of biotin deficiency, namely, dermatitis, glossitis, loss of appetite, and nausea. Raw egg white contains a glycoprotein, **avidin**, which tightly binds biotin and prevents its absorption from the intestine. However, with a normal diet, it has been estimated that 20 eggs per day would be required to induce a deficiency syndrome. Thus, inclusion of an occasional raw egg in the diet does not lead to biotin deficiency.

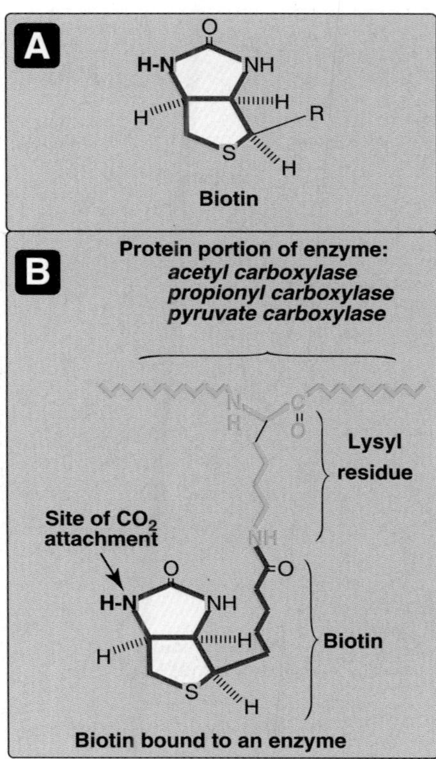

Figure 28.16
A. Structure of biotin. B. Biotin covalently bound to a lysyl residue of a biotin-dependent enzyme.

X. PANTOTHENIC ACID

Pantothenic acid is a component of coenzyme A, which functions in the transfer of acyl groups (Figure 28.17). Coenzyme A contains a thiol group that carries acyl compounds as activated thiol esters. Examples of such structures are succinyl CoA, fatty acyl CoA, and acetyl CoA. Pantothenic acid is also a component of *fatty acid synthase* (see p. 182). Eggs, liver, and yeast are the most important sources of pantothenic acid, although the vitamin is widely distributed. Pantothenic acid deficiency is not well characterized in humans, and no RDA has been established.

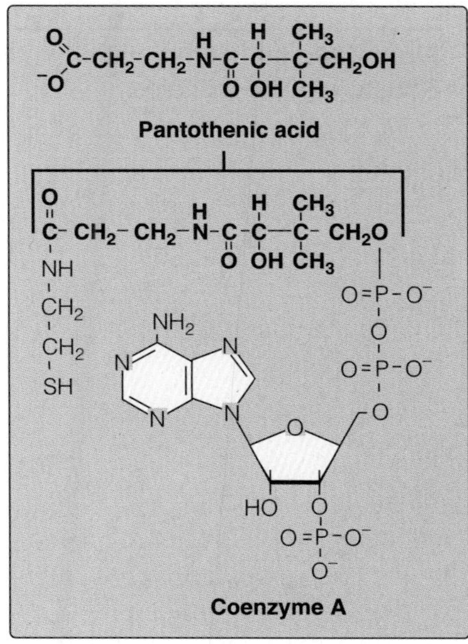

Figure 28.17
Structure of coenzyme A.

XI. VITAMIN A

The **retinoids**, a family of molecules that are related to retinol (vitamin A), are essential for vision, reproduction, growth, and maintenance of epithelial tissues. **Retinoic acid**, derived from oxidation of dietary **retinol**, mediates most of the actions of the retinoids, except for vision, which depends on **retinal**, the aldehyde derivative of retinol.

A. Structure of vitamin A

Vitamin A is often used as a collective term for several related biologically active molecules (Figure 28.18). The term retinoids includes both natural and synthetic forms of vitamin A that may or may not show vitamin A activity.

1. **Retinol:** A primary alcohol containing a β-ionone ring with an unsaturated side chain, retinol is found in animal tissues as a retinyl ester with long-chain fatty acids.

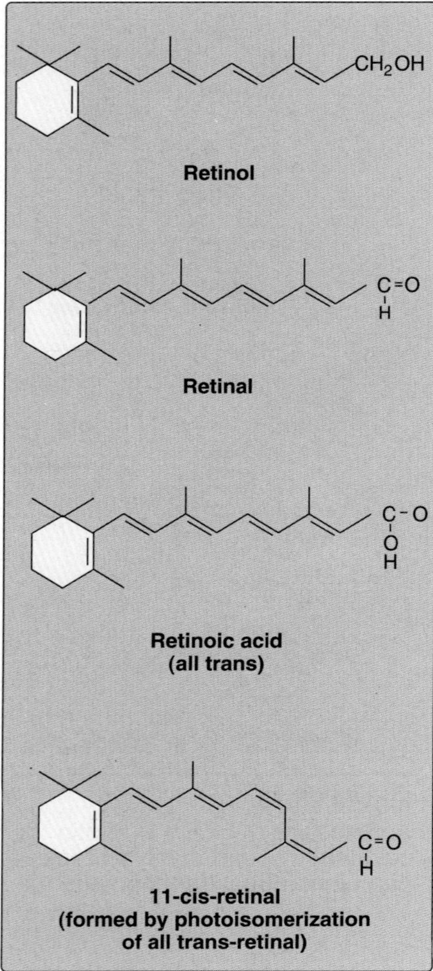

Figure 28.18
Structure of the retinoids.

2. **Retinal:** This is the aldehyde derived from the oxidation of retinol. Retinal and retinol can readily be interconverted.

3. **Retinoic acid:** This is the acid derived from the oxidation of retinal. Retinoic acid cannot be reduced in the bod, and, therefore, cannot give rise to either retinal or retinol.

4. **β-carotene:** Plant foods contain β-carotene, which can be oxidatively cleaved in the intestine to yield two molecules of retinal. In humans, the conversion is inefficient, and the vitamin A activity of β-carotene is only about one sixth that of retinol.

B. Absorption and transport of vitamin A

1. **Transport to the liver:** Retinol esters present in the diet are hydrolyzed in the intestinal mucosa, releasing retinol and free fatty acids (Figure 28.19). Retinol derived from esters and from the cleavage and reduction of carotenes is reesterified to long-chain fatty acids in the intestinal mucosa and secreted as a component of chylomicrons into the lymphatic system (see Figure 28.19). Retinol esters contained in chylomicrons are taken up by, and stored in, the liver.

2. **Release from the liver:** When needed, retinol is released from the liver and transported to extrahepatic tissues by the **plasma retinol-binding protein** (RBP). The retinol–RBP complex attaches to specific receptors on the surface of the cells of peripheral tissues, permitting retinol to enter. Many tissues contain a **cellular retinol-binding protein** that carries retinol to sites in the nucleus where the vitamin acts in a manner analogous to steroid hormones.

C. Mechanism of action of vitamin A

Retinoic acid binds with high-affinity to specific receptor proteins present in the nucleus of target tissues, such as epithelial cells (Figure 28.20). The activated retinoic acid–receptor complex interacts with nuclear chromatin to stimulate retinoid-specific RNA synthesis, resulting in the production of specific proteins that mediate several physiologic functions. For example, retinoids control the expression of the keratin gene in most epithelial tissues of the body. The specific retinoic acid–receptor proteins are part of the superfamily of transcriptional regulators that includes the steroid and thyroid hormones and 1,25-dihydroxycholecalciferol, all of which function in a similar way.

D. Functions of vitamin A

1. **Visual cycle:** Vitamin A is a component of the visual pigments of rod and cone cells. **Rhodopsin**, the visual pigment of the rod cells in the retina, consists of **11-cis retinal** specifically bound to the protein **opsin**. When rhodopsin is exposed to light, a series of photochemical isomerizations occurs, which results in the bleaching of the visual pigment and release of all trans retinal and opsin. This process triggers a nerve impulse that is transmitted by the optic nerve to the brain. Regeneration of rhodopsin requires isomerization of all trans retinal back to 11-cis retinal. Trans retinal, after being released from rhodopsin, is isomerized to 11-cis retinal, which spontaneously combines with opsin to form rhodopsin,

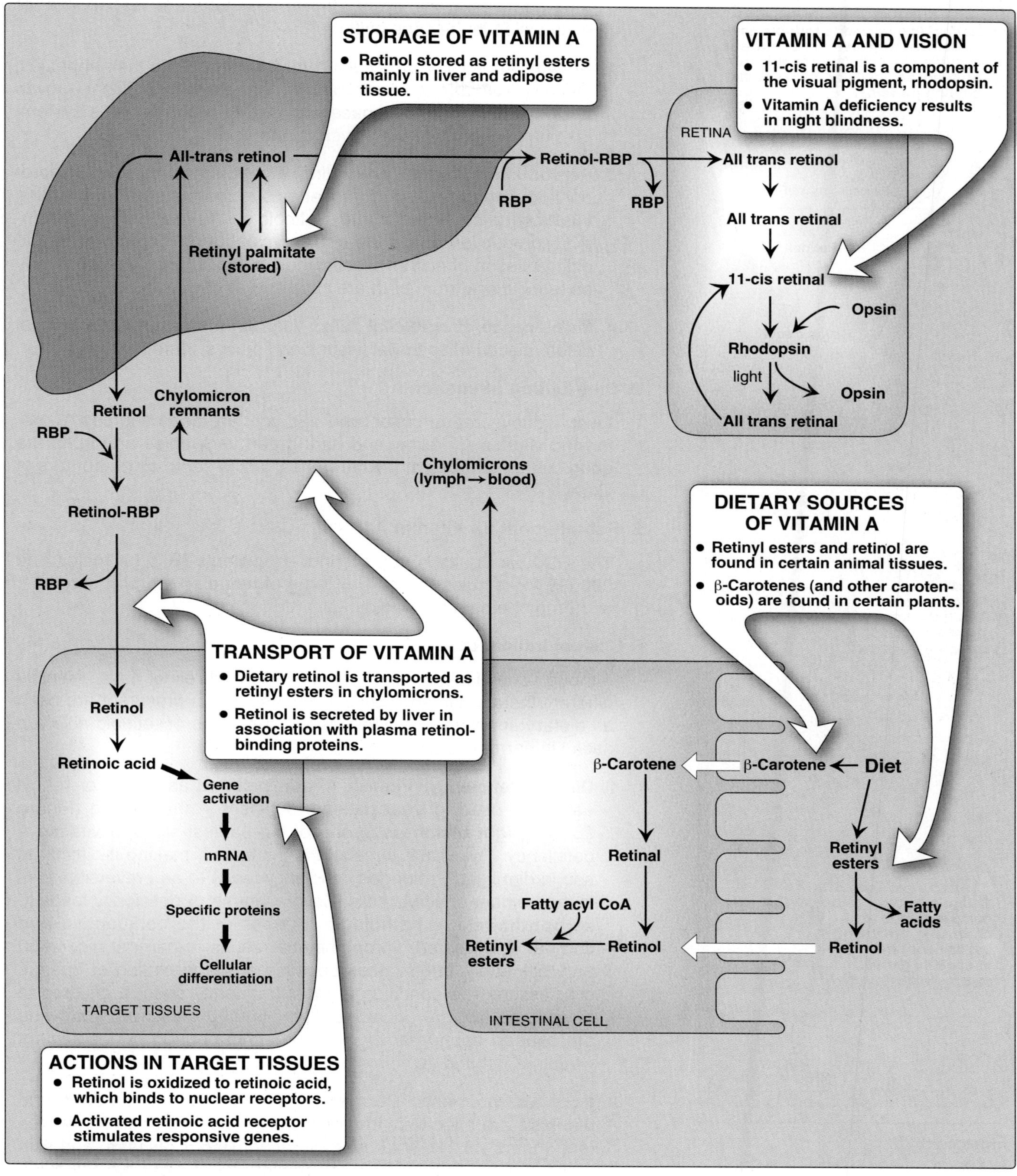

Figure 28.19
Absorption, transport, and storage of vitamin A and its derivatives. RBP = retinol-binding protein.

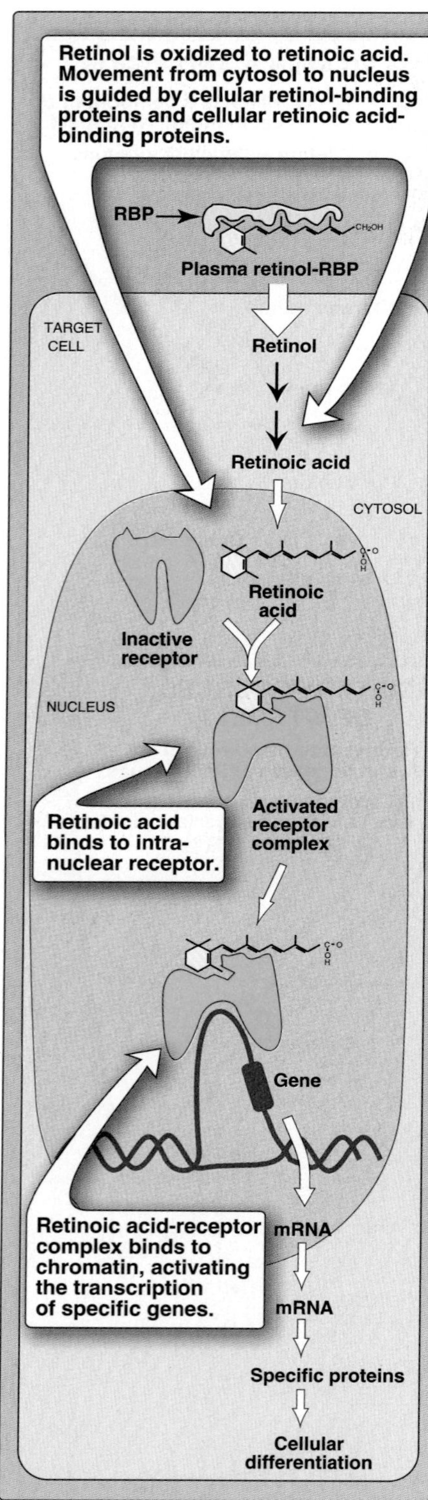

Figure 28.20
Action of retinoids (RBP = retinol-binding protein).

thus completing the cycle. Similar reactions are responsible for color vision in the cone cells.

2. **Growth:** Animals deprived of vitamin A initially lose their appetites, possibly because of keratinization of the taste buds. Bone growth is slow and fails to keep pace with growth of the nervous system, leading to central nervous system damage.

3. **Reproduction:** Retinol and retinal are essential for normal reproduction, supporting spermatogenesis in the male and preventing fetal resorption in the female. Retinoic acid is inactive in maintaining reproduction and in the visual cycle, but promotes growth and differentiation of epithelial cells; thus, animals given vitamin A only as retinoic acid from birth are blind and sterile.

4. **Maintenance of epithelial cells:** Vitamin A is essential for normal differentiation of epithelial tissues and mucus secretion.

D. Distribution of vitamin A

Liver, kidney, cream, butter, and egg yolk are good sources of preformed vitamin A. Yellow and dark green vegetables and fruits are good dietary sources of the carotenes, which serve as precursors of vitamin A.

E. Requirement for vitamin A

The RDA for adults is 1000 retinol equivalents (RE) for males and 800 RE for females. One RE = 1 mg of retinol, 6 mg of β-carotene, or 12 mg of other carotenoids.

F. Clinical indications

Although chemically related, retinoic acid and retinol have distinctly different therapeutic applications. Retinol and its precursor are used as dietary supplements, whereas various forms of retinoic acid are useful in dermatology.

1. **Dietary deficiency:** Vitamin A, administered as retinol or retinyl esters, is used to treat patients deficient in the vitamin (Figure 28.21). **Night blindness** is one of the earliest signs of vitamin A deficiency. The visual threshold is increased, making it difficult to see in dim light. Prolonged deficiency leads to an irreversible loss in the number of visual cells. Severe vitamin A deficiency leads to **xerophthalmia**, a pathologic dryness of the conjunctiva and cornea. If untreated, xerophthalmia results in corneal ulceration and, ultimately, in blindness because of the formation of opaque scar tissue. The condition is most frequently seen in children in developing tropical countries. Over 500,000 children worldwide are blinded each year by xerophthalmia caused by insufficient vitamin A in the diet.

2. **Acne and psoriasis:** Dermatologic problems such as acne and psoriasis are effectively treated with retinoic acid or its derivatives (see Figure 28.21). Mild cases of acne, **Darier disease**, and skin aging are treated with topical application of **tretinoin** (all trans retinoic acid), as well as benzoyl peroxide and antibiotics. [Note: Tretinoin is too toxic for systemic administration and is confined to topical application.] In patients with severe recalcitrant cystic acne

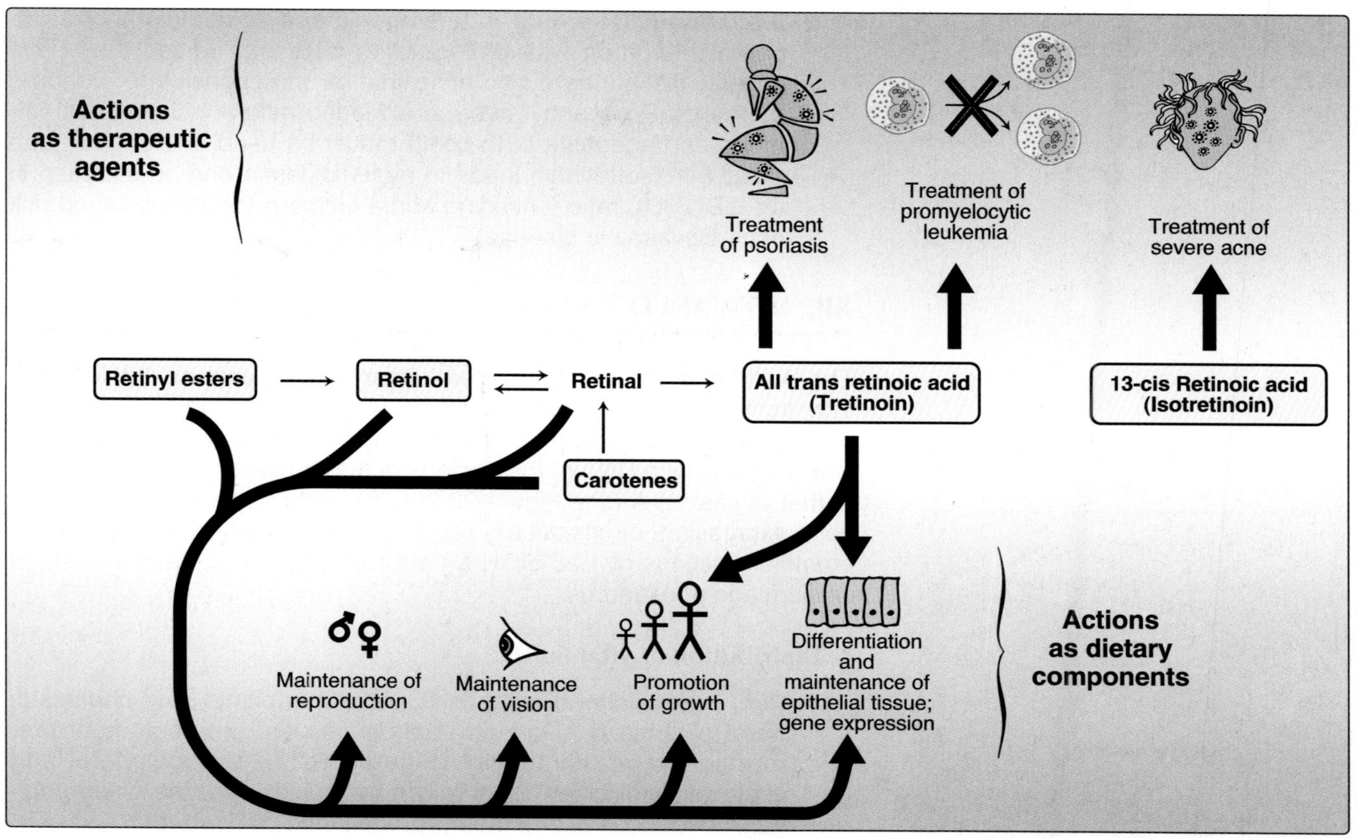

Figure 28.21
Summary of actions of retinoids. Compounds in boxes are available as dietary components or as pharmacologic agents.

unresponsive to conventional therapies, the drug of choice is **isotretinoin** (13-cis retinoic acid) administered orally.

3. **Prevention of chronic disease:** Populations consuming diets high in β-carotene show decreased incidence of heart disease and lung and skin cancer (see Figure 28.21). Consumption of foods rich in β-carotene is also associated with reduced risk of cataracts and macular degeneration. However, in clinical trials, β-carotene supplementation not only did not decrease the incidence of lung cancer, but actually increased cancer in individuals who smoke. Subjects in a clinical trial who received high doses of β-carotene unexpectedly had increased death due to heart disease.

G. Toxicity of retinoids

1. **Vitamin A:** Excessive intake of vitamin A produces a toxic syndrome called **hypervitaminosis A**. Amounts exceeding 7.5 mg/day of retinol should be avoided. Early signs of chronic hypervitaminosis A are reflected in the skin, which becomes dry and pruritic, the liver, which becomes enlarged and can become cirrhotic, and in the nervous system, where a rise in intracranial pressure may mimic the symptoms of a brain tumor. Pregnant women particularly should not ingest excessive quantities of vitamin A because of its potential for causing congenital malformations in the developing fetus.

2. **Isotretinoin:** The drug is teratogenic and absolutely contraindicated in women with childbearing potential unless they have severe, disfiguring cystic acne that is unresponsive to standard therapies. Pregnancy must be excluded before initiation of treatment, and adequate birth control must be used. Prolonged treatment with isotretinoin leads to hyperlipidemia and an increase in the LDL/HDL ratio, providing some concern for an increased risk of cardiovascular disease.

XII. VITAMIN D

The D vitamins are a group of sterols that have a hormone-like function. The active molecule, 1,25-dihydroxycholecalciferol (1,25 diOH D_3), binds to intracellular receptor proteins. The 1,25-diOH D_3-receptor complex interacts with DNA in the nucleus of target cells in a manner similar to that of vitamin A (see Figure 28.20), and either selectively stimulates gene expression, or specifically represses gene transcription. The most prominent actions of 1,25-diOH D_3 are to regulate the plasma levels of calcium and phosphorus.

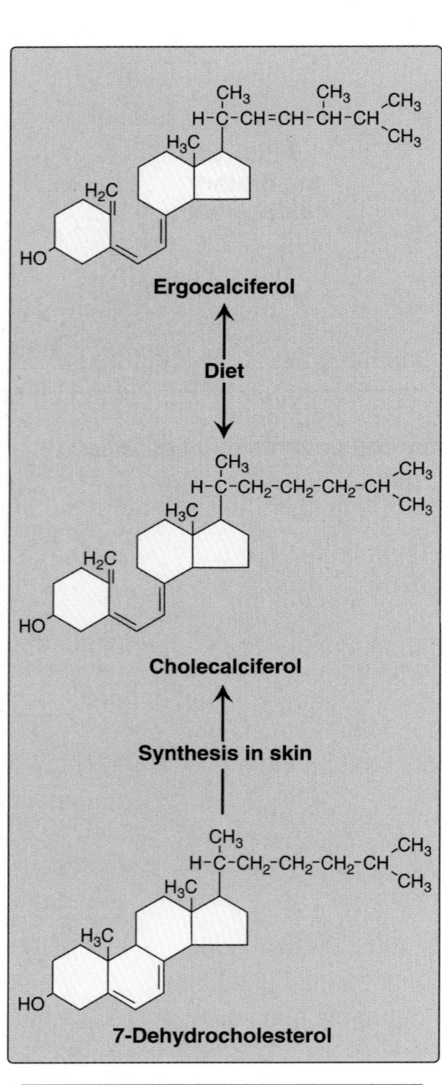

Figure 28.22
Sources of vitamin D.

A. Distribution of vitamin D

1. **Diet: Ergocalciferol** (vitamin D_2), found in plants, and **cholecalciferol** (vitamin D_3), found in animal tissues, are sources of preformed vitamin D activity (Figure 28.22). Ergocalciferol and cholecalciferol differ chemically only in the presence of an additional double bond and methyl group in the plant sterol.

2. **Endogenous vitamin precursor:** 7-Dehydrocholesterol, an intermediate in cholesterol synthesis, is converted to cholecalciferol in the dermis and epidermis of humans exposed to sunlight. Preformed vitamin D is a dietary requirement only in individuals with limited exposure to sunlight.

B. Metabolism of vitamin D

1. **Formation of 1,25-diOH D_3:** Vitamins D_2 and D_3 are not biologically active, but are converted in vivo to the active form of the D vitamin by two sequential hydroxylation reactions (Figure 28.23). The first hydroxylation occurs at the 25-position, and is catalyzed by a specific hydroxylase in the liver. The product of the reaction, 25-hydroxycholecalciferol (25-OH D_3), is the predominant form of vitamin D in the plasma and the major storage form of the vitamin. 25-OH D_3 is further hydroxylated at the one position by a specific *25-hydroxycholecalciferol 1-hydroxylase* found primarily in the kidney, resulting in the formation of 1,25-dihydroxycholecalciferol (1,25-diOH D_3). [Note: This hydroxylase, as well as the liver *25-hydroxylase*, employ cytochrome P450, molecular oxygen, and NADPH.]

2. **Regulation of 25-hydroxycholecalciferol 1-hydroxylase:** 1,25-diOH D_3 is the most potent vitamin D metabolite. Its formation is tightly regulated by the level of plasma phosphate and calcium ions (Figure 28.24). *25-Hydroxycholecalciferol 1-hydroxylase* activity is increased directly by low plasma phosphate or indirectly by low plasma calcium, which triggers the release of parathyroid hormone

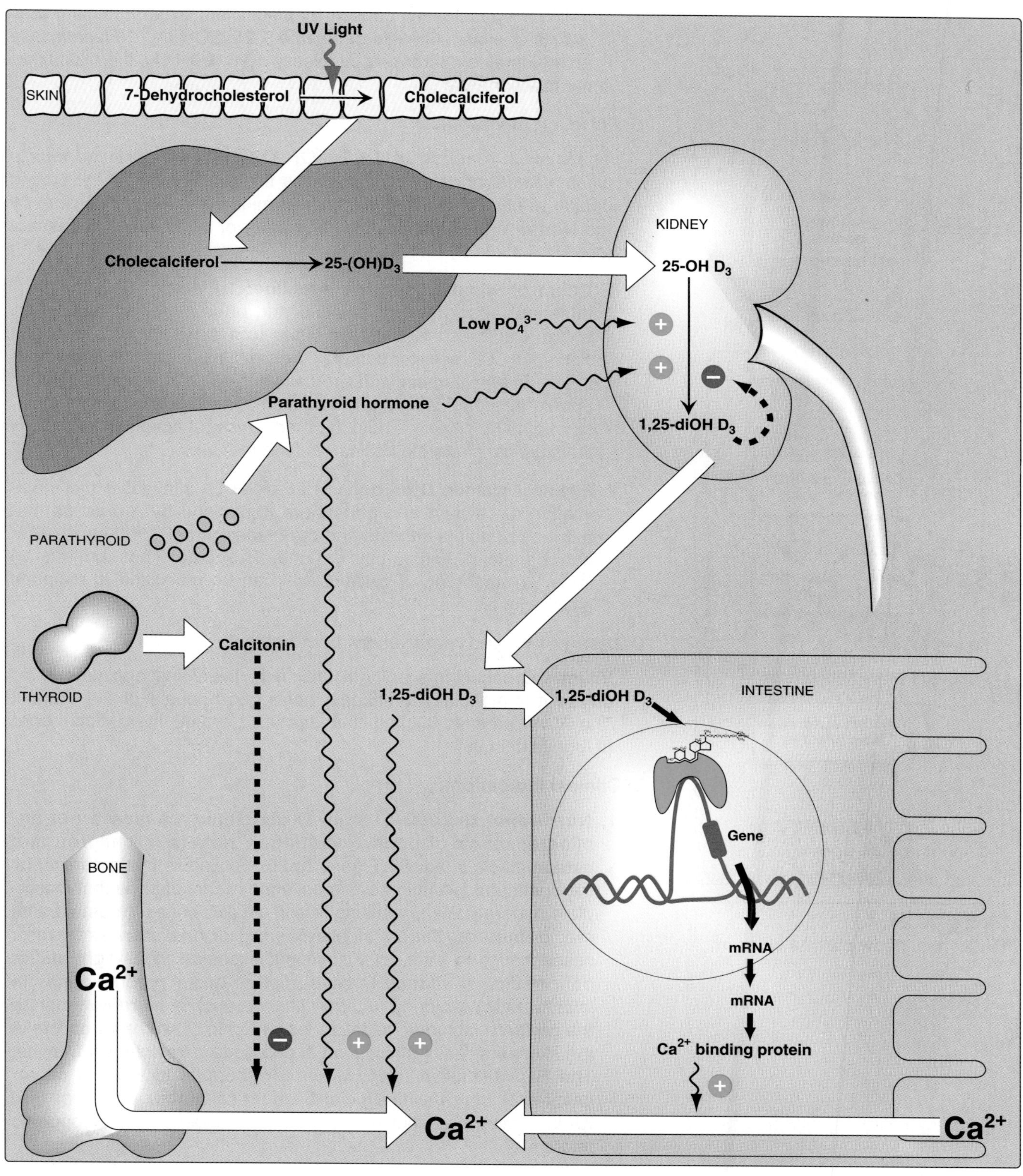

Figure 28.23
Metabolism and actions of vitamin D.

(PTH). Hypocalcemia caused by insufficient dietary calcium thus results in elevated levels of plasma 1,25 diOH D_3. *1-Hydroxylase* activity is also decreased by excess 1,25 diOH D_3, the product of the reaction.

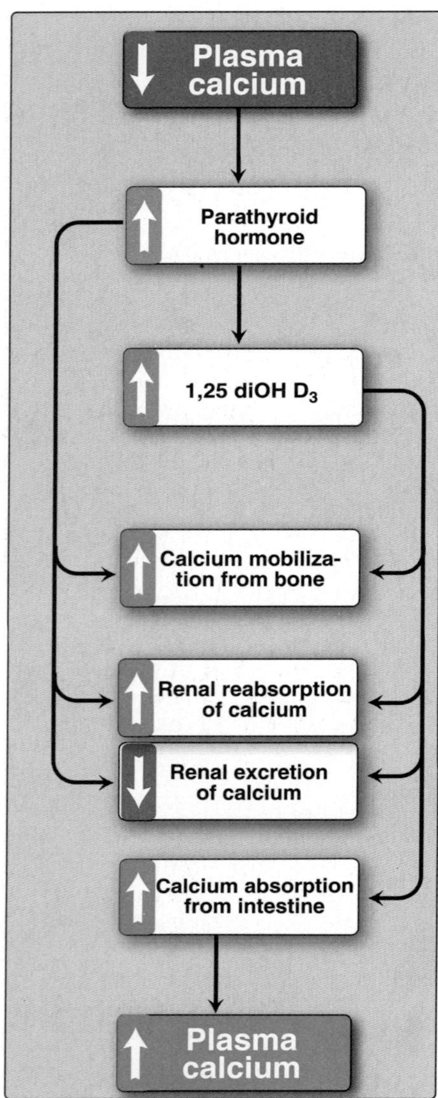

Figure 28.24
Response to low plasma calcium.

C. Function of vitamin D

The overall function of 1,25-diOH D3 is to maintain adequate plasma levels of calcium. It performs this function by: 1) increasing uptake of calcium by the intestine, 2) minimizing loss of calcium by the kidney, and 3) stimulating resorption of bone when necessary (see Figure 28.23).

1. **Effect of vitamin D on the intestine:** 1,25-diOH D_3 stimulates intestinal absorption of calcium and phosphate. 1,25-diOH D_3 enters the intestinal cell and binds to a cytosolic receptor. The 1,25-diOH D_3-receptor complex then moves to the nucleus where it selectively interacts with the cellular DNA. As a result, calcium uptake is enhanced by an increased synthesis of a specific calcium-binding protein. Thus, the mechanism of action of 1,25-diOH D_3 is typical of steroid hormones (see p. 238).

2. **Effect of vitamin D on bone:** 1,25-diOH D_3 stimulates the mobilization of calcium and phosphate from bone by a process that requires protein synthesis and the presence of PTH. The result is an increase in plasma calcium and phosphate. Thus, bone is an important reservoir of calcium that can be mobilized to maintain plasma levels.

D. Distribution and requirement of vitamin D

Vitamin D occurs naturally in fatty fish, liver, and egg yolk. Milk, unless it is artificially fortified, is not a good source of the vitamin. The RDA for adults is 5 mg cholecalciferol, or 200 international units (IU) of vitamin D.

E. Clinical indications

1. **Nutritional rickets:** Vitamin D deficiency causes a net demineralization of bone, resulting in **rickets** in children and **osteomalacia** in adults (Figure 28.25). Rickets is characterized by the continued formation of the collagen matrix of bone, but incomplete mineralization, resulting in soft, pliable bones. In osteomalacia, demineralization of preexisting bones increases their susceptibility to fracture. Insufficient exposure to daylight and/or deficiencies in vitamin D consumption occur predominantly in infants and the elderly. Vitamin D deficiency is more common in the northern latitudes, because less vitamin D synthesis occurs in the skin as a result of reduced exposure to ultraviolet light. [Note: The RDA of 200 IU/day (which corresponds to 5 μg of cholecalciferol) may be insufficient, because higher doses of 800 IU/day have been shown to reduce the incidence of osteoporotic fractures.]

2. **Renal rickets (renal osteodystrophy):** This disorder results from chronic renal failure and, thus, the decreased ability to form the active form of the vitamin. 1,25-diOH cholecalciferol (calcitriol) administration is effective replacement therapy.

3. **Hypoparathyroidism:** Lack of parathyroid hormone causes hypocalcemia and hyperphosphatemia. These patients may be treated with any form of vitamin D, together with parathyroid hormone.

F. Toxicity of vitamin D

Vitamin D is the most toxic of all vitamins. Like all fat-soluble vitamins, vitamin D can be stored in the body and is only slowly metabolized. High doses (100,000 IU for weeks or months) can cause loss of appetite, nausea, thirst, and stupor. Enhanced calcium absorption and bone resorption results in hypercalcemia, which can lead to deposition of calcium in many organs, particularly the arteries and kidneys.

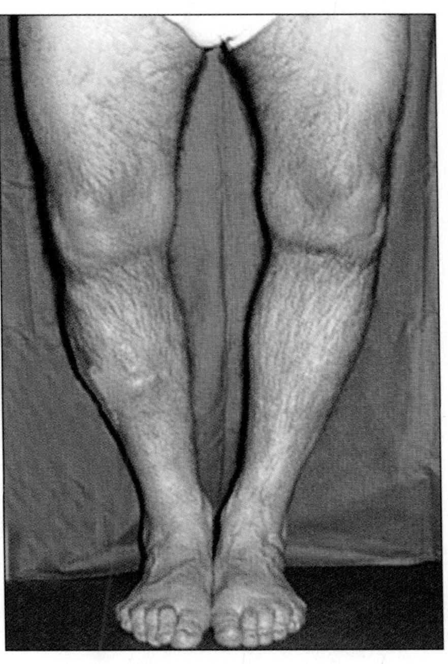

Figure 28.25
Bowed legs of middle-aged man with osteomalacia, a nutritional vitamin D deficiency which results in malformation of the skeleton.

XIII. VITAMIN K

The principal role of vitamin K is in the post-translational modification of various blood clotting factors, in which it serves as a coenzyme in the carboxylation of certain glutamic acid residues present in these proteins. Vitamin K exists in several forms, for example, in plants as **phylloquinone** (or **vitamin K_1**), and in intestinal bacterial flora as **menaquinone** (or **vitamin K_2**). For therapy, a synthetic derivative of vitamin K, **menadione**, is available.

A. Function of vitamin K

1. **Formation of γ-carboxyglutamate:** Vitamin K is required in the hepatic synthesis of prothrombin and blood clotting factors II, VII, IX, and X. These proteins are synthesized as inactive precursor molecules. Formation of the clotting factors requires the vitamin K-dependent carboxylation of glutamic acid residues (Figure 28.26). This forms a mature clotting factor that contains γ-carboxyglutamate (Gla) and is capable of subsequent activation. The reaction requires O_2, CO_2, and the hydroquinone form of vitamin K. The formation of Gla is sensitive to inhibition by **dicumarol**, an anticoagulant occurring naturally in spoiled sweet clover, and by **warfarin**, a synthetic analog of vitamin K.

2. **Interaction of prothrombin with platelets:** The Gla residues of prothrombin are good chelators of positively charged calcium ions, because of the two adjacent, negatively charged carboxylate groups. The prothrombin–calcium complex is then able to bind to phospholipids essential for blood clotting on the surface of platelets. Attachment to the platelet increases the rate at which the proteolytic conversion of prothrombin to thrombin can occur (Figure 28.27).

3. **Role of γ-carboxyglutamate residues in other proteins:** Gla is also present in other proteins (for example, osteocalcin of bone) unrelated to the clotting process. However, the physiologic role of these proteins and the function of vitamin K in their synthesis is not yet understood.

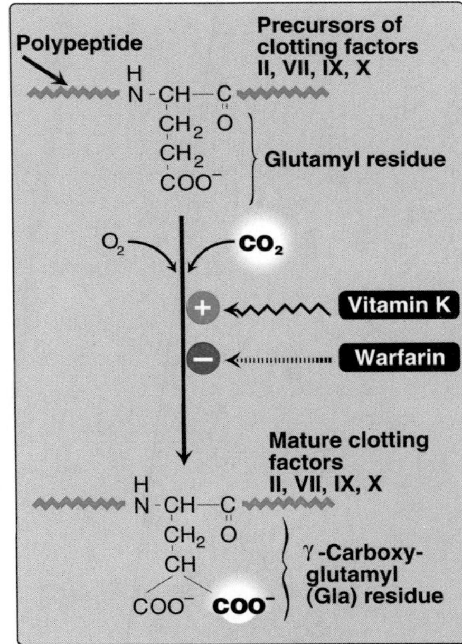

Figure 28.26
Carboxylation of glutamate to form γ-carboxyglutamate (Gla).

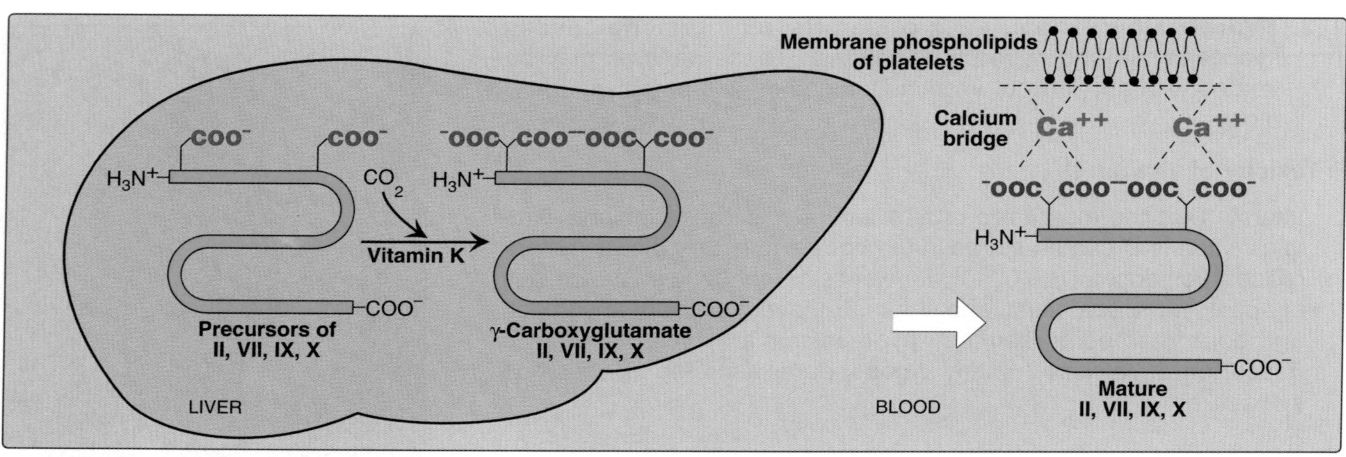

Figure 28.27
Role of vitamin K in blood coagulation.

B. Distribution and requirement of vitamin K

Vitamin K is found in cabbage, cauliflower, spinach, egg yolk, and liver. There is also extensive synthesis of the vitamin by the bacteria in the gut. There is no RDA for vitamin K, but 70 to 140 mg/day is recommended as an adequate level. The lower level assumes one half of the estimated requirement comes from bacterial synthesis, whereas the upper figure assumes no bacterial synthesis.

C. Clinical indications

1. **Deficiency of vitamin K:** A true vitamin K deficiency is unusual because adequate amounts are generally produced by intestinal bacteria or obtained from the diet. If the bacterial population in the gut is decreased, for example by antibiotics, the amount of endogenously formed vitamin is depressed, and can lead to **hypoprothrombinemia** in the marginally malnourished individual (for example, a debilitated geriatric patient). This condition may require supplementation with vitamin K to correct the bleeding tendency. In addition, certain second generation cephalosporins (for example, cefoperazone, cefamandole, and moxalactam) cause hypoprothrombinemia, apparently by a warfarin-like mechanism. Consequently, their use in treatment is usually supplemented with vitamin K.

2. **Deficiency of vitamin K in the newborn:** Newborns have sterile intestines and cannot initially synthesize vitamin K. Because human milk provides only about one fifth of the daily requirement for vitamin K, it is recommended that all newborns receive a single intramuscular dose of vitamin K as prophylaxis against hemorrhagic disease.

D Toxicity of vitamin K

Prolonged administration of large doses of vitamin K can produce hemolytic anemia and jaundice in the infant, due to toxic effects on the membrane of red blood cells.

XIV. VITAMIN E

The E vitamins consist of eight naturally occurring tocopherols, of which **α-tocopherol** is the most active (Figure 28.28). The primary function of vitamin E is as an antioxidant in prevention of the nonenzymic oxidation of cell components (for example, polyunsaturated fatty acids) by molecular oxygen and free radicals.

A. Distribution and requirements of vitamin E

Vegetable oils are rich sources of vitamin E, whereas liver and eggs contain moderate amounts. The RDA for α-tocopherol is 10 mg for men and 8 mg for women. Vitamin E requirement increases as the intake of polyunsaturated fatty acid increases.

B. Deficiency of vitamin E

Vitamin E deficiency is almost entirely restricted to premature infants. When observed in adults, it is usually associated with defective lipid absorption or transport. The signs of human vitamin E deficiency include sensitivity of erythrocytes to peroxide, and the appearance of abnormal cellular membranes.

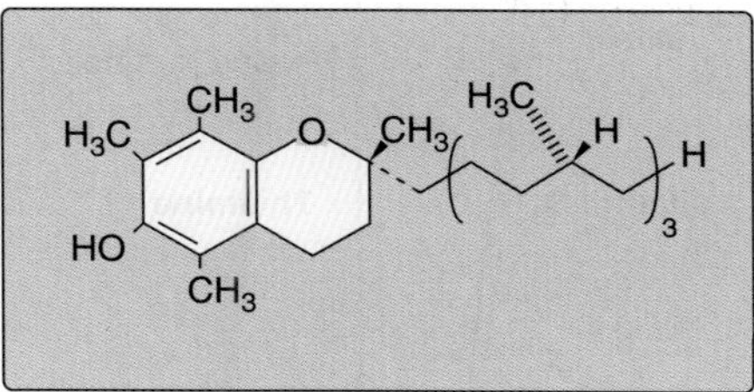

Figure 28.28
Structure of vitamin E.

C. Clinical indications

Vitamin E is not recommended for the prevention of chronic disease, such as coronary heart disease or cancer. Clinical trials using vitamin E supplementation have been uniformly disappointing. For example, subjects in the Alpha-Tocopherol, Beta Carotene Cancer Prevention Study trial who received high doses of vitamin E, not only lacked cadiovascular benefit but also had an increased incidence of stroke.

D. Toxicity of vitamin E

Vitamin E is the least toxic of the fat-soluble vitamins, and no toxicity has been observed at doses of 300 mg/day.

XV. VITAMIN SUPPLEMENTS

Because the potential benefits outweigh the possibilities of harm, many experts recommend a daily multivitamin that does not exceed the RDA of it component vitamins. Multivitamins ensure an adequate intake for those vitamins—folic acid, vitamin B_6, vitamin B_{12}, and vitamin D—that are most likely to be deficient. However, the the evidence is insufficient to recommend for or against the use of supplements of vitamins A, C, or E; multivitamins with folic acid; or antioxidant combinations for the prevention of cancer or cardiovascular disease. Most experts recommend against the use of β-carotene supplements, either alone or in combination, for the prevention of cancer or cardiovascular disease.

XVI. CHAPTER SUMMARY

The vitamins are summarized in Figure 28.29.

VITAMIN	OTHER NAMES	ACTIVE FORM	FUNCTION
Folic acid	—	Tetrahydro-folic acid	Transfer one-carbon units; Synthesis of methionine, purines, and thymine
Vitamin B_{12}	Cobalamin	Methylcobalamin Deoxyadenosyl cobalamin	Cofactor for reactions: Homocysteine → Methionine Methylmalonyl CoA → Succinyl CoA
Vitamin C	Ascorbic acid	Ascorbic acid	Antioxidant Cofactor for hydroxylation reactions, for example: In procollagen: Proline → hydroxyproline Lysine → hydroxylysine
Vitamin B_6	Pridoxine Pyridoxamine Pyridoxal	Pyridoxal phosphate	Cofactor for enzymes, particularly in amino acid metabolism
Vitamin B_1	Thiamine	Thiamine pyrophosphate	Cofactor of enzymes catalyzing: Pyruvate → acetyl CoA α-Ketoglutarate → Succinyl CoA Ribose 5-P xylulose 5-P → Sedoheptulose 7-P + Glyceraldehyde 3-P
Niacin	Nicotinic acid Nicotinamide	NAD^+, $NADP^+$	Electron transfer
Vitamin B_2	Riboflavin	FMN, FAD	Electron transfer
Biotin	—	Enzyme-bound biotin	Carboxylation reactions
Pantothenic acid	—	Coenzyme A	Acyl carrier **WATER-SOLUBLE**
			FAT-SOLUBLE
Vitamin A	Retinol Retinal Retinoic acid β-Carotene	Retinol Retinal Retinoic acid	Maintenance of reproduction Vision Promotion of growth Differentiation and maintenance of epithelial tissues Gene expression
Vitamin D	Cholecalciferol Ergocalciferol	1,25-Dihydroxy-cholecalciferol	Calcium uptake
Vitamin K	Menadione Menaquinone Phylloquinone	Menadione Menaquinone Phylloquinon	γ-Carboxylation of glutamate residue in clotting and other proteins
Vitamin E	α-Tocopherol	Any of several tocopherol derivatives	Antioxidant

Figure 28.29 (continued on next page)
Summary of vitamins.

DEFICIENCY	SIGNS AND SYMPTOMS	TOXICITY	NOTES
Megaloblastic anemia Neural tube defects	Anemia Birth defects	None	Administration of high levels of folate can mask vitamin B_{12} deficiency.
Pernicious anemia Dementia Spinal degeneration	Megaloblastic anemia Neuropsychiatric symptoms	None	Pernicious anemia is treated with IM or high-dose oral vitamin B_{12}
Scurvy	Sore, spongy gums Loose teeth Poor wound healing	None	Benefits of supplementation not established in controlled trials
Rare	Glossitis Neuropathy	Yes	Deficiency can be induced by isoniazid Sensory neuropathy occurs at high doses
Beriberi Wernicke-Korsakoff syndrome (most common in alcoholics)	Tachycardia, vomiting, convulsions Apathy, loss of memory, eye movements	None	—
Pellagra	Dermatitis Diarrhea Dementia	None	High doses of niacin used to treat hyperlipidemia
Rare	Dermatitis Angular stomatitis	None	—
Rare	—	None	Consumption of large amounts of raw egg whites (which contains a protein, avidin, that binds biotin) can induce a biotin deficiency
Rare	—	None	—
			WATER-SOLUBLE
			FAT-SOLUBLE
Impotence Night blindness Retardation of growth Xerophthalmia	Increased visual threshold Dryness of cornea	Yes	β-Carotene not acutely toxic, but supplementation is not recommended Excess vitamin A can increase incidence of fractures
Rickets (in children) Osteomalacia (in adults)	Soft, pliable bones	Yes	Vitamin D is not a true vitamin because it can be synthesized in skin. Application of sunscreen lotions or presence of dark skin color decreases this synthesis.
Newborn Rare in adults	Bleeding	Rare	Vitamin K produced by intestinal bacteria. Vitamin K deficiency common in newborns Parenteral treatment with the vitamin K is recommended at birth
Rare	Red blood cell fragility leads to hemolytic anemia	None	Benefits of supplementation not established in controlled trials

Figure 28.29 (continued from previous page)
Summary of vitamins.

Study Questions

Choose the ONE correct answer

28.1 Which one of the following statements concerning vitamin B_{12} is correct?

A. The cofactor form is vitamin B_{12} itself.
B. It is involved in the transfer of amino groups.
C. It requires a specific glycoprotein for its absorption.
D. t is present in plant products
E. It's deficiency is most often caused by a lack of the vitamin in the diet.

Correct answer = C. Vitamin B_{12} requires intrinsic factor for its absorption. A deficiency of vitamin B_{12} is most often caused by a lack of intrinsic factor. However, high does of the vitamin, given orally, are sufficiently absorbed to serve as treatment for pernicious anemia. The cofactor forms are methycobalamine and deoxyadenosylcobalamin. Vitamin B_6, not vitamin B_{12}, is involved in the transfer of amino groups. B_{12} is found in food derived from animal sources.

28.2 Retinol:

A. can be enzymically formed from retinoic acid.
B. is transported from the intestine to the liver in chylomicrons.
C. is the light-absorbing portion of rhodopsin.
D. is phosphorylated and dephosphorylated during the visual cycle.
E. mediates most of the actions of the retinoids.

Correct answer = B. Retinyl esters are incorporated into chylomicrons. Retinoic acid cannot be reduced to retinol. Retinal, the aldehyde form of retinol, is the chromophore for rhodopsin. Retinal is photoisomerized during the visual cycle. Retinoic acid, not retinol, is the most important retinoid.

28.3 Which one of the following statements concerning vitamin D is correct?

A. Chronic renal failure requires the oral administration of 1,25-dihydroxycholecalciferol.
B. It is required in the diet of individuals exposed to sunlight.
C. 25-Hydroxycholecalciferol is the active form of the vitamin.
D. Vitamin D opposes the effect of parathyroid hormone.
E. A deficiency in vitamin D results in an increased secretion of calcitonin.

Correct answer = A. Renal failure results in the decreased ability to form the active form of the vitamin, which must be supplied. The vitamin is not required in individuals exposed to sunlight. 1,25-dihydroxycholecalciferol is the active form of the vitamin. Vitamin D and parathyroid hormone both increase serum calcium. A deficiency of vitamin D decreases the secretion of calcitonin.

28.4 Vitamin K:

A. plays an essential role in preventing thrombosis.
B. increases the coagulation time in newborn infants with hemorrhagic disease.
C. is present in high concentration in cow or breast milk.
D. is synthesized by intestinal bacteria.
E. is a water-soluble vitamin.

Correct answer = D. Vitamin K is essential for clot formation, decreases coagulation time, and is present in low concentrations in milk.

UNIT VI:
Storage and Expression of
Genetic Information

DNA Structure and Replication

29

I. OVERVIEW

Nucleic acids are required for the storage and expression of genetic information. There are two chemically distinct types of nucleic acids: deoxyribonucleic acid (DNA) and ribonucleic acid (RNA, see Chapter 30). DNA, the storehouse of genetic information, is present not only in chromosomes in the nucleus of eukaryotic organisms, but also in mitochondria and the chloroplasts of plants. Prokaryotic cells, which lack nuclei, have a single chromosome, but may also contain nonchromosomal DNA in the form of plasmids. The genetic information found in DNA is copied and transmitted to daughter cells through DNA replication. The DNA contained in a fertilized egg encodes the information that directs the development of an organism. This development may involve the production of billions of cells. Each cell is specialized, expressing only those functions that are required for it to perform its role in maintaining the organism. Therefore, DNA must be able to not only replicate precisely each time a cell divides, but also to have the information that it contains be selectively expressed. Transcription (RNA synthesis) is the first stage in the expression of genetic information (see Chapter 30). Next, the code contained in the nucleotide sequence of messenger RNA molecules is translated (protein synthesis, see Chapter 31), thus completing gene expression. This flow of information from DNA to RNA to protein is termed the "central dogma of molecular biology" (Figure 29.1), and is descriptive of all organisms, with the exception of some viruses that have RNA as the repository of their genetic information.

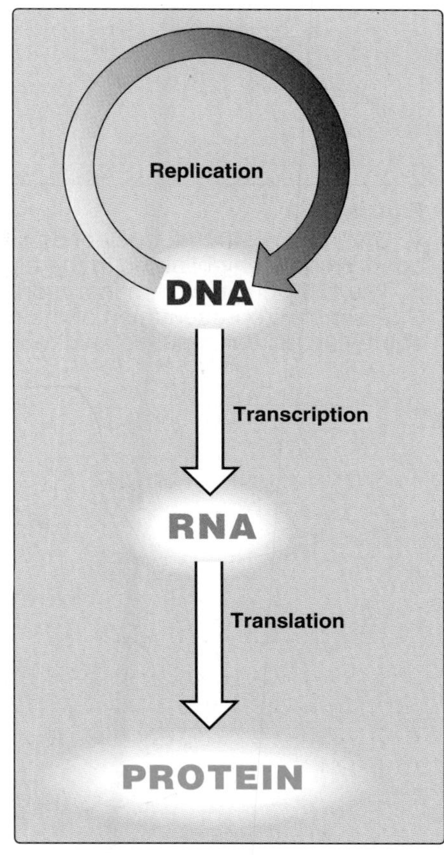

Figure 29.1
The "central dogma" of molecular biology.

II. STRUCTURE OF DNA

DNA is a polydeoxyribonucleotide that contains many monodeoxyribonucleotides covalently linked by **3'→5'–phosphodiester bonds**. With the exception of a few viruses that contain single-stranded DNA, DNA exists as a **double-stranded molecule**, in which the two strands wind around each other, forming a **double helix**. In eukaryotic cells, DNA is found associated with various types of proteins (known collectively as **nucleoprotein**) present in the nucleus, whereas in prokaryotes, the protein–DNA complex is present in the **nucleoid**.

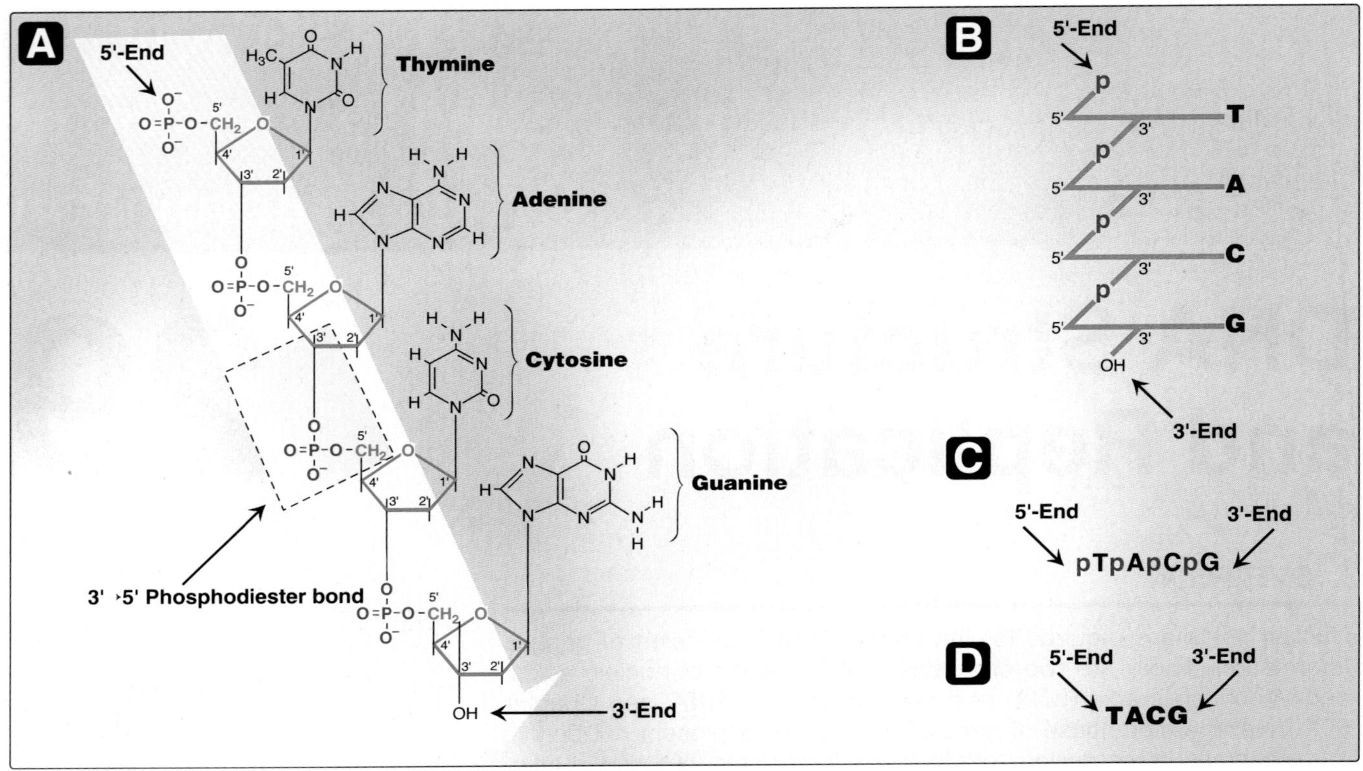

Figure 29.2
A. DNA chain with the nucleotide sequence shown written in the 5' → 3' direction. A 3' → 5' phosphodiester bond is shown highlighted in the blue box, and the deoxyribose-phosphate backbone is shaded in yellow. B. The DNA chain written in a more stylized form, emphasizing the ribose–phosphate backbone. C. A simpler representation of the nucleotide sequence. D. The simplest representation, with the abbreviations for the bases written in the conventional 5' → 3' direction.

A. 3'→5'-Phosphodiester bonds

Phosphodiester bonds join the 5'-hydroxyl group of the deoxypentose of one nucleotide to the 3'-hydroxyl group of the deoxypentose of an adjacent nucleotide through a phosphate group (Figure 29.2). The resulting long, unbranched chain has **polarity**, with both a 5'-end (the end with the free phosphate) and a 3'-end (the end with the free hydroxyl) that are not attached to other nucleotides. The bases located along the resulting deoxyribose–phosphate backbone are, by convention, always written in sequence from the 5'-end of the chain to the 3'-end. For example, the sequence of bases in the DNA shown in Figure 29.2 is read "thymine, adenine, cytosine, guanine" (5'-TACG-3'). Phosphodiester linkages between nucleotides (in DNA or RNA) can be cleaved hydrolytically by chemicals, or hydrolyzed enzymatically by a family of nucleases: **deoxyribonucleases** for DNA and **ribonucleases** for RNA. [Note: Nucleases that cleave the nucleotide chain at positions in the interior of the chain are called **endonucleases**. Those that cleave the chain only by removing individual nucleotides from one of the two ends are called **exonucleases**.]

B. Double helix

In the double helix, the two chains are coiled around a common axis called the **axis of symmetry**. The chains are paired in an **anti-parallel manner**, that is, the 5'-end of one strand is paired with the 3'-end of the other strand (Figure 29.3). In the DNA helix, the hydrophilic deoxyribose–phosphate backbone of each chain is on the outside of the molecule, whereas the hydrophobic bases are stacked inside. The overall structure resembles a twisted ladder. The spatial relationship between the two strands in the helix creates a **major (wide) groove** and a **minor (narrow) groove**. These grooves provide access for the binding of regulatory proteins to their specific recognition sequences along the DNA chain. [Note: Certain anti-cancer drugs, such as dactinomycin (actinomycin D), exert their cytotoxic effect by intercalating into the narrow groove of the DNA double helix, thus interfering with RNA and DNA synthesis.[1]]

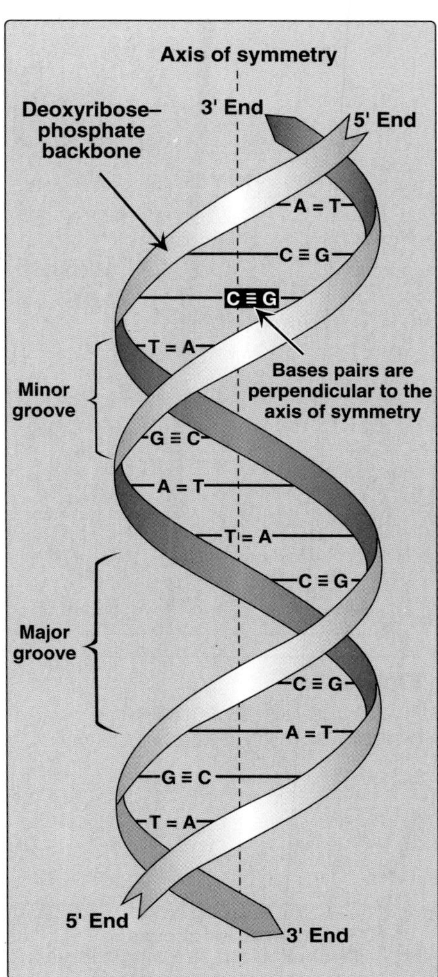

Figure 29.3
DNA double helix, illustrating some of its major structural features.

1. **Base pairing:** The bases of one strand of DNA are paired with the bases of the second strand, so that an adenine is always paired with a thymine and a cytosine is always paired with a guanine. [Note: The base pairs are perpendicular to the axis of the helix (see Figure 29.3).] Therefore, one polynucleotide chain of the DNA double helix is always the **complement** of the other. Given the sequence of bases on one chain, the sequence of bases on the complementary chain can be determined (Figure 29.4). [Note: The specific base pairing in DNA leads to **Chargaff's Rules**: In any sample of double-stranded DNA, the amount of adenine equals the amount of thymine, the amount of guanine equals the amount of cytosine, and the total amount of purines equals the total amount of pyrimidines.] The base pairs are held together by **hydrogen bonds**: two between A and T and three between G and C (Figure 29.5). These hydrogen bonds, plus the hydrophobic interactions between the stacked bases, stabilize the structure of the double helix.

2. **Separation of the two DNA strands in the double helix:** The two strands of the double helix separate when hydrogen bonds between the paired bases are disrupted. Disruption can occur in the laboratory if the pH of the DNA solution is altered so that the nucleotide bases ionize, or if the solution is heated. [Note: Phosphodiester bonds are not broken by such treatment.] When DNA is heated, the temperature at which one half of the helical structure is lost is defined as the **melting temperature (T_m)**. The loss of helical structure in DNA, called **denaturation**, can be monitored by measuring its absorbance at 260 nm. [Note: Single-stranded DNA has a higher relative absorbance at this wavelength than does double-stranded DNA.] Because there are three hydrogen bonds between G and C but only two between A and T, DNA that contains high concentrations of A and T denatures at a lower temperature than G- and C-rich DNA (Figure 29.6). Under appropriate conditions, complementary DNA strands can reform the double helix by the process called **renaturation** (or **reannealing**).

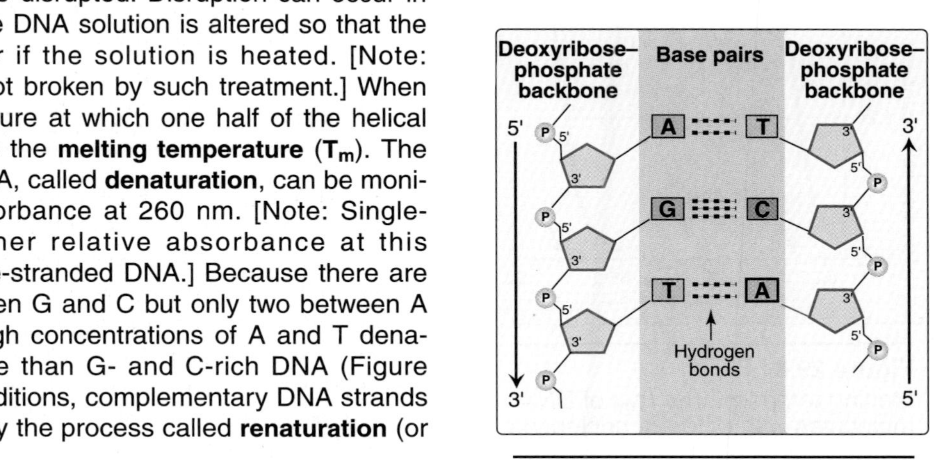

Figure 29.4
Two complementary DNA sequences.

[1]See Chapter 40 in ***Lippincott's Illustrated Reviews: Pharmacology*** (3rd Ed.) and Chapter 38 in (2nd Ed.) for a discussion of the anticancer drug, actinomycin D.

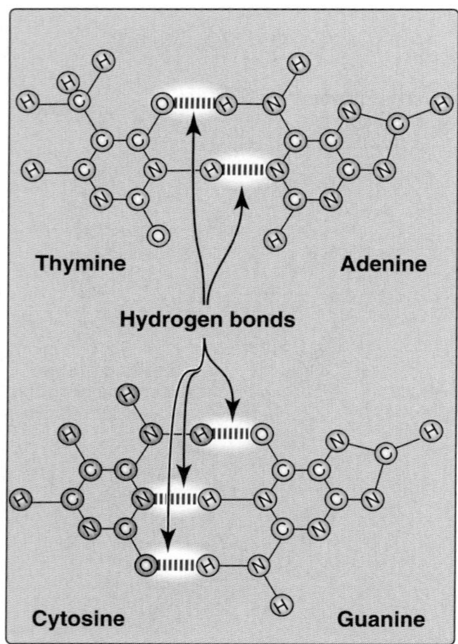

Figure 29.5
Hydrogen bonds between complementary bases.

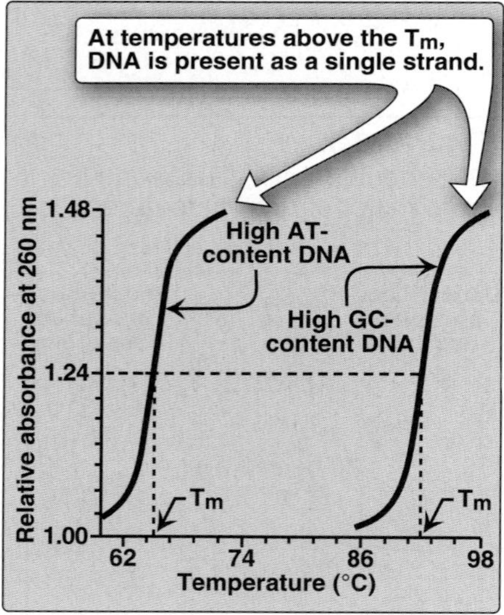

Figure 29.6
Melting temperatures (T_m) of DNA molecules with different nucleotide compositions. (At a wavelength of 260 nm, single-stranded DNA has a higher relative absorbance than does double-stranded DNA.)

3. **Structural forms of the double helix:** There are three major structural forms of DNA: the B form, described by Watson and Crick in 1953, the A form, and the Z form. The B form is a right-handed helix with ten residues per 360° turn of the helix, and with the planes of the bases perpendicular to the helical axis. Chromosomal DNA is thought to consist primarily of **B-DNA** (Figure 29.7 illustrates a space-filling model of B-DNA). The A form is produced by moderately dehydrating the B form. It is also a right-handed helix, but there are eleven base pairs per turn, and the planes of the base pairs are tilted 20° away from the perpendicular to the helical axis. The conformation found in DNA–RNA hybrids or RNA–RNA double-stranded regions is probably very close to the A form. **Z-DNA** is a left-handed helix that contains about twelve base pairs per turn (see Figure 29.7). [Note: The deoxyribose–phosphate backbone "zigzags," hence, the name "Z"-DNA.] Stretches of Z-DNA can occur naturally in regions of DNA that have a sequence of alternating purines and pyrimidines, for example, poly GC. Transitions between the helical forms of DNA may play an important role in regulating gene expression.

C. Circular DNA molecules

Each chromosome in the nucleus of a eukaryote contains one long linear molecule of double-stranded DNA, which is bound to a complex mixture of proteins to form **chromatin**. Eukaryotes have closed circular DNA molecules in their mitochondria, as do plant chloroplasts. A prokaryotic organism contains a single, double-stranded, supercoiled, circular chromosome. Each prokaryotic chromosome is associated with histone-like proteins (see p. 406) and RNA that can condense the DNA to form a **nucleoid**. In addition, most species of bacteria also contain small, circular, extrachromosomal DNA molecules called **plasmids**. Plasmid DNA carries genetic information, and undergoes replication that may or may not be synchronized to chromosomal division.[2] Plasmids may carry genes that convey antibiotic resistance to the host bacterium, and may facilitate the transfer of genetic information from one bacterium to another. [Note: The use of plasmids as vectors in recombinant DNA technology is described in Chapter 32.]

III. STEPS IN PROKARYOTIC DNA SYNTHESIS

When the two strands of the DNA double helix are separated, each can serve as a template for the replication of a new complementary strand. This produces two daughter molecules, each of which contains two DNA strands with an antiparallel orientation (see Figure 29.3). This process is called **semiconservative replication** because, although the parental duplex is separated into two halves (and, therefore, is not "conserved" as an entity), each of the individual parental strands remains intact in one of the two new duplexes (Figure 29.8). The enzymes involved in the DNA replication process are template-directed polymerases that can synthesize the complementary sequence of each strand with extraordinary fidelity. The reactions described in this section were first known from

[2]See p.116 in ***Lippincott's Illustrated Reviews: Microbiology*** for a discussion of plasmids.

studies of the bacterium Escherichia coli (E. coli), and the description given below refers to the process in that microorganism. DNA synthesis in higher organisms is less well understood, but involves the same types of mechanisms. In either case, initiation of DNA replication commits the cell to continue the process until the entire genome has been replicated.

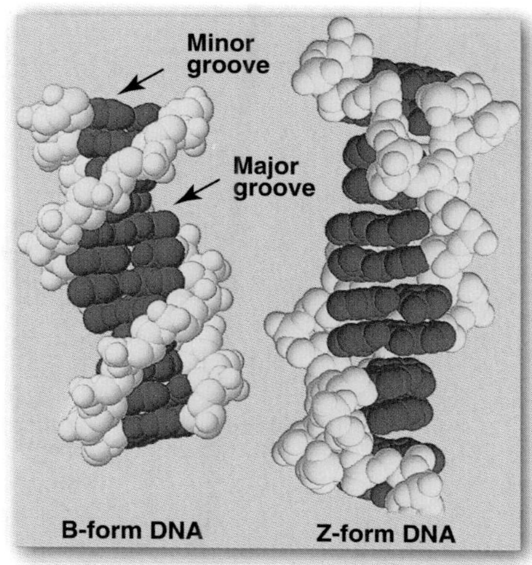

Figure 29.7
Structures of B-DNA and Z-DNA.

A. Separation of the two complementary DNA strands

In order for the two strands of the parental double helical DNA to be replicated, they must first separate (or "melt"), at least in a small region, because the polymerases use only single-stranded DNA as a template. In prokaryotic organisms, DNA replication begins at a single, unique nucleotide sequence—a site called the **origin of replication** (Figure 29.9 A). In eukaryotes, replication begins at multiple sites along the DNA helix (Figure 29.9 B). These sites include a short sequence composed almost exclusively of AT base pairs. [Note: This is referred to as a **consensus sequence**, because the order of nucleotides is essentially the same at each site.] Having multiple origins of replication provides a mechanism for rapidly replicating the great length of the eukaryotic DNA molecules.

B. Formation of the replication fork

As the two strands unwind and separate they form a "V" where active synthesis occurs. This region is called the **replication fork**. It moves along the DNA molecule as synthesis occurs. Replication of double-stranded DNA is **bidirectional**—that is, the replication forks move in both directions away from the origin (see Figure 29.9).

1. **Proteins required for DNA strand separation:** Initiation of DNA replication requires the recognition of the origin of replication and/or the replication fork by a group of proteins that form the **prepriming complex**. These proteins are responsible for maintaining the separation of the parental strands, and for unwinding the double helix ahead of the advancing replication fork. These proteins include the following:

   a. **DnaA protein:** Twenty to fifty monomers of dnaA protein bind to specific nucleotide sequences at the origin of replication, which is particularly rich in AT base pairs. This ATP-requiring process causes the double-stranded DNA to melt—that is, the strands separate, forming localized regions of single-stranded DNA.

   b. **Single-stranded DNA-binding (SSB) proteins:** Also called **helix-destabilizing proteins**, these bind only to single-stranded DNA (Figure 29.10). They bind **cooperatively**—that is, the binding of one molecule of SSB protein makes it easier for additional molecules of SSB protein to bind tightly to the DNA strand. The SSB proteins are not enzymes, but rather serve to shift the equilibrium between double- and single-stranded DNA in the direction of the single-stranded forms. These proteins not only keep the two strands of DNA separated in the area of the replication origin, thus providing the single-stranded template required by polymerases, but also protect the DNA from nucleases that cleave single-stranded DNA.

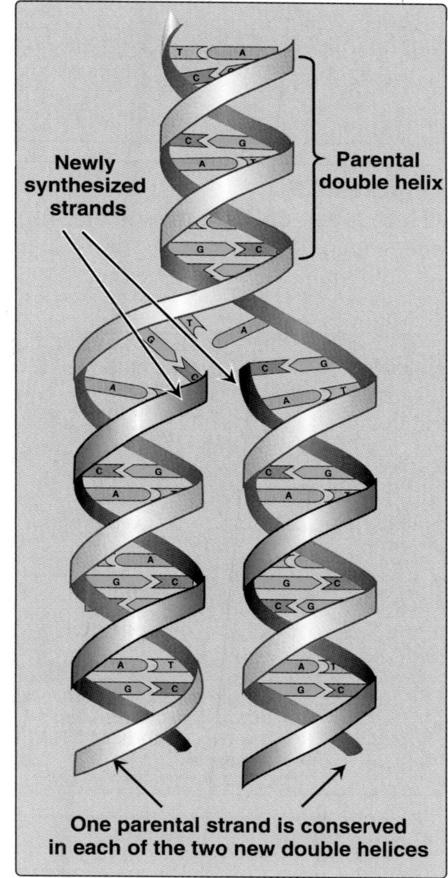

Figure 29.8
Semiconservative replication of DNA.

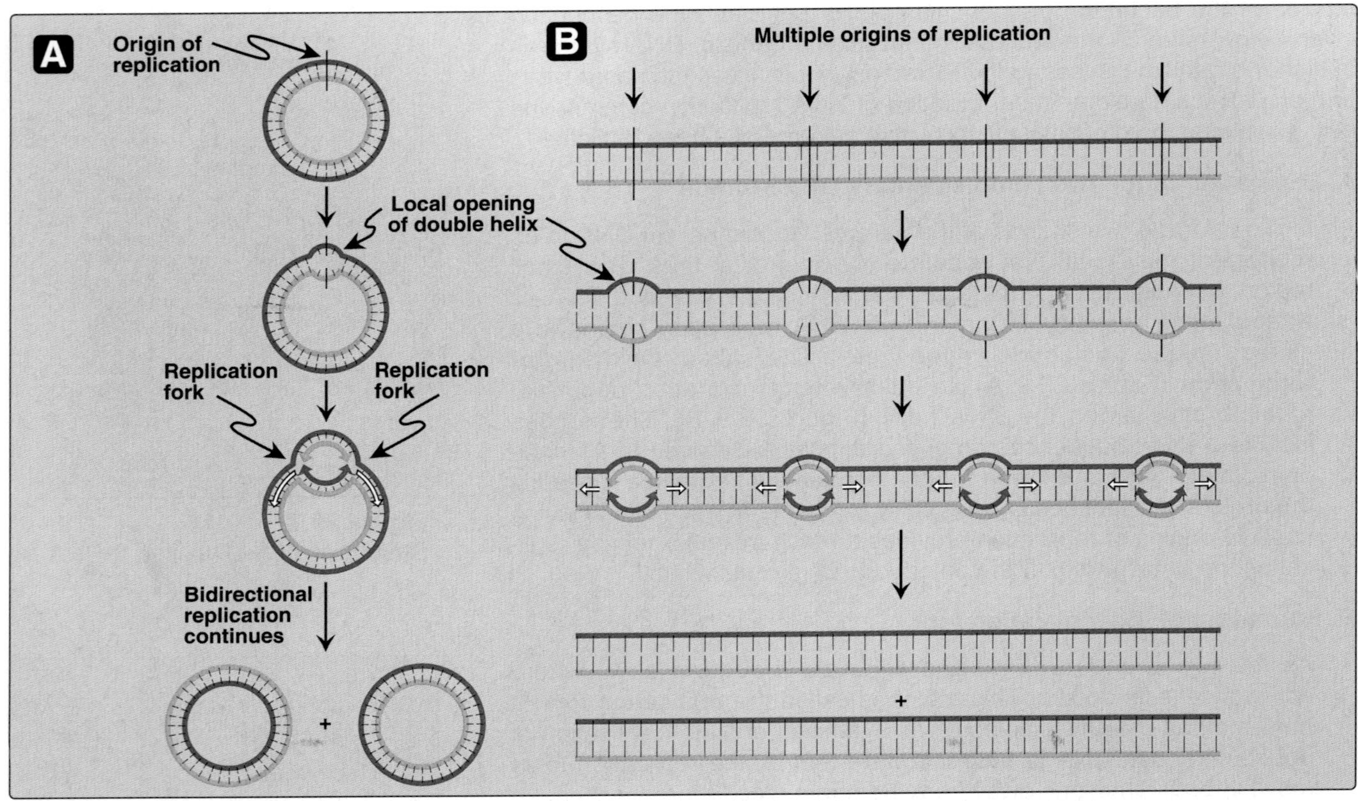

Figure 29.9
Replication of DNA: origins and replication forks. A. Small prokaryotic circular DNA. B. Very long eukaryotic DNA.

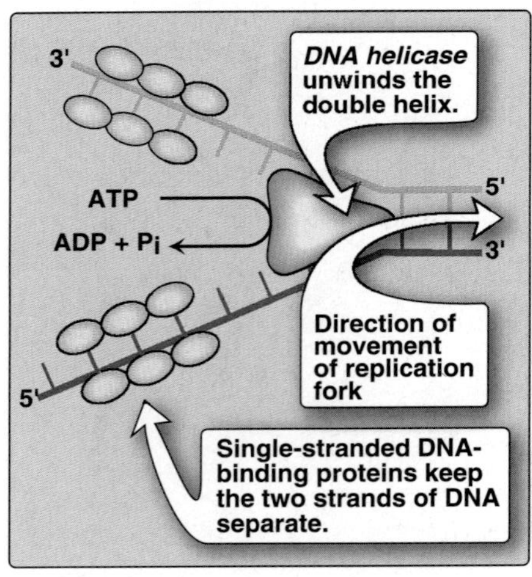

Figure 29.10
Proteins responsible for maintaining the separation of the parental strands and unwinding the double helix ahead of the advancing replication fork.

c. **DNA helicases:** These enzymes bind to single-stranded DNA near the replication fork, and then move into the neighboring double-stranded region, forcing the strands apart—in effect, unwinding the double helix. *Helicases* require energy provided by ATP. When the strands separate, SSB proteins bind, preventing reformation of the double helix (see Figure 29.10).

2. **Solving the problem of supercoils:** As the two strands of the double helix are separated, a problem is encountered, namely, the appearance of **positive supercoils** (also called **supertwists**) in the region of DNA ahead of the replication fork (Figure 29.11). The accumulating positive supercoils interfere with further unwinding of the double helix. [Note: Supercoiling can be demonstrated by tightly grasping one end of a telephone cord while twisting the other end. If the cord is twisted in the direction of tightening the coils, the cord will wrap around itself in space to form positive supercoils. If the cord is twisted in the direction of loosening the coils, the cord will wrap around itself in the opposite direction to form **negative supercoils**.] To solve this problem, there is a group of enzymes called **DNA topoisomerases**, which are responsible for removing supercoils in the helix.

   a. **Type I DNA topoisomerases** reversibly cut a single strand of the double helix. They have both **nuclease (strand-cutting)** and **ligase (strand-resealing)** activities. They do not require ATP, but rather appear to store the energy from the phosphodiester bond they cleave, reusing the energy to reseal the strand (Figure 29.12). Each time a transient "nick" is created in one DNA strand, the intact DNA strand is passed through the break before it is resealed, thus relieving ("relaxing") accumulated supercoils. *Type I topoisomerases* relax negative supercoils (that is, those that contain fewer turns of the helix than relaxed DNA) in E. coli, and both negative and positive supercoils (that is, those that contain fewer or more turns of the helix than relaxed DNA) in eukaryotic cells.

b. **Type II DNA topoisomerases** bind tightly to the DNA double helix and make transient breaks in both strands. The enzyme then causes a second stretch of the DNA double helix to pass through the break and, finally, reseals the break (Figure 29.13). As a result, both negative and positive supercoils can be relieved. *Type II DNA topoisomerases* are also required in both prokaryotes and eukaryotes for the separation of interlocked molecules of DNA following chromosomal replication. [Note: **Anticancer agents**, such as **etoposide**,[3] target human *topoisomerase II*.] *DNA gyrase*, a *type II topoisomerase* found in E. coli, has the unusual property of being able to introduce negative supercoils into relaxed circular DNA using energy from the hydrolysis of ATP. This facilitates the future replication of DNA because the negative supercoils neutralize the positive supercoils introduced during opening of the double helix. It also aids in the transient strand separation required during transcription (see p. 416). [Note: Bacterial *DNA gyrase* is the unique target of a group of **antimicrobial agents** called quinolones, for example, **ciprofloxacin**.[4]]

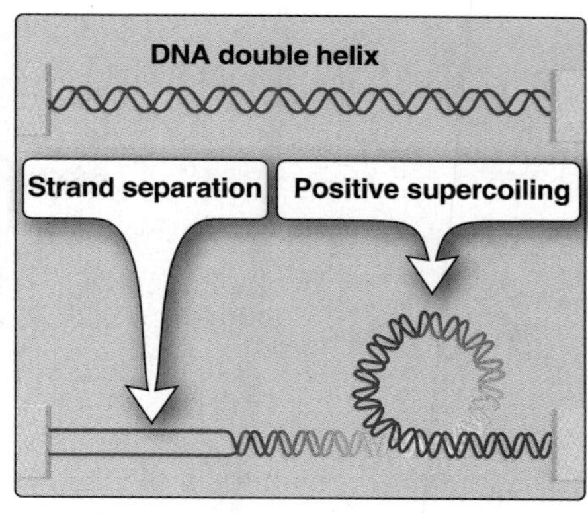

Figure 29.11
Positive supercoiling resulting from DNA strand separation.

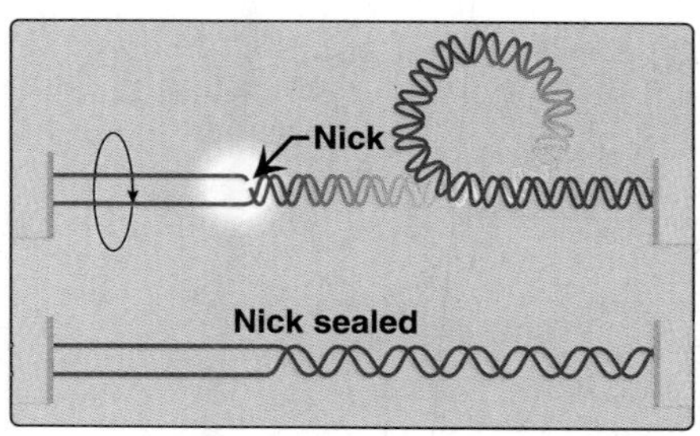

Figure 29.12
Action of type I DNA topoisomerases.

C. Direction of DNA replication

The DNA polymerases responsible for copying the DNA templates are only able to "read" the parental nucleotide sequences in the 3'→5' direction, and they synthesize the new DNA strands in the 5'→3' (antiparallel) direction. Therefore, beginning with one parental double helix, the two newly synthesized stretches of nucleotide chains must grow in opposite directions—one in the 5'→3' direction toward the replication fork and one in the 5'→3' direction away from the replication fork (Figure 29.14). This feat is accomplished by a slightly different mechanism on each strand.

1. **Leading strand:** The strand that is being copied in the direction of the advancing replication fork is called the leading strand and is **synthesized almost continuously**.

2. **Lagging strand:** The strand that is being copied in the direction away from the replication fork is **synthesized discontinuously**, with small fragments of DNA being copied near the replication fork. These short stretches of discontinuous DNA, termed **Okazaki fragments**, are eventually joined to become a single, continuous strand. The new strand of DNA produced by this mechanism is termed the lagging strand.

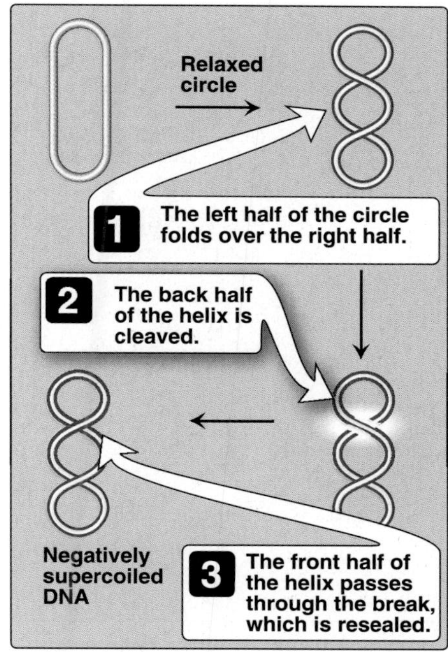

Figure 29.13
Action of *type II DNA topoisomerase*.

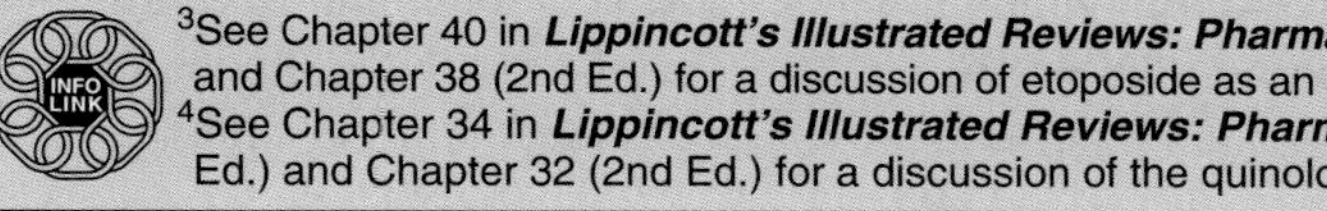

[3]See Chapter 40 in ***Lippincott's Illustrated Reviews: Pharmacology*** (3rd Ed.) and Chapter 38 (2nd Ed.) for a discussion of etoposide as an anticancer agent.
[4]See Chapter 34 in ***Lippincott's Illustrated Reviews: Pharmacology*** (3rd Ed.) and Chapter 32 (2nd Ed.) for a discussion of the quinolones.

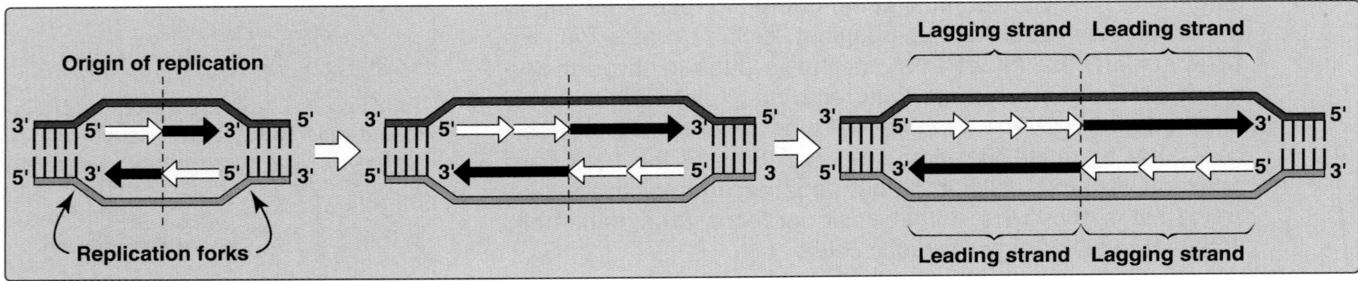

Figure 29.14
Discontinuous synthesis of DNA.

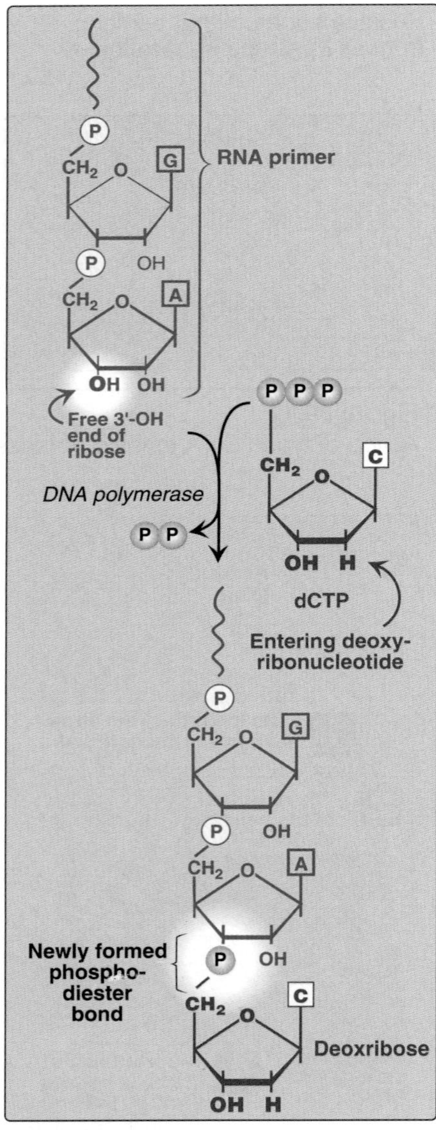

Figure 29.15
Use of an RNA primer to initiate DNA synthesis.

D. RNA primer

DNA polymerases cannot initiate synthesis of a complementary strand of DNA on a totally single-stranded template. Rather, they require an RNA primer—that is, a short, double-stranded region consisting of RNA base-paired to the DNA template, with a free hydroxyl group on the 3'-end of the RNA strand (Figure 29.15). This hydroxyl group serves as the first acceptor of a nucleotide by action of DNA polymerase. In de novo DNA synthesis, that free 3'-hydroxyl is provided by the short stretch of RNA, rather than DNA.

1. **Primase:** A specific **RNA polymerase**, called *primase*, synthesizes the short stretches of RNA (approximately ten nucleotides long) that are complementary and antiparallel to the DNA template. In the resulting **hybrid duplex**, the U in RNA pairs with A in DNA. As shown in Figure 29.16, these short RNA sequences are constantly being synthesized at the replication fork on the lagging strand, but only one RNA sequence at the origin of replication is required on the leading strand. The building blocks for this process are **5'-ribonucleoside triphosphates**, and pyrophosphate is released as each phosphodiester bond is made. [Note: The RNA primer is later removed as described on p. 402.]

2. **Primosome:** Prior to the beginning of RNA primer synthesis on the lagging strand, a prepriming complex of several proteins is assembled and binds to the single strand of DNA, displacing some of the single-stranded DNA-binding proteins. This protein complex, plus *primase*, is called the primosome. It initiates Okazaki fragment formation by moving along the template for the lagging strand in the 5'→3' direction, periodically recognizing specific sequences of nucleotides that direct it to create an RNA primer that is synthesized in the 5'→3' direction (antiparallel to the DNA template chain).

E. Chain elongation

Prokaryotic and eukaryotic DNA polymerases elongate a new DNA strand by adding deoxyribonucleotides, one at a time, to the 3'-end of the growing chain (see Figure 29.16). The sequence of nucleotides that are added is dictated by the base sequence of the template strand with which the incoming nucleotides are paired.

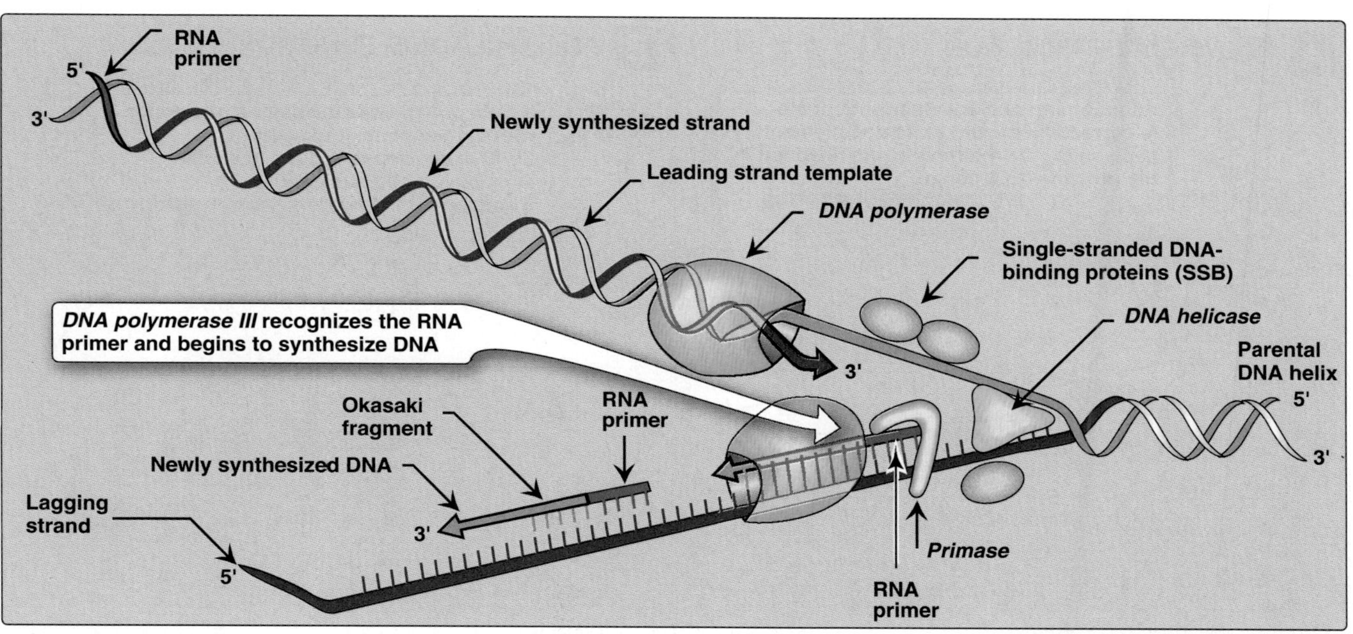

Figure 29.16
Elongation of the leading and lagging strands.

1. **DNA polymerase III:** DNA chain elongation is catalyzed by *DNA polymerase III*. Using the 3'-hydroxyl group of the RNA primer as the acceptor of the first deoxyribonucleotide, *DNA polymerase III* begins to add nucleotides along the single-stranded template that specifies the sequence of bases in the newly synthesized chain. *DNA polymerase III* is a highly "processive" enzyme—that is, it remains bound to the template strand as it moves along, and does not have to diffuse away and rebind before adding each new nucleotide. The new strand grows in the 5'→3' direction, **antiparallel** to the parental strand (see Figure 29.16). The nucleotide building blocks are **5'-deoxyribonucleoside triphosphates**. Pyrophosphate (PP_i) is released when each new nucleotide is added to the growing chain (see Figure 29.15). [Note: The further hydrolysis of pyrophosphate to two phosphates means that a total of two high-energy bonds are used to drive the addition of each deoxynucleotide.] All four deoxyribonucleoside triphosphates (dATP, dTTP, dCTP, and dGTP) must be present for DNA elongation to occur. If one of the four is in short supply, DNA synthesis stops when that nucleotide is depleted.

2. **Proofreading of newly synthesized DNA:** It is highly important for the survival of an organism that the nucleotide sequence of DNA be replicated with as few errors as possible. Misreading of the template sequence could result in deleterious, perhaps lethal, mutations. To ensure replication fidelity, *DNA polymerase III* has, in addition to its *5'→3' polymerase* activity, a "proofreading" activity (*3'→5' exonuclease*, Figure 29.17). As each nucleotide is added to the chain, *DNA polymerase III* checks to make certain the added

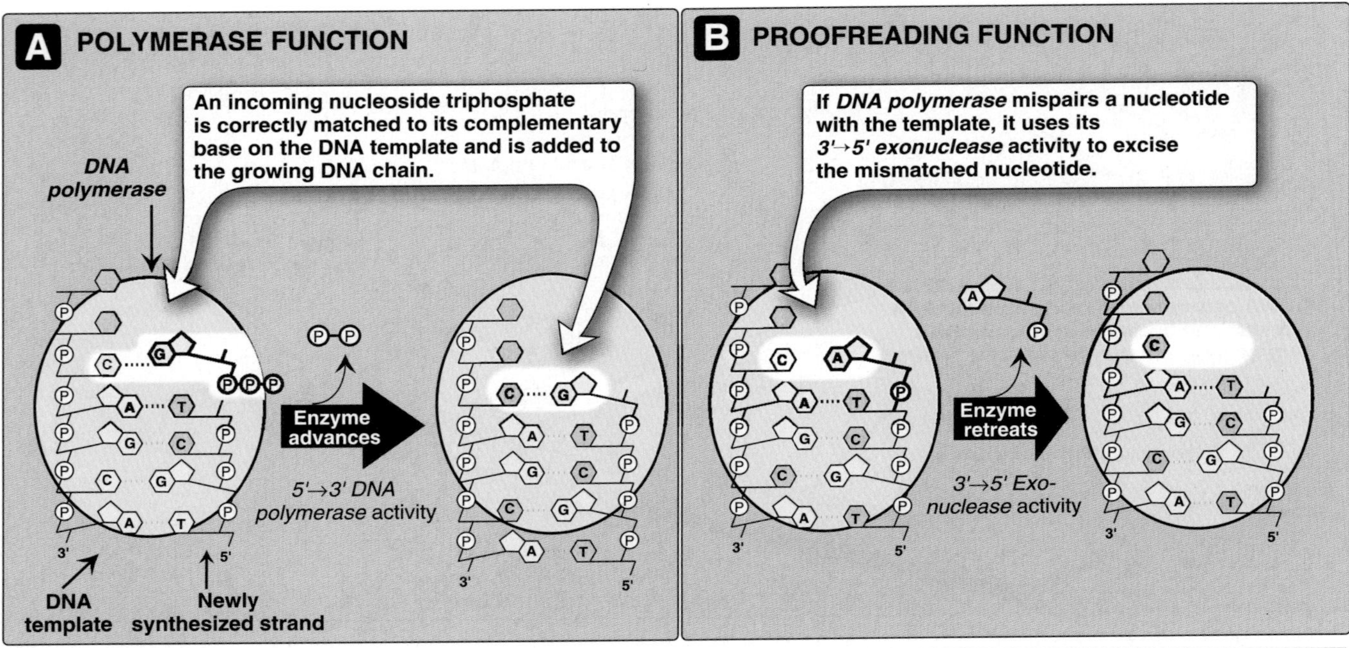

Figure 29.17
3'→5'-Exonuclease activity enables *DNA polymerase III* to "proofread" the newly synthesized DNA strand.

nucleotide is, in fact, correctly matched to its complementary base on the template. If it is not, the *3'→5' exonuclease* activity edits the mistake. [Note: The enzyme requires an improperly base-paired 3'-hydroxy terminus and, therefore, does not degrade correctly paired nucleotide sequences.] For example, if the template base is cytosine and the enzyme mistakenly inserts an adenine instead of a guanine into the new chain, the *3'→5' exonuclease* hydrolytically removes the misplaced nucleotide. The *5'→3' polymerase* then replaces it with the correct nucleotide containing guanine (see Figure 29.17). [Note: The proofreading activity requires an *exonuclease* that moves in the 3'→5' direction, not 5'→3' like the *polymerase* activity. This is because the excision must be done in the **reverse direction** from that of synthesis.]

F. Excision of RNA primers and their replacement by DNA

DNA polymerase III continues to synthesize DNA on the lagging strand until it is blocked by proximity to an RNA primer. When this occurs, the RNA is excised and the gap filled by *DNA polymerase I*.

1. **5'→3' Exonuclease activity:** In addition to having the *5'→3' polymerase* activity that synthesizes DNA, and the *3'→5' exonuclease* activity that proofreads the newly synthesized DNA chain like *DNA polymerase III*, *DNA polymerase I* also has a *5'→3' exonuclease* activity that is able to hydrolytically remove the RNA primer. [Note: These activities are **exonucleases** because they remove one nucleotide at a time from the end of the DNA chain, rather than cleaving it internally as do the **endonucleases** (Figure 29.18).] First, *DNA polymerase I* locates the space ("nick") between the 3'-end of the DNA newly synthesized by *DNA polymerase III* and the 5'-end of the adjacent RNA primer. Next, *DNA*

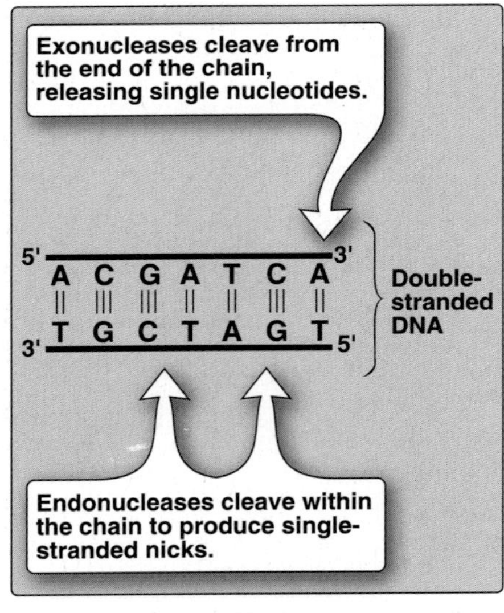

Figure 29.18
Endonuclease versus *exonuclease* activity.

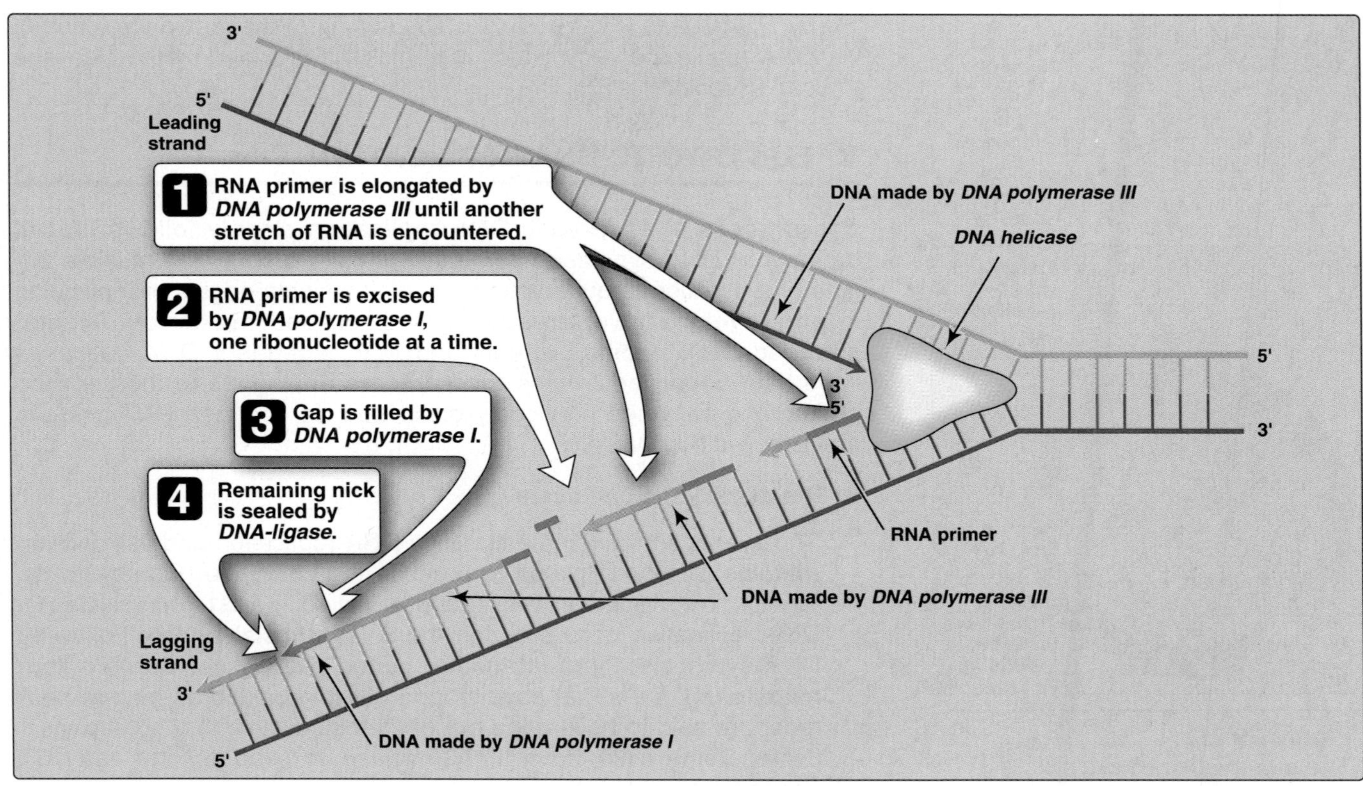

Figure 29.19
Removal of RNA primer and filling of the resulting "gaps" by *DNA polymerase I*.

polymerase I hydrolytically removes the RNA nucleotides "ahead" of itself, moving in the 5'→3' direction (**5'→3' exonuclease activity**). As it removes the RNA, *DNA polymerase I* replaces it with deoxyribonucleotides, synthesizing DNA in the 5'→3' direction (**5'→3' polymerase activity**). As it synthesizes the DNA, it also "proofreads" the new chain using **3'→5' exonuclease activity**. This removal/synthesis/proofreading continues, one nucleotide at a time, until the RNA is totally degraded and the gap is filled with DNA (Figure 29.19).

2. **Differences between 5'→3' and 3'→5' exonucleases:** The *5'→3' exonuclease* activity of *DNA polymerase I* differs from the *3'→5' exonuclease* used by both *DNA polymerase I* and *III* in two important ways. First, 5'→3' *exonuclease* can remove one nucleotide at a time from a region of DNA that is properly base-paired. The nucleotides it removes can be either ribonucleotides or deoxyribonucleotides. Second, 5'→3' *exonuclease* can also remove groups of altered nucleotides in the 5'→3' direction, removing from one to ten nucleotides at a time. This ability is important in the repair of some types of damaged DNA.

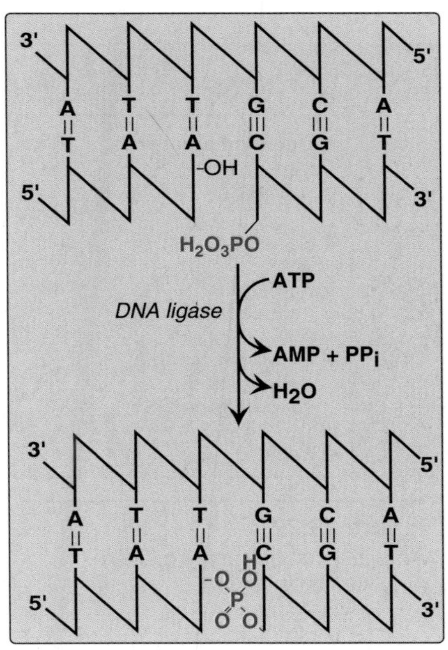

Figure 29.20
Formation of a phosphodiester bond by *DNA ligase*.

G. DNA ligase

The final phosphodiester linkage between the 5'-phosphate group on the DNA chain synthesized by *DNA polymerase III* and the 3'-hydroxyl group on the chain made by *DNA polymerase I* is catalyzed by *DNA ligase* (Figure 29.20). The joining of these two stretches of DNA requires energy, which in humans is provided by the cleavage of ATP to AMP + PP_i.

V. EUKARYOTIC DNA REPLICATION

The process of eukaryotic DNA replication closely follows that of prokaryotic DNA synthesis. Some differences, such as the multiple origins of replication in eukaryotic cells versus single origins of replication in prokaryotes, have already been discussed. Eukaryotic single-stranded DNA-binding proteins and ATP-dependent *DNA helicases* have been identified, whose functions are analogous to those of the prokaryotic enzymes previously discussed. In contrast, RNA primers are removed by *RNase H*.

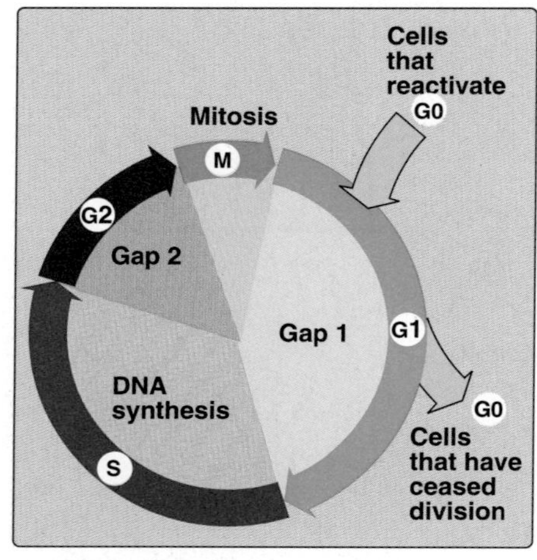

Figure 29.21
The eukaryotic cell cycle.

A. The eukaryotic cell cycle

The events surrounding eukaryotic DNA replication and cell division (**mitosis**) are coordinated to produce the cell cycle (Figure 29.21). The period preceding replication is called the **G1 phase (Gap1)**. DNA replication occurs during the **S (synthesis) phase**. Following DNA synthesis, there is another period (**G2 phase**, **Gap2**) before **mitosis (M)**. Cells that have stopped dividing, such as mature neurons, are said to have gone out of the cell cycle into the **G0 phase**. [Note: Some cells leave the **G0 phase** and reenter the early **G1 phase** to resume division.]

B. Eukaryotic DNA polymerases

At least five classes of eukaryotic DNA polymerases have been identified and categorized on the basis of molecular weight, cellular location, sensitivity to inhibitors, and the templates or substrates on which they act. They are designated by Greek letters rather than Roman numerals (Figure 29.22).

POLYMERASE	FUNCTION	PROOF-READING
Pol α	• Contains primase • Initiates DNA synthesis	–
Pol β	• Repair	–
Pol γ	• Replicates mitochondrial DNA	+
Pol δ	• Elongates leading strands and Okazai fragments	+
Pol ε	• Repair	+

Figure 29.22
Activities of eukaryotic DNA polymerases (pols).

1. **Pol α and pol δ:** *Pol α* is a multisubunit enzyme. One subunit has *primase* activity, which initiates strand synthesis on the leading strand and at the beginning of each Okazaki fragment on the lagging strand. The *primase* subunit synthesizes a short **RNA primer** that is extended by the *pol α 5'→3' polymerase* activity, which adds a short piece of DNA. *Pol δ* is then recruited to complete DNA synthesis on the leading strand and elongate each Okazaki fragment, using *3'→5' exonuclease* activity to **proofread** the newly synthesized DNA.

2. **Pol β, pol ε, and pol γ:** *Pol β* and *pol ε* are involved in carrying out DNA repair (see below). *Pol γ* replicates mitochondrial DNA.

C. Telomerase

Eukaryotic cells face a special problem in replicating the ends of their linear DNA molecules. Following removal of the RNA primer from the extreme 5'–end of the lagging strand, there is no way to fill in the remaining gap with DNA. To solve this problem, and to protect the ends of the chromosomes from attack by nucleases, noncoding sequences of DNA complexed with proteins are found at these ends. Called **telomeres**, their DNA consists of a repetitive sequence of T's and G's (T_xG_y, where x and y are usually in the range of one to four), base-paired to a complementary chain of A's and C's. The TG strand is longer than its complement, leaving a region of single-stranded DNA at the 3'–end of the double helix that is a few hundred nucleotides long. The single-stranded region folds back on itself, forming a structure that is stabilized by protein. This complex protects the ends of the chromosomes. In cells undergoing the aging process (**senescence**), the ends of their chromosomes get slightly shorter with each cell division until the telomeres are gone, and DNA essential for cell function is degraded—a phenomenon related to cellular aging and death. Cells that do not age (for example, germ-line cells and cancer cells) contain an enzyme called *telomerase* that is responsible for replacing these lost ends. *Telomerase* is a special kind of *reverse transcriptase* (see below) that carries its own RNA molecule of about 150 nucleotides long. In that RNA are copies of the A/C sequence that is complimentary to the T/G repeat sequence. The RNA base-pairs with the terminal nucleotides at the single-stranded 3'-end of the DNA (Figure 29.23). The RNA then serves as a template for extending the DNA strand. Once the next repeat sequence is complete, *telomerase* RNA is translocated to the newly synthesized end of the DNA, where it again hydrogen bonds, and the process is repeated.

D. Reverse transcriptase

A **retrovirus**, such as **human immunodeficiency virus** (**HIV**), carries its genome in the form of single-stranded RNA molecules. Following infection of a host cell, the viral enzyme, *reverse transcriptase*, uses the RNA as a template for the synthesis of viral DNA, which then becomes integrated into host chromosomes.[5] Like all the other enzymes that synthesize nucleic acids, *reverse transcriptase* moves along the template in the 3'→5' direction, synthesizing the DNA product in the 5'→3' direction. The lack of proofreading by *reverse transcriptase* provides an explanation for the high mutation rate of such viruses. [Note: In an attempt to prevent HIV infection from progressing to **acquired immune deficiency syndrome** (**AIDS**), patients are commonly treated with nucleoside and/or non-nucleoside inhibitors of *reverse transcriptase* accompanied by *protease* inhibitors (which target another HIV maturation enzyme), thus producing a mixture ("cocktail") that requires the virus to develop multiple resistance in order to continue to replicate.[6]]

E. Inhibition of DNA synthesis by nucleoside analogs

DNA chain growth can be blocked by the incorporation of certain nucleoside analogs that have been modified in the sugar portion of the nucleoside (Figure 29.24).[7] For example, removal of the hydroxyl group from the 3'-carbon of the deoxyribose ring as in **2',3'-dideoxyinosine** (**ddI**), or conversion of the deoxyribose to another

[5]See p. 361 in ***Lippincott's Illustrated Reviews: Microbiology*** for a discussion of retroviruses.
[6]See Chapter 39 in ***Lippincott's Illustrated Reviews: Pharmacology*** (3rd Ed.) and Chapter 37 (2nd Ed.) for a discussion of HIV therapy
[7]See Chapter 40 in ***Lippincott's Illustrated Reviews: Pharmacology*** (3rd Ed.) or Chapter 38 (2nd Ed.) for a discussion of nucleoside analogs.

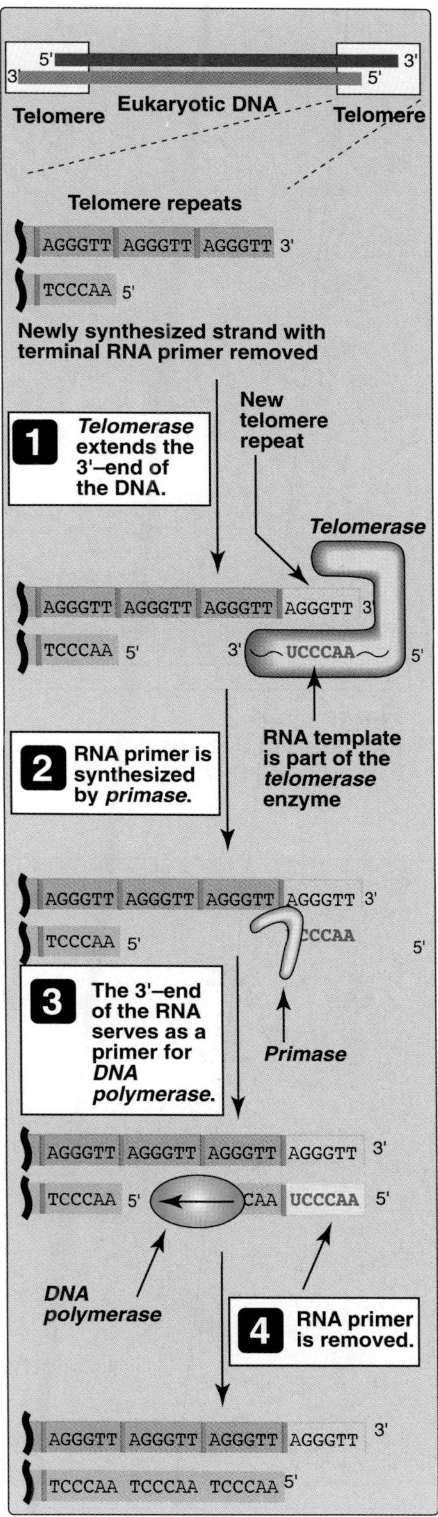

Figure 29.23
Mechanism of action of *telomerase*.

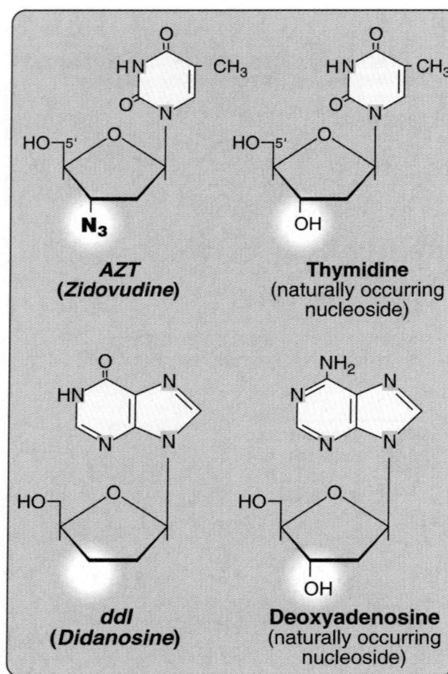

Figure 29.24
Examples of nucleoside analogs that lack a 3'-hydroxyl group.

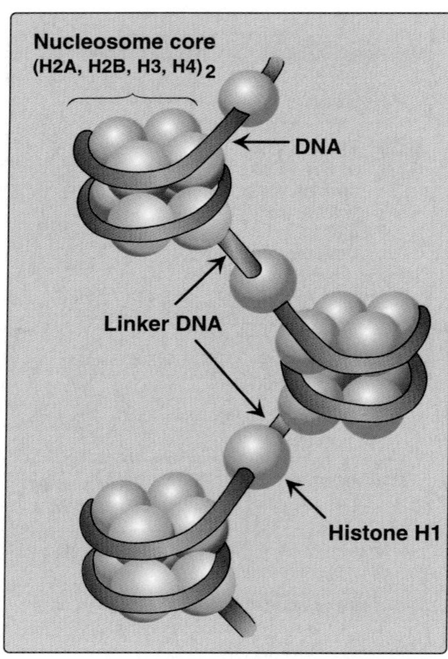

Figure 29.25
Organization of human DNA, illustrating the structure of nucleosomes.

sugar as in arabinose, prevents further chain elongation. By blocking DNA replication, these compounds slow the division of rapidly growing cells and viruses. For example, **cytosine arabinoside** (**cytarabine**, **araC**) has been used in anticancer chemotherapy, whereas **adenine arabinoside** (**vidarabine**, **araA**) is an antiviral agent. Chemically modifying the sugar moiety, as seen in **zidovudine** (**AZT**), accomplishes the same goal of termination of DNA chain elongation. [Note: These drugs are generally supplied as nucleosides, which are then converted to the active nucleotides by cellular "**salvage**" **enzymes** (see p. 294).]

VI. ORGANIZATION OF EUKARYOTIC DNA

A typical human cell contains 46 chromosomes, whose total DNA is approximately one meter long! It is difficult to imagine how such a large amount of genetic material can be effectively packaged into a volume the size of a cell nucleus so that it can be efficiently replicated and its genetic information expressed. To do so requires the interaction of DNA with a large number of proteins, each of which performs a specific function in the ordered packaging of these long molecules of DNA. Eukaryotic DNA is associated with tightly bound basic proteins, called histones. These serve to order the DNA into basic structural units, called nucleosomes, that resemble beads on a string. Nucleosomes are further arranged into increasingly more complex structures that organize and condense the long DNA molecules into chromosomes that can be segregated during cell division.

A. Histones and the formation of nucleosomes

There are five classes of histones, designated H1, H2A, H2B, H3, and H4. These small proteins are positively charged at physiologic pH as a result of their high content of lysine and arginine. Because of their positive charge, they form ionic bonds with negatively charged DNA. Histones, along with positively charged ions such as Mg^{++}, help neutralize the negatively charged DNA phosphate groups.

1. Nucleosomes: Two molecules each of H2A, H2B, H3, and H4 form the structural core of the individual nucleosome "beads." Around this core, a segment of the DNA double helix is wound nearly twice, forming a negatively supertwisted helix (Figure 29.25). [Note: The N-terminal ends of these histones can be acetylated, methylated, or phosphorylated. These reversible modifications can influence how tightly the histones bind to the DNA, thereby affecting the expression of specific genes.] Neighboring nucleosomes are joined by "linker" DNA approximately fifty base pairs long. Histone H1, of which there are several related species, is not found in the nucleosome core, but instead binds to the linker DNA chain between the nucleosome beads. H1 is the most tissue-specific and species-specific of the histones. It facillitates the packing of nucleosomes into the more compact structures.

2. **Higher levels of organization:** Nucleosomes can be packed more tightly to form a **polynucleosome** (also called a **nucleofilament**). This structure assumes the shape of a coil, often referred to as a **30-nm fiber**. The fiber is organized into loops that are anchored by

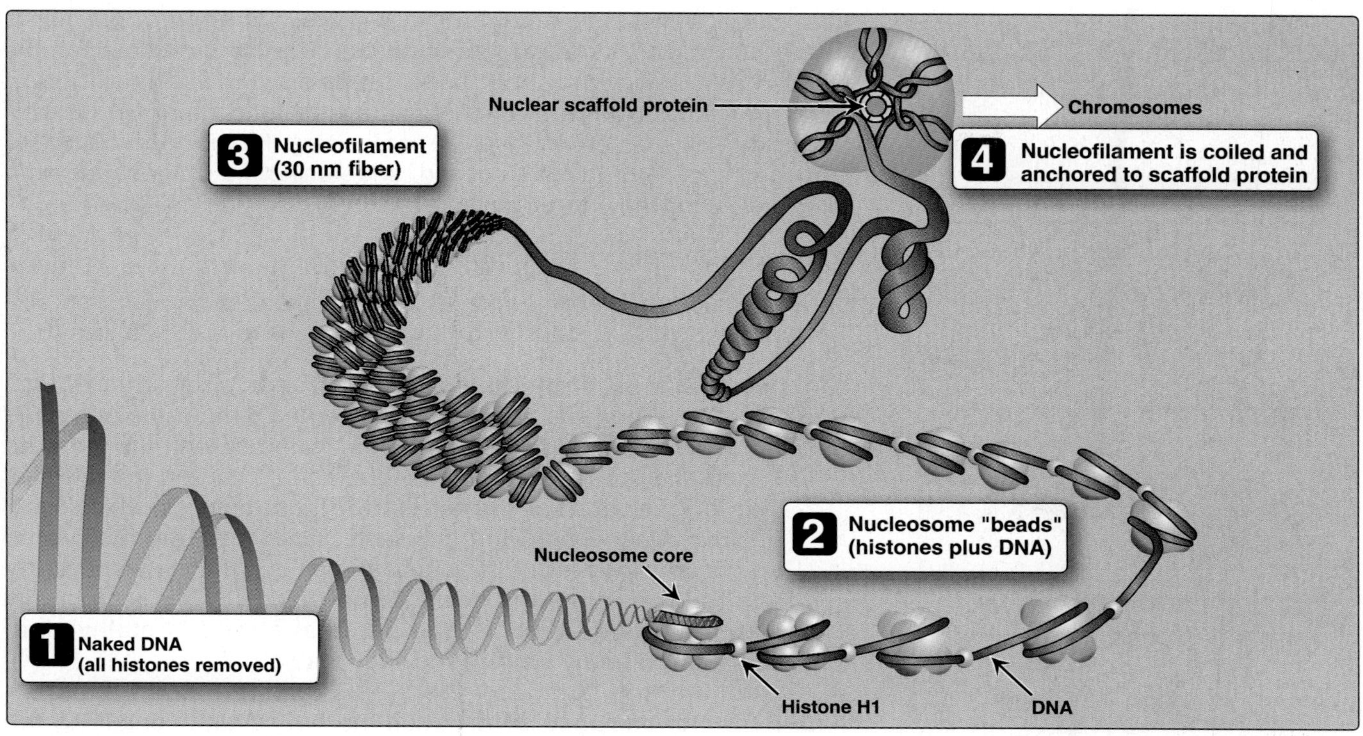

Figure 29.26
Structural organization of eukaryotic DNA.

a **nuclear scaffold** containing several proteins. Additional levels of organization lead to the final chromosomal structure (Figure 29.26).

B. Fate of nucleosomes during DNA replication

In order to replicate, the highly structured and constrained chromatin must be relaxed. Although the nucleosomes are displaced, dissociation of the nucleosome core from the DNA is incomplete, with all the parental histones remaining loosely associated with only one of the parental DNA strands. Synthesis of new histones occurs simultaneously with DNA replication, and nucleosomes containing the newly synthesized histones associate with only one of the new daughter helices. Therefore, the parental histone octamers are conserved.

VII. DNA REPAIR

Despite the elaborate proofreading system employed during DNA synthesis, mismatches—including incorrect base-pairing or insertion of one to a few extra nucleotides—can occur. In addition, DNA is constantly being subjected to environmental insults that cause the alteration or removal of nucleotide bases. The damaging agents can be either chemicals, for example, nitrous acid, or radiation, for example, ultraviolet light, which can fuse two pyrimidines adjacent to each other in the DNA, and high-energy radiation, which can cause double-strand breaks. Bases are also altered or lost spontaneously from mammalian DNA at a rate of many thousands per cell per day. If the damage is not repaired, a perma-

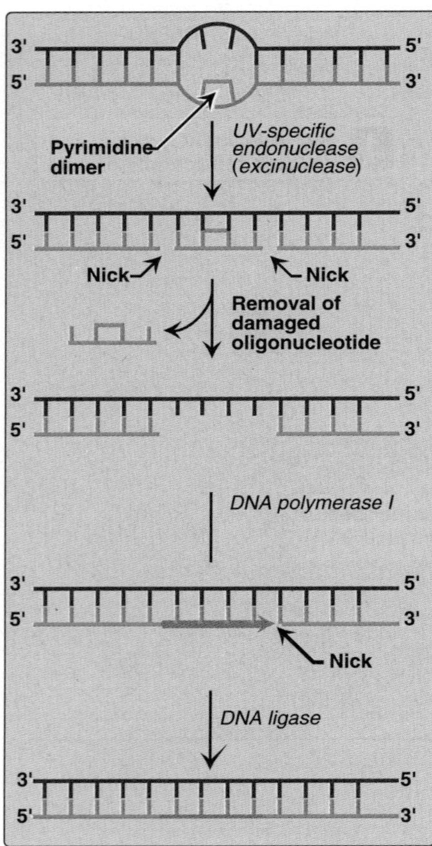

Figure 29.27
Excision repair of pyrimidine dimers in E. coli DNA.

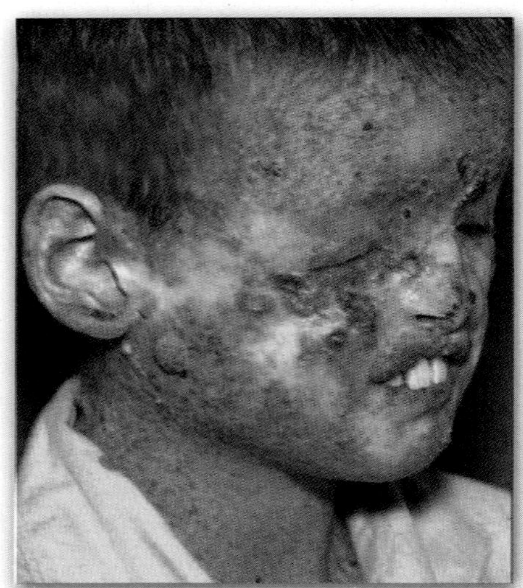

Figure 29.28
Patient with xeroderma pigmentosum.

nent mutation may be introduced that can result in any of a number of deleterious effects, including loss of control over the proliferation of the mutated cell, leading to cancer. Luckily, cells are remarkably efficient at repairing the mismatches and other damage done to their DNA. Most of the repair enzymes are involved in recognizing the lesion, excising the damaged section of the DNA strand, and—using the sister strand as a template—filling the gap left by the excision of the abnormal DNA.

A. Strand-directed mismatch repair system

Sometimes replication errors escape the proofreading function during DNA synthesis, causing a mismatch of one to several bases.

1. **Identification of the mismatched strand:** When a mismatch occurs, the proteins that are to identify and remove the mispaired nucleotide(s) must be able to discriminate between the template strand and the newly synthesized strand containing the mistake. This is done based on the fact that GATC sequences, which occur approximately once every thousand nucleotides, are methylated on the adenine residue. This methylation is not done immediately after synthesis, so the newly synthesized DNA is temporarily hemimethylated (that is, the parental strand is methylated, whereas the newly synthesized strand is not).

2. **Repair of damaged DNA:** When the new strand containing the mismatch is identified, an endonuclease nicks the mismatched strand, and the mismatched base(s) is/are removed. The gap left by removal of the mismatched nucleotide(s) is filled, using the sister strand as a template, by a *5'→3' DNA polymerase* (*DNA polymerase I* in E. coli). The 3'-hydroxyl of the newly synthesized DNA is spliced to the 5'-phosphate of the remaining stretch of the original DNA strand by *DNA ligase* (see p. 403). [Note: A defect in mismatch repair in humans has been shown to cause **hereditary nonpolyposis colon cancer** (**HNPCC**), one of the most common inherited cancers.]

B. Repair of damage caused by ultraviolet light

Exposure of a cell to ultraviolet light can result in the covalent joining of two adjacent pyrimidines (usually thymines), producing a dimer. These **thymine dimers** prevent *DNA polymerase* from replicating the DNA strand beyond the site of dimer formation. Thymine dimers are excised in bacteria as illustrated in Figure 29.27. A similar pathway is present in humans.

1. **Recognition and excision of dimers by UV-specific endonuclease:** First, a *UV-specific endonuclease* (called *uvrABC excinuclease*) recognizes the dimer, and cleaves the damaged strand at phosphodiester bonds on both the 5'-side and 3'-side of the dimer. The damaged oligonucleotide is released, leaving a gap in the DNA strand that formerly contained the dimer. This gap is filled in using the same repair system described above.

2. **Ultraviolet radiation and cancer:** Pyrimidine dimers can be formed in the skin cells of humans exposed to unfiltered sunlight.

In the rare genetic disease **xeroderma pigmentosum**, the cells cannot repair the damaged DNA, resulting in extensive accumulation of mutations and, consequently, skin cancers (Figure 29.28). The most common form of this disease is caused by the absence of the *UV-specific excinuclease*.

C. Correction of base alterations (base excision repair)

The bases of DNA can be altered, either spontaneously, as is the case with cytosine, which slowly undergoes deamination (the loss of its amino group) to form uracil, or by the action of deaminating or alkylating compounds. For example, **nitrous acid**, which is formed by the cell from precursors, such as the nitrosamines, nitrites, and nitrates, is a potent compound that deaminates cytosine, adenine, and guanine. Bases can also be lost spontaneously. For example, approximately 10,000 purine bases are lost this way per cell per day. Lesions involving base alterations or loss can be corrected by the following mechanisms (Figure 29.29).

1. **Removal of abnormal bases:** Abnormal bases, such as uracil, which can occur in DNA either by deamination of cytosine or improper incorporation of dUTP instead of dTTP during DNA synthesis, are recognized by specific glycosylases that hydrolytically cleave them from the deoxyribose–phosphate backbone of the strand. This leaves an **apyrimidinic site** (or **apurinic**, if a purine was removed), referred to as an **AP-site**.

2. **Recognition and repair of an AP-site:** Specific *AP-endonucleases* recognize that a base is missing and initiate the process of excision and gap filling by making an endonucleolytic cut just to the 5' side of the AP-site. A *deoxyribose-phosphate lyase* removes the single, empty, sugar-phosphate residue. *DNA polymerase* (*pol I* in E. coli) and *DNA ligase* complete the repair process.

D. Repair of double-strand breaks

High-energy radiation or oxidative free radicals (see p. 145) can cause double-strand breaks in DNA, which are potentially lethal to the cell. Double-strand breaks also occur naturally during gene rearrangements. Double-strand DNA breaks cannot be corrected by the previously described strategy of excising the damage on one strand and using the remaining strand as a template for replacing the missing nucleotide(s). Instead, double-strand breaks are repaired by one of two systems. The first is **nonhomologous end-joining repair**, in which the ends of two DNA fragments are brought together by a group of proteins that effect their religation. This system does not require that the two DNA sequences have any sequence homology. However, this mechanism of repair, which is the main repair mechanism in humans, is error prone and mutagenic. Defects in this repair system are associated with a predisposition to **cancer** and **immunodeficiency syndromes**. The second repair system, **homologous recombination repair**, uses the enzymes that normally perform genetic recombination between homologous chromosomes during meiosis. This system is used predominantly by the lower eukaryotes to repair double-strand breaks.

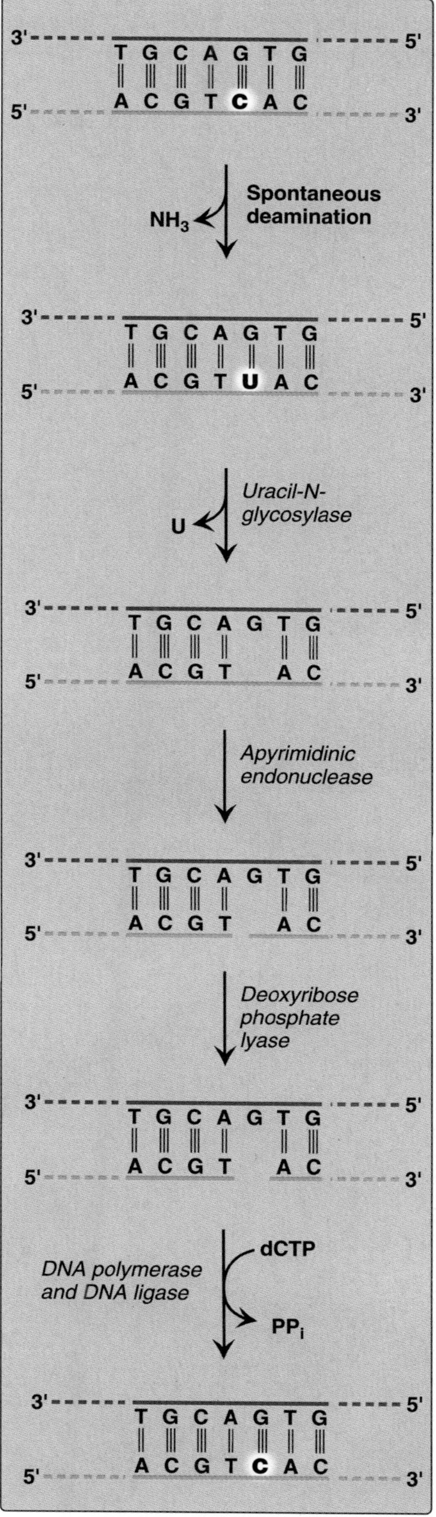

Figure 29.29
Correction of base alterations.

VIII. CHAPTER SUMMARY

DNA contains many monodeoxyribonucleotides covalently linked by **3'→5'–phosphodiester bonds**. The resulting long, unbranched chain has **polarity**, with both a 5'-end and a 3'-end. The sequence of nucleotides is read 5' to 3'. DNA exists as a **double-stranded molecule**, in which the two chains are paired in an **antiparallel manner**, and wind around each other, forming a **double helix**. **Adenine** pairs with **thymine** and **cytosine** pairs with **guanine**. Each strand of the double helix serves as a **template** for constructing a **complementary** daughter strand (**semiconservative replication**). DNA replication begins at the **origin of replication**. The strands are separated locally, forming two **replication forks**. Replication of double-stranded DNA is **bidirectional**. **Helicase** unwinds the double helix. As the two strands of the double helix are separated, **positive supercoils** are produced in the region of DNA ahead of the replication fork. **DNA topoisomerases types I and II** remove supercoils. DNA polymerases **synthesize** new DNA strands only in the **5'→3' direction**. Therefore, one of the newly synthesized stretches of nucleotide chains must grow in the 5'→3' direction toward the replication fork (**leading strand**), and one in the 5'→3' direction away from the replication fork (**lagging strand**). DNA polymerases require a **primer**. The primer for de novo DNA synthesis is a short stretch of **RNA** synthesized by **primase**. The leading strand only needs one RNA primer, whereas the lagging strand needs many. E. coli DNA chain elongation is catalyzed by **DNA polymerase III**, using **5'–deoxyribonucleoside triphosphates** as substrates. The enzyme "**proofreads**" the newly synthesized DNA, removing terminal mismatched nucleotides with its **3'→5' exonuclease** activity. RNA primers are removed by **DNA polymerase I**, using its **5'→3' exonuclease** activity. The resulting gaps are filled in by this enzyme. The final phosphodiester linkage is catalyzed by **DNA ligase**. There are at least five classes of **eukaryotic DNA polymerases**. **Pol α** is a multisubunit enzyme, one subunit of which is a **primase**. **Pol α 5'→3' polymerase** activity adds a short piece of DNA to the RNA primer. **Pol δ** completes DNA synthesis on the leading strand and elongates each lagging strand fragment using **3'→5' exonuclease** activity to **proofread**. **Pol β** and **pol ε** are involved with DNA "repair", and **pol γ** replicates mitochondrial DNA. **Nucleoside analogs** containing modified sugars can be used to block DNA chain growth. They are useful in anticancer and antiviral chemotherapy. **Telomeres** are stretches of **highly repetitive DNA** found at the ends of linear chromosomes. As cells divide and age, these sequences are shortened, contributing to cell death. In cells that do not age (for example, germ-line and cancer cells) the enzyme **telomerase** replaces the telomeres. There are five classes of **positively charged histone proteins**. Two each of histones H2A, H2B, H3, and H4 form a structural core around which DNA is wrapped creating a **nucleosome**. The DNA connecting the nucleosomes, called **linker DNA**, is bound to histone H1. Nucleosomes can be packed more tightly to form a **nucleofilament**. Additional levels of organization create a **chromosome**. Exposure of a cell to **ultraviolet light** can cause pyrimidine (usually thymine) dimers. **Thymine dimers** are removed by **UV-specific endonuclease** (**uvrABC excinulease**), and the resulting gap is filled by **DNA polymerase I**. In eukaryotes, a deficiency of UV-specific excinulease causes **xeroderma pigmentosum**. **Abnormal** or **mismatched bases** can be removed by a similar mechanism.

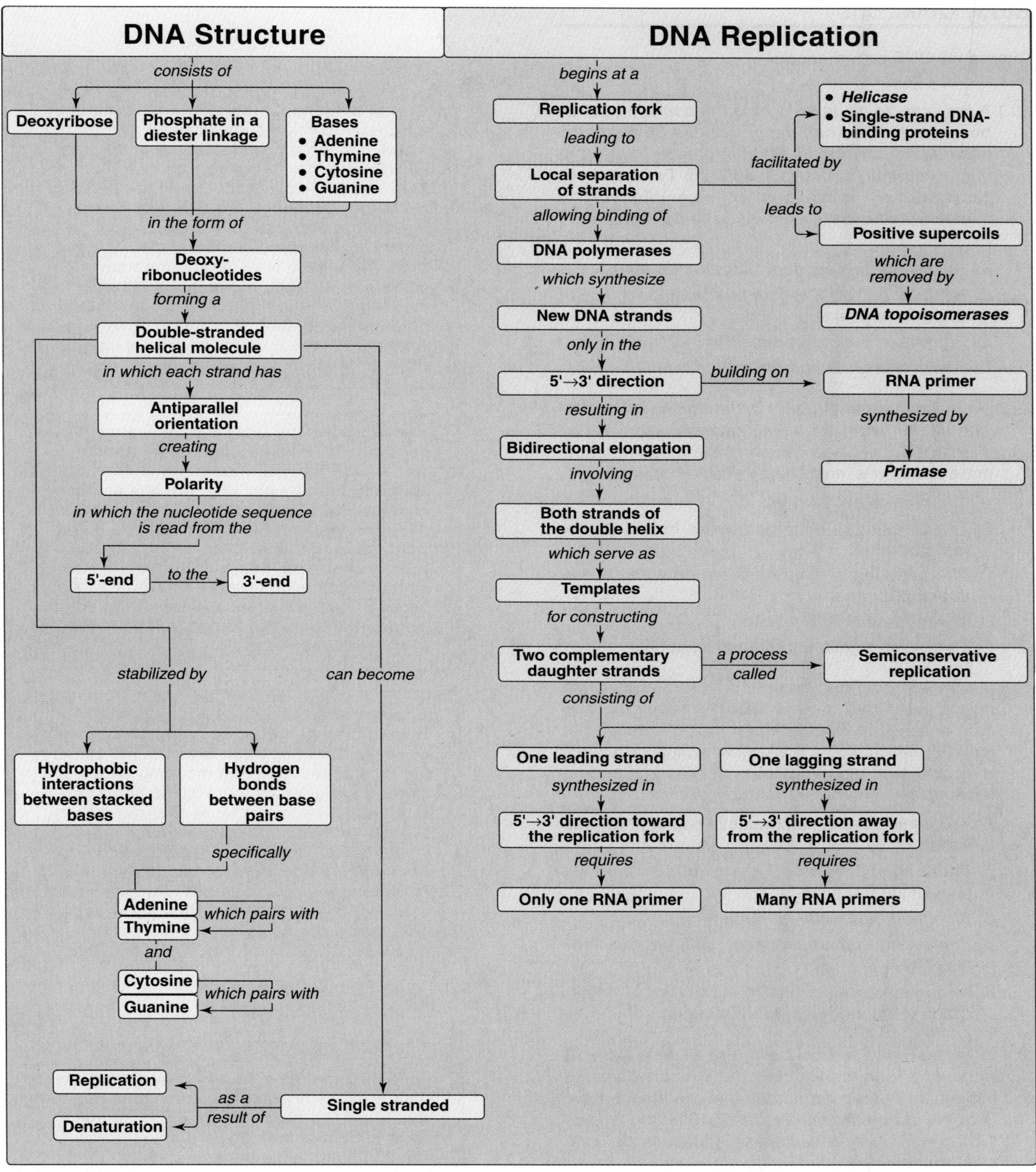

Figure 29.30
Key concept map for DNA structure and replication.

Study Questions

Choose the ONE correct answer

29.1 A ten-year-old girl is brought to the dermatologist by her parents. She has many freckles on her face, neck, arms, and hands, and the parents report that she is unusually sensitive to sunlight. Two basal cell carcinomas are identified on her face. Which of the following processes is most like to be defective in this patient?

A. Removal of primers from Okazaki fragments
B. Removal of mismatched bases from the 3' end of Okazaki fragments
C. Removal of pyrimidine dimers from DNA
D. Removal of uracil from DNA

Correct answer = C. The sensitivity to sunlight, extensive freckling on parts of the body exposed to the sun, and presence of skin cancer indicates that the patient most likely suffers from xeroderma pigmentosum. These patients are usually deficient in the UV-specific excinuclease that is used to remove pyrimidine dimers during the repair of ultraviolet-damaged DNA. None of the other choices have any relationship to sunlight. Mismatched bases from the 3' end of Okazaki fragments are removed by the proofreading function of DNA polymerases. Uracil is removed from damaged DNA molecules by a specific glycosylase.

29.2 An eight-year-old girl with cystic fibrosis is treated with ciprofloxacin for a Pseudomonas aeruginosa infection in her lungs. Which of the following enzymatic activities is most directly affected by this drug?

A. The synthesis of RNA primers
B. The breaking of hydrogen bonds in front of the replication fork
C. The breaking and subsequent rejoining of the DNA backbone
D. The removal of RNA primers
E. The joining together of Okazaki fragments

Correct answer = C. Fluoroquinolones, such as ciprofloxacin, inhibit bacterial DNA gyrase—a type II DNA topoisomerase. This enzyme catalyzes the transient breaking and rejoining of the phosphodiester bonds of the DNA backbone, to allow the removal of positive supercoils during DNA replication. The other enzyme activities mentioned are not affected. Primase synthesizes RNA primers, helicase breaks hydrogen bonds in front of the replication fork, DNA polymerase I removes RNA primers, and DNA ligase joins Okazaki fragments.

29.3 Didanosine (dideoxyinosine, ddI) is a nucleoside analog sometimes used to treat HIV infections. This drug is converted metabolically to 2',3'-dideoxyATP (ddATP), which blocks DNA chain elongation when it is incorporated into viral DNA synthesized by reverse transcriptase. Why does DNA synthesis stop?

A. The analog becomes covalently bound to reverse transcriptase, thus inactivating the enzyme.
B. There is no 3'-hydroxyl group to form the next phosphodiester bond.
C. Proofreading is inhibited.
D. The analog cannot hydrogen bond to the RNA template.
E. Incorporation of the analog initiates rapid degradation of the newly synthesized strand.

Correct answer = B. The DNA chain cannot continue to grow if it does not have a 3'-hydroxyl group, because this group is required to attack the phosphate of the next incoming deoxynucleotide during the formation of the phosphodiester bond. The analog forms hydrogen bonds perfectly well with the template as it is being incorporated. Reverse transcriptase has no proofreading ability and does not covalently bind to its substrates. The defective new strand will eventually be degraded, but this is not an especially rapid process.

29.4 While studying the structure of a small gene that was recently sequenced during the Human Genome Project, an investigator notices that one strand of the DNA molecule contains 20 A's, 25 G's, 30 C's, and 22 T's. How many of each base is found in the complete double-stranded molecule?

A. A = 40, G = 50, C = 60, T = 44
B. A = 44, G = 60, C = 50, T = 40
C. A = 45, G = 45, C = 52, T = 52
D. A = 50, G = 47, C = 50, T = 47
E. A = 42, G = 55, C = 55, T = 42

Correct answer = E. The two DNA strands are complementary to each other, with A base-paired with T, and G base-paired with C. So, for example, the 20 A's on the first strand would be paired with 20 T's on the second strand, the 25 G's on the first strand would be paired with 25 C's on the second strand, and so forth. When these are all added together, the correct numbers of each base are indicated in choice E. Notice that, in the correct answer, A = T and G = C.

RNA Structure and Synthesis

30

I. OVERVIEW

The genetic master plan of an organism is contained in the sequence of deoxyribonucleotides that constitute the DNA. However, it is through the ribonucleic acid (RNA)—the "working copies" of the DNA—that the master plan is expressed (Figure 30.1). The copying process, during which a DNA strand serves as a template, is called **transcription**. Following their synthesis, messenger RNAs are translated into sequences of amino acids (polypeptide chains or proteins). Ribosomal RNAs, transfer RNAs, and additional small RNA molecules perform specialized structural and regulatory functions and are not translated. A central feature of transcription is that it is highly selective. For example, many transcripts are made of some regions of the DNA. In other regions, few or no transcripts are made. This selectivity is due, at least in part, to signals embedded in the nucleotide sequence of the DNA. These signals instruct the *RNA polymerase* where to start, how often to start, and where to stop transcription. A variety of regulatory proteins is also involved in this selection process. The biochemical differentiation of an organism's tissues is ultimately a result of the selectivity of the transcription process. Another important feature of transcription is that many RNA transcripts that initially are faithful copies of one of the two DNA strands may undergo various modifications, such as terminal additions, base modifications, trimming, and internal segment removal, followed by splicing, which convert the inactive primary transcript into a functional molecule.

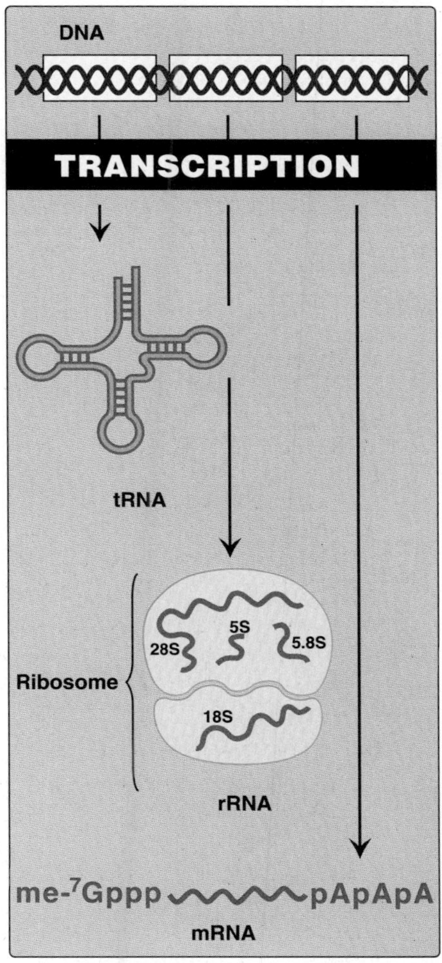

Figure 30.1
Expression of genetic information by transcription. [Note: RNAs shown are eukaryotic.] me-^{7}Gppp = 7-methylguanosine triphosphate "cap," described on p. 414. AAA = poly-A tail, described on p. 414.

II. STRUCTURE OF RNA

There are three major types of RNA that participate in the process of protein synthesis: **ribosomal RNA** (**rRNA**), **transfer RNA** (**tRNA**), and **messenger RNA** (**mRNA**). Like DNA, these three types of RNA are unbranched polymeric molecules composed of mononucleotides joined together by phosphodiester bonds (see p. 392). However, they differ as a group from DNA in several ways, for example, they are considerably smaller than DNA, and they contain **ribose** instead of deoxyribose and **uracil** instead of thymine. Unlike DNA, most RNAs exist as single strands that are capable of folding into complex structures. The three major types of RNA also differ from each other in size, function, and special structural modifications. [Note: In eukaryotes, small RNA molecules found in the nucleus (snRNAs) perform specialized functions as described on p. 424.]

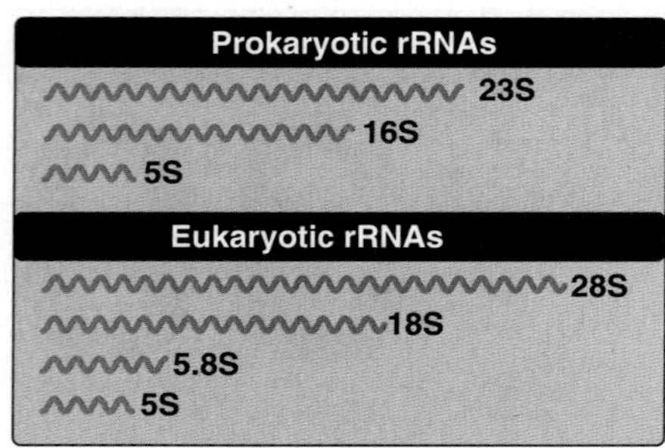

Figure 30.2
Prokaryotic and eukaryotic rRNAs.

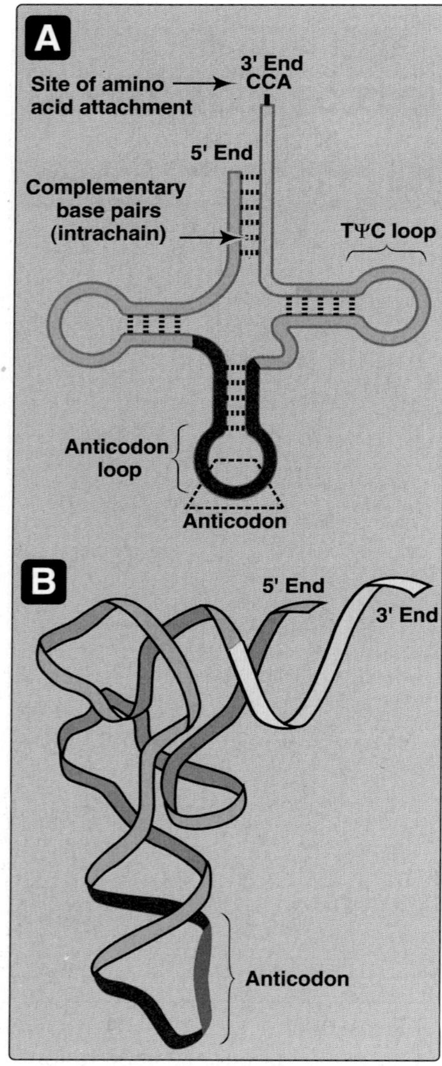

Figure 30.3
A. Characteristic tRNA structure.
B. Folded tRNA structure found in cells.

A. Ribosomal RNA

Ribosomal RNAs (rRNAs) are found in association with several proteins as components of the **ribosomes**—the complex structures that serve as the sites for protein synthesis (see p. 433). There are three distinct size species of rRNA (23S, 16S, and 5S) in prokaryotic cells (Figure 30.2). In the eukaryotic cytosol, there are four rRNA size species (28S, 18S, 5.8S, and 5S). [Note: "S" is the **Svedberg unit**, which is related to the molecular weight and shape of the compound.] Together, rRNAs make up eighty percent of the total RNA in the cell.

B. Transfer RNA

Transfer RNAs (tRNAs), the smallest of the three major species of RNA molecules (4S), have between 74 and 95 nucleotide residues. There is at least one specific type of tRNA molecule for each of the twenty amino acids commonly found in proteins. Together, tRNAs make up about fifteen percent of the total RNA in the cell. The tRNA molecules contain **unusual bases** (for example, pseudouracil, see Figure 22.2, p. 290) and have **extensive intrachain base-pairing** (Figure 30.3). Each tRNA serves as an "adaptor" molecule that carries its specific amino acid—covalently attached to its 3'-end—to the site of protein synthesis. There it recognizes the genetic code word on an mRNA, which specifies the addition of its amino acid to the growing peptide chain (see p. 429).

C. Messenger RNA

Messenger RNA (mRNA) comprises only about five percent of the RNA in the cell, yet is by far the most heterogeneous type of RNA in size (500 to 6000 nucleotides) and base sequence. The mRNA carries genetic information from the nuclear DNA to the cytosol, where it is used as the template for protein synthesis. Special structural characteristics of **eukaryotic mRNA** (but not prokaryotic) include a long sequence of adenine nucleotides (a "**poly-A tail**") on the 3'-end of the RNA chain, plus a "**cap**" on the 5'-end consisting of a molecule of **7-methylguanosine** attached "backward" (5'→5') through a triphosphate linkage as shown in Figure 30.4. [Note: The mechanisms for modifying mRNA to create these special structural characteristics are discussed on p. 422.]

III. TRANSCRIPTION OF PROKARYOTIC GENES

The structure of *RNA polymerase*, the signals that control transcription, and the varieties of modification that RNA transcripts can undergo differ among organisms, and particularly from prokaryotes to eukaryotes. Therefore, in this chapter, the discussions of prokaryotic and eukaryotic transcription are presented separately.

A. Properties of prokaryotic RNA polymerase

In bacteria, one species of *RNA polymerase* synthesizes all of the RNA except for the short RNA primers needed for DNA replication (RNA primers are synthesized by a specialized enzyme, *primase*, see p. 400). *RNA polymerase* is a multisubunit enzyme that recognizes a nucleotide sequence (the **promoter region**) at the beginning of a length of DNA that is to be transcribed. It next makes a complementary RNA copy of the DNA template strand, and then recognizes the end of the DNA sequence to be transcribed (the **termination region**). RNA is synthesized from its 5'-end to its 3'-end, antiparallel to its DNA template strand (see p. 395). The template is copied as it is in DNA synthesis, in which a G on the DNA specifies a C in the RNA, a C specifies a G, a T on the DNA template specifies an A in the RNA, but an A on the template specifies a U (instead of a T) on the RNA (Figure 30.5). A **transcription unit** extends from the promoter to the termination region, and the product of the process of transcription by *RNA polymerase* is termed the **primary transcript**. Transcription by *RNA polymerase* involves a core enzyme and several auxillary proteins:

1. **Core enzyme:** Four of the enzyme's peptide subunits, **2α**, **1β**, and **1β'**, are responsible for the *5'→3' RNA polymerase* activity, and are referred to as the core enzyme (Figure 30.6). However, this enzyme lacks specificity, that is, it cannot recognize the promoter region on the DNA template.

2. **Holoenzyme:** The **σ subunit** ("**sigma factor**") enables *RNA polymerase* to recognize promoter regions on the DNA. The σ subunit plus the core enzyme make up the holoenzyme. [Note: Different σ factors recognize different groups of genes.]

3. **Termination factor:** Some regions on the DNA that signal the termination of transcription are recognized by the *RNA polymerase* itself. Others are recognized by specific termination factors, an example of which is the **rho** (ρ) **factor** of E. coli.

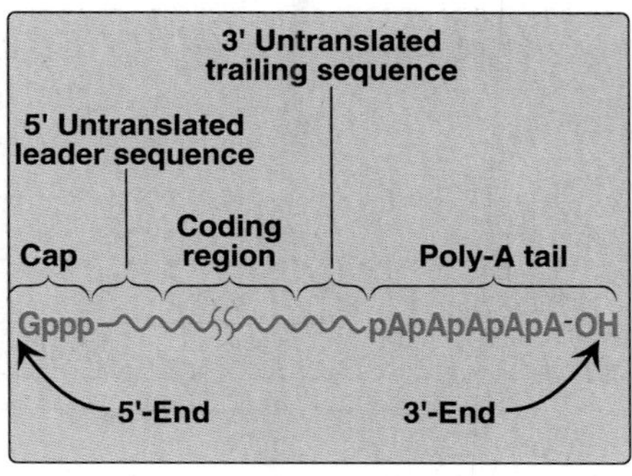

Figure 30.4
Structure of eukaryotic messenger RNA.

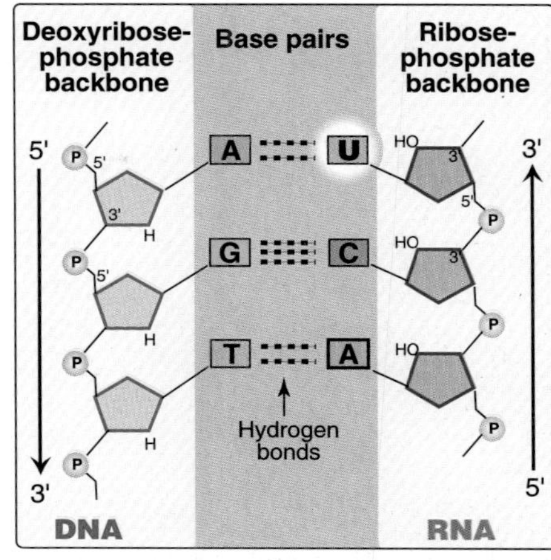

Figure 30.5
Antiparallel, complementary base pairs between DNA and RNA.

B. Steps in RNA synthesis

The process of transcription of a typical gene of E. coli can be divided into three phases: initiation, elongation, and termination. [Note: Within the DNA molecule, regions of both strands can serve as templates for specific RNA molecules. However, only one of the two DNA strands serves as a template within a specific stretch of double helix.]

1. **Initiation:** Initiation of transcription involves the binding of the *RNA polymerase* holoenzyme to a region on the DNA that determines the specificity of transcription of that particular gene. That DNA sequence is known as the **promoter region** (Figure 30.7). Characteristic "**consensus**" **nucleotide sequences** of the prokaryotic promoter region are highly **conserved**, that is, many different promoters contain some very similar or identical sequences. Those that are recognized by prokaryotic *RNA polymerase* σ factors include:

   a. **Pribnow box:** This is a stretch of six nucleotides (5'-TATAAT-3') **centered** about eight to ten nucleotides to the left of the transcription start site that codes for the initial base of the mRNA (see Figure 30.7). [Note: The regulatory sequences that control transcription are, by convention, designated by the 5'→3' nucleotide sequence on the nontemplate strand. A base in the promoter region is assigned a negative number if it occurs prior to ("upstream" of) the transcription start site. Therefore, the Pribnow box is centered approximately around base –9. The first base at the transcription start site is assigned a position of +1. There is no base designated "0."]

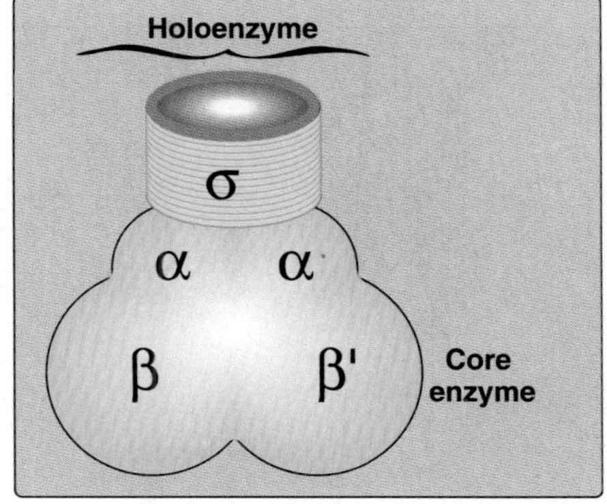

Figure 30.6
Prokaryotic *RNA polymerase*.

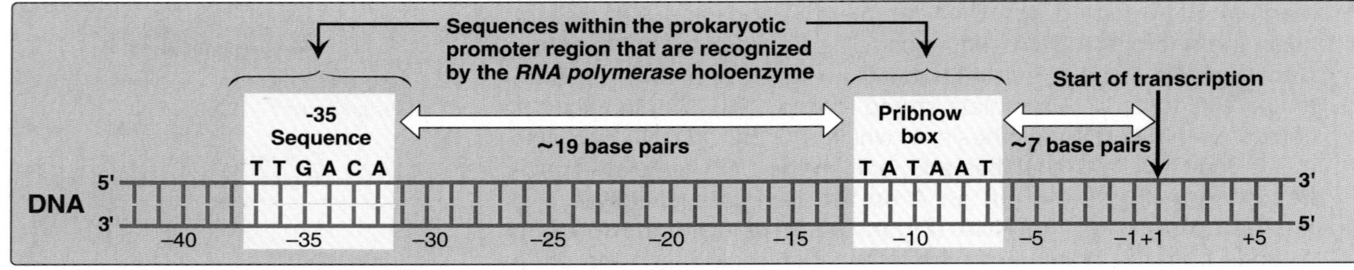

Figure 30.7
Structure of the prokaryotic promoter region.

b. **–35 sequence:** A second consensus nucleotide sequence (5'–TTGACA–3'), is centered about 35 bases to the left of the transcription start site (see Figure 30.7). [Note: A mutation in either the Pribnow box or the –35 sequence can affect the transcription of the gene controlled by the mutant promoter.]

2. **Elongation:** Once the promoter region has been recognized by the holoenzyme, *RNA polymerase* begins to synthesize a transcript of the DNA sequence (usually beginning with a purine), and the σ subunit is released. Unlike *DNA polymerase*, *RNA polymerase* does not require a primer and has no known endonuclease or exonuclease activity. It, therefore, has no ability to repair mistakes in the RNA, as does *DNA polymerase* during DNA synthesis (see p. 407). *RNA polymerase* uses **ribonucleoside triphosphates**, and releases pyrophosphate each time a nucleotide is added to the growing chain. [Note: As in DNA synthesis, two high-energy bonds are thus used for the addition of each nucleotide.] The binding of the enzyme to the DNA template results in a local unwinding of the DNA helix (Figure 30.8). [Note: This process can generate supercoils that can be relaxed by *DNA topoisomerases I* and *II* (see p. 398).]

3. **Termination:** The process of elongation of the RNA chain continues until a termination signal is reached. An additional protein, ρ **(rho) factor**, may be required for the release of the RNA product (**ρ-dependent termination**). Alternatively, the tetrameric *RNA polymerase* can, in some instances, recognize termination regions on the DNA template (**ρ-independent termination**).

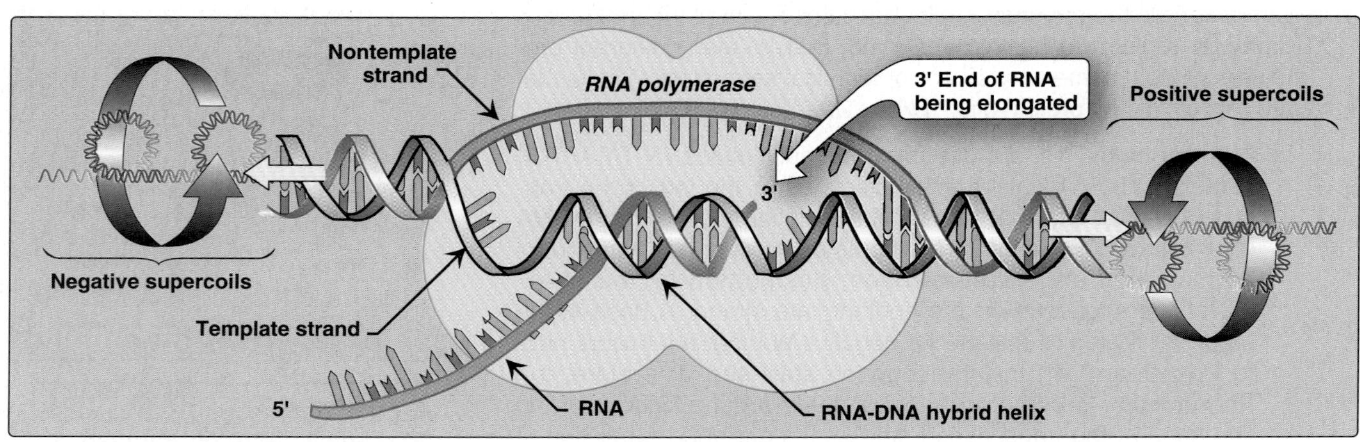

Figure 30.8
Local unwinding of DNA caused by *RNA polymerase*.

a. **Rho-dependent termination** requires the participation of an additional protein, ρ **factor**. This factor binds to a C-rich region near the 3'-end of the newly synthesized RNA, and migrates along behind the *RNA polymerase* in the 5'→3' direction until the termination site is reached. [Note: Rho factor has *ATP-dependent RNA–DNA helicase* activity that hydrolyzes ATP, and uses the energy to unwind the 3'-end of the transcript from the template. This facilitates the movement of the protein along the RNA/DNA duplex.] At the termination site, ρ factor displaces the DNA template strand, facilitating the dissociation of the RNA molecule.

b. **Rho-independent termination** requires that the newly synthesized RNA have two important structural features. First, the RNA transcript must be able to form a **stable hairpin turn** that slows down the progress of *RNA polymerase* and causes it to pause temporarily. The hairpin turn of the RNA is complementary to a region of the DNA template near the termination region that exhibits two-fold symmetry as a result of the presence of a **palindrome**. [Note: A palindrome is a region of double-stranded DNA in which each of the two strands contain stretches that have the same nucleotide sequence when read in the same (for example, 5'→3') direction (Figure 30.9 A).] Near the base of the stem of the hairpin, a sequence occurs that is rich in G and C. This stabilizes the secondary structure of the hairpin. Next, beyond the hairpin turn, the RNA transcript contains a string of U's. The bonding of U's to the corresponding DNA template A's is weak (see p. 395). This

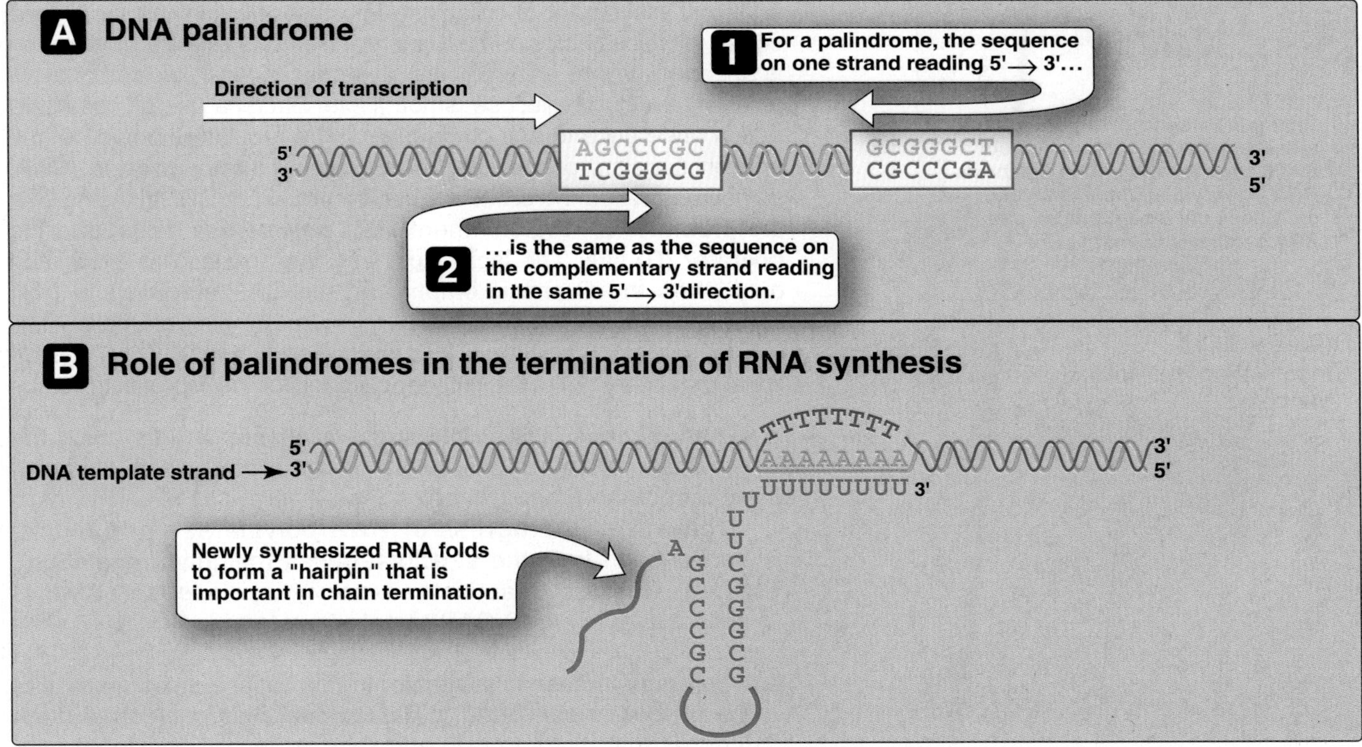

Figure 30.9
Rho-independent termination of transcription. A. An example of a palindrome in double-stranded DNA. B. A transcribed DNA palindrome codes for RNA that can form a hairpin turn.

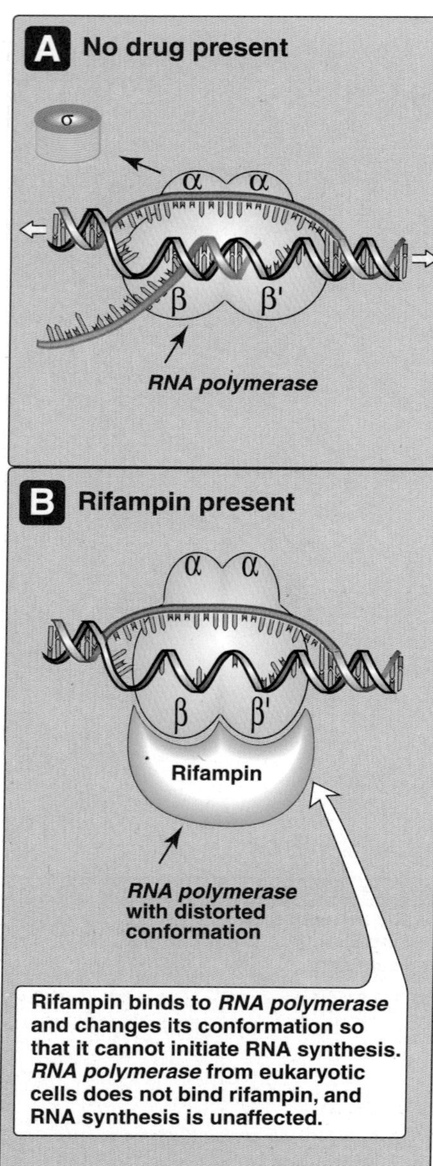

Figure 30.10
Inactivation of *RNA polymerase* by rifampin.

facilitates the separation of the newly synthesized RNA from its DNA template, as the double helix "zips up" behind the *RNA polymerase* (Figure 30.9 B).

4. **Action of antibiotics:** Some antibiotics prevent bacterial cell growth by inhibiting RNA synthesis. For example, rifampin inhibits the initiation of transcription by binding to the β-subunit of prokaryotic *RNA polymerase*, thus interfering with the formation of the first phosphodiester bond (Figure 30.10). Rifampin is useful in the treatment of tuberculosis.[1] Dactinomycin (known to biochemists as actinomycin D) was the first antibiotic to find therapeutic application in tumor chemotherapy.[2] It binds to the DNA template and interferes with the movement of *RNA polymerase* along the DNA.

C. Transcription from bacterial operons

In bacteria, the structural genes that code for the enzymes of a metabolic pathway are often found grouped together on the chromosome together with the regulatory genes that determine their transcription as a single long piece of mRNA. The genes are, thus, **coordinately expressed**. This entire package is referred to as an operon. One of the best understood examples is the **lactose operon** of E. coli, which illustrates both positive and negative regulation (Figure 30.11).

1. **The lactose operon:** The lactose (lac) operon codes for three enzymes involved in the catabolism of the sugar lactose: the **lacZ gene** codes for *β-galactosidase*, which hydroyzes lactose to galactose and glucose; the **lacY gene** codes for a *permease* that facilitates the movement of galactose into the cell; and the **lacA gene** codes for *thiogalactoside transacetylase* whose physiologic function is unknown. These enzymes are all produced when lactose is available to the cell but glucose is not. [Note: Bacteria use glucose as a fuel in preference to any other sugar.] The regulatory portion of the operon consists of the **catabolite gene activator protein (CAP**, sometimes called the cAMP regulatory protein or CRP) binding site; the **promoter (P) region** where *RNA polymerase* binds; and the **operator (O) site**. The lacZ, lacY, and lacA genes are expressed only when the O site is empty and the CAP binding site (just upstream of the P region) is bound by a complex of cyclic AMP (cAMP, see p. 93) and the CAP protein (see Figure 30.11). An additional regulatory gene, the **lacI gene**, codes for the repressor protein,

   a. **When glucose is the only sugar available:** In this case, the repressor protein binds to the operator site, which is downstream of the promotor region (see Figure 30.11A). This interferes with the progress of *RNA polymerase* and blocks transcription from the structural genes (**negative regulation**). [Note: Glucose deactivates *adenylyl cyclase*, so cAMP is absent and no cAMP-CAP complex can form.]

   b. **When only lactose is available:** In this case, a small amount of the lactose is converted to **allolactose**. This compound is an **inducer** that binds to the repressor protein, changing its con-

[1]See Chapter 35 in ***Lippincott's Illustrated Reviews: Pharmacology*** (3rd Ed.) and Chapter 33 (2nd Ed.) for a discussion of rifampin in the treatment of tuberculosis.
[2]See Chapter 40 in ***Lippincott's Illustrated Reviews: Pharmacology*** (3rd Ed.) and Chapter 38 (2nd Ed.) for a discussion of dactinomycin in treating cancer.

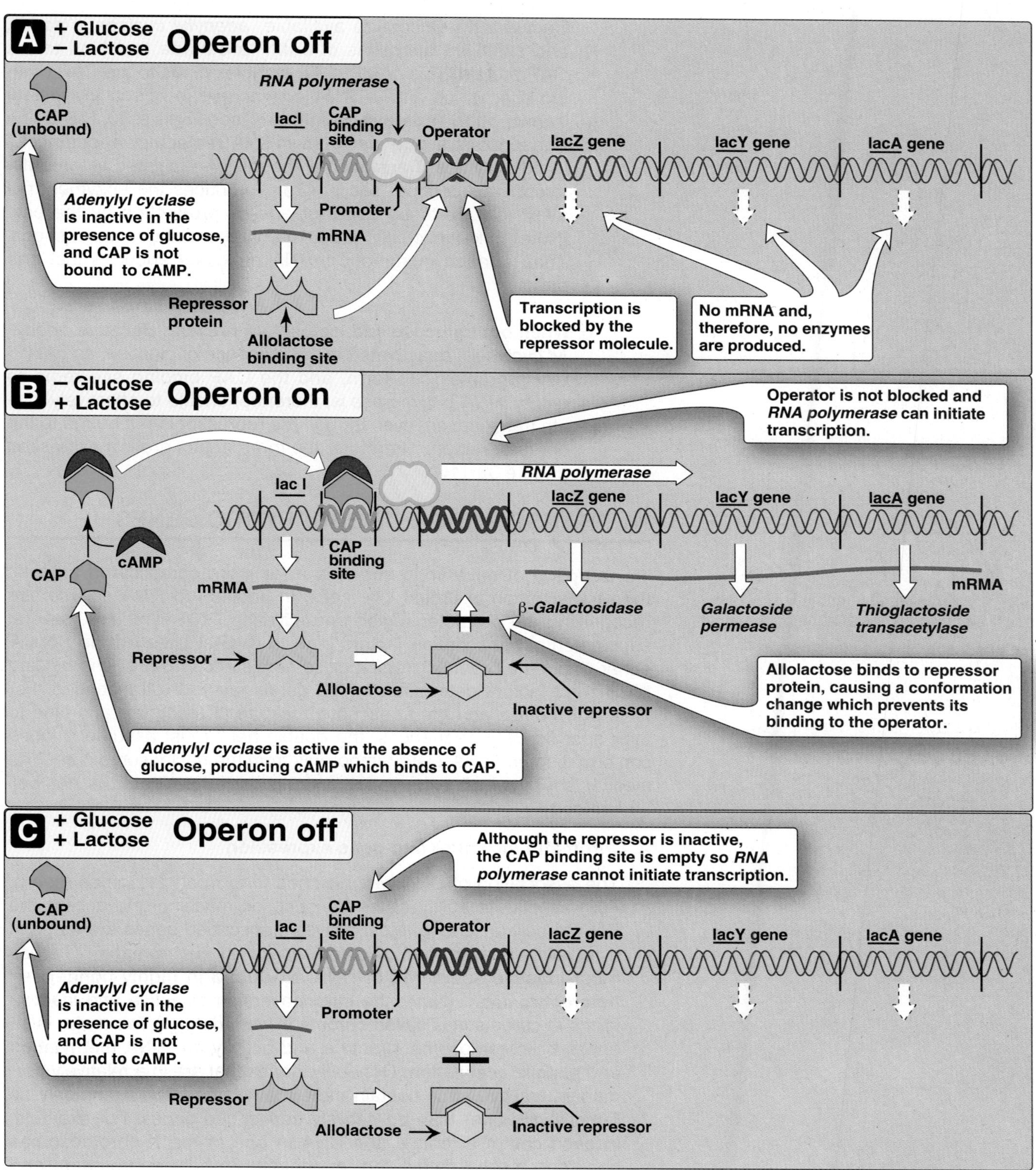

Figure 30.11
The lactose operon of E. coli.

formation so that it can no longer bind to the operator. Because no glucose is available, *adenylyl cyclase* is active, and sufficient quantities of cAMP are made that bind to the CAP protein. This cAMP–CAP complex binds to the CAP binding site, which allows *RNA polymerase* to effectively initiate transcription (**positive regulation**, see Figure 30.11B). The transcript is a **polycistronic** mRNA molecule, encoding all three enzymes. Translation of the mRNA—initiated at three different start codons (see p. 436)—produces the enzymes that allow lactose to be used for energy production by the cell. [Note: Eukaryotic cells produce only monocistronic messages. That is, each eukaryotic mRNA molecule encodes just one protein.]

c. **When both glucose and lactose are present:** Because *adenylyl cyclase* is deactivated in the presence of glucose, no cAMP-CAP complex can form, and the CAP binding site remains empty. *RNA polymerase* is, therefore, unable to effectively initiate transcription, even though the repressor is not bound to the operator region. Therefore, the three genes are not expressed (Figure 30.11C).

IV. TRANSCRIPTION OF EUKARYOTIC GENES

Transcription of eukaryotic genes is a far more complicated process than transcription in prokaryotic cells. In addition to *RNA polymerase* recognizing the promoter region and initiating RNA synthesis, several supplemental transcription factors bind to distinct sites on the DNA—either within the promoter region or some distance from it. The binding of different factors determines which genes are to be transcribed. For *RNA polymerase* and the transcription factors to recognize and bind to their specific DNA sequences, the double helix must assume a loose conformation and dissociate temporarily from the nucleosome core. The mechanism by which eukaryotic transcription is terminated is not well understood.

A. Chromatin structure and gene expression

The association of DNA with histones to form nucleosomes (see p. 406) affects the ability of the transcription machinery to access the DNA to be transcribed. Most actively transcribed genes are found in a relatively relaxed form of chromatin called **euchromatin**, whereas most inactive segments of DNA are found in highly condensed **heterochromatin**. [Note: the interconversion of active and inactive forms of chromatin is called **chromatin remodeling**.] Two major influences on chromosome structure and activity are **DNA methylation** and **histone acetylation**. Generally, genes that are in a relatively permanent inactive state contain more methylated DNA (commonly as 5-methylcytosine) than do actively transcribed genes. For example, in each cell of a female, the DNA in one of the X chromosomes becomes highly methylated, condenses to heterochromatin, and turns off. In contrast, when histones become acetylated, the structure of the chromatin becomes looser, DNA becomes more accessible to the transcription machinery, and genes are actively transcribed.

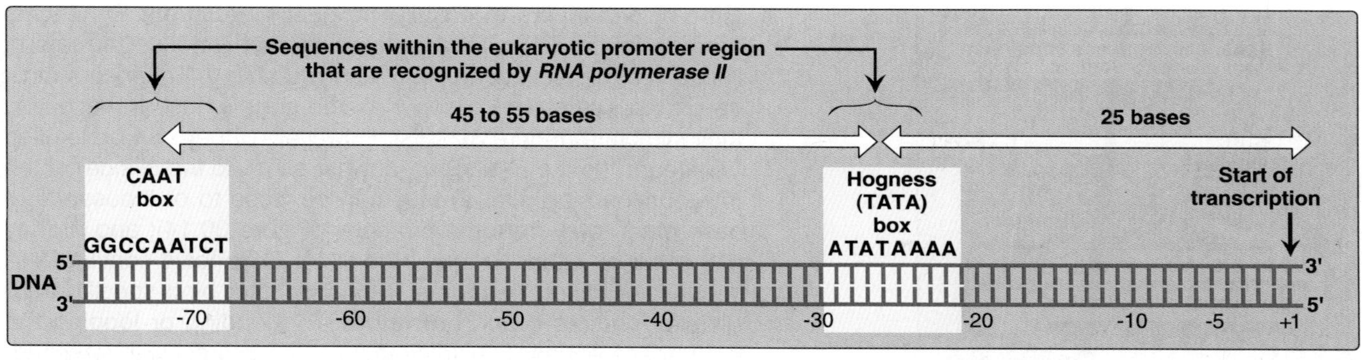

Figure 30.12
Eukaryotic gene promoter concensus sequences.

B. Nuclear RNA polymerases of eukaryotic cells

There are three distinct classes of *RNA polymerase* in the nucleus of eukaryotic cells. All are large enzymes with multiple subunits. Each class of *RNA polymerase* recognizes particular types of genes.

1. **RNA polymerase I:** This enzyme synthesizes the precursor of the **large ribosomal RNAs** (28S, 18S, and 5.8S) in the **nucleolus**. [Note: mRNA and tRNA are synthesized in the **nucleoplasm**.]

2. **RNA polymerase II:** This enzyme synthesizes the precursors of **messenger RNAs** that are subsequently translated to produce proteins. *Polymerase II* also synthesizes certain **small nuclear RNAs** (snRNA, see p. 424), and is used by some viruses to produce viral RNA.

   a. **Promoters for class II genes:** A sequence of DNA nucleotides that is almost identical to that of the Pribnow box is usually found centered about 25 nucleotides upstream of the initial base of the transcription start site for an mRNA molecule. This consensus sequence is called the **TATA** or **Hogness box** (Figure 30.12). Between seventy and eighty nucleotides upstream from the transcription start site is often found a second consensus sequence, known as the **CAAT box** (see Figure 30.12). In addition, many promoters contain a GC box (GGGCGG). Because these DNA sequences are on the same molecule of DNA as the genes being transcribed, they are called **cis-acting genetic elements**. They serve as binding sites for proteins called **general transcription factors**, which, in turn, interact with each other and with *RNA polymerase* II (Figure 30.13A).These are required for simple transcription of most class II genes (that is, those genes that are transcribed by *RNA polymerase* II). [Note: Because transcription factors, encoded by genes on different chromosomes and synthesized in the cytosol, can diffuse through the cell to their points of action (which may be on different chromosomes), they are called **trans-acting elements**. They can either stimulate or inhibit transcription of particular genes.]

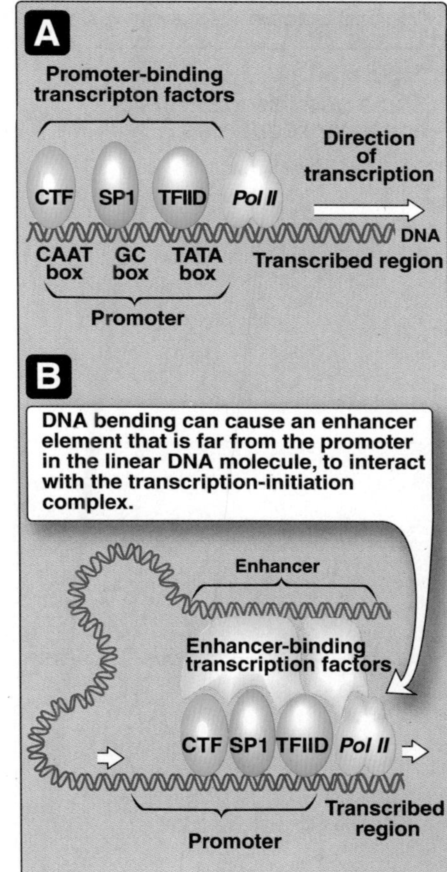

Figure 30.13
A. Eukaryotic general transcription factors bound to the promoter. CTF, SP1, and TFIID are general transcription factors. B. Enhancer stimulation of *RNA polymerase II*.

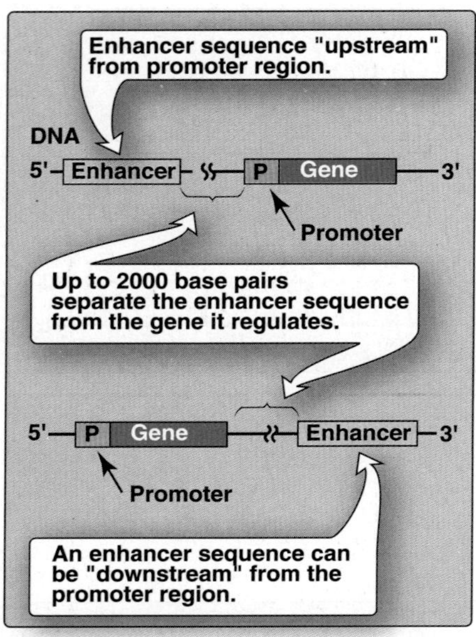

Figure 30.14
Some possible locations of enhancer sequences.

b. **Role of enhancers in eukaryotic gene regulation:** Enhancers are special cis-acting DNA sequences that increase the rate of initiation of transcription by *RNA polymerase II*. Enhancers must be on the same chromosome as the gene whose transcription they stimulate (Figure 30.13B). However, 1) they can be located "upstream" (to the 5'-side) or "downstream" (to the 3'-side) of the transcription start site; 2) they can be close to or thousands of base pairs away from the promoter (Figure 30.14); and 3) they can occur on either strand of the DNA. Enhancers contain DNA sequences called "**response elements**" that bind specific transcription factors called **activators**. By bending or looping the DNA, these enhancer-binding factors can interact with transcription factors bound to a promoter and with *RNA polymerase II*, thereby stimulating transcription (see Figure 30.13B). [Note: **Silencers** are similar to enhancers, in that they act over long distances to reduce the level of gene expression.]

c. **Inhibitors of RNA polymerase II:** This enzyme is inhibited by **α-amanitin**—a potent toxin produced by the poisonous mushroom Amanita phalloides (sometimes called "death cap" or "destroying angel"—it is said to taste delicious!). α-Amanitin forms a tight complex with the *polymerase*, thereby inhibiting mRNA synthesis and, ultimately, protein synthesis.

3. **RNA polymerase III:** This enzyme produces the small RNAs, including **tRNAs**, the **small 5S ribosomal RNA**, and some **snRNAs**.

B. Mitochondrial RNA polymerase

Mitochondria contain a single *RNA polymerase* that resembles bacterial *RNA polymerase* more closely than it does the eukaryotic enzyme.

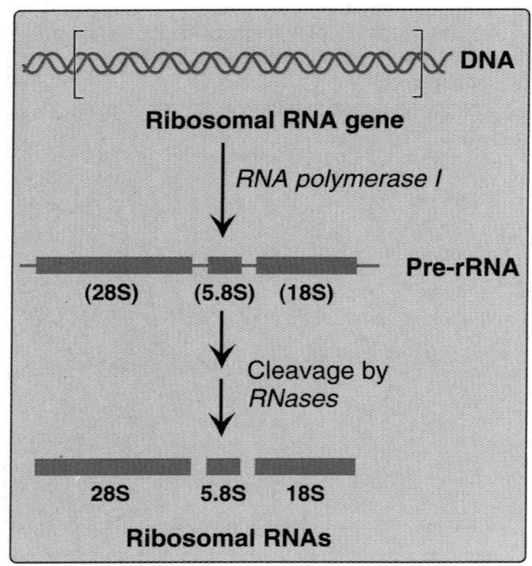

Figure 30.15
Posttranscriptional processing of eukaryotic ribosomal RNA by ribonucleases.

V. POSTTRANSCRIPTIONAL MODIFICATION OF RNA

A **primary transcript** is a linear copy of a transcriptional unit—the segment of DNA between specific initiation and termination sequences. The primary transcripts of both prokaryotic and eukaryotic tRNAs and rRNAs are post-transcriptionally modified by cleavage of the original transcripts by ribonucleases. tRNAs are then further modified to help give each species its unique identity. In contrast, prokaryotic mRNA is generally identical to its primary transcript, whereas eukaryotic mRNA is extensively modified posttranscriptionally.

A. Ribosomal RNA

Ribosomal RNAs of both prokaryotic and eukaryotic cells are synthesized from long precursor molecules called **preribosomal RNAs**. The 23S, 16S, and 5S ribosomal RNAs of prokaryotes are produced from a single RNA precursor molecule, as are the 28S, 18S, and 5.8S rRNAs of eukaryotes (Figure 30.15). [Note: Eukaryotic 5S rRNA is synthesized by *RNA polymerase III* and modified separately.] The preribosomal RNAs are cleaved by ribonucleases to yield intermediate-sized pieces of rRNA, which are further "trimmed"

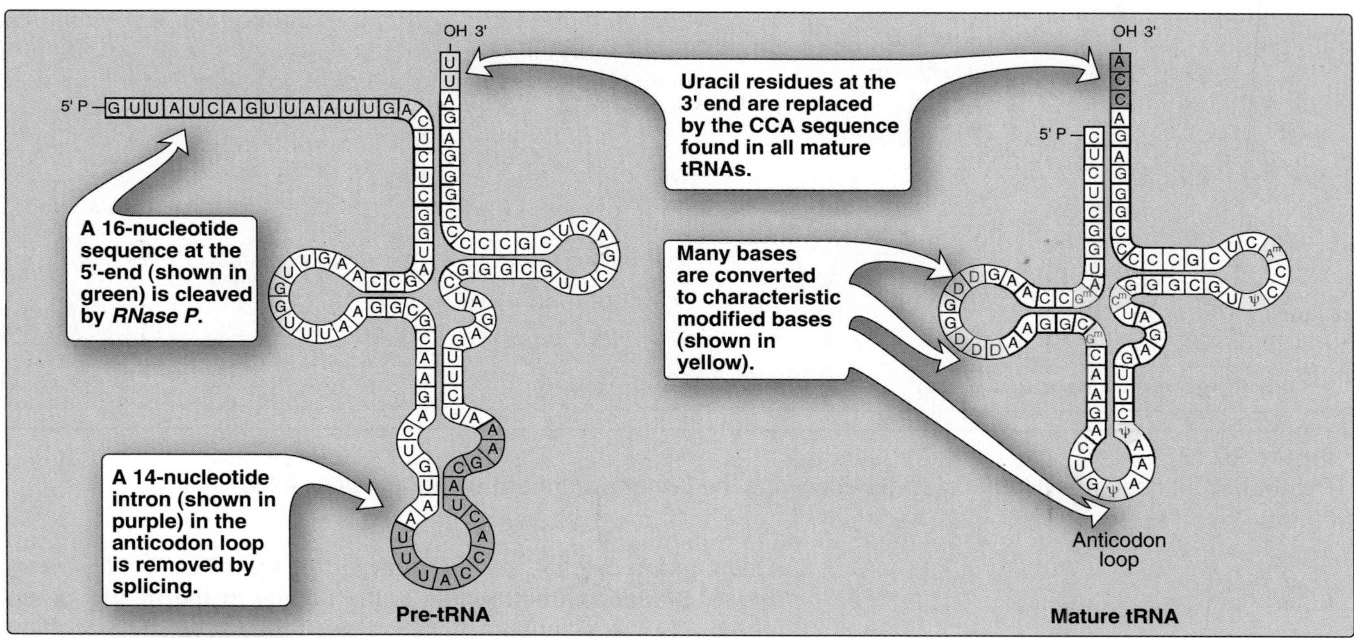

Figure 30.16
A. Primary tRNA transcript. B. Functional tRNA after posttranscriptional modification. Modified bases include D (dihydrouridine), ψ (pseudouridine), and ^{m}, which means that the base has been methylated.

to produce the required ribosomal RNA species. [Note: Some of the proteins destined to become components of the ribosome associate with the rRNA precursor prior to and during its posttranscriptional modification in the nucleolus.]

B. Transfer RNA

Both eukaryotic and prokaryotic transfer RNAs are also made from longer precursor molecules that must be modified (Figure 30.16). An intron (see below) must be removed from the anticodon loop, and sequences at both the 5'- and the 3'-ends of the molecule must be trimmed. Other posttranscriptional modifications include addition of a –CCA sequence by *nucleotidyltransferase* to the 3'-terminal end of tRNAs, and modification of bases at specific positions to produce **"unusual bases"** (see p. 290).

C. Eukaryotic messenger RNA

The RNA molecule synthesized by *RNA polymerase II* (the **primary transcript**) contains the sequences that are found in cytosolic mRNA. The collection of all the precursor molecules for mRNA is known **heterogeneous nuclear RNA** (hnRNA). The primary transcripts are extensively modified in the nucleus after transcription. These modifications usually include:

1. **5' "Capping":** This process is the first of the processing reactions for hnRNA (Figure 30.17). The cap is a **7-methyl-guanosine** attached "backward" to the 5'-terminal end of the mRNA, forming an unusual 5'→5' triphosphate linkage. The addition of the guanosine triphosphate part of the cap is catalyzed by the nuclear enzyme *guanylyltransferase*. Methylation of this terminal guanine

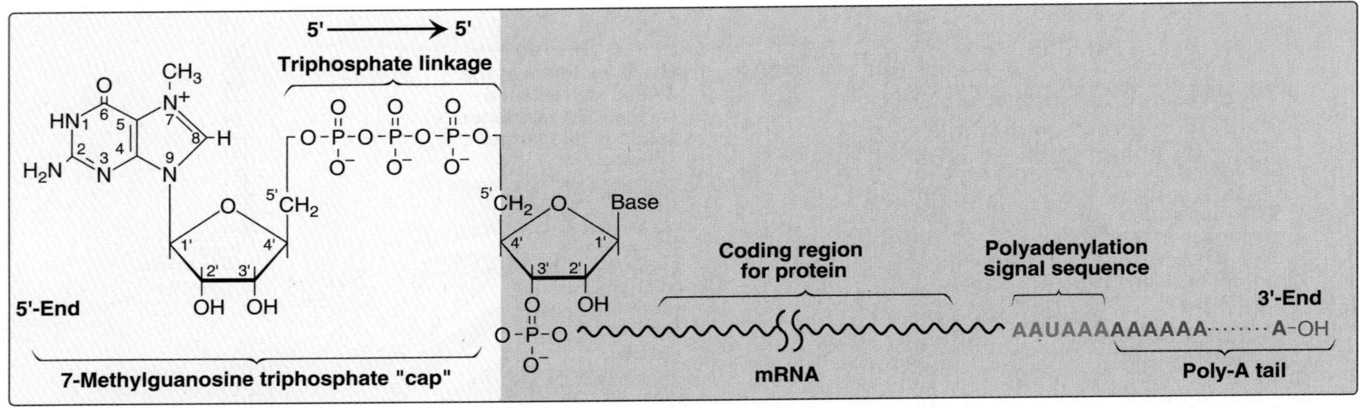

Figure 30.17
Posttranscriptional modification of mRNA showing the 7-methylguanosine cap and poly-A tail.

occurs in the cytosol, and is catalyzed by *guanine-7-methyltransferase*. **S-adenosylmethionine** is the source of the methyl group (see p. 261). Additional methylation steps may occur. The addition of this 7-methylguanosine "cap" permits the initiation of translation (see p. 435), and helps stabilize the mRNA. Eukaryotic mRNAs lacking the cap are not efficiently translated.

2. **Addition of a poly-A tail:** Most eukaryotic mRNAs (with several notable exceptions, including those coding for the histones and some interferons) have a chain of 40 to 200 adenine nucleotides attached to the 3'-end (see Figure 30.17). This poly-A tail is not transcribed from the DNA, but rather is added after transcription by the nuclear enzyme, *polyadenylate polymerase*. A consensus sequence, called the **polyadenylation signal sequence** (AAUAAA), found near the 3'-end of the RNA molecule, signals that a poly-A tail is to be added to the mRNA. These tails help stabilize the mRNAs and facilitate their exit from the nucleus. After the mRNA enters the cytosol, the poly-A tail is gradually shortened.

3. **Removal of introns:** Maturation of eukaryotic mRNA usually involves the removal of RNA sequences, which do not code for protein (**introns**, or **intervening sequences**) from the primary transcript. The remaining coding sequences, the **exons**, are spliced together to form the mature mRNA. The molecular machine that accomplishes these tasks is known as the **spliceosome**. [Note: A few eukaryotic primary transcripts contain no introns. Others contain a few introns, whereas some, such as the primary transcripts for the α-chains of collagen, contain more than fifty intervening sequences that must be removed before mature mRNA is ready for translation.]

   a. **Role of small nuclear RNAs (snRNAs):** snRNAs, in association with proteins, form **small nuclear ribonucleoprotein particles** (**snRNPs**, or "snurps"). These facilitate the splicing of exon segments by forming base pairs with the consensus sequences at each end of the intron (Figure 30.18). [Note: **Systemic lupus**

erythematosus, an often fatal inflammatory disease, results from an autoimmune response in which the patient produces antibodies against host proteins, including snRNPs.]

b. **Mechanism of excision of introns:** The binding of snRNPs brings the sequences of the neighboring exons into the correct alignment for splicing. The 2'-OH group of an adenosine (A) residue (known as the branch site) in the intron attacks and forms a phosphodiester bond with the phosphate at the 5'-end of intron 1 (see Figure 30.18). The newly-freed 3'-OH of the upstream exon 1 then forms a phosphodiester bond with the 5'-end of the downstream exon 2. The excised intron is released as a "lariat" structure, which is degraded. [Note: The GU and AG sequences at the branch site are invariant.] After removal of all the introns, the mature mRNA molecules leave the nucleus by passing into the cytosol through pores in the nuclear membrane.

c. **Effect of splice site mutations:** Mutations at splice sites can lead to improper splicing and the production of aberrant proteins. It is estimated that fifteen percent of all genetic diseases are a result of mutations that affect RNA splicing. For example, mutations that cause the incorrect splicing of β-globin mRNA are responsible for some cases of **β thalassemia**—a disease in which the production of the β-globin protein is defective (see p. 38).

4. **Alternative splicing of mRNA molecules:** The pre-mRNA molecules from some genes can be spliced in two or more alternative ways in different tissues. This produces multiple variations of the mRN, and, therefore, of its protein product (Figure 30.19). This appears to be a mechanism for producing a diverse set of proteins from a limited set of genes. For example, different types of muscle cells all produce the same primary transcript from the tropomyosin gene. However, different patterns of splicing in the different cell types produce a family of tissue-specific tropomyosin protein molecules.

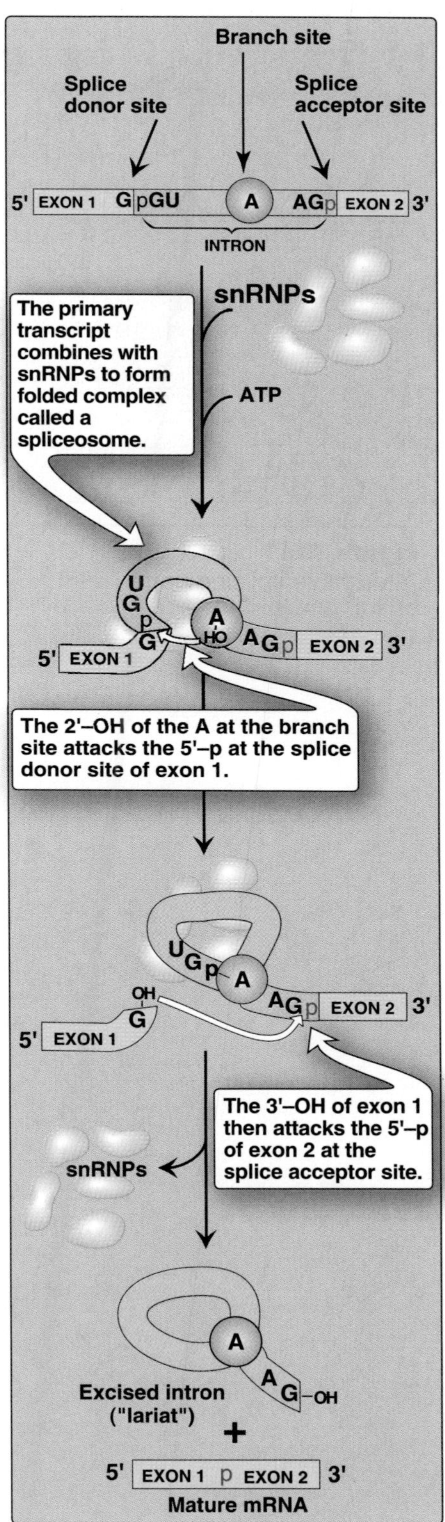

Figure 30.18
Removal of introns. snRNP = small nuclear ribonucleoprotein particle.

VI. CHAPTER SUMMARY

There are three major types of RNA that participate in the process of protein synthesis: **ribosomal RNA (rRNA)**, **transfer RNA (tRNA)**, and **messenger RNA (mRNA)**. They are unbranched polymers of nucleotides, but differ from DNA by containing **ribose** instead of deoxyribose and **uracil** instead of thymine. **rRNA** is a component of the **ribosomes**. **tRNA** serves as an "adaptor" molecule that carries a specific amino acid to the site of protein synthesis. **mRNA** carries genetic information from the nuclear DNA to the cytosol, where it is used as the template for protein synthesis. The process of RNA synthesis is called **transcription**, and its substrates are **ribonucleoside triphosphates**. The enzyme that synthesizes RNA is **RNA polymerase**, which is a multisubunit enzyme. In **prokaryotic cells**, the **core enzyme** has four subunits—

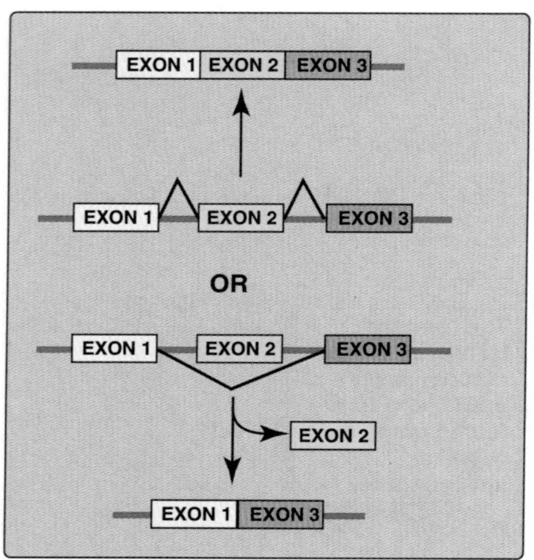

Figure 30.19
Alternative splicing patterns in eukaryotic mRNA.

2α, 1 β, and **1 β'**, and possesses **5'→3' polymerase activity** that **elongates** the growing RNA strand. This enzyme requires an additional subunit—**sigma (σ) factor**—that recognizes the nucleotide sequence (**promoter region**) at the beginning of a length of DNA that is to be transcribed. This region contains characteristic **consensus nucleotide sequences** that are highly conserved and include the **Pribnow box** and the **–35 sequence**. Another protein—**rho** (ρ) **factor**—is required for **termination** of transcription of some genes. A bacterial **operon** is a group of structural genes that code for the enzymes of a metabolic pathway, and are grouped together on the chromosome along with the regulatory genes that determine their transcription. The genes are thus **coordinately expressed**. The **lactose** (**lac**) **operon** of E. coli is one of the best understood. It codes for the enzymes needed to metabolize lactose when it is the only available sugar substrate. There are three distinct classes of RNA polymerase in the nucleus of **eukaryotic cells**. **RNA polymerase I** synthesizes the precursor of **large rRNAs** in the **nucleolus**. **RNA polymerase II** synthesizes the precursors for **mRNAs**, and **RNA polymerase III** produces the precursors of **tRNAs** and some other small RNAs in the **nucleoplasm**. **Promoters** for **class II genes** contain consensus sequences, such as the **TATA** or **Hogness box**, the **CAAT box**, and the **GC box**. They serve as binding sites for proteins called **general transcription factors**, which, in turn, interact with each other and with *RNA polymerase II*. **Enhancers** are DNA sequences that increase the rate of initiation of transcription by binding to specific transcription factors called **activators.** A **primary transcript** is a linear copy of a **transcriptional unit**—the segment of DNA between specific initiation and termination sequences. The primary transcripts of both prokaryotic and eukaryotic tRNAs and rRNAs are **posttranscriptionally modified** by cleavage of the original transcripts by ribonucleases. **rRNAs** of both prokaryotic and eukaryotic cells are synthesized from long precursor molecules called **preribosomal RNAs**. These precursors are cleaved and trimmed by ribonucleases, producing the three largest rRNAs. (Eukaryotic 5S rRNA is synthesized by RNA polymerase III instead of I, and is modified separately.) Prokaryotic mRNA is generally identical to its primary transcript, whereas **eukaryotic mRNA** is extensively modified posttranscriptionally. For example, a **7-methyl-guanosine "cap"** is attached to the 5'-terminal end of the mRNA through a triphosphate linkage. A **long poly-A tail**—not transcribed from the DNA—is attached to the 3'-end of most mRNAs. Many eukaryotic mRNAs also contain **intervening sequences** (**introns**) that must be removed to make the mRNA functional. Their removal requires **small nuclear RNAs**. Prokaryotic and eukaryotic **tRNAs** are also made from longer precursor molecules. These must have an **intron** removed, and the 5'- and 3'-ends of the molecule are trimmed by ribonuclease. A **3' –CCA sequence** is added (if not already present), and bases at specific positions are modified, producing **"unusual" bases**.

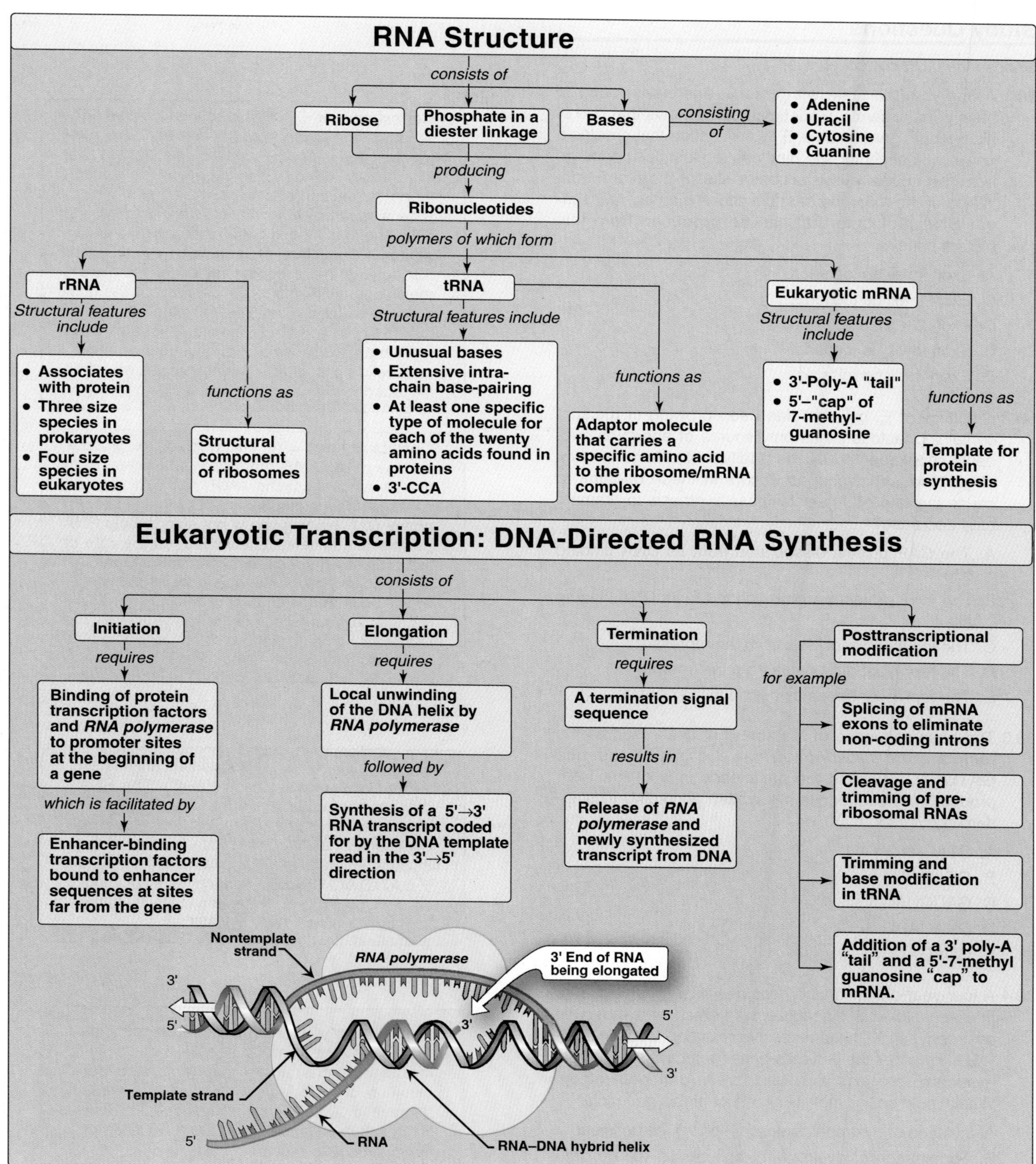

Figure 30.20
Key concept map for RNA structure and synthesis.

Study Questions

Choose the ONE correct answer

30.1 A one-year-old male with chronic anemia is found to have β-thalassemia. Genetic analysis shows that one of his β-globin genes has a G to A mutation that creates a new splice acceptor site nineteen nucleotides upstream from the normal splice acceptor site of the first intron. Which of the following best describes the new messenger RNA molecule that can be produced from this mutant gene?

A. Exon 1 will be too short.
B. Exon 1 will be too long.
C. Exon 2 will be too short.
D. Exon 2 will be too long.
E. Exon 2 will be missing.

Correct answer = D. Because the mutation adds an additional splice acceptor site (the 3'-end) of intron 1 upstream, the nineteen nucleotides that are usually found at the 3'-end of the excised intron 1 lariat can remain behind as part of exon 2 as a result of aberrant splicing. Exon 2 can therefore, have these extra nineteen nucleotides at its 5'-end. The presence of these extra nucleotides in the coding region of the mutant messenger RNA molecule will prevent the ribosome from translating the message into a normal β-globin protein molecule. Those mRNAs for which the normal splice site is used to remove the first intron, will be normal, and their translation will produce normal β-globin protein.

30.2 A culture of E. coli that has been growing in medium containing lactose as its only source of energy is suddenly supplemented by the addition of a large amount of glucose. What change occurs in these bacteria to cause the rate of β-galactosidase synthesis to dramatically decrease?

A. The CAP protein dissociates from its DNA binding site.
B. The CAP protein becomes bound to its DNA binding site.
C. The inducer dissociates from the repressor.
D. The repressor dissocates from the operator.
E. The repressor becomes bound to the operator.

Correct answer = A. The addition of glucose causes cyclic AMP production to decrease. In the absence of cyclic AMP, the CAP protein cannot remain bound to its DNA binding site. An empty CAP binding site is not able to help RNA polymerase initiate transcription, so the rate of transcription decreases. Lower mRNA production results in decreased β-galactosidase synthesis. Because lactose is still present, the inducer (allolactose) remains bound to the repressor, which continues to be unable to bind to the operator.

30.3 The base sequence of the strand of DNA used as the template for transcription has the base sequence GATCTAC. What is the base sequence of the RNA product? (All sequences are written according to standard convention.)

A. CTAGATG
B. GTAGATC
C. GAUCUAC
D. CUAGAUG
E. GUAGAUC

Correct answer = E. All sequences are written in the standard convention (5'→3'). The RNA product has a sequence that is complementary to the sequence of the template strand of DNA. Uracil (U) is found in RNA in place of the thymine (T) in DNA. Thus, the DNA template 5'-GATCTAC-3' would produce the RNA product 3'-CUAGAUG-5' or, written correctly in the standard direction, 5'-GUAGAUC-3'.

30.4 A four-year-old child who becomes easily tired and has trouble walking is diagnosed with Duchenne muscular dystrophy, an X-linked recessive disorder. Genetic analysis shows that the patient's gene for the muscle protein dystrophin contains a mutation in its promoter region. What would be the most likely effect of this mutation?

A. Initiation of dystrophin transcription will be deficient.
B. Termination of dystrophin transcription will be deficient.
C. Capping of dystrophin mRNA will be defective.
D. Splicing of dystrophin mRNA will be defective.
E. Tailing of dystrophin mRNA will be defective.

Correct answer = A. Mutations in the promoter prevent formation of the RNA polymerase II transcription complex, and the initiation of mRNA synthesis will be greatly decreased. A deficiency of dystrophin mRNA will result in a deficiency in the production of the dystrophin protein.

Protein Synthesis

31

I. OVERVIEW

Genetic information, stored in the chromosomes and transmitted to daughter cells through DNA replication, is expressed through transcription to RNA and, in the case of mRNA, subsequent **translation** into polypeptide chains (Figure 31.1). The pathway of protein synthesis is called translation because the "language" of the nucleotide sequence on the mRNA is translated into the language of an amino acid sequence. The process of translation requires a genetic code, through which the information contained in the nucleic acid sequence is expressed to produce a specific sequence of amino acids. Any alteration in the nucleic acid sequence may result in an improper amino acid being inserted into the polypeptide chain, potentially causing disease or even death of the organism. Many polypeptide chains are covalently modified following their synthesis to activate them, alter their activities, or target them to their final intracellular or extracellular destinations.

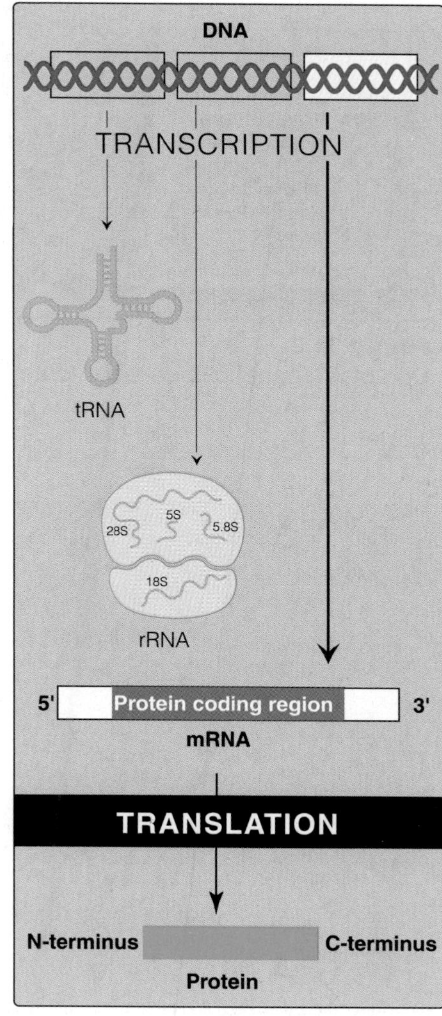

Figure 31.1
Protein synthesis or "translation."

II. THE GENETIC CODE

The genetic code is a dictionary that identifies the correspondence between a sequence of nucleotide bases and a sequence of amino acids. Each individual word in the code is composed of three nucleotide bases. These genetic words are called codons.

A. Codons

Codons are usually presented in the messenger RNA language of adenine (A), guanine (G), cytosine (C), and uracil (U). Their nucleotide sequences are always written from the 5'-end to the 3'-end. The four nucleotide bases are used to produce the three-base codons. There are, therefore, 64 different combinations of bases, taken three at a time as shown in Figure 31.2.

1. **How to translate a codon:** This table (or "dictionary") can be used to translate any codon sequence and, thus, to determine which amino acids are coded for by an mRNA sequence. For example, the codon 5'–AUG–3' codes for methionine (see Figure 31.2). Sixty-one of the 64 codons code for the twenty common amino acids.

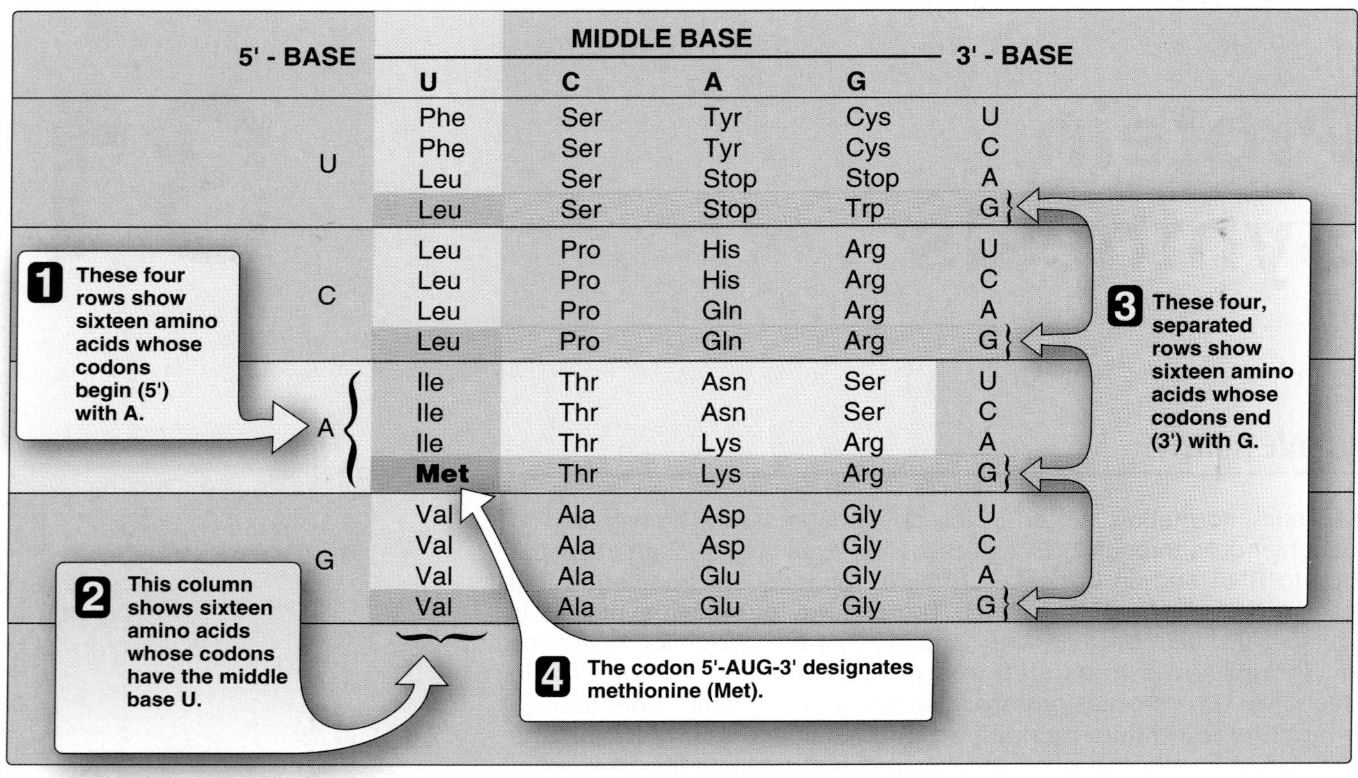

Figure 31.2
Use of the genetic code table to translate the codon AUG.

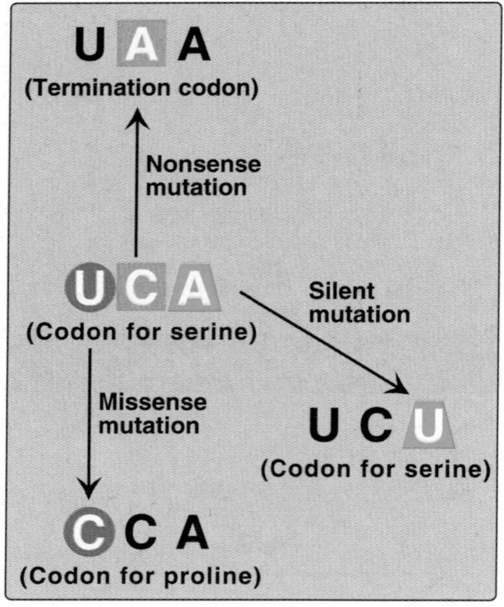

Figure 31.3
Possible effects of changing a single nucleotide base in the coding region of an mRNA chain.

2. **Termination ("stop" or "nonsense") codons:** Three of the codons, UAG, UGA, and UAA, do not code for amino acids, but rather are termination codons. When one of these codons appears in an mRNA sequence, it signals that synthesis of the peptide chain coded for by that mRNA is completed.

B. Characteristics of the genetic code

Usage of the genetic code is remarkably consistent throughout all living organisms. It is assumed that once the standard genetic code evolved in primitive organisms, any mutation that altered the manner in which the code was translated would have caused the alteration of most, if not all, protein sequences, which would certainly have been lethal. Adoption of the genetic code has been described as "an accident frozen in time." Characteristics of the genetic code include the following:

1. **Specificity:** The genetic code is specific (unambiguous), that is, a specific codon always codes for the same amino acid.

2. **Universality:** The genetic code is virtually universal, that is, the specificity of the genetic code has been conserved from very early stages of evolution, with only slight differences in the manner in which the code is translated. [Note: An exception occurs in mitochondria, in which a few codons have different meanings than those shown in Figure 31.2.]

3. **Redundancy:** The genetic code is redundant (sometimes called **degenerate**). Although each codon corresponds to a single amino acid, a given amino acid may have more than one triplet coding for it. For example, arginine is specified by six different codons (see Figure 31.2).

4. **Nonoverlapping and commaless:** The genetic code is nonoverlapping and commaless, that is, the code is read from a fixed starting point as a continuous sequence of bases, taken three at a time. For example, ABCDEFGHIJKL is read as ABC/DEF/GHI/JKL without any "punctuation" between the codons.

C. Consequences of altering the nucleotide sequence:

Changing a single nucleotide base on the mRNA chain (a "**point mutation**") can lead to any one of three results (Figure 31.3):

1. **Silent mutation:** The codon containing the changed base may code for the same amino acid. For example, if the serine codon UCA is given a different third base—U—to become UCU, it still codes for serine. Therefore, this is termed a "silent" mutation.

2. **Missense mutation:** The codon containing the changed base may code for a different amino acid. For example, if the serine codon UCA is given a different first base—C—to become CCA, it will code for a different amino acid, in this case, proline. This substitution of an incorrect amino acid is called a "missense" mutation.

3. **Nonsense mutation:** The codon containing the changed base may become a **termination codon**. For example, if the serine codon UCA is given a different second base—A—to become UAA, the new codon causes termination of translation at that point. The creation of a termination codon at an inappropriate place is called a "nonsense" mutation.

4. **Other mutations:** These can alter the amount or structure of the protein produced by translation.

   a. **Trinucleotide repeat expansion:** Occasionally, a sequence of three bases that is repeated in tandem will become amplified in number, so that too many copies of the triplet occur. If this occurs within the coding region of a gene, the protein will contain many extra copies of one amino acid. For example, amplification of the CAG codon leads to the insertion of many extra glutamine residues in the **huntingtin protein**, causing the neurogenerative disorder, **Huntington disease** (Figure 31.4). The additional glutamines result in unstable proteins that cause the accumulation of protein aggregates. If the trinucleotide repeat expansion occurs in the untranslated portion of a gene, the result can be a decrease in the amount of protein produced as seen, for example, in **fragile X syndrome** and **myotonic dystrophy**.

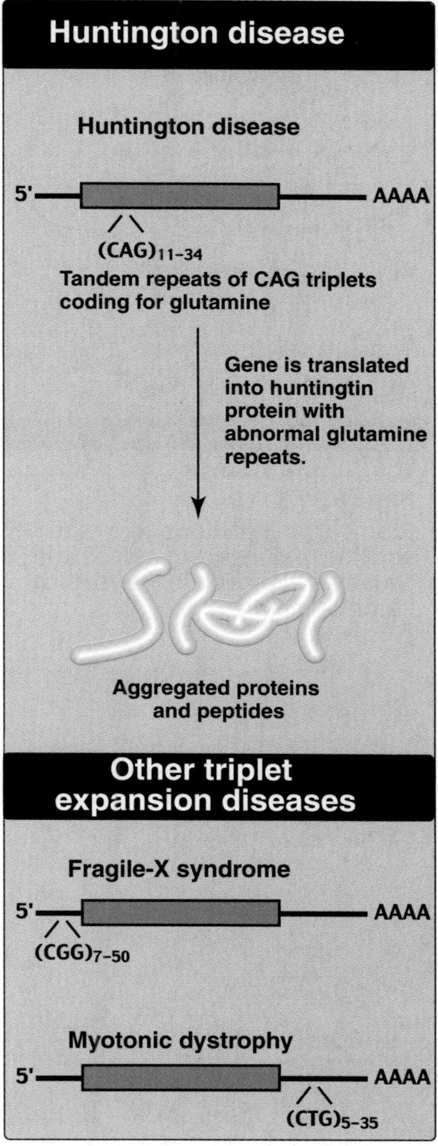

Figure 31.4
Role of tandem triplet repeats in mRNA causing Huntington disease and other gene expansion diseases.

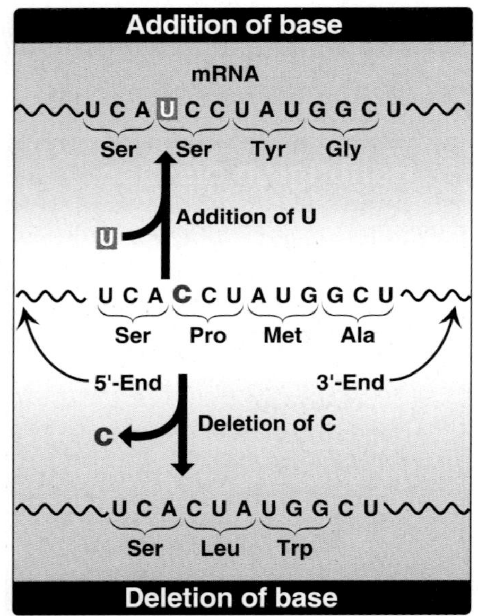

Figure 31.5
Frame-shift mutations as a result of addition or deletion of a base can cause an alteration in the reading frame of mRNA.

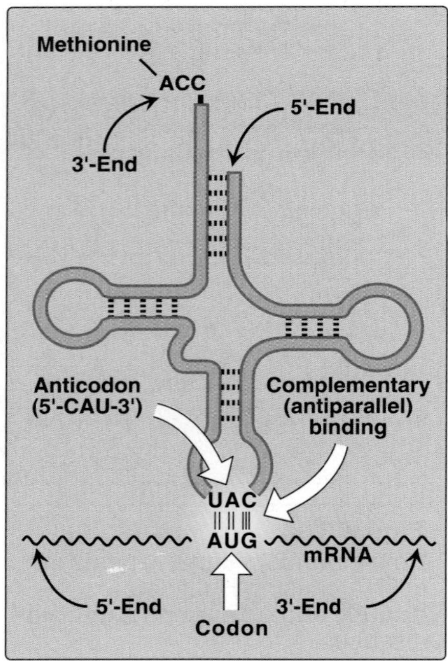

Figure 31.6
Complementary, antiparallel binding of the anticodon for methionyl-tRNA (CAU) to the mRNA codon for methionine (AUG).

b. **Splice site mutations:** Mutations at splice sites (see p. 425) can alter the way in which introns are removed from pre-mRNA molecules, producing aberrant proteins.

c. **Frame-shift mutations:** If one or two nucleotides are either deleted from or added to the interior of a message sequence, a frame-shift mutation occurs and the reading frame is altered. The resulting amino acid sequence may become radically different from this point on (Figure 31.5). [Note: If three nucleotides are added, a new amino acid is added to the peptide or, if three nucleotides are deleted, an amino acid is lost. In these instances, the reading frame is not affected.]

III. COMPONENTS REQUIRED FOR TRANSLATION

A large number of components are required for the synthesis of a polypeptide chain. These include all the amino acids that are found in the finished product, the mRNA to be translated, tRNAs, functional ribosomes, energy sources, and enzymes, as well as protein factors needed for initiation, elongation, and termination of the polypeptide chain.

A. Amino acids

All the amino acids that eventually appear in the finished protein must be present at the time of protein synthesis. [Note: If one amino acid is missing (for example, if the diet does not contain an essential amino acid), that amino acid is in limited supply in the cell, and translation, therefore, stops at the codon specifying that amino acid. This demonstrates the importance of having all the essential amino acids in sufficient quantities in the diet to ensure continued protein synthesis (see p. 259 for a discussion of the essential amino acids).]

B. Transfer RNA (tRNA)

At least one specific type of tRNA is required per amino acid. In humans, there are at least fifty species of tRNA, whereas bacteria contain thirty to forty species. Because there are only twenty different amino acids commonly carried by tRNAs, some amino acids have more than one specific tRNA molecule. This is particularly true of those amino acids that are coded for by several codons.

1. **Amino acid attachment site:** Each tRNA molecule has an attachment site for a specific amino acid at its 3'-end (Figure 31.6). The carboxyl group of the amino acid is in an ester linkage with the 3'-hydroxyl of the ribose moiety of the adenosine nucleotide at the 3'-end of the tRNA. [Note: When a tRNA has a covalently attached amino acid, it is said to be charged; when tRNA is not bound to an amino acid, it is described as being uncharged.] The amino acid that is attached to the tRNA molecule is said to be activated.

2. **Anticodon:** Each tRNA molecule also contains a three-base nucleotide sequence—the anticodon—that recognizes a specific codon on the mRNA (see Figure 31.6). This codon specifies the insertion into the growing peptide chain of the amino acid carried

by that tRNA. [Note: Because of their ability to both carry a specific amino acid and to recognize the codon for that amino acid, tRNAs are known as **adaptor molecules**.]

C. Aminoacyl-tRNA synthetases

This family of enzymes is required for attachment of amino acids to their corresponding tRNAs. Each member of this family recognizes a specific amino acid and the tRNAs that correspond to that amino acid. These enzymes, thus, implement the genetic code because they act as molecular dictionaries that can read both the three-letter code of nucleic acids and the twenty-letter code of amino acids. Each *aminoacyl-tRNA synthetase* catalyzes a two-step reaction that results in the covalent attachment of the carboxyl group of an amino acid to the 3'-end of its corresponding tRNA. The overall reaction requires ATP, which is cleaved to AMP and PP_i (Figure 31.7). The extreme specificity of the *synthetase* in recognizing both the amino acid and and its specific tRNA is largely responsible for the high fidelity of translation of the genetic message. In addition, the synthetases have a "proofreading" or "editing" activity that can remove mischarged amino acids from the tRNA molecule.

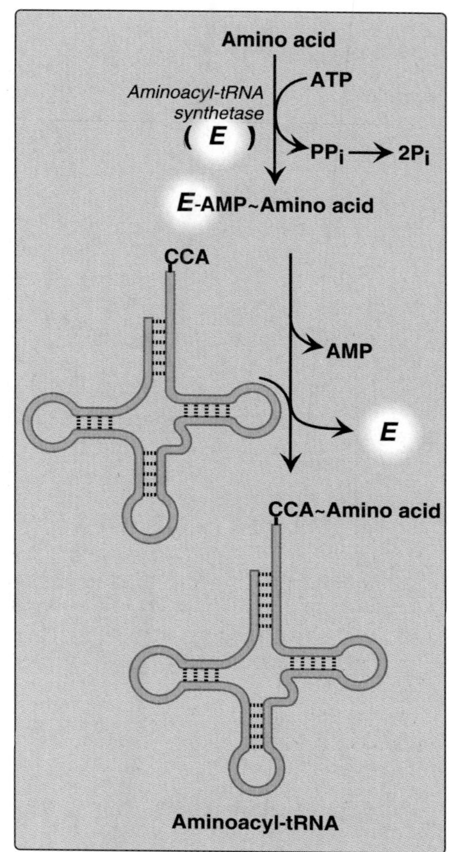

Figure 31.7
Attachment of a specific amino acid to its corresponding tRNA by *aminoacyl-tRNA synthetase* (*E*).

D. Messenger RNA (mRNA)

The specific mRNA required as a template for the synthesis of the desired polypeptide chain must be present.

E. Functionally competent ribosomes

Ribosomes are large complexes of protein and rRNA (Figure 31.8). They consist of two subunits—one large and one small—whose relative sizes are generally given in terms of their sedimentation coefficients, or S (Svedberg) values. [Note: Because the S values are determined both by shape as well as molecular mass, their numeric values are not strictly additive. For example, the prokaryotic 50S and 30S ribosomal subunits together form a ribosome with an S value of 70. The eukaryotic 60S and 40S subunits form an 80S ribosome.] Prokaryotic and eukaryotic ribosomes are similar in structure, and serve the same function, namely, as the "factories" in which the synthesis of proteins occurs.

1. **Ribosomal RNA (rRNA):** As discussed on p. 414, prokaryotic ribosomes contain three molecules of rRNA, whereas eukaryotic ribosomes contain four molecules of rRNA (see Figure 31.8). The rRNAs have extensive regions of secondary structure arising from the base-pairing of complementary sequences of nucleotides in different portions of the molecule. The formation of intramolecular, double-stranded regions is comparable to that found in tRNA.

2. **Ribosomal proteins:** Ribosomal proteins are present in considerably greater numbers in eukaryotic ribosomes than in prokaryotic ribosomes. These proteins play a number of roles in the structure and function of the ribosome and its interactions with other components of the translation system.

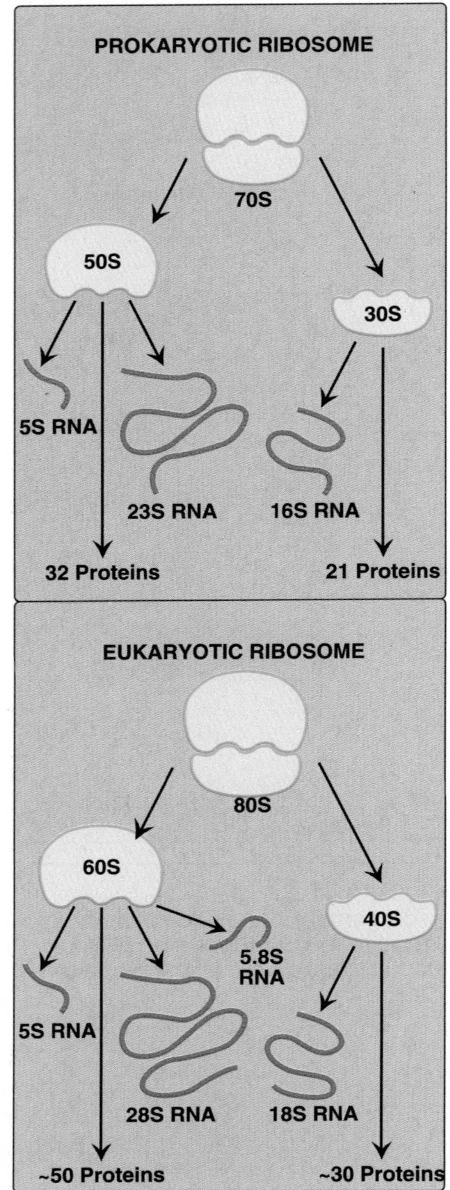

Figure 31.8
Ribosomal composition. (The number of proteins in the eukaryotic ribosomal subunits varies somewhat from species to species.)

3. **A, P, and E sites on the ribosome:** The ribosome has three binding sites for tRNA molecules—the A, P, and E sites—each of which extends over both subunits. Together, they cover three neighboring codons. During translation, the A site binds an incoming aminoacyl-tRNA as directed by the codon currently occupying this site. This codon specifies the next amino acid to be added to the growing peptide chain. The P site codon is occupied by peptidyl-tRNA. This tRNA carries the chain of amino acids that has already been synthesized. The E site is occupied by the empty tRNA as it is about to exit the ribosome. (See Figure 31.13, p. 438 for an illustration of the role of the A, P, and E sites in translation.)

4. **Cellular location of ribosomes:** In eukaryotic cells, the ribosomes are either "free" in the cytosol or are in close association with the endoplasmic reticulum (which is then known as the "rough" ER or RER). The RER–associated ribosomes are responsible for synthesizing proteins that are to be exported from the cell, as well as those that are destined to become integrated into plasma, ER, or Golgi membranes, or incorporated into lysosomes (see p. 167 for an overview of the latter process). [Note: Mitochondria contain their own set of ribosomes and their own unique, circular DNA.]

F. Protein factors

Initiation, elongation, and termination (or release) factors are required for peptide synthesis. Some of these protein factors perform a catalytic function, whereas others appear to stabilize the synthetic machinery.

G. ATP and GTP are required as sources of energy

Cleavage of four high-energy bonds is required for the addition of one amino acid to the growing polypeptide chain: two from ATP in the *aminoacyl-tRNA synthetase* reaction—one in the removal of pyrophosphate (PP_i), and one in the subsequent hydrolysis of the PP_i to inorganic phosphate by *pyrophosphatase*—and two from GTPs (one for binding the aminoacyl-tRNA to the A site and one for the translocation step (see Figure 31.13, p. 438). [Note: Additional ATP and GTP molecules are required for initiation and termination of polypeptide chain synthesis.]

IV. CODON RECOGNITION BY tRNA

Recognition of a particular codon in an mRNA sequence is accomplished by the anticodon sequence of the tRNA (see Figure 31.6). Some tRNAs recognize more than one codon for a given amino acid.

A. Antiparallel binding between codon and anticodon

Binding of the tRNA anticodon to the mRNA codon follows the rules of complementary and antiparallel binding, that is, the mRNA codon is "read" 5'→3' by an anticodon pairing in the "flipped" (3'→5') orientation (Figure 31.9). [Note: When writing the sequences of both codons and anticodons, the nucleotide sequence must ALWAYS be listed in the 5'→3' order.]

B. Wobble hypothesis

The mechanism by which tRNAs can recognize more than one codon for a specific amino acid is described by the "wobble" hypothesis in which the base at the 5'-end of the anticodon (the "first" base of the anticodon) is not as spatially defined as the other two bases. Movement of that first base allows nontraditional base-pairing with the 3'-base of the codon (the "last" base of the codon). This movement is called "wobble" and allows a single tRNA to recognize more than one codon. Examples of these flexible pairings are shown in Figure 31.9. The result of wobbling is that there need not be 61 tRNA species to read the 61 codons coding for amino acids.

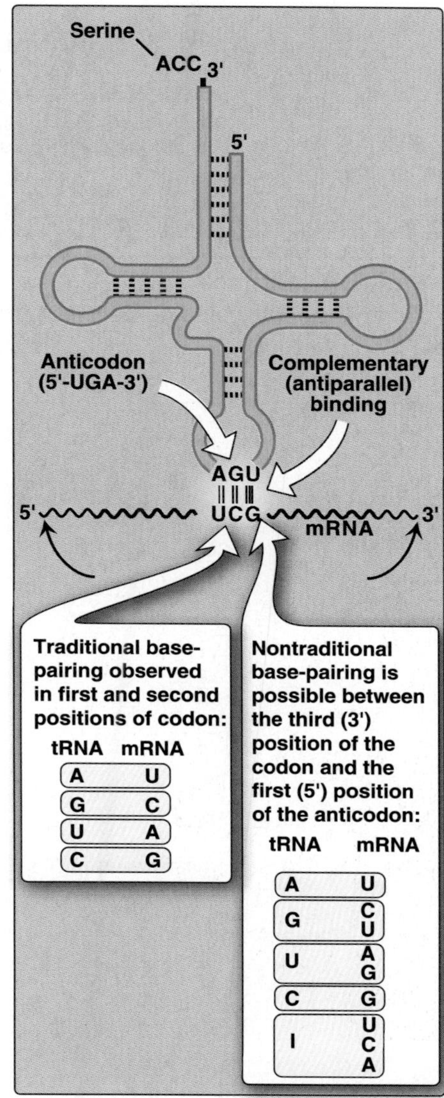

Figure 31.9
Wobble: Nontraditional base-pairing between the 5'-nucleotide (first nucleotide) of the anticodon with the 3'-nucleotide (last nucleotide) of the codon. I = inosine.

V. STEPS IN PROTEIN SYNTHESIS

The pathway of protein synthesis translates the three-letter alphabet of nucleotide sequences on mRNA into the twenty-letter alphabet of amino acids that constitute proteins. The mRNA is translated from its 5'-end to its 3'-end, producing a protein synthesized from its amino-terminal end to its carboxyl-terminal end. Prokaryotic mRNAs often have several coding regions, that is, they are **polycistronic** (see p. 420). Each coding region has its own initiation codon and produces a separate species of polypeptide. In contrast, each eukaryotic mRNA codes for only one polypeptide chain, that is, it is **monocistronic**. The process of translation is divided into three separate steps: initiation, elongation, and termination. The polypeptide chains produced may be modified by posttranslational modification. Eukaryotic protein synthesis resembles that of prokaryotes in most details. [Note: Individual differences are mentioned in the text.]

A. Initiation

Initiation of protein synthesis involves the assembly of the components of the translation system before peptide bond formation occurs. These components include the two ribosomal subunits, the mRNA to be translated, the aminoacyl-tRNA specified by the first codon in the message, GTP (which provides energy for the process), and initiation factors that facilitate the assembly of this initiation complex (see Figure 31.13). [Note: In prokaryotes, three initiation factors are known (IF-1, IF-2, and IF-3), whereas in eukaryotes, there are at least ten (designated eIF to indicate eukaryotic origin).] There are two mechanisms by which the ribosome recognizes the nucleotide sequence that initiates translation:

1. **Shine-Dalgarno sequence:** In E. coli, a purine-rich sequence of nucleotide bases (for example, 5'-UAAGGAGG-3'), known as the Shine-Dalgarno sequence, is located six to ten bases upstream of the AUG codon on the mRNA molecule—that is, near its 5'-end. The 16S ribosomal RNA component of the 30S ribosomal subunit has a nucleotide sequence near its 3'-end that is complementary to all or part of the Shine-Dalgarno sequence. Therefore, the mRNA 5'-end and the 3'-end of the 16S ribosomal RNA can form complementary base pairs, thus facilitating the binding and positioning of the mRNA on the 30S ribosomal subunit (Figure 31.10). [Note: Eukaryotic messages do not have Shine-Dalgarno

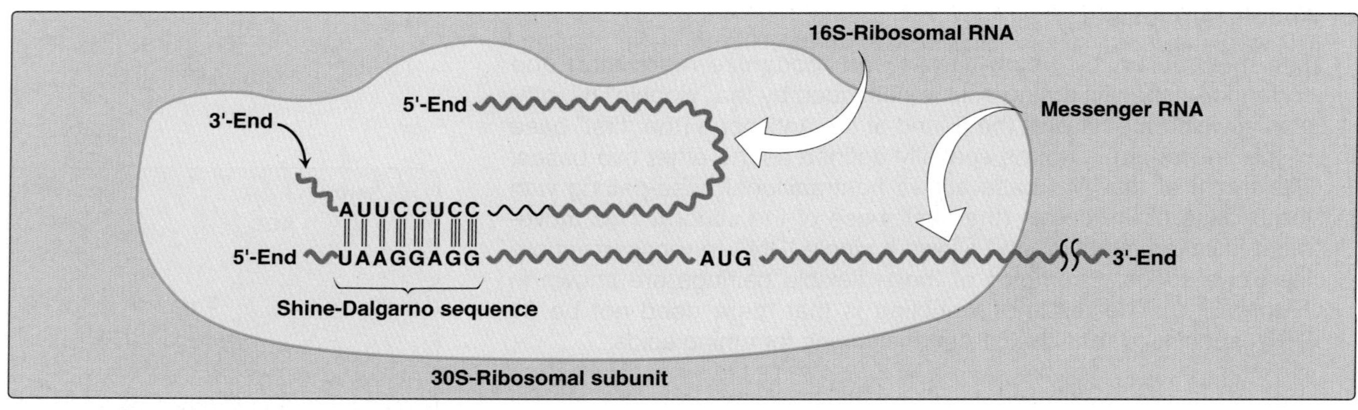

Figure 31.10
Complementary binding between prokaryotic mRNA Shine-Dalgarno sequence and 16S rRNA.

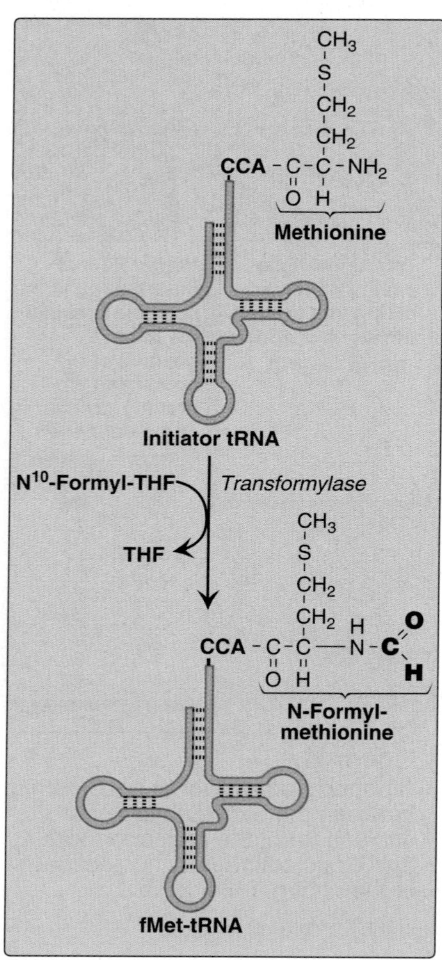

Figure 31.11
Generation of the initiator N-formyl-methionyl-tRNA (fMet-tRNA).

sequences. In eukaryotes, the 40S ribosomal subunit binds to the cap structure at the 5'-end of the mRNA and moves down the mRNA until it encounters the initiator AUG codon.]

2. **Initiation codon:** The codon AUG at the beginning of the message is recognized by a special **initiator tRNA** that enters the ribosomal P site. This recognition is facilitated by IF-2 in E. coli and several eIFs in humans. [Note: Only the initiator tRNA goes to the P site—all other charged tRNAs enter at the A site.] In bacteria and in mitochondria, the initiator tRNA carries an N-formylated methionine (Figure 31.11). The formyl group is added to the methionine after that amino acid is attached to the initiator tRNA by the enzyme *transformylase*, which uses N^{10}-formyl tetrahydrofolate (see p. 265) as the carbon donor. In eukaryotes, the initiator tRNA carries a methionine that is not formylated. [Note: In both prokaryotic cells and mitochondria, this N-terminal methionine is usually removed before the protein is completed.]

B. Elongation

Elongation of the polypeptide chain involves the addition of amino acids to the carboxyl end of the growing chain. During elongation, the ribosome moves from the 5'-end to the 3'-end of the mRNA that is being translated. Delivery of the aminoacyl-tRNA whose codon appears next on the mRNA template in the ribosomal A site is facilitated in E. coli by elongation factors EF-Tu and EF-Ts and requires GTP. [Note: In eukaryotes, comparable elongation factors are designated eEF.] The formation of the peptide bonds is catalyzed by *peptidyltransferase*, an activity intrinsic to the 23S rRNA found in the 50S ribosomal subunit (Figure 31.12). [Note: Because this rRNA catalyzes the reaction, it is referred to as a **ribozyme**.] After the peptide bond has been formed, the ribosome advances three nucleotides toward the 3'-end of the mRNA. This process is known as **translocation** and, in E. coli, requires the participation of EF-G and GTP (eukaryotic cells have similar requirements). This causes movement of the uncharged tRNA into the ribosomal E site (before being released) and movement of the peptidyl-tRNA into the P site. The steps in protein synthesis in the prokaryotic bacterium E. coli are summarized in detail in Figure 31.13.

C. Termination

Termination occurs when one of the three termination codons moves into the A site. These codons are recognized in E. coli by **release factors**: RF-1, which recognizes the termination codons UAA and UAG; RF-2, which recognizes UGA and UAA; and RF-3, which binds GTP and stimulates the activity of RF-1 and RF-2. These factors cause the newly synthesized protein to be released from the ribosomal complex, and, at the same time, cause the dissociation of the ribosome from the mRNA (see Figure 31.13). [Note: Eukaryotes have a single release factor, eRF, that also binds GTP.] The newly synthesized polypeptide may undergo further modification as described below, and the ribosomal subunits, mRNA, tRNA, and protein factors can be recycled and used to synthesize another polypeptide. Some inhibitors of the process of protein synthesis are illustrated in Figure 31.13. In addition, **ricin** (from castor beans) is a very potent toxin that exerts its effects by removing an adenine from 28S ribosomal RNA, thus inhibiting eukaryotic ribosomes.

D. Polysomes

Translation begins at the 5'-end of the mRNA, with the ribosome proceeding along the RNA molecule. Because of the length of most mRNAs, more than one ribosome at a time can generally translate a message (Figure 31.14). Such a complex of one mRNA and a number of ribosomes is called a polysome or **polyribosome**.

E. Protein targeting

Although most protein synthesis occurs in the cytoplasm of eukaryotic cells, many proteins are destined to perform their functions within specific cellular organelles. Such proteins usually contain amino acid sequences that direct these proteins to their final locations. For example, nuclear proteins contain a "**nuclear localization signal**," whereas mitochondrial proteins have a "**mitochondrial entry sequence**."

F. Regulation of translation

Although gene expression is most commonly regulated at the transcriptional level, the rate of protein synthesis is also sometimes regulated. For example, heme stimulates overall translation by preventing the phosphorylation of eukaryotic initiation factor eIF-2, which is only active in its unphosphorylated form. The translation of some messenger RNA molecules is regulated by the binding of regulatory proteins, which sometimes block translation (for example, of ferritin mRNA), and sometimes stabilize the mRNA to extend its lifetime (for example, of transferrin receptor mRNA). [Note: In the presence of adequate supplies of iron, these regulatory proteins dissociate from the mRNA molecules, causing the rate of ferritin synthesis to increase and the rate of transferrin receptor synthesis to decrease.]

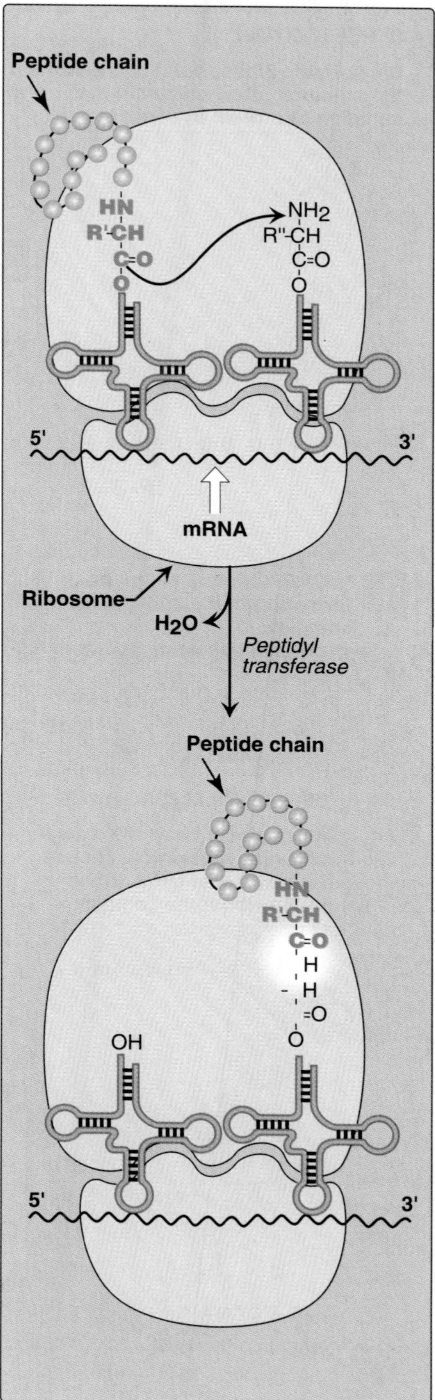

Figure 31.12
Formation of a peptide bond.

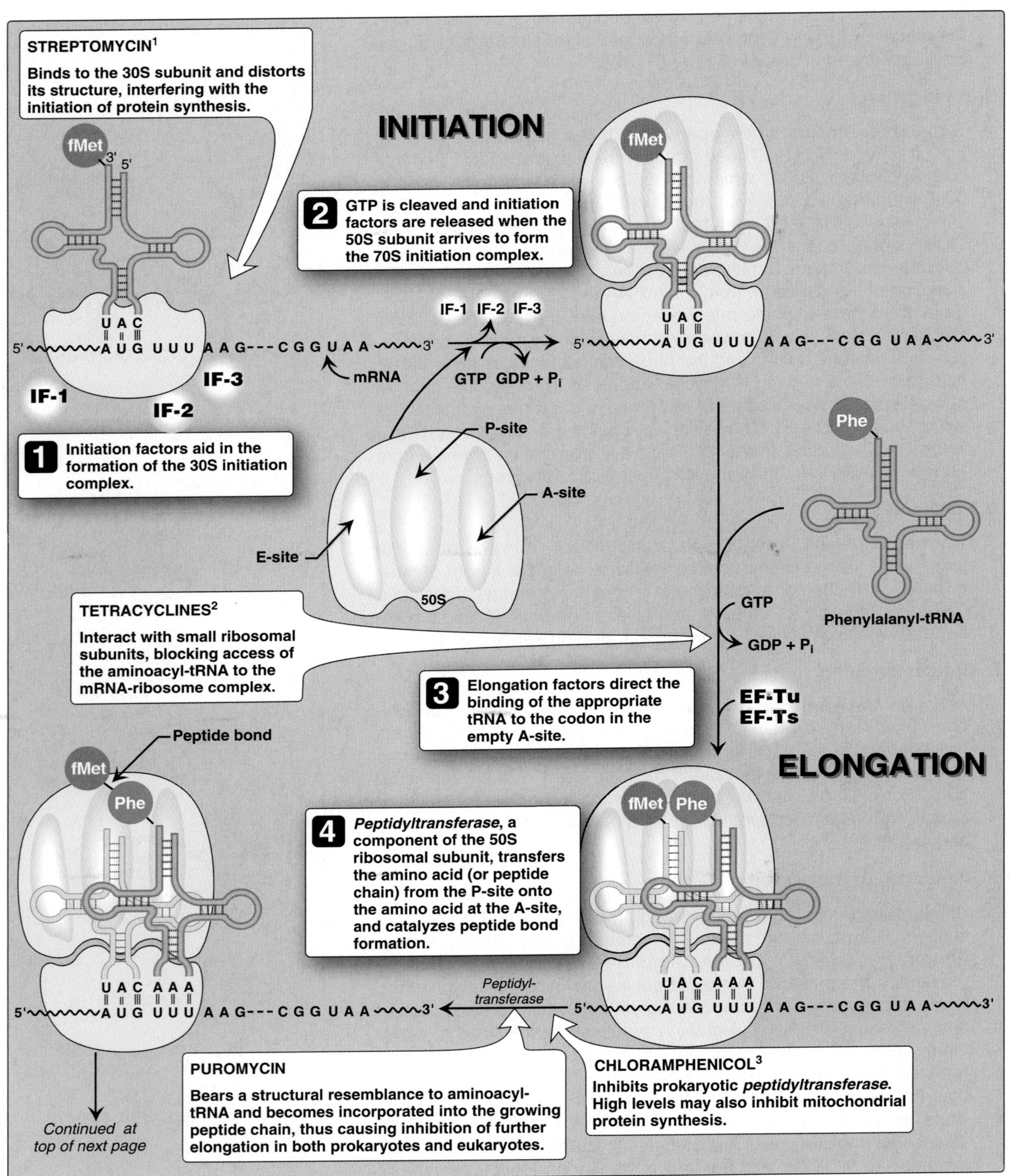

Figure 31.13
Steps in prokaryotic protein synthesis (translation). *(Continued on the next page)*

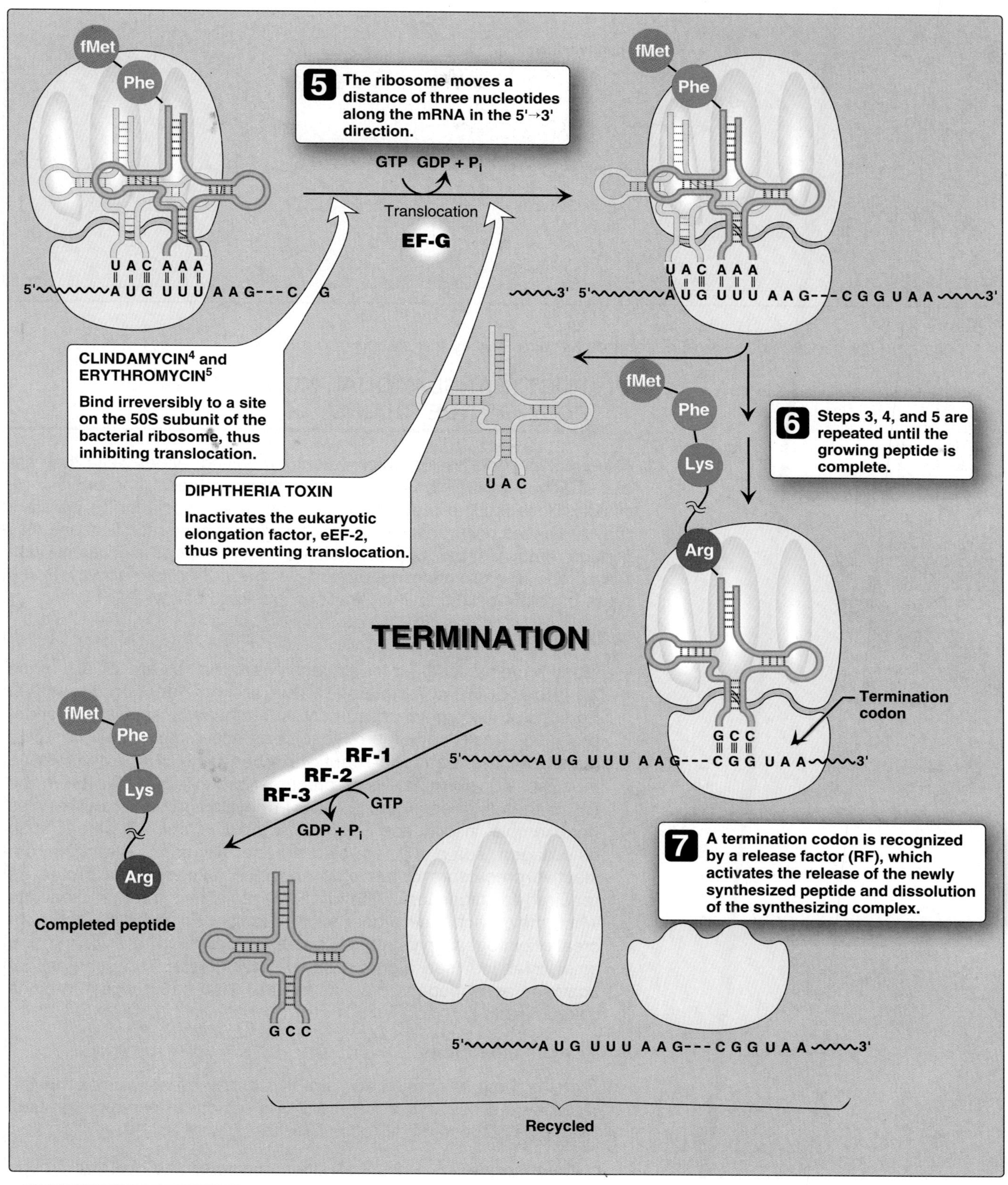

Figure 31.13 *(Continued)*

INFO LINK [1-5]See Chapter 33 in ***Lippincott's Illustrated Reviews: Pharmacology*** (3rd Ed.) and Chapter 31 (2nd Ed.) for a more detailed discussion of antibiotics that inhibit bacterial protein synthesis.

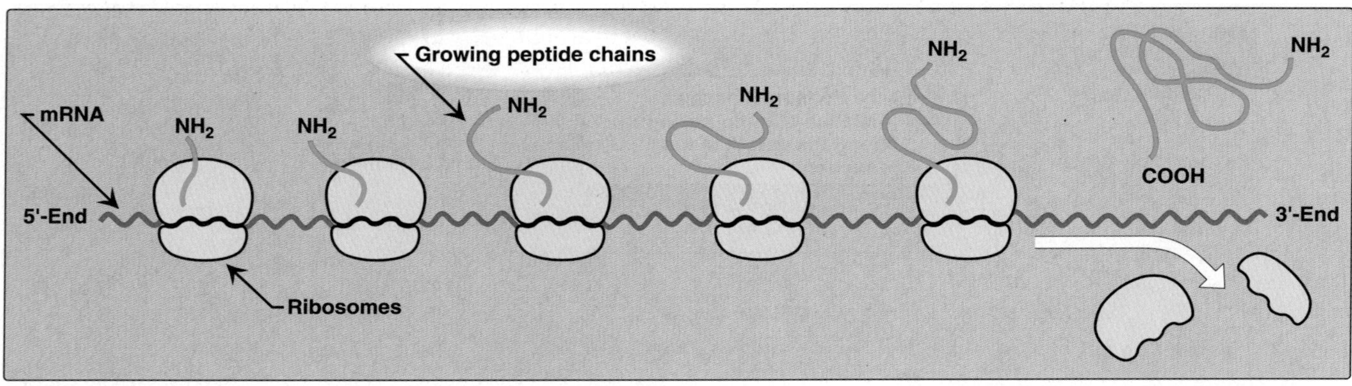

Figure 31.14
A polyribosome consists of several ribosomes simultaneously translating one mRNA.

VI. POSTTRANSLATIONAL MODIFICATION OF POLYPEPTIDE CHAINS

Many polypeptide chains are covalently modified, either while they are still attached to the ribosome or after their synthesis has been completed. Because the modifications occur after translation is initiated, they are called posttranslational modifications. These modifications may include removal of part of the translated sequence, or the covalent addition of one or more chemical groups required for protein activity. Some types of posttranslational modifications are listed below.

A. Trimming

Many proteins destined for secretion from the cell are initially made as large, precursor molecules that are not functionally active. Portions of the protein chain must be removed by specialized endoproteases, resulting in the release of an active molecule. The cellular site of the cleavage reaction depends on the protein to be modified. For example, some precursor proteins are cleaved in the ER or the Golgi apparatus, others in developing secretory vesicles (for example, insulin, see Figure 23.4, p. 307), and still others, such as collagen (see p. 47), are cleaved after secretion. **Zymogens** are inactive precursors of secreted enzymes (including the proteases required for digestion). They become activated through cleavage when they reach their proper sites of action. For example, the pancreatic zymogen, trypsinogen, becomes activated to *trypsin* in the small intestine (see Figure 19.5, p. 247). [Note: The synthesis of enzymes as zymogens protects the cell from being digested by its own products.]

B. Covalent alterations

Proteins, both enzymatic and structural, may be activated or inactivated by the covalent attachment of a variety of chemical groups. Examples of these modifications include (Figure 31.15):

1. **Phosphorylation:** Phosphorylation occurs on the hydroxyl groups of serine, threonine, or, less frequently, tyrosine residues in a pro-

tein. This phosphorylation is catalyzed by one of a family of protein kinases and may be reversed by the action of cellular protein phosphatases. The phosphorylation may increase or decrease the functional activity of the protein. Several examples of these phosphorylation reactions have been previously discussed (for example, see Chapter 11, p. 123 for the regulation of synthesis and degradation of glycogen).

2. **Glycosylation:** Many of the proteins that are destined to become part of a plasma membrane or lysosome or to be secreted from the cell have carbohydrate chains attached to serine or threonine hydroxyl groups (O-linked) or the amide nitrogen of asparagine (N-linked). The stepwise addition of sugars occurs in the endoplasmic reticulum and the Golgi apparatus. The process of producing such glycoproteins was discussed on p. 156. Sometimes glycosylation is used to target proteins to specific organelles. For example, enzymes destined to be incorporated into lysosomes are modified by the addition of mannose-6-phosphate residues (see p. 166).

3. **Hydroxylation:** Proline and lysine residues of the α-chains of collagen are extensively hydroxylated in the endoplasmic reticulum. A discussion of this process was presented on p. 47.

4. **Other covalent modifications:** These may be required for the functional activity of a protein. For example, additional **carboxyl groups** can be added to glutamate residues by vitamin K–dependent carboxylation (see p. 387). The resulting γ-carboxyglutamate resides are esssential for the activity of several of the blood-clotting proteins. Attachment of lipids, such as **farnesyl groups**, can help anchor proteins in membranes. In addition, many proteins are acetylated postranslationally.

C. Protein degradation

Proteins that are defective or destined for rapid turnover are often marked for destruction by **ubiquitination**—the attachment of a small, highly conserved protein, called **ubiquitin** (see p .244). Proteins marked in this way are rapidly degraded by a cellular component known as the "**proteasome**", which is a complex, ATP-dependent, proteolytic system located in the cytosol.

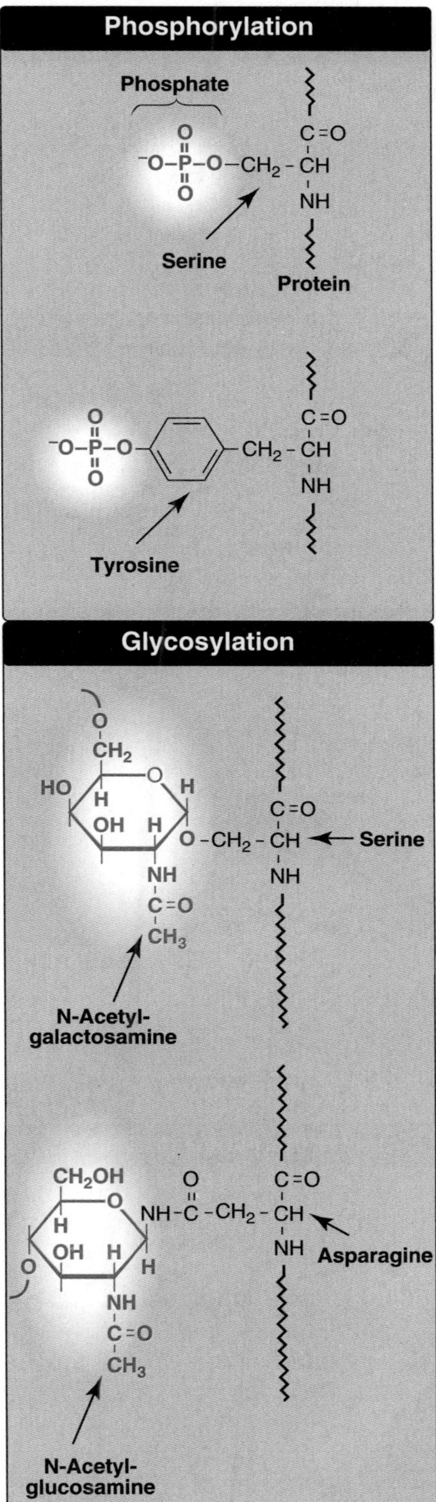

Figure 31.15 *(continued on next page)* Posttranslation modifications of some amino acid residues.

VII. CHAPTER SUMMARY

Codons are composed of three nucleotide bases usually presented in the mRNA language of **A**, **G**, **C**, and **U**. They are always written **5'→3'**. Of the 64 possible three-base combinations, 61 code for the twenty common amino acids and three signal termination of protein synthesis (**translation**). Altering the nucleotide sequence in a codon can cause **silent mutations** (the altered codon also codes for the original amino acid), **missense mutations** (the altered codon codes for a different amino acid), or **nonsense mutations** (the altered codon is a termination

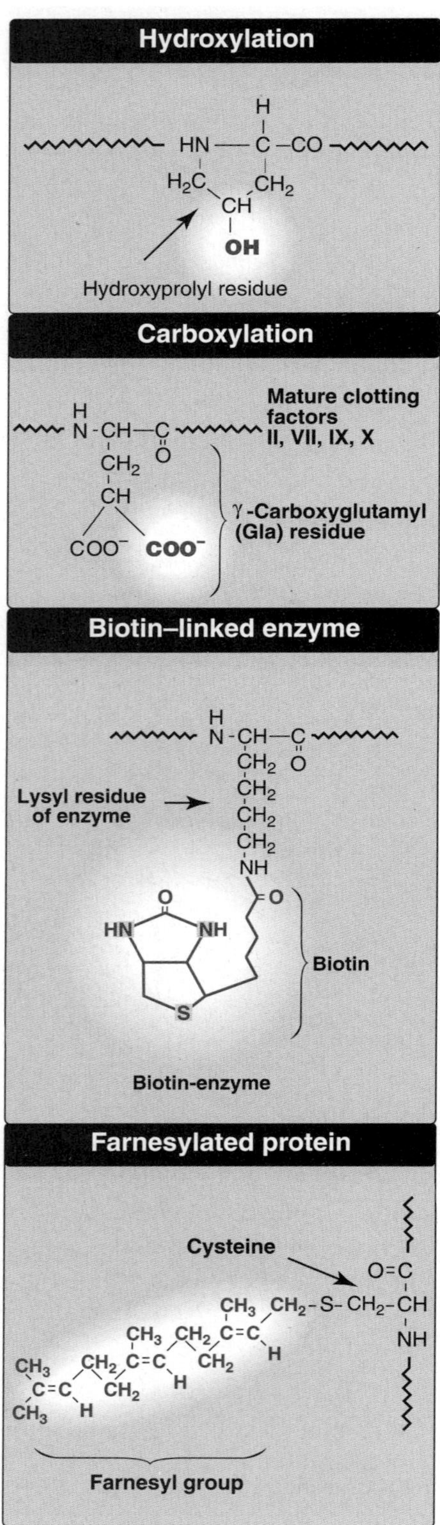

Figure 31.15 *(continued)*
Posttranslation modifications of some amino acid residues.

codon). Characteristics of the genetic code include **specificity**, **universality**, and **redundancy**, and it is **nonoverlapping** and **commaless**. Requirements for protein synthesis include all the **amino acids** that eventually appear in the finished protein, at least one specific type of **tRNA** for each amino acid, one **aminoacyl-tRNA synthetase** for each amino acid, the **mRNA** coding for the protein to be synthesized, fully competent **ribosomes**, **protein factors** needed for initiation, elongation, and termination of protein synthesis, and **ATP** and **GTP** as energy sources. tRNA has an attachment site for a specific amino acid at its 3'-end, and an **anticodon** region that can recognize the codon specifying the amino acid the tRNA is carryiing. **Ribosomes** are large complexes of protein and rRNA. They consist of **two subunits**. Each ribosome has three binding sites for tRNA molecules—the A, P, and E sites that cover three neighboring codons. The **A site** codon binds an **incoming aminoacyl-tRNA**, the **P site** codon is occupied by **peptidyl-tRNA**, and the **E site** is occupied by the **empty tRNA** as it is about to exit the ribosome. Recognition of an mRNA codon is accomplished by the tRNA **anticodon**. The anticodon binds to the codon following the rules of **complementarity** and **antiparallel binding**. (When writing the sequences of both codons and anticodons, the nucleotide sequence must ALWAYS be listed in the 5'→3' order.) The "**wobble**" **hypothesis** states that the first (5') base of the anticodon is not as spatially defined as the other two bases. Movement of that first base allows nontraditional base-pairing with the last (3') base of the codon, thus allowing a single tRNA to recognize more than one codon for a specific amino acid. **Initiation of protein synthesis:** The components of the translation system are assembled, and mRNA associates with the small ribosomal subunit. The process requires **initiation factors**. In **prokaryotes**, a purine-rich region of the mRNA (the **Shine-Dalgarno sequence**) base-pairs with a complementary sequence on 16S rRNA, resulting in the positioning of the mRNA so that translation can begin. The **5'-cap** on **eukaryotic mRNA** is used to position that structure on the ribosome. The **initiation codon** is **5'–AUG–3'**. **Elongation:** The polypeptide chain is elongated by the addition of amino acids to the carboxyl end of its growing chain. The process requires **elongation factors**. The formation of the peptide bond is catalyzed by **peptidyltransferase**, which is an activity intrinsic to the ribosomal 23S rRNA. Following peptide bond formation, the ribosome advances along the mRNA in the **5'→3' direction** to the next codon (**translocation**). Because of the length of most mRNAs, more than one ribosome at a time can translate a message, forming a **polysome**. **Termination:** Termination begins when one of the three termination codons moves into the A site. These codons are recognized by **release factors**. The newly synthesized protein is released from the ribosomal complex, and the ribosome is disassociated from the mRNA. Numerous **antibiotics** interfere with the process of protein synthesis. Many polypeptide chains are covalently modified after translation. Such modifications include **trimming** excess amino acids, **phosphorylation** which may activate or inactivate the protein, **glycosylation**, which targets a protein to become part of a plasma membrane or lysosome or be secreted from the cell, or **hydroxylation** such as that seen in collagen. Proteins that are defective or destined for rapid turnover are marked for destruction by the attachment of a small, highly conserved protein called **ubiquitin**. Proteins marked in this way are rapidly degraded by a cellular component known as the **proteasome**.

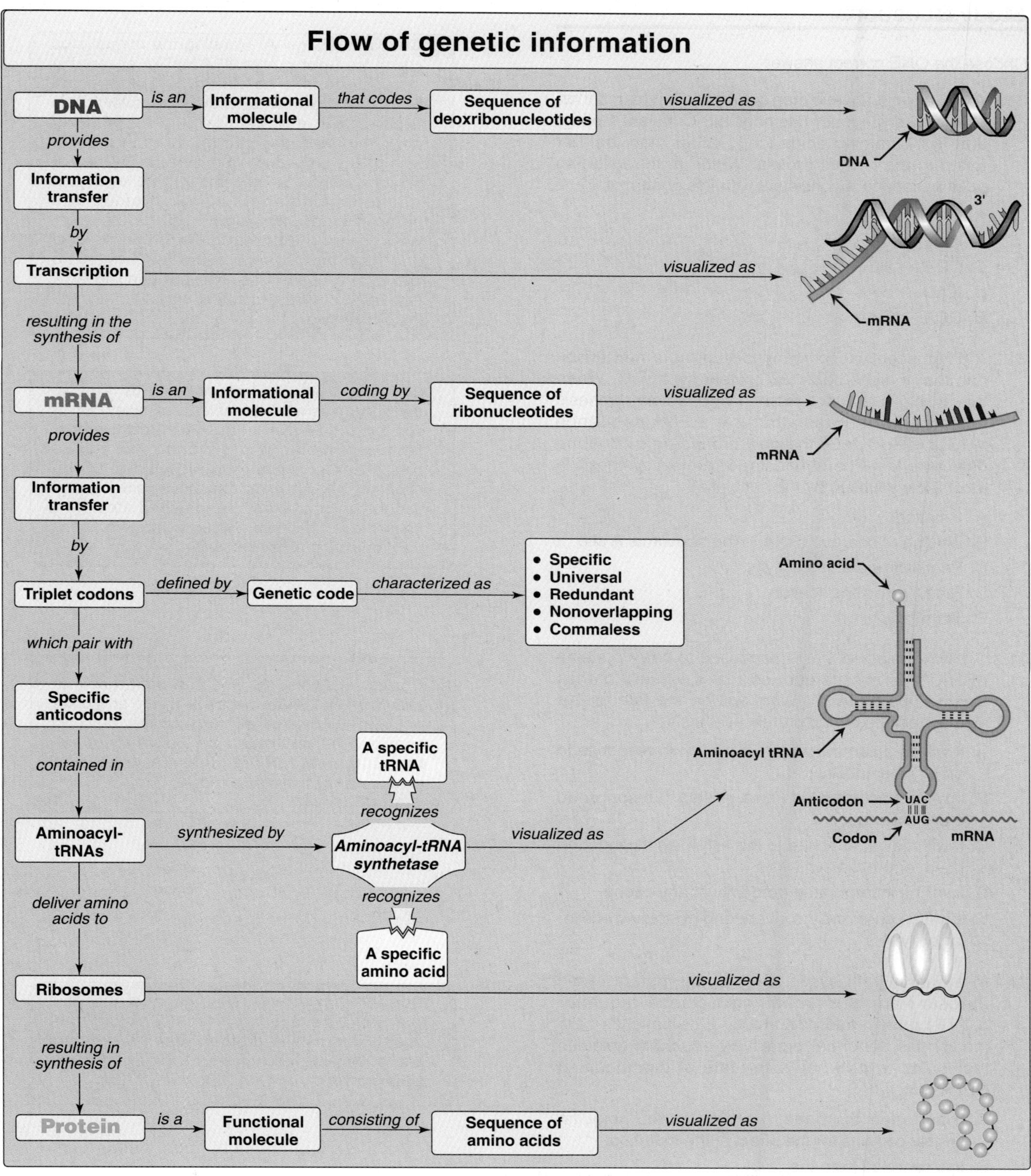

Figure 31.16
Key concept map for protein synthesis.

Study Questions

Choose the ONE correct answer

31.1 A 20-year-old anemic man is found to have an abnormal form of β-globin (Hemoglobin Constant Spring) that is 172 amino acids long, rather than the 141 found in the normal protein. Which of the following point mutations is consistent with this abnormality?

A. UAA → CAA
B. UAA → UAG
C. CGA → UGA
D. GAU → GAC
E. GCA → GAA

Correct answer = A. Mutating the normal stop codon for β-globin from UAA to CAA causes the ribosome to insert a glutamine at that point. Thus, it will continue extending the protein chain until it comes upon the next stop codon further down the message, resulting in an abnormally long protein. A change from UAA to UAG would simply change one stop codon for another and would have no effect on the protein. The replacement of CGA (arginine) with UGA (stop) would cause the protein to be too short. GAU and GAC both encode aspartate and would cause no change in the protein. Changing GCA (alanine) to GAA (glutamate) would not change the size of the protein product.

31.2 A pharmaceutical company is studying a new antibiotic that inhibits bacterial protein synthesis. When this antibiotic is added to an <u>in vitro</u> protein synthesis system that is translating the mRNA sequence AUGUUUUUUUAG, the only product formed is the dipeptide fMet-Phe. What step in protein synthesis is most likely inhibited by the antibiotic?

A. Initiation
B. Binding of charged tRNA to the ribosomal A site
C. Peptidyltransferase activity
D. Ribosomal translocation
E. Termination

Correct answer = D. Because fMet-Phe is made, the ribosomes must be able to complete initiation, bind Phe-tRNA to the A-site, and use peptidyltransferase activity to form the first peptide bond. Because the ribosome is not able to proceed any further, ribosomal movement (translocation) is most likely the inhibited step. The ribosome is, therefore, frozen before it reaches the stop codon of this message.

31.3 A tRNA molecule that is supposed to carry cysteine ($tRNA^{cys}$) is mischarged, so that it actually carries alanine (ala-$tRNA^{cys}$). What will be the fate of this alanine residue during protein synthesis?

A. It will be incorporated into a protein in response to an alanine codon.
B. It will be incorporated into a protein in response to a cysteine codon.
C. It will remain attached to the tRNA, as it cannot be used for protein synthesis.
D. It will be incorporated randomly at any codon.
E. It will be chemically converted to cysteine by cellular enzymes.

Correct answer = B. Once an amino acid is attached to a tRNA molecule, only the anticodon of that tRNA determines the specificity of incorporation. The mischarged alanine will, therefore, be incorporated in the protein at a position determined by a cysteine codon.

31.4 In a patient with cystic fibrosis, the mutant cystic fibrosis transmembrane conductance regulator (CFTR) protein folds incorrectly. The patient's cells modify this abnormal protein by attaching ubiquitin molecules to it. What is the fate of this modified CFTR protein?

A. It performs its normal function, as the ubiquitin largely corrects for the effect of the mutation.
B. It is secreted from the cell.
C. It is placed into storage vesicles.
D. It is degraded by the proteasome.
E. It is repaired by cellular enzymes.

Correct answer = D. Ubiquitination usually marks old, damaged, or misfolded proteins for destruction by the proteasome. There is no known cellular mechanism for repair of damaged proteins.

Biotechnology and Human Disease

32

I. OVERVIEW

In the past, efforts to understand genes and their expression have been confounded by the immense size and complexity of human DNA. The human genome contains DNA with approximately three billion (10^9) base pairs that encode 30,000 to 40,000 genes located on 23 pairs of chromosomes. It is now possible to determine the nucleotide sequence of long stretches of DNA, and essentially the entire sequence of the human genome is now known. This effort (called the Human Genome Project) was made possible by several techniques that have already contributed to our understanding of many genetic diseases (Figure 32.1). These include, first, the discovery of restriction endonucleases that permit the dissection of huge DNA molecules into defined fragments. Second, the development of cloning techniques, providing a mechanism for amplification of specific nucleotide sequences. Finally, the ability to synthesize specific probes, which has allowed the identification and manipulation of nucleotide sequences of interest. These and other experimental approaches have permitted the identification of both normal and mutant nucleotide sequences in DNA. This knowledge has led to the development of methods for the prenatal diagnosis of genetic diseases, and initial successes in the treatment of patients by gene therapy.

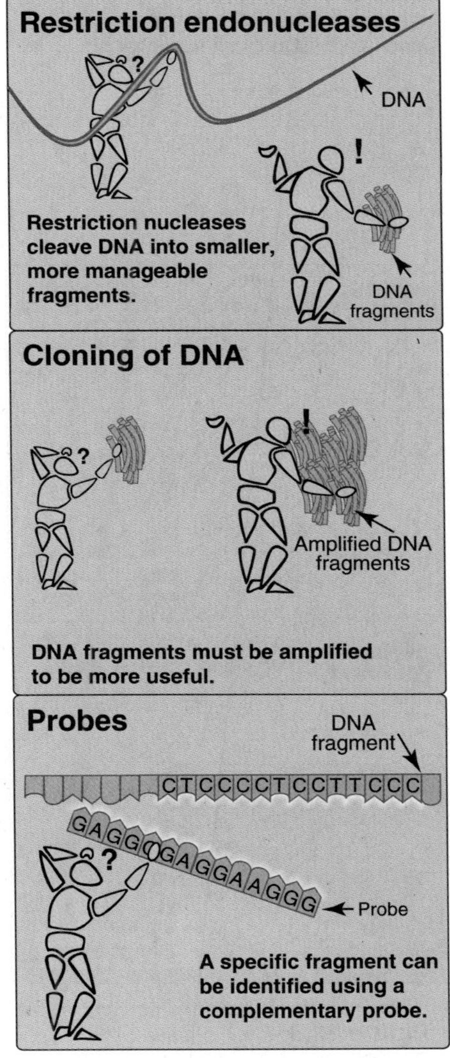

Figure 32.1
Three techniques that facilitate analysis of human DNA.

II. RESTRICTION ENDONUCLEASES

One of the major obstacles to molecular analysis of genomic DNA is the immense size of the molecules involved. The discovery of a special group of bacterial enzymes, called **restriction endonucleases** (**restriction enzymes**), which cleave double-stranded DNA into smaller, more manageable fragments, has opened the way for DNA analysis. Because each enzyme cleaves DNA at a specific nucleotide sequence, restriction enzymes are used experimentally to obtain precisely defined DNA segments called **restriction fragments**.

A. Specificity of restriction endonucleases

Restriction endonucleases recognize short stretches of DNA (generally four or six base pairs) that contain specific nucleotide sequences. These sequences, which differ for each restriction endonuclease, are **palindromes**, that is, they exhibit two-fold rotational symmetry (Figure

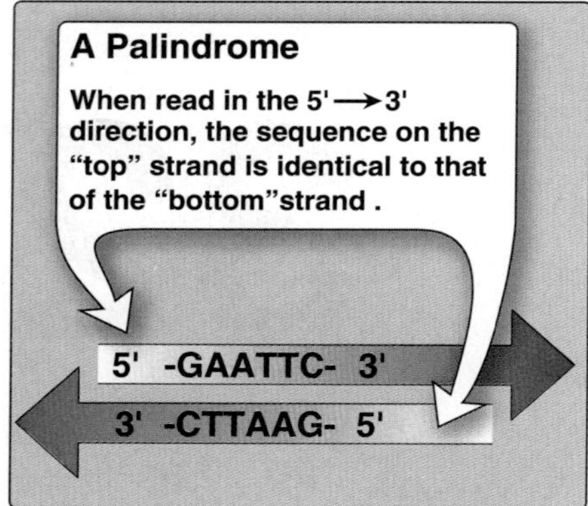

Figure 32.2
Recognition sequence of restriction endonuclease *Eco*RI shows two-fold rotational symmetry.

32.2). This means that, within a short region of the double helix, the nucleotide sequence on the "top" strand, read 5'→3', is identical to that of the "bottom" strand, also read in the 5'→3' direction. Therefore, if you turn the page upside down—that is, rotate it 180 degrees around its **axis of symmetry**—the structure remains the same.

B. Nomenclature

A restriction enzyme is named according to the organism from which it was isolated. The first letter of the name is from the genus of the bacterium. The next two letters are from the name of the species. An additional subscript letter indicates the type or strain, and a final number is appended to indicate the order in which the enzyme was discovered in that particular organism. For example, *Hae*III is the third restriction endonuclease isolated from the bacterium Haemophilus aegyptius.

C. "Sticky" and "blunt" ends

Restriction enzymes cleave DNA so as to produce a 3'-hydroxyl group on one end and a 5'-phosphate group on the other. Some restriction endonucleases, such as *Taq*I, form staggered cuts that produce "sticky" or cohesive ends—that is, the resulting DNA fragments have single-stranded sequences that are complementary to each other (Figure 32.3). Other restriction endonucleases, such as *Hae*III, cleave in the middle of their recognition sequence—that is, at the axis of symmetry (see Figures 32.2 and 32.3)—and produce fragments that have "blunt" ends that do not form hydrogen bonds with each other. Using the enzyme *DNA ligase* (see p. 403), sticky ends of a DNA fragment of interest can be covalently joined with other DNA fragments that have sticky ends produced by cleavage with the same restriction endonuclease (Figure 32.4). [Note: Another ligase, encoded by bacteriophage T4, can covalently join blunt-ended fragments.] The hybrid combination of two fragments is called a **recombinant DNA molecule**.

D. Restriction sites

A DNA sequence that is recognized by a restriction enzyme is called a restriction site. These sites are recognized by restriction endonucleases that cleave DNA into fragments of different sizes. For example, an enzyme that recognizes a specific four base-pair sequence produces many cuts in the DNA molecule. In contrast, an enzyme requiring a unique sequence of six base pairs produces fewer cuts and, hence, longer pieces. Hundreds of these enzymes, having different cleavage specificities (varying in both nucleotide sequences and length of recognition sites), are commercially available as analytic reagents.

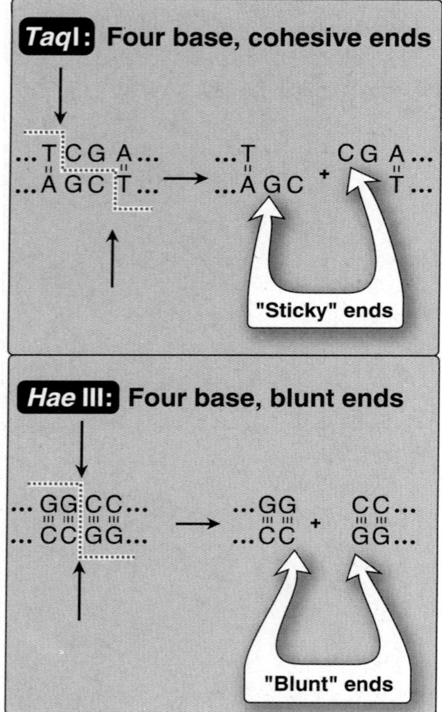

Figure 32.3
Specificity of *Taq*I and *Hae*III restriction endonucleases.

III. DNA CLONING

Introduction of a foreign DNA molecule into a replicating cell permits the amplification (that is, production of many copies) of the DNA. In some cases, a single DNA fragment can be isolated and purified prior to cloning. More commonly, to clone a nucleotide sequence of interest, the total cellular DNA is first cleaved with a specific restriction enzyme, creating hundreds of thousand s of fragments. Therefore, individual fragments cannot be isolated. Instead, each of the resulting DNA fragments is joined to a DNA vector molecule (referred to as a **cloning vector**) to form a hybrid molecule. Each hybrid recombinant DNA molecule conveys its inserted DNA fragment into a single host cell, for example, a bacterium, where it is replicated (or "**amplified**"). As the host cell multiplies, it forms a clone in which every bacterium carries copies of the same inserted DNA fragment, hence, the name "**cloning**." The cloned DNA is eventually released from its vector by cleavage (using the appropriate restriction endonuclease) and is isolated. By this mechanism, many identical copies of the DNA of interest can be produced. [Note: An alternative to cloning—the **polymerase chain reaction**—is described on p. 459.]

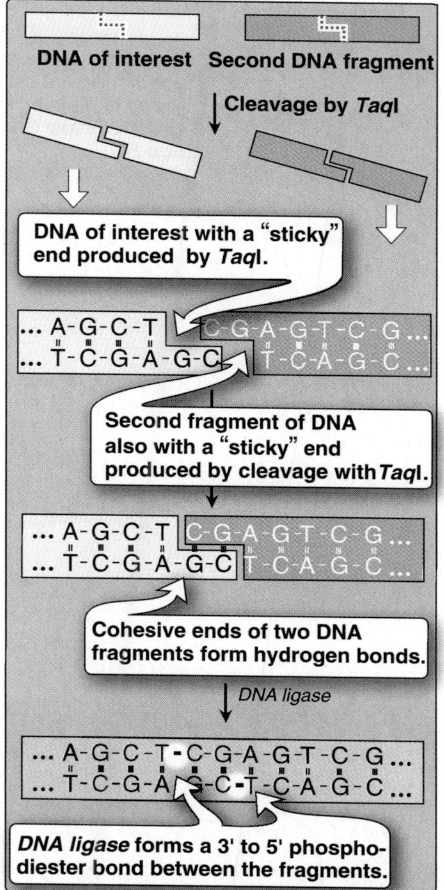

Figure 32.4
Formation of recombinant DNA from restriction fragments with "sticky"ends.

A. Vectors

A vector is a molecule of DNA to which the fragment of DNA to be cloned is joined. Essential properties of a vector include: 1) it must be capable of autonomous replication within a host cell, 2) it must contain at least one specific nucleotide sequence recognized by a restriction endonuclease, and 3) it must carry at least one gene that confers the ability to select for the vector, such as an antibiotic resistance gene. Commonly used vectors include **plasmids** and **bacterial and animal viruses**.

1. **Prokaryotic plasmids:** Prokaryotic organisms contain single, large, circular chromosomes. In addition, most species of bacteria also normally contain small, circular, extrachromosomal DNA molecules called plasmids[1] (Figure 32.5). Plasmid DNA undergoes replication that may or may not be synchronized to chromosomal division. Plasmids may carry genes that convey antibiotic resistance to the host bacterium, and may facilitate the transfer of genetic information from one bacterium to another. [Note: Bacteria are grown in the presence of antibiotics, thus selecting for cells containing the hybrid plasmids, which provide antibiotic resistance.] Plasmids can be readily isolated from bacterial cells, their circular DNA cleaved at specific sites by restriction endonucleases, and foreign DNA inserted into the circle. The hybrid plasmid can be reintroduced into a bacterium, and large numbers of copies of the plasmid containing the foreign DNA produced (Figure 32.6). [Note: The experiment is conducted to favor only one DNA fragment being inserted into each plasmid and only one plasmid being taken up by each bacterium.]

2. **Other vectors:** The development of improved vectors that can more efficiently accommodate large DNA segments, or express the passenger genes in different cell types, is an ongoing endeavor of molecular genetics research. In addition to the prokaryotic plasmids described above, bacteriophage lambda (λ), yeast artificial chromosomes (YACs), and mammalian viruses (retroviruses, for example) are currently in wide use as cloning vectors.

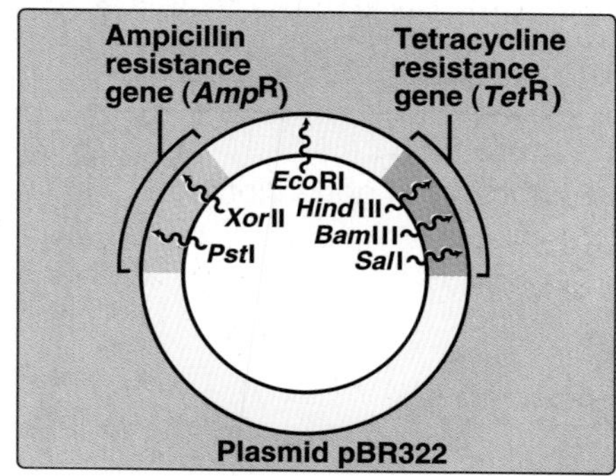

Figure 32.5
A restriction map of plasmid pBR322 indicating the positions of its antibiotic resistance genes and the sites of nucleotide sequences recognized by specific restriction endonucleases.

B. DNA libraries

A DNA library is a collection of cloned restriction fragments of the DNA of an organism. Two kinds of libraries will be discussed: genomic libraries and cDNA libraries. Genomic libraries ideally con-

[1]See p. 116 in ***Lippincott's Illustrated Reviews: Microbiology*** for a discussion of plasmids.

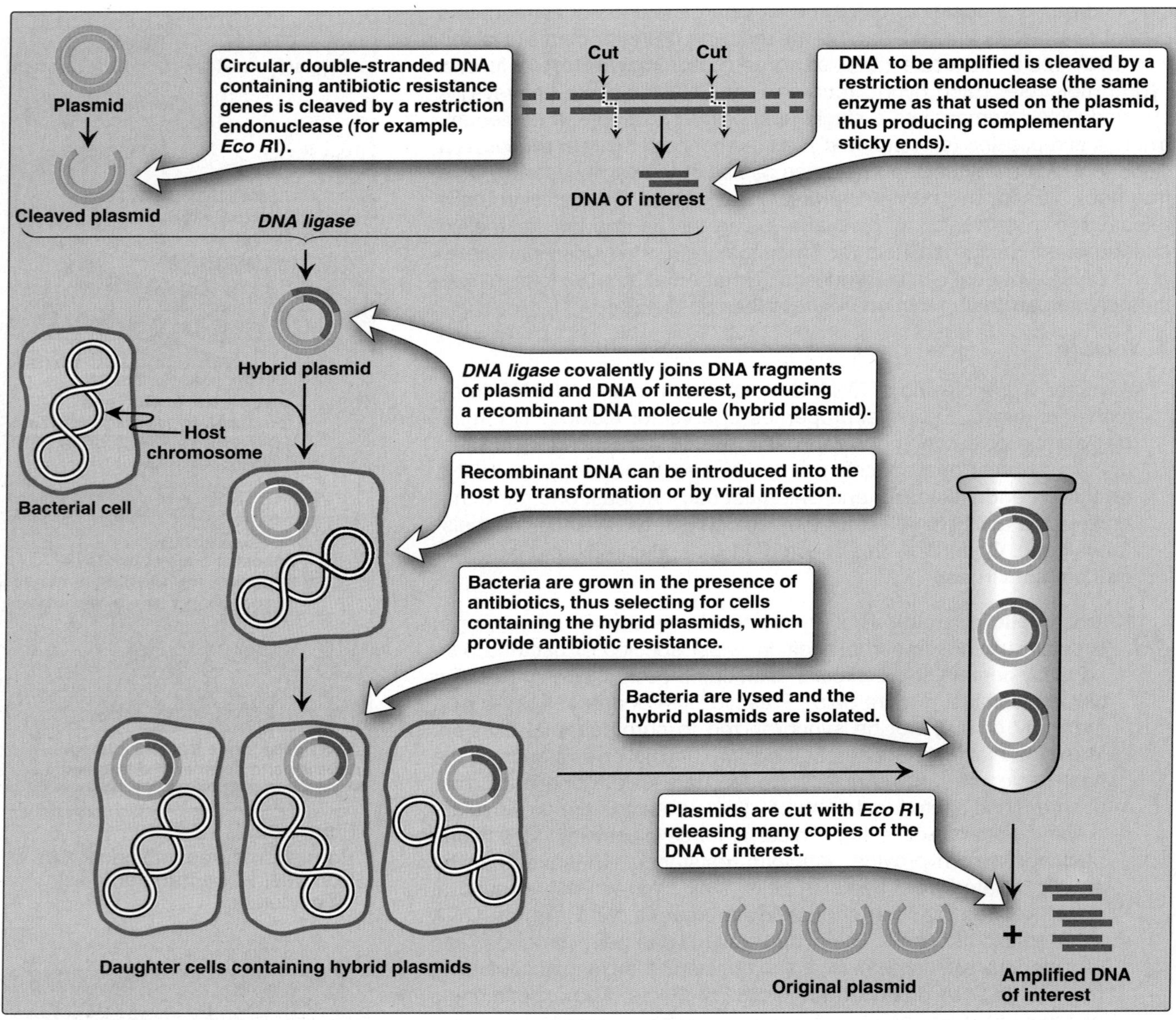

Figure 32.6
Summary of gene cloning.

tain a copy of every DNA nucleotide sequence in the genome. In contrast, cDNA libraries contain those DNA sequences that appear as mRNA molecules, and these differ from one cell type to another. [Note: Cloned cDNAs lack introns and the control regions of the genes, whereas these are present in genomic libraries.]

1. **Genomic DNA libraries:** A genomic library is the collection of fragments of double-stranded DNA obtained by digestion of the total DNA of the organism with a restriction endonuclease and subsequent ligation to an appropriate vector. The recombinant DNA molecules are replicated within host bacteria. The amplified DNA fragments represent the entire genome of the organism and are called a genomic library. Regardless of the restriction enzyme

used, the chances are rather good that the gene of interest contains more than one restriction site recognized by that enzyme. If this is the case, and if the digestion is allowed to go to completion, the gene of interest is fragmented—that is, it is not contained in any one clone in the library. To avoid this usually undesirable result, a **partial digestion** is performed in which either the amount or the time of action of the enzyme is limited. This results in cleavage occuring at only a fraction of the restriction sites on any one DNA molecule, thus producing fragments of about 20,000 base pairs. Enzymes that cut very frequently (that is, enzymes that recognize four base pair sequences) are generally used for this purpose so that the result is an almost random collection of fragments. This ensures a high degree of probability that the gene of interest is contained, intact, in some fragment.

2. **Complementary DNA (cDNA) libraries:** If a gene of interest is expressed at a very high level in a particular tissue, it is likely that the mRNA corresponding to that gene is also present at high concentrations in the cell. For example, reticulocyte mRNA is composed largely of molecules encoding the α-globin and β-globin chains of hemoglobin. This mRNA can be used as a template to make a complementary double-stranded DNA (cDNA) molecule using the enzyme *reverse transcriptase* (Figure 32.7). The resulting cDNA is thus a double-stranded copy of mRNA. cDNA can be amplified by cloning or by the polymerase chain reaction. It can be used as a probe to locate the gene that coded for the original mRNA (or fragments of the gene) in mixtures containing many unrelated DNA fragments. If the mRNA used as a template is a mixture of many different species, the resulting cDNAs are heterogeneous. These mixtures can be cloned to form a cDNA library. Because cDNA has no intervening sequences, it can be cloned into an **expression vector** for the synthesis of eukaryotic proteins by bacteria (Figure 32.8). These special plasmids contain a bacterial promoter for transcription of the cDNA, and a Shine-Dalgarno sequence (see p. 435) that allows the bacterial ribosome to initiate translation of the resulting mRNA molecule.

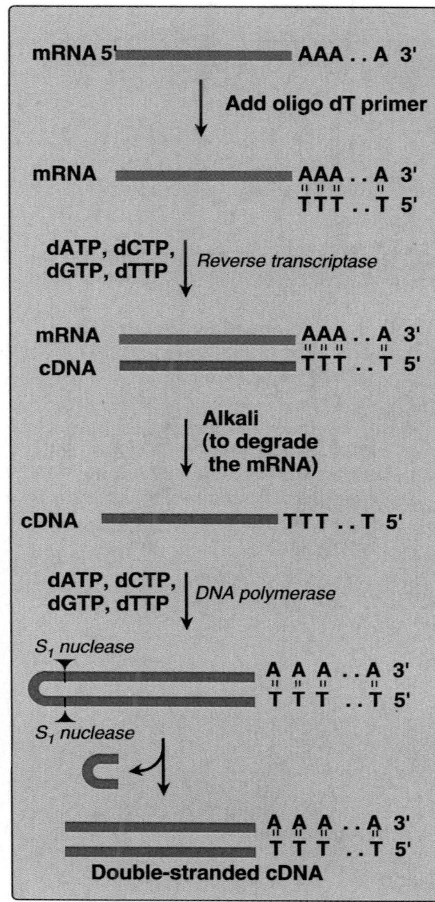

Figure 32.7
Synthesis of cDNA from mRNA using *reverse transcriptase.*

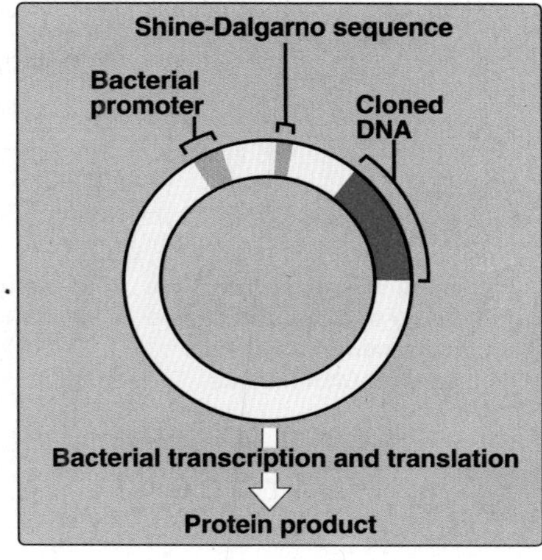

Figure 32.8
An expression vector.

C. Sequencing of cloned DNA fragments

The base sequence of DNA fragments that have been cloned and purified can be determined in the laboratory. The original procedure for this purpose was the **Sanger dideoxy method** illustrated in Figure 32.9. The single-stranded DNA to be sequenced is used as the template for DNA synthesis by *DNA polymerase.* A radioactive primer complementary to the 3'-end of the target DNA is added, along with the four deoxyribonucleoside triphosphates (dNTPs). The sample is divided into four reaction tubes, and a small amount of one of the four dideoxyribonucleoside triphosphates (ddNTPs) is added to each tube. Because it contains no 3' hydroxyl group, incorporation of a ddNTP into a newly synthesized strand terminates its elongation at that point. The products of this reaction then consist of a mixture of DNA strands of different lengths, each terminating at a specific base. Separation of the various DNA products by size using polyacrylamide gel electrophoresis, followed by autoradiography, yields a pattern of bands from which the DNA base sequence can be read. The **Human Genome Project** used highly automated variations of this technique to determine the base sequence of essentially the entire human genome.

IV. PROBES

Cleavage of large DNA molecules by restriction endonucleases produces a bewildering array of fragments. How can a specific gene or DNA sequence of interest be picked out of the mixture of thousands or even millions of irrelevant DNA fragments? The answer lies in the use of a **probe**—a single-stranded piece of DNA, labeled with a radioisotope, such as ^{32}P, or with a non-radioactive probe, such as biotin. The nucleotide sequence of a probe is complementary to the DNA of interest, called the **target DNA**. Probes are used to identify which clone of a library or which band on a gel contains the target DNA.

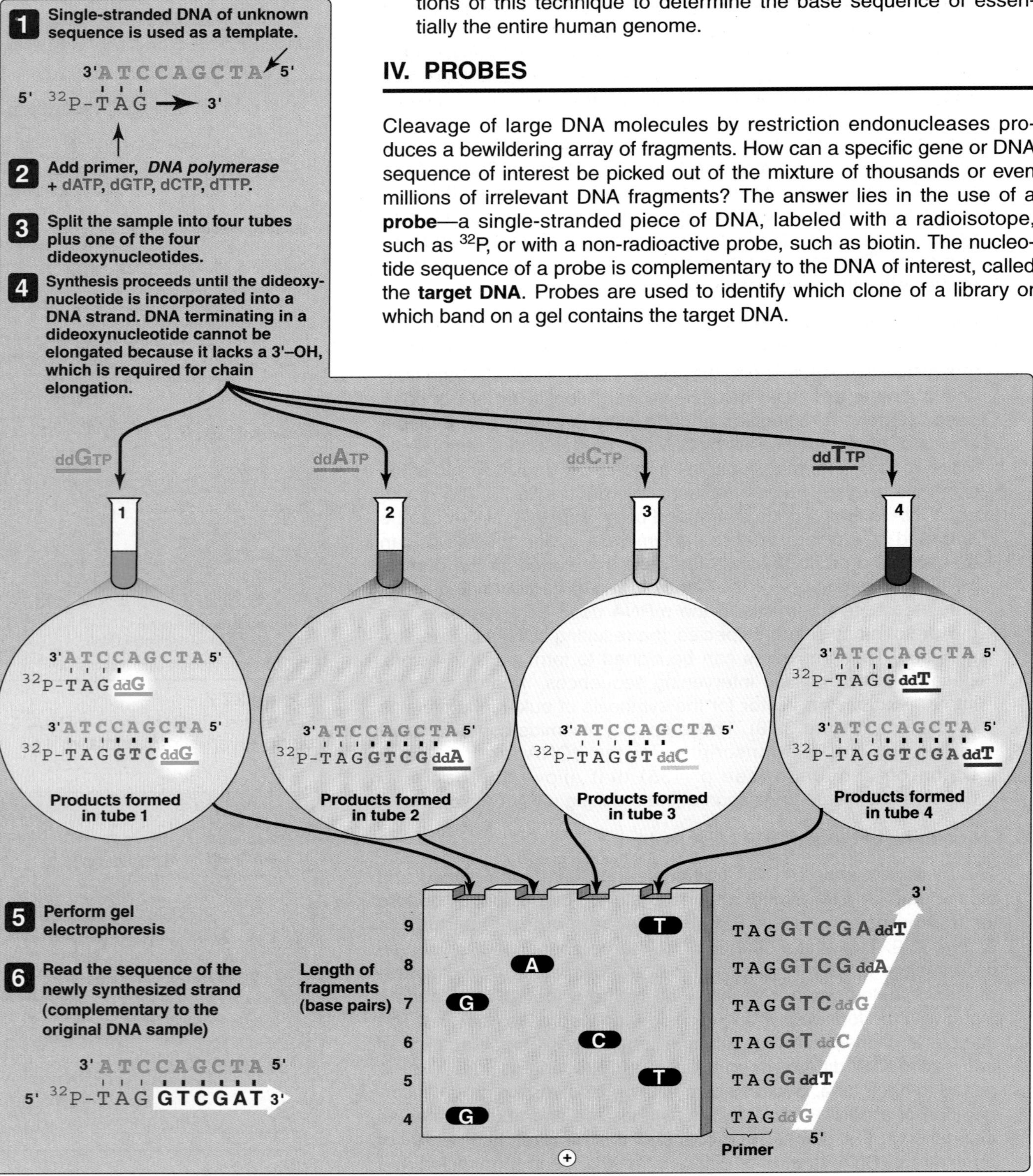

Figure 32.9
DNA sequencing by the Sanger method.

A. Hybridization of a probe to DNA fragments

The utility of probes hinges on the phenomenon of hybridization in which a single-stranded sequence of a target DNA binds to a probe containing a complementary nucleotide sequence. Single-stranded DNA, produced by alkaline denaturation of double-stranded DNA, is first bound to a solid support, such as a nitrocellulose membrane. The immobilized DNA strands are prevented from self-annealing, but are available for hybridization to an exogenous, single-stranded, radiolabeled DNA probe. The extent of hybridization is measured by the retention of radioactivity on the membrane. Excess probe molecules that do not hybridize are removed by washing the filter and, therefore, do not interfere.

B. Synthetic oligonucleotide probes

If the sequence of all or part of the target DNA is known, single-stranded oligonucleotide probes of twenty to thirty nucleotides can be synthesized that are complementary to a small region of the gene of interest. If the sequence of the gene is unknown, the amino acid sequence of the protein—that is, the gene product—may be used to construct a probe. Short, single-stranded DNA sequences (fifteen to thirty nucleotides) are synthesized, using the genetic code as a guide. Because of the degeneracy of the genetic code, it is necessary to synthesize several oligonucleotides. [Note: Oligonucleotides can be used to detect single base changes in the sequence to which they are complementary. In contrast, cDNA probes contain many thousands of bases, and their binding to a target DNA with a single base change is unaffected.]

1. **Detecting the β^S-globin mutation:** Figure 32.10 shows how an **allele-specific oligonucleotide** (**ASO**) **probe** can be used to detect the presence of the sickle cell mutation in the β-globin gene. DNA, isolated from white blood cells, is denatured into single strands. An oligonucleotide is constructed that is complementary to the portion of the mutant globin gene coding for the amino-terminal sequence of the β-globin protein. DNA isolated from a heterozygous individual (sickle cell trait) or a homozygous patient (sickle cell disease) contains a nucleotide sequence that is complementary to the probe. Thus, a double-stranded hybrid forms that can be detected by electrophoresis. In contrast, DNA obtained from normal individuals is not complementary at the sixth nucleotide triplet (coding for glutamate in normal individuals but for valine in patients with the β^S-gene) and, therefore, does not form a hybrid (see Figure 32.10). Use of a pair of such ASOs (one specific for the normal allele and one specific for the mutant allele) allows one to distinguish the DNA from all three possible genotypes—homozygous normal, heterozygous, and homozygous mutant (Figure 32.11.)

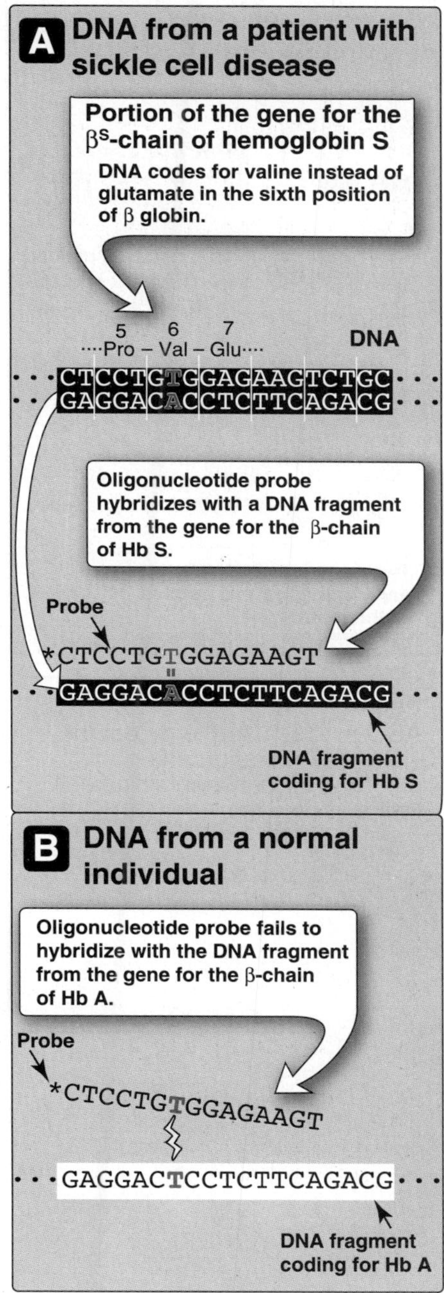

Figure 32.10
Oligonucleotide probe detects Hb S allele. [Note: * indicates ^{32}P radioactive label.]

C. Biotinylated probes

Because the disposal of radioactive waste is becoming increasingly expensive, nonradioactive probes have been developed. One of the most successful is based on the vitamin biotin (see p. 379), which can be chemically coupled to the nucleotides used to synthesize the probe. Biotin was chosen because it binds very tenaciously to

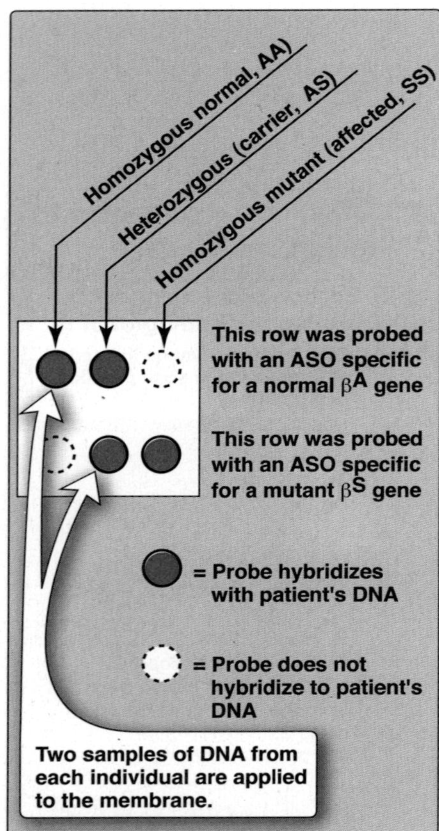

Figure 32.11
ASO probes used to detect the sickle cell mutation and differentiate between sickle cell trait and disease.

avidin—a readily available protein contained in chicken egg whites. Avidin can be attached to a fluorescent dye detectable optically with great sensitivity. Thus, a DNA fragment (displayed, for example, by gel electrophoresis) that hybridizes with the biotinylated probe can be made visible by immersing the gel in a solution of dye-coupled avidin. After washing away the excess avidin, the DNA fragment that binds the probe is fluorescent.

D. Antibodies

Sometimes no amino acid sequence information is available to guide the synthesis of a probe for direct detection of the DNA of interest. In this case, a gene can be identified indirectly by cloning cDNA in a vector that allows the cloned cDNA to be transcribed and translated. A labeled antibody is used to identify which bacterial colony produces the protein and, therefore, contains the cDNA of interest.

V. SOUTHERN BLOTTING

Southern blotting is a technique that can detect mutations in DNA. It combines the use of restriction enzymes and DNA probes.

A. Experimental procedure

This method, named after its inventor, Edward Southern, involves the following steps (Figure 32.12). First, DNA is extracted from cells, for example, a patient's leukocytes. Second, the DNA is cleaved into many fragments using a restriction enzyme. Third, the resulting fragments are separated on the basis of size by electrophoresis. [Note: The large fragments move more slowly than the smaller fragments. Therefore, the lengths of the fragments, usually expressed as the number of base pairs, can be calculated from comparison of the position of the band relative to standard fragments of known size.] The DNA fragments in the gel are denatured and transferred to a nitrocellulose membrane for analysis. If the original DNA represents the individual's entire genome, the enzymic digest contains a million or more fragments. The gene of interest is on only one (or a few if the gene itself was fragmented) of these pieces of DNA. If all the DNA segments were visualized by a nonspecific technique, they would appear as an unresolved blur of overlapping bands. To avoid this, the last step in Southern blotting uses a probe to identify the DNA fragments of interest. The patterns observed on Southern blot analysis depend both on the specific restriction endonuclease and on the probe used to visualize the restriction fragments. [Note: Variants of the Southern blot have been facetiously named "northern" (electrophoresis of mRNA followed by hybridization with a specific probe), and "western" (electrophoresis of protein followed by detection with an antibody directed against the protein of interest), neither of which relate to anyone's name or to points of the compass.]

B. Detection of mutations

The presence of a mutation affecting a restriction site causes the pattern of bands to differ from those seen with a normal gene. For

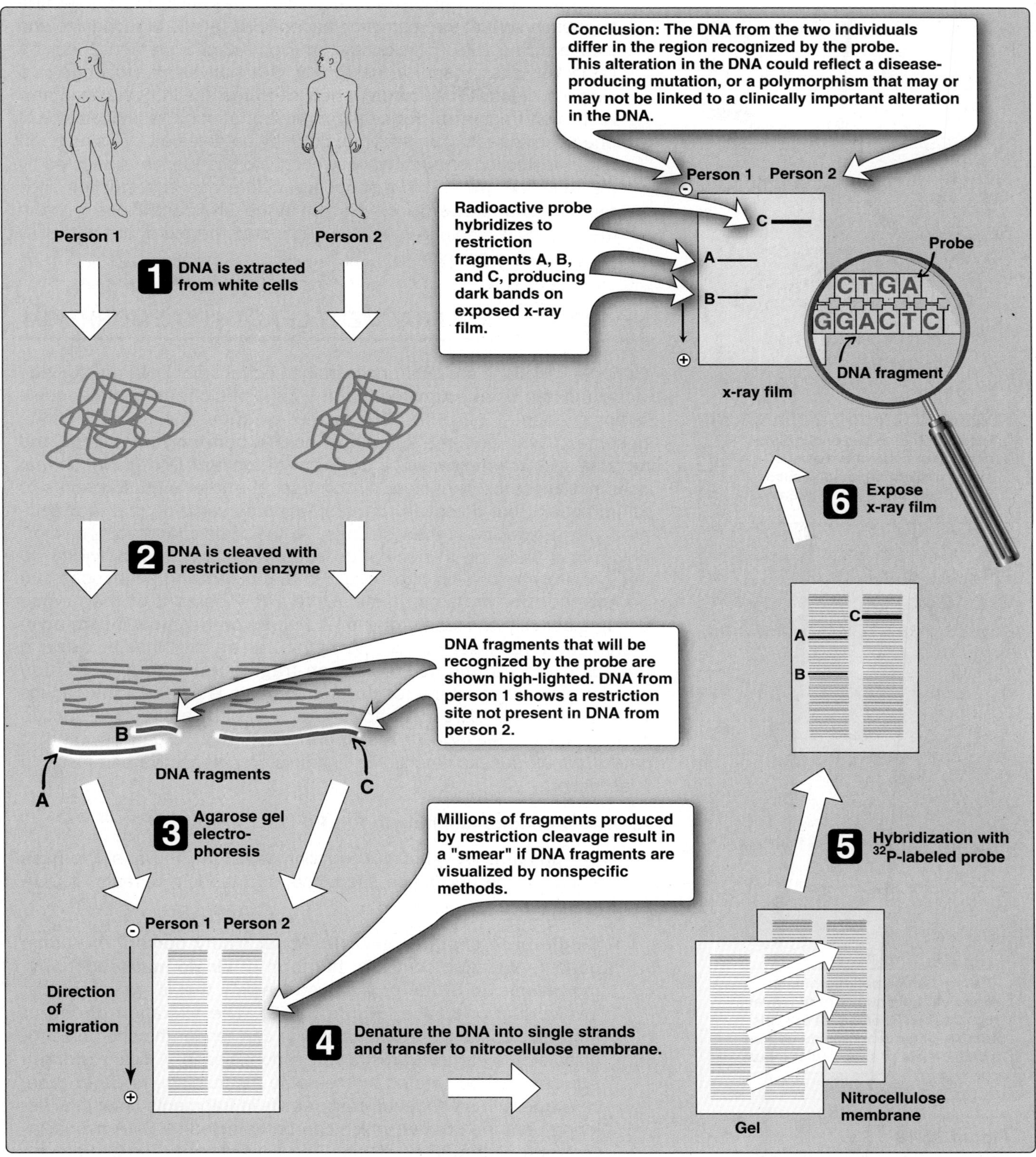

Figure 33.12
Southern blotting procedure.

example, a change in one nucleotide may alter the nucleotide sequence so that the restriction endonuclease fails to recognize and cleave at that site (for example, in Figure 32.12, person 2 lacks a restriction site present in person 1). Alternatively, the change in a single nucleotide may create a new cleavage site that results in new restriction fragments. A mutation may not affect a restriction site of one specific restriction enzyme, but may be revealed by using a different restriction enzyme whose recognition sequence is affected by the mutation. [Note: Most sequence differences at restriction sites represent normal variations present in the DNA, and those found in the noncoding regions are often silent and, therefore, are generally not clinically significant.]

VI. RESTRICTION FRAGMENT LENGTH POLYMORPHISM

Genome variations are differences in the sequence of DNA among individuals. It has been estimated that the genomes of non-related people differ at about 1 of 1500 DNA bases, or about 0.1 percent of the genome. These genome variations include both polymorphisms and mutations. A **polymorphism** is a clinically harmless DNA variation that does not affect the phenotype. In contrast, the term **mutation** refers to an infrequent, but potentially harmful, genome variation that is associated with a specific human disease. At the molecular level, polymorphism is a variation in nucleotide sequence from one individual to another. Polymorphisms often occur in the intervening sequences that do not code for proteins. [Note: Only a few percent of the human genome actually encodes proteins.] A **restriction fragment length polymorphism** (**RFLP**) is a genetic variant that can be examined by cleaving the DNA into fragments (restriction fragments) with a restriction enzyme. The length of the restriction fragments is altered if the genetic variant alters the DNA so as to create or abolish a site of restriction endonuclease cleavage (a **restriction site**). RFLPs can be used to detect human genetic defects, for example, in prospective parents or in fetal tissue.

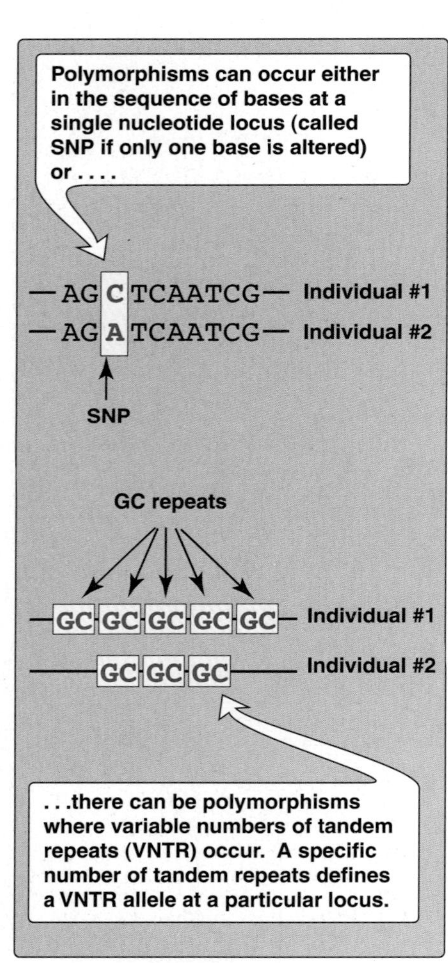

Figure 32.13
Common forms of genetic polymorphism. SNP = single nucleotide polymorphism.

A. DNA variations resulting in RFLPs

Two types of DNA variation commonly result in RFLPs: single base changes in the nucleotide sequence, and tandem repeats of DNA sequences.

1. **Single base changes in DNA:** About ninety percent of human genome variation comes in the form of **single nucleotide polymorphisms**, or **SNPs** (pronounced "snips"), that is, variations that involve just one base (Figure 32.13). The alteration of one or more nucleotides at a restriction site can render the site unrecognizable by a particular restriction endonuclease. A new restriction site can also be created by the same mechanism. In either case, cleavage with an endonuclease results in fragments of lengths differing from the normal, which can be detected by DNA hybridization (see Figure 32.12). [Note: The altered restriction site can be either at the site of a disease-causing mutation or at a site some distance from the mutation.]

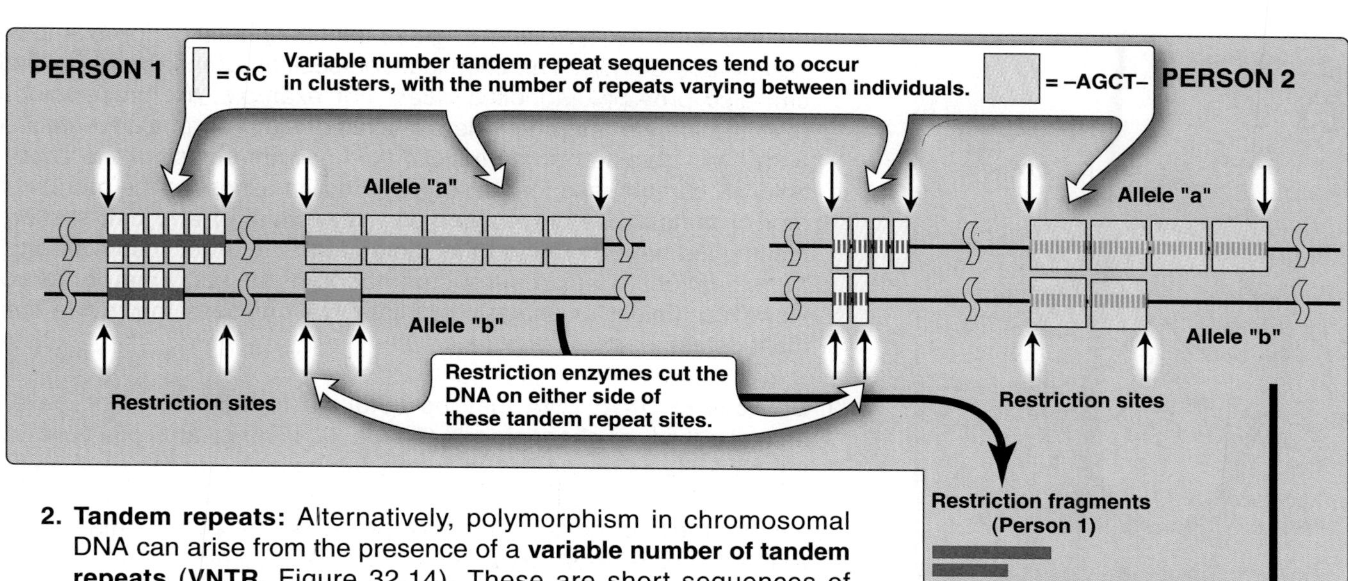

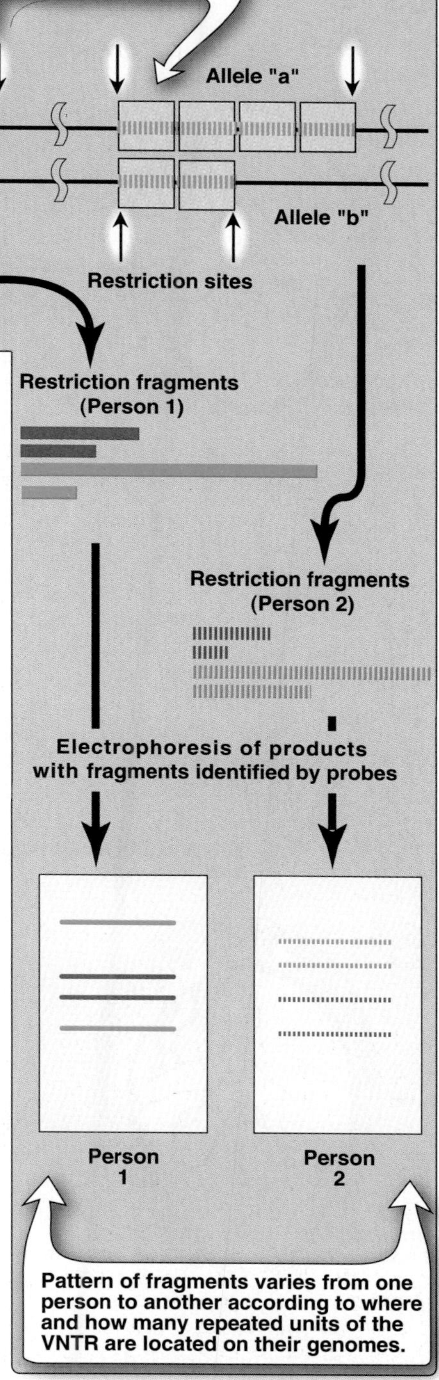

Figure 32.14
Restriction fragment length polymorphism (RFLP) of variable number tandem repeats (VNTR). For each person, a pair of homologous chromosomes is shown.

2. **Tandem repeats:** Alternatively, polymorphism in chromosomal DNA can arise from the presence of a **variable number of tandem repeats** (**VNTR**, Figure 32.14). These are short sequences of DNA at scattered locations in the genome, repeated in tandem (like freight cars of a train). The number of these repeat units varies from person to person, but is unique for any given individual and, therefore, serves as a molecular fingerprint. Cleavage by restriction enzymes yields fragments that vary in length depending on how many repeated segments are contained in the fragment. Variation in the number of tandem repeats can lead to polymorphisms. Many different VNTR loci have been identified, and are extremely useful for DNA fingerprint analysis, such as in forensic and paternity identity cases. It is important to emphasize that these polymorphisms, whether SNPs or VNTRs, are simply markers, and which, in most cases, have no known effect on the structure or rate of production of any particular protein.

B. Tracing chromosomes from parent to offspring

If the DNA of an individual has gained a restriction site by base substitution, then enzymc cleavage yields at least one additional fragment. Conversely, if a mutation results in loss of a restriction site, fewer fragments are produced by enzymic cleavage. An individual who is heterozygous for a polymorphism has a sequence variation in the DNA of one chromosome, and not in the DNA of the companion chromosome. In such individuals, each chromosome can be traced from parent to offspring by determining the presence or absence of the polymorphism.

C. Prenatal diagnosis

Families with a history of severe genetic disease, such as an affected previous child or near relative, may wish to determine the presence of the disorder in a developing fetus. Prenatal diagnosis allows for an informed reproductive choice if the fetus is affected.

1. **Methods available:** The available diagnostic methods vary in sensitivity and specificity. Visualization of the fetus, for example, by ultrasound or fiberoptic devices (fetoscopy), is useful only if the

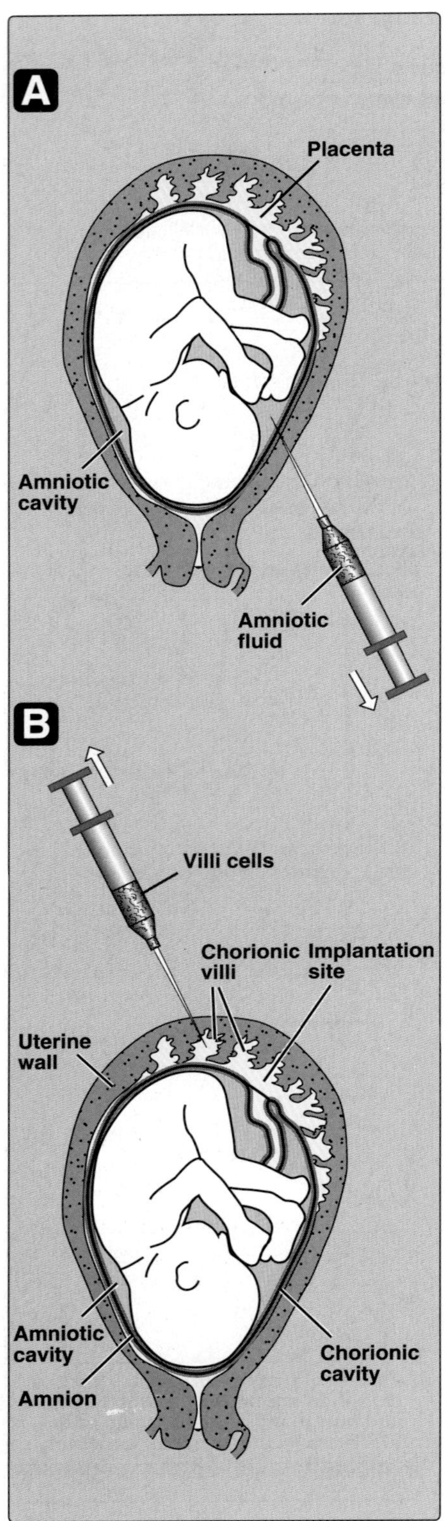

Figure 32.15
Sampling of fetal cells. A. Amniotic fluid; B. Chorionic villus.

genetic abnormality results in gross anatomic defects, for example, neural tube defects. The chemical composition of the amniotic fluid can also provide diagnostic clues. For example, the presence of high levels of α-fetoprotein is associated with neural tube defects. Fetal cells obtained from amniotic fluid or from biopsy of the chorionic villi can be used for **karyotyping**, which assesses the morphology of metaphase chromosomes. New staining and cell sorting techniques have permitted the rapid identification of trisomies and translocations that produce chromosomes of abnormal lengths. However, molecular analysis of fetal DNA promises to provide the most detailed genetic picture.

2. **Sources of DNA:** DNA may be obtained from white blood cells, amniotic fluid, or chorionic villi (Figure 32.15). For amniotic fluid, in the past, it was necessary to culture the cells in order to have sufficient DNA for analysis. This took two to three weeks to grow a sufficient number of cells. The development of the polymerase chain reaction (PCR, see below) has dramatically shortened the time needed for a DNA analysis.

3. **Direct diagnosis of sickle cell disease using RFLPs:** The genetic disorders of hemoglobin are the most common genetic diseases in humans. At present, there is no satisfactory treatment for most of these disorders, and prenatal diagnosis is the only available method for limiting the number of afflicted individuals. In the case of sickle cell disease (Figure 32.16), the mutation that gives rise to the disease is actually one and the same as the mutation that gives rise to the polymorphism. Direct detection by RFLPs of diseases that result from point mutations is at present limited to only a few genetic diseases.

   a. **Early efforts to diagnose sickle cell anemia:** Prenatal diagnosis of hemoglobinopathies has in the past involved the determination of the amount and kinds of hemoglobin synthesized in red cells obtained from fetal blood. For example, the presence of hemoglobin S in hemolysates indicated sickle cell anemia. However, the invasive procedures to obtain fetal blood have a high mortality rate (approximately five percent), and diagnosis cannot be carried out until late in the second trimester of pregnancy when HbS begins to be produced (see p. 35).

   b. **RFLP analysis:** Sickle cell anemia is an example of a genetic disease caused by a point mutation (see p. 36). The sequence altered by the mutation abolishes the recognition site of the restriction endonuclease *Mst*II that recognizes the nucleotide sequence CCTNAGG (where N is any nucleotide, see Figure 32.16). Thus, the A to T mutation within a codon of the β^S-globin gene eliminates a cleavage site for the enzyme. Normal DNA digested with *Mst*II yields a 1.15 kb fragment, whereas a 1.35 kb fragment is generated from the β^S gene as a result of the loss of one *Mst*II cleavage site. Diagnostic techniques for analyzing fetal DNA (from amniotic cells or chorionic villus sampling), rather than fetal blood, have proved valuable because they provide safe, early detection of sickle cell anemia, as well as other genetic diseases.

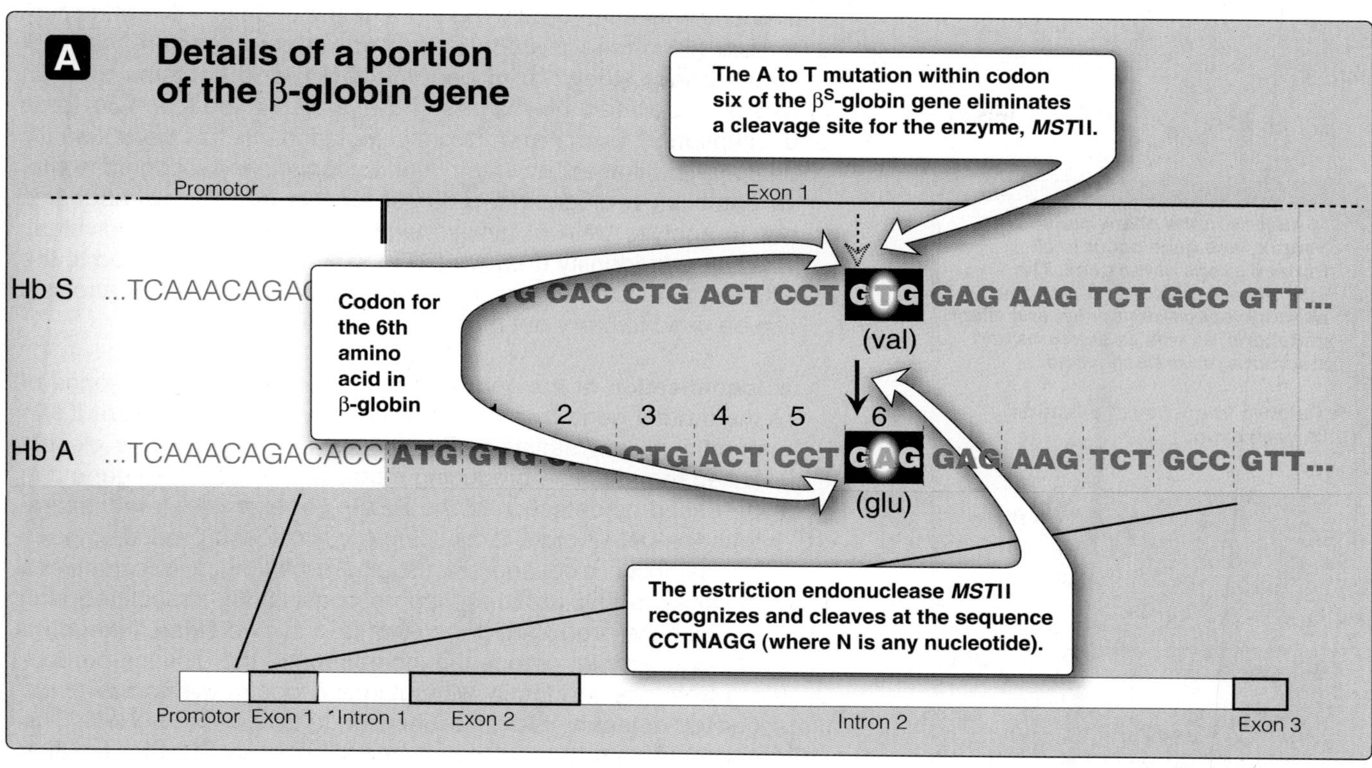

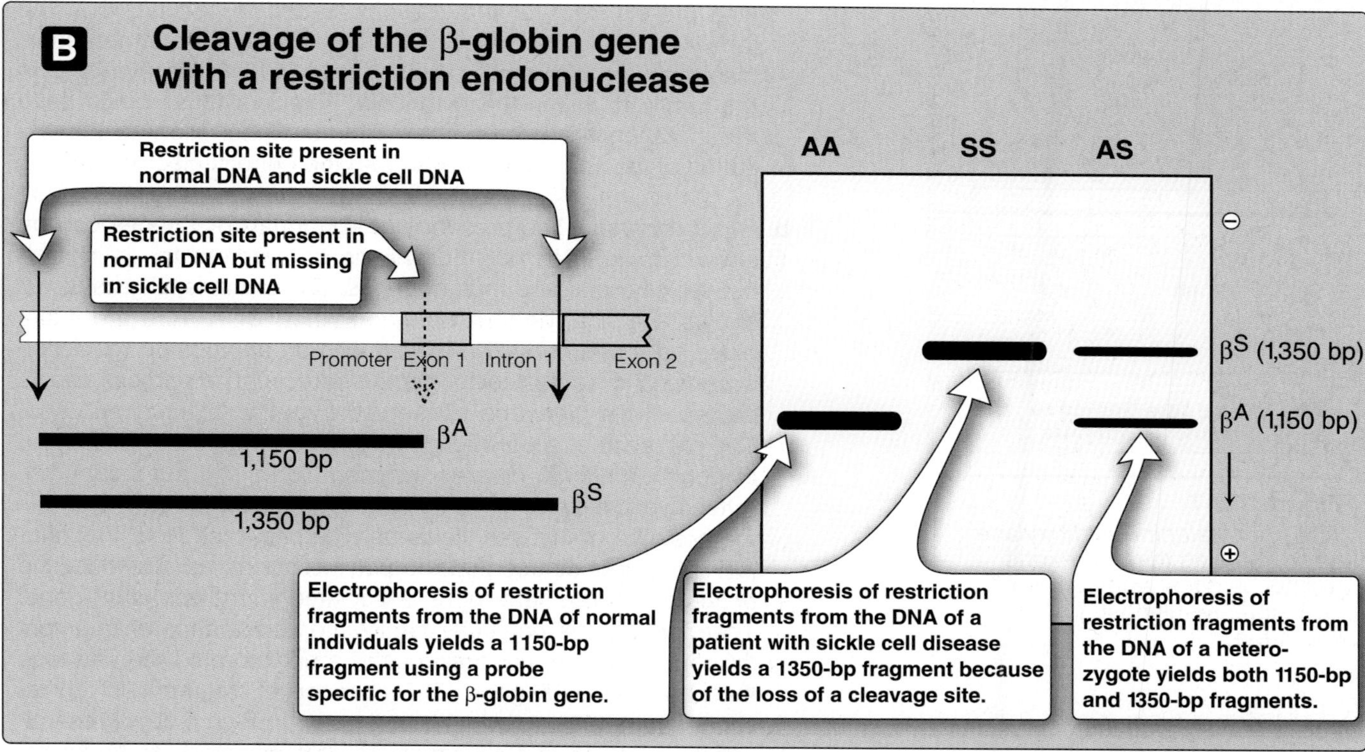

Figure 32.16
Detection of β^S-globin mutation. bp = Base pair.

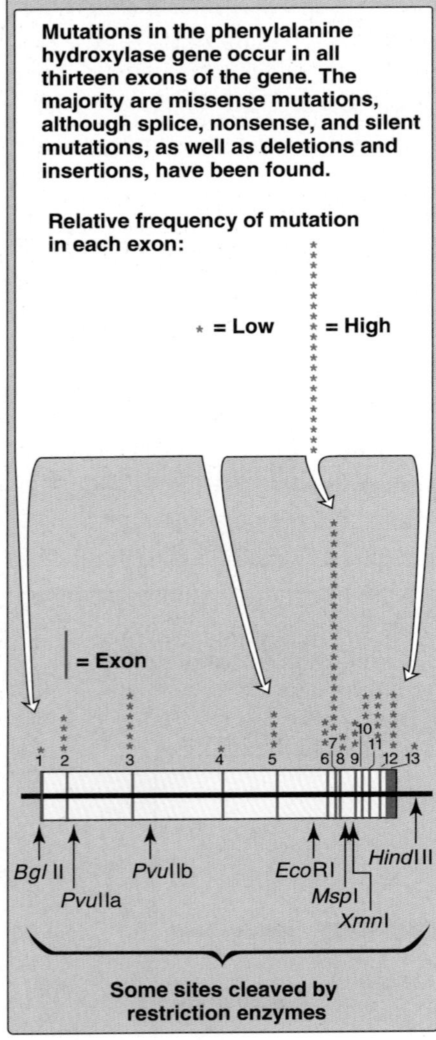

Figure 32.17
The *phenylalanine hydroxylase* gene showing thirteen exons, restriction sites, and some of the mutations resulting in phenylketonuria.

4. Indirect, prenatal diagnosis of phenylketonuria using RFLPs: The *phenylalanine hydroxylase* (*PAH*) gene, deficient in phenylketonuria (PKU, see p. 268), is located on chromosome twelve. It spans about ninety kb of genomic DNA, and contains thirteen exons separated by introns (Figure 32.17; see p. 424 for a description of exons and introns). Mutations in this gene usually do not directly affect any restriction endonuclease recognition site. To establish a diagnostic protocol for this genetic disease, one has to analyze DNA of family members of the afflicted individual. The key is to identify markers (RFLPs) that are tightly linked to the disease trait. Once these markers are identified, RFLP analysis can be used to carry out prenatal diagnosis.

a. Identification of the gene: One can determine the presence of the mutant gene by identifying the polymorphism marker if two conditions are satisfied. First, if the polymorphism is closely linked to a disease-producing mutation, the defective gene can be traced by detection of the RFLP. For example, if one examines the DNA from a family carrying a disease-producing gene by restriction cleavage and Southern blotting, it is sometimes possible to find an RFLP that is consistently associated with the disease-producing gene (that is, it shows **close linkage**). It is then possible to trace the inheritance of the disease-producing DNA within a family without knowledge of the nature of the genetic defect or its precise location in the genome. [Note: The polymorphism may be known from the study of other families with the disorder, or may be discovered to be unique in the family under investigation.] Second, for autosomal recessive disorders, an affected individual would ideally be available in the family to aid in the diagnosis. This individual would have the mutation present on both chromosomes, allowing identification of the RFLP associated with the genetic disorder.

b. RFLP analysis: The presence of abnormal *PAH* genes can be shown using DNA polymorphisms as markers to distinguish between normal and mutant genes. For example, Figure 32.18 shows a typical pattern obtained when DNA from the white blood cells of a family is cleaved with an appropriate restriction enzyme and subjected to electrophoresis. The vertical arrows represent the cleavage sites for the restriction enzyme used. The presence of a polymorphic site (see p. 454 for a discussion of polymorphisms) creates fragment "b" in the autoradiogram (after hybridization with a labeled PAH-cDNA probe), whereas the absence of this site yields only fragment "a." Note that subject II-2 demonstrates that the polymorphism, as shown by the presence of fragment "b," is associated with the mutant gene. Therefore, in this particular family, the appearance of fragment "b" corresponds to the presence of a polymorphic site that tags the abnormal *PAH* gene. The absence of fragment "b" corresponds to having only the normal gene. In Figure 32.18, examination of fetal DNA shows that the fetus inherited two abnormal genes from its parents and, therefore, has PKU.

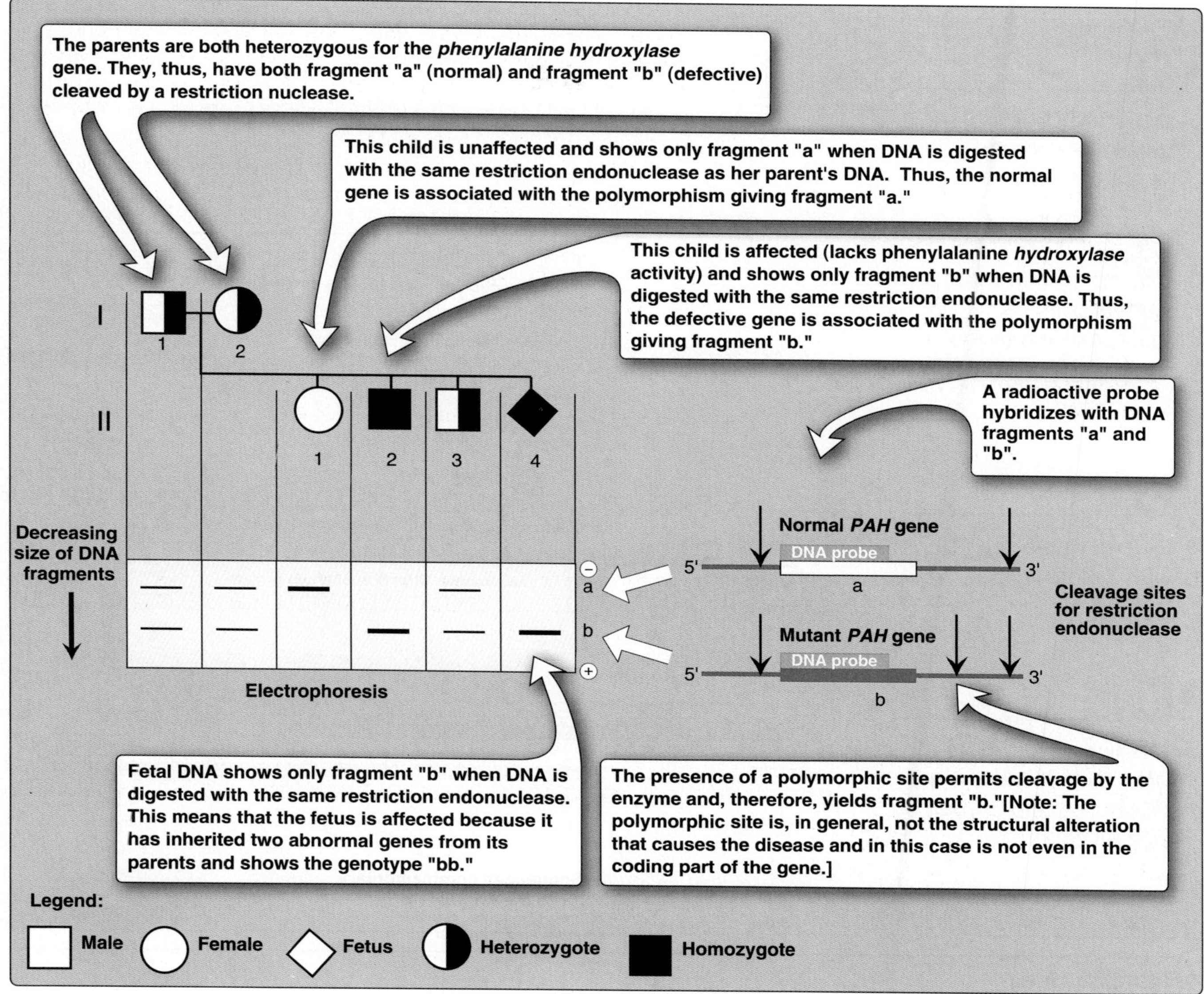

Figure 32.18
Analysis of restriction fragment length polymorphism in a family with a child affected by phenylketonuria. The molecular defect in the *phenylalanine hydroxylase* (*PAH*) gene in the family is not known. The family wanted to know if the current pregnancy would be affected by phenylketonuria.

c. **Value of screening:** DNA-based screening is useful not only in determining if an unborn fetus is affected, but also in detecting carriers of the PKU gene. PKU, like many inborn errors of amino acid metabolism, is inherited as an autosomal recessive trait. It is important to identify heterozygotes for future family planning.

VII. POLYMERASE CHAIN REACTION

The polymerase chain reaction (PCR) is a test tube method for amplifying a selected DNA sequence that does not rely on the biologic cloning method described on p. 446. PCR permits the synthesis of millions of

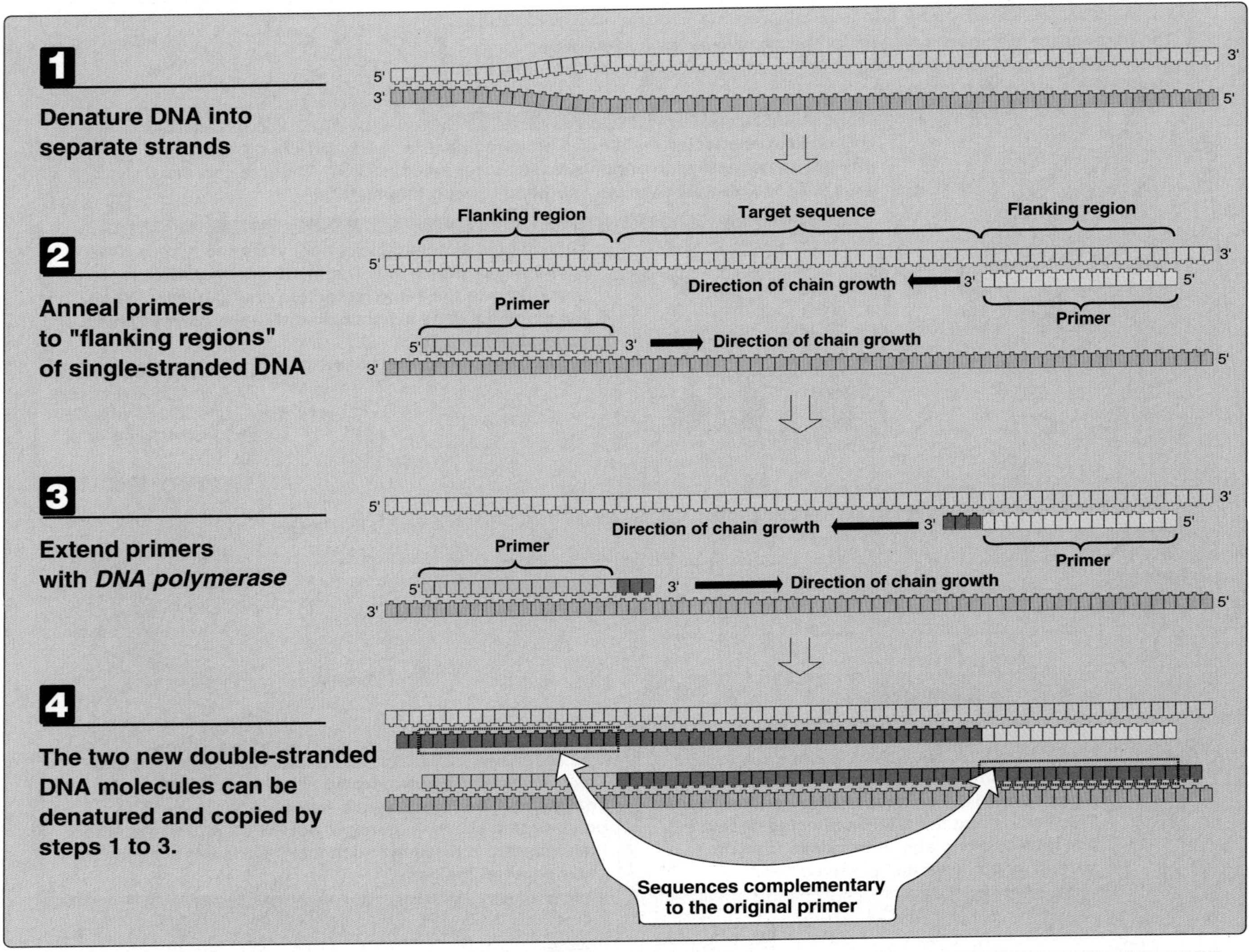

Figure 32.19
Steps in one cycle of the polymerase chain reaction.

copies of a specific nucleotide sequence in a few hours. It can amplify the sequence, even when the targeted sequence makes up less than one part in a million of the total initial sample. The method can be used to amplify DNA sequences from any source—bacterial, viral, plant, or animal. The steps in PCR are summarized in Figures 32.19 and 32.20.

A. Steps of a PCR

PCR uses *DNA polymerase* to repetitively amplify targeted portions of DNA. Each cycle of amplification doubles the amount of DNA in the sample, leading to an exponential increase in DNA with repeated cycles of amplification. The amplified DNA sequence can then be analyzed by gel electrophoresis, Southern hybridization, or direct sequence determination.

1. **Primer construction:** It is not necessary to know the nucleotide sequence of the target DNA in the PCR method. However, it is necessary to know the nucleotide sequence of short segments on each side of the target DNA. These stretches, called **flanking sequences**, bracket the DNA sequence of interest. The nucleotide sequences of the flanking regions are used to construct two, single-stranded oligonucleotides, usually 20 to 35 nucleotides long, which are complementary to the respective flanking sequences. The 3'-hydroxyl end of each primer points toward the target sequence (see Figure 32.19). These synthetic oligonucleotides function as primers in PCR reactions.

2. **Denature the DNA:** The DNA to be amplified is heated to separate the double-stranded target DNA into single strands.

3. **Annealing of primers to single-stranded DNA:** The separated strands are cooled and allowed to anneal to the two primers (one for each strand).

4. **Chain extension:** *DNA polymerase* and deoxyribonucleoside triphosphates (in excess) are added to the mixture to initiate the synthesis of two new chains complementary to the original DNA chains. *DNA polymerase* adds nucleotides to the 3'-hydroxyl end of the primer, and strand growth extends across the target DNA, making complementary copies of the target. [Note: PCR products can be several thousand base pairs long.] At the completion of one cycle of replication, the reaction mixture is heated again to denature the DNA strands (of which there are now four). Each DNA strand binds a complementary primer, and the cycle of chain extension is repeated. By using a heat-stable DNA polymerase (for example, *Taq polymerase*) from a bacterium that normally lives at high temperatures (a thermophilic bacterium), the polymerase is not denatured and, therefore, does not have to be added at each successive cycle. Typically twenty to thirty cycles are run during this process, amplifying the DNA by a million-fold to a billion-fold. [Note: Each extension product of the primer includes a sequence complementary to the primer at the 5' end of the target sequence (see Figure 32.19). Thus, each newly synthesized polynucleotide can act as a template for the successive cycles (see Figure 32.20). This leads to an exponential increase in the amount of target DNA with each cycle, hence, the name "polymerase chain reaction."]

B. Advantages of PCR

The major advantages of PCR over cloning as a mechanism for amplifying a specific DNA sequence are sensitivity and speed. DNA sequences present in only trace amounts can be amplified to become the predominant sequence. PCR is so sensitive that DNA sequences present in an individual cell can be amplified and studied. Isolating and amplifying a specific DNA sequence by PCR is faster and less technically difficult than traditional cloning methods using recombinant DNA techniques.

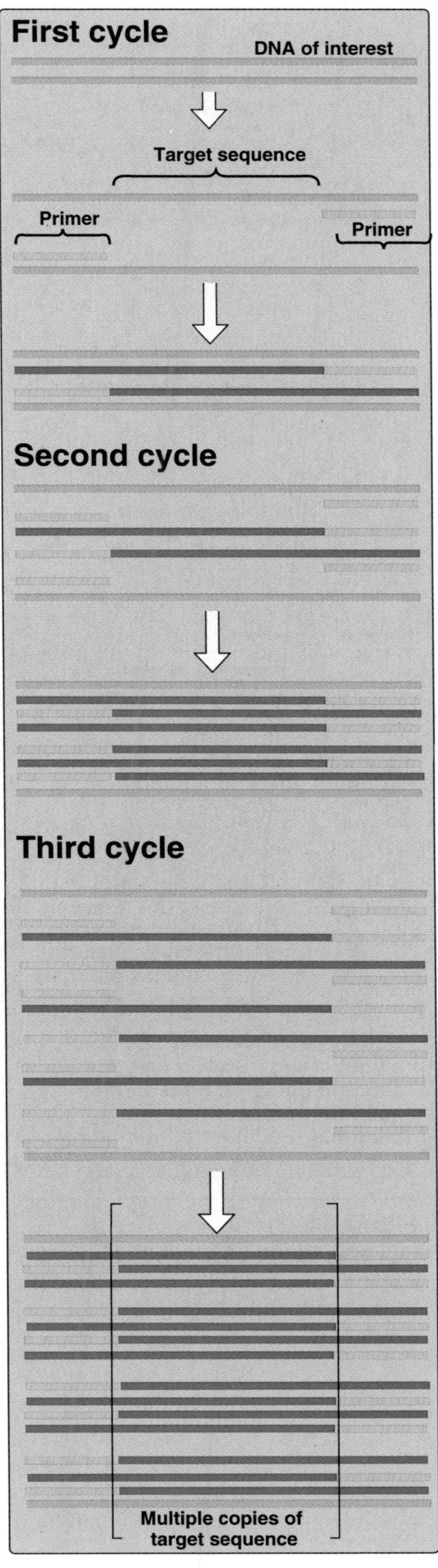

Figure 32.20
Multiple cycles of polymerase chain reaction.

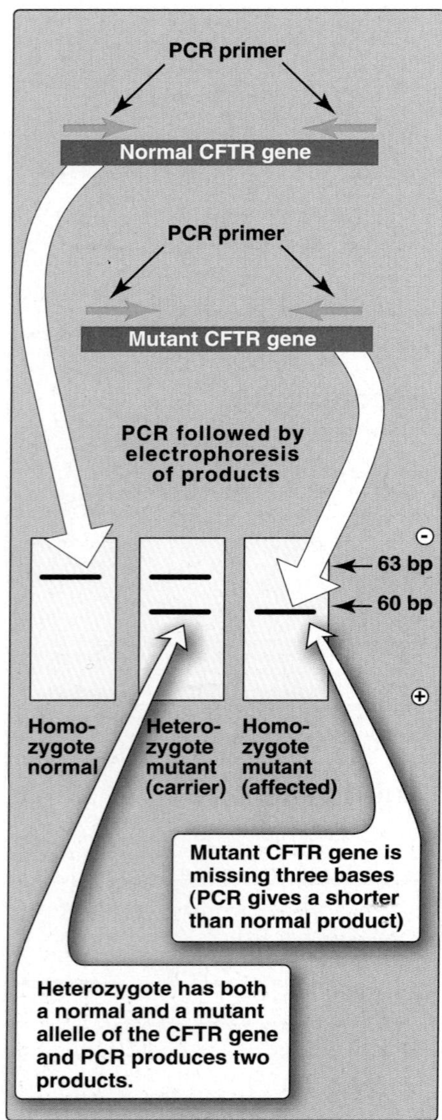

Figure 32.21
Genetic testing for cystic fibrosis using PCR. CFTR = cystic fibrosis transmembrane regulator.

C. Applications

PCR has become a very common tool for a large number of applications. These include:

1. **Comparison of a normal cloned gene with an uncloned mutant form of the gene:** PCR allows the synthesis of mutant DNA in sufficient quantities for a sequencing protocol without laboriously cloning the altered DNA.

2. **Detection of low–abundance nucleic acid sequences:** For example, viruses that have a long latency period, such as HIV, are difficult to detect at the early stage of infection using conventional methods. PCR offers a rapid and sensitive method for detecting viral DNA sequences even when only a small proportion of cells is harboring the virus.

3. **Forensic analysis of DNA samples:** DNA fingerprinting by means of PCR has revolutionized the analysis of evidence from crime scenes. DNA isolated from a single human hair, a tiny spot of blood, or a sample of semen is sufficient to determine whether the sample comes from a specific individual. The DNA markers analyzed for such fingerprinting are most commonly **short tandem repeat polymorphisms** (**STRs**). These are very similar to the VNTRs described previously (see p. 455), but are smaller in size. [Note: Verification of paternity uses the same techniques.]

4. **Prenatal diagnosis and carrier detection of cystic fibrosis:** Cystic fibrosis is an autosomal recessive genetic disease resulting from mutations in the **cystic fibrosis transmembrane regulator** (**CFTR**) gene. The most common mutation is a three-base deletion that results in the loss of a phenylalanine residue from the CFTR protein. Because the mutant allele is three bases shorter than the normal allele, it is possible to distinguish them from each other by the size of the PCR products obtained by amplifying that portion of the DNA. Figure 32.21 illustrates how the results of such a PCR test can distinguish between homozygous normal, heterozygous (carriers), and homozygous mutant (affected) individuals.

VIII. ANALYSIS OF GENE EXPRESSION

The tools of biotechnology not only allow the study of gene structure, but also provide ways of analyzing the products of gene expression—mRNA and proteins.

A. Determination of mRNA levels

Messenger RNA levels are usually determined by the hybridization of labeled probes to either mRNA itself or to cDNA produced from mRNA.

1. **Northern blots:** Northern blots are very similar to Southern blots (see Figure 32.12, p. 453), except that the original sample contains a mixture of **mRNA molecules** that are separated by electrophoresis, then transferred to a membrane and hybridized to a

radioactive probe. The bands obtained by autoradiography give a measure of the amount and size of particular mRNA molecules in the sample.

2. **Microarrays:** DNA microarrays contain thousands of immobilized DNA sequences organized in an area no larger than a microscope slide. These microarrays are used to analyze a sample for the presence of gene variations or mutations (**genotyping**), or to determine the patterns of mRNA production (**gene expression anlaysis**), analyzing thousands of genes at the same time. For genotyping analysis, the cellular sample is genomic DNA. For expression analysis, the population of mRNA molecules from a particular cell type is converted to cDNA and labeled with a fluorescent tag (Figure 32.22). This mixture is then exposed to a gene chip, which is a glass slide or membrane containing thousands of tiny spots of DNA, each corresponding to a different gene. The amount of fluorescence bound to each spot is a measure of the amount of that particular mRNA in the sample. DNA microarrays are often used to determine the differing patterns of gene expression in two different types of cell—for example, normal and cancer cells (see Figure 32.22). [Note: Physicians hope to one day be able to tailor particular treatment regimens to each cancer patient, based on the specific microarray expression patterns exhibited by that patient's individual tumor.]

B. Analysis of proteins

The kinds and amounts of proteins in cells do not always directly correspond to the amounts of mRNA present. Some mRNAs are translated more efficiently than others, and some proteins undergo posttranslational modifications by adding sugars or lipids, or both. Thus, the genome contains about 40,000 genes, but a typical cell produces hundreds of thousands of distinct proteins. When investigating one, or a limited number of gene products, it is convenient to use labeled antibodies to detect and quantify specific proteins. However, when analyzing the abundance and interactions of large numbers of cellular proteins (called **proteomics**, see below), automated methods employing two-dimensional gel electrophoresis, mass spectrometry, multidimensional liquid chromatography, and bioinformatics are employed.

1. **Enzyme-Linked Immunosorbent Assays (ELISAs):** These assays are performed in the wells of a plastic microtiter dish. The antigen (protein) is bound to the plastic of the dish. The probe used consists of an antibody specific for the particular protein to be measured. The antibody is covalently bound to an ezyme, which will produce a colored product when exposed to its substrate. The amount of color produced can be used to determine the amount of protein (or antibody) in the sample to be tested.

2. **Western blots:** Western blots (also called **immunoblots**) are similar to Southern blots, except that protein molecules in the sample are separated by electrophoresis and blotted to a membrane. The probe is a labeled antibody, which produces a band at the location of its antigen.

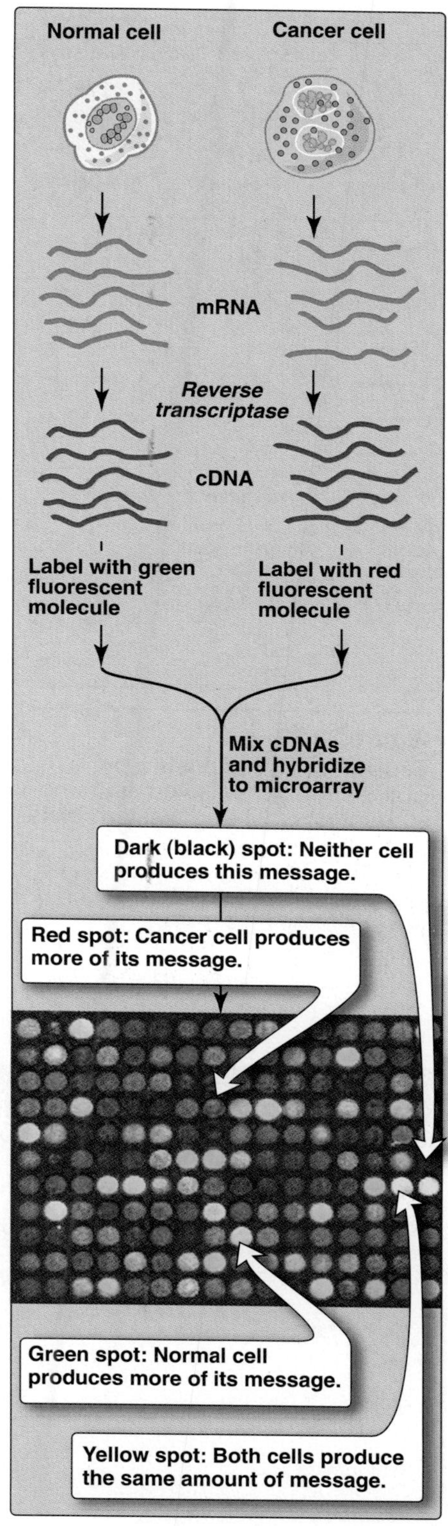

Figure 32.22
Microarray analysis of gene expression.

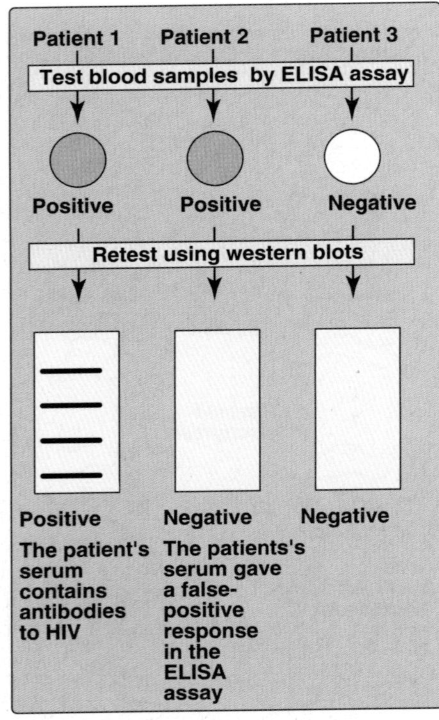

Figure 32.23
Testing for HIV exposure by ELISAs (enzyme–linked immunosorbent assays) and western blots.

3. **Detecting exposure to HIV:** ELISA assays and western blots are commonly used to detect exposure to HIV by measuring the amount of anti-HIV antibodies present in a patient's blood sample. ELISA assays are used as the primary screening tool, because they are very sensitive. These assays sometimes give false-positives, however, so western blots, which are more specific, are often used as a confirmatory test (Figure 32.23). [Note: ELISA and western blots can only detect HIV exposure after anti-HIV antibodies appear in the bloodstream. PCR-based testing for HIV is more useful in the first few months after exposure.]

4. **Proteomics** involves the study of all proteins expressed by a genome, including their relative abundance, distribution, posttranslational modifications, functions, and interactions with other macromolecules. The approximately 40,000 genes of the human genone translate into hundreds of thousands of proteins when alternate splicing and post-translational modifications are considered. While a genome remains unchanged, the amounts and types of proteins in any particular cell change dramatically as genes are turned on and off. Proteomics offers the potential of identifying new disease markers and drug targets. Figure 32.24 compares some of the analytic techniques discussed in this chapter.

IX. GENE THERAPY

The goal of gene therapy is to insert the normal, cloned DNA for a gene into the somatic cells of a patient who is defective in that gene as a result of some disease-causing mutation. The DNA must become permanently integrated into the patient's chromosomes in such a way as to be properly expressed to produce the correct protein. For example, patients with **severe combined immunodeficiency disease** (**SCID**) have an immune deficiency as a result of mutations in either the *adenosine deaminase* gene (p. 299) or a gene coding for an interleukin receptor subunit (**X-linked severe combined immunodeficiency**, **SCID-X1**).

TECHNIQUE	SAMPLE ANALYZED	GEL USED	PURPOSE
Southern blot	DNA	Yes	Detects DNA changes
Northern blot	RNA	Yes	Measures mRNA amounts
Western blot	Protein	Yes	Measures protein amounts
ASO	DNA	No	Detects DNA mutations
Microarray	RNA or cDNA	No	Measures many mRNA levels at once
ELISA	Proteins or antibodies	No	Detects proteins (antigens) or antibodies
Proteomics	Proteins	Yes	Measures abundance, distribution, posttranslational modifications, functions, and interactions of cellular proteins

Figure 32.24
Techniques used to analyze DNA, RNA, and proteins. ASO = allele-specific oligonucleotides.
ELISA = enzyme-linked immunosorbent assays.

Patients with both kinds of SCIDs have been successfully treated by incorporating functional copies of the appropriate gene into their cells (Figure 32.25). [Note: This is often called "**gene replacement therapy**."] Although gene therapy is an attractive therapeutic strategy for individuals with inherited diseases, the method is not without risks. For example, retrovirus-mediated gene transfer was able to correct X-linked severe combined immunodeficiency (SCID-X1) in nine of ten patients. However, leukemias developed in several of the patients, presumably because of activation of a hematopoietic oncogene. Clearly, gene therapy is a work in progress.

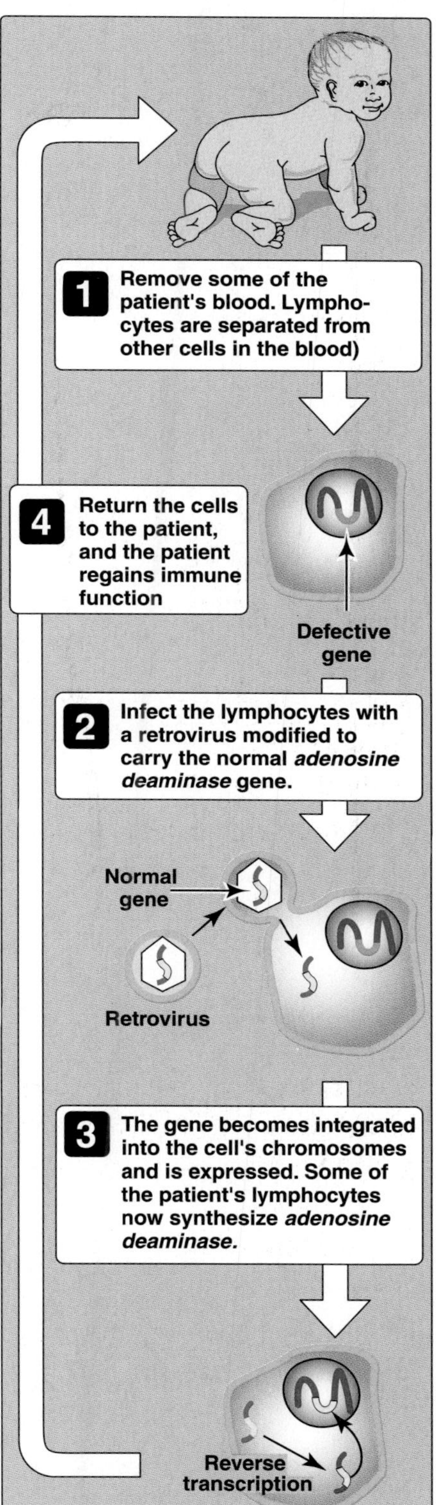

Figure 32.25
Gene therapy for a patient with severe combined immunodeficiency caused by *adenosine deaminase* deficiency.

X. TRANSGENIC ANIMALS

Transgenic animals can be produced by injecting a cloned gene into the fertilized egg. If the gene becomes successfully integrated into a chromosome, it will be present in the germline of the resulting animal, and can be passed along from generation to generation. A giant mouse called "Supermouse" was produced in this way by injecting the gene for rat growth hormone into a fertilized mouse egg. In a similar way, transgenic goats and cows can now be designed that produce human hormones in their milk. Sometimes, rather than introducing a functional gene into a transgenic mouse, a mutant gene is used to replace the normal copies of that gene in the cells of the mouse. This can be used to produce a colony of "**knockout mice**" that are deficient in a particular enzyme. Such animals can then serve as models for the study of a corresponding human disease. For example, transgenic mice carrying mutant copies of the dystrophin gene serve as animal models for the study of muscular dystrophy.

XI. CHAPTER SUMMARY

Restriction endonucleases are bacterial enzymes that cleave double-stranded DNA into smaller fragments. Each enzyme cleaves DNA at a specific four to six base-long nucleotide sequence, producing DNA segments called **restriction fragments**. The sequences that are recognized are **palindromic**. These enzymes form either **staggered cuts** (**sticky ends**) or **blunt end cuts** on the DNA. A DNA sequence that is recognized by a restriction enzyme is called a **restriction site**. **Bacterial DNA ligases** can anneal two DNA fragments from different sources if they have been cut by the same restriction endonuclease. This hybrid combination of two fragments is called a **recombinant DNA molecule**. Introduction of a foreign DNA molecule into a replicating cell permits the **amplification** (production of many copies) of the DNA—a process called **cloning**. A **vector** is a molecule of DNA to which the fragment of DNA to be joined is attached. Vectors must be capable of **autonomous replication** within the host cell, and must contain at least one specific nucleotide sequence recognized by a restriction endonuclease. It must also carry at least one gene that confers the ability to select for the vector, such as an **antibiotic resistance gene**. Prokaryotic organisms normally contain small, circular, extrachromosomal DNA molecules called **plasmids** that can serve as **vectors**. They can be readily isolated from the bacterium, annealed with the DNA of interest, and reintroduced into the bacterium which will replicate, thus making multiple copies of the **hybrid plasmid**. A **DNA library** is a collection of cloned restriction fragments of the DNA of an organism. **A genomic library** is a collection of fragments of double-stranded DNA obtained by digestion of the total DNA of the organism with a restriction endonuclease and subsequent ligation to an appropriate vector. It ideally contains a copy of every DNA nucleotide sequence in the genome. In contrast, **cDNA (complementary DNA) libraries** contain only those DNA sequences that are complementary to mRNA molecules present in a cell, and differ from one cell type to another. Because cDNA has no intervening sequences, it can be cloned into an **expression vector** for the synthesis of eukaryotic proteins by bacteria. Cloned, then purified, fragments of DNA can be sequenced, for example, using the **Sanger dideoxy method**. A **probe** is a single-stranded piece of DNA (usually labeled with a radioisotope, such as ^{32}P, or another recognizable compound, such as biotin) which has a nucleotide sequence complementary to the DNA molecule of interest (**target DNA**). Probes can be used to identify which clone of a library or which band on a gel contains the target DNA. **Southern blotting** is a technique that can be used to detect specific genes present in DNA. The DNA is cleaved using a **restriction endonuclease**, the pieces are separated by **gel electrophoresis**, and then are denatured and transferred to a **nitrocellulose membrane** for analysis. The fragment of interest is detected using a **probe**. The human genome contains may thousands of **polymorphisms** that do not affect the structure or function of the individual. A **polymorphic gene** is one in which the variant alleles are common enough to be useful as genetic markers. A **restriction fragment length polymorphism** (**RFLP**) is a genetic variant that can be examined by cleaving the DNA into **restriction fragments** using a **restriction enzyme**. A **base substitution** of one or more nucleotides at a restriction site can render the site unrecognizable by a particular restriction endonuclease. A new restriction site also can be created by the same mechanism. In either case, cleavage with endonuclease results in fragments of lengths differing from the normal that can be detected by DNA hybridization. This technique can be used to diagnose genetic diseases early in the gestation of a fetus. The **polymerase chain reaction** (**PCR**), a test tube method for **amplifying** a selected DNA sequence, does not rely on the biologic cloning method. PCR permits the synthesis of millions of copies of a specific nucleotide sequence in a few hours. It can amplify the sequence, even when the targeted sequence makes up less than one part in a million of the total initial sample. The method can be used to amplify DNA sequences from any source. **Applications of the PCR technique** include: 1) efficient comparison of a normal cloned gene with an uncloned mutant form of the gene, 2) detection of low-abundance nucleic acid sequences, 3) forensic analysis of DNA samples, and 4) prenatal diagnosis and carrier detection, for example, of cystic fibrosis. The **products of gene expression** (mRNA and proteins) can be measured by techniques such as the following. **Northern blots** are very similar to Southern blots except that the original sample contains a mixture of **mRNA** molecules that are separated by electrophoresis, then hybridized to a radioactive probe. **Microarrays** are used to determine the differing patterns of gene expression in two different types of cells—for example, normal and cancer cells. **Enzyme-linked immunosorbent assays** (**ELISAs**) and **western blots** (**immunoblots**) are used to detect specific proteins.

Study Questions

Choose the ONE correct answer

32.1 *Hind*III is a restriction endonuclease commonly used to cut human DNA into pieces before inserting it into a plasmid. Which of the following is most likely to be the recognition sequence for this enzyme?

A. AAGGAA
B. AAGAAG
C. AAGTTC
D. AAGCTT
E. AAGAGA

Correct answer = D. The vast majority of restriction endonucleases recognize palindromes, and AAGCTT is the only palindrome among the choices. Because the sequence of only one DNA strand is given, one must determine the base sequence of the complementary strand. To be a palindrome, both strands must have the same sequence when read in the 5'→3' direction. Thus, the complement of 5'AAGCTT3' is also 5'AAGCTT3'.

32.2 An Ashkenazi Jewish couple brings their six-month-old son to you for evaluation of listlessness, poor head control, and a fixed gaze. You determine that he has Tay-Sachs disease, an autosomal recessive disorder. The couple also has a daughter. The diagram below shows this family's pedigree, along with Southern blots of an RFLP very closely linked to the hexosaminidase A gene. Which of the statements below is most accurate with respect to the daughter?

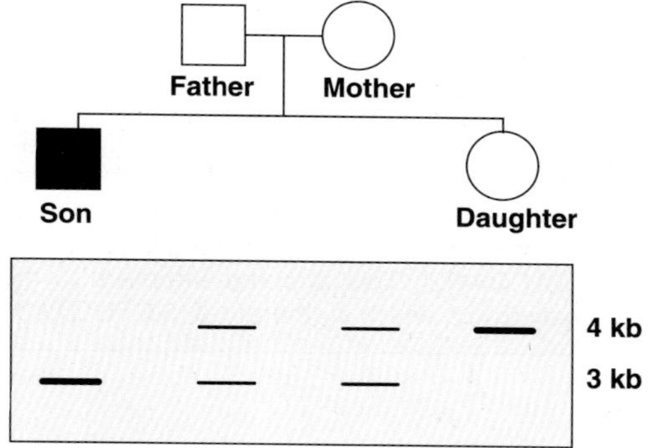

A. She has a 25 percent chance of having Tay-Sachs disease.
B. She has a 50 percent chance of having Tay-Sachs disease.
C. She has Tay-Sachs disease.
D. She is a carrier for Tay-Sachs disease.
E. She is homozygous normal.

Correct answer = E. Both the father and mother must be carriers for this disease. The son must have inherited a mutant allele from each parent. Because he shows only the 3 kb band on the Southern blot, the mutant allele for this disease must be linked to the 3 kb band for both parents. The normal allele must be linked to the 4 kb band in both parents. Because the daughter inherited the 4 kb band from both parents, she must be homozygous normal for the hexosaminidase A gene.

32.3 A physician would like to determine the global patterns of gene expression in two different types of tumor cells in order to develop the most appropriate form of chemotherapy for each patient. Which of the following techniques would be most appropriate for this purpose?

A. Southern blot
B. Northern blot
C. Western blot
D. ELISA
E. Microarray

Correct answer = E. Microarray analysis allows the determination of mRNA production (thus, gene expression) from thousands of genes at once. A northern blot only measures mRNA production from one gene at a time. Western blots and ELISAs measure protein production (also gene expression), but only from one gene at a time. Southern blots are used to analyze DNA, not gene expression.

32.4 A pharmaceutical company wants to produce a transgenic goat that will secrete human growth hormone into its milk. Which of the following tests would be most appropriate to apply to milk samples in order to identify a goat meeting this requirement?

A. ELISA
B. Northern blot
C. Southern blot
D. Dot blot using allele specific oligonucleotide probes
E. RFLP analysis

Correct answer = A. ELISA assays are commonly used to detect production of proteins. Northern blots could detect mRNA production, but would not guarantee that the protein was made. Southern blots, ASO probes, and RFLP analysis could be used to study the DNA inserted into the goat, but would not give information about protein production.

32.5 A researcher has cloned and isolated a small fragment of single-stranded DNA. She sequences this fragment using the Sanger dideoxy method. The results from her sequencing gel are shown below. What is the sequence of her original single-stranded fragment?

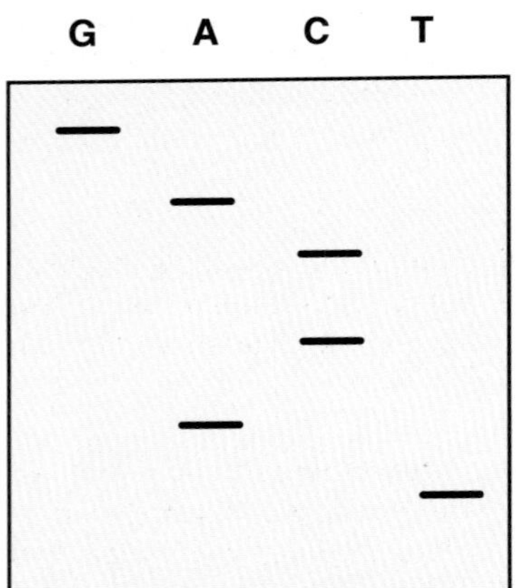

A. TACCAG
B. CTGGTA
C. ATGGTC
D. GACCAT
E. TCCAAG

Correct answer = B. The sequence of the new strand of DNA synthesized using the Sanger dideoxy method may be determined by reading the bands from the bottom to the top of the gel as 5'TACCAG3'. This is complementary to the original strand, which is, therefore, 5'CTGGTA3'.

Summary of Key Biochemical Facts

33

Chapter 1: Amino Acids

Overview of amino acids:	AMINO ACID STRUCTURE (p. 1)
• Number of common amino acids	• **Twenty common amino acids** are found in human proteins (coded for by DNA).
• Structural characteristics of amino acids	• Every amino acid has a **carboxyl group**, and every one except proline has an **amino group** (proline has an imino group) attached to the **α-carbon**. Both groups are charged at physiologic pH ($-COO^-$ and $-NH_3^+$).
• Types of side chains • Amino acids named according to side chain	• Individual amino acids have distinctive **side chains** (R-groups) that are classified as **nonpolar** (alanine, glycine, isoleucine, leucine, methionine, phenylalanine, proline, tryptophan, valine), **uncharged polar** (asparagine, cysteine, glutamine, serine, threonine, tyrosine), **acidic** (aspartate, glutamate), or **basic** (arginine, histidine, lysine).
• Buffering capacity	• All free amino acids plus charged amino acids in peptide chains can serve as **buffers**.
• Location of nonpolar amino acids in proteins • Location of polar amino acids in proteins	• Amino acids with **nonpolar** (hydrophobic) R-groups are generally found in the **interior** of proteins that function in an aqueous environment, and on the surface of proteins (such as membrane proteins) that interact with lipids. Amino acids with **polar** side chains are generally found on the **outside** of proteins that function in an aqueous environment, and in the interior of membrane-associated proteins.
• Covalent bond between amino acids in a peptide chain	• Amino acids are attached to each other by **peptide bonds**. All α-amino and α-carboxyl groups participate in these bonds except the terminal groups on the protein. These are charged at physiologic pH. Sequences of amino acids in a peptide are read from the amino-terminal end to the carboxy-terminal end.
• Additional types of bonds made between amino acids	• **Nonpolar** amino acids participate in **hydrophobic interactions**. **Polar** amino acids form **hydrogen bonds**. Two cysteine residues can form a **disulfide bond**, producing cystine. **Acidic** side chains are negatively charged, and **basic** side chains are generally positively charged at physiologic pH. All can form **ionic bonds**.
• Stereoisomers of amino acids	• D- and L-amino acids are **stereoisomers** of each other. Only **L-amino acids** are found in human proteins. D-amino acids are found in some antibiotics and in bacterial cell walls.
• Use of the Henderson-Hasselbalch equation	• The **Henderson-Hasselbalch equation** can be used to calculate the quantitative relationship between the concentration of a weak acid and its conjugate base.

Chapter 2: Structure of Proteins

Primary structure:

- Definition of primary structure
- Structure of peptide bonds
- Characteristics of peptide bonds

PRIMARY STRUCTURE (p. 13)

- The **primary structure** of a protein is defined as the **linear sequence** of its **amino acids**.
- **Peptide bonds** join the individual amino acids, attaching the **α–amino group** of one amino acid to the **α-carboxyl group** of another. Prolonged exposure to a strong acid or base at elevated temperatures is required to hydrolyze these bonds nonenzymatically.
- Peptide bonds have a **partial double-bond character** (**rigid** and **planar**). They are generally in the **trans configuration**. They are **polar**, and can **hydrogen bond** to each other or to other polar compounds.

Secondary structure:

- Definition
- Examples of secondary structural elements
- Bond stabilizing secondary structure
- Motifs

SECONDARY STRUCTURE (p. 16)

- The **secondary structure** of a protein is generally defined as regular arrangements of amino acids that are located near to each other in the linear sequence. Examples of such elements are the **α–helix**, **β–sheet**, and **β–bend**. Some secondary structure is not regular, but rather is considered **non-repetitive** (**loop** and **coil**).
- Secondary structural elements are stabilized by extensive **hydrogen bonding**.
- **Supersecondary structures** (**motifs**) are produced by packing side chains from adjacent secondary structural elements close to each other.

Tertiary structure:

- Definition of domains
- Definition of tertiary structure.
- Bonds stabilizing tertiary structure
- Role of chaperones

TERTIARY STRUCTURE (p. 18)

- **Domains** are the fundamental **functional** and **three-dimensional structural units** of a polypeptide. They are formed from **combinations of motifs**.
- **Tertiary structure** refers to the folding of the domains and their final arrangement in the polypeptide. Tertiary structure is stabilized by **disulfide bonds**, **hydrophobic interactions**, **hydrogen bonds**, and **ionic bonds**.
- A specialized group of proteins, named **chaperones**, is required for the proper folding of many species of proteins.

Quaternary structure:

- Definition of quaternary structure
- Bonds involved

QUATERNARY STRUCTURE (p. 20)

- Proteins consisting of **more than one polypeptide** chain have **quaternary structure**. The polypeptides are held together by noncovalent bonds.

Denaturation and misfolding:

- Definition of denaturation
- Cause of protein misfolding
- Role of amyloid protein in disease
- Role of prion protein in disease

DENATURATION AND MISFOLDING (p. 21)

- Proteins can be **denatured**, that is, **unfolded** and **disorganized**, making them nonfunctional. **Denaturing agents** include heat, organic solvents, mechanical mixing, strong acids or bases, detergents, and ions of heavy metals such as lead and mercury.
- **Protein misfolding** is most commonly caused by a **gene mutation**, which produces an altered protein. An example of this is the **amyloid protein** that spontaneously aggregates in many degenerative diseases, for example, **Alzheimer disease**.
- The **prion protein** (**PrP**) is an infectious protein that converts noninfectious PrP into the infectious form, which precipitates. PrP is implicated as the causative agent of the **transmissible spongiform encephalopathies**, including **Creutzfeld-Jakob disease**.

Chapter 3: Globular Proteins

Globular hemeproteins:	GLOBULAR HEMEPROTEINS (p. 25)
• Definition and examples of hemeproteins	• Hemeproteins are a group of specialized proteins that contain **heme** as a tightly bound prosthetic group. Examples of hemeproteins include cytochromes, catalase, hemoglobin, and myoglobin.
• Structure of a heme group	• Heme is a complex of **protoporphyrin IX** and **ferrous iron** (Fe^{2+}).The iron is held in the center of the heme molecule by bonds to the four nitrogens of the porphyrin ring. The Fe^{2+} can then form two additional bonds, including one to molecular oxygen.
• Structure, location, and function of myoglobin (Mb)	• **Myoglobin** (Mb) consists of one heme group bound to a single polypeptide. It is present in **heart** and **skeletal muscle**. Mb functions both as a reservoir for oxygen, and as an oxygen carrier within the muscle cell.
• Structure, location, and function of hemoglobin (Hb)	• **Hemoglobin** (Hb) consists of four polypeptides, each of which binds a heme group. Hb is found exclusively in **red blood cells** (RBCs), within which it transports oxygen from the lungs to the capillaries of the tissues.
• Taut versus relaxed forms of Hb	• Hemoglobin can exist in two forms. The **deoxy form** of Hb is called the **taut** (**T**) **form**, and the **oxygenated form** of Hb is called the **relaxed** (**R**) **form**. The R form is the **high oxygen-affinity form** of hemoglobin.
• Number of O_2 that Mb and Hb can bind • Shapes of the oxygen dissociation curves for Hb and Mb	• **Mb** binds only **one molecule of oxygen** (O_2) because it contains only one heme group. **Hb** can bind **four molecules** of oxygen. The **oxygen-dissociation curve** (a plot of degree of saturation, Y, measured at different partial pressures of oxygen, pO_2) is **hyperbolic** for **Mb** and **sigmoidal** for **Hb**. This indicates that the four subunits of Hb **cooperate** in binding oxygen.
• Compounds affecting the ability of Hb to bind oxygen reversibly	• The ability of Hb to bind oxygen reversibly is affected by the **pO_2**, the **pH** of the environment, the **pCO_2**, and the availability of **2,3-bisphosphoglycerate (2,3-BPG)**. For example, the release of O_2 from Hb is enhanced when the pH is lowered or the pCO_2 is increased (the **Bohr effect**), as is seen in **exercising muscle**. Here, the oxygen-dissociation curve of Hb is shifted to the right. To cope long-term with the effects of **chronic hypoxia** or **anemia**, the concentration of **2,3-BPG** in **RBCs** increases. **2,3-BPG** binds to the Hb and decreases its oxygen affinity, and it, therefore, also shifts the oxygen-dissociation curve to the right.
• Mechanism of carbon monoxide toxicity ("CO poisoning")	• **Carbon monoxide** (**CO**) binds tightly to the Hb iron. It stabilizes the **R form** of Hb and, thus, prevents release of O_2 to the tissues. **CO toxicity** is, in large part, a result of tissue hypoxia. CO poisoning is treated with **100 percent oxygen therapy**, which facilitates the dissociation of CO from the Hb, allowing more O_2 to be bound to Hb.
• Peptide subunits found in HbA, HbF, and HbA_2	• The most plentiful adult Hb is **HbA**, consisting of two α and two β chains ($\alpha_2\beta_2$). Fetal Hb (**HbF**) is $\alpha_2\gamma_2$, and **HbA_2**, a minor adult Hb, is $\alpha_2\delta_2$.
• Significance of HbA_{1c}	• Under physiologic conditions, HbA is slowly and nonenzymically **glycosylated**, becoming **HbA_{1c}**, the amount of which is dependent on the plasma concentration of glucose. Increased amounts of HbA_{1c} are found in the RBCs of patients with **diabetes mellitus**, because their HbA has contact with higher glucose concentrations in the blood during the 120 day life span of the RBC.
Hemoglobinopathies:	**HEMOGLOBINOPATHIES (p. 35)**
• Definition of hemoglobinopathies	• Hemoglobinopathies are disorders caused either by production of a **structurally abnormal Hb molecule** (seen in sickle cell disease), synthesis of **insufficient amounts** of normal Hb subunits (seen in thalassemias), or, rarely, both.

- The cause of sickle cell disease
- Symptoms of sickle cell disease
- Conditions that favor deoxy-Hb (and, therefore, that cause crises)

- **Sickle cell disease** (**HbS disease**) is caused by a **point mutation** in both genes coding for the β-chain that results in a valine rather than a glutamate at position six. The mutation leads to a polymerization (and therefore decreased solubility) of the deoxy form of Hb, which causes distortion of the RBC membrane. The misshapen, rigid RBCs occlude capillaries, leading to a shortened RBC life span (hence, **anemia**), **tissue anoxia**, and **painful "crises."** Thus anything that favors the deoxygenated form of Hb (for example, low pO_2, increased pCO_2, decreased pH, or an increased concentration of 2,3-BPG) can precipitate a crisis.

- The causes of α- and β-thalassemia

- **Thalassemias** are hereditary hemolytic diseases in which an imbalance occurs in the synthesis of either α or β globin chains. Each thalassemia can be classified as either a disorder in which no globin chains are produced (α^o- or β^o-thalassemia), or in which some chains are synthesized but at a reduced rate (α^+- or β^+-thalassemia). [Note: In β-thalassemias, synthesis of β globin chains is decreased or absent, whereas in α-thalassemias, synthesis of α globin chains is decreased or absent.]

Chapter 4: Fibrous Proteins

Collagen:

COLLAGEN (p. 43)

- Structure of collagen

- A typical collagen molecule is a long, stiff, **extracellular** structure in which three polypeptides (referred to as "**α-chains**," each 1000 amino acids in length) are wound around one another in a rope-like **triple-helix**. The chains are held together by **hydrogen bonds**. Variations in the amino acid sequence of the α-chains result in collagen molecules with slightly different properties.

- Types of collagen and their functions

- **Type I, II, and III collagens** are **fibrillar**, and are found in tendon, skin, bone, cornea, cartilage, vitreous body, and blood vessels. **Types IX and XII** are **fibril-associated**, and are found in cartilage, tendon, and ligaments. **Type IV and VII** form **networks** in basement membrane and beneath stratified squamous epithelia.

- Amino acid sequence of collagen

- Collagen is rich in **glycine** and **proline**. The glycine residues are part of a repeating sequence, **–Gly–X–Y–**, where **X** is frequently **proline** and **Y** is often **hydroxyproline** or **hydroxylysine**.

- Posttranslational modifications of collagen
- Disease caused by ascorbic acid deficiency

- **Hydroxyproline** and **hydroxylysine** result from the hydroxylation by specific hydroxylases of proline and lysine residues after their incorporation into α-chains. The enzymes require **ascorbic acid** as a cofactor. [Note: An **ascorbic acid deficiency** results in **scurvy**.] The hydroxyl group of the **hydroxylysine** residues of collagen may be enzymatically **glycosylated** (most commonly, **glucose** and **galactose** are added sequentially to the triple helix).

- Steps in type 1 collagen synthesis

- The precursors of collagen α-chains are formed in **fibroblasts**, **osteoblasts**, and **chondroblasts**, and travel via the endoplasmic reticulum and Golgi to the **extracellular matrix**. There, the **N-terminal** and **C-terminal propeptides** are removed by **procollagen peptidases**. In some collagens (for example, type 1), the collagen molecules self-assemble into **fibrils** in which the adjacent triple helices are arranged in a staggered pattern, each overlapping its neighbor by a length approximately three-quarters of a molecule. The triple helices are then **cross-linked**, giving the fibrillar array great tensile strength.

- Cause and symptoms of Ehlers-Danlos syndrome
- Cause and symptoms of osteogenesis imperfecta

- **Collagen diseases: Ehlers-Danlos syndrome** is a heterogeneous group of generalized tissue disorders that results from inheritable defects in the metabolism of fibrillar collagen molecules. Symptoms can include stretchy skin, loose joints, and vascular problems. **Osteogenesis imperfecta**, also known as **brittle bone syndrome**, is also a heterogeneous group of inherited disorders involving mutations in the collagen genes themselves that is distinguished by bones that easily bend and fracture. Retarded wound healing, and a rotated and twisted spine, leading to a humped-back appearance, are common features of the disease.

Elastin:

- Function and location of elastin
- Cause of Marfan syndrome

ELASTIN (p. 49)

- Elastin is a **connective tissue protein** with **rubber-like properties**. Elastic fibers composed of elastin and glycoprotein microfibrils, such as fibrillin, are found in the **lungs**, the walls of **large arteries**, and **elastic ligaments**. [Note: Mutations in the fibrillin gene are responsible for **Marfan syndrome**.]

- Structure of tropoelastin

- Elastin is synthesized from a precursor, **tropoelastin**, that is rich in **proline** and **lysine**, but contains only a little hydroxyproline and no hydroxylysine. In the **extracellular matrix,** tropoelastin is converted to elastin.

- Function of α1-antitrypsin
- Effect of α1-antitrypsin deficiency, its treatment

- **α1-Antitrypsin (a1-AT)**, found in blood and other body fluids, **inhibits** a number of **proteolytic enzymes**,including **elastase**. Most α1-AT is produced by the **liver**. In the normal lung, activated and degenerating neutrophils release elastase, which is inactivated by α1-AT. If this inhibitor is absent, lung tissue is destroyed and cannot regenerate, leading to **emphysema**. A deficiency of α1-AT can be reversed by weekly intravenous administration of α1-AT.

Chapter 5: Enzymes

Properties of enzymes:

- Definition of an enzyme
- Definition of an active site

PROPERTIES OF ENZYMES (p. 53)

- Enzymes are **protein catalysts** that increase the velocity of a chemical reaction, and are not consumed during the reaction they catalyze. Each enzyme has an **active site** where the substrate binds and is converted to product.

- Definition of a cofactor, coenzyme, and prosthetic group

- Some enzymes require **cofactors** for activity. These can be metal ions or organic molecules called **coenzymes** that are often derivatives of **vitamins**. Tightly bound coenzymes are called **prosthetic groups**.

How enzymes work:

- How enzymes increase the velocity of a reaction

HOW ENZYMES WORK (p. 55)

- Enzymes provide an **alternate reaction pathway** with a **lower free energy of activation**. They do not change the free energies of the reactants or products and, therefore, do not change the equilibrium of the reaction.

- Factors affecting reaction velocity

- **Substrate concentrations**, **temperature**, and **pH** are among the factors that can affect the reaction velocity of enzymes.

- Shapes of the kinetics curves for simple and allosteric enzymes

- Enzymes following **Michaelis-Menten kinetics** show **hyperbolic curves** when the initial reaction velocity (v_o) of the reaction is plotted against substrate concentration. In contrast, **allosteric enzymes** generally show **sigmoidal curves**.

Inhibition of enzyme activity:

- Definition of reversible and irreversible inhibitors

INHIBITION OF ENZYME ACTIVITY (p. 60)

- Any substance that can **diminish the velocity** of an enzyme-catalyzed reaction is called an **inhibitor**. **Reversible inhibitors** bind to enzymes through noncovalent bonds, and can be diluted away from the enzyme. **Irreversible inhibitors** do not dissociate from the enzyme.

- Properties of a competitive inhibitor

- **Competitive inhibitors** are **reversible**. Their inhibition can be overcome by adding additional substrate They act by **increasing the apparent K_m** of the enzyme for a given substrate, but do not affect the V_{max} of the enzyme.

- Properties of a noncompetitive inhibitor

- **Noncompetitive inhibitors** are **reversible**. Their inhibition cannot be overcome by adding additional substrate. They act by decreasing the V_{max} of the enzyme for a given substrate, but do not affect the K_m of the enzyme.

- Examples of enzyme inhibitors that can be used as drugs

- **Enzyme inhibitors** can be used as **drugs**, inhibiting either intracellular or extracellular reactions. For example, the **β-lactam antibiotics**, such as penicillin and amoxicillin, act by inhibiting one or more of the enzymes of bacterial cell wall synthesis.

Regulation of enzyme activity:

- How the rates of enzymes can be regulated

- Definition of allosteric enzymes and their function

REGULATION OF ENZYME ACTIVITY (p. 62)

- The **rates** of most enzymes are responsive to **changes in substrate concentration** because the intracellular level of many substrates is in the range of the K_m. In addition, some enzymes with specialized regulatory functions respond to **allosteric effectors** , to **covalent modification** (for example, by phosphorylation/dephosphorylation), or they show altered rates of enzyme synthesis when physiologic conditions are changed.

- **Allosteric enzymes** are **multi-subunit** proteins that are **regulated** by molecules called **effectors**. The effectors bind noncovalently at a site other than the active site, and can alter the affinity of the enzyme for its substrate, modify the maximal catalytic activity of the enzyme, or both. Allosteric enzymes frequently catalyze the **committed step** early in a pathway.

Enzymes in clinical diagnosis:

- Reasons for the presence of enzymes in the plasma

- Definition of isoenzymes
- How isozymes are used to diagnose tissue damage

ENZYMES IN CLINICAL DIAGNOSIS (p. 64)

- **Enzymes** can normally be found in the **plasma** either because they were **specifically secreted** to fulfill a function in the blood, or because they were **released by dead or damaged cells**. Many diseases that cause tissue damage result in an increased release of intracellular enzymes into the plasma. The activities of many of these enzymes (for example, creatine kinase, lactate dehydrogenase, and alanine aminotransferase) are routinely determined for **diagnostic purposes** in diseases of the heart, liver, skeletal muscle, and other tissues.

- **Isoenzymes** (also called isozymes) are enzymes that **catalyze the same reaction**. However, they do not necessarily have the same physical properties, because of genetically determined differences in amino acid sequence. For this reason, isoenzymes may contain different numbers of charged amino acids, and may be separated from each other by **electrophoresis**. Different organs frequently contain characteristic proportions of different isoenzymes. The pattern of isoenzymes found in the plasma may, therefore, serve as a means of identifying the site of tissue damage.

Chapter 6: Bioenergetics and Oxidative Phosphorylation

Free energy:

- Definition of enthalpy and entropy
- Definition of free energy

- Significance of negative and positive ΔGs
- Value of ΔG at equilibrium

- Result of coupling reactions with large positive ΔG°s with hydrolysis of adenosine triphosphate

FREE ENERGY (p. 69)

- **Enthalpy** (a measure of the change in **heat content** of the reactants and products) and **entropy** (a measure of the change in the **randomness** or **disorder** of reactants and products) determine the direction and extent to which a chemical reaction will proceed. When combined mathematically, they can be used to define a third quantity, **free energy**, which predicts the direction in which a reaction will spontaneously proceed.

- If the change in **free energy** (**ΔG**) is **negative** (that is, the product has a lower free energy than the substrate), the **reaction goes spontaneously**. If **ΔG is positive**, the reaction **does not go spontaneously**. If **ΔG = 0**, the reactants are in **equilibrium**. The change in free energy of the forward reaction ($A \rightarrow B$) is equal in magnitude but opposite in sign to that of the backward reaction ($B \rightarrow A$).

- The **standard free energy changes** (**ΔG°s**) are **additive** in any sequence of consecutive reactions. Therefore, reactions or processes that have a large, positive ΔG are made possible by **coupling** with hydrolysis of **adenosine triphosphate**, which has a large, negative ΔG°.

Electron transport chain:

- Types of components found in the electron transport chain
- Location of the chain
- Cytochrome that can bind oxygen

ELECTRON TRANSPORT CHAIN (p. 73)

- The reduced coenzymes **NADH** and **$FADH_2$** each donate a pair of electrons to a specialized set of electron carriers, consisting of **FMN**, **coenzyme Q**, and a series of **cytochromes**, collectively called the **electron transport chain**. This pathway is present in the **inner mitochondrial membrane**, and is the final common pathway by which electrons derived from different fuels of the body flow to oxygen. The terminal cyctochrome, **cytochrome a+ a_3**, is the only cytochrome able to bind oxygen.

Oxidative phosphorylation:

- Mechanism by which electron transport creates an electrical and pH gradient across the inner mitochondrial membrane.
- Mechanism by which these gradients are dissipated
- Results of uncoupling electron transport and phosphorylation
- Examples of uncouplers
- Cause of Leber hereditary optic neuropathy

OXIDATIVE PHOSPHORYLATION (p. 77)

- **Electron transport** is **coupled** to the **transport of protons** (H^+) across the inner mitochondrial membrane from the matrix to the intermembrane space. This process creates an **electrical gradient** and a **pH gradient** across the **inner mitochondrial membrane**. After protons have been transferred to the cytosolic side of this membrane, they can reenter the mitochondrial matrix by passing through a channel in the **ATP synthase complex**, resulting in the synthesis of ATP from ADP + Pi, and at the same time dissipating the pH and electrical gradients. **Electron transport** and **phosphorylation** are thus said to be **tightly coupled**.

- These processes can be **uncoupled** by **uncoupling proteins** found in the inner mitochondrial membrane, and by synthetic compounds, such as **2,4-dinitrophenol** and **aspirin**, all of which increase the permeability of the inner mitochondrial membrane to protons. The energy produced by the transport of electrons is released as **heat**, rather than being used to synthesize ATP.

- **Mutations in mtDNA** are responsible for some cases of **mitochondrial diseases**, such as **Leber hereditary optic neuropathy**.

Chapter 7: Introduction to Carbohydrates

Carbohydrate classification and structure:

- Definition of aldose and ketose sugars
- Type of bond that links carbohydrates
- Definition of isomer, epimer, and enantiomer
- Definition of anomeric carbon
- Definition of a reducing sugar
- Definition of N- and O-glycosides

CARBOHYDRATE CLASSIFICATION AND STRUCTURE (p. 83)

- **Monosaccharides** (simple sugars) containing an aldehyde group are called **aldoses** and those with a keto group are called **ketoses. Disaccharides**, **oligosaccharides**, and **polysaccharides** consist of monosaccharides linked by **glycosidic bonds**.

- Compounds with the same chemical formula are called **isomers**. If two monosaccharide isomers differ in configuration around one specific carbon atom (with the exception of the carbonyl carbon) they are defined as **epimers** of each other. If a pair of sugars are mirror images of each other (**enantiomers**), the two members of the pair are designated as **D-** and **L-sugars**.

- When a sugar cyclizes, an **anomeric carbon** is created from the aldehyde group of an aldose or keto group of a ketose. This carbon can have two configurations, α and β. If the oxygen on the anomeric carbon is not attached to any other structure, that sugar is a **reducing sugar**. A sugar with its anomeric carbon linked to another structure is called a **glycosyl residue**. Sugars can be attached either to a $-NH_2$ or an $-OH$ group, producing **N-** and **O-glycosides**.

Digestion of carbohydrates:

- Enzymes that degrade starch

DIGESTION OF CARBOHYDRATES (p. 85)

- The principal sites of dietary carbohydrate digestion are the **mouth** and **intestinal lumen**. **Salivary α-amylase** acts on **dietary starch** (glycogen, amylose, amylopectin), producing **oligosaccharides**. **Pancreatic α-amylase** continues the process of starch digestion.

- Source and kinds of disaccharidases

- The final digestive processes occur at the **mucosal lining** of the **small intestine**. Several **disaccharidases** [for example, **lactase** (**β-galactosidase**), **sucrase**, **maltase**, and **isomaltase**] produce monosaccharides (glucose, galactose, and fructose). These enzymes are secreted by and remain associated with the **luminal side** of the **brush border membranes** of **intestinal mucosal cells**. Absorption of the monosaccharides requires specific transporters.

- Effect on the body of a deficiency of carbohydrate degradation
- Most common of these deficiencies

- If carbohydrate degradation is deficient as a result of heredity, intestinal disease, malnutrition, or drugs that injure the mucosa of the small intestine, undigested carbohydrate will pass into the large intestine, where it can cause **osmotic diarrhea**. Bacterial fermentation of the compounds produces large volumes of CO_2 and H_2 gas, causing abdominal cramps, diarrhea, and flatulence. **Lactose intolerance** caused by a lack of lactase is by far the most common of these deficiencies.

Chapter 8: Glycolysis

Introduction to metabolism:

- Definition of catabolic and anabolic pathways

INTRODUCTION TO METABOLISM (p. 89)

- Most pathways can be classified as either **catabolic** (they **degrade** complex molecules to a few simple products, such as CO_2, NH_3, and water) or **anabolic** (they synthesize complex end-products from simple precursors). **Catabolic reactions** also **capture chemical energy** in the form of ATP from the degradation of energy-rich molecules. **Anabolic reactions require energy**, which is generally provided by the breakdown of ATP.

Regulation of metabolism:

- Examples of intracellular regulatory signals that control the metabolic rates of pathways

REGULATION OF METABOLISM (p. 92)

- The rate of a metabolic pathway can respond to **regulatory signals** that arise from **within the cell**, such as the availability of **substrates**, **product inhibition**, or alterations in the levels of **allosteric activators or inhibitors**. These intracellular signals typically elicit rapid responses.

- Examples of chemical signals that aid communication between cells

- Signaling **between cells** provides for the integration of metabolism. The most important route of this communication is **chemical signaling** between cells, for example, by **hormones** or **neurotransmitters**.

- Function of second messenger molecules

- **Second messenger molecules** convey the intent of a chemical signal (hormone or neurotransmitter) to appropriate intracellular responders.

- Summary of the cAMP second messenger system

- **Adenylyl cyclase** is a membrane-bound enzyme that **synthesizes cyclic AMP** (**cAMP**) in response to chemical signals, such as the hormones **glucagon** and **epinephrine**. Following binding of a hormone to its **cell-surface receptor**, a GTP-dependent regulatory protein (**G-protein**) is activated that, in turn, activates adenylyl cyclase. The cAMP that is synthesized activates a **protein kinase**, which phosphorylates a cadre of enzymes, causing their activation or deactivation. Phosphorylation is reversed by **protein phosphatases**. cAMP is inactivated by conversion to AMP, catalyzed by **cAMP phosphodiesterase**.

Glycolysis:

- Definition of aerobic and anaerobic glycolysis

GLYCOLYSIS (p. 95)

- **Aerobic glycolysis**, in which pyruvate is the end-product, occurs in cells with **mitochondria** and an adequate supply of **oxygen**. **Anaerobic glycolysis**, in which lactic acid is the end-product, occurs in cells that **lack mitochondria** or in cells **deprived of sufficient oxygen**.

- Mechanism by which glucose is transported into cells, and tissue-specific examples

- Glucose is transported across membranes by one of at least fourteen **glucose transporter isoforms** (**GLUTs**). **GLUT-1** is abundant in **erythrocytes** and **brain**, **GLUT-4** (which is **insulin-dependent**) is found in **muscle** and **adipose tissue**, and **GLUT-2** is found in **liver**

- Energy used/generated of the two phases of glycolysis

- The conversion of glucose to pyruvate occurs in two stages: an **energy investment phase** in which phosphorylated intermediates are synthesized at the expense of ATP, and an **energy generation phase**, in which ATP is produced.

- Regulated enzymes in the energy investment phase of glycolysis, and the compounds that increase or decrease their activity

- In the **energy investment phase**, glucose is phosphorylated by **hexokinase** (found in **most tissues**) or **glucokinase** (a hexokinase found in **liver cells** and the **β cells** of the pancreas). **Hexokinase** has a **high affinity** (**low K_m**) and a **small V_{max}** for glucose, and is **inhibited** by **glucose 6-phosphate**. **Glucokinase** has a **large K_m** and a **large V_{max}** for glucose. It is indirectly **inhibited** by **fructose 6-phosphate** and **activated** by **glucose**, and the **transcription** of the glucokinase gene is **enhanced by insulin**. Glucose 6-phosphate is isomerized to fructose 6-phosphate, which is phosphorylated to **fructose 1,6-bisphosphate** by **phosphofructokinase**. This enzyme is allosterically **inhibited by ATP** and **citrate**, and **activated** by **AMP**. **Fructose 2,6-bisphosphate**, whose synthesis is **activated by insulin**, is the **most potent allosteric activator** of this enzyme. A total of **two ATP are used** during this phase of glycolysis.

- Regulated enzyme in the energy generation phase of glycolysis, and the compounds that regulate it

- **Fructose 1,6-bisphosphate** is cleaved to form two trioses that are further metabolized by the glycolytic pathway, forming pyruvate. During their interconversions, **four ATP** and **two NADH** are produced from ADP and NAD^+. The final step in pyruvate synthesis from phosphoenolpyruvate is catalyzed by **pyruvate kinase**. This enzyme is **allosterically activated** by **fructose 1,6-bisphosphate**, **hormonally activated** by **insulin**, and **inhibited** by **glucagon** via the **cAMP pathway**.

- Effect of pyruvate kinase deficiency

- **Pyruvate kinase deficiency** accounts for 95 percent of all inherited defects in glycolytic enzymes. It is restricted to **erythrocytes**, and causes mild to severe **chronic hemolytic anemia**. Altered kinetics (for example, increased K_m, decreased V_{max}, etc.) most often account for the enzyme deficiency.

- Tissues using anaerobic glycolysis

- In **anaerobic glycolysis**, NADH is reoxidized to NAD^+ by the **conversion of pyruvate to lactic acid**. This occurs in cells such as **erythrocytes** that have few or no mitochondria, and in tissues such as **exercising muscle**, where production of NADH exceeds the oxidative capacity of the respiratory chain.

- Causes of lactic acidosis

- Elevated concentrations of lactate in the plasma (**lactic acidosis**) occur when there is a **collapse of the circulatory system**, or when an individual is in **shock**.

Alternate fates of pyruvate:

- Compounds other than lactate to which pyruvate can be converted

ALTERNATE FATES OF PYRUVATE (p. 103)

- Pyruvate can be **oxidatively decarboxylated** by **pyruvate dehydrogenase**, producing **acetyl CoA**—a major fuel for the tricarboxylic acid cycle (TCA cycle) and the building block for fatty acid synthesis. Pyruvate can be **carboxylated** to **oxaloacetate** (a TCA cycle intermediate) by **pyruvate carboxylase**. Pyruvate can be **reduced** by microorganisms to **ethanol** by **pyruvate decarboxylase**.

Chapter 9: Tricarboxylic Acid Cycle

Reactions of the TCA cycle:

- Enzyme that oxidatively decarboxylates pyruvate, its coenzymes, activators, and inhibitors

REACTIONS OF THE TRICARBOXYLIC ACID CYCLE (p. 107)

- **Pyruvate** is **oxidatively decarboxylated** by **pyruvate dehydrogenase complex** producing **acetyl CoA**, which is the major fuel for the tricarboxylic acid cycle (TCA cycle). The irreversible set of reactions catalyzed by this enzyme complex requires five coenzymes: **thiamine pyrophosphate**, **lipoic acid**, **coenzyme A** (which contains the vitamin pantothenic acid), **FAD**, and **NAD^+**. The reaction is **activated** by **NAD^+**, **coenzyme A**, and **pyruvate**, and **inhibited** by **ATP**, **acetyl CoA**, and **NADH**.

- Most common biologic cause of congenital lactic acidosis
- Biochemical mechanism of arsenic poisoning

- **Pyruvate dehydrogenase deficiency** is the most common biochemical cause of **congenital lactic acidosis**. Because the deficiency deprives the brain of acetyl CoA, the CNS is particularly affected, with profound **psychomotor retardation** and **death** occurring in most patients. A **ketogenic diet** may be of benefit is some cases. The deficiency is **X-linked dominant**. **Arsenic poisoning** causes inactivation of pyruvate dehydrogenase by binding to lipoic acid.

- Enzymes synthesizing citrate and isocitrate, their substrates, activators and inhibitors

- **Citrate** is synthesized from **oxaloacetate** (OAA) and **acetyl CoA** by **citrate synthase**. This enzyme is allosterically activated by ADP, and inhibited by ATP, NADH, succinyl CoA, and fatty acyl CoA derivatives. Citrate is isomerized to **isocitrate** by **aconitase**, an enzyme that is targeted by the rat poison, **fluoroacetate**.

- TCA cycle enzyme synthesizing α-ketoglutarate, its products, inhibitors, and activators
- Enzyme synthesizing succinyl CoA, and its additional products, activators, and inhibitors

- **Isocitrate** is oxidized and decarboxylated by **isocitrate dehydrogenase** to **α-ketoglutarate**, producing CO_2 and **NADH**. The enzyme is **inhibited** by **ATP** and **NADH**, and **activated** by **ADP** and Ca^{++}.

- **α-Ketoglutarate** is oxidatively decarboxylated to **succinyl CoA** by the **α-ketoglutarate dehydrogenase complex**, producing CO_2 and **NADH**. The enzyme is very similar to pyruvate dehydrogenase and uses the same coenzymes. α-Ketoglutarate dehydrogenase complex is **activated** by **calcium**, and **inhibited** by **ATP**, **GTP**, **NADH**, and **succinyl CoA**. It is not regulated by phosphorylation/dephosphorylation.

- Enzymes synthesizing succinate, fumarate, malate, and oxaloacetate
- The above reactions that produce ATP, $FADH_2$, or NADH
- Total number of ATPs produced per acetyl CoA entering the TCA cycle

- **Succinyl CoA** is cleaved by **succinate thiokinase** (also called **succinyl CoA synthetase**), producing **succinate** and **ATP** (or **GTP**). This is an example of **substrate-level phosphorylation**. **Succinate** is oxidized to **fumarate** by **succinate dehydrogenase**, producing $FADH_2$. The enzyme is **inhibited** by **oxaloacetate**. **Fumarate** is hydrated to **malate** by **fumarase** (**fumarate hydratase**), and **malate** is oxidized to **oxaloacetate** by **malate dehydrogenase**, producing **NADH**.

- **Three NADH**, **one** $FADH_2$, and one **ATP** (or **GTP**) are produced by one round of the TCA cycle. Oxidation of the NADHs and $FADH_2$ by the electron transport chain yields approximately eleven ATPs, making **twelve** the total number of ATPs produced.

Chapter 10: Gluconeogenesis

Reactions unique to gluconeogenesis:

- Examples of gluconeogenic precursors

REACTIONS UNIQUE TO GLUCONEOGENESIS (p. 115)

- **Gluconeogenic precursors** include all the **intermediates of glycolysis** and the **tricarboxylic acid cycle**, **glycerol** released during the hydrolysis of triacylglycerols in adipose tissue, **lactate** released into the blood by cells that lack mitochondria and by exercising skeletal muscle, and **α-ketoacids** derived from the metabolism of glucogenic amino acids.

- Glycolytic enzymes that are physiologically irreversible

- Seven of the reactions of glycolysis are reversible and are used for gluconeogenesis in the liver and kidneys. Three reactions are **physiologically irreversible** and must be circumvented. These reactions are catalyzed by the glycolytic enzymes **pyruvate kinase**, **phosphofructokinase**, and **hexokinase**.

- Enzymes required to reverse the pyruvate kinase reaction

- **Pyruvate** is converted to **phosphoenolpyruvate** (**PEP**) by **pyruvate carboxylase** and **PEP carboxykinase**. The carboxylase requires **biotin** and **ATP**, and is allosterically **activated** by **acetyl CoA**. PEP carboxykinase, which requires **GTP**, is the rate-limiting step in gluconeogenesis. The transcription of its mRNA is increased by glucagon and decreased by insulin.

- Enzyme required to reverse the phosphofructokinase reaction, its activator/inhibitors

- **Fructose 1,6-bisphosphate** is converted to **fructose 1-phosphate** by **fructose 1,6-bisphosphatase**. This enzyme is **inhibited** by elevated levels of **AMP** and **activated** by elevated levels of **ATP**. The enzyme is also **inhibited** by **fructose 2,6-bisphosphate**, the primary allosteric activator of glycolysis.

- Result of a deficiency in the enzyme required to reverse the hexokinase reaction

- **Glucose 6-phosphate** is converted to **glucose** by **glucose 6-phosphatase**. This enzyme activity is required for the final step in glycogen degradation, as well as gluconeogenesis. A deficiency of this enzyme results in **type Ia glycogen storage disease**.

Chapter 11: Glycogen Metabolism

Structure and function of glycogen

- Primary locations and functions of glycogen
- Structure of glycogen
- Types of bonds

STRUCTURE AND FUNCTION OF GLYCOGEN (p. 123)

- The **main stores** of glycogen in the body are found in **skeletal muscle**, where they serve as a **fuel reserve** for the synthesis of ATP during **muscle contraction**, and in the **liver**, where glycogen is used to **maintain the blood glucose** concentration, particularly during the **early stages of a fast**.
- Glycogen is a **highly branched** polymer of **α-D-glucose**. The primary glycosidic bond is an **α(1→4)-linkage**. After about eight to ten glucosyl residues, there is a **branch** containing an **α(1→6)-linkage**.

Synthesis of glycogen:

- Form of glucose used to synthesize glycogen
- Enzyme that adds glucose to the ends of glycogen chains
- Mechanism by which branches are formed

SYNTHESIS OF GLYCOGEN (GLYCOGENESIS) (p. 124)

- **UDP-glucose**, the **building block** of glycogen, is synthesized from **glucose 1-phosphate** and **UTP** by **UDP-glucose pyrophosphorylase**. Glucose from UDP-glucose is transferred to the non-reducing ends of glycogen chains by **glycogen synthase**, which makes **α(1→4)-linkages**.
- **Branches** are formed by **glucosyl-α(1→4)→α(1→6)-transferase**, which transfers a chain of five to eight glucosyl residues from the nonreducing end of the glycogen chain (**breaking** an **α(1→4)-linkage**), and attaches it with an **α(1→6)-linkage** to another residue in the chain.

Degradation of glycogen:

- Enzyme that cleaves α(1→4)-bonds
- Product produced by that enzyme
- Enzymes and mechanisms required to remove a branch
- Products produced by these enzymes
- Final products and their functions from glycogen degradation in the muscle and liver
- Cause and result of glycogen storage disease type Ia (Von Gierke disease)

DEGRADATION OF GLYCOGEN (GLYCOGENOLYSIS) (p. 126)

- **Glycogen phosphorylase** cleaves the **α(1→4)-bonds** between glucosyl residues at the **nonreducing ends** of the glycogen chains, producing **glucose 1-phosphate**. It requires **pyridoxyl phosphate** as a coenzyme. This sequential degradation continues until four glucosyl units remain on each chain before a branch point. The resulting structure is called a **limit dextrin**.
- **Oligo–α(1→4)→α(1→4)-glucan transferase**, common name, **glucosyl-α(1→4)-α(1→4)-transferase**, removes the outer three of the four glucosyl residues attached at a branch and transfers them to the non-reducing end of another chain, where they can be converted to **glucose 1-phosphate** by **glycogen phosphorylase**. Next, the remaining single glucose residue attached in an **α(1→6)-linkage** is removed hydrolytically by the **amylo-α(1→6)-glucosidase** activity, releasing **free glucose**.
- **Glucose 1-phosphate** is converted to **glucose 6-phosphate** by **phosphoglucomutase**. In the **muscle**, glucose 6-phosphate enters glycolysis. In the **liver**, the phosphate is removed by **glucose 6-phosphatase**, releasing **free glucose** that can be used to **maintain blood glucose** levels at the beginning of a fast.
- A **deficiency** of **glucose 6-phosphatase** causes **glycogen storage disease type Ia** (**Von Gierke disease**). This disease results in an inability of the liver to provide free glucose to the body during a fast. It affects both glycogen degradation and the last step in gluconeogenesis, and causes **severe fasting hypoglycemia**.

Regulation of glycogen synthesis/degradation:

- Allosteric activators and inhibitors of glycogen synthesis and degradation
- Mechanism and effect of calcium on glycogen degradation in muscle
- Effects of insulin and glucagon on glycogen synthesis and degradation

REGULATION OF GLYCOGEN SYNTHESIS AND DEGRADATION (p. 129)

- **Glycogen synthase** and **glycogen phosphorylase** are **allosterically regulated**. In the well-fed state, **glycogen synthase** is **activated** by **glucose 6-phosphate**, but **glycogen phosphorylase** is **inhibited** by **glucose 6-phosphate**, as well as by **ATP**. In the **liver**, **glucose** also serves as an allosteric **inhibitor** of **glycogen phosphorylase**.

- **Ca^{2+}** is released from the **sarcoplasmic reticulum** during exercise. It activates **phosphorylase kinase** in the muscle by binding to the enzyme's **calmodulin** subunit. This allows the enzyme to activate **glycogen phosphorylase**, thereby causing glycogen degradation.

- Glycogen synthesis and degradation are regulated by the same hormonal signals, namely, an **elevated insulin** level results in overall **increased glycogen synthesis** and **decreased glycogen degradation**, whereas an **elevated glucagon** (or **epinephrine**) level causes **increased glycogen degradation** and **decreased glycogen synthesis**. Key enzymes are phosphorylated by a family of **protein kinases**, some of which are **cAMP-dependent** (a compound increased by **glucagon** and **epinephrine**). Phosphate groups are removed by **protein phosphatase** (activated when **insulin** levels are elevated).

Chapter 12: Metabolism of Monosaccharides and Disaccharides

Fructose metabolism:

- Major source of fructose
- Enzymes required for fructose to enter intermediary metabolism,
- Cause and treatment of hereditary fructose intolerance
- Pathway for entry of mannose into intermediary metabolism
- Enzymes required to convert glucose to fructose via sorbitol and their locations in the body
- Pathology resulting from elevated sorbitol in diabetics

FRUCTOSE METABOLISM (p. 135)

- The major source of fructose is **sucrose**, which, when cleaved, releases equimolar amounts of **fructose** and **glucose**. Entry of fructose into cells is **insulin-independent**.

- Fructose is first phosphorylated to **fructose 1-phosphate** by **fructokinase**, and then cleaved by **aldolase B** to **dihydroxyacetone phosphate** and **glyceraldehyde**. These enzymes are found in the **liver**, **kidney**, and **small intestinal mucosa**.

- A deficiency of fructokinase causes a benign condition, but a **deficiency** of **aldolase B** causes **hereditary fructose intolerance**, in which severe hypoglycemia and liver damage can lead to death if the amount of fructose (and, therefore, sucrose) in the diet are not severely limited.

- **Mannose**, an important component of **glycoproteins**, is phosphorylated by **hexokinase** to **mannose 6-phosphate**, which is reversibly isomerized to **fructose 6-phosphate** by **phosphomannose isomerase**.

- **Glucose** can be reduced to **sorbitol** (**glucitol**) by **aldose reductase** in many tissues including the **lens**, **retina**, **Schwann cells**, **liver**, **kidney**, **ovaries**, and **seminal vesicles**. In cells of the **liver**, **ovaries**, **sperm**, and **seminal vesicles**, a second enzyme, **sorbitol dehydrogenase**, can oxidize sorbitol to produce **fructose**. **Hyperglycemia** results in the accumulation of sorbitol in those cells lacking sorbitol dehydrogenase. The resulting **osmotic events** cause **cell swelling**, and contribute to the **cataract formation**, **peripheral neuropathy**, **nephropathy**, and **retinopathy** seen in **diabetes**.

Galactose metabolism:

- Major dietary source of galactose
- Enzymes required for the conversion of galactose to UDP-galactose

GALACTOSE METABOLISM (p. 138)

- The major dietary source of galactose is **lactose**. The entry of galactose into cells is not insulin-dependent.

- Galactose is first phosphorylated by **galactokinase**, which produces **galactose 1-phosphate**. This compound is converted to **UDP-galactose** by **galactose 1-phosphate uridyltransferase**, with the nucleotide supplied by UDP-glucose.

- Enzyme deficiency causing classic galactosemia, and its pathology

- **A deficiency** of **uridyltransferase** causes **classic galactosemia**. Galactose 1-phosphate accumulates, causing **phosphate sequestration**, and excess galactose is converted to **galactitol** by **aldose reductase**. This causes **liver damage**, **severe retardation**, and **cataracts**. **Treatment** requires removal of galactose (and, therefore, lactose) from the diet.

- Enzyme required to interconvert UDP-galactose and UDP-glucose

- In order for **UDP-galactose** to enter the mainstream of glucose metabolism, it must first be converted to **UDP-glucose** by **UDP-hexose 4-epimerase**. This enzyme can also be used to produce UDP-galactose from UDP-glucose when the former is required for the synthesis of structural carbohydrates.

Lactose synthesis:

LACTOSE SYNTHESIS (p. 140)

- Structural components of lactose

- **Lactose** is a **disaccharide** that consists of **galactose** and **glucose**. Milk and other **dairy products** are the dietary sources of lactose.

- Structure and location of the enzyme that synthesizes lactose
- Functions of the enzyme's subunits

- Lactose is synthesized by **lactase synthase** from **UDP-galactose** and **glucose** in the **lactating mammary gland**. The enzyme has two subunits, **protein A** (which is a **galactosyl transferase** found in most cells where it synthesizes **N-acetyllactosamine**) and **protein B** (**α-lactalbumin**, which is found only in the lactating mammary glands, and whose synthesis is stimulated by the peptide hormone, **prolactin**). When both subunits are present, the transferase produces lactose.

Chapter 13: Pentose Phosphate Pathway and NADPH

Pentose phosphate pathway:
- Summary of the pathway
- Reduced coenzymes produced by the pathway

PENTOSE PHOSPHATE PATHWAY (p. 143)

- Also called the **hexose monophosphate shunt**, or **6-phosphogluconate pathway**, the pentose phosphate pathway is found in all cells. It consists of two irreversible oxidative reactions followed by a series of reversible sugar–phosphate interconversions. **No ATP** is directly consumed or produced in the cycle, and **two NADPH are produced** for each glucose 6-phosphate entering the oxidative part of the pathway.

- Most important tissues requiring the pathway
- Regulated enzyme in the pathway and its inhibitor

- The **oxidative portion** of the pentose phosphate pathway is particularly important in **liver** and **mammary glands**, which are active in the biosynthesis of fatty acids, in the **adrenal cortex**, which is active in the NADPH-dependent synthesis of steroids, and in **erythrocytes**, which require NADPH to keep glutathione reduced. **Glucose 6-phosphate** is irreversibly converted to **ribulose 5-phosphate**, and **two NADPH are produced**. The regulated step is **glucose 6-phosphate dehydrogenase** (**G6PD**), which is strongly **inhibited** by **NADPH**.

- Sugars participating in reversible, nonoxidative reactions of the pathway
- Product of these reactions that is required for nucleotide synthesis

- **Reversible nonoxidate reactions** interconvert three-carbon (glyceraldehyde 3-phosphate), four-carbon (erythrose 4-phosphate), five-carbon (ribulose 5-, ribose 5-, and xylulose 5-phosphate), six-carbon (fructose 6-phosphate), and seven-carbon (sedoheptulose 7-phosphate) sugars. This part of the pathway is the source of **ribose 5-phosphate** required for nucleotide and nucleic acid synthesis. Because the reactions are reversible, they can be entered from fructose 6-phosphate and glyceraldehyde 3-phosphate (glycolytic intermediates) if ribose is needed and glucose 6-phosphate dehydrogenase is inhibited.

Uses of NADPH:

USES OF NADPH (p. 145)

- Role of NADPH in fatty acid and steroid synthesis
- Role of NADPH in the conversion of hydrogen peroxide to water

- NADPH is a source of reducing equivalents in **reductive biosynthesis**, such as the production of fatty acids and steroids. It is also required for the **reduction of hydrogen peroxide**, providing the reducing equivalents required by **glutathione** (**GSH**). GSH is used by **glutathione peroxidase** to reduce peroxide to water. The oxidized glutathione is reduced by **glutathione reductase**, using NADPH as the source of electrons.

- Functions of the cytochrome P450 system

- NADPH provides reducing equivalents for the **cytochrome P450 monooxygenase system**, which is used in the **hydroxylation of steroids** to produce steroid hormones, **bile acid synthesis** by the liver, and **activation of vitamin D**. The system also **detoxifies** foreign compounds such as drugs and varied pollutants, including carcinogens, pesticides, and petroleum products.

- Product of NADPH oxidase reaction
- Enzyme deficiency causing chronic granulomatosis

- NADPH provides the reducing equivalents for **phagocytes** in the process of eliminating invading microorganisms. **NADPH oxidase** uses molecular oxygen and NADPH to produce **superoxide radicals**, which in turn can be converted to peroxide, hypochlorous acid, and hydroxyl radicals. **Myeloperoxidase** is an important enzyme in this pathway. A genetic defect in NADPH oxidase causes **chronic granulomatosis**, a disease characterized by severe, persistent, chronic pyogenic infections.

- Functions of nitric oxide

- NADPH is required for the synthesis of **nitric oxide** (**NO**), an important molecule that causes **vasodilation** by relaxing vascular smooth muscle, acts as a kind of **neurotransmitter**, **prevents platelet aggregation**, and helps mediate **macrophage bactericidal activity**.

Glucose 6-P dehydrogenase deficiency:

- Cell type most affected by the deficiency
- Reason that cell type is most affected
- Result of the enzyme deficiency

GLUCOSE 6-P DEHYDROGENASE (G6PD) DEFICIENCY (p. 149)

- This deficiency is a **genetic disease** characterized by **hemolytic anemia**. G6PD deficiency impairs the ability of the cell to form the NADPH that is essential for the maintenance of the reduced glutathione pool. The cells most affected are the **red blood cells** because they do not have additional sources of NADPH. **Free radicals** and **peroxides** formed within the cells cannot be neutralized, causing denaturation of protein (hemoglobin, forming Heinz bodies) and membrane proteins. The cells become rigid, and they are removed by the reticuloendothelial system of the spleen and liver.

- Types of compounds that cause difficulty for people with G6PD deficiency
- Classes of G6PD deficiency and their relative severity

- **Hemolytic anemia** can be caused by the production of free radicals and peroxides, following the taking of **oxidant drugs**, ingestion of **fava beans**, or **severe infections**. Babies with G6PD deficiency may experience **neonatal jaundice** appearing one to four days after birth.
- The degree of severity of the anemia depends on the location of the mutation in the G6PD gene. **Class I mutations** are the **most severe** (for example, **G6PD Mediterranean**). They are often associated with **chronic nonspherocytic anemia**. **Class III mutations** (for example, **G6PD A$^-$**) have a **more moderate form** of the disease.

Chapter 14: Glycosaminoglycans and Glycoproteins

Glycosaminoglycans:

- Composition and charge of glycosaminoglycans

GLYCOSAMINOGLYCANS (p. 155)

- Glycosaminoglycans are **long**, **negatively charged**, **unbranched**, **heteropolysaccharide chains** generally composed of a **repeating disaccharide unit** [acidic sugar–amino sugar]$_n$. The **amino sugar** is either **D-glucosamine** or **D-galactosamine** in which the amino group is usually acetylated, thus eliminating its positive charge. The amino sugar may also be sulfated on carbon four or six or on a nonacetylated nitrogen. The **acidic sugar** is either **D-glucuronic acid** or its carbon-five epimer, **L-iduronic acid**.

- Functions of glycosaminoglycans

- These compounds **bind large amounts of water**, thereby producing the gel-like matrix that forms the basis of the body's **ground substance.** The viscous, lubricating properties of **mucous secretions** are also a result of the presence of glycosaminoglycans, which led to the original naming of these compounds as **mucopolysaccharides.** As essential components of cell surfaces, glycosaminoglycans play an important role in mediating **cell-cell signaling** and **adhesion**.

- Six major classes of glycosaminoglycans

- There are **six major classes** of glycosaminoglycans. These include **chondroitin 4- and 6-sulfates**, **keratan sulfate**, **dermatan sulfate**, **heparin**, **heparan sulfate**, and **hyaluronic acid**.

- Structure of proteoglycan monomers
- Structure of proteoglycan aggregates

- All of the glycosaminoglycans, except hyaluronic acid, are found as components of **proteoglycan monomers**, which consist of a **core protein** to which the linear glycosaminoglycan chains are covalently attached. The proteoglycan monomers associate with a molecule of **hyaluronic acid** to form **proteoglycan aggregates**.

- Cellular location of glycosaminoglycan synthesis
- Steps in glycosaminoglycan synthesis

- Glycosaminoglycans are synthesized in the **endoplasmic reticulum** and the **Golgi**. The polysaccharide chains are elongated by the sequential addition of alternating acidic and amino sugars, donated by their **UDP-derivatives**. The last step in synthesis is sulfation of some of the amino sugars. The source of the sulfate is **3'-phosphoadenosyl-5'-phosphosulfate**.

- Mechanism of degradation of glycosaminoglycans
- Symptoms and examples of mucopolysaccharidoses

- Glycosaminoglycans are **degraded** by **lysosomal hydrolases**. They are first broken down to **oligosaccharides**, which are degraded sequentially by hydrolases from the non-reducing end of each chain. A **deficiency** of one of the hydrolases results in a **mucopolysaccharidosis**. These are hereditary disorders in which glycosaminoglycans accumulate in tissues, causing symptoms such as **skeletal** and **extracellular matrix deformities** and **mental retardation**. Examples of these genetic diseases include **Hunter** and **Hurler syndromes**.

Glycoproteins:

GLYCOPROTEINS (p. 163)

- Structure of glycoproteins

- Glycoproteins are **proteins** to which **oligosaccharides** are covalently attached. They differ from the proteoglycans in that the length of the glycoprotein's carbohydrate chain is relatively short (usually two to ten sugar residues long, although they can be longer). The carbohydrate units of glycoproteins do not have serial repeats as do glycosaminoglycans. The oligosaccharides are attached to proteins by **N-** or **O-glycosidic bonds**.

- Functions of glycoproteins

- **Membrane-bound** glycoproteins participate in a broad range of cellular phenomena, including **cell surface recognition** (by other cells, hormones, viruses), **cell surface antigenicity** (such as the blood group antigens), and as components of the **extracellular matrix** and of the **mucins** of the gastrointestinal and urogenital tracts, where they act as protective biologic lubricants. In addition, almost all of the globular proteins present in human plasma are glycoproteins.

- Site of glycoprotein synthesis
- Intermediate required for N-linked glycoprotein synthesis
- Cause of I-cell disease

- Glycoproteins are **synthesized** in the **endoplasmic reticulum** and the **Golgi**. The precursors of the carbohydrate components of glycoproteins are **sugar nucleotides**. **O-linked glycoproteins** are synthesized by the sequential transfer of sugars from their nucleotide carriers to the protein. **N-linked glycoproteins** contain varying amounts of **mannose**. They also require **dolichol**, an intermediate carrier of the growing oligosaccharide chain. A **deficiency** in the phosphorylation of mannose residues in N-linked glycoprotein pre-enzymes destined for the lysosomes results in **I-cell disease**.

- Site of glycoprotein degradation
- Cause of oligosaccaridoses

- Glycoproteins are degraded in lysosomes by acid hydrolases. A deficiency of one of these enzymes results in a **glycoprotein storage disease** (**oligosaccharidosis**), resulting in accumulation of partially degraded structures in the lysosome.

Chapter 15: Metabolism of Dietary Lipids

Digestion of dietary lipids:

DIGESTION OF DIETARY LIPIDS (p. 171)

- Dietary lipids

- Dietary lipids consist primarily of **triacylglycerol**, with some **cholesterol**, **cholesteryl esters**, **phospholipids**, and free (nonesterified) **fatty acids**.

- Populations for whom acid-stable lipases in the stomach are important for digestion

- Digestion of dietary lipids begins in the **stomach**, where acid-stable lipases (**lingual** and **gastric lipases**) begin removing primarily medium-chain-length fatty acids from triacylglycerol. These enzymes are important in **neonates**, and in individuals with **pancreatic insufficiency**, such as those with **cystic fibrosis**.

- Requirements for lipid emulsification

- In the duodenum, the mixture of lipids is emulsified by **peristalsis**, using **bile salts** as the detergent.

- Pancreatic enzymes required for dietary lipid degradation
- Hormone causing these enzymes' release

- In the duodenum, dietary lipids are degraded by **pancreatic enzymes**: triacylglycerol by **pancreatic lipase**, phospholipids by **phospholipase A_2** and **lysophospholipase**, and cholesteryl esters by **cholesterol esterase**. Enzyme release from the pancreas is controlled by **cholecystokinin**, produced by cells in the intestinal mucosa.

Absorption and secretion of dietary lipids:

- Role of mixed micelles
- Activated form of a fatty acid
- Components of a chylomicron
- Fates of chylomicron components
- Causes and effects of type I and type III hyperlipoproteinemias

ABSORPTION, SECRETION, AND USE OF DIETARY LIPIDS (p. 174)

- The products of lipid digestion—**free fatty acids**, **2-monoacylglycerol**, and **cholesterol**—plus **bile salts**, form **mixed micelles** that are able to cross the unstirred water layer on the surface of the brush border membrane. Individual lipids enter the intestinal mucosal cell cytosol.

- The mixture of lipids moves to the endoplasmic reticulum, where fatty acyl CoA synthetase converts free fatty acids into their activated CoA derivatives. **Fatty acyl CoAs** are then used to produce triacylglycerols, cholesteryl esters, and phospholipids. These, together with the fat-soluble vitamins (A, D, E, and K) and a single protein (**apolipoprotein B-48**), form a **chylomicron**, which is secreted into the **lymphatic system** and carried to the **blood**.

- **Triacylglycerol** in chylomicrons is degraded to **free fatty acids** and **glycerol** by **lipoprotein lipase**, synthesized primarily by the adipocytes and fibroblasts. A deficiency of this enzyme or its coenzyme, apo C-II, causes massive chylomicronemia (**type I hyperlipoproteinemia**). Free fatty acids can be taken up directly, or be carried by **serum albumin** until the fatty acid is absorbed by a cell. Glycerol is metabolized by the liver. **Chylomicron remnants** containing little remaining triacylglycerol, but still carrying **dietary cholesterol**, are absorbed by the **liver**. If removal of the remnants is defective, they accumulate in the plasma (**type III hyperlipoproteinemia**).

Chapter 16: Fatty Acid and Triacylglycerol Metabolism

Structure of fatty acids:

- Structural characteristics of fatty acids
- Cause of Refsum disease
- Essential fatty acids

STRUCTURE OF FATTY ACIDS (p. 179)

- Generally a linear hydrocarbon chain with a terminal carboxyl group, a fatty acid can be **saturated** or **unsaturated**. Branched-chain **phytanic acid** is found in dairy products. An inability to degrade phytanic acid causes its accumulation in plasma and tissues (**Refsum disease**).

- Two fatty acids are essential (must be obtained from the diet): **linoleic** and **linolenic acids**.

<u>De</u> <u>novo</u> synthesis of fatty acids:

- Major tissue site of fatty acid synthesis
- Source of building blocks, energy, reducing equivalents
- Location of the pathway in cells
- Regulated enzyme and its activators and inhibitors

<u>DE</u> <u>NOVO</u> SYNTHESIS OF FATTY ACIDS (p. 180)

- Most fatty acids are synthesized in the **liver** following a meal containing excess carbohydrate and protein. Fatty acids are also synthesized in **lactating mammary glands** and, to a lesser extent, in adipose and kidney.

- Carbons used to synthesize fatty acids are provided by **acetyl CoA**, energy is provided by **ATP**, and reducing equivalents are provided by **NADPH**.

- Fatty acids are synthesized in the **cytosol**. **Citrate** carries two-carbon acetyl units from the mitochondrial matrix to the cytosol.

- The regulated step in fatty acid synthesis (acetyl CoA $\rightarrow$ malonyl CoA) is catalyzed by **acetyl CoA carboxylase**, which requires **biotin**. **Citrate** is the allosteric **activator**, and **long-chain fatty acyl CoA** is the **inhibitor**. The enzyme can also be activated in the presence of **insulin** and inactivated in the presence of **epinephrine** or **glucagon**.

- End-product of the pathway
- Sites of further fatty acid elongation and desaturation
- Form in which fatty acids are stored

- The rest of the steps in fatty acid synthesis are catalyzed by the **fatty acid synthase complex**, which produces **palmitoyl CoA** from acetyl CoA and malonyl CoA, with NADPH as the source of reducing equivalents.
- Palmitoyl can be further elongated in the endoplasmic reticulum and the mitochondria, and desaturated by mixed function oxidases in the endoplasmic reticulum.
- Fatty acids are stored as components of triacylglycerol in adipose tissue.

Fatty acid degradation:

- Regulation of triacylglycerol degradation in adipocytes
- Carrier of fatty acids in the blood
- Tissues that cannot use fatty acids as fuel, and why
- Fate of the glycerol backbone
- Cellular location of fatty acid degradation
- Components of the shuttle required for fatty acid transport into the mitochondria
- Result of shuttle component deficiencies in muscle and liver.
- Examples of causes of carnitine deficiency
- End-products of β-oxidation
- Characteristics of MCAD deficiency
- Products of odd-number chain oxidation
- Causes of methylmalonic aciduria

FATTY ACID DEGRADATION (p. 187)

- When lipids are required by the body for energy, adipose cell **hormone-sensitive lipase** (**activated** by **epinephrine**, and **inhibited** by **insulin**) initiates degradation of stored triacylglycerol.
- Fatty acids are carried by **serum albumin** to the liver and to peripheral tissues, where oxidation of the lipids provides energy. (Cells, such as red blood cells, with few or no mitochondria cannot oxidize fatty acids, nor can the brain, because long-chain fatty acids do not cross the blood-brain barrier.)
- The **glycerol** backbone of the degraded triacylglycerol is carried by the blood to the **liver**, where it serves as an important **gluconeogenic precursor**.
- Fatty acid degradation (**β-oxidation**) occurs in **mitochondria**. The **carnitine shuttle** is required to transport fatty acids from the cytosol to the mitochondria. Enzymes required are **carnitine palmitoyltransferases I** (CPT I, cytosolic side of inner mitochondrial membrane) and **II** (CPT II, an enzyme of the inner mitochondrial membrane). CPT I is **inhibited** by **malonyl CoA**. This prevents fatty acids that are being synthesized in the cytosol from malonyl CoA from being transported into the mitochondria where they would be degraded.
- Genetic **CPT II deficiency** in **cardiac** and **skeletal muscle** causes **cardiomyopathy**, and **myoglobinemia** and **weakness** following exercise. Genetic **CPT I deficiency** affects the **liver**, where an inability to use long-chain fatty acids for energy during a fast can cause severe hypoglycemia. **Carnitine deficiency** can be caused, for example, by malnutrition, liver disease, strict vegetarianism, and hemodialysis.
- Once in the mitochondria, fatty acids are oxidized, producing acetyl CoAs, NADHs, and $FADH_2$s. The first step in the β-oxidation pathway is catalyzed by one of a family of four acyl CoA dehydrogenases that each has a specificity for either short-, medium-, long-, or very-long-chain fatty acids. **Medium-chain fatty acyl CoA dehydrogenase** (**MCAD**) **deficiency** is one of the most common inborn errors of metabolism. It causes a decrease in fatty acid oxidation, resulting in severe hypoglycemia. Treatment includes a carbohydrate-rich diet.
- Oxidation of fatty acids with an odd number of carbons proceeds two carbons at a time (producing acetyl CoA) until the last three carbons (**propionyl CoA**). This compound is converted to **methylmalonyl CoA** (a reaction requiring **biotin**), which is then converted to **succinyl CoA** by methylmalonyl CoA mutase (requiring **vitamin B_{12}**). A genetic error in the mutase or vitamin B_{12} deficiency causes **methylmalonic acidemia** and **aciduria**.

Ketone bodies:

- Names of the ketone bodies
- Fate of the ketone bodies

KETONE BODIES (p. 193)

- Liver mitochondria can convert acetyl CoA derived from fatty acid oxidation into the ketone bodies, **acetoacetate** and **β-hydroxybutyrate**. (**Acetone**, a nonmetabolizable ketone body, is produced spontaneously from acetoacetate in the blood.) Peripheral tissues possessing mitochondria can oxidize β-hydroxybutyrate to acetoacetate, which can be reconverted to acetyl CoA, thus producing energy for the cell.

• Tissues that can use ketone bodies for fuels	• Unlike fatty acids, ketone bodies can be used by the **brain** (but not by cells, such as red blood cells, that lack mitochondria) and, thus, are important fuels during a fast. The liver lacks the ability to degrade ketone bodies, and so synthesizes them specifically for the peripheral tissues.
• Definition of ketoacidosis and example of a disease where it occurs	• **Ketoacidosis** occurs when the rate of formation of ketone bodies is greater than the rate of use, as seen in cases of uncontrolled, **insulin-dependent diabetes mellitus**.

Chapter 17: Complex Lipid Metabolism

Phospholipids:	**PHOSPHOLIPIDS (p. 199)**
• General structure of a phospholipid	• Phospholipids are **polar**, **ionic** compounds composed of an **alcohol** (for example, **choline**, **ethanolamine**, **serine**, and **glycerol**) attached by a phosphodiester bridge to either **diacylglycerol** or to **sphingosine**.
• Functions of the phospholipid	• Phospholipids are the predominant lipids of **cell membranes**. They also function as a reservoir of **intracellular messengers** and as **anchors** for some proteins to cell membranes. Non–membrane-bound phospholipids serve as components of **lung surfactant** and **bile**.
• Definition and example of a glycerophospholipid	• Phospholipids that contain glycerol are called **glycerophospholipids** or **phosphoglycerides**. All contain **phosphatidic acid**, the simplest glycerophospholipid. When an alcohol, such as choline, is esterified to phosphatidic acid, the product is **phosphatidylcholine**.
• Structure and function of cardiolipin	• **Cardiolipin** contains two molecules of phosphatidic acid esterified through their phosphate groups to an additional molecule of glycerol. This is the only human glycerophospholipid that is **antigenic**. It is an important component of the **inner mitochondrial membrane**.
• Structures and tissue locations of two important plasmalogens	• **Plasmalogens** are glycerophospholipids that have the fatty acid at carbon 1 of the glycerol backbone attached by an **ether**, rather than an ester linkage. **Phosphatidalethanolamine** (abundant in nerve tissue) and **phosphatidalcholine** (abundant in heart muscle) are the two quantitatively significant plasmalogens.
• Structural components and function of sphingomyelin	• The alcohol **sphingosine** attached to a long-chain fatty acid produces a **ceramide**. Addition of a **phosphorylcholine** produces the phospholipid **sphingomyelin**, which is the only significant sphingophospholipid in humans. It is an important constituent of **myelin**.
• Source of choline and ethanolamine used for phospholipid synthesis	• **Phosphatidylethanolamine** (**PE**) and **phosphatidylcholine** (**PC**) are the most abundant phospholipids in most eukaryotic cells. The primary route of their synthesis uses choline and ethanolamine obtained either from the **diet** or from the **turnover** of the body's phospholipids. Because the amount of choline the body makes is insufficient for its need, **choline** is an **essential dietary nutrient**.
• Phospholipid that is the major component of lung surfactant, and the syndrome caused by its deficiency	• **Dipalmitoylphosphatidylcholine** (DPPC, also called **dipalmitoyllecithin**, DPPL) is the major lipid component of **lung surfactant**. It is made and secreted by type II granular pneumocytes. Insufficient surfactant production causes **respiratory distress syndrome**, which can occur in preterm infants or adults whose surfactant-producing pneumocytes have been damaged or destroyed.
• Fatty acid for which phosphatidylinositol (PI) serves as a reservoir • Role PI plays in signal transmission across cell membranes	• **Phosphatidylinositol** (**PI**) serves as a reservoir for **arachidonic acid** in membranes. The phosphorylation of membrane-bound PI produces **phosphatidylinositol 4,5-bisphosphate** (**PIP_2**). This compound is degraded by **phospholipase C** in response to the binding of a variety of neurotransmitters, hormones, and growth factors to membrane receptors. The products of this degradation, **inositol 1,4,5-trisphosphate** (**IP_3**) and **diacylglycerol** mediate the mobilization of intracellular **calcium** and the activation of **protein kinase C**, which act synergistically to evoke specific cellular responses.

- Cause of paroxysmal nocturnal hemoglobinuria

- Specific proteins can be covalently attached via a carbohydrate bridge to membrane-bound PI (**glycosylphosphatidylinositol**, or **GPI**). This allows GPI-anchored proteins rapid lateral mobility on the surface of the plasma membrane. A deficiency in the synthesis of GPI in hematopoietic cells results in a hemolytic disease, **paroxysmal nocturnal hemoglobinuria**.

- Enzymes involved in phospholipid degradation
- Cause of Niemann-Pick disease

- The **degradation** of phosphoglycerides is performed by **phospholipases** found in all tissues and pancreatic juice. **Sphingomyelin** is degraded to a ceramide plus phosphorylcholine by the lysosomal enzyme **sphingomyelinase**. A deficiency in sphingomyelinase causes **Niemann-Pick disease**, which can cause rapid and progressive neurodegeneration in infants.

Glycolipids:

GLYCOLIPIDS (p. 207)

- Structures of cerebrosides, globosides, gangliosides, and sulfatides

- Almost all glycolipids are derivatives of ceramides to which carbohydrates have been attached (**glycosphingolipids**). When one sugar molecule is added to the ceramide, a **cerebroside** is produced. If an oligosaccharide is added, a **globoside** is produced; if an acidic N-acetylneuraminic acid molecule is added, a **ganglioside** is produced; if a cerebroside is sulfated, a **sulfoglycosphingolipid** (**sulfatide**) is produced.

- Predominant locations of glycolipids

- Glycolipids are found predominantly in cell membranes of the **brain** and **peripheral nervous tissue**, with high concentrations in the **myelin sheath**. Glycolipids are very **antigenic**.

- Sites of glycolipid synthesis and degradation

- Glycolipids are synthesized in the **endoplasmic reticulum** and **Golgi**. They are degraded in the **lysosomes** by hydrolytic enzymes that sequentially remove groups from the glycolipid in the reverse order from which they were added during synthesis ("last on, first off").

- Cause of sphingolipidoses

- **Sphingolipidoses** are genetic lipid storage diseases. A specific **lysosomal hydrolytic enzyme** is deficient in each disorder, resulting in the accumulation of a characteristic sphingolipid. These diseases are all autosomal recessive except **Fabry disease**, which is X-linked. The incidence of these diseases is low in most populations, except for **Gaucher** and **Tay-Sachs diseases**, which show a high incidence in Ashkenazi Jews.

Prostaglandins and related compounds:

PROSTAGLANDINS AND RELATED COMPOUNDS (p. 211)

- Compounds known as eicosanoids

- **Prostaglandins** (**PG**), **thromboxanes** (**TX**), and **leukotrienes** (**LT**) are collectively known as **eicosanoids**. Produced in very small amounts in almost all tissues, they act locally, and have an extremely short half-life.

- Dietary precursor
- Enzyme that releases arachidonic acid from membrane phospholipid
- Enzymes that catalyze the first step in prostaglandin and leukotriene synthesis

- The dietary precursor of the eicosanoids is the essential fatty acid, **linoleic acid**. It is elongated and desaturated to **arachidonic acid**, the immediate precursor of prostaglandins, which is stored in the membrane as a component of a phospholipid—generally phosphatidylinositol (PI). Arachidonic acid is released from PI by **phospholipase** A_2.

- Synthesis of **prostaglandins** and **thromboxanes** begins with the oxidative cyclization of free arachidonic acid to yield PGH_2 by **prostaglandin endoperoxide synthase**—a microsomal protein that has two catalytic activities: **fatty acid cyclooxygenase** (**COX**) and peroxidase. There are two isozymes of the synthase: **COX-1** and **COX-2**. **Leukotrienes** are produced by the **5-lipoxygenase** pathway.

- Drugs that inhibit eicosanoid synthesis

- **Cortisol** inhibits phospholipase A_2 and COX-2. Non-steroidal antiinflammatory drugs, such as **aspirin**, inhibit both COX-1 and COX-2, whereas **celecoxib** inhibits COX-2. 5-Lipoxygenase inhibitors are used in the treatment of asthma.

Chapter 18: Cholesterol and Steroid Metabolism

Structure of cholesterol:

- Overall structure of cholesterol
- Structure of a cholesterol ester

STRUCTURE OF CHOLESTEROL (p. 217)

- Cholesterol is a very **hydrophobic** compound. It consists of four fused hydrocarbon rings (A, B, C, and D) plus an eight-membered, branched hydrocarbon chain attached to the D ring. Cholesterol has a single hydroxyl group—located at carbon 3 of the A ring—to which a fatty acid can be attached, producing a **cholesteryl ester**.

Synthesis of cholesterol:

- Tissues synthesizing most of the body's cholesterol
- Sources of carbons, reducing equivalents, and energy sources
- Cellular site of cholesterol synthesisr
- Rate-limiting enzyme in cholesterol synthesis
- Mechanisms by which this enzyme is regulated

SYNTHESIS OF CHOLESTEROL (p. 218)

- Cholesterol is synthesized by virtually all tissues in humans, although **liver**, **intestine**, **adrenal cortex**, and **reproductive tissues** make the largest contribution to the body's cholesterol pool.
- As with fatty acids, all the carbon atoms in cholesterol are provided by **acetate**, and **NADPH** provides the reducing equivalents. The pathway is driven by hydrolysis of the high-energy thioester bond of acetyl CoA and the terminal phosphate bond of ATP.
- Cholesterol is synthesized in the **cytoplasm**, with enzymes in both the **cytosol** and the membrane of the endoplasmic reticulum.
- The **rate-limiting step** in cholesterol synthesis is cytosolic **HMG CoA reductase**, which produces **mevalonic acid** from hydroxymethylglutaryl CoA (HMG CoA). The enzyme is regulated by a number of mechanisms: 1) **Expression of the HMG CoA reductase gene** is controlled by a transcription factor that is activated when cholesterol levels are low, resulting in increased enzyme and, therefore, more cholesterol synthesis. 2) HMG CoA reductase activity is controlled covalently through the actions of a **protein kinase** (**inactivates** the enzyme) and a **protein phosphatase** (**activates** the enzyme). 3) **Insulin activates** HMG CoA reductase, whereas **glucagon deactivates** it. 4) Drugs such as **lovastatin** and **mevastatin** are **competitive inhibitors** of HMG CoA reductase, and are used to decrease plasma cholesterol in patients with hypercholesterolemia.

Degradation of cholesterol:

- Mechanisms of cholesterol disposal

DEGRADATION OF CHOLESTEROL (p. 222)

- The ring structure of cholesterol can not be metabolized in humans. Cholesterol can be eliminated from the body either by **conversion to bile salts** or by **secretion into the bile**. Intestinal bacteria can reduce cholesterol to **coprostanol** and **cholestanol**, which together with cholesterol make up the bulk of **neutral fecal sterols**.

Bile acids and bile salts:

- Most important organic components of bile
- Names, structural characteristics, and functions of bile acids
- Tissue where bile acids are synthesized, and the regulated step

BILE ACIDS AND BILE SALTS (p. 222)

- **Bile salts** and **phosphatidylcholine** are quantitatively the most important organic components of bile. Bile salts are **conjugated bile acids**.
- The **primary bile acids**, **cholic** or **chenodeoxycholic acids,** contain two or three alcohol groups, respectively. Both have a shortened side chain that terminates in a carboxyl group. These structures are **amphipathic**, and can serve as **emulsifying agents**.
- **Bile acids** are synthesized in the **liver**. The rate-limiting step is catalyzed by **cholesterol-7-α-hydroxylase**, which is **activated** by **cholesterol** and **inhibited** by **bile acids**.

- How primary bile salts are formed and their names
- Names and mechanism of production of the secondary bile acids

- Before the bile acids leave the liver, they are conjugated to a molecule of either **glycine** or **taurine**, producing the **primary bile salts**: **glycocholic** or **taurocholic acids**, and **glycochenodeoxycholic** or **taurochenodeoxycholic acids**. Bile salts are more amphipathic than bile acids and, therefore, are more effective emulsifiers. In the intestine, bacteria can remove the glycine and taurine, and can remove a hydroxyl group, producing the **secondary bile acids—deoxycholic** and **lithocholic acids**.

- Definition of the enterohepatic circulation

- Bile is secreted into the intestine, and more than 95 percent of the bile acids and salts are efficiently reabsorbed. They are actively transported from the intestinal mucosal cells into the portal blood, where they are carried by albumin back to the liver (**enterohepatic circulation**). In the liver, the primary and secondary bile acids are reconverted to bile salts, and secreted into the bile.

- Major causes of cholelithiasis

- If more cholesterol enters the bile than can be solubilized by the available bile salts and phosphatidylcholine, **cholesterol gallstone disease** (**cholelithiasis**) can occur. This is generally caused by gross malabsorption of bile acids from the intestine, obstruction of the biliary tract, or severe hepatic dysfunction, leading to abnormalities in bile or bile salt production.

Plasma lipoproteins:

PLASMA LIPOPROTEINS (p. 225)

- Major classes of lipoproteins
- Functions of these particles

- The plasma lipoproteins include **chylomicrons**, **very-low-density lipoproteins** (**VLDL**), **low-density lipoproteins** (LDL), and **high-density lipoproteins** (**HDL**). They keep lipids (primarily, **triacylglycerol** and **cholesteryl esters**) soluble as they transport them in the plasma, and provide an efficient mechanism for transporting their lipid contents between tissues.

- Structural components of the lipoproteins

- Lipoproteins are composed of a **neutral lipid core** (containing **triacylglycerol**, **cholesteryl esters**, or both) surrounded by a shell of amphipathic **apolipoproteins**, **phospholipid**, and **nonesterified cholesterol**.

- Tissue where chylomicrons are synthesized
- Types and sources of the molecules that comprise a functional chylomicron
- Activator of lipoprotein lipase
- Products of lipoprotein lipase and their fates
- Disease caused by a deficiency of lipoprotein lipase or apo C-II
- Fate of the chylomicron

- **Chylomicrons** are assembled in **intestinal mucosal cells** from dietary lipids (primarily **triacylglycerol**), plus additional lipids synthesized in these cells. Each nascent chylomicron particle has one molecule of **apolipoprotein B-48** (**apo B-48**). They are released from the cells into the lymphatic system and travel to the blood, where they receive **apo C-II** and **apo E** from **HDLs**. This makes the chylomicrons functional.

- Apo C-II activates **lipoprotein lipase**, which degrades the chylomicron's **triacylglycerol** to fatty acids and glycerol. The **fatty acids** that are released are **stored** (in the **adipose**) or used for **energy** (by the **muscle**). The **glycerol** is metabolized by the **liver**. Patients with a **deficiency** of **lipoprotein lipase** or **apo C-II** show a dramatic accumulation of chylomicrons in the plasma (**type 1 hyperlipoproteinemia**, **familial lipoprotein lipase deficiency**, or **hypertriacylglycerolemia**).

- After most of the triacylglycerol is removed, apo C-II is returned to the HDL, and the **chylomicron remnant**—carrying most of the **dietary cholesterol**—binds to a **receptor** on the **liver** that recognizes **apo E**. The particle is **endocytosed** and its contents degraded by **lysosomal enzymes**.

- Site of production, particle composition, and function of VLDLs

- **Nascent VLDLs** are produced in the **liver**, and are composed predominantly of **triacylglycerol**. They contain a single molecule of **apo B-100**. Like nascent chylomicrons, VLDLs receive **apo C-II** and **apo E** from **HDLs** in the plasma. VLDLs **carry triacylglycerol** from the liver to the **peripheral tissues**, where **lipoprotein lipase** degrades the lipid.

- Modifications of VLDL in the plasma
- Composition and fate of LDLs
- Disease caused by a deficiency of LDL receptors

- As triacylglycerol is removed from the VLDL, the particle receives **cholesteryl esters** from **HDL**. This process is accomplished by **cholesteryl ester transfer protein**. Eventually, VLDL in the **plasma** is **converted to LDL**—a much smaller, denser particle. Apo CII and apo E are returned to HDLs, but the LDL retains **apo B-100**, which is recognized by **receptors** on **peripheral tissues** and the **liver**. LDLs undergo **receptor-mediated endocytosis**, and their contents are degraded in the **lysosomes**. A **deficiency of functional LDL receptors** causes **type II hyperlipidemia** (**familial hypercholesterolemia**). The endocytosed cholesterol **inhibits HMG CoA reductase** and **decreases synthesis of LDL receptors**. Some of it can also be esterified by **acyl CoA:cholesterol acyltransferase** and stored.

- Site of HDL synthesis
- Functions of HDLs in the body

- **HDLs** are synthesized by the **liver** and **intestine**. They have a number of functions, including: 1) serving as a **circulating reservoir of apo C-II** and **apo E** for chylomicrons and VLDL; 2) removing **unesterified cholesterol** from cell surfaces and other lipoproteins and **esterifying it** using **phosphatidylcholine:cholesterol acyl transferase**, a liver-synthesized plasma enzyme that is activated by **apo A-1**; and 3) delivering these cholesterol esters to the liver ("**reverse cholesterol transport**").

Steroid hormones:

STEROID HORMONES (p. 235)

- Classes of steroid hormones
- Sites of steroid hormone synthesis

- **Cholesterol** is the precursor of all classes of steroid hormones (**glucocorticoids**, **mineralocorticoids**, and the **sex hormones**—the **androgens**, **estrogens**, and **progestins**). **Synthesis**, using primarily **mixed-function oxidases**, occurs in the **adrenal cortex** (cortisol, aldosterone, and androgens), **ovaries** and **placenta** (estrogens and progestins), and **testes** (testosterone).

- Mechanism of action of steroid hormones

- Each steroid hormone diffuses across the plasma membrane of its target cell and binds to a specific **cytosolic** or **nuclear receptor**. These **receptor–ligand complexes** accumulate in the nucleus, dimerize, and **bind** to specific regulatory DNA sequences (**hormone-response elements**) in association with coactivator proteins, thereby causing **promoter activation** and **increased transcription** of targeted genes.

- Tissues and their hormones that control the secretion of the steroid hormones

- Steroid hormones are secreted from their tissues of origin in response to hormonal signals. **Corticotropin-releasing hormone** produced by the **hypothalamus** stimulates the **pituitary** to secrete **adrenocorticotropic hormone**, which stimulates the **middle layer** of the **adrenal cortex** to secrete **cortisol**. The **hypothalamus** also secretes **gonadotropin-releasing hormone**, which stimulates the **anterior pituitary** to release: 1) **luteinizing hormone** (which stimulates the **testes** to produce **testosterone**, and the **ovaries** to produce **estrogens** and **progesterone**), and 2) **follicle-stimulating hormone**, which regulates the growth of **ovarian follicles**, and stimulates **testicular spermatogenesis**). The secretion of **aldosterone** from the **outer layer** of the **adrenal cortex** is induced by the hormone **angiotensin II**, and by a decrease in the **plasma Na^+/K^+ ratio**.

Chapter 19: Amino Acids: Disposal of Nitrogen

Overall nitrogen metabolism:

OVERALL NITROGEN METABOLISM (p. 243)

- Definition of amino acid pool
- Sources and fates of amino acids in the body

- The **amino acid pool** is defined as all the free amino acids in cells and extracellular fluid.

- These free amino acids are obtained from the **degradation of dietary protein**, the constant **turnover of body protein**, and the **synthesis of nonessential amino acids**. Free amino acids are consumed by **synthesis of body protein** and **metabolism of their carbon skeletons**.

- Definition of protein turnover

- In a healthy adult, the rate of protein synthesis is just sufficient to replace the protein that is degraded. This process is called **protein turnover**.

- Two major enzyme systems responsible for degrading damaged or unneeded proteins

- Two major enzyme systems are responsible for degrading damaged or unneeded proteins. The first is the **ubiquitin-proteasome mechanism**, in which intracellular proteins destined for degradation are covalently tagged with ubiquitin. They are then recognized and degraded by the proteasome. **Lysosomal enzymes** primarily degrade extracellular proteins.

- Location of polar and nonpolar R-groups in proteins found in either an aqueous or hydrophobic environment

- Amino acids with **nonpolar** (hydrophobic) R-groups are generally found in the **interior** of proteins that function in an aqueous environment, and on the surface of proteins (such as membrane proteins) that interact with lipids. Amino acids with **polar** side chains are generally found on the **outside** of proteins that function in an aqueous environment, and in the interior of membrane-associated proteins.

Digestion of dietary proteins:

DIGESTION OF DIETARY PROTEINS (p. 245)

- Role of the stomach in dietary protein degradation

- In the **stomach**, **hydrochloric acid** denatures dietary proteins, making them more susceptible to proteases. ***Pepsin***, an enzyme secreted in zymogen form by the serous cells of the stomach, releases peptides and a few free amino acids from dietary proteins.

- Roles of the pancreas and small intestine in dietary protein degradation

- In the **small intestine**, proteases released by the **pancreas** as zymogens become active. Each has a different specificity for the amino acid R-groups adjacent to the susceptible peptide bond. Examples of these enzymes are **trypsin**, **chymotrypsin**, **elastase**, and **carboxypeptidase A** and **B**. The resulting **oligopeptides** are cleaved by **aminopeptidase** found on the **luminal surface of the intestine**. Free amino acids and dipeptides are then absorbed by the intestinal epithelial cells.

- Cause of cystinuria

- At least seven different systems are known for transporting amino acids into cells. In the inherited disorder **cystinuria**, the carrier system responsible for reabsorption of the amino acids **cysteine**, **ornithine**, **arginine**, and **lysine** in the proximal convoluted tubule of the kidney is defective. The inability to reabsorb **cystine** leads to **kidney stones**.

Removal of nitrogen from amino acids:

REMOVAL OF NITROGEN FROM AMINO ACIDS (p. 247)

- Two most important enzymes involved in transamination, their coenzyme, and diagnostic value

- Amino groups are funneled to glutamate from all amino acids except lysine and threonine. The enzymes are **aminotransferases**, and they are reversible. The two most important of these enzymes are **alanine aminotransferase** (**ALT**) and **aspartate aminotransferase** (**AST**). Aminotransferases require **pyridoxal phosphate** as a coenzyme. The presence of elevated levels of aminotransferases in the plasma can be used to diagnose **liver disease**.

- Enzyme used to oxidatively deaminate glutamate

- Glutamate can be oxidatively deaminated in the liver by **glutamate dehydrogenase**, liberating **free ammonia** that can be used to make urea.

- Two major mechanisms for ammonia transport in the blood

- **Ammonia transport** in the blood from the peripheral tissues to the liver occurs by two major mechanisms: **glutamine** can be synthesized from glutamate and ammonia (glutamine synthetase) or pyruvate can be transaminated to **alanine**. In the liver, the ammonia group is removed from glutamine by glutaminase and from alanine by transamination.

Urea cycle:

UREA CYCLE (p. 251)

- Sources of the two nitrogen atoms in urea

- A portion of the free ammonia is excreted in the urine, but most is used in the synthesis of **urea** by the liver, which is quantitatively the most important route for disposing of nitrogen from the body. One nitrogen of the urea molecule is supplied by free **NH_3** and one by **aspartate**.

- Rate-limiting step in the urea cycle

- **Carbamoyl phosphate synthetase I** produces carbamoyl phosphate in the mitochondria from CO_2, **NH_3**, and two ATP molecules. The enzyme, which has an absolute requirement for its **positive allosteric effector**, **N-acetylglutamate**, is the **rate-limiting** step in the cycle.

- Components of the urea cycle

- Carbamoyl phosphate and **ornithine** combine to form **citruline**, which is transported out of the mitochondrion. **Aspartic acid** (the source of the **second N** in urea) and citruline combine to form **argininosuccinate**, which is converted to **arginine**. Arginase cleaves the arginine, releasing urea and ornithine.

- Causes and symptoms of hyperammonemia

- When liver function is compromised, as a result of genetic defects in one of the urea cycle enzymes or to liver disease, **hyperammonemia** (**ammonia intoxication**) can occur. Symptoms include tremors, slurring of speech, somnolence, vomiting, cerebral edema, and blurring of vision. All inherited deficiencies of urea cycle enzymes cause mental retardation.

Chapter 20: Amino Acids Degradation and Synthesis

Catabolism of amino acid carbon skeletons:

CATABOLISM OF AMINO ACID CARBON SKELETONS (p. 260)

- Definition of glucogenic amino acids, and examples of the products made from them

- Amino acids whose catabolism yields pyruvate or one of the intermediates of the TCA cycle, are termed **glucogenic**. They can give rise to the net formation of glucose or glycogen in the liver and glycogen in the muscle. The glucogenic amino acids that form **α-ketoglutarate** are glutamine, glutamate, proline, arginine, and histidine; those that form **pyruvate** are alanine, serine, glycine, cysteine, and threonine; those that form **fumarate** are phenylalanine and tyrosine; those that form **succinyl CoA** are methionine, valine, isoleucine, and threonine; and the amino acids aspartate and asparagine form **oxaloacetate**.

- Definition of ketogenic amino acids, and examples of the products made from them

- Amino acids whose catabolism yields either **acetoacetate** or one of its precursors, **acetyl CoA** or **acetoacetyl CoA**, are termed **ketogenic**. Tyrosine, phenylalanine, tryptophan, and isoleucine are both ketogenic and glucogenic. Leucine and lysine are solely ketogenic.

Biosynthesis of nonessential amino acids:

BIOSYNTHESIS OF NONESSENTIAL AMINO ACIDS (p. 265)

- Non-essential amino acids and the sources of their carbons

- **Nonessential** amino acids can be synthesized from metabolic intermediates, or from the carbon skeletons of essential amino acids. Nonessential amino acids include **alanine**, **aspartate**, and **glutamate** (made by transamination of α-keto acids), **glutamine** and **asparagine** (made by amidation of glutamate and aspartate), **proline** (made from glutamate), **cysteine** (made from methionine and serine), **serine** (made from 3-phosphoglycerate), **glycine** (made from serine), and **tyrosine** (made from phenylalanine). **Essential amino acids** need to be obtained from the **diet**.

Metabolic defects:

METABOLIC DEFECTS IN AMINO ACID METABOLISM (p. 266)

- Enzyme deficiency, diagnosis, and treatment of phenylketonuria

- **Phenylketonuria** (**PKU**) is caused by a **deficiency** of **phenylalanine hydroxylase**—the enzyme that converts phenylalanine to tyrosine. **Hyperphenylalaninemia** may also be caused by deficiencies in the enzymes that synthesize or reduce the hydroxylase's coenzyme, **tetrahydrobiopterin**. Untreated patients with PKU suffer from **mental retardation**, failure to walk or talk, seizures, hyperactivity, tremor, microcephaly, and failure to grow. A blood test administered 48 hours after a newborn has started ingesting protein can be used to diagnose the disease. Treatment involves controlling dietary phenylalanine. Note that tyrosine becomes an essential dietary component for people with PKU.

- Enzyme deficiency, diagnosis, and treatment of maple syrup urine disease

- **Maple syrup urine disease** (**MSUD**) is a recessive disorder in which there is a partial or complete **deficiency** in **branched-chain α-ketoacid dehydrogenase**—an enzyme that decarboxylates **leucine**, **isoleucine**, and **valine**. These amino acids and their corresponding α-keto acids accumulate in the blood, causing a toxic effect that interferes with brain function. Symptoms include feeding problems, vomiting, dehydration, severe metabolic acidosis, and a characteristic smell of the urine. If untreated, the disease leads to **mental retardation**, physical disabilities, and death. Diagnosis is based on a blood sample within 24 hours of birth. Treatment of MSUD involves a synthetic formula that contains limited amounts of leucine, isoleucine, and valine.

• Examples of other important genetic diseases associated with amino acid metabolism	• Other important genetic diseases associated with amino acid metabolism include **albinism**, **homocystinuria**, **methylmalonyl CoA mutase deficiency**, **alkaptonuria**, **histidinemia**, and **cystathioninuria**.

Chapter 21: Conversion of Amino Acids to Specialized Products

Porphyrin metabolism:	**PORPHYRIN METABOLISM (p. 275)**
• Structure and use of porphyrins	• Porphyrins are cyclic compounds that readily bind metal ions—usually Fe^{2+} or Fe^{3+}. The most prevalent metalloporphyrin in humans is **heme**, which is found in **hemoglobin**, **myoglobin**, **cytochromes**, and the enzymes **catalase** and **tryptophan pyrrolase**.
• Tissues that synthesize heme, and the sources of porphyrin's carbon and nitrogen	• The major sites of **heme biosynthesis** are the **liver** (where the rate of synthesis is highly variable) and the erythrocyte-producing cells of the **bone marrow** (where the rate is generally constant). All the carbon and nitrogen atoms are provided by **glycine** and **succinyl CoA**.
• Committed step in heme synthesis, its coenzyme, and inhibitor	• The **committed step** in heme synthesis is the formation of **δ-aminolevulinic acid** (**ALA**). The reaction, which requires **pyridoxal phosphate** as a coenzyme, is catalyzed by ALA synthase. The reaction is **inhibited** by **hemin** (the oxidized form of heme that accumulates in the cell when it is being under-used). The conversion of protoporphyrin IX to heme, catalyzed by ferrochelatase, is **inhibited** by **lead**.
• Definition of porphyrias, their modes of genetic inheritance, and their treatment	• **Porphyrias** are caused by inherited (or occasionally acquired) **defects in heme synthesis**, resulting in the accumulation and increased excretion of porphyrins or porphyrin precursors. Porphyrias are classified as **erythropoietic** or **hepatic**, depending where the enzyme deficiency occurs. With the exception of congenital erythropoietic porphyria, which is a genetically recessive disease, all the porphyrias are inherited as **autosomal dominant disorders**. All porphyrias result in a decreased synthesis of heme and, therefore, ALA synthase is derepressed. The severity of symptoms of the porphyrias can be diminished by intravenous injections of hemin. Because some porphyrias result in **photosensitivity**, avoidance of sunlight is helpful.
Degradation of heme:	**DEGRADATION OF HEME (p. 281)**
• Degradation pathway for heme	• Heme is degraded by **macrophages** to **biliverdin** (green), which is reduced to **bilirubin** (red-orange). Bilirubin complexed with albumin is carried via the blood to the **liver**, where the bilirubin's solubility is increased by the addition of two molecules of glucuronic acid. **Bilirubin diglucuronide** is transported into the bile canaliculi, where it is first hydrolyzed and reduced by bacteria to yield **urobilinogen**, then oxidized by intestinal bacteria to **stercobilin** (brown). A portion of the urobilinogen is transported by the blood to the kidney, where it is converted to **urobilin** (yellow) and eliminated in the urine.
• Definition and causes of jaundice	• **Jaundice** (**icterus**) refers to the yellow color of the skin, nail beds, and sclerae caused by deposition of **bilirubin**, secondary to increased bilirubin levels in the blood. There are three major forms of jaundice: **hemolytic jaundice**, caused by massive lysis of red blood cells, releasing more heme than can be handled by the reticuloendothelial system; **obstructive jaundice**, resulting from obstruction of the bile duct; and **hepatocellular jaundice**, caused by damage to liver cells that decreases the liver's ability to take up and conjugate bilirubin. In addition, **neonatal jaundice** is caused by the low activity of hepatic glucuronylation of bilirubin, especially in premature infants.

Other nitrogen-containing compounds:

- Examples of other N-containing compounds, and the amino acids from which they are synthesized

OTHER NITROGEN-CONTAINING COMPOUNDS (p. 283)

- Other important nitrogen-containing compounds made from amino acids include the **catecholamines** (dopamine, norepinephrine, and epinephrine), which are synthesized from tyrosine; **creatine**, which is synthesized from arginine and glycine; **histamine**, which is synthesized from histidine; and **serotonin**, which is synthesized from tryptophan.

Chapter 22: Nucleotide Metabolism

Nucleotide structure:

- Structural components of nucleotides
- Sugars, purines, and pyrimidines found in DNA and RNA

NUCLEOTIDE STRUCTURE (p. 289)

- **Nucleotides** are composed of a **nitrogen-containing base**, a **pentose monosaccharide** (**ribose** or **deoxyribose**), and one, two, or three **phosphate groups**. The nitrogen-containing bases belong to two families of compounds: the **purines** (**adenine**, "**A**," and **guanine,** "**G**") and the **pyrimidines** (**cytosine**, "**C**," **uracil**, "**U**," and **thymine**, "**T**"). [Note: A base plus a pentose produces a **nucleoside**.] Deoxyribonucleic acid (DNA) contains deoxyribose and A, G, C, and T. Ribonucleic acid (RNA) contains ribose, and A, G, C, and U.

Synthesis of purine nucleotides:

- Sources of atoms for the purine ring
- Role of PRPP in nucleotide synthesis
- Enzyme that synthesizes PRPP, and its regulation
- Regulated step in purine synthesis
- End-products of the pathway of purine synthesis
- Enzymes required for the production of nucleoside diphosphtes and triphosphates, and their specificities
- Clinical uses of the sulfonamides, trimethoprim, and methotrexate
- Enzymes required for the salvage of purine bases

SYNTHESIS OF PURINE NUCLEOTIDES (p. 291)

- The atoms of a purine are contributed by amino acids (**aspartic acid**, **glutamine**, and **glycine**), CO_2, and **N^{10}-formyl tetrahydrofolic acid**.
- **5-phosphoribosyl-1-pyrophosphate** (**PRPP**) is an "activated pentose" that participates in the synthesis of purines and pyrimidines nucleotides and in the salvage of purine bases. It donates the ribose–phosphate unit found in nucleotides. PRPP is produced by **PRPP synthetase**, an enzyme that is **activated** by **inorganic phosphate** and **inhibited** by purine nucleoside diphosphates and triphosphates (**purine nucleotides**—the end-products of this pathway).
- Synthesis of 5'phosphoribosylamine from PRPP and glutamine is catalized by **glutamine:phosphoribosyl pyrophosphate amidotransferase**. This enzyme is **inhibited** by the **purine 5'-nucleotides**, AMP, GMP, and IMP—the end-products of the pathway. This is the **committed step in** purine nucleotide biosynthesis.
- The end-product of this pathway is **inosine monophosphate** (**IMP**), the "parent" purine nucleotide that contains the base, **hypoxanthine**. IMP is converted to **AMP** and **GMP**.
- Nucleoside diphosphates (NDP) are synthesized from the corresponding nucleoside monophosphates (NMP) by **base-specific nucleoside monophosphate kinases**. NDPs and nucleoside triphosphates (NTP) are interconverted by **nucleoside diphosphate kinase**—an enzyme that, unlike the monophosphate kinases, has **broad specificity**.
- Some synthetic inhibitors of purine synthesis (for example, the **sulfonamides** or **trimethoprim**), inhibit the growth of **rapidly dividing microorganisms** without interfering with human cell functions. Other purine synthesis inhibitors, such as structural analogs of folic acid (for example, **methotrexate**), are used pharmacologically to control the spread of **cancer** by interfering with the synthesis of nucleotides and, therefore, of DNA and RNA.
- Purines that result from the normal turnover of cellular nucleic acids can be reconverted into nucleoside triphosphates and used by the body. Thus, they are "**salvaged**" instead of being degraded to uric acid. **PRPP** is the source of the ribose-phosphate, and the reactions are catalyzed by adenine phosphoribosyltransferase, and hypoxanthine-guanine phosphoribosyltransferase (HPRT).

- Cause and symptoms of Lesch-Nyhan syndrome

- A **deficiency of HPRT** results in the X-linked, recessive, inherited disorder, **Lesch-Nyhan syndrome**. Decreased salvage of hypoxanthine and guanine result in large amounts of **circulating uric acid**, causing **gout**. Symptoms also include neurologic features, such as **self-mutilation** and involuntary movements.

Synthesis of deoxyribonucleotides:

- Enzyme used to produce deoxyribonucleotides from ribonucleotides, its cofactor, and its regulation

SYNTHESIS OF DEOXYRIBONUCLEOTIDES (p. 295)

- All deoxyribonucleotides (used to synthesize DNA) are synthesized from ribonucleotides by the enzyme **ribonucleotide reductase**, which requires **thioredoxin** as a cofactor. This enzyme is highly regulated, for example, it is strongly **inhibited by dATP**—a compound that is overproduced in bone marrow cells in individuals with **adenosine deaminase deficiency** (see below).

Degradation of purine nucleotides:

- Degradation and fate of dietary nucleic acids
- Overview and end-product of purine nucleotide degradation
- Causes of primary and secondary hyperuricemia (gout)
- Result of *adenosine deaminase* deficiency

DEGRADATION OF PURINE NUCLEOTIDES (p. 296)

- Degradation of **dietary nucleic acids** occurs in the **small intestine**, where a family of **pancreatic enzymes** hydrolyze the nucleotides to nucleosides and free bases. Dietary purines are generally converted to uric acid, and dietary pyrimidines are degraded to small compounds by the intestinal mucosal cells.

- Purine nucleotides are converted to **uric acid** by a pathway that removes amino groups and ribose 1-phosphate from the nucleotide, then uses xanthine oxidase to oxidize the carbon rings to uric acid. **Allopurinol**, a drug that inhibits xanthine oxidase, is used to treat gout.

- High levels of uric acid in the blood can cause gout. **Primary gout** is caused by a genetic defect resulting in the overproduction or underexcretion of uric acid. **Secondary hyperuricemia** is caused by a variety of disorders and lifestyles, for example, in patients with chronic renal insufficiency, those who have myeloproliferative disorders, or those who consume excessive amounts of alcohol or purine-rich foods. Secondary gout can also be an adverse effect of metabolic diseases, such as von Gierke disease or fructose intolerance.

- **Adenosine deaminase deficiency** results in an accumulation of adenosine, which is converted to its ribonucleotide or deoxyribonucleotide forms by cellular kinases. As dATP levels rise, they inhibit ribonucleotide reductase, thus preventing the production of deoxyribonucleotides, so that the cell cannot produce DNA and divide. This causes **severe combined immunodeficiency disease**, involving a lack of T cells and B cells.

Pyrimidine synthesis and degradation:

- Sources of the atoms in the pyrimidine ring
- Committed step in pyrimidine synthesis
- End-product of pyrimidine base synthesis
- Cause of orotic aciduria
- Enzyme that synthesizes dTMP from dUMP and anticancer drugs that affect this reaction, and source of CTP

PYRIMIDINE SYNTHESIS AND DEGRADATION (p. 299)

- The sources of the atoms in the pyrimidine ring are **glutamine**, **CO_2**, and **aspartic acid**.

- The **committed step** of this pathway is the synthesis of **carbamoyl phosphate** from **glutamine** and **CO_2**, catalyzed by **carbamoyl phosphate synthetase II**. This enzyme is **inhibited** by **UTP** and **activated** by **ATP** and **PRPP**.

- The **end-product** of pyrimidine base synthesis is **orotic acid**, which is converted to the nucleotide **OMP** by the addition of ribose 6-phosphate (donated by **PRPP**). OMP is then converted to UMP, which is phosphorylated to UTP. UTP is then aminated to form CTP. A **deficiency** of the enzyme complex (UMP synthase) that converts orotic acid to UMP causes **orotic aciduria**.

- dUMP is converted to **dTMP** by **thymidylate synthase**, which utilizes **N^5,N^{10}-methylene tetrahydrofolate** as the source of the methyl group. **Thymine analogs** such as **5-flurouracil**, and **dihydrofolate reductase inhibitors**, such as **methotrexate**, are used as **anticancer drugs** because they prevent the production of dTMP and, therefore, stop DNA synthesis.

- Degradation products of pyrimidines

- Pyrimidines are degraded to highly soluble structures such as β-alanine and β-aminoisobutyrate, which can serve as precursors of **acetyl CoA** and **succinyl CoA**, respectively.

Chapter 23: Metabolic Effects of Insulin and Glucagon

Insulin:

- Site of synthesis and secretion
- Enzyme that degrades insulin and its source
- Half-life of insulin
- Compounds that stimulate insulin secretion
- One that decreases insulin secretion
- Pathways that are stimulated or inhibited by insulin
- Mechanism of insulin action

INSULIN (p. 305)

- **Insulin** is a polypeptide hormone produced by the **β cells** of the **islets of Langerhans** of the **pancreas**. Its **synthesis** involves two inactive precursors, **preproinsulin** and **proinsulin**, which are subsequently cleaved to form the active hormone. Insulin is stored in the **cytosol** in **granules** that are released by **exocytosis**. Insulin is degraded by the enzyme **insulinase** produced primarily by the **liver**. Insulin has a plasma half-life of approximately six minutes.
- A rise in **blood glucose** is the most important signal for **increased insulin secretion**. Plasma **amino acid levels** and the intestinal peptide **secretin** also stimulate insulin secretion. Its synthesis and release are **decreased by epinephrine**, which is secreted in response to stress, trauma, or extreme exercise.
- Insulin **increases glucose uptake** and the **synthesis of glycogen, protein, and triacylglycerol**. It **decreases triacylglycerol** and **glycogen degradation**. These actions are mediated by the binding of insulin to the **insulin receptor**, which initiates a cascade of cell-signaling responses, including phosphorylation of a family of proteins called **insulin receptor substrate (IRS) proteins**.

Glucagon:

- Site of synthesis
- The counterregulatory hormones
- Pathways increased in the presence of glucagon
- Compounds stimulating and inhibiting glucagon secretion
- Mechanism of glucagon action

GLUCAGON (p. 311)

- Glucagon is a polypeptide hormone secreted by the **α cells** of the pancreatic islets. Glucagon, along with epinephrine, cortisol, and growth hormone (the "**counterregulatory hormones**"), opposes many of the actions of insulin.
- Glucagon acts to **maintain blood glucose** during periods of potential hypoglycemia. Glucagon **increases glycogenolysis, gluconeogenesis, ketogenesis**, and **uptake of amino acids**.
- Glucagon **secretion** is **stimulated** by **low blood glucose**, **amino acids**, and **epinephrine**. Its secretion is **inhibited** by **elevated blood sugar** and by **insulin**.
- Glucagon binds to **high-affinity receptors** of **hepatocytes**. This binding results in the activation of **adenylate cyclase**, which produces the second messenger, **cyclic AMP**. Subsequent activation of **cAMP-dependent protein kinase** results in the phosphorylation-mediated activation or inhibition of key regulatory enzymes involved in carbohydrate and lipid metabolism.

Hypoglycemia:

- Characteristics of hypoglycemia
- Three major types of hypoglycemia

HYPOGLYCEMIA (p. 312)

- **Hypoglycemia** is characterized by: 1) central nervous system symptoms, including confusion, aberrant behavior, or coma; 2) a simultaneous blood glucose level equal to or less than 40 mg/dl; and 3) symptoms that resolve within minutes following the administration of glucose. Hypoglycemia most commonly occurs in patients receiving insulin treatment with tight control.
- Hypoglycemia may be divided into three groups: 1) **insulin-induced**, 2) **postprandial** (sometimes called reactive hypoglycemia), and 3) **fasting hypoglycemia**.

- Effect of ethanol on blood glucose levels

- The consumption and subsequent metabolism of **ethanol** inhibits gluconeogenesis, leading to hypoglycemia in individuals with depleted stores of glycogen. Alcohol consumption can also increase the risk for hypoglycemia in patients using insulin.

Chapter 24: The Feed/Fast Cycle

The fed state:

THE FED STATE (p. 319)

- Four mechanisms of control of the flow of intermediates through metabolic pathways

- The flow of intermediates through metabolic pathways is controlled by four mechanisms: 1) the availability of substrates; 2) allosteric activation and inhibition of enzymes; 3) covalent modification of enzymes; and 4) induction-repression of enzyme synthesis. In the fed state, these regulatory mechanisms ensure that available nutrients are captured as **glycogen**, **triacylglycerol**, and **protein**.

- Definition of absorptive state and compounds found in the plasma during this period

- The **absorptive state** is the two- to four-hour period after ingestion of a normal meal. During this interval, transient increases in plasma glucose, amino acids, and triacylglycerols occur, the last primarily as components of chylomicrons synthesized by the intestinal mucosal cells.

- Pancreatic hormones seen during fed state

- The **pancreas** responds to elevated levels of glucose and amino acids with an **increased secretion of insulin** and a **drop in the release of glucagon** by the **islets of Langerhans**. The elevated insulin to glucagon ratio and the ready availability of circulating substrates make the two to four hours after ingestion of a meal into an **anabolic period**.

- Major compounds synthesized/used by liver, adipose, muscle, and brain during fed state

- During this absorptive period, virtually all tissues use **glucose** as a fuel. In addition, the **liver** replenishes its **glycogen** stores, replaces any needed **hepatic proteins**, and increases **triacylglycerol** synthesis. The latter are packaged in **very-low-density glycoproteins**, which are exported to the peripheral tissues. The **adipose** increases **triacylglycerol** synthesis and storage, whereas the **muscle** increases **protein** synthesis to replace protein degraded since the previous meal. In the well-fed state, the **brain** uses glucose exclusively as a fuel.

The fasting state:

THE FASTING STATE (p. 327)

- Definition of catabolic period, and the plasma hormones seen in this state

- In the **absence of food**, plasma levels of glucose, amino acids, and triacylglycerols fall, triggering a **decline** in **insulin secretion** and an **increase in glucagon** and **epinephrine release**. The decreased insulin to glucagon ratio, and the decreased availability of circulating substrates, make the period of nutrient deprivation a **catabolic period**.

- Major compounds synthesized/used by liver, adipose, muscle, and brain during a fasting state

- To accomplish these goals, the **liver** degrades **glycogen** and initiates **gluconeogenesis**, using **increased fatty acid oxidation** both as a source of energy and to supply the acetyl CoA building blocks for **ketone body synthesis**. The **adipose** degrades stored **triacylglycerols**, thus providing **fatty acids** and **glycerol** to the liver. The **muscle** can also use **fatty acids** as fuel, as well as **ketone bodies** supplied by the liver. Muscle **protein** is **degraded** to supply **amino acids** for the liver to use in gluconeogenesis. The **brain** can use both **glucose** and **ketone bodies** as fuels.

Chapter 25: Diabetes Mellitus

Type 1 diabetes:

TYPE 1 DIABETES (p. 334)

- Defect in diabetes mellitus
- Possible health consequences of having diabetes

- **Diabetes mellitus** is a heterogeneous group of syndromes characterized by an **elevation of fasting blood glucose** that is caused by a relative or absolute deficiency in insulin. Diabetes is the leading cause of **adult blindness** and **amputation** and a major cause of **renal failure**, **heart attack**, and **stroke**. The disease can be classified into two groups, type 1 and type 2.

- Cause of type 1 diabetes mellitus

- Persons with **type 1** diabetes constitute approximately ten percent of the diabetics in the United States. The disease is characterized by an **absolute deficiency of insulin** caused by an **autoimmune attack** on the **β cells of the pancreas**. This destruction requires a **stimulus from the environment** (such as a viral infection) and a **genetic determinant** that allows the β cell to be recognized as "non-self."

- Metabolic abnormalities associated with type 1 diabetes mellitus

- The **metabolic abnormalities** of type 1 diabetes mellitus include **hyperglycemia**, **ketoacidosis**, and **hypertriglyceridemia**. They result from a deficiency of insulin and a relative excess of glucagon.

- Drug treatment for type 1 diabetes mellitus

- Persons with type 1 diabetes must rely on **exogenous insulin**, injected subcutaneously, to control the hyperglycemia and ketoacidosis.

Type 2 diabetes:

TYPE 2 DIABETES (p. 340)

- Definition of insulin resistance

- **Type 2** diabetes has a strong **genetic** component. It results from a combination of **insulin resistance** and **dysfunctional β cells**. Insulin resistance is the decreased ability of target tissues, such as liver, adipose tissue, and muscle, to respond properly to normal circulating concentrations of insulin.

- Most common cause of insulin resistance

- **Obesity** is the most common cause of insulin resistance. However, the majority of people with obesity and insulin resistance do not become diabetic. In the absence of a defect in β cell function, **non-diabetic obese individuals** can compensate for insulin resistant with **elevated levels of insulin**. Insulin resistance alone will not lead to type 2 diabetes. Rather, type 2 diabetes develops in insulin-resistant individuals who also show impaired β-cell function.

- Why metabolic alterations in type 2 are milder than those seen in type 1 diabetes

- The **metabolic alterations** observed in type 2 diabetes are **milder** than those described for the insulin-dependent form of the disease, in part, because insulin secretion in type 2 diabetes—although not adequate—restrains ketogenesis and blunts the development of diabetic ketoacidosis.

- Chronic complications of diabetes

- Available treatments for diabetes moderate the hyperglycemia, but fail to completely normalize metabolism. The long-standing elevation of blood glucose causes the **chronic complications of diabetes**—**premature atherosclerosis**, **retinopathy**, **nephropathy**, and **neuropathy**.

Chapter 26: Obesity

Assessment of obesity:

ASSESSMENT OF OBESITY (p. 347)

- Fundamental cause of obesity

- **Obesity**—the accumulation of excess body fat—results when energy intake exceeds energy expenditure.

- Equation for the body mass index (BMI)
- BMI ranges for overweight and obese

- The **body mass index** (**BMI**) is a measure of body fat. It is calculated in both men and women as the (weight in kg)/(height in inches^2). Individuals with a BMI between 25.1 and 29 are considered **overweight**, and those with a BMI greater than 30 are defined as **obese**.

- Differences between the location of android versus gynoid fat, and their association with diseases

- Excess fat can be located in the central **abdominal area** (android, upper body obesity). This fat is associated with a **greater risk for hypertension, insulin resistance, diabetes, dyslipidemia,** and **coronary heart disease**. That distributed in the **lower extremities** (gynoid, lower body obesity) is **relatively benign**, healthwise.

- Effect of obesity on adipocyte size and number

- Most **obesity** is thought to involve an **increaase** in both the **number** and **size** of **adipocytes**.

Body weight recognition:

- Role of a predetermined body weight set point
- Role of genetics in obesity
- Types of environmental and lifestyle factors that impact on obesity

BODY WEIGHT REGULATION (p. 349)

- The body attempts to add adipose tissue when the body weight falls below a **predetermined weight set point**, and to lose weight when the body weight is higher than the set point.
- **Genetic mechanisms** play a primary role in determining body weight. Obesity behaves as a **complex polygenic disease**, involving interactions between multiple genes and the environment.
- **Environmental factors** such as the ready availability of palatable, energy-dense foods, and the increasingly **sedentary lifestyles** encouraged by TV watching, autos, computer usage, and energy-sparing devices in the workplace and at home, decrease physical activity and enhance the tendency to gain weight.

Molecules that influence obesity:

- Compounds that signal the hypothalamus to influence appetite and energy expenditure

MOLECUES THAT INFLUENCE OBESITY (p. 350)

- The **hypothalamus** releases peptides that regulate appetite and energy consumption in response to afferent signals from other tissues, such as from the pancreas (**insulin**), stomach (**ghrelin**), intestine (**cholecystokinin**), peripheral nervous system (**norepinephrine**), and adipose tissue (**leptin**, **adiponectin**, and **resistin**).

Metabolic changes observed in obesity:

- Disorders associated with syndrome X (insulin resistance syndrome, metabolic syndrome)

METABOLIC CHANGES OBSERVED IN OBESITY (p. 351)

- **Abdominal obesity** is associated with glucose intolerance, insulin resistance, hyperinsulinemia, dyslipidemia (low HDL and VLDL), and hypertension. This cluster of metabolic abnormalities is known as **syndrome X**, the **insulin resistance syndrome**, or the **metabolic syndrome**. Individuals with this syndrome have a significantly increased risk for developing diabetes mellitus and cardiovascular disease.

Weight reduction:

- Definition of negative energy balance, and its role in weight reduction
- Two weight-loss medications currently approved by the FDA, and modes of action

WEIGHT REDUCTION (p. 352)

- Weight reduction is achieved with a **negative energy balance**, that is, **decreasing caloric intake** and/or **increasing energy expenditure**. In addition, two weight-loss medications are currently approved by the FDA for obese patients: **sibutramine**, which is an appetite supressant that inhibits the reuptake of both serotonin and norepinephrine, and **orlistat**, which inhibits gastric and pancreatic lipases, thus decreasing the breakdown of dietary fat into smaller molecules. **Surgical procedures** designed to reduce food consumption are an option for the severely obese patient who has not responded to other treatments.

Chapter 27: Nutrition

Dietary reference intakes:

- Definition of Dietary Reference intake
- Definition of Estimated Average Requirement
- Definition of Recommended Daily Allowance
- Definition of Adequate Intake
- Definition of Tolerable Upper Intake Level

DIETARY REFERENCE INTAKES (p. 355)

- **Dietary Reference intakes** (**DRIs**) are estimates of the amounts of nutrients required to prevent deficiencies and maintain optimal health. DRIs consist of the following four dietary reference standards. **Estimated Average Requirement** (**EAR**) is the average daily nutrient intake level estimated to meet the requirement of one half the healthy individuals in a particular life stage and gender group. The **Recommended Dietary Allowance** (**RDA**) is the average daily dietary intake level that is sufficient to meet the nutrient requirements of nearly all (97 to 98 percent) individuals. **Adequate Intake** (**AI**) is set instead of an RDA if sufficient scientific evidence is not available to calculate the RDA. The **Tolerable Upper Intake Level** (**UL**) is the highest average daily nutrient intake level that is likely to pose no risk of adverse health effects to almost all individuals in the general population.

Energy requirements in humans:

- Three major energy-requiring processes occurring in the body
- Definition of Acceptable Macronutrient Distribution Ranges
- Percent of total calories adults should consume as fats, carbohydrates, and protein

Dietary fats:

- Relative effects of high serum LDLs versus HDLs on cardiovascular disease
- Relationship between dietary saturated fat and plasma and LDL cholesterol
- Effect of n-6 and n-3 polyunsaturated fatty acids on plasma LDLs and HDLs and on heart disease

Dietary carbohydrates:

- Functions of dietary carbohydrates

Dietary protein:

- Function of dietary protein
- Definition of protein quality
- Definition of positive and negative protein balance
- Health situations leading to positive and negative protein balance
- Cause of kwashiorkor
- Cause of marasmus

ENERGY REQUIREMENTS IN HUMANS (p. 356)

- The energy generated by the metabolism of **macronutrients** is used for three energy-requiring processes that occur in the body: **resting metabolic rate**, **thermic effect of food**, and **physical activity**.

- **Acceptable Macronutrient Distribution Ranges** (**AMDR**) are defined as a range of intakes for a particular macronutrient that is associated with reduced risk of chronic disease while providing adequate amounts of essential nutrients. Adults should consume **45 to 65 percent** of their **total calories** from **carbohydrates**, **20 to 35 percent** from **fat**, and **10 to 35 percent** from **protein**.

DIETARY FATS (p. 358)

- Elevated levels of total cholesterol or LDL cholesterol result in increased risk for **cardiovascular disease**. In contrast, high levels of HDL cholesterol have been associated with a decreased risk for heart disease.

- Consumption of **saturated fats** is strongly associated with high levels of total plasma cholesterol and LDL cholesterol. When substituted for saturated fatty acids in the diet, **monounsaturated fats** lower both total plasma cholesterol and LDL cholesterol, but increase HDLs.

- Consumption of fats containing **n-6 polyunsaturated fatty acids** lowers plasma LDLs, but HDLs, which protect against coronary heart disease, are also lowered. Dietary **n-3 polyunsaturated fats** have little effect on plasma HDL or LDL levels, but they suppress cardiac arrhythmias and reduce serum triacylglycerols, decrease the tendency to thrombosis, and substantially reduce the risk of cardiovascular mortality.

DIETARY CARBOHYDRATES (p. 363)

- **Carbohydrates** provide **energy** and **fiber** to the diet. When they are consumed as part of a diet in which caloric intake is equal to energy expenditure, they do not promote obesity.

DIETARY PROTEIN (p. 365)

- Dietary **protein** provides **essential amino acids**. The **quality of a protein** is a measure of its ability to provide the essential amino acids required for tissue maintenance. Proteins from animal sources, in general, have a higher quality protein than that derived from plants. However, proteins from different plant sources may be combined in such a way that the result is equivalent in nutritional value to animal protein.

- **Positive nitrogen balance** occurs when nitrogen intake exceeds nitrogen excretion. It is observed in situations in which tissue growth occurs, for example, in children, pregnancy, or during recovery from an emaciating illness. **Negative nitrogen balance** occurs when nitrogen losses are greater than nitrogen intake. It is associated with inadequate dietary protein, lack of an essential amino acid, or during physiologic stresses, such as trauma, burns, illness, or surgery.

- **Kwashiorkor** is caused by inadequate intake of protein. **Marasmus** Marasmus occurs when calorie deprivation is relatively greater than the reduction in protein.

Chapter 28: Vitamins

Water-soluble vitamins:

WATER-SOLUBLE VITAMINS (p. 372)

- Folic acid: active form, function, and results of deficiency

- **Folic acid's** active form is **tetrahydrofolic acid**. Its function is to **transfer one-carbon units** in the synthesis of methionine, purines, and thymine. **Deficiency** of this vitamin results in **megaloblastic anemia** and **neural tube defects** at birth. There is no known toxicity of this vitamin, but administration of high levels of folate can mask vitamin B_{12} deficiency.

- Vitamin B_{12}: active forms, function, and results of deficiency

- **Vitamin B_{12}** (**cobalamin**) has as its active forms, **methylcobalamin** and **deoxyadenosylcobalamin**. It serves as a cofactor for the conversion of homocysteine to methionine, and methylmalonyl CoA to succinyl CoA. A **deficiency** of cobalamin results in **pernicious** (**megaloblastic**) **anemia**, **dementia**, and **spinal degeneration**. The anemia is treated with IM or high oral doses of vitamin B_{12}. There is no known toxicity for this vitamin.

- Vitamin C: chemical name, function, and results of deficiency

- **Vitamin C** (**ascorbic acid**) functions as an **antioxidant** and as a cofactor for **hydroxylation reactions** in procollagen. A **deficiency** of vitamin C results in **scurvy**, a disease characterized by **sore, spongy gums**, **loose teeth**, and **poor wound healing**. There is no known toxicity for this vitamin.

- Vitamin B_6: other names, active form, function, and results of deficiency

- **Vitamin B_6** (**pyridoxine**, **pyridoxamine**, and **pyridoxal**) has the active form, **pyridoxal phosphate**. It functions as a cofactor for enzymes, particularly in amino acid metabolism. **Deficiency** of this vitamin is rare, but causes **glossitis** and **neuropathy**. The deficiency can be induced by **isoniazid**, which causes sensory neuropathy at high doses.

- Vitamin B_1: other name, active form, function, and results of deficiency

- **Vitamin B_1** (**thiamine**) has the active form, **thiamine pyrophosphate**. It is a cofactor of enzymes catalyzing the conversion of pyruvate to acetyl CoA, α-ketoglutarate to succinyl CoA, and the transketolase reactions in the pentose phosphate pathway. A **deficiency** of thiamine causes **beriberi**, with symptoms of **tachycardia**, **vomiting**, and **convulsions**. In **Wernicke-Korsakoff syndrome** (most common in alcoholics), individuals suffer from **apathy**, **loss of memory**, and **eye movements**. There is no known toxicity for this vitamin.

- Niacin: other names, active forms, function, and results of deficiency

- **Niacin** (**nicotinic acid**, **nicotinamide**) has the active forms **NAD^+** and **NADPH**. It functions in electron transfer. A **deficiency** of niacin causes **pellagra**, which is characterized by **dermatitis**, **diarrhea**, and **dementia**. There is no known toxicity for this vitamin. High doses of niacin are used to treat hyperlipidemia.

- Vitamin B_2: other name, active forms, function, and results of deficiency

- **Vitamin B_2** (**riboflavin**) has the active forms **FAD** and **FMN**. It functions in electron transfer. A **deficiency** of riboflavin is rare, but it causes **dermatitis** and **angular stomatitis**. There is no known toxicity.

- Biotin: active form and function

- **Biotin** is active when covalently attached to a **carboxylase**, participating in carboxylation reactions. A deficiency of biotin is rare, and it has no known toxicity.

- Pantothenic acid: active form and function

- **Pantothenic acid** has the active form **coenzyme A**. It functions as an acyl carrier. A deficiency of pantothenic acid is rare, and it has no known toxicity.

Fat-soluble vitamins:

FAT-SOLUBLE VITAMINS (p. 379)

- Vitamin A: other names, active forms, functions, and results of deficiency

- **Vitamin A** (**retinol**, **retinal**, **retinoic acid**—the three active forms of vitamin A, and **β-carotene**) function in the maintenance of reproduction, vision, promotion of growth, differentiation and maintenance of epithelial tissues, and gene expression. A **deficiency** of vitamin A results in **impotence**, **night blindness, retardation of growth**, and **xerophthalmia**. Large amounts of vitamin A are **toxic** and can result in an **increased incidence of fractures**.

- Vitamin D: other names, active forms, function, and results of deficiency

- **Vitamin D** (**cholecalciferol**, **ergocalciferol**) has its active form as **1,25-dihydroxylcholecalciferol**. It is responsible for **calcium uptake**, and a **deficiency** of the vitamin results in **rickets** (in children) and **osteomalacia** (in adults). The symptoms of both syndromes are soft, pliable bones. High levels of vitamin D are toxic.

- Vitamin K: other names, function, and result of deficiency

- **Vitamin K** (**menadione**, **menaquinone**, and **phyloquinone**) are responsible for the **γ-carboxylation of glutamate residues** in clotting factors and other proteins. A **deficiency** of vitamin K is seen in newborns but is rare in adults; it causes bleeding. The vitamin has little toxicity.

- Vitamin E: other name, active forms, function, and result of deficiency

- **Vitamin E** (**α-tocopherol**) has as its active form any of several tocopherol derivatives. It functions as an **antioxidant**. Vitamin E **deficiency** is rarely seen, but can lead to red blood cell fragility that leads to **hemolytic anemia**. It has no known toxicity.

Chapter 29: DNA Structure and Replication

Structure of DNA:

- Bonds linking nucleotides
- Definition of polarity
- Sequence in which nucleotides are read
- Overall structure of DNA
- Definition of axis of symmetry
- Definition of antiparallel arrangements of chains
- Bases that are paired in DNA
- Bonds that hold the base pairs together
- Where circular DNA is found

STRUCTURE OF DNA (p. 393)

- DNA contains many monodeoxyribonucleotides covalently linked by **3',5'–phosphodiester bonds**. The resulting long, unbranched chain has **polarity**, with both a 5'-end and a 3'-end that are not attached to other nucleotides. The sequence of nucleotides is read 5' to 3'.

- With the exception of a few viruses that contain single-stranded DNA, DNA exists as a **double-stranded molecule**, in which the two strands wind around each other forming a **double helix**. In the double helix, the two chains are coiled around a common axis called the **axis of symmetry**. The chains are paired in an **antiparallel manner**, that is, the 5'-end of one strand is paired with the 3'-end of the other strand.

- The bases of one strand are paired with the bases of the second strand so that an **adenine** is always paired with a **thymine**, and a **cytosine** is always paired with a **guanine**. Therefore, one polynucleotide chain of the DNA double helix is always the **complement** of the other. Base pairs are held together by **hydrogen bonds.**

- **Eukaryotic chromosomes** consist of one long, **linear molecule of DNA** bound to a complex of proteins to form **chromatin**. **Circular DNA** is found in **eukaryotic mitochondria**, **prokaryotic chromosomes**, and extrachromasomal **plasmids.**

Prokaryotic DNA synthesis:

- Definition of semiconservative replication
- Direction in which the replication forks move from the origin of replication
- Names and functions of members of the prepriming complex

PROKARYOTIC DNA SYNTHESIS (p. 396)

- Each strand of the double helix serves as a **template** for constructing a **complementary** daughter strand. The resulting duplex contains one parental and one daughter strand, and the mode of replication is, thus, **semiconservative**.

- DNA replication begins at the **origin of replication** (one in prokaryotes, multiple in eukaryotes). The strands are separated locally, forming two **replication forks**. Replication of double-stranded DNA is **bidirectional**.

- A group of proteins form the **prepriming complex**. They recognize the origin of replication (**dnaA protein**), maintain the separation of the parental strands (**single-stranded DNA-binding proteins**), and unwind the double helix ahead of the advancing replication fork (**helicase**).

- Cause of positive supercoils and enzymes that can relax them
- Topoisomerases that are targeted by therapeutic drugs

- As the two strands of the double helix are separated, **positive supercoils** are produced in the region of DNA ahead of the replication fork. These interfere with further unwinding of the double helix. **DNA topoisomerases Types I and II** remove supercoils. Human topoisomerase II is targeted by **anticancer agents**, such as **etoposide**, and **DNA gyrase** (a Type II topoisomerase found in E. coli that can introduce negative supercoils) is targeted by the **antimicrobial quinolones**.

- Direction in which DNA polymerases "read" the template and the direction in which they synthesize DNA

- DNA polymerases are only able to "**read**" the parental nucleotide template sequences in the **3'→5' direction** and **synthesize** the new DNA strands in the **5'→3'** (**antiparallel**) **direction**. Therefore, beginning with one parental double helix, the two newly synthesized stretches of nucleotide chains must grow in opposite directions—one in the 5'→3' direction toward the replication fork (**leading strand**), and one in the 5'→3' direction away from the replication fork (**lagging strand**). The lagging strand is synthesized **discontinuously**.

- The function of a primer
- Enzyme that synthesizes the primer in de novo DNA synthesis

- DNA polymerases require a **primer**—a short, double-stranded region with a free hydroxyl group on the 3'-end of the shorter strand. The primer for de novo DNA synthesis is a short stretch of **RNA** synthesized by an RNA polymerase called **primase**. The leading strand only needs one RNA primer, whereas the lagging strand needs many.

- Enzymes that catalyze DNA chain elongation, proofreading, removal of RNA primers, filling gaps, and making the final phosphodiester bond

- DNA chain elongation is catalyzed by **DNA polymerase III** using **5'-deoxyribonucleoside triphosphates** as substrates. The enzyme "**proofreads**" the newly synthesized DNA, removing terminal mismatched nucleotides with its **3'→5' exonuclease** activity.

- RNA primers are removed by **DNA polymerase I** using its **5'→3' exonuclease** activity. The resulting gaps are filled in by this enzyme which can also proofread. The final phosphodiester linkage is catalyzed by **DNA ligase**.

Eukaryotic DNA replication:
- The functions of DNA polymerases α, β, γ, δ, and ε

EUKARYOTIC DNA REPLICATION (p. 404)

- There are at least five classes of **eukaryotic DNA polymerases**. **Pol α** is a multisubunit enzyme, one subunit of which performs the primase function. **Pol α 5'→3' polymerase** activity adds a short piece of DNA to the RNA primer. **Pol δ** completes DNA synthesis on the leading strand and elongates each lagging strand fragment, using **3'→5' exonuclease** activity to **proofread** the newly synthesized DNA. **Pol β** and **pol ε** are involved in carrying out DNA "repair," and **pol γ** replicates mitochondrial DNA.

- Definition of telomeres
- Function of telomerase

- **Telomeres** are stretches of **highly repetitive DNA** found at the ends of linear chromosomes. As cells divide and age, these sequences are shortened, contributing to cell death. In cells that do not age (for example, germline and cancer cells) the enzyme **telomerase** replaces the telomeres, thus extending the life of the cell.

- Function of reverse transcriptase

- **Retroviruses**, such as the **human immunodeficiency virus**, carry their genomes in the form of single-stranded RNA molecules. They use **reverse transcriptase** to make a DNA copy of their RNA and can integrate the copy into host cells.

- Function and usefulness of nucleoside analogs

- **Nucleoside analogs** that have been modified in the sugar portion of the nucleoside can be used to block DNA chain growth. They are useful in anticancer and antiviral chemotherapy.

Organization of eukaryotic DNA:
- Role of histones
- Definition of nucleosome

ORGANIZATION OF EUKARYOTIC DNA (p. 406)

- There are five classes of **histones**, which are **positively charged** small proteins that form **ionic bonds** with **negatively charged DNA**. Two each of histones H2A, H2B, H3, and H4 form a structural core around which DNA is wrapped creating a **nucleosome**. The DNA connecting the nucleosomes is called **linker DNA**, and is bound to histone H1.

- Levels of chromosomal organization

- Nucleosomes can be packed more tightly to form a **polynucleosome** (also called a **nucleofilament**), which is organized into loops that are anchored by a **nuclear scaffold** containing several proteins. Additional levels of organization create a **chromosome**.

DNA repair:

- Production of thymine dimers
- Mechanism of excision of thymine dimers
- Cause of xeroderma pigmentosum

- Causes of base alterations
- Mechanism of abnormal base removal
- Mechanism of recognition and repair of an AP-site

DNA REPAIR (p. 407)

- Exposure of a cell to **ultraviolet light** can cause covalent joining of two adjacent pyrimidines (usually thymines), producing a dimer. These **thymine dimers** prevent DNA polymerase from replicating the DNA strand beyond the site of dimer formation. These are removed by **UV-specific endonuclease** (**uvrABC excinulease**), and the resulting gap is filled by **DNA polymerase I**. In eukaryotes, a deficiency of UV-specific excinulease causes **xeroderma pigmentosum**, a rare disease in which cells cannot repair DNA damaged by UV.

- The bases of DNA can be altered spontaneously or by the action of deaminating or alkylating compounds. **Abnormal bases** are recognized by specific glycosylases that hydrolytically cleave them from the deoxyribose–phosphate backbone of the strand. This leaves an **apyrimidinic** or **apurinic site** (**AP-site**). Specific **AP-endonucleases** make a nick at the 5'-side of the AP-site. **Deoxyribose-phosphate lyase** removes the single empty sugar–phosphate residue. **DNA polymerase** and **DNA ligase** complete the repair process.

Chapter 30: RNA Synthesis

Structure of RNA:

- Three major types of RNA

- Functions of the three major types of RNA

STRUCTURE OF RNA (p. 413)

- There are three major types of RNA that participate in the process of protein synthesis: **ribosomal RNA** (**rRNA**), **transfer RNA** (**tRNA**), and **messenger RNA** (**mRNA**). They are unbranched polymers of nucleotides, but differ from DNA by containing **ribose** instead of deoxyribose and **uracil** instead of thymine.

- **rRNA** is found in association with several proteins as a component of the **ribosomes**. **Prokaryotic cells** have **three** and **eukaryotic cells** have **four** distinct size species. **tRNA** serves as an “adaptor” molecule that carries a specific amino acid to the site of protein synthesis. There is at least one specific type of tRNA molecule for each of the common twenty amino acids. **mRNA** carries genetic information from the nuclear DNA to the cytosol, where it is used as the template for protein synthesis.

Transcription of prokaryotic genes:

- Structure of RNA polymerase
- Functions of σ and ρ factors

- Steps in transcription
- Substrates for RNA polymerase

TRANSCRIPTION OF PROKARYOTIC GENES (p. 414)

- The process of RNA synthesis is called **transcription**. The enzyme that synthesizes RNA is **RNA polymerase**, which is a multisubunit enzyme. The **core enzyme** has four subunits—**2** α, **1** β, and **1** β', and possesses **5'→3' polymerase activity**. The enzyme requires an additional subunit—**sigma** (σ) **factor**—that recognizes the nucleotide sequence (**promoter region**) at the beginning of a length of DNA that is to be transcribed. Another protein—**rho** (ρ) **factor**—is required for **termination** of transcription of some genes.

- **Initiation** of transcription involves binding of the RNA polymerase to the promoter region. This sequence contains characteristic **consensus nucleotide sequences** that are highly conserved. These include the **Pribnow box** and the **–35 sequence**. **Elongation** involves **RNA polymerase** copying one strand of the DNA double helix, pairing C's with G's and A's (on the DNA template) with U's on the RNA transcript. Substrates are **ribonucleoside triphosphates**. **Termination** may be accomplished by the RNA polymerase alone, or may require ρ factor.

- Definition of an operon
- Function of the lactose operon

- A bacterial **operon** is a group of structural genes that code for the enzymes of a metabolic pathway, which are often found grouped together on the chromosome along with the regulatory genes that determine their transcription as a single long piece of mRNA. The genes are thus **coordinately expressed**. The **lactose (lac) operon** of E. coli is one of the best understood. It codes for the enzymes needed to metabolize lactose when it is the only available sugar substrate.

Transcription of eukaryotic genes:
- Functions of the three classes of RNA polymerase
- Characteristics of the promoter regions
- Function of enhancers

TRANSCRIPTION OF EUKARYOTIC GENES (p. 420)

- There are three distinct classes of RNA polymerase in the nucleus of eukaryotic cells. **RNA polymerase I** synthesizes the precursor of **large rRNAs** in the **nucleolus**. **RNA polymerase II** synthesizes the precursors for **mRNAs**, and **RNA polymerase III** produces the precursors of **tRNAs** and some other small RNAs in the **nucleoplasm**.

- **Promoters** for **class II genes** contain **consensus sequences**, such as the **TATA** or **Hogness box**, the **CAAT box**, and the **GC box**. They serve as binding sites for proteins called **general transcription factors**, which, in turn, interact with each other and with RNA polymerase II. **Enhancers** are DNA sequences that increase the rate of initiation of transcription by binding to specific transcription factors called **activators.**

Posttranscriptional modification of RNA:
- Definition of a primary transcript
- Posttranscriptional modification of rRNAs
- Posttranscriptional modification of tRNAs
- Posttranscriptional modification of mRNAs

POSTTRANSCRIPTIONAL MODIFICATION OF RNA (p. 422)

- A **primary transcript** is a linear copy of a **transcriptional unit**—the segment of DNA between specific initiation and termination sequences. The primary transcripts of both prokaryotic and eukaryotic tRNAs and rRNAs are posttranscriptionally modified through cleavage of the original transcripts by ribonucleases.

- **rRNAs** of both prokaryotic and eukaryotic cells are synthesized from long precursor molecules called **preribosomal RNAs**. These precursors are cleaved and trimmed by ribonucleases, producing the three largest rRNAs. (Eukaryotic 5S rRNA is synthesized by RNA polymerase II instead of I, and is modified separately.)

- Prokaryotic and eukaryotic **tRNAs** are also made from longer precursor molecules. These must have an **intervening sequence (intron)** removed, and the 5'- and 3'-ends of the molecule are trimmed by ribonuclease. A **3' –CCA sequence** is added and bases at specific positions are modified, producing **"unusual" bases**.

- Prokaryotic mRNA is generally identical to its primary transcript, whereas **eukaryotic mRNA** is extensively modified posttranscriptionally. For example, a **7-methyl-guanosine "cap"** is attached to the 5'–terminal end of the mRNA through a triphosphate linkage by guanylyltransferase. A **long poly-A tail**—not transcribed from the DNA—is attached to the 3'-end of most mRNAs. Many eukaryotic mRNAs also contain **introns** that must be removed to make the mRNA functional. Their removal requires **small nuclear RNAs**.

Chapter 31: Protein Synthesis

The genetic code:
- Definition of a codon
- Types of mutations caused by altering the nucleotide sequence in a codon

THE GENETIC CODE (p. 429)

- **Codons** are composed of three nucleotide bases usually presented in the mRNA language of **A**, **G**, **C**, and **U**. They are always written **5'→3'**. Of the 64 possible three-base combinations, 61 code for the twenty common amino acids, and three signal termination of protein synthesis (**translation**).

- Altering the nucleotide sequence in a codon can cause **silent mutations** (the altered codon also codes for the original amino acid), **missense mutations** (the altered codon codes for a different amino acid), or **nonsense mutations** (the altered codon is a termination codon).

- Characteristics of the genetic code

- Characteristics of the genetic code include **specificity**, **universality**, and **redundancy**, and it is **nonoverlapping** and **commaless**.

Components required for translation:

- Components required for translation
- Important sites on tRNA required for translation
- Structure of ribosomes
- Function of the A, P, and E binding sites

COMPONENTS REQUIRED FOR TRANSLATION (p. 432)

- Requirements include: all the **amino acids** that eventually appear in the finished protein, at least one specific type of **tRNA** for each amino acid, one **aminoacyl-tRNA synthetase** for each amino acid, the **mRNA** coding for the protein to be synthesized, fully competent **ribosomes**, **protein factors** needed for initiation, elongation, and termination of protein synthesis, and **ATP** and **GTP** as energy sources.
- tRNA has an attachment site for a specific amino acid at its 3'-end, and an **anticodon** region that can recognize the codon specifying the amino acid the tRNA is carrying.
- **Ribosomes** are large complexes of protein and rRNA. They consist of **two subunits**. Each ribosome has three binding sites for tRNA molecules, the A, P, and E sites that cover three neighboring codons. The **A site** codon binds an **incoming aminoacyl-tRNA**, the **P site** codon is occupied by **peptidyl-tRNA**, and the **E site** is occupied by the **empty tRNA** as it is about to exit the ribosome.

Codon recognition by tRNA:

- Rules for codon/anticodon binding
- Direction in which nucleotide sequences are listed
- Explanation of the "wobble" hypothesis

CODON RECOGNITION BY tRNA (p. 434)

- Recognition of an mRNA codon is accomplished by the tRNA **anticodon**. The anticodon binds to the codon following the rules of **complementarity** and **antiparallel binding**. (When writing the sequences of both codons and anticodons, the nucleotide sequence must ALWAYS be listed in the 5'→3' order.)
- The "**wobble**" **hypothesis** states that the first (5') base of the anticodon is not as spatially defined as the other two bases. Movement of that first base allows nontraditional base-pairing with the last (3') base of the codon, thus allowing a single tRNA to recognize more than one codon for a specific amino acid.

Steps in protein synthesis:

- Mechanism of initiation of protein synthesis
- Mechanism of elongation during protein synthesis
- Definition of a polysome
- Mechanism of termination of protein synthesis

STEPS IN PROTEIN SYNTHESIS (p. 435)

- **Initiation:** The components of the translation system are assembled, and mRNA associates with the small ribosomal subunit. The process requires **initiation factors**. In **prokaryotes**,a purine-rich region (the **Shine-Dalgarno sequence**) of the mRNA base-pairs with a complementary sequence on 16S rRNA, resulting in the positioning of the mRNA so that translation can begin. The **5'-cap** on **eukaryotic mRNA** is used to position that structure on the ribosome. The **initiation codon** is **5'–AUG–3'**.
- **Elongation:** The polypeptide chain is elongated in the **5'→3' direction** by the addition of amino acids to the carboxyl end of its growing chain. The process requires **elongation factors**. The formation of the peptide bond is catalyzed by **peptidyltransferase**, which is an activity intrinsic to the ribosomal 23S rRNA. Following peptide bond formation, the ribosome advances to the next codon (**translocation**). Because of the length of most mRNAs, more than one ribosome at a time can translate a message, forming a **polysome**.
- **Termination:** Termination begins when one of the three termination codons moves into the A site. These codons are recognized by **release factors**. The newly synthesized protein is released from the ribosomal complex, and the ribosome is dissociated from the mRNA. Numerous **antibiotics** interfere with the process of protein synthesis.

Posttranslational modification:

- Examples of posttranslational modification
- Mechanism for degrading defective proteins

POSTTRANSLATIONAL MODIFICATION OF POLYPEPTIDE CHAINS (p. 440)

- Many polypeptide chains are covalently modified after translation. Such modifications include **trimming** excess amino acids, **phosphorylation** which may activate or inactivate the protein, **glycosylation** which targets a protein to become part of a plasma membrane or lysosome or be secreted from the cell, or **hydroxylation** such as that seen in collagen.
- Proteins that are defective or destined for rapid turnover are marked for destruction by the attachment of a small, highly conserved protein called **ubiquitin**. Proteins marked in this way are rapidly degraded by a cellular component known as the **proteasome**.

Chapter 32: Biotechnology and Human Disease

Restriction endonucleases:

- Function of restriction endonucleases
- Type of sequences recognized by these enzymes
- Definition of a restriction site
- Mechanism for producing a recombinant DNA molecule

RESTRICTION ENDONUCLEASES (p. 445)

- **Restriction endonucleases** are bacterial enzymes that cleave double-stranded DNA into smaller fragments. Each enzyme cleaves DNA at a specific four- to six-base long nucleotide sequence, producing DNA segments called **restriction fragments**. The sequences that are recognized are **palindromic**. These enzymes form either **staggered cuts** (**sticky ends**) or **blunt end cuts** on the DNA. A DNA sequence that is recognized by a restriction enzyme is called a **restriction site**.
- **Bacterial DNA ligases** can anneal two DNA fragments from different sources if they have been cut by the same restriction endonuclease. This hybrid combination of two fragments is called a **recombinant DNA molecule**.

DNA cloning:

- Definition of a vector
- Requirements for a vector to be functional
- Definition of a plasmid
- Function in amplification of DNA
- Difference between a genomic DNA library and cDNA library
- Common method of DNA sequencing

DNA CLONING (p. 446)

- Introduction of a foreign DNA molecule into a replicating cell permits the **amplification** (production of many copies) of the DNA—a process called **cloning**. A **vector** is a molecule of DNA to which the fragment of DNA to be cloned is joined. Vectors must be capable of **autonomous replication** within the host cell, and must contain at least one specific nucleotide sequence recognized by a restriction endonuclease. It must also carry at least one gene that confers the ability to select for the vector, such as an **antibiotic resistance gene**.
- Prokaryotic organisms normally contain small, circular, extrachromosomal DNA molecules called **plasmids** that can serve as vectors. They can be readily isolated from the bacterium, annealed with the DNA of interest, and reintroduced into the bacterium which will replicate, thus making multiple copies of the hybrid plasmid.
- A **DNA library** is a collection of cloned restriction fragments of the DNA of an organism. **A genomic library** is a collection of fragments of double-stranded DNA obtained by digestion of the total DNA of the organism with a restriction endonuclease and subsequent ligation to an appropriate vector. It ideally contains a copy of every DNA nucleotide sequence in the genome. In contrast, **cDNA** (**complementary DNA**) **libraries** contain only those DNA sequences that are complementary to mRNA molecules present in a cell, and differ from one cell type to another. Because cDNA has no intervening sequences, it can be cloned into an **expression vector** for the synthesis of eukaryotic proteins by bacteria.
- Cloned, purified fragments of DNA can be sequenced using the **Sanger dideoxy method**.

Probes:

- Definition and function of a probe
- Definition and function of Southern blotting

PROBES (p. 450)

- A **probe** is a single-stranded piece of DNA, usually labeled with a radioisotope such, as ^{32}P, that has a nucleotide sequence complementary to the DNA molecule of interest (**target DNA**). Probes can be used to identify which clone of a library or which band on a gel contains the target DNA.
- **Southern blotting** is a technique that can be used to detect specific genes present in DNA. The DNA is cleaved using a **restriction endonuclease**, the pieces are separated by **gel electrophoresis** and then transferred to a **nitrocellulose membrane** for analysis. The fragment of interest is detected using a **probe**.

Restriction fragment length polymorphism:

- Definition of a polymorphic gene
- Definition of a restriction fragment length polymorphism
- Potential effect of a mutation at a restriction site on the cleavage of DNA by a restriction endonuclease

RESTRICTION FRAGMENT LENGTH POLYMORPHISM (p. 454)

- The human genome contains many thousands of **polymorphisms** that do not affect the structure or function of the individual. A **polymorphic gene** is one in which the variant alleles are common enough to be useful as genetic markers. A **restriction fragment length polymorphism** (**RFLP**) is a genetic variant that can be examined by cleaving the DNA into **restriction fragments** using a **restriction enzyme**.
- A **mutation** of one or more nucleotides at a restriction site can render the site unrecognizable by a particular restriction endonuclease. A new restriction site also can be created by the same mechanism. In either case, cleavage with endonuclease results in fragments of lengths differing from the normal that can be detected by DNA hybridization. This technique can be used to diagnose genetic diseases early in the gestation of a fetus.

Polymerase chain reaction:

- Function of the polymerase chain reaction
- Examples of applications of the PCR technique

POLYMERASE CHAIN REACTION (p. 459)

- The **polymerase chain reaction** (**PCR**) is a test tube method for amplifying a selected DNA sequence, and does not rely on the biologic cloning method. PCR permits the synthesis of millions of copies of a specific nucleotide sequence in a few hours. It can amplify the sequence, even when the targeted sequence makes up less than one part in a million of the total initial sample. The method can be used to amplify DNA sequences from any source.
- **Applications** of the PCR technique include: 1) efficient comparison of a normal cloned gene with an uncloned mutant form of the gene, 2) detection of low-abundance nucleic acid sequences, 3) forensic analysis of DNA samples, and 4) prenatal diagnosis and carrier detection, for example, of cystic fibrosis.

Analysis of gene expression

- Techniques that measure mRNA production
- Techniques that measure protein synthesis

ANALYSIS OF GENE EXPRESSION (p. 462)

- The products of gene expression (mRNA and proteins) can be measured by techniques such as the following. **Northern blots** are very similar to Southern blots except that the original sample contains a mixture of **mRNA** molecules that are separated by electrophoresis, then hybridized to a radioactive probe. **Microarrays** are used to determine the differing patterns of gene expression in two different types of cells—for example, normal and cancer cells. **Enzyme-linked immunosorbent assays** (**ELISAs**) and **western blots** (**immunoblots**) are used to detect specific proteins.

Index

Page numbers in **bold** indicate main discussion of topic. Page numbers followed by f denote figures. "*See*" cross-references direct the reader to the synonymous term. "*See also*" cross-references direct the reader to related topics. [Note: Positional and configurational designations in chemical names (for example, "3-", "α-", "N-", "D-") are ignored in alphabetizing.]

Lippincott's Illustrated Reviews: Biochemistry, 3rd Edition
by Pamela C. Champe and Richard A. Harvey.

C

D

E

H

I

R

S

U

Figure Sources

Figure 2.12: Modified from Garrett, R. H. and Grisham, C. M. *Biochemistry.* Saunders College Publishing, 1995. Figure 6.36, p. 193.

Figure 2.13 (panel C, top): Abdulla, S. Basic mechanisms of protein folding disease. Nature Publishing Group.

Figure 3.1A: Illustration: Irving Geis. Rights owned by Howard Hughes Medical Institute. Not to be used without permission

Figure 3.20: Photo curtesy of Photodyne Incorporated, Hartland, WI.

Figure 3.21: Corbis

Figure 4.3: Electron micrograph of collagen: Natural Toxin Research Center. Texas A&M University-Kingsville. Collagen molecule modified from Mathews, C. K., van Holde, K. E. and Ahern, K. G. Biochemistry. 3rd Ed., Addison Wesley Longman, Inc. 2000. Figure 6.13, p.175.

Figure 4.4: modified from Yurchenco, P. D., Birk, D. E., and Mecham, R. P., eds. (1994) Extracellular Matrix Assembly and Structure, Academic Press, San Diego, California

Figure 4.8: Kronauer and Buhler, Images in Clinical Medicine, The New England Journal of Medicine, June 15, 1995 , Vol. 332, No. 24, p.1611.

Figure 4.10: Photo from Web site Derma.de

Figure 4.11: Jorde, L. B., Carey, J. C., Bamshad, M. J. and White, R. L. Medical Genetics, 2nd Ed. http://medgen.genetics.utah.edu/index.htm

Question 4.3: from Berge LN, Marton V, Tranebjaerg L, Kearney MS, Kiserud T, Oian P. Prenatal diagnosis of osteogenesis imperfecta. Acta Obstetricia et Gynecologica Scandinavica. 74(4):321-3, 1995 Apr.

Figure 13.11: The Crookston Collection, University of Toronto.

Figure 17.13: Urbana Atlas of Pathology, University of Illinois College of Medicine at urbana-champaign. Image number 26.

Figure 17.19: Interactive Case Study Companion to Robbins Pathologic Basis of Disease.

Figure 18.12: Custom Medical Stock Photo.

Figure 20.20: Success in MRCOphth. http://www.mrcophth.com/iriscases/albinism.html

Figure 20.22 (top): Rubin, E. and Farber, J. L. Pathology (2nd edition). J. B. Lippinoctt. 194.Figure 6-30, p. 244.

Figure 20.22 (bottom): Gilbert-Barness, E. and Barness L. *Metabolic Disease*. Eaton Publishing, 2000. Figure 15, p42.

Figure 21.6: Rich, M. W. Porphyria cutanea tarda, Postgraduate Medicine, 105: 208-214 (1999).

Figure 21.5: Department of Dermatlogy, University of Pittsburgh. http://www.upmc.edu/dermatology/MedStudentInfo/introLecture/enlarged/vespct.htm.

Figure 21.10: Custom Medical Stock Photo, Inc.

Figure 21.13: Phototake.

Figure 22.16: Wuthrich, D. A., and Lebowitz. Tophaceous gout. Images in clinical Medicine. N Engl J Med, 332:646, 1995.

Figure 22.17: WebMD Inc. http://www.samed.com/sam/forms/index.htm

Figure 22.18 : Corbis.

Figure 23.2: Childs, G. http://www.cytochemistry.net/.

Figure 24.4: Phototake.

Figure 26.6: Corbis.

Figure 27.20: Joseph E. Armstrong Illinois State University.

Figure 28.4: Matthews, J. H. Queen's University Department of Medicine, Division of Hematology/Oncology, Kingston, Canada.

Figure 29.7: Nolan, J., Department of Biochemistry, Tulane University, New Orleans, LA.

Figure 23.13: Modified from Cryer, P. E., Fisher, J. N. and Shamoon, H. Hypoglycemia. Diabetes Care 17:734-753, 1994.